T0313914

Power System Modeling, Computation, and Control

Power System Modeling, Computation, and Control

Joe H. Chow
Rensselaer Polytechnic Institute
Troy, NY, USA

Juan J. Sanchez-Gasca
GE Energy Consulting
Schenectady, NY, USA

This edition first published 2020

Registered Offices
John Wiley & Sons, Inc., 111 River Street, Hoboken, NJ 07030, USA
John Wiley & Sons Ltd, The Atrium, Southern Gate, Chichester, West Sussex, PO19 8SQ, UK

Editorial Office
The Atrium, Southern Gate, Chichester, West Sussex, PO19 8SQ, UK

For details of our global editorial offices, customer services, and more information about Wiley products visit us at www.wiley.com.

Wiley also publishes its books in a variety of electronic formats and by print-on-demand. Some content that appears in standard print versions of this book may not be available in other formats.

Library of Congress Cataloging-in-Publication data applied for

Hardback : 9781119546870

Cover Design: Wiley
Cover Images: Courtesy of New York Independent System Operator, Inc.,
Courtesy of Nicholas W. Miller at Hickory Ledge,
Courtesy of New York Power Authority

Set in 10/12pt Warnock by SPi Global, Pondicherry, India
Printed and bound in Singapore by Markono Print Media Pte Ltd

10 9 8 7 6 5 4 3 2 1

To
Doris and Peggy

Contents

Preface

The aim of this book is to make advanced power system modeling, analysis, and control design more accessible to graduate students and power engineers. The materials in this book have been used in teaching a two-course sequence on power system modeling, computation, and control for graduate power students at Rensselaer Polytechnic Institute (RPI). Initially, Dr. Prabha Kundur's book [6] was used as the textbook. Subsequently, with the rearrangement of the course materials and the addition of more recent topics, such as flexible AC transmission systems (FACTS) and wind turbine-generators, it was decided that there were sufficient new materials to write this book.

This book reflects the preferences of the authors as to what topics are important to cover, how much material should be included, and how the material should be explained. The authors, each with 40-plus years of experience working on power systems, were colleagues at General Electric Company for many years. In some senses, our approach is also shaped by the other engineers who have worked with us, both inside and outside our organizations.

In covering the topics, we provide the fundamental models and equations, with complexities that are appropriate for graduate power students who have taken courses on power system analysis with symmetrical components, and control systems with advanced calculus and matrix analysis. Most of the topics are illustrated with numerical examples, which are a distinct feature of this book. The power system data used are either actual or typical parameters. The illustrative examples are generally of two types. There are examples that require calculations using the expressions given in the text; these examples may involve setting up system matrices and calculating eigenvalues. Some other examples require simulating certain power system dynamics beyond the capability of manual computation. For these examples, data files are provided which can be used directly by the Power System Toolbox (PST)©, a MATLAB®-based software package that was first developed in 1990. For models not included in PST, Simulink® models are provided, for example, wind turbine-generators. The computer code for the examples is available on the Wiley website for this textbook. Of course, students can use any other power system simulation software of their choice, requiring only some input data format conversion. For instructors, a solution manual and computer files for the problems are available.

The second feature is the organization of the materials. In addition to being a modeling and control book, this text is also a systems and computation book. As such, the first part of the text is on system and stability concepts, which can be delivered using only the electromechanical model for synchronous machines. Afterward, detailed models of power equipment are provided. To go full circle, the synchronous machine chapter shows how the detailed 6-state 2-axis generator model can be systematically reduced to the 2-state electromechanical model. As each control function is introduced, it is folded back into stability and linear model analysis. Such repeated reinforcement of fundamental concepts, with more advanced models added every time, has worked out well for the students at RPI.

As this is also a text on power system control, the third feature is that the book describes the details of five swing-mode damping control design examples, including power system stabilizers, high-voltage DC modulation control, static var compensators, and thyristor-controlled series compensators. In each design, a different input signal is used because of the characteristics of the control systems and the siting of the controller. Also, a dual-input power system stabilizer is demonstrated. Several similar damping control designs using complementary signals have been included as problems for students to try their design skills.

The text also contains quite a number of observations from our engineering practice. For example, the state estimator in a control center indicated that one transformer was receiving reactive power and another transformer was supplying reactive power. Then, the control room engineering manager explained that these two auto-transformers, connected in parallel, were made by different vendors with slightly different tap ratios. This rather unique circulating reactive power situation is posed as a problem for Chapter 2.

As another example, the first author, early on in his career, was working on a system planning study with a data file provided by a utility company. The computer simulation was unstable even before the disturbance was applied. An experienced engineer sat with him for two hours, going through the parameters of every piece of equipment in the data file and pointing out that many parameters were outside of normal parameter ranges. Once the data was adjusted, the initial simulation was stable. This example illustrates that proper model data should be used in power system analysis, which is emphasized throughout this book.

The book is divided into three parts. Part I, on systems concepts, consists of Chapters 2 to 6. The topics include power flow solution methods, voltage stability, dynamic simulation, and direct stability analysis. Part II, on synchronous machine models and their control systems, consists of Chapters 7 to 12. The topics include the systematic development of steady-state and dynamic models of synchronous machines, excitation systems, power system stabilizers, and turbine-generators. Part III, consisting of Chapters 13 to 16, deals with advanced topics including high-voltage direct current systems, FACTS, wind power generation, and model reduction.

Parts I and II, without Chapters 11 and 12, are suitable for a one semester course, whereas, Part III, together with Chapters 11 and 12, and supplemented by lectures on synchrophasor measurements and other advanced control design methods, can be covered as a second semester course. For schools with quarter or trimester systems, each part can be covered in a quarter.

We would like to acknowledge the many individuals who taught us and worked with us on the topics in this book. At General Electric, they include Jim Winkelman, Dale Swann, Don Ewart, Bill Price, Einar Larsen, Harold Javid, and the late Tom Younkins. Outside of General Electric, they include Professors Petar Kokotović, M. A. Pai, and Peter Sauer at the University of Illinois, Urbana-Champaign, Professor Fernando Alvarado at the

University of Wisconsin, Madison, the late Aty Edris of Electric Power Research Institute, Bruce Fardanesh of New York Power Authority, and Dave Bertagnolli of ISO-New England. Most of all, we are indebted to Graham Rogers of Ontario Hydro and Cherry Tree Scientific Software, who developed many advanced functions in the Power System Toolbox.

We would also like to thank many RPI PhD students whose work appears in various parts of this book. They include Ranjit Date (coherency), Kwok Cheung (Power System Toolbox), Glauco Taranto (power system damping control), Jaewon Chang (TCSC), George Boukarim (power system stabilizers), Xuan Wei (FACTS), Xia Jiang (UPFC/IPFC), Xinghao Fang (B2B STATCOM), Aranya Chakrabortty (energy functions), Felipe Wilches-Bernal (wind turbines), and, particularly, Luigi Vanfretti (synchrophasors) and Scott Ghiocel (voltage stability), who also developed solutions to many examples and problems. Luigi also taught from the draft manuscript.

We are grateful to the research support from many organizations, including General Electric Company, US Department of Energy, Tokyo Electric Power Company, National Science Foundation, Electric Power Research Institute, New York State Energy Research and Development Administration, and New York Power Authority. The first author is especially grateful for the support provided by the NSF/DOE CURENT Engineering Research Center, directed by Professors Kevin Tomsovic and Yilu Liu of the University of Tennessee, Knoxville.

We would also like to thank the Wiley editors Peter Mitchell and Michelle Dunckley for their assistance, and Dr. Barbara Lewis of the RPI Center for Global Communication and Design for providing comments on the writing style of the book.

Finally, we would like to thank our family for their love and patience during the preparation of this manuscript: Joe's wife Doris, daughters Amy and Tammy, and sisters Yvonne and Wanda, and Juan's wife Peggy. We are extremely grateful to them.

Joe H. Chow
Troy, New York
Juan J. Sanchez-Gasca
Schenectady, New York

About the Companion Website

The companion website for this book is at:

www.wiley.com/go/chow/power-system-modeling

The website includes:

- Examples, lecture slides & problems

Scan this QR code to visit the companion website.

1 Introduction

1.1 Electrification

The construction of power grids and their associated infrastructures all over the world was recognized by the US National Academy of Engineering as the top engineering accomplishment in the 20th century for the advancement of human well-being.[1] Electrification drives industrialization and increases human productivity in developed countries, significantly improving the quality of life in underdeveloped countries. Although power generation and transmission in the USA started with direct current (DC), power transmission had relied exclusively on alternating current (AC) for over 50 years until a few early high-voltage DC systems were installed. Power systems will continue to use these two modes of generation and transmission, especially with the increased integration of renewable resources using power electronic interfaces and with the emergence of smart grids.

The topics in this book cover a slice of these enormously complex, continuously operating electric power systems, namely the dynamic aspects of their components and how these components interact with each other in the prevailing central station model of power generation. It is expected that these topics will provide relevant background to prepare power engineers to deal with future power systems in which most households will be equipped with solar panels and every town will have a few wind turbines.

The objective of this introductory chapter is to provide some high-level notions of power systems that will be helpful in the later chapters. The central generating station model is useful in understanding the test systems for illustrating the main concepts and techniques presented in this book. The discussion on time-scales of power system dynamics will be helpful in understanding the assumptions used on models and in control designs.

---

1 US National Academy of Engineering Report, *Greatest Engineering Achievements of the 20th Century.* available online: http://www.greatachievements.org/.

Power System Modeling, Computation, and Control, First Edition. Joe H. Chow and Juan J. Sanchez-Gasca.

Companion website: www.wiley.com/go/chow/power-system-modeling

1.2 Generation, Transmission, and Distribution Systems

1.2.1 Central Generating Station Model

Early power systems started as a local generator supplying a neighborhood of customers. These generators were noisy and dirty, and unpopular in the relatively well-to-do communities. Before long, large efficient generators were built in relatively remote locations, using both hydraulic and steam turbines. These new generators could deliver power to large load centers using long-distance transmission lines. Figure 1.1 shows the essential structure of a modern power system. Large generating stations directly supply power to a high-voltage transmission system (230–765 kV). The power is then supplied to a medium voltage subtransmission system (69–138 kV), which is then further stepped down to lower voltage distribution systems (2–35 kV) and eventually to customers at household voltage levels. The loads shown in Figure 1.1 represent the power consumption by households, office buildings, commercial centers, and factories. A distribution system is typically a radial network like the roots of a tree that have no connections to other parts of the network, that is, they do not form loops.

This central generating station model is the basis of a traditional vertically integrated utility company in which a utility company has a monopoly of serving loads in a region and is also responsible for building generators to secure the energy supply and transmission systems for bringing the supply to the distribution systems. The transmission system is needed as the generating stations are not necessarily in the company's service area. As the transmission systems of neighboring power companies become intertwined, power pools are formed, either in a single state (like New York) or multiple states (like New England), to oversee the security of such power transfer across long distances.

Recently many US power grids, under FERC Order 888 to promote wholesale energy competition through open access,[2] have been deregulated, thus separating the traditional vertically integrated utility companies into independent generation, transmission, and distribution operators. Still, the central generating station model remains unchanged, as the distribution operators, on behalf of the loads, have to secure supplies from the large generating units, except now the energy clearing prices are settled on an open electricity market, instead of within a single company. Thus power system dynamics consideration and analysis methods, such as those discussed in this book, are still valid in the deregulation environment.

Power systems are complex because they are interconnected and constantly being expanded. The US eastern power system consists of an interconnection of the power grids in all the states east of the Rocky Mountains and northeast of Texas, as well as several Canadian provinces. Models as large as 50,000 buses have been built such that the dynamics of the systems can be studied in great detail.

Given such complexity, a reader may ask what is the best way to understand the operation of such a large system. Fortunately, the design of power systems is hierarchical. The power from the remote large generators will be supplied on high-voltage transmission systems, not low-voltage distribution systems. For example, the loads served from Bus 2 in Figure 1.1 are supplied by Generators A and B. If Generator A is out for service, the power would come from Generator B by rerouting the power flow on the transmission

2 See Federal Energy Regulatory Commission (FERC) website: http://www.ferc.gov/legal/maj-ord-reg/land-docs/order888.asp.

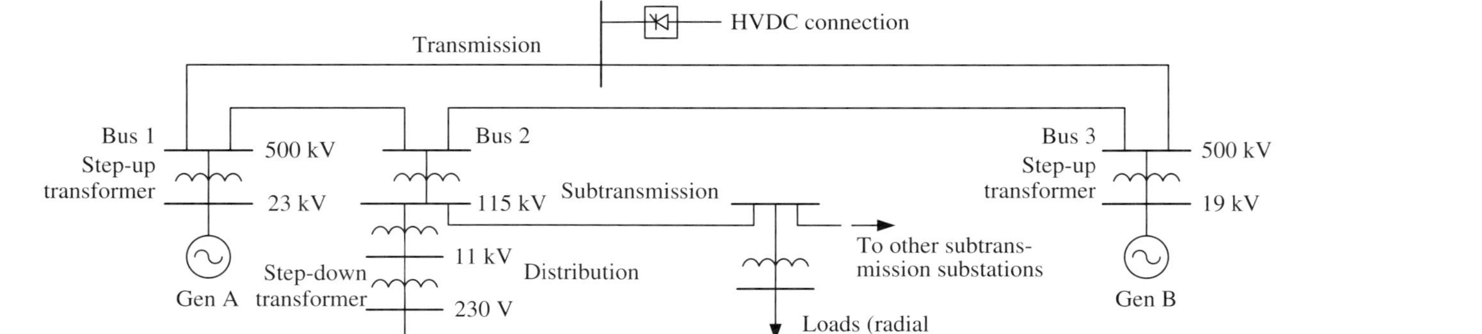

Figure 1.1 Central generation centric transmission and distribution network structure.

system. Some of the power flow may find a path through the subtransmission systems, but the amount should be small, as the effective impedance through a lower-voltage path is much higher. If this is not the case, some severe consequences such as voltage collapse may occur. Thus in the analysis of the stability of a generator or a group of generators subject to a fault, conceptually one can analyze a Single-Machine Infinite-Bus (SMIB) system to investigate critical clearing time, and a two-area system for the damping of interarea swing modes. Loads can be represented by a constant-impedance model before considering a more complex non-conforming model. A reader will find that there are many examples in the text using these two systems. Once mastered, these concepts can be applied to understand the results of power flow solutions or time responses obtained from dynamic simulations of very large realistic power systems.

1.2.2 Renewable Generation

Since the 1990s, renewable generation has been expanding at a rapid pace in many countries. Due to technology improvement, cost reduction, and government incentives for green energy, first wind-turbine generators (WTGs) were installed, which were followed by photo-voltaic (PV) systems. Modern WTGs are mostly 1–3 MW land-based units using doubly-fed induction generator (DFIG) technology and 4 MW and above off-shore units based mostly on full converter-based technology. PV plants are much smaller in power output, ranging from several kilowatts for units on the rooftops of single-family homes to much larger megawatt units, such as the 32-MW solar power plant on Long Island.[3]

As the percentage of renewable source generation becomes higher, the traditional central generating station model is being disrupted. First, the amount of generator inertias will become lower as some of the less efficient fossil generating units are displaced. PV plants have no inertia and the wind-turbine rotor and generator inertias are mostly hidden from the grid by the power electronic interface of the WTGs. Second, depending on the grid code, reactive power supply can be a challenge. On the other hand, the power electronics interface on renewable resources may be used productively for control.

The disruption to the distribution systems by renewable resources, mostly rooftop PV systems, may be more severe. Traditional distribution systems are designed with uni-directional power flowing from distribution transformers to homes. However, with high concentration, rooftop units can have the capability of supplying power back to a distribution transformer on sunny days. Such bi-directional flow complicates the design of relays and reactive power compensation. In addition, electric vehicle charging can double or triple the power consumption of a household, requiring increased transformer ratings and reactive power compensation capacity.

WTGs, because of their larger sizes and installation as wind farms, will have an impact on power system dynamics. Chapter 15 of this book discusses the modeling of WTGs. However, PV systems, because of their smaller size, faster dynamics, and limited impact on the transmission systems, will not be covered here.

3 See http://www.bnl.gov/SET/LISF.php.

1.2.3 Smart Grids

A further deviation from the central generation model is the establishment of smaller grid operating models that can take on a variety of forms, with designations like community grid and micro-grid. The main concept of these smart grid models is some form of self supply (local generation), via PV systems, energy storage, and small conventional diesel units, coupled to demand response or control, utilizing loads such as heating, ventilation, and air-conditioning (HVAC) systems. In normal operating conditions, these smart grids are supplied by the main power grid. When energy prices are high, these grids will reduce their intake of external power. In case of an emergency disrupting the main grid, these smart grids can operate as a standalone system supporting only the essential consumption. Smart grid models are beyond the scope of this book. Interested readers can find relevant information in many magazine articles and textbooks [1].

1.3 Time Scales

The complexity of a power system lies not only in having large interconnected control regions covering sometime immense areas, but also in the time scales of dynamics, measurements, control, and operating regimes. Figure 1.2 is a summary of the time spans of power system components and operation.

1.3.1 Dynamic Phenomena

The dynamic phenomena part of Figure 1.2 originates from [2] and can be found in various forms in many power system textbooks [3]. Lightning and switching surges are uncontrolled micro-second phenomena on transmission and distribution systems [4]. Stator transients and subsynchronous resonance are millisecond phenomena in synchronous generators. An analytical treatment of stator transients can be found in [3]. Subsynchronous oscillations are covered in Chapter 12 on turbine generators. Transient stability denoting the dynamics determined by the swing equations of synchronous generators interacting through the transmission network covers a time span from fractions of a second to tens of seconds. Governor dynamics and load frequency regulation involving the control of turbine valves and generator setpoints are on a time scale of seconds to minutes. Boiler dynamics involving steam generation will involve an even longer time span. In this text, such dynamics which require an understanding of thermal cycles will not be discussed. Here the effect of steam flow is represented by turbine time constants of several seconds. Thus this textbook covers dynamics that are important in the range of fractions of a second to 20–30 seconds, as is the case with books such as [3, 5–7].

1.3.2 Measurements and Data

Power system data consist of both analog measurements, such as 3-phase voltages and currents, and digital statuses of circuit breakers and switches. With the advent of inexpensive and yet powerful microprocessors, much traditional equipment based on analog measurements has been converted to use digital data. For example, digital relays with sampling rates of 2.8 and 5.6 kHz (for 60 Hz systems) are commonplace. Substation

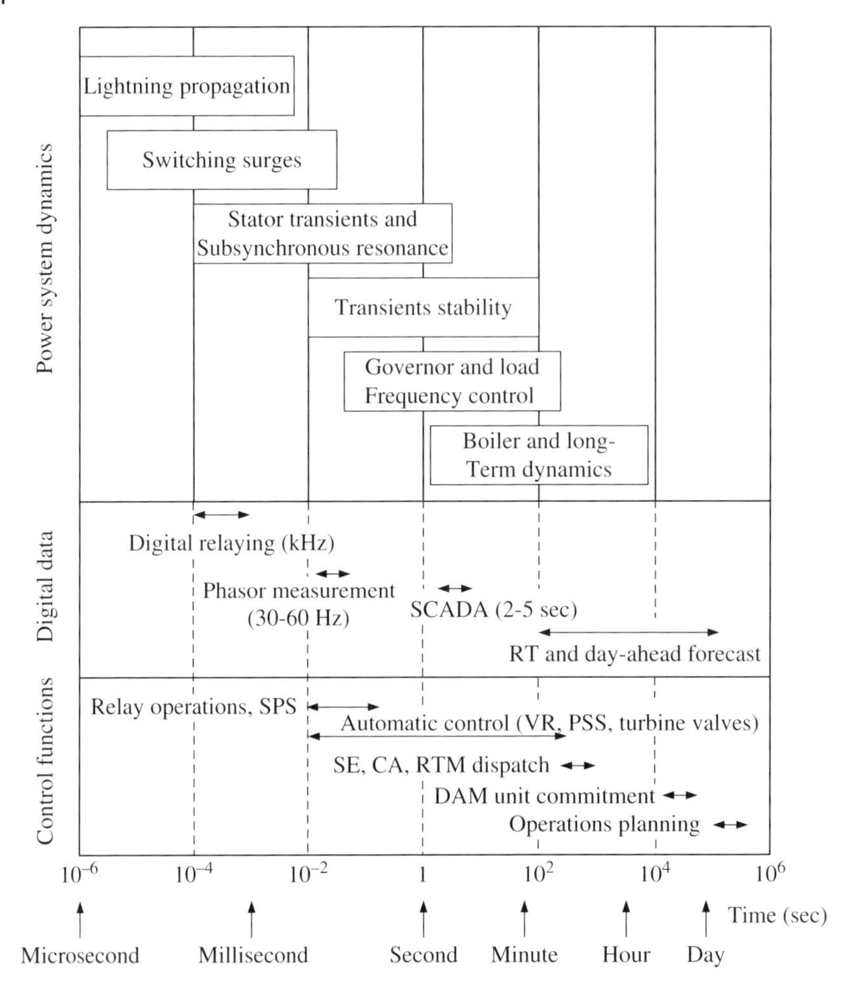

Figure 1.2 Power system time scales for dynamics, data, and control functions. SPS, special protection schemes; VR, voltage regulator; PSS, power system stabilizer; SE, state estimation; CA, contingency analysis; RTM, real-time market; DAM, day-ahead market. The dynamics part is from [3].

SCADA data (voltage, current, and power) are sampled every 2–5 seconds and sent via the internet for use in power system control centers. With the increased penetration of wind and solar power plants, some control centers are supplementing their operations with real-time and day-ahead wind and insolation forecast data obtained from complex atmospheric computer simulation models.

Perhaps the data most relevant for power system dynamics and control are voltage and current phasor data obtained from phasor measurement units (PMUs), which provide digital data at 30/60 samples per second for 60 Hz systems (50 samples per second for 50 Hz systems) [8]. Machine swing dynamics can be readily visible in PMU data. When the PMUs covering a wide region are synchronized via the GPS timing signal, the propagation of a disturbance across a power system is clearly observable from the synchrophasor data. Because a chapter on synchrophasor has already been provided in [3], the topic is not repeated here.

1.3.3 Control Functions and System Operation

Controls in power systems can be broadly grouped into two types. The first type is fast closed-loop control that needs to be implemented automatically. These controls include for active power control, frequency regulation by turbine valve governing and high-voltage direct current (HVDC) system power regulation [9], for reactive power control, voltage control by excitation systems and a flexible AC transmission system (FACTS) [10], and for damping control, supplementary signals from power system stabilizers. These topics are covered in this text.

Power system operation in control centers occurs at a much slower time scale and is also automated, except that additional discrete control inputs are computed separately. For example, real-time dispatch based on the state estimator solution and contingency analysis computes new generator setpoints every 5–10 minutes, which are then issued to the generators. Energy bids in the day-ahead market are normally submitted 12–36 hours ahead so that an independent system operator can secure and schedule generators for reliable operation. In addition, operation engineers will determine the scheduling of seasonal maintenance of transmission lines and generators to ensure normal operations will not be disrupted. A reader can find a good coverage of these topics in [11] and control center operating manuals on the websites of independent system operators.[4]

1.4 Organization of the Book

This book is organized into three parts. Part I consists of Chapters 2 to 6. These chapters cover power flow, voltage stability, and dynamic power system models based on the classical model of synchronous machines. Stability concepts, methods for power system simulation, and linear model generation are discussed. This part of the book may also be suitable for an advanced undergraduate course for students who have taken the introductory power system analysis course [12].

Part II consists of Chapters 7 to 12. They contain detailed materials on synchronous machine modeling, excitation systems and voltage regulators, power system stabilizer and damping control design, load models, and turbine-governor models and frequency control. These chapters start with a description of the equipment and its operating principles. They are then followed by modeling and analysis: steady-state operating characteristics, dynamic models, initialization (steady-state operating point computation), and dynamic simulation. With a good understanding of these models, a user of commercial power system simulation programs can literally "think" through the dynamics computation.

Part III consists of the last four chapters and can be considered as advanced topics, with contents on HVDC systems, FACTS, wind energy, and power system model reduction. At the Rensselaer Polytechnic Institute, these chapters plus the synchrophasor topic are offered as a second graduate course on power system dynamics. These chapters can also be used to supplement individualized courses on these topics, with the intention of showing how the models and techniques are used in a dynamic simulation context.

4 For example, the New York power system operation manuals can be found on http://www.nyiso.com.

Examples and problems are an important part of this book and serve two purposes. Some of them serve the academic purpose of illustrating analytical expressions with numerical values. Some others are motivated by practical systems. These problems can be quite interesting and perplexing at first. Furthermore, they illustrate that even seemingly complex situations can be solved by some straightforward calculations.

As this is a book on power system computation, students in a course using this book are expected to perform dynamic simulations. Those students without access to commercial power system software tools can use several free MATLAB®-based tools, such as the Power System Toolbox [13]. The computer files to run the examples can be downloaded from the Wiley website for this book.

Part I

System Concepts

2

Steady-State Power Flow

2.1 Introduction

Steady-state power flow, also known as loadflow, is the determination of the steady-state flow of active and reactive power from the generators to the loads in a positive-sequence network. A power flow solution represents an equilibrium operating condition of the power network. The assumptions made in computing the power flow solution include the following:

- The generator bus voltages are fixed at desired levels as they are supported by adequate reactive power supplied from the generators.
- All generator active power outputs are fixed at desired levels, except for one generator, which is known as the swing or slack bus, whose generation is also needed to make up for transmission losses.
- The active and reactive power consumptions of loads are known and given.

Power flow solutions are used in many applications, including ensuring all line flows below thermal limits, voltage stability studies (Chapter 3), and the initialization of dynamic simulation to study the impact of disturbances (Chapter 4). For large interconnected power systems, it is not unusual for a power engineer to solve power flow solutions for systems with more than 50,000 buses. Commercial power flow software packages using the Newton-Raphson method with optimized sparse factorization methods can readily handle these large data sets.

This chapter discusses the power flow formulation and solution process. The topics include calculating transmission-line active and reactive power flow between the buses, setting up the admittance matrix of a power network, formulating the power flow problem, and the Newton-Raphson algorithm for solving nonlinear power flow equations. It also discusses more advanced topics on using sparse factorization to minimize the storage and computation effort of the Newton-Raphson method, and performing power flow for multiple power control regions with specified interface flows, which is commonly known as multi-area power flow.

Power System Modeling, Computation, and Control, First Edition. Joe H. Chow and Juan J. Sanchez-Gasca.

Companion website: www.wiley.com/go/chow/power-system-modeling

2.2 Power Network Elements and Admittance Matrix

As the power system transmission and distribution network is responsible for delivering the power supply from the generators to the loads, the computation of the power flow in a large power network requires efficient organization of the power network by setting up an admittance matrix. This power network admittance matrix relating the bus voltages and line currents can be developed sequentially by adding one branch (either a transmission line or a transformer) at a time, as the branch data is being read in serially by a computer from a power flow data file or database.

2.2.1 Transmission Lines

In power flow formulation, a transmission line connecting two buses is represented by a π-equivalent [12], as shown in Figure 2.1, connecting Buses i and j. This transmission line model, obtained from considering conductor types and tower configurations, is defined by three parameters: resistance R_{ij}, reactance $X_{ij} = \omega L_{ij}$, and line susceptance (charging) $B_{ij} = \omega C_{ij}$, where L_{ij} is the line inductance, C_{ij} is the line capacitance, $\omega = 2\pi f$, and f is the nominal system frequency in Hz.[1] As shown in Figure 2.1, the voltage phasors at Buses i and j are denoted by $\tilde{V}_i$ and $\tilde{V}_j$, respectively. The current flowing from Bus i into the line is denoted by the current phasor $\tilde{I}_{ij}$ and that flowing from Bus j into the line by $\tilde{I}_{ji}$.

The series admittance is given by

$$Y_{ij} = 1/Z_{ij} = 1/(R_{ij} + jX_{ij}) \tag{2.1}$$

and the shunt susceptance at each bus is $B_{ij}/2$. Thus the current $\tilde{I}_{ij}$ flowing from Bus i into the line is given by

$$\begin{aligned} \tilde{I}_{ij} &= \text{current in series reactance } + \text{ current in shunt susceptance} \\ &= Y_{ij}(\tilde{V}_i - \tilde{V}_j) + \tilde{V}_i \cdot j\frac{B_{ij}}{2} = \left(Y_{ij} + j\frac{B_{ij}}{2}\right)\tilde{V}_i - Y_{ij}\tilde{V}_j \end{aligned} \tag{2.2}$$

Similarly, the current $\tilde{I}_{ji}$ is given by

$$\tilde{I}_{ji} = \left(Y_{ij} + j\frac{B_{ij}}{2}\right)\tilde{V}_j - Y_{ij}\tilde{V}_i \tag{2.3}$$

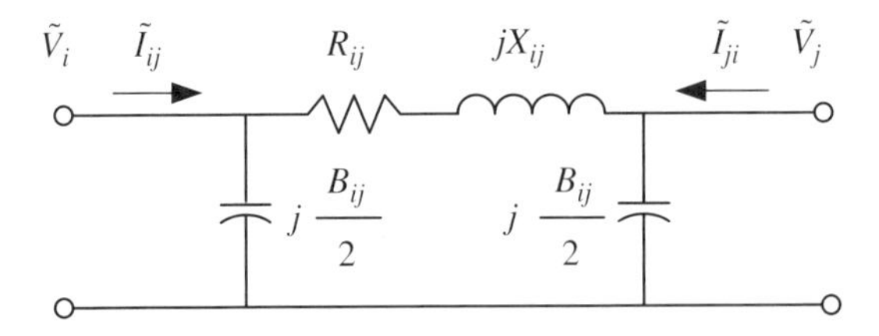

Figure 2.1 Line parameters, bus voltages, and line currents in a π-equivalent transmission line.

1 Note that all three parameters are in general positive. An exception is when the transmission line is used to represent a series capacitor, in which case $X_{ij} < 0$. For a long transmission line going over different terrains, it may be modeled with multiple π sections.

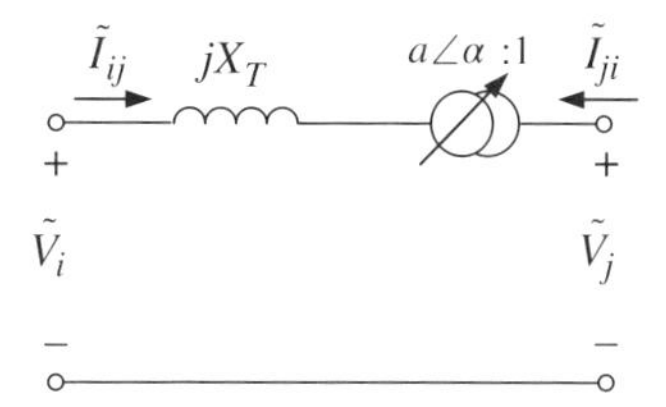

Figure 2.2 Transformer parameters, bus voltages, and line currents.

These two equations can be combined in matrix form as

$$\begin{bmatrix} Y_{ij} + j\frac{B_{ij}}{2} & -Y_{ij} \\ -Y_{ij} & Y_{ij} + j\frac{B_{ij}}{2} \end{bmatrix} \begin{bmatrix} \tilde{V}_i \\ \tilde{V}_j \end{bmatrix} = \begin{bmatrix} \tilde{I}_{ij} \\ \tilde{I}_{ji} \end{bmatrix} \quad \Rightarrow \quad Y\tilde{V} = \tilde{I} \tag{2.4}$$

where Y is known as the admittance matrix, $\tilde{V}$ is the vector of voltage phasors, and $\tilde{I}$ is the current flow vector. Note that Y is symmetric.

Recall that a transmission line can also be represented by a T-equivalent model [12]. However, the T-equivalent model has a node in the middle of the transmission line and thus would require a 3×3 admittance matrix for its representation. Hence it is less suitable for use in large power systems.

2.2.2 Transformers

Power transformers are primarily used in power systems to interconnect transmission and distribution systems at different voltage levels. Some phase-shifting transformers [14], consisting of a shunt transformer and a series transformer, are also used to change the phase of the current flow so that the power flow on a transmission line can be controlled to the extent allowed by the physical equipment.

A transformer model connecting Buses i and j is shown in Figure 2.2. It consists of a transformer reactance X_T, and a tap ratio a for a tap-changing transformer or a phase α for a phase-shifting transformer.[2] In general the resistance of a transformer is small and thus neglected. However, transformers with nonzero resistance can be readily accommodated. For shunt transformers, X_T is the leakage reactance, which is typically 12–15% on the transformer rating. The derivation of the equivalent circuit of a power transformer is discussed in Appendix 2.A.

The bus voltages and line currents in a transformer branch are related by

$$\begin{bmatrix} Y_T & -\tilde{a}Y_T \\ -\tilde{a}^* Y_T & a^2 Y_T \end{bmatrix} \begin{bmatrix} \tilde{V}_i \\ \tilde{V}_j \end{bmatrix} = \begin{bmatrix} \tilde{I}_{ij} \\ \tilde{I}_{ji} \end{bmatrix} \quad \Rightarrow \quad Y\tilde{V} = \tilde{I} \tag{2.5}$$

where $Y_T = 1/(jX_T)$ and

$$\tilde{\alpha} = ae^{j\alpha} = a\angle\alpha \tag{2.6}$$

If $\alpha \neq 0$, Y is no longer symmetric.

2 Power system software allows a user to put the transformer tap on either side of the transformer, depending on the physical equipment.

2.2.3 Per Unit Representation

In a practical power system with various voltage levels, the formulation of the voltage–current equations (2.4) and (2.5) may be cumbersome if absolute values of voltages, currents, and impedances are used. Thus to reduce such complexity, the power industry practice is to normalize the voltages with respect to their base values and the power outputs and impedances with respect to a system power rating. The resulting variable and parameter values are then converted to per unit (pu) and are used in power flow data sets.

A per-unit system is defined by three base quantities, namely, the base voltage, power, and frequency:

- The base voltage V_{base}, nominally in kV, is not a single value. It is defined by the nominal voltage value of a substation or transmission line. For example, an actual voltage value of 510 kV on a 500 kV substation is normalized to $510/500 = 1.02$ pu.
- The base power (VA_{base}) is defined in terms of VA, kVA, or MVA, usually in convenient orders of the power of 10. In a Single-Machine Infinite-Bus (SMIB) system, the generator MVA base can be taken to be the system MVA base.
- The base frequency can be given by f_{base} in Hz (cycles per second) or ω_{base} in rad/s (radians per second). The nominal system frequency is mostly 50 or 60 Hz, or 314 or 377 rad/s as calculated from $\omega_{\text{base}} = 2\pi f_{\text{base}}$.

There are also two derived base quantities:

- The base current (I_{base}) is computed as

$$I_{\text{base}} = \frac{\text{VA}_{\text{base}}}{V_{\text{base}}} \tag{2.7}$$

- The base impedance (Z_{base}) is computed as

$$Z_{\text{base}} = \frac{V_{\text{base}}}{I_{\text{base}}} = \frac{(\text{kV}_{\text{base}})^2}{\text{MVA}_{\text{base}}} \tag{2.8}$$

  Consequently, the base admittance is $Y_{\text{base}} = 1/Z_{\text{base}}$.

If a power system model has several voltage levels, then I_{base} and Z_{base} will be different for each voltage level. In particular, I_{base} will be smaller and Z_{base} will be larger at higher voltage levels given the same base power. In a power flow data set, it is common to provide the pu values of branch impedances on the system MVA base at relevant voltage levels without requiring the computer program to calculate the pu impedance values.

The per-unitized process can be applied to both single-phase and 3-phase systems, with the appropriate base quantities shown in Table 2.1.

2.2.4 Building the Network Admittance Matrix

Using the 2×2 Y matrix entries in (2.4) and (2.5), the Y matrix for the overall power system can be built progressively as each of the branches is processed from the branch data set. Initially the Y matrix contains all zeros, if a full storage matrix format is used, or is empty, if the sparse matrix storage format is used.

As a new branch connecting Buses i and j is processed, Y_{ij} is added to the (i, j) and (j, i) entries, and $jB_{ij}/2$ is added to the (i, i) and (j, j) entries of the overall Y matrix, with

Table 2.1 Base quantities of single-phase and 3-phase systems.

Single-phase	Three-phase
MVA_{base} (1ϕ)	MVA_{base} (3ϕ)
kV_{base} (L-N)	kV_{base} (L-L)
$I_{\text{base}} = \text{VA}_{\text{base}}/V_{\text{base}}$	$I_{\text{base}} = \text{VA}_{\text{base}}/(\sqrt{3}V_{\text{base}})$
$Z_{\text{base}} = (\text{kV}_{\text{base}})^2/\text{MVA}_{\text{base}}$	$Z_{\text{base}} = (\text{kV}_{\text{base}})^2/\text{MVA}_{\text{base}}$

appropriate scaling factor modifications for off-nominal transformer taps. The addition operation is used as there may be parallel branches between these two buses. As a result, the network admittance matrix Y has the following characteristics:

1) If Buses i and j are not directly connected, then the (i, j) entry of Y is either zero or empty in sparse storage format.
2) The (i, j) entries of Y are the sum of the admittances of all the direct connections between Buses i and j.
3) The self (i, i) entries of Y is the sum of the admittances of direct connections of Bus i to all the other buses, and all the shunt components (line susceptance, shunt capacitors, and shunt reactors) representable as linear, passive circuit elements.

There are two ways to store the complex Y matrix:

1) As a full complex $N_B \times N_B$ matrix, where N_B is the number of buses: This is not practical for large power networks as the Y matrix will contain many zero entries, which take up a large amount of unnecessary storage.
2) As a sparse matrix, in which each non-zero (i, j)th entry of Y is stored as a three-tuple (i, j, Y_{ij}): This is the preferred storage format for large systems. A 1,000-bus system with an average of four connections per bus will result in a sparse matrix with approximately 5,000 non-zero entries, instead of a million entries in the full-storage format.

Example 2.1: Building an admittance matrix

Consider a radial five-bus section of a power system shown in Figure 2.3, in which there are no injections (that is, loads) at Buses 2, 3, and 4. Line 1-2 is a generator step-up transformer,[3] Line 2-3 is modeled with line charging, Line 3-4 represents a series capacitor compensation,[4] and Line 4-5 is a short line with line charging neglected. Bus 5 is connected to other buses, which are not shown here but are represented by a current injection $\tilde{I}_5$. The generator output is modeled as a current injection $\tilde{I}_1$. Find the admittance matrix of this five-bus section.

3 Normally a generator step-up transformer has a fixed ratio, as voltage regulation is performed by the generator excitation system. Here the adjustable tap is used as an example.

4 The admittance Y_{34} is the inverse of the line impedance Z_{34}. For a series capacitor, the imaginary part of Z_{34} is negative.

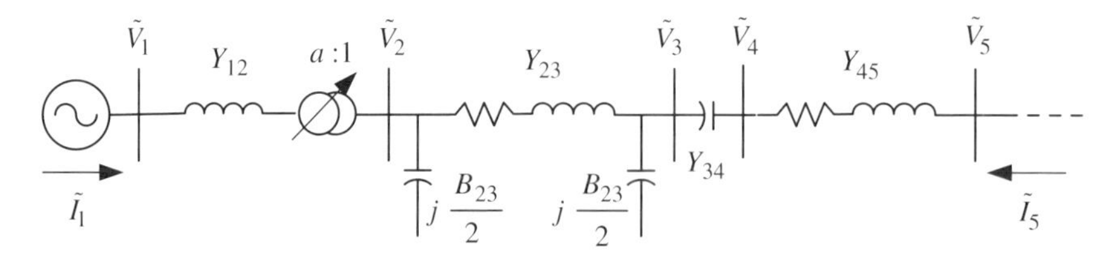

Figure 2.3 Five-bus system example.

Solutions: The network voltage–current equation $Y\tilde{V} = \tilde{I}$ is given by, following the earlier discussion,

$$\begin{bmatrix} Y_{11} & -aY_{12} & 0 & 0 & 0 \\ -aY_{12} & Y_{22} & -Y_{23} & 0 & 0 \\ 0 & -Y_{23} & Y_{33} & -Y_{34} & 0 \\ 0 & 0 & -Y_{34} & Y_{44} & -Y_{45} \\ 0 & 0 & 0 & -Y_{45} & Y_{55} \end{bmatrix} \begin{bmatrix} \tilde{V}_1 \\ \tilde{V}_2 \\ \tilde{V}_3 \\ \tilde{V}_4 \\ \tilde{V}_5 \end{bmatrix} = \begin{bmatrix} \tilde{I}_1 \\ \tilde{I}_2 \\ \tilde{I}_3 \\ \tilde{I}_4 \\ \tilde{I}_5 \end{bmatrix} = \begin{bmatrix} \tilde{I}_1 \\ 0 \\ 0 \\ 0 \\ \tilde{I}_5 \end{bmatrix} \tag{2.9}$$

where the direction of current injection $\tilde{I}$ is leaving the node (bus). Because there are no parallel branches, the non-diagonal entries of Y are the same as those corresponding values shown in Figure 2.3. The diagonal entries of Y are the sum of the branch admittances and the line susceptances, given by

$$Y_{11} = Y_{12}, \quad Y_{22} = aY_{12} + Y_{23} + j\frac{B_{23}}{2}, \quad Y_{33} = Y_{23} + j\frac{B_{23}}{2} + jY_{34}$$
$$Y_{44} = Y_{34} + Y_{45}, \quad Y_{55} = Y_{45} \tag{2.10}$$

In addition, there are no injections from Buses 2, 3, and 4, and thus the current vector $\tilde{I}$ has only two non-zero values. ■

2.3 Active and Reactive Power Flow Calculations

From the network equation $Y\tilde{V} = \tilde{I}$, the current injections $\tilde{I}$ can be directly computed if the bus voltage $\tilde{V}$ at each of the buses is known. This is indeed the case in dynamic simulation when $\tilde{V}$ is the internal voltage of a generator. For non-source buses, the load is converted into an admittance (for a constant-impedance load) and added to the corresponding diagonal entry of the overall admittance matrix. This process is discussed in Chapter 4.

The network equation $Y\tilde{V} = \tilde{I}$, however, cannot be used directly for steady-state power flow calculation in which the loads are specified as fixed active and reactive power consumption $P + jQ$, and not as fixed current injections. Expressing $P + jQ = -\tilde{V}\tilde{I}^*$ and replacing the right-hand side of (2.4) results in the expression[5]

$$Y\tilde{V} = -\text{vec}\left(\frac{P_i + jQ_i}{\tilde{V}_i}\right)^* = -\text{vec}\left(\frac{(P_i + jQ_i)^*}{\tilde{V}_i^*}\right) = \tilde{I} \tag{2.11}$$

where the "vec" notation means a vector formed by the entry shown in the parentheses.

5 The "vec" notation denotes the stacking of individual components into a vector.

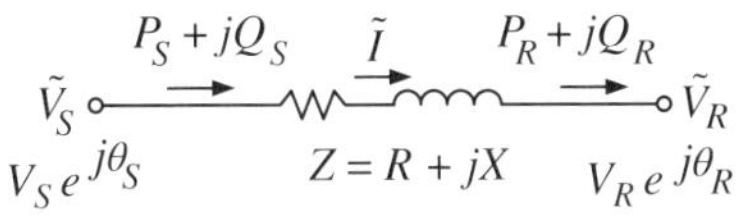

$P_S + jQ_S$ $\tilde{I}$ $P_R + jQ_R$

$\tilde{V}_S$ $\tilde{V}_R$

$V_S e^{j\theta_S}$ $Z = R + jX$ $V_R e^{j\theta_R}$

Figure 2.4 Single branch showing active and reactive power at the sending and receiving buses.

Note that the right-hand-side of the equation carries a negative sign because P and Q are load power consumption, and $\tilde{I}$ is current injection leaving the bus into the power network.

To solve the power flow, it is necessary to compute the power flow between two connected buses, as shown in Figure 2.4, where the subscript "S" denotes the sending end, and the subscript "R" denotes the receiving end. Note that this designation is arbitrary as it is acceptable for P_S to be negative.

The line current is, with $\varphi = \tan^{-1}(X/R)$ (which is close to 90° if $R \ll X$),

$$\tilde{I} = \frac{\tilde{V}_S - \tilde{V}_R}{Z} = \frac{V_S e^{j\theta_S} - V_R e^{j\theta_R}}{|Z| e^{j\varphi}} = \frac{V_S}{|Z|} e^{j(\theta_S - \varphi)} - \frac{V_R}{|Z|} e^{j(\theta_R - \varphi)} \tag{2.12}$$

The apparent or complex power S_S flowing from the sending end into the transmission line is

$$\begin{aligned} S_S = P_S + jQ_S = \tilde{V}_S \tilde{I}^* &= V_S e^{j\theta_S} \frac{V_S}{|Z|} e^{-j(\theta_S - \varphi)} - V_S e^{j\theta_S} \frac{V_R}{|Z|} e^{-j(\theta_R - \varphi)} \\ &= \frac{V_S^2}{|Z|} e^{j\varphi} - \frac{V_S V_R}{|Z|} e^{j(\theta_S - \theta_R + \varphi)} \end{aligned} \tag{2.13}$$

Thus the active and reactive power can be obtained, respectively, as

$$\begin{aligned} P_S = \text{Real}(S_S) &= \frac{V_S^2}{|Z|} \cos\varphi - \frac{V_S V_R}{|Z|} \cos(\theta_S - \theta_R + \varphi) \\ &= \frac{V_S^2}{|Z|} \cos\varphi + \frac{V_S V_R}{|Z|} \sin\left(\theta_S - \theta_R + \varphi - \frac{\pi}{2}\right) \end{aligned} \tag{2.14}$$

$$\begin{aligned} Q_S = \text{Imag}(S_S) &= \frac{V_S^2}{|Z|} \sin\varphi - \frac{V_S V_R}{|Z|} \sin(\theta_S - \theta_R + \varphi) \\ &= \frac{V_S^2}{|Z|} \sin\varphi - \frac{V_S V_R}{|Z|} \cos\left(\theta_S - \theta_R + \varphi - \frac{\pi}{2}\right) \end{aligned} \tag{2.15}$$

Similarly, the expressions for the power flow (P_R,Q_R) going to the receiving end can be derived. Define the active power losses as $P_S - P_R$ (due to R) and the reactive power losses as $Q_S - Q_R$ (due to X). Both active and reactive power losses will be higher for higher line active power flow P_S.

To obtain some additional insights on the expressions (2.14) and (2.15), assume that R is small. Hence $\varphi \simeq \pi/2$ and $|Z| \simeq X$. Then (2.14) and (2.15) simplify, respectively, to

$$P_S \simeq \frac{V_S V_R}{X} \sin(\theta_S - \theta_R), \quad Q_S \simeq \frac{V_S^2}{X} - \frac{V_S V_R}{X} \cos(\theta_S - \theta_R) \tag{2.16}$$

From the P_S expression in (2.16), the active power flow P_S is positive if $\theta_S - \theta_R > 0$ (assuming an inductive line with $X > 0$), that is, active power flows from a higher bus

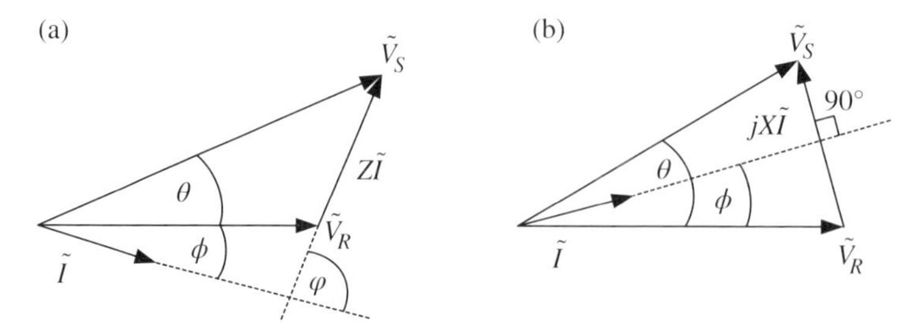

Figure 2.5 V–I phasor diagrams: (a) $V_S > V_R$ and (b) $V_S = V_R$.

voltage angle to a lower bus voltage angle. From the Q_S expression in (2.16), if $V_S > V_R$, then

$$\frac{V_S^2 - V_S V_R \cos(\theta_S - \theta_R)}{X} > 0 \tag{2.17}$$

as the cosine function is at most equal to 1. Thus reactive power flows from a higher bus voltage magnitude to a lower bus voltage magnitude. Of course, both buses may be supplying reactive power into the line to support active power transfer.

In visualizing power flow, it is useful to develop a graphical representation of the voltage and current phasors, as shown in Figure 2.5. Figure 2.5.a represents a case with non-zero resistance $R \neq 0$ such that $\varphi < 90°$, where $\theta = \theta_S - \theta_R$ is the load angle and ϕ is the power factor angle. Note that $\tilde{I}$ lags $\tilde{V}_R$, that is, the load at the receiving end is inductive. Also the sending end voltage magnitude is higher, and thus the sending end bus is supplying reactive power to the receiving end bus. Figure 2.5.b is a special case in which $V_R = V_S$ in magnitude and $R = 0$. In this case, $\tilde{I} \perp jX\tilde{I}$, $\phi = \theta/2$, and $Q_S = -Q_R = XI^2/2$, that is, equal var (reactive power) contributions are injected into the line from both buses.

Example 2.2: Power flow calculation

By looking at the bus voltage magnitudes and angles, one can get a feel of which directions the active power and reactive power are flowing in a power system. For a 345 kV transmission line with the parameters $R = 0.004$ pu, $X = 0.04$ pu, and $B = 0.68$ pu connecting Buses 1 and 2, calculate P_i and Q_i, $i = 1, 2$, and the active and reactive power losses, for several cases of $\tilde{V}_1$ and $\tilde{V}_2$ given in Table 2.2. Take P_1 and Q_1 as the power leaving Bus 1, and P_2 and Q_2 as the power going into Bus 2. Compare the power flow for the different cases.

Solutions: The current leaving Bus 1 is given by

$$\tilde{I}_1 = \frac{\tilde{V}_1 - \tilde{V}_2}{R + jX} + j\frac{B}{2}\tilde{V}_1 \tag{2.18}$$

and the current arriving at Bus 2 is given by

$$\tilde{I}_2 = \frac{\tilde{V}_1 - \tilde{V}_2}{R + jX} - j\frac{B}{2}\tilde{V}_2 \tag{2.19}$$

Table 2.2 Various cases of bus voltages for Example 2.2.

Case	$\tilde{V}_1$ (pu)	$\tilde{V}_2$ (pu)
a	$1.03\angle 8^\circ$	$1.00\angle 0^\circ$
b	$1.00\angle 8^\circ$	$1.00\angle 0^\circ$
c	$1.05\angle 8^\circ$	$1.00\angle 0^\circ$
d	$1.03\angle 16^\circ$	$1.00\angle 0^\circ$

Table 2.3 Solutions for Example 2.2 (all quantities in pu).

Case	P_1	Q_1	P_2	Q_2	P_{loss}	Q_{loss}
a	3.6495	0.2974	3.5977	0.4796	0.0519	−0.1822
b	3.4690	−0.4436	3.4208	−0.2454	0.0482	−0.1982
c	3.7724	0.8159	3.7156	0.9630	0.0568	−0.1471
d	7.2026	0.6890	7.0029	−0.6078	0.1998	1.2968

From these expressions, the power flow on the transmission line can be readily computed and is shown in Table 2.3.

Several conclusions can be drawn from Table 2.3:

1) The active power transmitted is mostly dependent on the angular separation between the two buses, with the active power flowing from the bus with a larger angle.
2) The active power losses P_{loss} is a nonlinear function of the amount of power transferred, increasing quadratically as P is increased.
3) In Case a, reactive power is provided by Bus 1 but a higher amount arrives at Bus 2, as the line charging is supplying additional reactive power (note that $Q_{loss} < 0$). In Case b, as the voltage on Bus 1 is reduced to the Bus 2 voltage, the reactive power flow is reversed. On the other hand, in Case c, as the voltage on Bus 1 is further increased, Bus 1 is supplying more reactive power to Bus 2.
4) In Case d, as the active power flow is increased, both Buses 1 and 2 have to supply reactive power, as the reactive power losses have increased significantly.
5) Line charging contributes additional reactive power (Q_{loss} more negative) as the voltage at Bus 1 is increased.

■

2.4 Power Flow Formulation

The objective of a power flow is to determine the bus voltage and line current phasors in a power network to satisfy a given set of generation, load, and voltage conditions. At each bus, there are four quantities: the bus voltage magnitude and angle denoted by $\tilde{V} = Ve^{j\theta}$, and the bus generation and/or load: $P + jQ$. At a bus with zero injections,

Table 2.4 Power flow bus types and variables.

Bus type	V	θ	P	Q
PQ (load bus)	Computed	Computed	Specified	Specified
PV (generator)	Specified	Computed	Specified	Computed
Swing (slack) bus	Specified	$\theta = 0°$	Computed	Computed

$P = Q = 0$. In general, two of the four quantities are specified depending on the type of buses, and the other two quantities can be computed if the power flow problem is formulated consistently.

There are two main types of buses. A generator bus supplies a certain amount of active power and maintains the bus voltage magnitude at a certain desired value, and any bus that is not a generator bus is represented as a load bus. A bus with zero-injection is also known as a connection bus. Its voltage cannot be controlled, thus it has the same characteristic as a load bus.

Table 2.4 shows the various bus types and the known or specified quantities associated with them. Some explanations of the formulation in Table 2.4 are given below:

1) The P and Q values of a load bus are specified. Thus a load bus is also known as a PQ bus. With no means of varying the reactive power, its voltage cannot be controlled.
2) The P and V values of a generator bus are specified, as it is assumed that the turbine can provide the active power P and the generator excitation system can supply sufficient reactive power output Q to support the specified voltage at the generator bus. Thus a generator bus is also known as a PV bus.
3) The flow of active power in the power system is primarily determined by the relative bus voltage angles. As a result, one of the bus voltage angles should be fixed. Otherwise, there will be an infinite number of solutions for the angles so that no iterative algorithm will likely be able to converge. For convenience, the angle of one generator bus is fixed, which is defaulted to zero (or any arbitrary fixed value).
4) The bus voltage magnitude and angle of the swing bus are specified, but not its active power output P. This "slackness" is necessary for the swing generator to supply part of the total load and compensate for the active power losses due to the line resistance. Such losses are dependent on the voltage and current solutions and thus are not known ahead of time. Thus the swing bus is also known as the slack bus.
5) Each generator has an upper reactive power limit Q_{max} and a lower reactive power limit Q_{min}, which are functions of the generator P, Q capability curve (see Section 7.9). If the reactive power output Q at a generator bus in a converged solution is outside the maximum or minimum limit, then Q is set to the appropriate Q limit Q_ℓ and the PV bus becomes a PQ bus with $Q = Q_\ell$. The power flow solution is then repeated until no other generator Q limits are violated.

Power flow formulation involving high-voltage direct current (HVDC) systems is discussed in Chapter 13 and flexible AC transmission systems (FACTS) in Chapter 14.

2.5 Newton-Raphson Method

From Table 2.4, for a power system with N_L load buses and N_G generator buses, the solution variable set consists of N_L unknown bus voltage magnitude at the load buses, and $N_B - 1$ unknown bus voltage angle at all the buses except the swing bus, where $N_B = N_L + N_G$. These $N_B + N_L - 1$ unknown variables are solved from $N_B - 1$ active power injection balance equations at all N_B buses except the swing bus, and N_L reactive power injection balance equations at the N_L load buses.

This section describes the Newton-Raphson (NR) method for solving nonlinear power flow equations [16]. Note that the power flow equations used in this section do not include the effect of transformers, which is discussed in the next section.

Before discussing the NR method, it should be mentioned that the Gauss-Seidel (GS) method [17] was one of the first numerical methods proposed for iteratively solving power flow equations using the network equations. It is not commonly used in commercial power flow software anymore because its rate of convergence is only linear. However, some power flow packages may use the GS method to initiate the NR method. In addition, the Gauss method, which does not perform immediate updating as in the GS method, is readily amenable for parallel computing and thus is being considered for implementation on computers with massively parallel processors [18]. A reader interested to review the GS method is referred to [17] or power system analysis textbooks such as [12].

2.5.1 General Procedure

The NR method uses partial derivative information to iteratively solve for a set of nonlinear equations

$$f(x) = b \tag{2.20}$$

where $x \in R^N$ is the vector of unknowns, $f(x) \in R^N$ is the nonlinear vector function of x, and $b \in R^N$ is a constant vector

$$x = \begin{bmatrix} x_1 \\ \vdots \\ x_N \end{bmatrix}, \quad f = \begin{bmatrix} f_1 \\ \vdots \\ f_N \end{bmatrix}, \quad b = \begin{bmatrix} b_1 \\ \vdots \\ b_N \end{bmatrix} \tag{2.21}$$

such that (2.20) is a system of N nonlinear equations in N unknowns.

Suppose an initial guess $x^{(0)}$ of the solution is provided and the true solution is $x^{(0)} + \Delta x$, where

$$\Delta x = \begin{bmatrix} \Delta x_1 \\ \vdots \\ \Delta x_N \end{bmatrix} \tag{2.22}$$

Considering Δx as a perturbation, (2.20) can be expanded about $x^{(0)}$ as

$$f(x^{(0)} + \Delta x) = f(x^{(0)}) + \left.\frac{\partial f}{\partial x}\right|_{x=x^{(0)}} \Delta x + \frac{1}{2}\Delta x^T F|_{x=x^{(0)}} \Delta x + \text{ higher order terms} = b \tag{2.23}$$

where

$$\frac{\partial f}{\partial x} = J = \begin{bmatrix} J_{11} & J_{12} & \cdots & J_{1N} \\ \vdots & \vdots & \vdots & \vdots \\ J_{N1} & J_{N2} & \cdots & J_{NN} \end{bmatrix}, \quad J_{ij} = \frac{\partial f_i}{\partial x_j} \tag{2.24}$$

is the first derivative of f with respect to x and is known as the Jacobian matrix, and F is the second derivative and is known as the Hessian matrix.

If $\Delta x_1, \Delta x_2, \ldots, \Delta x_N$ are small, then $\Delta x_i \Delta x_j$ and higher order terms can be neglected. Thus the expansion (2.23) can be approximated by

$$f(x^{(0)}) + J(x^{(0)})\Delta x^{(1)} = b \quad \Rightarrow \quad J(x^{(0)})\Delta x^{(1)} = b - f(x^{(0)}) = \varepsilon_{\text{NR}}(x^{(0)}) \tag{2.25}$$

where ε_{NR} is the mismatch vector. For numerical stability, it is not recommended to solve for the linear system of equations (2.25) as $\Delta x^{(1)} = J^{-1}\varepsilon_{\text{NR}}(x^{(0)})$ by first computing J^{-1}, the inverse of J. In addition, in many applications, such as power systems, the Jacobian J is a sparse matrix, with many zero entries. The inverse J^{-1}, however, is a full matrix.

Instead, the proper way to solve (2.25) is to first perform an LU decomposition of $J = LU$, where L is a lower triangular matrix, and U is an upper triangular matrix [19]. Furthermore, the diagonal elements of L can be set to unity. More details on LU decomposition and sparse factorization are given in Appendix 2.B. Then at iteration k, with the $(k-1)$ iterate of x as $x^{(k-1)}$, the update equation

$$LU\Delta x^{(k)} = L\Delta y = \varepsilon_{\text{NR}}(x^{(k-1)}) \tag{2.26}$$

is solved in two stages:

1) Forward elimination: solve for Δy from $L\Delta y = \varepsilon_{\text{NR}}(x^{(k-1)})$.
2) Back substitution: solve for the update $\Delta x^{(k)}$ from $U\Delta x^{(k)} = \Delta y$.

Both steps are fast because L and U are triangular matrices.

At iteration k, the solution is updated as

$$x^{(k)} = x^{(k-1)} + \Delta x^{(k)} \tag{2.27}$$

The NR iteration continues until convergence is achieved when

$$|x^{(k)} - x^{(k-1)}| < \delta \tag{2.28}$$

where δ is a user-specified tolerance.

Near the solution, the convergence of the NR method is quadratic, as the higher order terms are very small, thus the method is very efficient. On the other hand, if the initial condition is poorly selected, then the NR method may not converge or may converge to an undesirable solution. The selection of initial conditions for solving power flow equations is discussed later.

2.5.2 NR Solution of Power Flow Equations

To apply the NR algorithm to the power flow problem, it is necessary to develop the $(N_B - 1) + N_L$ nonlinear power flow equations in the form of (2.20).

For each bus i, except for the swing bus, the power injection from the node into the system is

$$S_i = P_i + jQ_i = \tilde{V}_i \tilde{I}_i^* \tag{2.29}$$

From $Y\tilde{V} = \tilde{I}$ (2.4), it follows that (2.29) can be rewritten as

$$\sum_{j=1}^{N_B} Y_{ij}\tilde{V}_j = \tilde{I}_i \quad \Rightarrow \quad S_i = \tilde{V}_i\left(\sum_{j=1}^{N_B} Y_{ij}^* \tilde{V}_j^*\right) \tag{2.30}$$

This expression is expanded into

$$P_i + jQ_i = V_i e^{j\theta_i}\left(\sum_{j=1}^{N_B} |Y_{ij}| e^{-j\varphi_{ij}} V_j e^{-j\theta_j}\right) = V_i\left(\sum_{j=1}^{N_B} |Y_{ij}| V_j e^{j(\theta_i-\theta_j-\varphi_{ij})}\right) \tag{2.31}$$

that is,

$$P_i = V_i \sum_{j=1}^{N_B} |Y_{ij}| V_j \cos(\theta_i - \theta_j - \varphi_{ij}), \quad i = 1, 2, ..., N_B - 1 \tag{2.32}$$

$$Q_i = V_i \sum_{j=1}^{N_B} |Y_{ij}| V_j \sin(\theta_i - \theta_j - \varphi_{ij}), \quad i = 1, 2, ..., N_L \tag{2.33}$$

which are specified and φ_{ij} is the impedance angle of Line i-j. Equations (2.32) and (2.33) can be written in the form of $f(x) = b$ (2.20) as

$$f(x) = \begin{bmatrix} V_1 \sum_{j=1}^{N_B} |Y_{1j}| V_j \cos(\theta_1 - \theta_j - \varphi_{1j}) \\ V_2 \sum_{j=1}^{N_B} |Y_{2j}| V_j \cos(\theta_2 - \theta_j - \varphi_{2j}) \\ \vdots \\ V_{N_B-1} \sum_{j=1}^{N_B} |Y_{(N_B-1)j}| V_j \cos(\theta_{N_B-1} - \theta_j - \varphi_{(N_B-1)j}) \\ V_1 \sum_{j=1}^{N_B} |Y_{1j}| V_j \sin(\theta_1 - \theta_j - \varphi_{1j}) \\ \vdots \\ V_{N_L} \sum_{j=1}^{N_B} |Y_{N_L j}| V_j \sin(\theta_{N_L} - \theta_j - \varphi_{N_L j}) \end{bmatrix}, \quad b = \begin{bmatrix} P_1 \\ P_2 \\ \vdots \\ P_{N_B-1} \\ Q_1 \\ \vdots \\ Q_{N_L} \end{bmatrix} \tag{2.34}$$

where the unknown vector x is

$$x = \left[\theta_1 \; \theta_2 \cdots \theta_{N_B-1} \; V_1 \; V_2 \cdots V_{N_L}\right]^T \tag{2.35}$$

There are two options to set up the initial value $x^{(0)}$:

1) Cold start, in which all the unknown bus voltage magnitude values are set to 1 and all the unknown angle values are set to zero, as indicated in Table 2.5.
2) Warm start, in which the solution to a problem with a similar load condition is used as $x^{(0)}$.

At iteration k, $\Delta x^{(k)}$ is solved from

$$J(x^{(k-1)})\Delta x^{(k)} = f(x^{(k-1)}) - b = \varepsilon_{\mathrm{NR}}(x^{(k-1)}) \tag{2.36}$$

where

$$J(x^{(k-1)}) = \left.\begin{bmatrix} \frac{\partial P_1}{\partial \theta_1} & \frac{\partial P_1}{\partial \theta_2} & \cdots & \frac{\partial P_1}{\partial \theta_{N_B-1}} & \frac{\partial P_1}{\partial V_1} & \cdots & \frac{\partial P_1}{\partial V_{N_L}} \\ \frac{\partial P_2}{\partial \theta_1} & \frac{\partial P_2}{\partial \theta_2} & \cdots & \frac{\partial P_2}{\partial \theta_{N_B-1}} & \frac{\partial P_2}{\partial V_1} & \cdots & \frac{\partial P_2}{\partial V_{N_L}} \\ & & & \vdots & & & \\ \frac{\partial Q_1}{\partial \theta_1} & \frac{\partial Q_1}{\partial \theta_2} & \cdots & \frac{\partial Q_1}{\partial \theta_{N_B-1}} & \frac{\partial Q_1}{\partial V_1} & \cdots & \frac{\partial Q_1}{\partial V_{N_L}} \\ & & & \vdots & & & \end{bmatrix}\right|_{x=x^{(k-1)}} \tag{2.37}$$

Table 2.5 Initial voltage values used in the NR method for cold start.

Bus types	Voltage Phasors	Fixed values	Unknowns	Initial values
PQ buses	$V_i e^{j\theta_i}$	P_i, Q_i	$2{:}V_i, \theta_i$	$1e^{j0^\circ}$
PV buses	$V_i e^{j\theta_i}$	P_i, V_i	$1{:}\theta_i$	$V_i e^{j0^\circ}$
Swing bus	$V_{N_B} e^{j0^\circ}$	$V_{N_B}, \theta_{N_B} (= 0)$	None	$V_{N_B} e^{j0^\circ}$

in which

$$\frac{\partial P_i}{\partial \theta_i} = -V_i \sum_{j=1}^{N_B} |Y_{ij}| V_j \sin(\theta_i - \theta_j - \varphi_{ij}), \quad i = 1, 2, ..., N_B - 1 \tag{2.38}$$

$$\frac{\partial P_i}{\partial \theta_j} = -V_i |Y_{ij}| V_j \sin(\theta_i - \theta_j - \varphi_{ij}), \quad i = 1, 2, ..., N_B - 1, \quad j = 1, 2, ..., N_B - 1, \ j \neq i \tag{2.39}$$

$$\frac{\partial P_i}{\partial V_i} = \sum_{j=1}^{N} [|Y_{ij}| V_j \cos(\theta_i - \theta_j - \varphi_{ij})] - V_i |Y_{ii}| \cos \varphi_{ij}, \quad i = 1, 2, ..., N_L \tag{2.40}$$

$$\frac{\partial P_i}{\partial V_j} = V_i |Y_{ij}| \cos(\theta_i - \theta_j - \varphi_{ij}), \quad i = 1, 2, ..., N_L, \quad j = 1, 2, ..., N_L - 1, \ j \neq i \tag{2.41}$$

Similar expressions can be obtained for the partial derivatives of Q_i.

Then the solution vector is updated as

$$x^{(k)} = x^{(k-1)} + \Delta x^{(k)} \tag{2.42}$$

The NR algorithm converges if the increment $|\Delta x^{(k)}|$ is less than a desired value δ. A flow chart of the NR method as an inner loop of a power flow algorithm is shown in Figure 2.6. The transformer and shunt controls are discussed in the next section.

The NR is efficient as its rate of convergence is quadratic near the solution. The NR method allows a user to decelerate the algorithm by using a stepsize $\gamma < 1$ [16], such that the update equation becomes $x^{(k)} = x^{(k-1)} + \gamma \Delta x^{(k)}$. This decelerating feature is mostly used to prevent divergence in case the initial condition is not well chosen.

In each iteration of the NR method, the Q limits at the generator buses need to be checked:

1) If $Q_i^{(k)} > Q_{\max}$, set $Q_i^{(k)} = Q_{\max}$ and change the PV bus to a PQ bus.
2) If $Q_i^{(k)} < Q_{\min}$, set $Q_i^{(k)} = Q_{\min}$ and change the PV bus to a PQ bus.

Once a bus type has been adjusted and the Q limit used as the reactive power load, the power flow solution is recomputed. The Q limits of on the PV buses in the resulting power flow are checked again to ensure there are no more limit violations before proceeding to the next iteration.

At convergence, the Jacobian J is the sensitivity matrix whose entries can be used to study the impact of changing the loads on the power flow solution. For example, if a

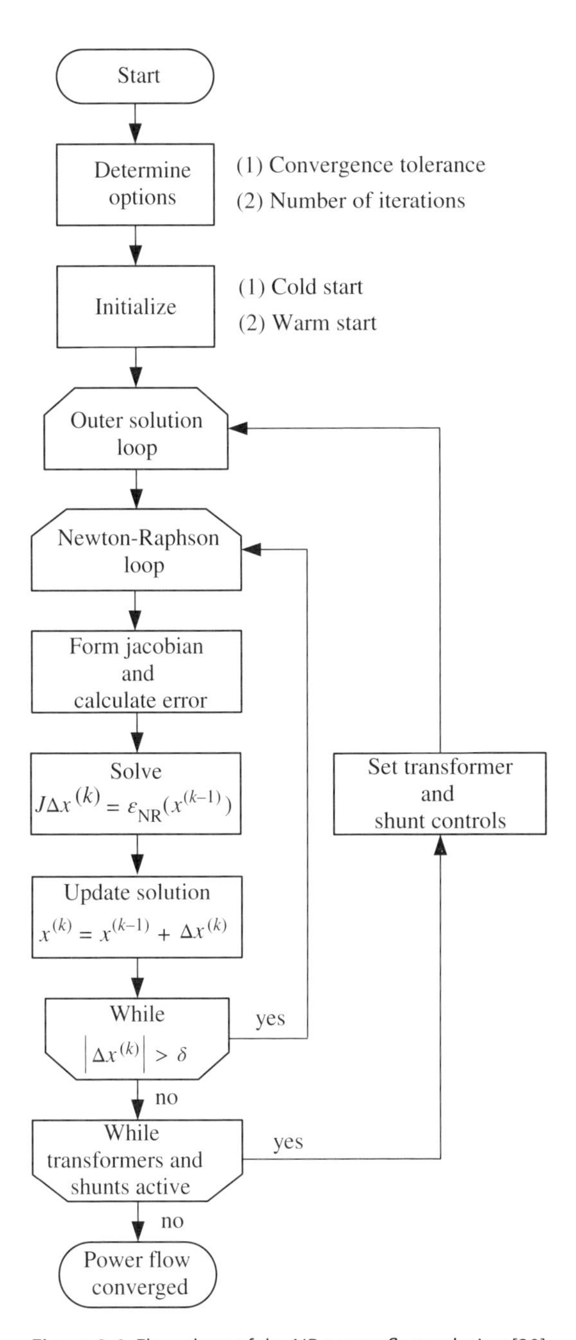

Figure 2.6 Flow chart of the NR power flow solution [20].

small change ($\Delta P_i, \Delta P_j = -\Delta P_i$) on a pair of loads is introduced, the changes in $\Delta\theta$ and ΔV can be solved from

$$J \begin{bmatrix} \Delta\theta \\ \Delta V \end{bmatrix} = \begin{bmatrix} \Delta P \\ 0 \end{bmatrix} \tag{2.43}$$

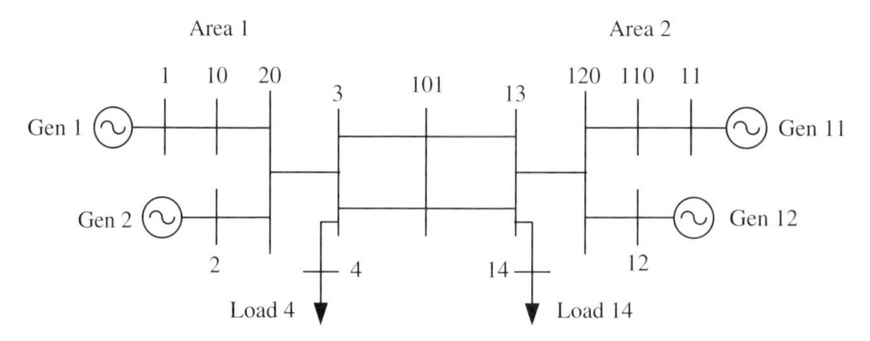

Figure 2.7 Two-area, 4-machine system [6, 21].

in which ΔP contains the two nonzero load changes in the i and j entries. From the $(\Delta\theta, \Delta V)$ solution, the changes in line flows can be computed.

Example 2.3: Power flow calculation

The data of the 2-area system shown in Figure 2.7 can be found in Appendix 2.C and in the file *data2a.m*. The bus and branch data are on 100 MVA base. Compute the power flow solution for the following scenarios. The reference voltages on the generators (1,2,11,12) are set to (1.03,1.01,1.03,1.01) pu, respectively. Use 10^{-6} as the tolerance for convergence for the power flow solution.

1) Set the load at Bus 4 to be $12.76 + j1.0$ pu and the load at Bus 14 to be $14.76 + j1.0$ pu. Compute the power flow.
2) For Scenario 1, note that the bus voltage magnitude on Bus 101 is less than 1.0 pu. Suppose there are several 0.5 pu shunt capacitor banks available at Bus 101. What is the minimum number of capacitor banks needed to be switched on to increase the Bus 101 voltage to above 1.0 pu?
3) Continue from Scenario 2 with the capacitor bank(s) in place, reducing the maximum reactive power output of Generator 2 to 2.5 pu. Compute the power flow solution.

Solutions: Several selected bus voltage magnitudes and angles, and generation reactive power outputs of the power flow solutions for all three scenarios are shown in Tables 2.6–2.8. In all three scenarios, the loads are fed through Buses 3 and 13. Thus their bus angles are lower than the generator bus angles. Furthermore, about 0.85 pu active power flows from Bus 3 to Bus 13, and as a result the angle at Bus 3 is higher than that at Bus 13. The angle varies slightly between these three scenarios as they are not significantly affected by changes in reactive power support.

The bus voltage magnitudes show more variations. For Scenario 1, the power flow solution shows that the voltage magnitudes of the load buses and the transmission buses are all less than 1 pu.[6] For Scenario 2, with the addition of 3 pu B shunt (six capacitor banks), the voltage magnitude of Bus 101 rises to 1.0236 pu. Note that the voltage magnitudes at Buses 3 and 13 also increase, and the generator reactive power outputs are reduced. For Scenario 3, the output of Generator 2 is reduced from 3.24 pu to 2.50 pu.

6 The power flow solution of Scenario 1 is not an acceptable practical operating condition as some of the bus voltages are below 0.95 pu.

Table 2.6 Bus voltage magnitudes in pu.

Buses	2	3	4	13	14	101
Scenario 1	1.0100	0.8889	0.8803	0.8764	0.8665	0.8899
Scenario 2	1.0100	0.9352	0.9272	0.9245	0.9155	1.0236
Scenario 3	0.9414	0.8740	0.8651	0.9094	0.9002	0.9811

Table 2.7 Bus voltage angles in degrees.

Buses	1	2	3	11	12	13	101
Scenario 1	0	−10.45	−26.92	−7.77	−18.28	−34.99	−30.99
Scenario 2	0	−10.13	−25.81	−6.11	−16.42	−32.30	−29.62
Scenario 3	0	−10.22	−28.10	−9.03	−19.39	−35.52	−32.47

Table 2.8 Reactive power in pu supplied by Generators 1 and 2.

Generators	1	2
Scenario 1	2.9211	4.8232
Scenario 2	2.2492	3.2393
Scenario 3	3.7362	2.5000

This reduction is compensated by increasing the reactive power output of Generator 1. Note that the voltage at Generator 2 is reduced from 1.04 to 0.9414 pu. The voltage magnitudes on Buses 3, 13, and 101 are also reduced. ■

2.6 Advanced Power Flow Features

The power flow formulation of a power network can also include additional features useful in practical applications. The following discussion in this section highlights some of these features.

2.6.1 Load Bus Voltage Regulation

Consider Bus j connected to a variable tap transformer as shown in Figure 2.2, where the tap a (with $\alpha = 0$) can be adjusted so that V_j (a PQ bus) is at a desired value $V_{j\text{ref}}$ or within a range of $(V_{\text{min}}, V_{\text{max}})$. If the bus voltage is to be kept at $V_{j\text{ref}}$ on Bus j, the unknowns in the power flow solution for Bus j are now a and θ_j instead of V_j and θ_j. Thus the solution vector still contains $2N_B - N_G - 1$ unknowns

$$x = \begin{bmatrix} \theta_1 \cdots \theta_{N-1} \; V_1 \cdots V_{N-N_g-1} \; a \end{bmatrix}^T \tag{2.44}$$

where V_j is removed. Accordingly, the column in the power flow Jacobian J corresponding to V_j is removed and a new column

$$\left[\partial P_1/\partial a \cdots \partial Q_1/\partial a \cdots \right]^T \tag{2.45}$$

is added to be the last column of J. Note that $\partial P_i/\partial a$ and $\partial Q_i/\partial a$ are all zero except for the bus (Bus i) connected to Bus j via the transformer.

In the NR method to solve for (2.44), a is normally taken to be a continuous variable. In practice, transformer taps are discrete, typically with a fixed number of discrete steps up and down from the nominal point. Thus at the completion of the power flow solution, the resulting continuous value of a needs to be set to the nearest discrete tap position. The power flow problem is solved again with a fixed at the final discrete-step position. As a result, the voltage magnitude on Bus j may be at a value slightly different from the desired value.

It is more common to use tap-changing transformers to maintain the Bus j voltage to within certain limits ($V_{\min}$, $V_{\max}$). These limits are defined as inputs in a power flow bus data set. In this case, the transformer is initially set to its nominal tap position. If the resulting power flow solution shows that V_j is outside the desired limits, tap a is adjusted with the intention of moving V_j to within the limits. For example, if V_j is above $V_{\max}$, then a is used to reduce the voltage to below $V_{\max}$. The power flow is solved again with the new tap position until this process converges. This is indicated as the outer loop in Figure 2.6.

Bus voltage control within ($V_{\min}$, $V_{\max}$) is also possible with switched shunt capacitor or reactor banks, which are in discrete blocks.[7] Initially all shunt banks are off and the power flow is solved. Suppose the voltage at Bus j is below $V_{\min}$ and Bus j has two capacitor banks of different sizes. The inverses of the sensitivity $\partial Q_j/\partial V_j$ can be used to determine whether one or both capacitor banks should be switched on. Again a new power flow is solved and repeated until V_j falls within the desired limits. If the voltage at Bus j is above $V_{\max}$, shunt reactors can be switched on if they are available on Bus j. The shunt bank adjustment is also a part of the outer-loop solution process.

2.6.2 Multi-area Power Flow

Typically a large power system consists of the interconnection of many separate power control areas, such as New York and New England. Each control area would solve its own power flow with its own generators supplying the load. In addition, power regions exchange active power for economic and security reasons. As a result, there are scheduled power exchanges between control areas.

Consider the 2-area power system in Figure 2.8 with generators and loads in both areas. Suppose Area 1 is scheduled to provide $P_{12\text{ref}}$ MW to Area 2 over the two tielines (the flow allocation on each tieline is not specified). It seems logical to set the swing bus in Area 1 and require the net difference between all the generation and load in Area 2 to be $-P_{12\text{ref}}$. If the power flow is solved in this manner, the flow from Area 1 to Area 2 will be different from $P_{12\text{ref}}$ because the swing bus in Area 1 also has to account for the losses in Area 2.

7 Exceptions are static var systems and static compensators, which are discussed in Chapter 14.

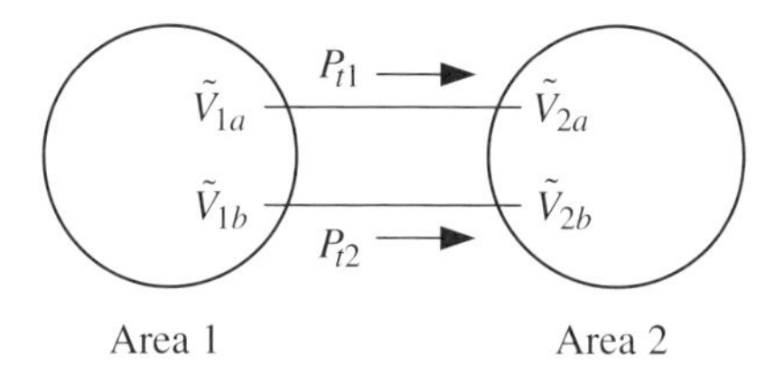

Figure 2.8 A 2-area power system.

A simple way to avoid this situation is to also assign a swing bus in Area 2 or, more specifically, a variable generation bus. The angle of this swing bus is not fixed, so it remains an unknown variable, as the angle of the swing bus in Area 1 is already fixed. The active power generation equation of this second swing bus is no longer needed. In its place, the desired combined tieline flow constraint

$$P_{12} = P_{t1} + P_{t2} = P_{12\text{ref}} \tag{2.46}$$

is added to power flow formulation. Thus the total number of unknowns and power flow equations is preserved.

For a power system with N_A areas, up to $N_A - 1$ tieline active power flow equations can be added to the power flow formulation. Note that the N_Ath interface flow constraint will be redundant. For every power interchange equation added, one of the generator buses in the affected area needs to be converted to a swing bus. In the case of $N_A - 1$ interface flow equations, one swing bus in each area is required, such that its active power generation equation is replaced by the interface flow constraint.

2.6.3 Active Line Power Flow Regulation

For Buses i and j connected by a variable phase shifting transformer as shown in Figure 2.2, the phase shift α (with $a = 1$) can be adjusted so that the active power flow P_{ij} on the branch can be controlled to a prespecified value $P_{ij\text{ref}}$. To enforce this additional constraint in the power flow formulation, the phase-shifter angle α will be added to the solution vector x

$$x = \left[\theta_1 \cdots \theta_{N_B-1}\ V_1 \cdots V_{N_L-1}\ \alpha\right]^T \tag{2.47}$$

As a result, the Jacobian matrix will grow by one column and one row.

Note that the phase-shifter tap position α for line power flow regulation is solved in power flow as a continuous variable. However, the tap position is discrete. Thus after a NR solution converges, the tap position is adjusted to the nearest discrete tap position and a new NR solution is obtained with all the tap positions fixed and without the line flow constraints.

Although phase-shifting transformers can change the active power flow on a transmission line in both directions, they are primarily used to reduce undesired "loop" flow of active power across multiple regions. Consider the 3-area power system shown in Figure 2.9, in which Area 1 intends to export $P_{12} = P_{t1} + P_{t2}$ MW to Area 2. However, as these three areas are interconnected, some of the export will go from Area 1 to Area 3, and then to Area 2. This unintended parallel flow is known as loop flow. Loop flow on

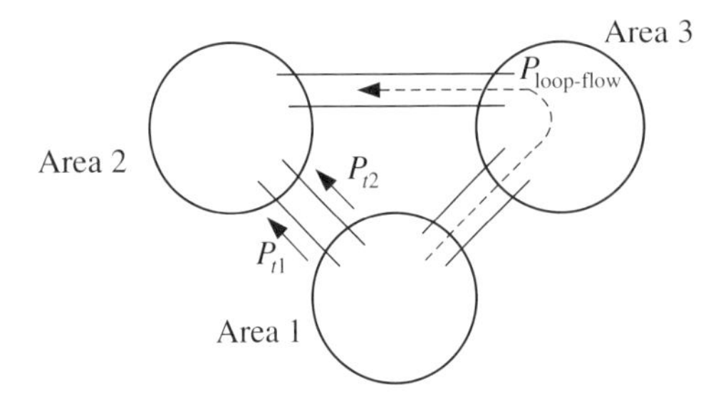

Figure 2.9 A 3-area power system.

unintended transfer paths is undesirable for economic and stability reasons.[8] If phase shifters are installed on the interface between Areas 1 and 2, the direct flow between these two areas can be adjusted to minimize the loop flow.

2.6.4 Dishonest Newton-Raphson Method

The Jacobian matrix J contains the search direction for the solution vector to converge. Its computation, however, can be expensive. On the other hand, J from the previous iteration may still contain useful convergence information so that it can be reused for the next iteration.

For this "dishonest" NR method, in the iterative solution of

$$J^{(k)} \begin{bmatrix} \Delta\theta^{(k)} \\ \Delta V^{(k)} \end{bmatrix} = \begin{bmatrix} \Delta P^{(k)} \\ \Delta Q^{(k)} \end{bmatrix} = \varepsilon_{\text{NR}}(x^{(k)}) \tag{2.48}$$

where $\varepsilon_{\text{NR}}(x)$ is the mismatch vector, instead of updating J at every iteration, J is updated at every other iteration. However, the mismatch $\varepsilon_{\text{NR}}(x)$ is updated at every iteration. Note that the LU decomposition of J can be saved, and thus no refactorization is needed.

The dishonest NR method will require more iterations than the regular NR method to converge, as it does not have strict quadratic convergence. Its advantage is that it requires less computation per iteration on average.

2.6.5 Fast Decoupled Loadflow

Another way to reduce the amount of computation in the NR method is to take advantage of the dependence of active and reactive power flows on the bus voltage angles and magnitudes [16]. Consider the NR method expressed as

$$J^{(k)} \begin{bmatrix} \Delta\theta^{(k)} \\ \Delta V^{(k)} \end{bmatrix} = \begin{bmatrix} \partial P^{(k)}/\partial\theta & \partial P^{(k)}/\partial V \\ \partial Q^{(k)}/\partial\theta & \partial Q^{(k)}/\partial V \end{bmatrix} \begin{bmatrix} \Delta\theta^{(k)} \\ \Delta V^{(k)} \end{bmatrix} = \begin{bmatrix} \Delta P^{(k)} \\ \Delta Q^{(k)} \end{bmatrix} \tag{2.49}$$

Because the active power injections P is largely determined by the bus voltage angle θ and the reactive power flow Q is largely determined by the bus voltage magnitude V, $J^{(k)}$ is block-diagonally dominant. Thus neglecting the off-diagonal blocks of $J^{(k)}$ results in the decoupled power flow

$$\begin{bmatrix} \partial P^{(k)}/\partial\theta & 0 \\ 0 & \partial Q^{(k)}/\partial V \end{bmatrix} \begin{bmatrix} \Delta\theta^{(k)} \\ \Delta V^{(k)} \end{bmatrix} = \begin{bmatrix} \Delta P^{(k)} \\ \Delta Q^{(k)} \end{bmatrix} \tag{2.50}$$

8 Loop flow can increase losses in other control regions.

Note that even though the increments $\Delta\theta^{(k)}$ and $\Delta V^{(k)}$ obtained from the solutions of

$$\frac{\partial P^{(k)}}{\partial \theta}\Delta\theta^{(k)} = \Delta P^{(k)}, \quad \frac{\partial Q^{(k)}}{\partial V}\Delta V^{(k)} = \Delta Q^{(k)} \tag{2.51}$$

may not be as good as those obtained from the fully coupled NR method, decoupled power flow will still converge to the right solution because the mismatch $(\Delta P^{(k)}, \Delta Q^{(k)})$ is computed without approximation. The decoupled power flow will achieve a quadratic convergence rate as long as the off-diagonal blocks are sufficiently small. In particular, this decoupled power flow works well when the line resistance to reactance ratio satisfies $R/X < 0.1$. Because of its faster computation time, this method is commonly known as the fast decoupled loadflow (FDLF).

A further simplification of the FDLF method is to observe that if the line resistances are neglected and the angular differences $|\theta_i - \theta_j|$ are assumed to be small, that is, $\sin(\theta_i - \theta_j) = 0$ and $\cos(\theta_i - \theta_j) = 1$, then

$$\frac{\partial P}{\partial \theta} = -\text{diag}(V) \cdot B \cdot \text{diag}(V), \quad \frac{\partial Q}{\partial V} = -\text{diag}(V) \cdot B \tag{2.52}$$

where $B = \text{Im}(Y)$ is the susceptance matrix, and the "diag" operation takes the entries of a vector and puts them into the diagonal entries of a matrix.

Thus at each iteration, the voltage angle and magnitude increments can be solved from

$$-B \cdot \text{diag}(V)\Delta\theta = (\text{diag}(V))^{-1}\Delta P, \quad -B\Delta V = (\text{diag}(V))^{-1}\Delta Q \tag{2.53}$$

in which the LU decomposition for B has to be computed only once. Convergence of this method is linear, but each iteration is very fast.

2.6.6 DC Power Flow

In power system applications requiring fast repeated power flow evaluations, such as screening of contingencies, all the bus voltages may be set to 1 pu and the reactive power flow calculation is skipped, assuming that sufficient reactive power support will be provided by the generators. This is known as a DC power flow [16], with the following assumptions:

1) All the bus voltage magnitudes are equal to unity.
2) All line resistances are neglected.
3) All angular differences $|\theta_i - \theta_j|$ are small.

Then the power flow equation (2.53) simplifies to

$$-B \cdot \theta = P \tag{2.54}$$

The DC power flow can also be used to initialize the NR method. It is common for unit commitment and optimal power flow programs to use the DC power flow formulation to reduce the dispatch complexity.

2.7 Summary and Notes

This chapter provides the power flow formulation of the steady-state operation of a power system and its solution using the Newton-Raphson method. It also describes advanced features such as multi-area power flow and tie-line flow control as well as computational savings from decoupled power flow and sparse factorization. Additional information can be found in the excellent power flow survey [16]. The text [23] also provides an extensive discussion on the state-of-the-art power flow computation, and the text [24] contains many power flow examples.

With modern computers, large power system data sets with more than 50,000 buses can be solved readily and quite rapidly. The methods covered here and in [16, 23] are based on serial computing, in which minimizing memory utilization and the number of floating point operations is important. Parallel computing is recognized as a means to speed up significantly the computing time. In a power flow solution, the updating of the Jacobian matrix can be readily computed in parallel. However, the NR solution is not amenable to parallel computing, except for the fact that in decoupled power flow computation, $\Delta\theta$ and ΔV (2.53) can be solved in parallel, a saving of roughly 50% computation time.

There is also active research on applying the Jacobi and Krylov parallel computing methods to the Gauss iteration of obtaining the power flow solution. A reader can consult [18] for further discussion.

Appendix 2.A Two-winding Transformer Model

This appendix develops a circuit model for two-winding transformers. In a two-winding transformer, the primary and secondary coils are coupled by a magnetic field via the iron core, as shown in Figure 2.10 [15]. The current in the primary coil (coil 1), the voltage across the primary coil, and the number of turns for the primary coil are denoted by i_1, v_1, and n_1, respectively. Similarly, these variables for the secondary coil (coil 2) are denoted by i_2, v_2, and n_2, respectively.

With coil 2 open, the magnetic flux produced by coil 1 is

$$\phi_{11} = \phi_{\ell 1} + \phi_{21} = n_1 i_1 (P_{\ell 1} + P_m) \tag{2.55}$$

where ϕ_{21} is the flux in the magnetic core, $\phi_{\ell 1}$ is the leakage flux, $P_{\ell 1}$ is the leakage permeance, and P_m is the magnetizing permeance.

Similarly, with coil 1 open, the magnetic flux produced by coil 2 is

$$\phi_{22} = \phi_{\ell 2} + \phi_{12} = n_2 i_2 (P_{\ell 2} + P_m) \tag{2.56}$$

where ϕ_{12} is the flux in the magnetic core, $\phi_{\ell 2}$ is the leakage flux, and $P_{\ell 2}$ is the leakage permeance.

With currents in both coils, using superposition, the flux ϕ_1 in coil 1 and the flux in ϕ_2 in coil 2 are given by, assuming no saturations,

$$\phi_1 = \phi_{11} + \phi_{12} = \phi_{\ell 1} + \phi_{21} + \phi_{12} \tag{2.57}$$

$$\phi_2 = \phi_{22} + \phi_{21} = \phi_{\ell 2} + \phi_{12} + \phi_{21} \tag{2.58}$$

Thus the flux linkages are

$$\psi_1 = n_1 \phi_1 = n_1 \phi_{11} + n_2 \phi_{12}, \quad \psi_2 = n_2 \phi_2 = n_2 \phi_{22} + n_2 \phi_{21} \tag{2.59}$$

where n_1 and n_2 are the number of turns of the winding on the primary and secondary coils, respectively.

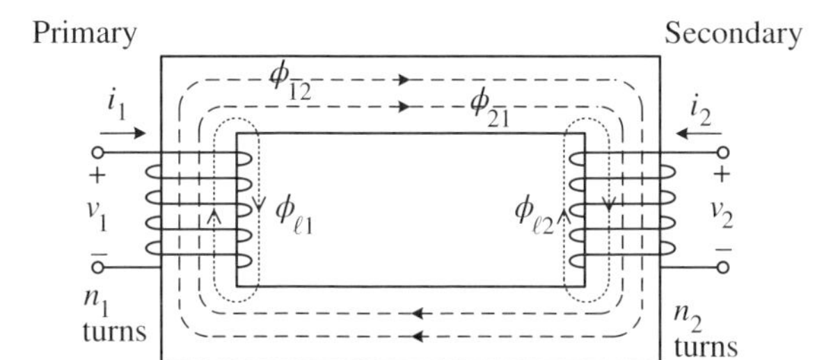

Figure 2.10 Magnetic circuit diagram of a two-winding transformer.

Define the self inductances as

$$n_1\phi_{11} = L_1 i_1, \quad n_2\phi_{22} = L_2 i_2 \tag{2.60}$$

and the mutual inductances as

$$n_2\phi_{21} = M_{21} i_1, \quad n_1\phi_{12} = M_{12} i_2, \quad M_{12} = M_{21} = M \tag{2.61}$$

such that

$$M = \frac{n_2\phi_{21}}{i_1} = \frac{n_1\phi_{12}}{i_2} = n_1 n_2 P_m \tag{2.62}$$

$$L_1 = \frac{n_1\phi_{11}}{i_1} = L_{\ell 1} + aM, \quad L_{\ell 1} = n_1^2 P_{\ell 1} \tag{2.63}$$

$$L_2 = \frac{n_1\phi_{22}}{i_2} = L_{\ell 2} + \frac{1}{a}M, \quad L_{\ell 2} = n_2^2 P_{\ell 2} \tag{2.64}$$

where $a = n_1/n_2$, and $L_{\ell 1}$ and $L_{\ell 2}$ are the leakage reactances of coils 1 and 2, respectively.

Applying Faraday's law of electromagnetic induction $v_i = d\psi_i/dt$ to the flux linkage expressions

$$\psi_1 = L_1 i_1 + M i_2, \quad \psi_2 = M i_1 + L_2 i_2 \tag{2.65}$$

and including resistances R_1 and R_2 to the coils, results in the circuit equations

$$v_1 = R_1 i_1 + L_1\frac{di_1}{dt} + M\frac{di_2}{dt} = R_1 i_1 + (L_{\ell 1} + aM)\frac{di_1}{dt} + M\frac{di_2}{dt} \tag{2.66}$$

$$v_2 = R_2 i_2 + M\frac{di_1}{dt} + L_2\frac{di_2}{dt} = R_2 i_2 + M\frac{di_1}{dt} + (L_{\ell 2} + \frac{1}{a}M)\frac{di_2}{dt} \tag{2.67}$$

Note that core losses due to hysteresis and eddy currents are neglected.

Scaling i_2 by a and multiplying the v_2 equation by a, (2.66) and (2.67) become, respectively,

$$v_1 = R_1 i_1 + (L_{\ell 1} + aM)\frac{di_1}{dt} + aM\frac{d(i_2/a)}{dt} \tag{2.68}$$

$$av_2 = a^2 R_2 \frac{i_2}{a} + aM\frac{di_1}{dt} + (a^2 L_{\ell 2} + aM)\frac{d(i_2/a)}{dt} \tag{2.69}$$

These equations can be represented by an equivalent circuit, shown in Figure 2.11.

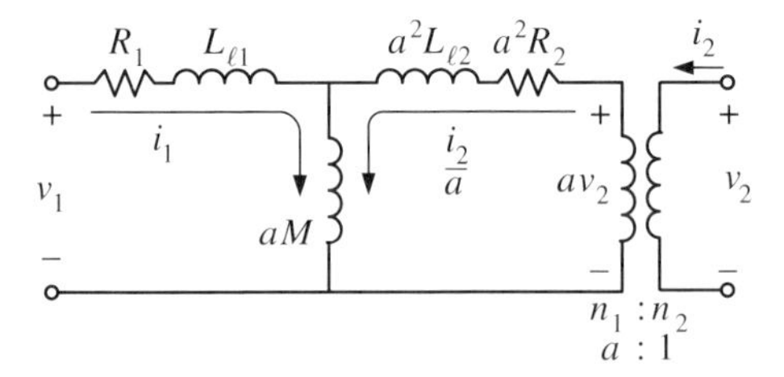

Figure 2.11 Equivalent circuit for a two-winding transformer.

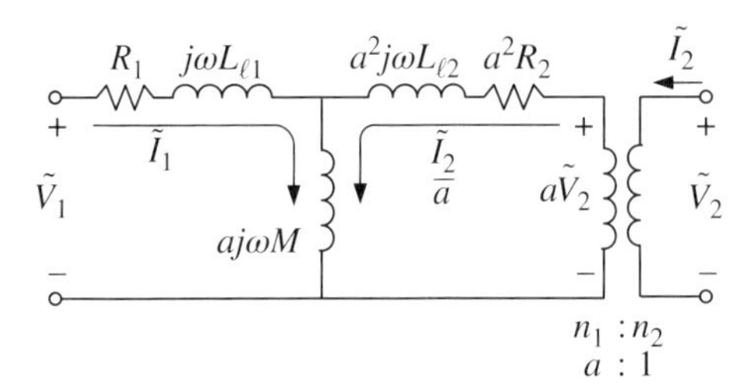

Figure 2.12 Sinusoidal steady-state equivalent circuit for a two-winding transformer.

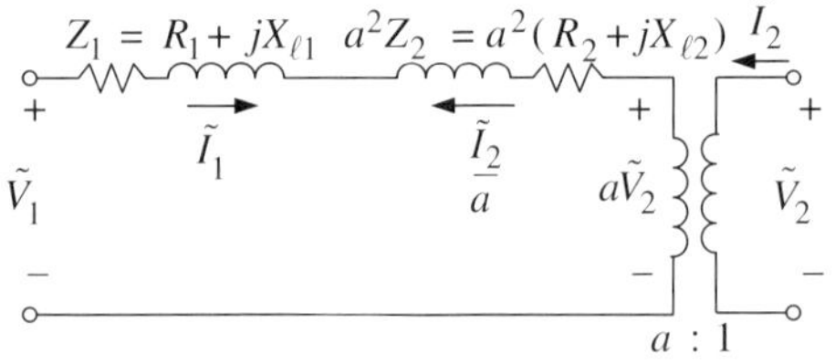

Figure 2.13 Equivalent circuit for a two-winding transformer with mutual reactance neglected.

For a transformer operating in sinusoidal steady-state at a frequency of ω, the Laplace transform can be applied to (2.68) and (2.69) by setting $s = j\omega$ to obtain

$$\tilde{V}_1 = R_1\tilde{I}_1 + (j\omega L_{\ell 1} + a(j\omega M))\tilde{I}_1 + a(j\omega M)\frac{\tilde{I}_2}{a} \tag{2.70}$$

$$a\tilde{V}_2 = a^2 R_2\frac{\tilde{I}_2}{a} + a(j\omega M)\tilde{I}_1 + (a^2(j\omega L_{\ell 2}) + a(j\omega M))\frac{\tilde{I}_2}{a} \tag{2.71}$$

These equations can be represented by the equivalent circuit shown in Figure 2.12.

The magnetizing reactance $X_m = \omega M$ is usually much larger than the leakage reactances $X_{\ell 1} = \omega L_{\ell 1}$ and $X_{\ell 2} = \omega L_{\ell 2}$. Thus if a load is connected across the terminals of coil 2, X_m would have a negligible contribution to the parallel combination of X_m and the load impedance in series with $X_{\ell 2}$. Thus X_m is normally neglected, in which case $\tilde{I}_1 = -\tilde{I}_2/a$. This simplified equivalent circuit is shown in Figure 2.13.

Thus the equivalent impedance of a transformer is given by

$$Z_T = Z_1 + a^2 Z_2 \tag{2.72}$$

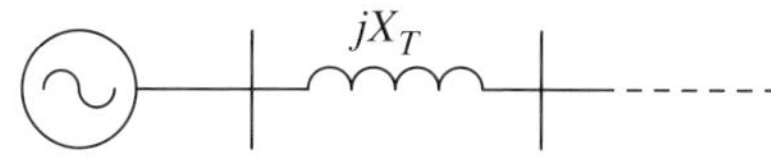

Figure 2.14 Generator connected via a transformer to a transmission system.

where the resistance is typically neglected. *Thus the transformer impedance consists of the leakage reactances of the primary and secondary sides.*

The transformer voltage and current equations in matrix form become

$$\begin{bmatrix} \tilde{I}_1 \\ \tilde{I}_2 \end{bmatrix} = \begin{bmatrix} Y_T & -aY_T \\ -aY_T & a^2Y_T \end{bmatrix} \begin{bmatrix} \tilde{V}_1 \\ \tilde{V}_2 \end{bmatrix} \tag{2.73}$$

where the transformer equivalent admittance is $Y_T = 1/Z_T$.

There are two main types of transformers: fixed tap and variable tap. In a fixed-tap transformer, $a = n_1/n_2$ is fixed. A variable tap transformer typically has 10 to 20 taps up and down from a nominal position of tap 0. When the voltage values are given in pu, the range of a is normally between 0.9 and 1.1. An *on-load (or under-load) tap-changing (OLTC/ULTC) transformer* regulates the voltage on one of its terminals by adjusting its taps. Deadbands and time delays are used to prevent frequent adjustment of taps (to reduce wear and tear). Note that for an accurate model, Z_T changes as a changes.

With resistance neglected,

$$Z_T = jX_T = j(X_{\ell 1} + a^2 X_{\ell 2}) \tag{2.74}$$

such that X_T is the leakage reactance, which is typically 12–15% of the transformer rating. Also this reactance needs to be per unitized with respect to the system base when entered as power flow branch data. From (2.74), X_T is dependent on the tap ratio. However, as a is normally close to 1, X_T is set to $X_{\ell 1} + X_{\ell 2}$.

Example 2.A.1 Generator transformer reactance

Consider a synchronous machine rated at 900 MVA connected to a transmission system via a step-up transformer, also rated at 900 MVA (Figure 2.14). The leakage reactance X_T is 0.12 pu (12%) on its own base (900 MVA). Compute the transformer reactance to be used in a power flow data set with a base of 100 MVA for the transmission system.

Solutions: The reactance X_T on the 100 MVA base is

$$X_T = 0.12 \times \frac{100}{900} = 0.0133 \text{ pu} \tag{2.75}$$

Note that a proper X_T value is needed to allow for the delivery of the generator active power.

■

Appendix 2.B LU Decomposition and Sparsity Methods

2.B.1 LU Decomposition

The solution of a system of linear equations written in the matrix form

$$Ax = b, \quad x, b \in R^N \tag{2.76}$$

is found numerically via the LU decomposition technique, as the inversion of the $N \times N$ matrix A is of the order of N^3 operations and thus is very ineffficient. In LU decomposition, the matrix A is factorized into

$$A = LU \tag{2.77}$$

where L is a lower triangular matrix with diagonal entries equal to 1, and U is an upper triangular matrix with nonzero diagonal entries. For example, the matrices L and U for a 4×4 A matrix has the following forms:

$$L = \begin{bmatrix} 1 & 0 & 0 & 0 \\ * & 1 & 0 & 0 \\ * & * & 1 & 0 \\ * & * & * & 1 \end{bmatrix}, \quad U = \begin{bmatrix} * & * & * & * \\ 0 & * & * & * \\ 0 & 0 & * & * \\ 0 & 0 & 0 & * \end{bmatrix} \tag{2.78}$$

where $*$ denotes an entry that may be nonzero.

The factors L and U are computed successively, proceeding one row and column at a time using Gaussian elimination. To illustrate, let the (1,1) entry a_{11} be nonzero and shown separately from A as

$$A = \begin{bmatrix} a_{11} & A_{12} \\ A_{21} & A_{22} \end{bmatrix} = A^{(1)} \tag{2.79}$$

where A_{12}, A_{21}, and A_{22} are submatrices of dimensions $1 \times (N-1)$, $(N-1) \times 1$, and $(N-1) \times (N-1)$, respectively. If a_{11} is zero, then the unknown vector x can be reordered such that the new a_{11} is nonzero.

Construct the lower-triangular matrix

$$M_1 = \begin{bmatrix} 1 & 0 \\ -\frac{A_{21}}{a_{11}} & I \end{bmatrix} \tag{2.80}$$

where I is an identity matrix of dimension $(N-1) \times (N-1)$. Then

$$M_1 A^{(1)} = \begin{bmatrix} 1 & 0 \\ -\frac{A_{21}}{a_{11}} & I \end{bmatrix} \begin{bmatrix} a_{11} & A_{12} \\ A_{21} & A_{22} \end{bmatrix} = \begin{bmatrix} a_{11} & A_{12} \\ 0 & A_{22} - \frac{A_{21}A_{12}}{a_{11}} \end{bmatrix} = \begin{bmatrix} a_{11} & A_{12} \\ 0 & A^{(2)} \end{bmatrix} \tag{2.81}$$

Next construct a lower triangular M_2 to make the first column of $A^{(2)}$ below the diagonal entry be zero. Continuing the process for $N-1$ times results in

$$(M_{N-1} \cdots M_2 M_1)A = U \quad \Rightarrow \quad L = (M_{N-1} \cdots M_2 M_1)^{-1} = M_1^{-1} M_2^{-1} \cdots M_{N-1}^{-1} \tag{2.82}$$

By construction L and U are, respectively, lower and upper triangular. Note that the inverses of M_i are obtained without computation, as

$$M_1^{-1} = \begin{bmatrix} 1 & 0 \\ \frac{A_{21}}{a_{11}} & I \end{bmatrix} \tag{2.83}$$

Furthermore, the multiplication of all the M_i^{-1} to form L is no more complicated than filling out the lower diagonal part of L one column at a time using the nonzero entries of M_i^{-1}. Thus L and U can be stored in the memory locations occupied by A.

2.B.2 Forward Elimination and Back Substitution

Using LU decomposition, the linear equation (2.76) can be expressed as

$$Ax = LUx = Ly = b \tag{2.84}$$

allowing x to be solved in two steps. In the forward elimination step, the entries of y are successively solved from $Ly = b$ starting from the first entry of y. To illustrate, consider the following example

$$\begin{bmatrix} 1 & 0 \\ \ell_{21} & 1 \end{bmatrix} \begin{bmatrix} y_1 \\ y_2 \end{bmatrix} = \begin{bmatrix} b_1 \\ b_2 \end{bmatrix} \tag{2.85}$$

Because L is lower diagonal, the entries of y are solved successively as

$$y_1 = b_1, \quad y_2 = b_2 - \ell_{21} y_1 \tag{2.86}$$

Note that no division is actually necessary as all diagonal entries of L are unity.

The second step is back substitution in which x is solved successively from $Ux = y$ starting from the last entry of x. Continuing the illustration, consider the example

$$\begin{bmatrix} u_{11} & u_{12} \\ 0 & u_{22} \end{bmatrix} \begin{bmatrix} x_1 \\ x_2 \end{bmatrix} = \begin{bmatrix} y_1 \\ y_2 \end{bmatrix} \tag{2.87}$$

The entries of x are solved successively as

$$x_2 = \frac{1}{u_{22}} y_2, \quad x_1 = \frac{1}{u_{11}} (y_1 - u_{12} x_2) \tag{2.88}$$

The solution procedure requires $u_{ii} \neq 0$. If u_{ii} is equal to zero or small (ill-conditioning), find the entry with the largest magnitude and permute x and b so that entry becomes the new u_{ii}. Computer packages for LU decomposition will carry out pivoting to ensure good numerical stability. For example, the MATLAB® function *[L,U] = lu(A)* uses pivoting to obtain the matrices L and U.

2.B.3 Sparse Factorization

In power system applications, the matrix A (which is the Jacobian matrix J) is usually very large and has many zero entries. Due to the nature of a power network, a bus (substation) is normally connected to three to four other buses with complex impedances. Thus independent of the size of the power system, the J matrix will have an average of four to five nonzero complex entries per row.

For a sparse matrix, one only needs to store its nonzero entries together with their row and column indices. If the L and U factors are also sparse, then the forward elimination and back substitution steps will be very efficient, as any multiplication operation with zero entries can be skipped. To keep the sparse structures of the L and U factors, Tinney [22] proposed an optimal ordering of the LU decomposition to minimize the introduction of nonzero entries, a process known as sparse factorization.

As an example, consider the matrix

$$A = \begin{bmatrix} a_{11} & A_{12} \\ A_{21} & A_{22} \end{bmatrix} = \left[\begin{array}{c|ccccc} a_{11} & a_{12} & a_{13} & a_{14} & a_{15} & a_{16} \\ \hline a_{21} & a_{22} & a_{23} & 0 & 0 & 0 \\ a_{31} & a_{32} & a_{33} & a_{34} & 0 & 0 \\ a_{41} & 0 & a_{43} & a_{44} & a_{45} & a_{46} \\ a_{51} & 0 & 0 & a_{54} & a_{55} & 0 \\ a_{61} & 0 & 0 & a_{64} & 0 & a_{66} \end{array}\right] \tag{2.89}$$

where the a_{ij} entries denote nonzero entries.

Following (2.81), $A^{(2)}$ becomes

$$A^{(2)} = A_{22} - \frac{A_{21}A_{12}}{a_{11}} \tag{2.90}$$

which is a full matrix with all nonzero entries, as A_{12} and A_{21} have all their entries nonzero.

The idea of sparse factorization is to permute the rows and columns of A such that the "fill-ins" in the LU decomposition process is minimal. "Fill-in" means a zero entry in a particular location becoming a nonzero entry during the factorization process. The idea basically says that one should start the factorization process with rows having the smallest number of nonzero entries to produce the smallest number of fill-ins.

In (2.89), because the second row of A has one of the smallest number of nonzero entries, the first and second rows and columns of A are exchanged to form A'

$$A' = \begin{bmatrix} a'_{11} & A'_{12} \\ A'_{21} & A'_{22} \end{bmatrix} = \left[\begin{array}{c|ccccc} a_{22} & a_{21} & a_{23} & 0 & 0 & 0 \\ \hline a_{12} & a_{11} & a_{13} & a_{14} & a_{15} & a_{16} \\ a_{32} & a_{31} & a_{33} & a_{34} & 0 & 0 \\ 0 & a_{41} & a_{43} & a_{44} & a_{45} & a_{46} \\ 0 & a_{51} & 0 & a_{54} & a_{55} & 0 \\ 0 & a_{61} & 0 & a_{64} & 0 & a_{66} \end{array}\right] \tag{2.91}$$

For (2.91), $A'_{22} - A'_{21}A'_{12}/a'_{11}$ has no additional fill-ins.

Sparse factorization typically uses first an optimal ordering procedure by checking on fill-ins without any numerical computation. This optimal ordering can be used in every NR iteration because the optimal ordering remains unchanged as the solution converges.

For power systems consisting of control regions with more transmission lines within the control regions than between control regions, the Jacobian matrix can be organized into a sparse matrix in block form as

$$\begin{bmatrix} A'_{11} & 0 & 0 & * \\ 0 & A'_{22} & 0 & * \\ 0 & 0 & A'_{33} & * \\ * & * & * & * \end{bmatrix} \tag{2.92}$$

where the A'_{ii} blocks, $i = 1, 2, 3$, are due to the internal connections within each control region. The boundary buses of these control regions form the (4, 4)th block of the matrix. Furthermore, the "*" entries in the last row and column are due to the tielines connecting the control regions. Sparse factorization can also be applied within each A_{ii}.

MATLAB® allows a user to store a sparse matrix with the data format (*i,j,value*), and its LU decomposition *[L,U]=lu(A)* automatically uses sparse factorization if *A* is sparse.

With sparse factorization, power flow solutions for power networks with tens of thousand of buses can be readily accommodated.

Appendix 2.C Power Flow and Dynamic Data for the 2-area, 4-machine System

The 2-area system (Figure 2.7) consists of Generators 1 and 2 in Area 1 and Generators 11 and 12 in Area 2. The data of this system used in this text is modified from the original 2-area, 4-machine system found in [82]. The major change is the addition of several buses to the system to facilitate the demonstration of some relevant power system concepts. The data in this appendix are contained in the file *data2a.m*.

The line parameters are given in Table 2.9. The bus voltage magnitudes and angles in Table 2.10 are the power flow solutions. As noted, the bus voltages on the two load buses (4 and 14) and the interconnection bus (101) are very low. Thus the two areas are weakly connected. Shunt capacitors can be added to the load buses to improve the voltage magnitude. To enhance the interarea stability, an additional line between Buses 3 and 13 can be added. Alternatively, controls such as a static var compensator can be installed on Bus 101 to improve transient and small-signal stability. Another method is to add an HVDC line between Buses 3 and 13. These controllers will be the subject of subsequent chapters. Note that the data here will be changed as needed in other chapters, with an updated data file provided in an appropriate example or problem folder.

Table 2.9 Two-area, 4-machine system line parameters.

From bus	To bus	*R* pu	*X* pu	*B* pu	Tap ratio	Tap phase	Tap max	Tap min	Tap size
1	10	0.0	0.0167	0.00	1.0	0	0	0	0
2	20	0.0	0.0167	0.00	1.0	0	0	0	0
3	4	0.0	0.005	0.00	1.0	0	1.2	0.8	0.05
3	20	0.001	0.010	0.175	1.0	0	0	0	0
3	101	0.011	0.110	0.1925	1.0	0	0	0	0
3	101	0.011	0.110	0.1925	1.0	0	0	0	0
10	20	0.0025	0.025	0.0437	1.0	0	0	0	0
11	110	0.0	0.0167	0.0	1.0	0	0	0	0
12	120	0.0	0.0167	0.0	1.0	0	0	0	0
13	14	0.0	0.005	0.0	1.0	0	1.2	0.8	0.05
13	101	0.011	0.11	0.1925	1.0	0	0	0	0
13	101	0.011	0.11	0.1925	1.0	0	0	0	0
13	120	0.001	0.01	0.0175	1.0	0	0	0	0
110	120	0.0025	0.025	0.0437	1.0	0	0	0	0

Table 2.10 Two-area, 4-machine system bus parameters (bus types: 1 for swing bus, 2 for generator bus, and 3 for load bus).

Bus number	V pu	θ deg.	P_g pu	Q_g pu	P_L pu	Q_L pu	G_{sh} pu	B_{sh} pu
1	1.03	0	7.00	1.61	0	0	0	0
2	1.01	−10.41	7.00	1.76	0	0	0	0
3	0.8858	−26.93	0	0	0	0	0	0
4	0.8766	−32.01	0	0	13.76	1.00	0	0
10	0.9888	−6.72	0	0	0	0	0	0
11	1.03	0.18	7.16	1.49	0	0	0	0
12	1.01	−10.29	7.00	1.39	0	0	0	0
13	0.8856	−26.83	0	0	0	0	0	0
14	0.8764	−31.92	0	0	13.76	1.00	0	0
20	0.9357	−17.51	0	0	0	0	0	0
101	0.8952	−26.94	0	0	0	0	0	0
110	0.9887	−6.56	0	0	0	0	0	0
120	0.9355	−17.40	0	0	0	0	0	0

Bus number	Bus type	Q_{gmax} pu	Q_{gmin} pu	Rated V kV	V_{max} pu	V_{min} pu
1	1	99.0	−1	22.0	1.1	0.9
2	2	99.0	−1	22.0	1.1	0.9
3	3	0	0	500.0	1.2	0.8
4	3	0	0	115.0	1.2	0.8
10	3	0	0	230.0	1.2	0.8
11	2	99.0	−1	22.0	1.1	0.9
12	2	99.0	−1	22.0	1.1	0.9
13	3	0	0	500.0	1.2	0.8
14	3	0	0	115.0	1.2	0.8
20	3	0	0	230.0	1.2	0.8
101	3	0	0	500.0	1.2	0.8
110	3	0	0	230.0	1.2	0.8
120	3	0	0	230.0	1.2	0.8

The four generators on Buses 1, 2, 11, and 12 have identical ratings and parameters. A subtransient model is used, with the parameters given in pu on the machine base shown in Table 2.11 (see Chapter 8 for the definition of the parameters).

The exciter data is given in Table 2.12. Block diagrams of different types of exciter models can be found in Chapter 9.

The turbine-governor models of the four generators are identical. The parameters given in pu on the machine base are shown in Table 2.13 (see Chapter 12 for the definition of the parameters).

Table 2.11 Two-area, 4-machine system generator parameters.

Machine base	900 MVA
Leakage reactance X_ℓ	0.200 pu
Resistance r_a	0.0025 pu
d-axis synchronous reactance X_d	1.8 pu
d-axis transient reactance X'_d	0.30 pu
d-axis subtransient reactance X''_d	0.25 pu
d-axis open-circuit time constant T'_{do}	8.00 sec
d-axis open-circuit subtransient time constant T''_{do}	0.03 sec
q-axis synchronous reactance X_q	1.7 pu
q-axis transient reactance X'_q	0.55 pu
q-axis subtransient reactance X''_q	0.25 pu
q-axis open-circuit time constant T'_{qo}	0.4 sec
q-axis open-circuit subtransient time constant T''_{qo}	0.05 sec
Inertia constant H	6.5 sec
Damping D	0 pu

Table 2.12 Two-area, 4-machine system excitation system parameters.

Generator **Exciter type**	**1** **1 (DC1A)**	**2** **3 (ST3A)**	**11** **0**	**12** **2 (AC2A)**
Input filter time constant T_R (sec)	0.01	0.01	0.01	0.01
Voltage regulator gain K_A	46	7.04	200	300
Voltage regulator time constant T_A (sec)	0.06	1.0	0	0.01
Voltage regulator time constant T_B (sec)	0	6.67	0	0
Voltage regulator time constant T_C (sec)	0	1.0	0	0
Maximum voltage regulator output $V_{R\max}$ (pu)	1.0	10	5.0	4.95
Minimum voltage regulator output $V_{R\min}$ (pu)	−0.9	−10	−5.0	−4.9
Maximum internal signal $V_{I\max}$ (pu)	0	0.2	0	1.0
Minimum internal signal $V_{I\min}$ (pu)	0.46	−0.2	0	1.33
First stage regulator gain K_J	3.1	200	0	3.05
Potential circuit gain coefficient K_p	0.33	4.37	0	0.279
Potential circuit phase angle θ_p	2.3°	20°	0	2.29°
Current circuit gain coefficient K_I	0.1	4.83	0	0.117
Potential source reactance X_L (pu)	0.1	0.09	0	0.1
Rectifier loading factor K_C	1.0	1.1	0	0.675
Maximum field voltage $E_{fd\max}$ (pu)	0	8.63	0	0
Inner loop feedback constant K_G	0	1.0	0	0
Maximum inner loop voltage feedback $V_{G\max}$ (pu)	0	6.53	0	0

Table 2.13 Two-area, 4-machine system turbine-governor model parameters.

Speed set point ω_{ref}	1 pu
Steady state gain $1/R$	25 pu
Maximum power order P_{max}	1 pu
Servo time constant T_s	0.1 sec
Governor time constant T_c	0.5 sec
Transient gain time constant T_3	0 sec
HP section time constant T_4	1.25 sec
Reheater time constant T_5	5.0 sec

Problems

2.1 (Admittance matrix) For Example 2.1, suppose that a new transmission line with series admittance Y_{25} and line charging B_{25} is added between Buses 2 and 5. Find the new admittance matrix and show all the diagonal entries.

2.2 (Admittance matrix) A five-bus power system is shown in Figure 2.15.

1) The line parameters are:
   Line 1-2: $(R, X, B) = (0.01, 0.1, 0.2)$ pu
   Line 2-3: $(R, X, B) = (0.006, 0.06, 0.12)$ pu
   In addition, there is a shunt capacitor with susceptance $j1.0$ pu located on Bus 2. Find Y_{22}, that is, the (2,2) entry of the admittance matrix Y corresponding to Bus 2 due to the transmission lines and the shunt capacitor. What other entries of the Y matrix can you also determine?
2) Set up the admittance matrix Y for the system, showing zero entries as "0" and nonzero entries as "*", with the voltage vector in the order of Bus 1 to Bus 5. Show how to re-order the Y matrix to minimize fill-ins in the sparse factorization process.

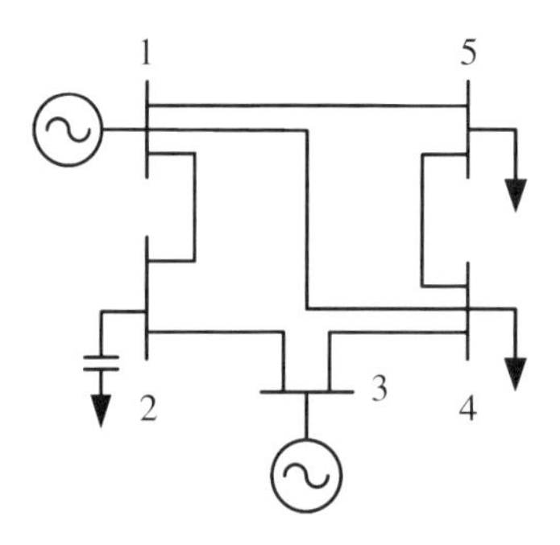

Figure 2.15 Power system for Problem 2.2.

2.3 (Power flow expressions) Derive the expressions for the power flow P_R and Q_R to the receiving-end bus in Figure 2.4. Also derive the expressions for the active and reactive power losses $P_S - P_R$ and $Q_R - Q_S$. Comment on these expressions.

2.4 (Power flow and losses) By examining the bus voltage magnitudes and angles, one can get a feel of which directions the active power and reactive power are flowing in a power system. For the transmission line shown in Figure 2.16 (the line charging is neglected), calculate P_i and Q_i, $i = 1, 2$, and the active and reactive power losses for several cases of $\tilde{V}_1$ and $\tilde{V}_2$ given in Table 2.14. Using Case a as the base case, discuss the changes in the other cases as compared to the base case.

$\tilde{V}_1$ $P_1 + jQ_1$ $P_2 + jQ_2$ $\tilde{V}_2$

$R = 0.01$ pu $X = 0.1$ pu

Figure 2.16 Two-bus system for Problem 2.4.

Table 2.14 Voltages for Problem 2.4.

Case	$\tilde{V}_1$ (pu)	$\tilde{V}_2$ (pu)
a	$1.03\angle 20°$	$1.00\angle 0°$
b	$0.97\angle 20°$	$1.00\angle 0°$
c	$1.03\angle 40°$	$1.00\angle 0°$
d	$1.05\angle 20°$	$1.00\angle 0°$

2.5 (Circulating reactive power) Two bus bars in the same substation but at different voltage levels are connected via two tap-changing transformers as shown in Figure 2.17. The transformer reactances and tap ratios are $X_1 = 0.04$ pu, $X_2 = 0.044$ pu, $n_1 = 1.01$, and $n_2 = 0.985$. (Note that these two transformers were made by different manufacturers having different leakage reactances and numbers of taps.) The high-side voltage is 1.02 pu and the power into the high-side bus is $P + jQ = 6 + j0$ pu (on a 100 MVA base). Find the low-side voltage and the power flow on each transformer from the high-side bus. Comment on the reactive power flow on these two transformers. (Hint: Start with an equation for balancing the currents into and out of the high-side bus.)

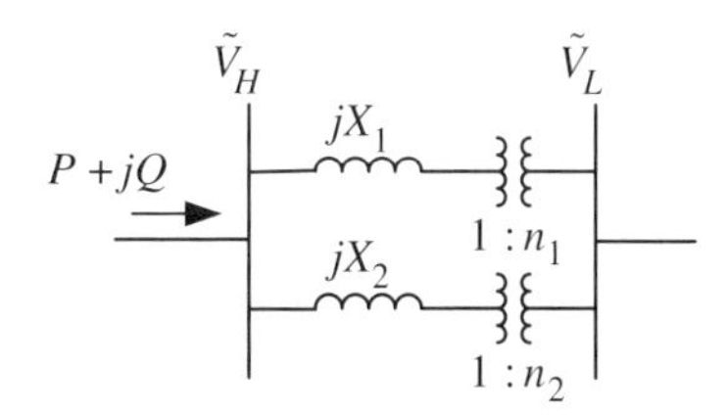

Figure 2.17 Two-bus system for Problem 2.5.

2.6 (Phase shifter) Based on the discussion in Section 2.6.3, develop the additional equations required to include a phase-shifter in the power flow formulation. Indicate the changes to the active and reactive flow power equations and the Jacobian matrix.

2.7 (Optimal ordering and sparse factorization) Reorder the matrix A

$$A = \begin{bmatrix} 2 & -0.05 & 0 & -1 & 0 \\ -0.5 & 4 & 0 & 0 & 0 \\ 0 & 0 & 3 & -1.5 & 0 \\ -1 & 0 & -0.2 & 2 & -1 \\ 0 & 0 & 0 & -1 & 4 \end{bmatrix} \tag{2.93}$$

into a block-diagonal form as suggested by [22] and Appendix 2.B, and compute the sparse LU factorization using the MATLAB® *lu* function.

2.8 (Jacobian matrix construction) The connections to Bus 2 in a large power system are shown in Figure 2.18.

1) Bus 1 is a PV (generator) bus, and Buses 2 and 3 are PQ (load) buses. The load on Bus 2 is $P_2 + jQ_2$. For simplicity, line resistances are neglected. The line reactances and capacitances are indicated on the figure. Derive the power flow expression at Bus 2 (that is, the power flow into Bus 2 from Buses 1 and 3 must be equal to the load at Bus 2). From this expression, derive the entries of the Jacobian matrix corresponding to Bus 2.
2) Given $X_{12} = 0.1$ pu, $X_{23} = 0.05$ pu, $B_{12} = 0.05$ pu, $B_{23} = 0.1$ pu, $P_2 = 2$ pu, and $Q_2 = 0.3$ pu, find the numerical values of these Jacobian entries at initialization when all load bus voltage magnitudes are assumed to be 1.0 pu and their angles are 0°.
3) What entries of the Jacobian derived in Part 2 will be neglected in a decoupled loadflow solution technique? Are they small?
4) Simplify the results of Part 1 if Bus 1 is the swing bus.

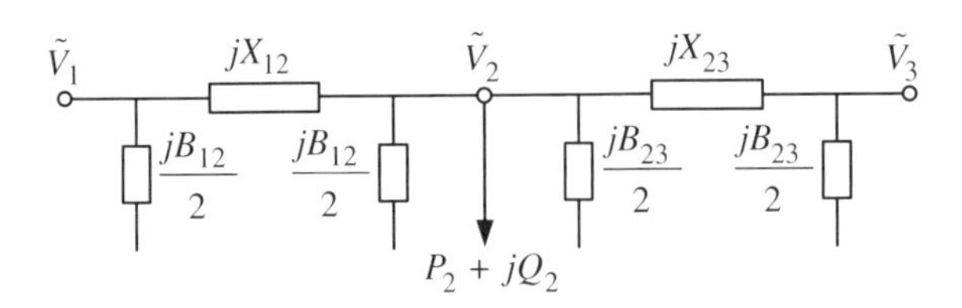

Figure 2.18 Figure for Problem 2.8.

2.9 (Unknown variables in power flow computation) For the 2-area system in Figure 2.7, the generator buses are PV buses with Generator 1 being the swing bus. All the other buses are PQ buses. Assume that there are no variable tap and phase-shifting transformers. The loads (at Buses 4 and 14) are given. List the unknown variables to be solved for this power flow.

2.10 (Power flow computation) Use the Power System Toolbox to run the power flow cases stated below. The system is the 2-area model in Figure 2.7 and the power flow data file is *data2a.m*, available in the problem folder. When you execute the *loadflow* function, use 1×10^{-6} as the tolerance and set the maximum iteration number to 50. You will have to change some of the loadflow data. Note that the system base is 100 MVA.

1) Set the load at Bus 4 to be $12.76 + j2.0$ pu and the load at Bus 14 to be $14.76 + j2.0$ pu. Solve the power flow.
2) Repeat Part 1, but now set the maximum reactive power output of Generator 2 at 5.5 pu.
3) Repeat Part 1, but decrease the load at Bus 4 to $11.76 + j2.0$ pu and increase the load at Bus 14 to $15.76 + j2.0$ pu.
4) Repeat Part 3 by installing a synchronous condenser on Bus 101.[9] In the power flow data, change Bus 101 to be a *PV* bus with its voltage set at 1.03 pu.

Show the power flow summary report for each case. Then make a table with each case as a column to show the generator active and reactive power output, voltages at the generator buses and at Buses 3, 13, and 101, and the active power flow from Bus 3 to Bus 13. What are your observations?

2.11 (Dishonest NR method) Repeat Part 1 of Problem 2.10 by updating the Jacobian at every other iteration. Compare the rates of convergence for the NR method and the dishonest NR method.

2.12 (Power flow computation) A four-bus system is given in Figure 2.19. Both generators are producing 400 MW, and both loads are consuming 400 MW and 100 MVar. The lines have no resistance, and their per unit reactance values are $X_{12} = 0.04$, $X_{23} = 0.12$, $X_{34} = 0.04$, and $X_{41} = 0.04$, all on a 100 MVA base.

1) Set up the line and bus data and solve the AC power flow using large reactive power limits for the generators. All generator bus voltage magnitude is set at 1.0 pu. Comment on the flow of active power from the generators to the loads.
2) Use the DC power flow method to compute the bus angles and compare them to the AC loadflow solution.

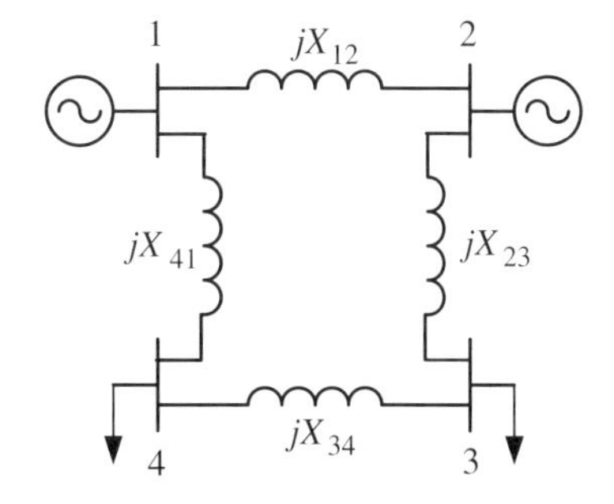

Figure 2.19 Four-bus system for Problem 2.12.

9 A synchronous condenser is a synchronous machine without a prime mover (turbine) and thus only provides reactive power.

2.13 (Multi-area power flow) Use the power flow expressions to assemble the 2-area tieline flow equation (2.46) for the system shown in Figure 2.8. Assume that the line parameters are (R_i, X_i, B_i), $i = 1, 2$. Take P_{12} as the total active power leaving Area 1.

2.14 (Multi-area power flow)

1) For the 3-area system shown in Figure 2.9, suppose Area 1 exports a total of 500 MW to the other two areas, and Area 2 imports 400 MW from the other two areas. Show how you can formulate a Newton-Raphson loadflow problem for this system.
2) Can this problem be formulated and solved if the tie flow from Area 1 to Area 2 is exactly 300 MW, the tie flow from Area 1 to Area 3 is 200 MW, and the tie flow from Area 3 to Area 2 is 100 MW? What happens if a phase-shifter (a rather powerful one) is available on one of the lines between Areas 1 and 2?

3

Steady-State Voltage Stability Analysis

3.1 Introduction

As evident from the power flow formulation in Chapter 2, active power transfer requires system bus voltages to be close to their nominal values. Thus a power system needs to have sufficient reactive power supply at appropriate locations, such as major power transfer paths, to support the bus voltages and maintain voltage stability. In particular, at high power transfer situations, the demand for reactive power would also be high. Although there are many definitions of voltage stability [25–27], voltage instability occurs when the voltages in a power system are much lower than 1.0 pu. It is, however, not possible to define a precise voltage level when voltage instability would occur because it is dependent on system operating conditions, the reactive power supply, and the network impedances and configurations. When an operating condition is voltage stable, reducing the power transfer will increase the bus voltages along the transfer path. If an operating condition is voltage unstable, reducing the power transfer will further reduce the voltage levels. As a result, the system loses its ability to recover the system voltage.

A power system rarely encounters voltage stability problems in normal operating conditions. Examples of causes of voltage instability due to insufficient reactive power support in critical locations include the following scenarios.

- The outage of a critical line carrying a significant amount of active power will have the dual impact of (1) losing the reactive power of the line provided through line charging and (2) forcing other lines in parallel to carry more active power flows, resulting in higher reactive power losses.
- A generator trip in a load area will result in (1) the loss of its reactive power output to maintain the voltages at the load buses and (2) additional inflow of power from areas external to the load area, causing lower voltage on the transfer path.
- After a severe disturbance, a generator temporarily increases its field current to its maximum short-term limit to provide additional voltage support. This ability, however, is thermally limited, and the field current will be reduced to its normal rating after a specified period of time, resulting in a significant reduction of reactive power output. Additional reactive power should be secured before the field current limiter times out.
- The load increases faster or to a higher than expected level, as switched capacitors may not be turned on in a timely manner, resulting in inadequate reactive power support.

Power System Modeling, Computation, and Control, First Edition. Joe H. Chow and Juan J. Sanchez-Gasca.

Companion website: www.wiley.com/go/chow/power-system-modeling

- A reversal of power flow may also cause voltage collapse. Reactive power compensations such as switched capacitor and reactor banks are dispatched to support active power flow in a given direction on certain transfer paths. In power systems with loop flows, a disruption in some part of the loop will result in a power reversal, creating new transmission paths without proper reactive power and voltage support.

This chapter starts by describing in Section 3.2 the salient features of two voltage stability incidents, each resulting in wide-spread blackout. These and other incidents have led to changes in planning and operating procedures for improving the reliability of power systems. In Section 3.3, reactive power required for power transmission on AC lines is analyzed using distributed parameter line models. Then in Section 3.4, the voltage stability analysis of a stiff voltage source supplying a load is discussed using a PV curve. Section 3.5 discusses the voltage stability analysis for large power systems. Sections 3.6 and 3.7 discuss methods for solving the power flow at the voltage collapse point. Other voltage stability topics including dynamic voltage regulation are considered in Sections 3.8 and 3.9. Section 3.10 describes several additional methods for assessing voltage stability.

3.2 Voltage Collapse Incidents

One of the first voltage stability incidents leading to a widespread blackout happened in the French power system on January 12, 1987 [28]. Subsequently many other voltage instability related blackouts have occurred in various countries. Two of these incidents are summarized in this section.

3.2.1 Tokyo, Japan: July 23, 1987

In the summers of 1986 and 1987, power consumption in the Tokyo Electric Power Company (TEPCO) system would dip 2000–3000 MW during lunch time. At 1 pm, the load would pick up rapidly when work resumed, as shown by the load profiles in Figure 3.1.a for several summer heavy-load days. On July 23, 1987, the pre-lunch power consumption was higher than the other heavy-load days. The post-lunch load increase was estimated at 400 MW/min, twice as fast as the previously fastest rate. As the system voltage started to drop, all available shunt capacitors were switched on. The problem was further compounded by constant power loads, such as air-conditioning equipment, whose current would increase as voltage decreased. As a result, the voltage around Tokyo dropped rapidly, as shown in Figure 3.1.b, eventually causing a blackout in the city of Tokyo shortly after 13:19. To reinforce reactive power supply and control, TEPCO has since taken many measures, including the installation of additional shunt capacitors, dynamic reactive power reserves, under-voltage load shedding controllers, and an on-line voltage security monitoring system [30]. No voltage collapse incidents have happened since then.

3.2.2 US Western Power System: July 2, 1996

In the July 2, 1996 event [31], the Western Electricity Coordinating Council (WECC) system loading was high. Several heavily loaded lines sagged into trees, creating

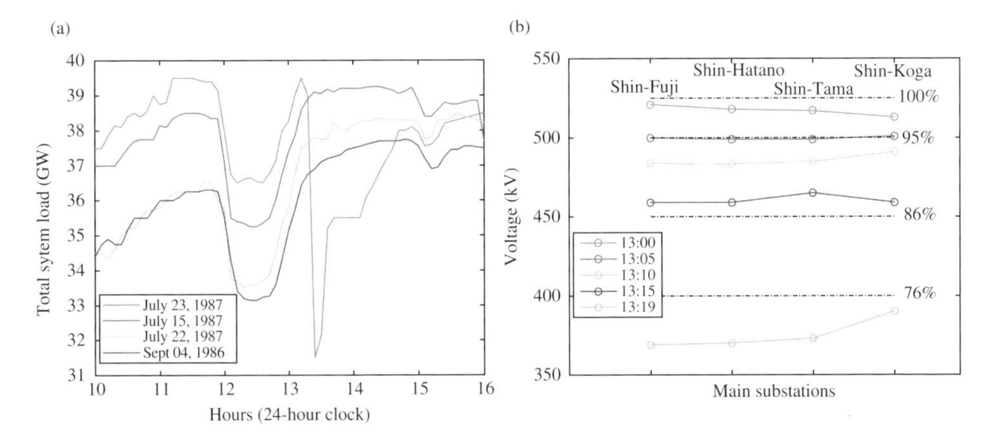

Figure 3.1 TEPCO system: (a) system load profile around noon time for several high-load days and (b) voltage drops at main substations with time progression indicated by different colors [29] (data courtesy of Teruo Ohno, TEPCO) (see color plate section).

short-circuit faults and causing the trip of several critical lines. The trip of the Jim Bridge generating unit in Wyoming resulted in a further drop in substation voltages, which was followed by a 0.25 Hz voltage oscillation with an amplitude of 10 kV peak-to-peak in the Malin Substation (near the Oregon–California border), as shown in Figure 3.2. A system voltage collapse occurred when a 230 kV line between Montana and Idaho tripped. In [31], it was noted that activating the two switched-capacitor banks at the Malin Substation would have been helpful. However, the voltage oscillation repeatedly pushed the voltage above 530 kV, so the arming of the capacitor installation was reset several times. The authors of [31] also mentioned the possibility that generator over-excitation limiter operations had probably occurred, although there was no definite confirmation. Such actions would have reduced the amount of reactive power output from generators. They also recommended several mechanisms to improve system voltage stability, such as state-of-the-art protective relaying, under-voltage load shedding, and online security assessment based on current operating conditions.

3.3 Reactive Power Consumption on Transmission Lines

In discussing the power flow solution, Chapter 2 stresses the importance of having the proper system voltage to support active power transfer. To understand the need for reactive power, it is appropriate to start by reviewing the role of reactive power in supporting active power transfer in transmission lines. Consider the long-line transmission line model shown in Figure 3.3 [12]. Let the current $\tilde{I}$ be flowing from the sending end, defined to be at $x = \ell$, to the receiving end, defined to be at $x = 0$, such that ℓ is the line length. Denote the sending end voltage as $\tilde{V}_S$ and the receiving end voltage as $\tilde{V}_R$. The line parameters operating at a frequency of ω rad/s are denoted by the series impedance per unit length $z = R + j\omega L$ and the shunt admittance per unit length $y = G + j\omega C$, where R, L, G, and C are the series resistance, series inductance, shunt conductance, and shunt susceptance per unit length, respectively.

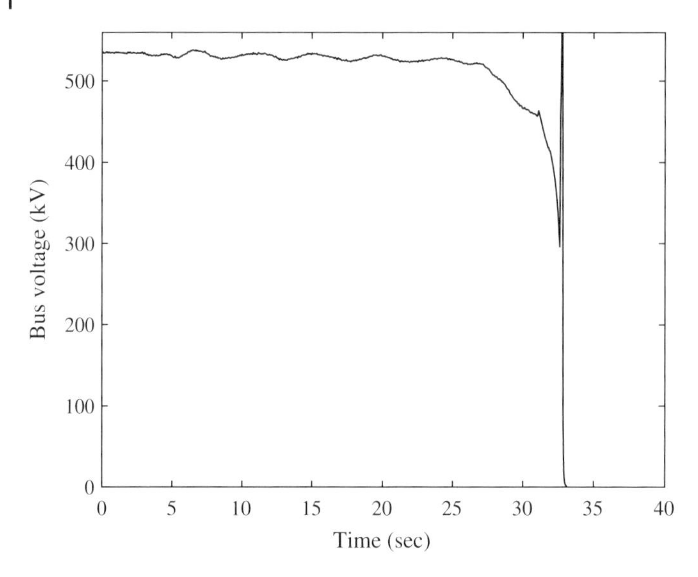

Figure 3.2 WECC voltage collapse incident: voltage response at the Malin substation on July 2, 1996 (data courtesy of Dmitry Kosterev, Bonneville Power Administration).

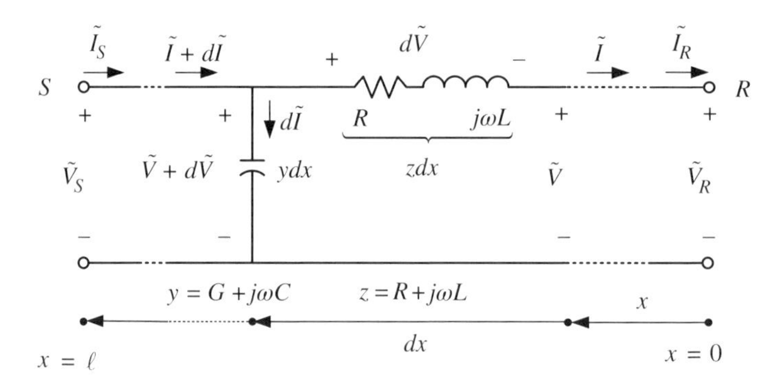

Figure 3.3 Model of a long transmission line as incremental sections.

Dividing a transmission line of length ℓ into sections with length dx, the incremental voltage phasor $d\tilde{V}$ and current phasor $d\tilde{I}$ can be represented by

$$d\tilde{V} = Iz \cdot dx, \quad d\tilde{I} = \tilde{V}y \cdot dx \tag{3.1}$$

These incremental variables can be combined to obtain the wave equations

$$\frac{d^2\tilde{V}}{dx^2} = z\frac{d\tilde{I}}{dx} = yz\tilde{V}, \quad \frac{d^2\tilde{I}}{dx^2} = y\frac{d\tilde{V}}{dx} = yz\tilde{I} \tag{3.2}$$

The solution to equation (3.2) consists of a forward travelling wave and a backward travelling wave, as determined by the boundary voltage and current conditions at the sending end and the receiving end. Defining the propagation constant as

$$\gamma = \sqrt{yz} = \alpha + j\beta \tag{3.3}$$

and the characteristic impedance and admittance as $Z_o = \sqrt{z/y}$ and $Y_o = \sqrt{y/z}$, respectively,[1] the voltage and current solutions, based on the receiving end voltage $\tilde{V}_R$ and current $\tilde{I}_R$, are given by

$$\tilde{V} = \frac{\tilde{V}_R + Z_o\tilde{I}_R}{2}e^{\gamma x} + \frac{\tilde{V}_R - Z_o\tilde{I}_R}{2}e^{-\gamma x} \tag{3.4}$$

$$\tilde{I} = \frac{\tilde{V}_R/Z_o + \tilde{I}_R}{2}e^{\gamma x} - \frac{\tilde{V}_R/Z_o - \tilde{I}_R}{2}e^{-\gamma x} \tag{3.5}$$

The $e^{\gamma x}$ term is the incident wave and the $e^{-\gamma x}$ term is the reflected wave. Using hyperbolic functions, (3.4) and (3.5) can be expressed as

$$\tilde{V} = \tilde{V}_R\cosh(\gamma x) + Z_o\tilde{I}_R\sinh(\gamma x), \quad \tilde{I} = \tilde{I}_R\cosh(\gamma x) + Y_o\tilde{V}_R\sinh(\gamma x) \tag{3.6}$$

At $x = \ell$, the sending voltage and current are given by

$$\tilde{V}_S = \tilde{V}_R\cosh(\gamma\ell) + Z_o\tilde{I}_R\sinh(\gamma\ell), \quad \tilde{I}_S = \tilde{I}_R\cosh(\gamma\ell) + Y_o\tilde{V}_R\sinh(\gamma\ell) \tag{3.7}$$

For a lossless transmission line, $\gamma = j\beta$, such that $\cosh(\gamma x) = \cos(\beta x)$ and $\sinh(\gamma x) = j\sin(\gamma x)$. Thus (3.6) reduces to

$$\tilde{V} = \tilde{V}_R\cos(\beta x) + jZ_o\tilde{I}_R\sin(\beta x), \quad \tilde{I} = \tilde{I}_R\cos(\beta x) + jY_o\tilde{V}_R\sin(\beta x) \tag{3.8}$$

As active power is delivered on the transmission line from the sending end to the receiving end, reactive power is consumed by the series inductance. This consumption, however, is partially offset by the reactive power injected by the shunt susceptance. Example 3.1 provides an assessment of reactive power consumption for different levels of power transfer.

Example 3.1: Voltage profile on a transmission line

Derive the voltage and current profiles on an overhead transmission line when the load is resistive for four cases given by (a) $Z_R = \infty$ (open circuit), (b) $Z_R = 2Z_o$, (c) $Z_R = Z_o$, and (d) $Z_R = 0.5Z_o$. Also compute the active and reactive power P_S and Q_S leaving the sending end and the active and reactive power P_R and Q_R arriving at the receiving end. Use the line parameters on p. 209 of [6]: the line length is 300 km, the nominal system voltage is 345 kV, the line is lossless with $\beta = 0.00129$ rad/m, and the surge impedance is $Z_o = 285$ ohms. Assume that the sending end voltage magnitude is $V_S = 345$ kV $= 1$ pu. Use the value of Z_o as the base impedance, such that the system base is $(345 \times 10^3)^2/285$ VA $= 417.6$ MVA.

Solutions: With a resistance as the load, the load is at unity power factor. The voltage and current relationship at the receiving (load) end is

$$\tilde{V}_R/\tilde{I}_R = kZ_o \quad \text{or} \quad \tilde{I}_R = \tilde{V}_R/(kZ_o) \tag{3.9}$$

for $k \to \infty$ and $k = 2, 1, 0.5$. Substitution of this relationship into (3.8) with $x = \ell$ results in

$$\tilde{V}_S = \tilde{V}_R\cos(\beta\ell) + j\tilde{V}_R/(k)\sin(\beta\ell) \quad \Rightarrow \quad \tilde{V}_R = \frac{\tilde{V}_S}{\cos(\beta\ell) + j(1/k)\sin(\beta\ell)} \tag{3.10}$$

---

1 The quantity Z_o is also known as the surge impedance.

Using these expressions of $\tilde{V}_R$ and $\tilde{I}_R$, the expressions in (3.8) become

$$\tilde{V}(x) = \frac{\tilde{V}_S}{\cos(\beta\ell) + j(1/k)\sin(\beta\ell)}(\cos(\beta x) + j(1/k)\sin(\beta x)) \tag{3.11}$$

$$\tilde{I}(x) = \frac{\tilde{V}_S}{\cos(\beta\ell) + j(1/k)\sin(\beta\ell)}(1/(kZ_o)\cos(\beta x) + jY_o\sin(\beta x)) \tag{3.12}$$

Using the sending end voltage as the reference, the phase of $\tilde{V}_S = 1$ pu is set to zero. The formulas (3.11) and (3.12) can be readily used to compute $\tilde{V}(x)$ and $\tilde{I}(x)$ for x from 0 to 300 km, with $k \to \infty$ and $k = 2, 1, 0.5$. The voltage and current profiles are shown in Figure 3.4. The voltage and current at the receiving end, the sending and receiving end active and reactive power, and the reactive power loss $Q_S - Q_R$ in the line are also given in Table 3.1.[2]

It is important to understand the implications of these four cases for voltage stability analysis:

1) $k \to \infty$: The load current $\tilde{I}_R = 0$, although $\tilde{I}_S \neq 0$. The current $\tilde{I}_S$ is in quadrature with $\tilde{V}_S$, thus there is no active power injected at the sending end. At the receiving end, there is an excess of reactive power from the line shunt capacitance, driving up the receiving end voltage.[3] At the sending end, the excess reactive power is injected out of the line.
2) $k = 2$: At this low loading level, reactive power is still generated by the transmission line, as seen in Q_S. Furthermore, the receiving end voltage is still higher than the sending end voltage.
3) $k = 1$: Under this condition, the reactive power required for the power transfer is exactly compensated by the transmission line. The sending and receiving reactive power flows are both zero. The voltage magnitude stays at 1 pu throughout the line. This fact can be readily shown analytically by setting $k = 1$ in (3.11). This power transfer level is known as the surge impedance loading.
4) $k = 0.5$: At this higher level of loading, the reactive power required for the power transfer has to be supplemented by Q_S from the sending end. Without voltage support at the receiving end, the receiving end voltage drops to 0.837 pu, which may cause voltage stability issues.

■

As shown in Chapter 2, in power flow and time simulation programs, the transmission line shown in Figure 3.3 is normally modeled by an equivalent π model, as shown in Figure 2.1. The series impedance Z and shunt admittance Y of this lumped-parameter model can be readily derived from (3.7) as

$$Z = Z_o\sinh(\gamma\ell) = R + jX, \quad Y = \frac{2}{Z_o}\tanh\frac{\gamma\ell}{2} = G + jB \tag{3.13}$$

---

2 In (3.12), $Z_o = 1$ for $\tilde{I}(x)$ to be in per unit.

3 When a transmission line is lightly loaded, the voltage on the line may be significantly higher than the nominal value. This is known as the Ferranti effect [4]. High voltages on transmission lines can cause insulation failures and result in faults. With the 3-phase conductors closely packed, underground cables have a higher shunt capacitance than overhead transmission lines and thus are more prone to the Ferranti effect, which is a concern particularly during system restoration as the load may be small. Thus cable systems are typically equipped with shunt reactors, which can be switched in to prevent high voltages.

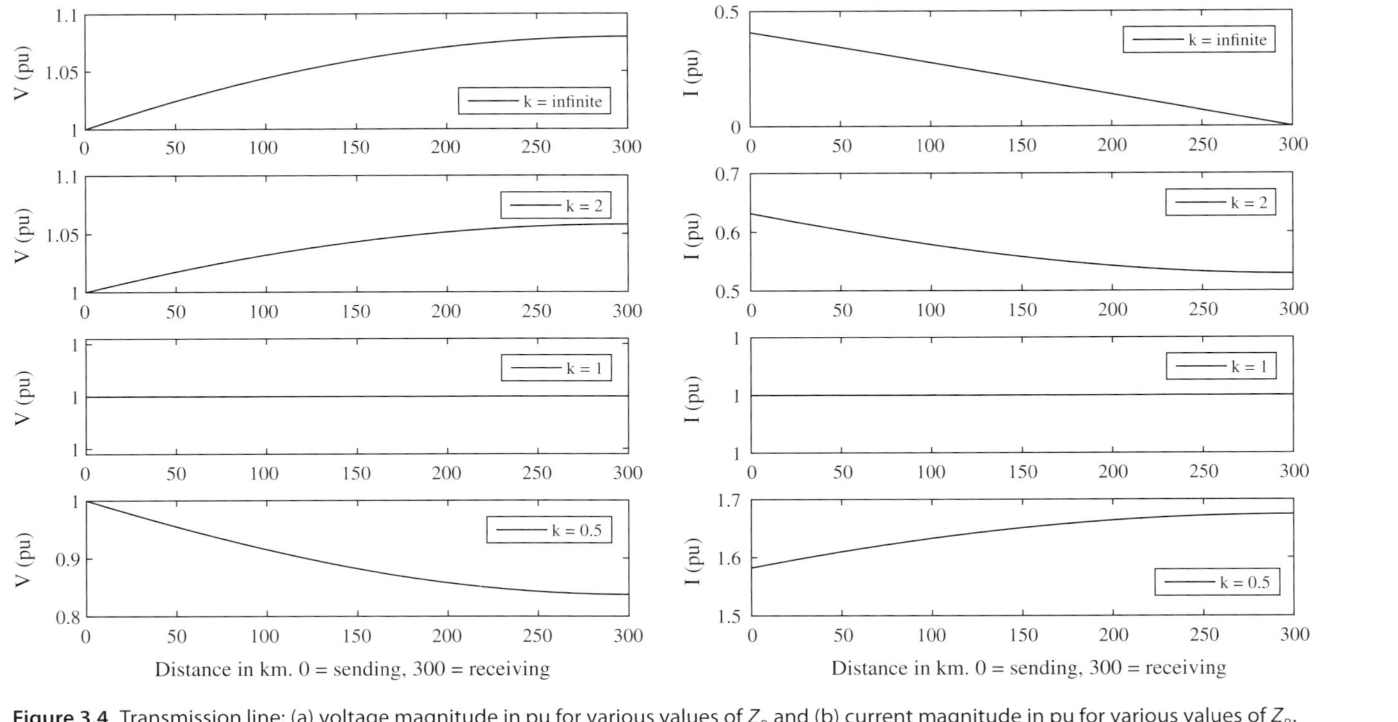

Figure 3.4 Transmission line: (a) voltage magnitude in pu for various values of Z_R and (b) current magnitude in pu for various values of Z_R.

Table 3.1 Voltage and flow quantities in pu at the sending and receiving ends of the transmission line.

k	$\tilde{I}_S$	$\tilde{V}_R$	$\tilde{I}_R$	P_S	Q_S	P_R	Q_R	Q_{loss}
∞	$0.4076\angle 90°$	$1.0799\angle 0°$	0	0	−0.4706	0	0	−0.4706
2	$0.6321\angle 27.67°$	$1.0581\angle -11.52°$	$0.5291\angle -11.52°$	0.5598	−2.935	0.5598	0	−0.2935
1	$1.0\angle 0°$	$1.0\angle -22.17°$	$1.0\angle -22.17°$	1.0	0	1.0	0	0
0.5	$1.5821\angle -27.67°$	$0.8370\angle -39.18°$	$1.6741\angle -39.18°$	1.4012	0.7346	1.4012	0	0.7346

Table 3.2 Transmission line parameters in pu on 100 MVA base (source: line parameters are from various US power system stability data sets; line lengths are estimated values; note that the actual line parameters would depend on the conductor structure and the terrain).

Line type	Voltage level (kV)	Length (miles)	R (pu)	X (pu)	B (pu)
Overhead	765	100	0.00044	0.01255	6.17
Overhead	500	150	0.0012	0.0297	0.9672
Overhead	345	100	0.0041	0.0423	0.6866
Overhead	230	100	0.01595	0.1081	0.2013
Underground cable	345	10	0.00067	0.00367	3.2300

In general, only R, X, and B are used to model transmission lines, as the shunt conductance G is negligible. The parameters of several real transmission lines of various lengths and voltage levels are shown in Table 3.2.

Note that as the voltage level increases, the pu values of the line resistance R and reactance X decrease, but that of the line charging value B increases. As a result, high-voltage transmission lines more readily facilitate the transfer of larger amount of power. In particular, transmitting the same amount of active power on a higher voltage level transmission line would require less reactive power support. As mentioned earlier, B is also larger for underground cables.

Example 3.2: Surge impedance loading

Find the surge impedance loading (SIL) of the 765 and 345 kV transmission lines in Table 3.2, neglecting the effect of the line resistance.

Solutions: The SIL is the level of power transfer through a transmission line in which the required reactive power is entirely supplied by line susceptance. It can be derived as

$$\text{SIL} = V_{\text{LL}}^2/Z_o \tag{3.14}$$

where V_{LL} is the line-to-line voltage in volts and Z_o is the characteristic impedance in ohms. Using per-unitized values, (3.14) simplifies to $\text{SIL} = 1/Z_o = 1/\sqrt{X/B}$ pu. Thus for the 765 kV line

$$Z_o = 0.04510 \text{ pu}, \quad \text{SIL} = 22.17 \text{ pu} = 2217 \text{ MW} \tag{3.15}$$

and for the 345 kV line

$$Z_o = 0.2482 \text{ pu}, \quad \text{SIL} = 4.029 \text{ pu} = 402.9 \text{ MW} \tag{3.16}$$

■

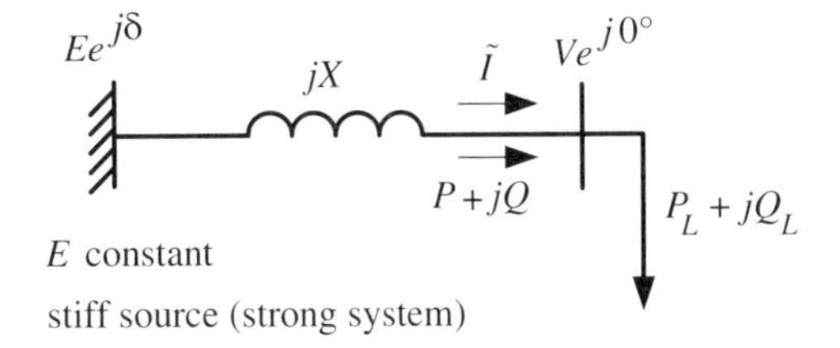

Figure 3.5 A stiff voltage source supplying power to a load.

3.4 Voltage Stability Analysis of a Radial Load System

From Figure 3.4, the voltage at the receiving end of the transmission line decreases as the active power flow is increased. For a power system to function properly, the system voltage must be maintained in a tight range about 1 pu. Low voltages at the receiving end will result in increased active and reactive power losses on the line. If the voltage is sufficiently low, induction machines will stall or overheat. This process can reinforce itself and lead to voltage collapse.

In this section, the discussion on the voltage variations profile on a transmission line as load changes will be expanded into an investigation on how much power can be transferred on a transmission line with limited reactive power support. The maximum power transfer limit will coincide with the voltage collapse point. Following the discourse in many voltage stability analysis textbooks [25–27], voltage stability analysis begins with the study of the stiff-source-to-load system shown in Figure 3.5.

In Figure 3.5, a load $P_L + jQ_L$ draws power from a stiff power source with a fixed voltage E via a transmission line with reactance X. The load bus voltage is denoted by V and its angle is fixed at zero. The source voltage angle is taken to be δ, at an appropriate value to deliver the active power to the load. This system is a good approximation of a radial load center supplied by a bulk power system, with many generating sources to hold the voltage constant. The voltage source E and the reactance X can be considered as a Thèvenin equivalent model. A resistance R can be added to the Thèvenin equivalent to represent active power transmission losses. Also the effect of the transmission line shunt susceptance can be included in the Thèvenin equivalent.

The normalization approach in [25] is used to develop a general result for the variation of the load bus voltage versus the active power transfer, commonly known as the PV-curve analysis.

Consider the short-circuit situation such that $P_L = Q_L = 0$. The short-circuit current and its magnitude are given by

$$\tilde{I}_{\text{SC}} = \frac{\tilde{E}}{jX} = -j\frac{Ee^{j\delta}}{X}, \quad I_{\text{SC}} = |\tilde{I}_{\text{SC}}| = \frac{E}{X} \tag{3.17}$$

The short-circuit power delivered by the source voltage and its magnitude are

$$S_{\text{SC}} = \tilde{E} \cdot \tilde{I}_{\text{SC}}^* = Ee^{j\delta} \cdot j\frac{Ee^{-j\delta}}{X} = j\frac{E^2}{X}, \quad |S_{\text{SC}}| = \frac{E^2}{X} \tag{3.18}$$

Note that the source power is purely reactive as there is no line resistance.

The expressions for the active and reactive power delivered from the source voltage to the load are given by (2.15) in Chapter 2 as

$$P = \frac{EV}{X}\sin\delta, \quad Q = \frac{EV\cos\delta}{X} - \frac{V^2}{X} \tag{3.19}$$

Introduce the normalized variables

$$p = \frac{P}{|S_{SC}|} = \frac{PX}{E^2}, \quad q = \frac{Q}{|S_{SC}|} = \frac{QX}{E^2}, \quad v = \frac{V}{E} \tag{3.20}$$

Substituting the variables (3.19) into (3.20), the normalized equations become

$$\frac{PX}{E^2} = \frac{V}{E}\sin\delta \quad \Rightarrow \quad p = v\sin\delta \tag{3.21}$$

$$\frac{QX}{E^2} = \frac{V}{E}\cos\delta - \frac{V^2}{E^2} \quad \Rightarrow \quad q = v\cos\delta - v^2 \tag{3.22}$$

Furthermore,

$$p = \sqrt{v^2\sin^2\delta} = \sqrt{v^2(1-\cos^2\delta)} = \sqrt{v^2 - (q+v^2)^2} \tag{3.23}$$

In voltage stability analysis, the objective is to assume certain properties of the load and compute the maximum amount of active power that can be supplied to the load. Equation (3.23) has three variables: p, q, and v. Consider the case of a unity power factor load such that $q = 0$, that is, the load is purely resistive and is denoted by the resistance R. Then (3.23) becomes

$$p = \sqrt{v^2 - v^4} \tag{3.24}$$

As v decreases from 1 to 0, p will increase from 0 but will eventually drop back to zero again. To find the maximum value of the power transmitted, the partial derivative of p with respect to v is set to zero:

$$\frac{\partial p}{\partial v} = \frac{1}{2}\frac{2v - 4v^3}{\sqrt{v^2 - v^4}} = 0 \tag{3.25}$$

The solution to (3.25) yields the critical voltage

$$2v^2 = 1 \quad \Rightarrow \quad v_{\text{crit}} = \frac{1}{\sqrt{2}} = 0.707 \tag{3.26}$$

Thus the maximum power transmitted to the load at unity power factor is

$$p_{\max} = \sqrt{v_{\text{crit}}^2 - v_{\text{crit}}^4} = 0.5 \tag{3.27}$$

The angle separation between the source and the load bus at maximum power is given by

$$\delta = \sin^{-1}\left(\frac{p_{\max}}{v_{\text{crit}}}\right) = \sin^{-1}\left(\frac{0.5}{1/\sqrt{2}}\right) = \sin^{-1}\left(\frac{1}{\sqrt{2}}\right) = 45^\circ \tag{3.28}$$

From (3.19) and (3.20), the maximum active power that can be delivered to the load is

$$P_{\max} = \frac{p_{\max}E^2}{X} = \frac{0.5E^2}{X} = \frac{v_{\text{crit}}^2}{X} = \frac{v_{\text{crit}}^2}{R} \tag{3.29}$$

that is, the maximum power is delivered when the resistive load is equal in value to the line reactance, that is, $R = X$. At maximum active power delivery, the current is

$$\tilde{I} = \frac{Ee^{j\delta}}{R + jX} = \frac{Ee^{j45^\circ}}{\sqrt{2}Xe^{j45^\circ}} = \frac{E}{\sqrt{2}X} = \frac{v_{\text{crit}}}{X} \tag{3.30}$$

Thus the reactive power supply from the stiff source is

$$Q_S = E \sin \delta \frac{E}{\sqrt{2}X} = \frac{E^2}{2X} \tag{3.31}$$

Summarizing, for a purely resistive load, the current, voltage, and active power for any value of the load resistance R with X fixed are given by

$$\frac{I}{I_{SC}} = \frac{I}{E/X} = \frac{1}{\sqrt{1 + (R/X)^2}} = \frac{X/R}{\sqrt{1 + (X/R)^2}} = \begin{cases} 1, & R = 0 \\ 0, & R \to \infty \end{cases} \tag{3.32}$$

$$\frac{V}{E} = \frac{R}{\sqrt{R^2 + X^2}} = \frac{1}{\sqrt{1 + (X/R)^2}} = \begin{cases} 0, & R = 0 \\ 1, & R \to \infty \end{cases} \tag{3.33}$$

$$\frac{P}{P_{\max}} = \frac{E^2R/(R^2 + X^2)}{E^2/(2X)} = \frac{2RX}{R^2 + X^2} = \frac{2X/R}{1 + (X/R)^2} = \begin{cases} 0, & R = 0 \\ 0, & R \to \infty \end{cases} \tag{3.34}$$

The asymptotic values when $R = 0$ (short circuit) and $R \to \infty$ (open circuit) are also shown on the right-hand side of these expressions. The variations of p, q, and v as functions of R are shown in Figure 3.6.

When $q \neq 0$, it is common to assume that the load has a fixed power factor as it varies. Thus (3.23) can be rewritten as

$$p = \sqrt{v^2 - (q + v^2)^2} = \sqrt{v^2 - (p \tan \phi + v^2)^2} \tag{3.35}$$

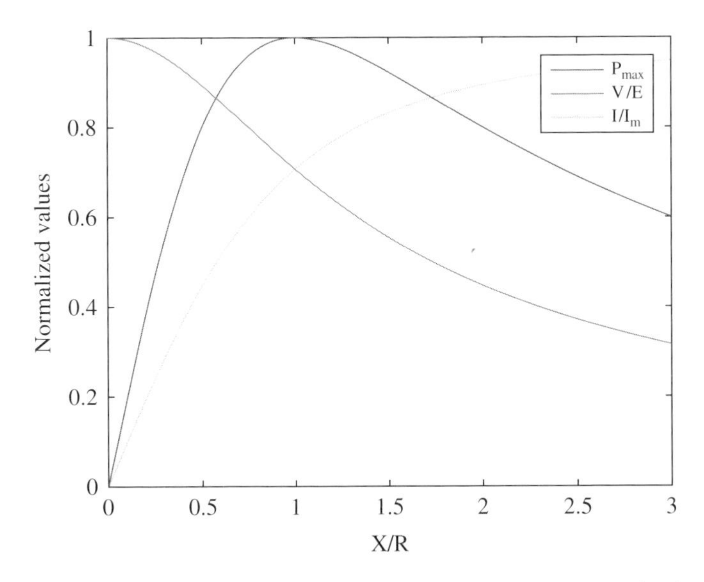

Figure 3.6 Voltage, current, and power profile for various resistive loads (see color plate section).

where ϕ is the power factor (pf) angle defined by

$$\phi = \cos^{-1}(\text{pf}), \quad \text{pf} = \frac{p}{p^2 + q^2} \tag{3.36}$$

It follows that the reactive power is

$$q = \frac{QX}{E^2} = \frac{PX}{E^2} \tan\phi = p\tan\phi \tag{3.37}$$

Equation (3.35) can be rewritten as a quadratic equation of v^2:

$$(v^2)^2 + (p^2\tan^2\phi - 1)v^2 + p^2(1 + \tan^2\phi) = 0 \tag{3.38}$$

For each pair of fixed p and ϕ, this equation yields two solutions for v^2. As p is increased from 0, both solutions will satisfy $v^2 \geq 0$. Thus v has two solutions, which are the positive roots of v^2. At the critical point of maximum power delivery, the two solutions of v are identical.[4]

The solutions to (3.38) for several different power factors can be computed numerically and are shown in Figure 3.7. These are the so-called power-voltage or *PV* curves. Note that the unity power factor curves in Figure 3.7 can also be obtained from the voltage and power profiles in Figure 3.6.

To compute the *PV* curves, it is more expedient to rewrite (3.38) as a quadratic equation for p

$$p^2(1 + \tan^2\phi) + 2\tan\phi v^2 p + v^4 - v^2 = 0 \tag{3.39}$$

The general solution to (3.39) is given by

$$p = \left(-v^2\tan\phi \pm v\sqrt{1 + \tan^2\phi - v^2}\right)/(1 + \tan^2\phi) \tag{3.40}$$

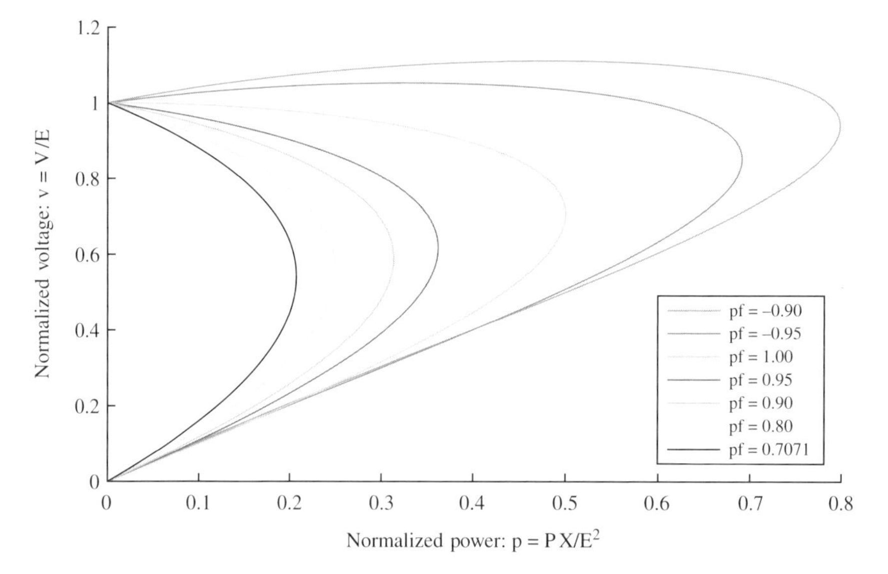

Figure 3.7 *PV* curves for various power factors (see color plate section).

4 Further increase in p will result in v^2 being a complex number, and hence there will be no admissible solutions for v.

For lagging power factor loads ($\tan\phi > 0$), the load bus voltage will be less than or equal to 1 pu as the load bus consumes reactive power. Thus varying v from 1 to 0 and computing p using the positive root of $\sqrt{1+\tan^2\phi - v^2}$, yields

$$p = \left(-v^2\tan\phi + v\sqrt{1+\tan^2\phi - v^2}\right)/(1+\tan^2\phi) \tag{3.41}$$

For leading power factor loads ($\tan\phi < 0$), the load bus voltage will be higher than 1 pu for light loads. At maximum voltage v_{max}, $v_{max}^2 = 1 + \tan^2\phi$. The computation of the PV curve will be divided into two segments: v from 1 to v_{max} using

$$p = \left(-v^2\tan\phi - v\sqrt{1+\tan^2\phi - v^2}\right)/(1+\tan^2\phi) \tag{3.42}$$

and v from v_{max} to 0 using (3.41).

For each of the PV curves in Figure 3.7, the maximum power transfer occurs at the voltage collapse point. The slopes at the voltage collapse points are vertical, indicating infinite sensitivities. At the lower part of the PV curves, the system is voltage unstable. Any decrease in power consumption will further decrease the load voltage.

The voltage collapse point is strongly dependent on the amount of reactive power consumed or provided by the load. Consider the unity power factor voltage collapse point (v_{crit}^1 and p_{crit}^1) as the reference. When $\tan\phi > 0$, that is, power factors are positive, the load also consumes reactive power. Thus the source needs to deliver reactive power to the load, reducing the ability of the transmission line to deliver active power. As a result the critical voltage and the critical power are both less than v_{crit}^1 and p_{crit}^1, respectively. On the other hand, when $\tan\phi < 0$, that is, power factors are negative, the load provides some reactive power. Thus the source needs to deliver less reactive power to the load, allowing the transmission line to deliver more active power. As a result the critical voltage and the critical power are both higher than v_{crit}^1 and p_{crit}^1, respectively.

3.4.1 Maximum Power Transfer

The analysis shown earlier assumes that the Thèvenin impedance is purely reactive. If the resistance of the Thèvenin impedance Z_{Thev} is nonzero, as shown in Figure 3.8, PV curves similar to those in Figure 3.7 can still be computed and plotted. In addition, there is a useful result from circuit theory to determine the maximum power transfer that is possible for a load with a given power factor. This result was elegantly derived in [32] (p. 40) and is presented as follows.

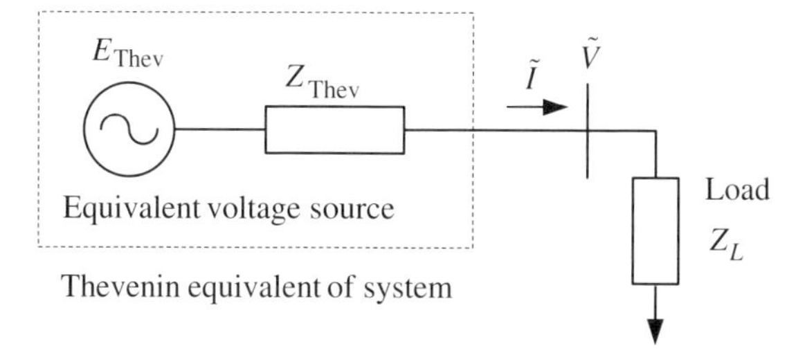

Figure 3.8 Thévenin equivalent model with nonzero line resistance.

The apparent power $S = P + jQ$ delivered to the load bus is given by

$$S = \tilde{V}\tilde{I}^* = \tilde{V}\left(\frac{\tilde{E}_{\text{Thev}} - \tilde{V}}{Z_{\text{Thev}}}\right)^* = \frac{E_{\text{Thev}}V}{|Z_{\text{Thev}}|}e^{j(\theta_V - \theta_E + \varphi)} - \frac{V^2}{|Z_{\text{Thev}}|}e^{j\varphi} \tag{3.43}$$

where E_{Thev}, V, and $|Z_{\text{Thev}}|$ are the magnitudes and θ_E, θ_V, and φ are the angles of the complex quantities $\tilde{E}_{\text{Thev}}$, $\tilde{V}$, and $\tilde{Z}_{\text{Thev}}$, respectively. Separate the real and imaginary parts of (3.43) to obtain two equations

$$\frac{E_{\text{Thev}}V}{|Z_{\text{Thev}}|}\cos(\theta_V - \theta_E + \varphi) = P + \frac{V^2}{|Z_{\text{Thev}}|}\cos\varphi \tag{3.44}$$

$$\frac{E_{\text{Thev}}V}{|Z_{\text{Thev}}|}\sin(\theta_V - \theta_E + \varphi) = Q + \frac{V^2}{|Z_{\text{Thev}}|}\sin\varphi \tag{3.45}$$

Squaring both sides of (3.44) and (3.45) and combining the resulting equations yields

$$\begin{aligned}\left(\frac{E_{\text{Thev}}V}{|Z_{\text{Thev}}|}\right)^2 &= \left(P + \frac{V^2}{|Z_{\text{Thev}}|}\cos\varphi\right)^2 + \left(Q + \frac{V^2}{|Z_{\text{Thev}}|}\sin\varphi\right)^2 \\ &= (P^2 + Q^2) + 2\frac{V^2}{|Z_{\text{Thev}}|}(P\cos\varphi + Q\sin\varphi) + \left(\frac{V^2}{|Z_{\text{Thev}}|}\right)^2\end{aligned} \tag{3.46}$$

Multiplying both sides of the equation by $|Z_{\text{Thev}}|^2$ and rearranging the resulting equation as a quadratic equation for V^2 leads to

$$(V^2)^2 + (2|Z_{\text{Thev}}|(P\cos\varphi + Q\sin\varphi) - E_{\text{Thev}}^2)V^2 + |Z_{\text{Thev}}|^2|S|^2 = 0 \tag{3.47}$$

The solution to this quadratic equation is

$$V^2 = \frac{(E_{\text{Thev}}^2 - 2|Z_{\text{Thev}}|(P\cos\varphi + Q\sin\varphi)) \pm \sqrt{D}}{2} \tag{3.48}$$

where

$$D = (E_{\text{Thev}}^2 - 2|Z_{\text{Thev}}|(P\cos\varphi + Q\sin\varphi))^2 - 4|Z_{\text{Thev}}|^2|S|^2 \tag{3.49}$$

At the voltage collapse point, the two solutions of V^2 will coalesce into the same solution, implying that $D = 0$. The resulting simplification of (3.49) when substituted into (3.48) yields the critical voltage as

$$V_{\text{crit}}^2 = |Z_{\text{Thev}}||S| \tag{3.50}$$

that is,

$$|Z_{\text{Thev}}| = \frac{V^2}{|S|} = |Z_L| \tag{3.51}$$

where Z_L is the load impedance. This derivation shows that the maximum power transfer occurs when $|Z_L| = |Z_{\text{Thev}}|$, which is also the voltage collapse point. Equation (3.29) is a special case of this property for unity power factor, such that the maximum power transfer occurs at $Z_L = R_L = X$. Similarly for maximum reactive power transfer, $Z_L = jX_L = jX$.

This result allows the monitoring of voltage stability by the closeness of the estimated load impedance to the Thèvenin equivalent impedance. The Voltage Instability Predictor (VIP) method proposed in [33] for real-time voltage stability margin calculation is based on this idea. For a load center having active power coming primarily on one transmission path, the VIP method relies on power system voltage and current measurements to construct a Thevénin equivalent for the supply source and to calculate the load impedance. Then the estimated $|Z_L|$ and $|Z_{\text{Thev}}|$ can be used to infer the voltage security margin.

3.5 Voltage Stability Analysis of Large Power Systems

Voltage stability analysis of a large power system typically follows the scenario that the power consumption increase in a load center is supplied by several generators in some pre-specified locations, as determined by the daily load forecast and the economic dispatch of the generators. As the power consumption continues to increase, the voltages of the load center buses will decrease until the critical voltage is reached, which corresponds to the maximum power transfer limit. For each of the load buses under investigation, an equivalent PV curve can be generated, as in the case of a stiff-source-to-load system.

This voltage stability investigation for a power system is typically performed via repeated power flow computations, and hence is referred to as the steady-state voltage stability analysis. It would be ideal if the Newton-Raphson (NR) power flow method could compute the power flow solution up to the voltage collapse point, such that the voltage stability margin for the current operating condition could be calculated. In practice, the linearized power flow Jacobian matrix J will become singular, that is, have less than full rank, at the voltage collapse point, causing difficulties in the convergence of the NR method.[5] The Gauss-Seidel method will have similar convergence problems at the voltage collapse point. The next example illustrates the singularity issue in a stiff-source-to-load system.

Example 3.3: Singularity of Jacobian matrix

Compute the Jacobian matrix J of the stiff-source-to-load system in Figure 3.9 and find $\det(J)$. Plot $\det(J)$ versus θ for the system with $E = 1$ pu and $X = 0.1$ pu, and for three different load power factors: 0.9 leading, unity, and 0.9 lagging. From the plot, determine the value of θ when $\det(J) = 0$.

Solutions: The active and reactive power equations at the load bus are

$$P = \frac{VE\sin\theta}{X} = P_L, \quad Q = \frac{VE\cos\theta}{X} - \frac{V^2}{X} = Q_L \tag{3.52}$$

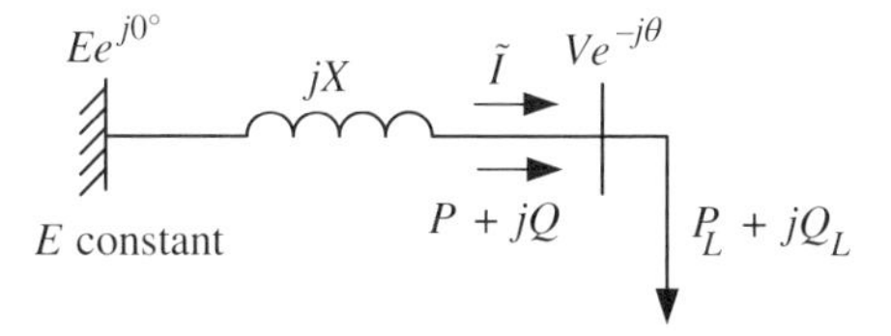

Figure 3.9 A stiff source supplying power to a load with source bus voltage angle at 0°.

5 When J is close to being singular, the region of convergence of the NR method is very small.

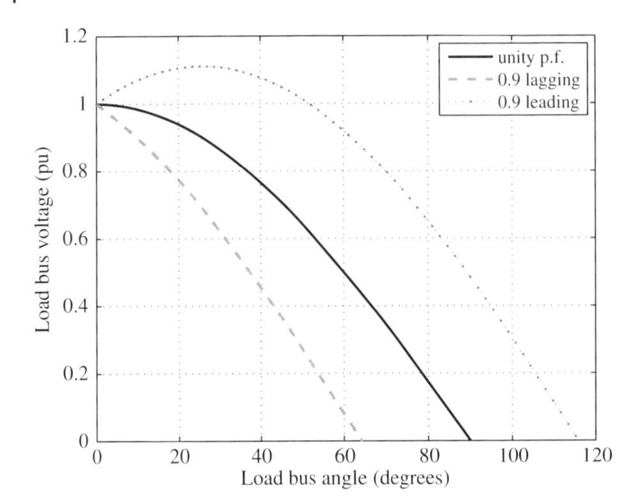

Figure 3.10 Variation of V versus θ (see color plate section).

Treating the stiff source as the swing bus and the load bus as a PQ bus, the Jacobian matrix obtained by taking the partial derivatives of these two equations with respect to the load bus variables θ and V is

$$J = \begin{bmatrix} \partial P/\partial\theta & \partial P/\partial V \\ \partial Q/\partial\theta & \partial Q/\partial V \end{bmatrix} = \frac{1}{X}\begin{bmatrix} VE\cos\theta & E\sin\theta \\ -VE\sin\theta & E\cos\theta - 2V \end{bmatrix} \tag{3.53}$$

The determinant of J is

$$\det(J) = (VE^2 - 2V^2E\cos\theta)/X^2 \tag{3.54}$$

Using the PV-curve results in Figure 3.7, the variations of V and P_L versus θ for several load power factors are shown in Figures 3.10 and 3.11, respectively. Figure 3.12 shows the variations of det(J) versus θ (dashed curves) for these power factors. In particular, det(J) = 0 at the voltage collapse points which, as expected, occur at maximum power transfer P_{max}. For 0.9 lagging power factor, det(J) = 0 at 32.9°, for unity power factor, $\theta = 45°$, and for 0.9 leading power factor, $\theta = 58.4°$. Note that none of the rows in J are identically zero at P_{max}. Thus the two rows of J are linearly dependent for det(J) = 0. The singularity of J at P_{max} is a manifestation of $\partial V/\partial P = 0$ at P_{max}, as shown in the PV curve in Figure 3.7. ■

From Chapter 2, the power flow incremental variables in the NR formulation are related by (2.46)

$$\begin{bmatrix} \Delta P \\ \Delta Q \end{bmatrix} = J \begin{bmatrix} \Delta\theta \\ \Delta V \end{bmatrix} = \begin{bmatrix} \partial P/\partial\theta & \partial P/\partial V \\ \partial Q/\partial\theta & \partial Q/\partial V \end{bmatrix} \begin{bmatrix} \Delta\theta \\ \Delta V \end{bmatrix} = \begin{bmatrix} J_{P\theta} & J_{PV} \\ J_{Q\theta} & J_{QV} \end{bmatrix} \begin{bmatrix} \Delta\theta \\ \Delta V \end{bmatrix} \tag{3.55}$$

where the matrix J is partitioned into four submatrices of the partial derivatives. As the load on one or more load buses is increased, leading to voltage collapse, the Jacobian matrix J will be close to a singular matrix, such that the rows in J are becoming linearly dependent, that is, J no longer has full rank. The rank of the $n \times n$ matrix J, where $n = N_B + N_L - 1$ with N_B the total number of buses and N_L the number of load buses, can be analyzed by computing the singular values [19] as

$$J = U\Sigma V^T \tag{3.56}$$

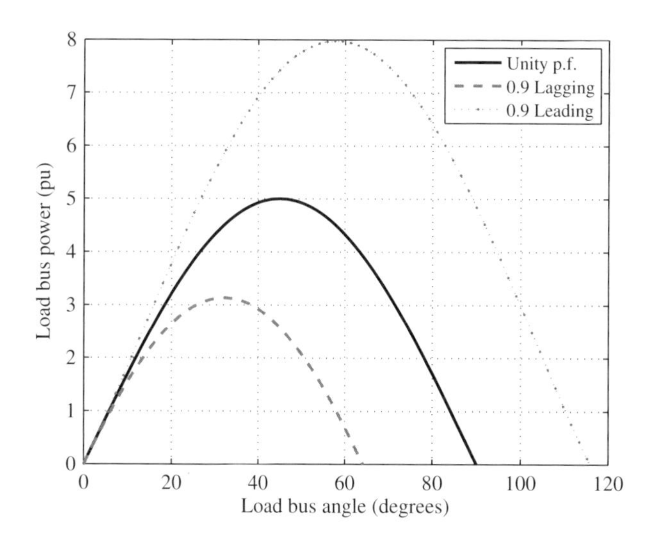

Figure 3.11 Variation of P_L versus θ (see color plate section).

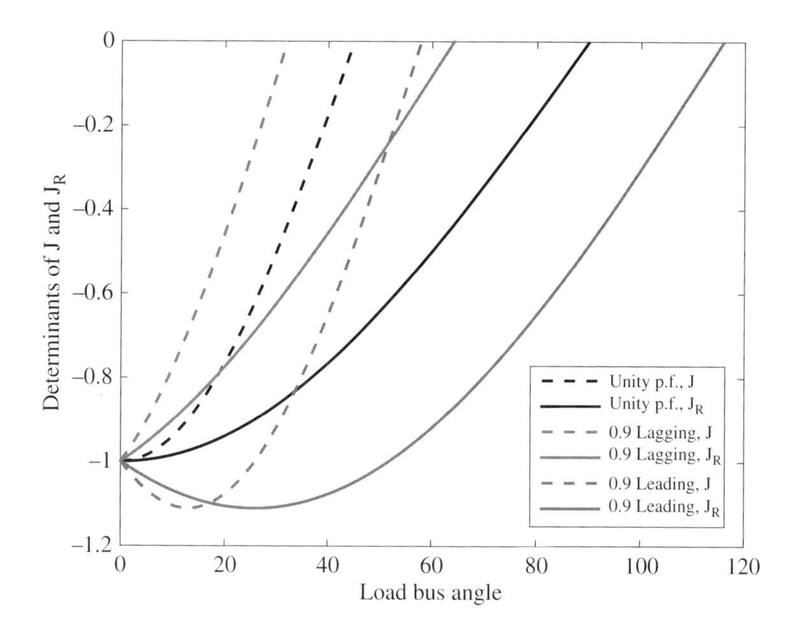

Figure 3.12 Determinants of J and J_R as a function of θ in degrees (see color plate section).

where the diagonal matrix

$$\Sigma = \text{diag}(\sigma_1, \sigma_2, \dots, \sigma_n) \tag{3.57}$$

contains the singular values $\sigma_1, \sigma_2, \dots, \sigma_n$ in descending order $\sigma_1 \geq \sigma_2 \cdots \geq \sigma_n \geq 0$, and

$$U = \left[u_1, \dots, u_n\right] \in R^{n\times n}, \quad V = \left[v_1, \dots, v_n\right] \in R^{n\times n} \tag{3.58}$$

are orthogonal matrices[6] where u_i are the left singular vectors and v_i are the right singular vectors. If any of the singular values of J is zero, then J is singular such that the solution to (3.55) cannot be computed. Thus one of the means to detect voltage stability is to check whether the smallest singular value σ_n of J satisfies $\sigma_n/\sigma_1 < \delta$, where δ is a small tolerance value. The singular vector corresponding to the smallest singular value can be used to find the load buses most susceptible to voltage collapse.

A singular value analysis of the Jacobian matrix for a larger size system is provided later in Example 3.6 (Section 3.7).

An alternative approach is to use eigenvalues and eigenvectors [34]. If ΔP is set to 0, (3.55) reduces to

$$\Delta Q = (J_{QV} - J_{Q\theta}J_{P\theta}^{-1}J_{PV})\Delta V = \bar{J}_{QV}\Delta V \tag{3.59}$$

This reduced Jacobian matrix $\bar{J}_{QV}$ directly relates the bus voltage magnitude and reactive power injection. Let the modal decomposition of $\bar{J}_{QV} \in R^{N_L \times N_L}$ be

$$H^T\bar{J}_{QV}\Xi = \Lambda = \begin{bmatrix} \lambda_1 & 0 & \cdot & 0 \\ 0 & \lambda_2 & \cdot & 0 \\ \cdot & \cdot & \cdot & \cdot \\ 0 & 0 & \cdot & \lambda_n \end{bmatrix} \tag{3.60}$$

where $H^T = \Xi^{-1}$,

$$H = \left[\eta_1, \dots, \eta_{N_L}\right] \in R^{N_L \times N_L}, \quad \Xi = \left[\xi_1, \dots, \xi_{N_L}\right] \in R^{N_L \times N_L} \tag{3.61}$$

in which η_i are the left eigenvectors and ξ_i are the right eigenvectors. If J is singular, $\bar{J}_{QV}$ would also be singular, such that one of the eigenvalues of $\bar{J}_{QV}$ is zero, indicating the voltage collapse point.

Additional voltage stability properties can be derived from the eigenvalues and eigenvectors [34]:

1) The system is voltage stable if all $\lambda_i > 0$, $i = 1, 2, \dots n$, with ΔQ taken as positive for reactive power injected into the bus.
2) The bus participation factor of Bus k to eigenvalue i

$$p_{ki} = \xi_{ik}\eta_{ik} \tag{3.62}$$

indicates the contribution of the ith eigenvalue to the voltage-reactive power sensitivity at Bus k. Using the participation factors corresponding to the small eigenvalues, the buses close to voltage instability can be identified.

3.6 Continuation Power Flow Method

As the load is increased in a voltage stability study, the Jacobian J will become ill-conditioned as the voltages on some load buses approach their critical values. In the NR method, the region of convergence becomes smaller when J is approaching rank deficiency. If the initial guess is outside the region of convergence, the NR method would not converge. To prevent the power flow solution from non-convergence, the

6 A matrix U is orthogonal if $U^{-1} = U^T$.

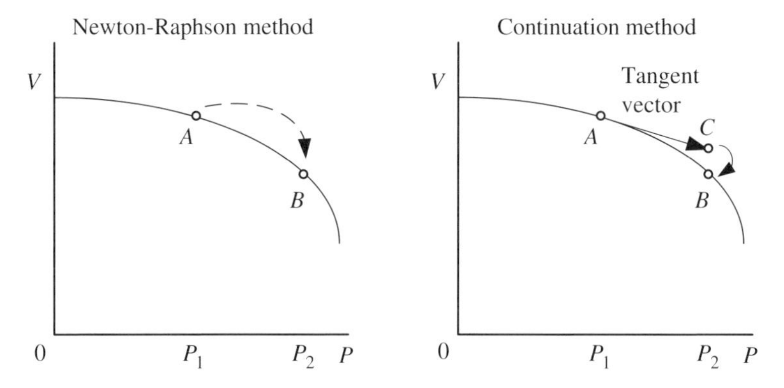

Figure 3.13 Illustration of NR and continuation methods.

load may have to be increased by very small increments so that the previous solution (at a lower loading level), when used as the initial guess, would be in the region of convergence for a higher loading level.

The *continuation power flow* technique [35] was introduced to overcome this slow solution process to reach the voltage collapse point. It consists of a predictor step and a corrector step (which is a regular NR power flow solution). The predictor step uses a tangent vector to get close to the new solution (with increased loading). Using a *PV* curve to denote the power flow solutions, Figure 3.13 shows how the NR method and the continuation method would solve for the solution at Point B starting from the solution at Point A. In the NR method, the solution at Point A is used as the initial guess for solving for the solution at Point B. In the continuation method, the solution at Point A is projected to the Point C, which is used as the initial condition to solve for Point B.

In the continuation method, a new load increase parameter λ is introduced to generate the tangent vector for the predictor step. The load at Bus i becomes $(P_{Li}(\lambda), Q_{Li}(\lambda))$ and the generation at Bus j becomes $P_{Gj}(\lambda)$. For example, at Bus i

$$P_{Li}(\lambda) = (I + \lambda\kappa_i)P_{Lo} \tag{3.63}$$

where κ_i is the load participation factor representing the proportional increase in load on Bus i. There is no need to specify the generator reactive power output $Q_{Gj}(\lambda)$, assuming that the generator reactive power limit is not reached. The parameter λ is expected to be within the range

$$0 < \lambda < \lambda_{\text{crit}} \tag{3.64}$$

where λ_{crit} is the load increase to the voltage collapse point and not known ahead of time.

From the power flow formulation, the error vector in the unknowns V_i and θ_i becomes, with λ added as a variable,

$$\varepsilon(x, \lambda) = \begin{bmatrix} V_1 \sum_{j=1}^{N} |Y_{1j}| V_j \cos(\theta_1 - \theta_j - \alpha_{1j}) - P_1(\lambda) \\ V_2 \sum_{j=1}^{N} |Y_{2j}| V_j \cos(\theta_2 - \theta_j - \alpha_{2j}) - P_2(\lambda) \\ \vdots \\ V_1 \sum_{j=1}^{N} |Y_{1j}| V_j \sin(\theta_1 - \theta_j - \alpha_{1j}) - Q_1(\lambda) \\ V_2 \sum_{j=1}^{N} |Y_{2j}| V_j \sin(\theta_2 - \theta_j - \alpha_{2j}) - Q_2(\lambda) \\ \vdots \end{bmatrix} \tag{3.65}$$

Setting the partial derivatives of ε with respect to θ, V, and λ to zero results in

$$J_\theta \Delta\theta + J_V \Delta V + J_\lambda \Delta\lambda = \begin{bmatrix} J_\theta & J_V & J_\lambda \end{bmatrix} \begin{bmatrix} \Delta\theta \\ \Delta V \\ \Delta\lambda \end{bmatrix} = \bar{J} t = 0 \tag{3.66}$$

where t is the tangent vector and the new Jacobian matrix $\bar{J}$ is of dimension $n \times (n+1)$. The matrix $\bar{J}$ has a one-dimensional null space, such that t is not unique. Thus to solve this equation properly, one of the entries of t needs to be fixed at a nonzero value (like 1). If the kth entry of t is set to unity, then define the row vector u_k with a dimension of $N_B + N_L$

$$u_k = [0 \ \ldots \ 1 \ \ldots \ 0] \tag{3.67}$$

where the only nonzero entry of u_k is at the k position. When u_k is added to $\bar{J}$ in (3.66), t can be solved from

$$\begin{bmatrix} J_\theta \ J_V \ J_\lambda \\ u_k \end{bmatrix} t = \begin{bmatrix} 0 \\ 1 \end{bmatrix} \tag{3.68}$$

This tangent vector t is then used to go from the solution at Point A in Figure 3.13 to a point near the solution at Point B.

A continuation power flow algorithm consisting of two steps in each iteration is described in the following procedure.

3.6.1 Continuation Power Flow Algorithm

1) **Predictor step**: From the existing power flow solution, use t to get a new starting point of the NR power flow

$$\begin{bmatrix} \theta^* \\ V^* \\ \lambda^* \end{bmatrix} = \begin{bmatrix} \theta^o \\ V^o \\ \lambda^o \end{bmatrix} + ht = \begin{bmatrix} \theta^o \\ V^o \\ \lambda^o \end{bmatrix} + h \begin{bmatrix} \Delta\theta \\ \Delta V \\ \Delta\lambda \end{bmatrix} \tag{3.69}$$

   where λ^o can be set to 0 and $h\Delta V$ should be negative, showing a voltage reduction. The stepsize h is between 0 and 1.
2) **Corrector step**: Use λ^* to compute the new load values $(P_{Li}(\lambda^*), Q_{Li}(\lambda^*))$, and the generation at Bus j becomes $P_{Gj}(\lambda^*)$ (3.63). Then use θ^* and V^* as the initial condition to obtain a new NR power flow solution.

The continuation algorithm is repeated until the voltage collapse point is reached. The continuation power flow, however, is still computationally intensive, as the NR convergence would be slow in the corrector step. An improved continuation power flow method using an arc-length parametrization method is described in [36].

Example 3.4: Continuation power flow

This example illustrates the continuation power flow method using the stiff-source-to-load model in Figure 3.5. Note that E and X are normalized to 1.0 pu. Let the power factor of the load be unity. Find the slope of the PV curve at $p = 0.4$ pu and discuss the initialization of the corrector step.

Solutions: At $p = 0.4$ pu $= P_o$, the power flow solution is $V = 0.8944$ pu and $\theta = 0.4636$ rad, which is marked by the point A in Figure 3.14. The load increase

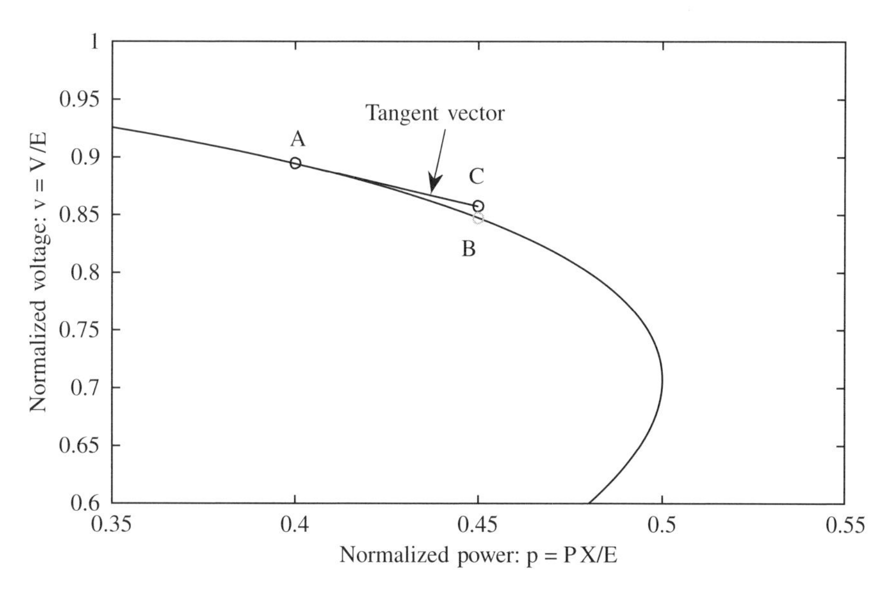

Figure 3.14 Tangent vector at $p = 0.4$ pu to find the initial guess of the power flow solution for $p = 0.45$ pu (see color plate section).

functions are $\Delta P = \lambda P_o$ and $\Delta Q = 0$ (unity power factor). Thus (3.68) becomes, by setting the third entry of u_k to unity,

$$\begin{bmatrix} (VE/X)\cos\theta & (E/X)\sin\theta & -P_o \\ -(VE/X)\sin\theta & (E/X)\cos\theta - 2V/X & 0 \\ 0 & 0 & 1 \end{bmatrix} \begin{bmatrix} \Delta\theta \\ \Delta V_L \\ \Delta\lambda \end{bmatrix} \tag{3.70}$$

$$= \begin{bmatrix} 0.8 & 0.4472 & -0.4 \\ -0.4 & -0.8944 & 0 \\ 0 & 0 & 1 \end{bmatrix} \begin{bmatrix} \Delta\theta \\ \Delta V \\ \Delta\lambda \end{bmatrix} = \begin{bmatrix} 0 \\ 0 \\ 1 \end{bmatrix} \tag{3.71}$$

yielding the solution $\Delta\theta = 0.6666$ rad, $\Delta V = -0.2981$ pu, and $\Delta\lambda = 1$. Thus the tangent vector of the PV curve at $P_o = 0.4$ pu has a slope of $\Delta V/P_o = -0.7453$, which is shown in Figure 3.14.

To use the result for the corrector step, a user has to decide on the stepsize h to use. Suppose $h = 0.125$ is chosen so that the new active power transfer level is $0.4(1 + 0.125) = 0.45$ pu. Then the initial guess for the NR power flow solution is

$$\theta_{\text{init}} = 0.4636 + 0.125 \times 0.6666 = 0.5469 \text{ rad}$$

$$V_{\text{init}} = 0.8944 + 0.125 \times (-0.2981) = 0.8571 \text{ pu} \tag{3.72}$$

shown as point C in Figure 3.14, which is much closer to the actual solution marked by point B, than using point A as the initial guess. ■

3.7 An *AQ*-Bus Method for Solving Power Flow

This section describes a power flow method to directly eliminate the singularity of the Jacobian matrix. Elimination of the singularity allows for a well-conditioned power flow solution even at the voltage collapse point. The central idea is to reformulate the power

Table 3.3 Power flow bus types.

Bus type	**Bus model**	**Fixed values**
Generator bus	*PV*	Active power generation, voltage magnitude
Load bus	*PQ*	Active power consumption, reactive power consumption
Swing bus	*AV*	Voltage angle, voltage magnitude
Load bus	*AQ*	Voltage angle, reactive power consumption

flow with the introduction of a new type of load bus, which is called an AQ bus (A stands for angle) [37]. A conventional power flow formulation uses three types of buses: PV buses, PQ buses, and the swing bus (Table 3.3). A swing bus can be considered as an AV bus because its angle is fixed and its voltage magnitude is specified. For an AQ bus, the bus voltage angle θ and the reactive power consumption Q are specified. When a load bus is modified to an AQ bus, the active power balance equation is no longer needed. Only the reactive power balance equation is kept. Furthermore, because θ at this bus is known, it is eliminated from the power flow solution vector consisting of bus voltage magnitudes of PQ buses and bus voltage angles of all the buses except for the swing bus. Thus the size of the resulting Jacobian matrix J_R is reduced by one. This J_R matrix is nonsingular at the voltage collapse point, and thus it avoids the singularity problem of the Jacobian matrix J from a conventional NR power flow method.

Before embarking on the general AQ-bus formulation, the singularity in Example 3.3 is resolved using the AQ-bus method.

Example 3.5: *AQ*-bus method

Suppose the angle of the load bus in Figure 3.9 is fixed at $-\theta$, so that the only unknown variable of the stiff-source-to-load system is V_L. Find the reduced Jacobian matrix J_R, assuming a constant power factor load. Plot J_R for $E = 1$ pu and $X = 0.1$ pu, and for three different load power factors: 0.9 leading, unity, and 0.9 lagging. From the plot, determine the value of θ when $\det(J_R) = J_R = 0$.

Solutions: If the load bus is taken as an AQ bus, then the separation angle θ can be specified instead of the active power equation at the load bus. If Q_L is fixed, then the reduced matrix J_R is simply the (2,2) entry of J (3.53). If the load is of constant power factor, i.e., $Q_L = P_L \tan\phi$, the reactive power equation can be rewritten as

$$Q_L = \frac{V_L E\cos\theta}{X} - \frac{V_L^2}{X} = -\frac{V_L E\sin\theta}{X}\tan\phi \tag{3.73}$$

that is,

$$0 = \frac{V_L E\cos\theta}{X} - \frac{V_L^2}{X} + \frac{V_L E\sin\theta}{X}\tan\phi \tag{3.74}$$

The reduced Jacobian is the partial derivative of (3.74) with respect to V_L

$$J_R = \frac{1}{X}(E\cos\theta - 2V_L + E\sin\theta\tan\phi) \tag{3.75}$$

The solid curves in Figure 3.12 show that det(J_R) becomes zero only when $V_L = 0$, for all three cases of 0.9 leading, unity, and 0.9 lagging power factor loads. The values of det(J_R) = 0 occur at $\theta = 64.66^\circ,\ 90^\circ,\ 115.9^\circ$, respectively, which are not the critical voltage points. Thus the NR iteration will readily converge at the critical voltage points. ■

3.7.1 Analytical Framework for the *AQ*-bus Method

This section develops the general framework of a power flow formulation including an AQ bus, and extends the method for steady-state voltage stability analysis allowing for load and generation increases on multiple buses and for constant power factor loads. The development follows the NR power flow formulation in Section 2.5.2.

Assume that the power system has N_G generator buses and N_L non-generator buses. The buses are ordered with the non-generator buses appearing before the generator buses. The last generator bus is designated as the swing bus. Assign a non-generator bus with increasing load as the first bus. This candidate bus for voltage collapse is designated as the AQ bus with $\theta_1 = \theta_1^\circ$ and Q_1 specified. Then the NR formulation (3.55) reduces to

$$J_R \begin{bmatrix} \Delta\theta_R \\ \Delta V \end{bmatrix} = \begin{bmatrix} J_{R11} & J_{R12} \\ J_{R21} & J_{R22} \end{bmatrix} \begin{bmatrix} \Delta\theta_R \\ \Delta V \end{bmatrix} = \begin{bmatrix} \Delta P_R \\ \Delta Q \end{bmatrix} \tag{3.76}$$

where, with $N_B = N_L + N_G$,

$$J_{R11} = J_{P\theta}(2, \ldots, N_B - 1; 2, \ldots, N_B - 1)\big|_{\theta_1=\theta_1^\circ} \tag{3.77}$$

$$J_{R12} = J_{PV}(2, \ldots, N_B - 1; 1, \ldots, N_L)\big|_{\theta_1=\theta_1^\circ} \tag{3.78}$$

$$J_{R21} = J_{Q\theta}(1, \ldots, N_L; 2, \ldots, N_B - 1)\Big|_{\theta_1=\theta_1^\circ} \tag{3.79}$$

$$J_{R22} = J_{QV}\Big|_{\theta_1=\theta_1^\circ} \tag{3.80}$$

In this notation, J_R is obtained simply by eliminating the first column and the first row from J and substituting the value of θ_1 by θ_1°. Also, the number of bus angle variables is reduced by one as

$$\Delta\theta_R = \begin{bmatrix} \Delta\theta_2 & \cdots & \Delta\theta_{N_B-1} \end{bmatrix}^T \tag{3.81}$$

The AQ bus active power flow equation is eliminated, such that ΔP_R is the vector of active power mismatches from Bus 2 to Bus ($N_B - 1$). The load P_1 on Bus 1 is no longer specified, but it can be computed after the solution converges using the power $f_{P_1}(\theta, V)$ flowing into Bus 1.

This reduced power flow formulation would not yield directly a specified P_1 on Bus 1 (AQ bus). This is, however, not a hindrance in voltage stability analysis. Instead of increasing P_1 on Bus 1 and not getting a converged solution, a user can keep increasing the angular separation between Bus 1 and the swing bus until the maximum power transfer point is reached. The reduced Jacobian J_R would not be singular at that point and the maximum power transfer can be readily computed.

3.7.2 *AQ*-Bus Formulation for Constant-Power-Factor Loads

In voltage stability analysis, loads are commonly assumed to be of constant-power-factor type. This section extends the NR iteration (3.76) by considering constant-power-factor load increases at multiple load buses to be supplied by generators at multiple locations.

Let Buses 1 to p be load buses with constant power factor $\cos\phi_\ell$, that is, $Q_\ell = P_\ell \tan\phi_\ell$ for $\ell = 1, \ldots, p$. The active power load increases at these load buses are scaled with respect to Bus 1, which is the AQ bus, according to

$$P_\ell - P_\ell^o = \alpha_\ell \left(P_1 - P_1^o\right), \quad \ell = 2, \ldots, p \tag{3.82}$$

and the increase in reactive power are similarly scaled as

$$Q_\ell - Q_\ell^o = \left(P_\ell - P_\ell^o\right) \tan\phi_\ell, \quad \ell = 2, \ldots, p \tag{3.83}$$

where the superscript "o" denotes the initial loading condition. The load increase is balanced by increases in outputs of generators on Buses $N_B - q$ to N_B, in which Bus N_B is the swing bus. The increases in the active power at these generators are scaled according to the swing bus

$$P_k - P_k^o = \beta_k \left(P_{N_B} - P_{N_B}^o\right), \quad k = N_B - q, \ldots, N_B - 1 \tag{3.84}$$

In a solved power flow solution, the active power injections at Buses 1 and N_B are computed as the power flow leaving these buses on the lines interconnecting them to the other buses. Thus in an AQ-bus formulation, the buses with increasing load and generation are accounted for by modifying the power flow injection equations such that, following the notation of (2.34),

$$f_{P\ell}(V,\theta) - \alpha_\ell f_{P1}(V,\theta) = P_\ell^o - \alpha_\ell P_1^o, \quad \ell = 2, \ldots, p \tag{3.85}$$

$$f_{Q\ell}(V,\theta) - f_{P\ell}(V,\theta)\tan\phi_\ell = Q_\ell^o - P_\ell^o \tan\phi_\ell, \quad \ell = 2, \ldots, p \tag{3.86}$$

$$f_{Pk}(V,\theta) - \beta_k f_{PN_B}(V,\theta) = P_k^o - \beta_k P_{N_B}^o, \quad k = N_B - q, \ldots, N_B - 1 \tag{3.87}$$

The other injection equations remain unchanged.

The partial derivatives of these injections with respect to θ and V need to be added to the reduced Jacobian matrix J_R as, with the indices ℓ and k denoting the ℓth and kth rows, respectively,

$$\begin{aligned}
&(\bar{J}_{R11})_\ell = (J_{R11})_\ell - \alpha_\ell \frac{\partial f_{P1}}{\partial \theta_R}, \quad (\bar{J}_{R12})_\ell = (J_{R12})_\ell - \alpha_\ell \frac{\partial f_{P1}}{\partial V} \\
&(\bar{J}_{R21})_\ell = (J_{R21})_\ell - \frac{\partial f_{P_\ell}}{\partial \theta_R}\tan\phi_\ell, \quad (\bar{J}_{R22})_\ell = (J_{R22})_\ell - \frac{\partial f_{P_\ell}}{\partial V}\tan\phi_\ell \\
&\qquad \ell = 2, \ldots, p
\end{aligned} \tag{3.88}$$

and

$$\begin{aligned}
&(\bar{J}_{R11})_k = (J_{R11})_k - \beta_k \frac{\partial f_{PN_B}}{\partial \theta_R}, \quad (\bar{J}_{R12})_k = (J_{R12})_k - \beta_k \frac{\partial f_{PN_B}}{\partial V} \\
&\qquad k = N_B - q, \ldots, N_B - 1
\end{aligned} \tag{3.89}$$

In this more general formulation of the AQ-bus power flow, the NR iteration becomes

$$\bar{J}_R \begin{bmatrix} \Delta\theta_R \\ \Delta V \end{bmatrix} = \begin{bmatrix} \Delta\bar{P} \\ \Delta\bar{Q} \end{bmatrix} \tag{3.90}$$

where

$$\bar{J}_R = \begin{bmatrix} \bar{J}_{R11} & \bar{J}_{R12} \\ \bar{J}_{R21} & \bar{J}_{R22} \end{bmatrix} \tag{3.91}$$

and the power mismatches $\Delta\bar{P}$ and $\Delta\bar{Q}$ reflect the required adjustments of active and reactive power after each iteration.

3.7.3 *AQ*-Bus Algorithm for Computing Voltage Stability Margins

Because $\bar{J}_R$ in (3.90) would not be singular at the maximum loadability point, fast and well-conditioned voltage stability margin calculation methods can be formulated. The following algorithm is the application of the *AQ*-bus method for contingency-based voltage stability analysis.

***AQ*-Bus Algorithm: using *AQ*-bus power flow with $\bar{J}_R$ to compute voltage stability margins**

1) From the current operating point (base case) with a power transfer of P_o, specify the load and generation increment schedule and the load composition (such as constant power factors).
2) Use a conventional power flow program with increasing loads until the NR algorithm no longer converges.
3) Starting from the last converged solution in Step 2, apply the *AQ*-bus power flow method (3.82)–(3.90) to extend the power flow solution by increasing the angle separation $(\theta_1 - \theta_N)$ between the *AQ* bus and the swing bus until the maximum power transfer P_{omax} is reached. Typically, the bus with the largest load increase will be selected to be the *AQ* bus. The base-case voltage stability margin is $P_{om} = P_{omax} - P_o$.
4) Specify a set of N_c contingencies to be analyzed and repeat Step 5 for all contingencies $i = 1, 2, \ldots, N_c$.
5) For contingency i, repeat Steps 2 and 3 for the post-contingency system to compute the maximum power transfer P_{imax} and the voltage stability margin $P_{im} = P_{imax} - P_o$.
6) The contingency-based voltage stability margin, measured as additional power delivered to the load until the maximum loadability point, is given by

$$P_m = \min_{i=0,\ldots,N_c} \{P_{im}\} \tag{3.92}$$

Note that for any of the contingencies in Step 5, if the *AQ*-bus algorithm for P_o fails to converge, that is, P_o is not a feasible solution, then the *AQ*-bus algorithm can be used to reduce P_o until a converged power flow solution is obtained. The new power flow solution would then be a voltage secure operating condition.

Also note in Steps 3 and 5 of the *AQ*-bus algorithm that all the adjustments and controls available in power flow can be used. For example, transformer taps can be adjusted to maintain voltages, and generators exceeding their reactive power capability can be changed to *PQ* buses. Both capabilities are important for finding the proper voltage stability limit.

The advantage of using a conventional power flow algorithm in Step 2 of the *AQ*-bus algorithm is that it will allow a user to select the *AQ* bus for Step 3. There are several ways to select the *AQ* bus: (1) use the bus with the largest load increase (as stated in Step 3 of the *AQ*-bus algorithm), (2) use the bus with the largest rate of decrease of the bus voltage magnitude, or (3) use the bus angle with the largest component in the singular vector of the smallest singular value of J from the last converged solution. Frequently all three will yield the same bus. If not, any of those choices is a suitable candidate for the AQ bus.

It is also possible to solve for voltage stability margins without updating J_R (3.76), in the so-called dishonest Newton's Method. This method can be useful when one wants to avoid changing the Jacobian matrix entries, but it will have slower convergence. The load increase condition (3.82), the generator increase condition (3.84), and the load power factor condition $Q_\ell = P_\ell \tan\phi_\ell$ are now enforced as fixed values after each power flow iteration has converged. See [38] for a more detailed discussion.

Example 3.6: *AQ*-Bus method applied to a 2-area system

This example will perform a voltage stability analysis of the 2-area system shown in Figure 2.7. Two cases are investigated.

1) Load 14 is increased at a constant power factor of 0.9 lagging whereas Load 4 is kept constant at $9.76 + j1$ pu. The load increase is supplied by Generator 1. Use the AQ-bus method to compute and plot the resulting PV curve, assuming that there are no limits on the generator reactive power output capability
2) Repeat the calculations if the maximum reactive power generation of Generator 2 is 3 pu.

Solutions: (1) The conventional power flow solution is shown as the black dashed line of the PV curve in Figure 3.15. It fails to converge when the active power of Load 14 is $P_{14} = 19.15$ pu, which occurs when the angle separation is $\theta_1 - \theta_{14} = 91.1°$. After this point, the AQ-bus algorithm is used to continue the power flow solution by further increasing the angle separation between Buses 1 and 14. The solution of the AQ-bus approach is shown as the solid black line of the PV curve in Figure 3.15. From the PV curve, the critical voltage is 0.8144 pu and the maximum active load power is 19.2 pu, with a power factor of 0.9 lagging. It is also illuminating to plot the load active power at Bus 14 versus the angle separation $\theta_1 - \theta_{14}$, shown as the solid black curve in Figure 3.16. Note that at maximum power transfer, $\theta_1 - \theta_{14} = 99.5°$.

At the maximum loadability point, the largest singular value of J is 423 and the two smallest singular values are 3.59 and 0.02. At the same operating point, the largest and smallest singular values of the $\bar{J}_R$ matrix are 423 and 2.49, respectively. Thus the iteration (3.90) does not exhibit any convergence problems.

At the point where the conventional power flow fails to converge, the smallest singular value of the Jacobian is 0.05 and its singular vector is given in Table 3.4. Note that the element of the singular vector with the largest magnitude corresponds to θ_{14} of the chosen AQ bus. Note that the components of the buses in Area 2 also have relatively high values.

(2) When a maximum reactive power generation of 3 pu is imposed for Generator 2, that is, if the reactive power generation of Generator 2 exceeds 3 pu, it will be changed into a PQ bus with $Q = 3$ pu. The resulting PV and $P\theta$ curves for the same load increase conditions are shown as the red curves in Figures 3.15 and 3.16. Figure 3.17 shows the reactive power plotted versus $\theta_1 - \theta_{14}$ for the var-limited case. Observe that the var

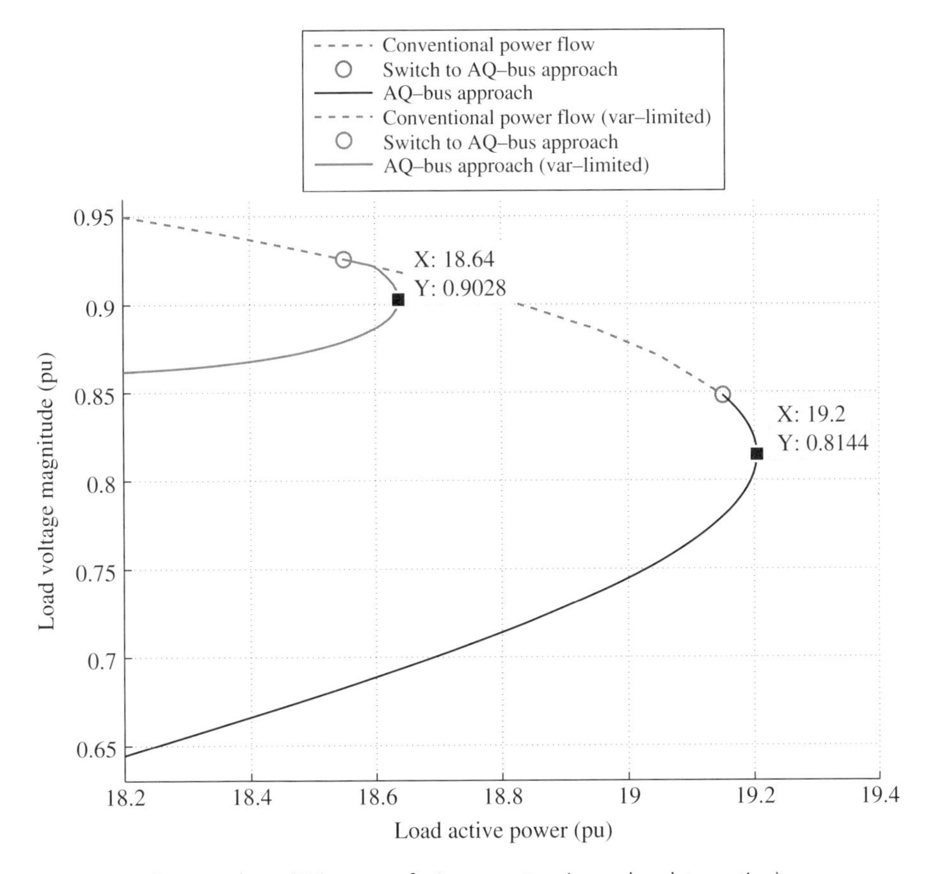

Figure 3.15 Power-voltage (*PV*) curves of a 2-area system (see color plate section).

limit on Generator 2 increases the reactive power burden on Generator 1, and the reactive power losses continue to increase after the point of maximum power transfer, even though the active power consumed by the load decreases. ■

To illustrate the contingency-based voltage stability analysis, an example from [37] is used. The test system is a simplified Northeast Power Coordinating Council (NPCC) system consisting of 148 buses and 48 generators [39]. The test scenario is to check the voltage stability margin of a base case to supply power to a load center. The resulting *PV* curve is shown in black in Figure 3.18. Then five contingency cases are evaluated, each contingency consisting of losing a line (without fault) near the load center. The resulting *PV* curves are also shown in Figure 3.18. In such analysis, the most limiting case will determine the voltage stability margin, which is about 9 pu (900 MW) for losing Line 8-73.

3.8 Power System Components Affecting Voltage Stability

This section describes two mechanisms that impact the voltage stability of a power system. The first consists of shunt capacitors which supply reactance power. Other components that also supply reactive power, such as generators and flexible AC transmission

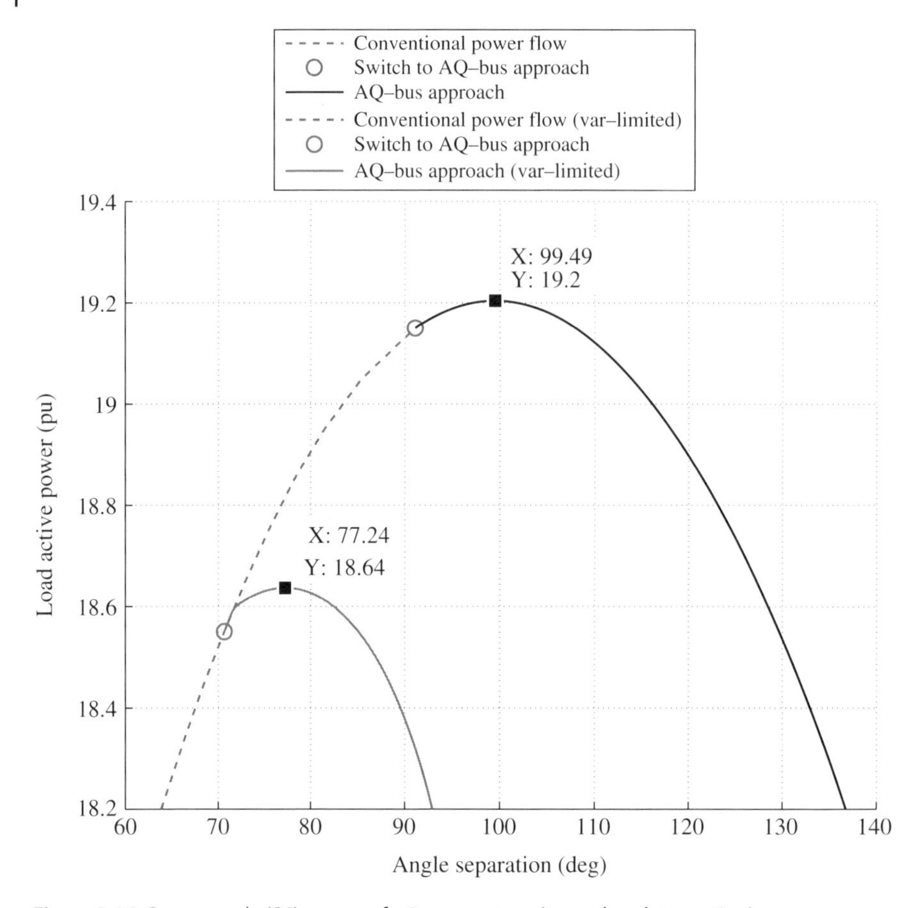

Figure 3.16 Power-angle ($P\theta$) curves of a 2-area system (see color plate section).

systems, will be discussed in later chapters. The second is tap-changing transformers, which affect the reactive power flow on a power system.

3.8.1 Shunt Reactive Power Supply

Shunt capacitors connected to buses requiring extra reactive power supply can be effective in enhancing the voltage stability margins. The reactive power generated by a capacitor of susceptance B is $Q = BV^2$, which is a squared function of the bus voltage V. The impact of a shunt capacitor on voltage stability can be readily seen using a PV-curve analysis.

Example 3.7: Impact of shunt capacitor on *PV* curve

For the system used in Example 3.3, suppose a 1 pu ($B = 1$ pu) capacitor is added to the load bus. The resulting system is shown in Figure 3.19. Plot the PV curve with and without the capacitor, using $\tilde{E} = 1.0\angle 0°$ pu, $X = 0.1$ pu, and $Q_L = 0$ pu (i.e., the load is of unity power factor).

Table 3.4 Singular vector corresponding to the smallest singular value of the conventional power flow Jacobian.

Singular vector component	Corresponding variable
0.025	θ_2
0.064	θ_3
0.075	θ_4
0.005	θ_{10}
0.329	θ_{11}
0.358	θ_{12}
0.416	θ_{13}
0.450	$\boldsymbol{\theta_{14}}$
0.031	θ_{20}
0.228	θ_{101}
0.332	θ_{110}
0.366	θ_{120}
0.085	V_3
0.086	V_4
0.021	V_{10}
0.117	V_{13}
0.125	V_{14}
0.048	V_{20}
0.172	V_{101}
0.024	V_{110}
0.062	V_{120}

Solutions: The PV curve for a unity-power-factor load without the shunt capacitor is similar to the one shown in Figure 3.7, and is shown as the dashed line in Figure 3.20, where the load bus voltage and active power in pu are used in the plot.

To find the impact of the shunt capacitor, it is more convenient to derive a new Thèvenin model including the shunt capacitor. With the load P_L disconnected in Figure 3.19, the short-circuit current at the load bus is $\tilde{E}/(jX)$ and the open-circuit voltage is

$$\tilde{E}' = \frac{1/(jB)}{jX + 1/(jB)}\tilde{E} = \frac{1}{1 - BX}\tilde{E} = 1.111\angle 0^\circ \tag{3.93}$$

Thus the new Thèvenin model including the shunt capacitor is given by the source voltage $\tilde{E}'$ with the reactance

$$X' = \frac{1}{j}\frac{\tilde{E}'}{\tilde{E}/(jX)} = \frac{X}{1 - BX} = 0.1111 \text{ pu} \tag{3.94}$$

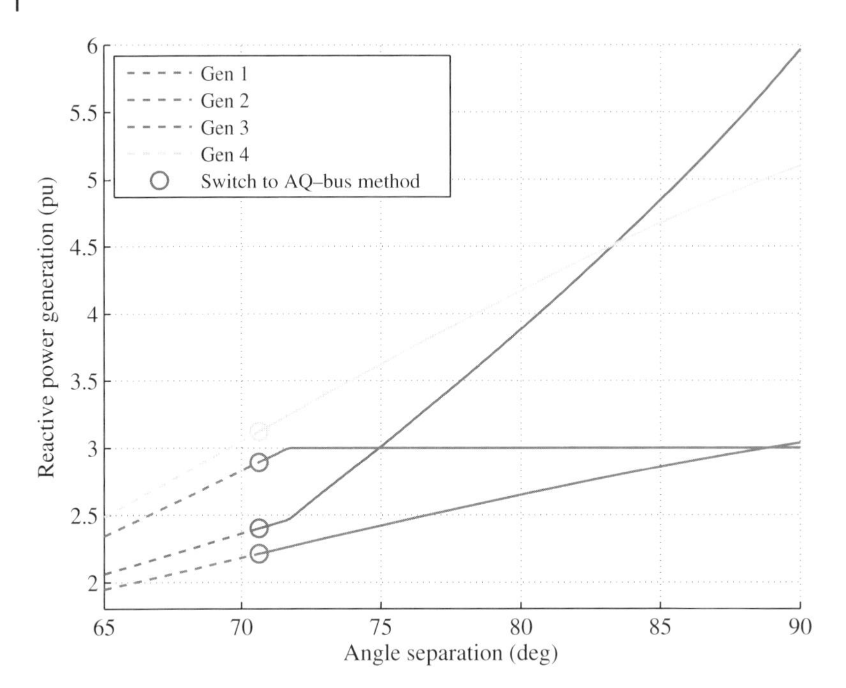

Figure 3.17 Reactive power output of generators in a 2-area system with a var limit (see color plate section).

Referring to (3.41) and (3.42) at unity power factor, the voltage V' and power P' of the PV curve with the shunt capacitor are given by

$$V' = vE' = 1.111v, \quad P' = p\frac{(E')^2}{X'} = 11.11p \tag{3.95}$$

where v and p are the normalized values associated with the PV curve without the shunt capacitors. This new PV curve with the shunt capacitor is shown as the solid curve in Figure 3.20. Note the power transfer enhancement provided by the shunt capacitors. ■

Shunt capacitors are normally of two types. The mechanically switched type is less expensive, but the switching is at discrete levels and can be slow. These mechanically switched shunt capacitors may range from a few MVar (on a distribution network) to hundreds of MVar (on a transmission network). To prevent over-voltage when switching on the capacitors, the shunt capacitors are typically installed as several smaller banks. As the reactive power generated by a capacitor of susceptance B is $Q = BV^2$, a shunt capacitor provides more reactive power if it is switched on while the voltage is still relatively high. The switching on of a shunt capacitor can be automated and triggered if the bus voltage falls below a certain level.

The electronically switched type, also known as a static var compensator (SVC), will be discussed in Chapter 14.

3.8.2 Under-Load Tap Changer

Although a variable-tap transformer acting as an under-load tap changer (ULTC) between the power supply and the load does not by itself supply reactive power, it can

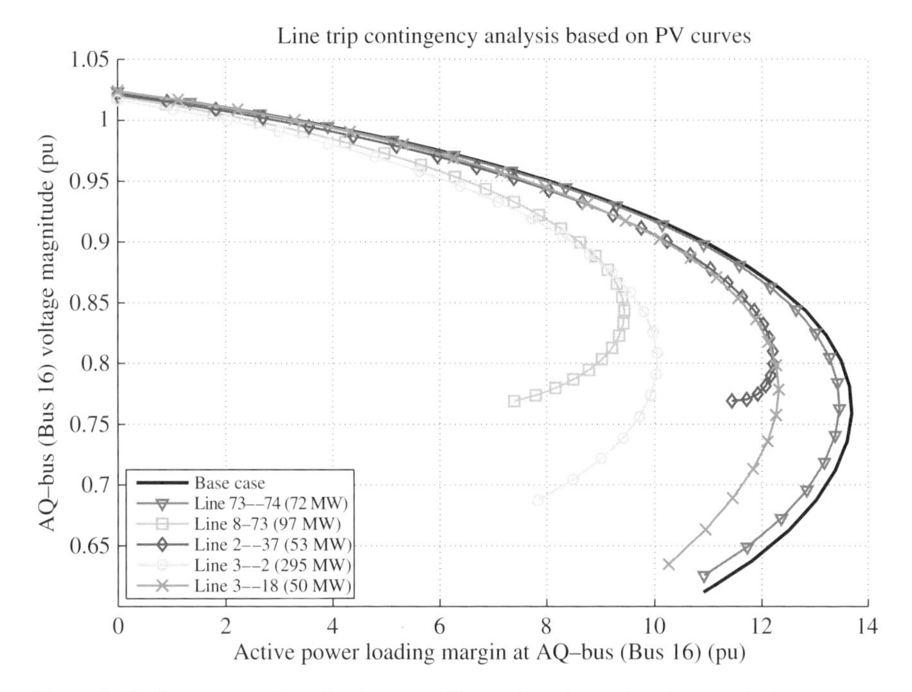

Figure 3.18 Contingency-based voltage stability analysis (see color plate section).

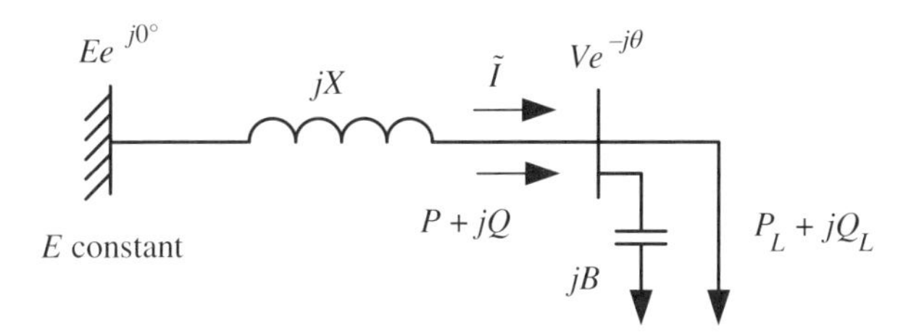

Figure 3.19 Stiff-source to load system with shunt capacitor at load bus.

be used to adjust the voltage of the load buses by more optimally utilizing the reactive power supply from the sources. Consider the simple system in Figure 3.21 in which a voltage source is supplying power via a transmission line to a load via a transformer. Let the source voltage be $Ee^{j\delta}$, the transmission line, transformer, and distribution line reactances be X_{ts}, X_T, and X_{ds}, respectively, the load consumption be $P_L + jQ_L$, and the tap ratio be denoted by a. For simplicity, the variation of the transformer reactance as a function of a is neglected.

If the load voltage V is low, the tap ratio a can be decreased so as to increase V. As an illustration, Figure 3.22 shows that the voltage V drops due to some disturbance. Then the action of an automatic under-load tap-changing transformer restores the load bus voltage V by reducing a one tap at a time until V falls into a desired range of operation, which is a deadband about a desired voltage. Note that a delay is implemented for such automatic tap-changing actions to avoid responding to noise or transient conditions.

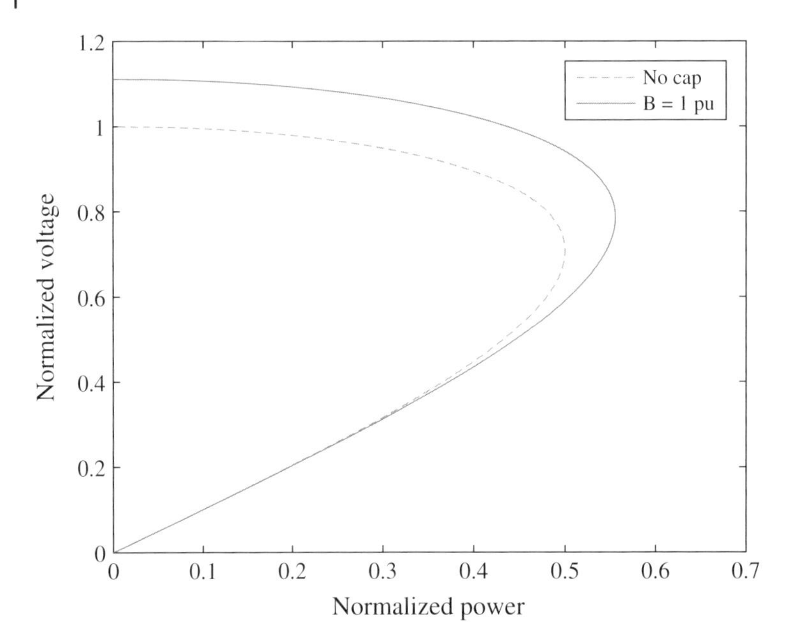

Figure 3.20 *PV* curves with and without shunt capacitor at the load bus (see color plate section).

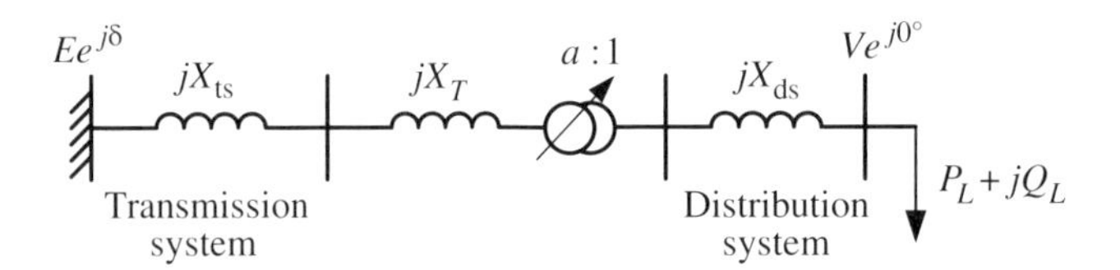

Figure 3.21 A transformer between a supply bus and a load bus.

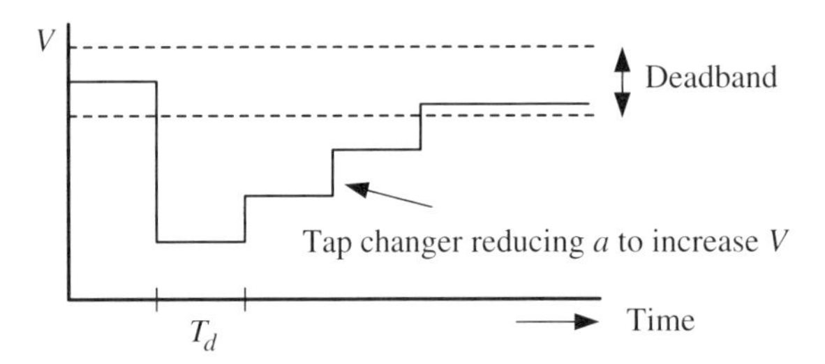

Figure 3.22 Transformer raising load bus voltage showing time delays: the delay time T_d is normally about 30–60 seconds.

Figure 3.22 illustrates an ideal situation of a tap changer affecting the voltage V. There are some further implications in tap-changer adjustments depending on the load model, as shown in the following example.

Example 3.8: Impact of tap-changing transformer on load-bus voltage

Consider a voltage source with a fixed voltage supplying a load through a transformer as shown in Figure 3.21. The system parameters are $E = 1.05$ pu, $X_{ts} = 0.1$ pu, and

Table 3.5 Table for Example 3.8: P_L, G, P_G, Q_G and V are all in pu.

	$P_L = 3.2426$			$G = 4$		
Tap ratio	P_G	Q_G	V	P_G	Q_G	V
0.95	3.2426	1.8201	0.9638	3.4564	2.2232	0.9296
1.00	3.2426	1.9456	0.9004	3.2426	1.9456	0.9004
1.05	3.2426	2.0998	0.8394	3.0378	1.7097	0.8715

$X_T = 0.05$ pu. The distribution system transformer reactance X_{ds} is neglected. Use a power flow program to compute the load-bus voltage V, and the real and reactive power $P_G + jQ_G$ output of the voltage source for three different values of $a = 0.95$, 1.0 and 1.05, given two types of loads: (a) constant power load with $P_L = 3.2426$ pu (unity power factor) and (b) constant impedance unity pf load given by the shunt conductance $G = 4$ pu. Comment on the results, that is, how V, the active power P_G, and the reactive power Q_G change when a deviates from 1.

Solutions: The results from the six cases are shown in Table 3.5, with columns 2–4 showing the results for constant power load and columns 5–7 showing the results for constant impedance load. In all cases, by reducing a, the load-bus voltage V increases, and vice versa. In the constant power load case, as V increases, the amount of reactive power supplied by the voltage source also decreases, indicating more efficient power transmission. However, in the constant impedance load case, as a is lowered, the increase in V is smaller. Also more load power is consumed as P_G increases and the demand of Q_G is higher. Thus in some cases when the load-bus voltage is low, the ULTC action is disabled to prevent the load area from drawing more active power and reactive power, which may cause a voltage instability situation [40]. ■

3.9 Hierarchical Voltage Control

Given all the choices of voltage components and given that reactive power cannot be readily transported for long distances, many power grid operators have chosen a hierarchical approach to coordinate voltage control and optimization over their control regions. A voltage control scheme first proposed by the French power system operator in response to its 1987 blackout [28] has been adopted by many other power grid operators. This hierarchical voltage control scheme is divided into three levels [41, 42] described as follows.

1) Primary voltage control: this level encompasses all automatic closed-loop voltage control functions, without requiring operator manual actions, including:
   a) Regulation of the generator terminal voltages using the generator excitation systems. The regulation can be made more effective by controlling the voltage on the high-side of the generator step-up transformer [25], also known as line-drop compensation (see Section 9.6).
   b) Use of available SVCs to regulate bus voltages on critical power transfer paths (see Section 14.2).

2) Secondary voltage control: this level establishes a "pilot" point in each zone consisting of many buses and a reference voltage V_p^{ref} for that pilot point. The pilot point tends to be at a central location in each zone. The error signal $V_p^{\text{ref}} - V_p$ is passed through a proportional-integral (PI) controller as a supplemental signal to the voltage regulators for adjusting their voltage reference setpoints to supply additional reactive power if needed.
3) Tertiary voltage control: this level executes security-constrained optimal power flow to set appropriate voltage levels over the entire system.

A number of variants of this three-level control strategy have been used. For example, some smaller power systems would combine the secondary and tertiary levels into a single process [26]. In larger systems, the zonal boundaries in secondary voltage control may have to be adjusted according to the operating condition.

3.10 Voltage Stability Margins and Indices

The outcome of a voltage stability analysis is normally an indicator of some stability margins or indices, allowing system planners and operators to secure additional reactive power supplies if the margin drops below a certain threshold or the index falls into an undesirable range. This section discusses some of the proposed methods, which are grouped into several categories for discussion purposes.

3.10.1 Voltage Stability Margins

For a particular operating condition, the voltage stability margin can be computed from the PV curve, assuming the load grows in a certain fashion, like constant power factor loads. Methods for such calculations include the VIP technique [33], the continuous power flow method [27], and the AQ-bus method [37]. Some of these techniques require a complete or partial data set including line parameters. Others require some online estimation of the Thévenin equivalent values of the voltage source and the line impedance [33, 41, 46] using SCADA data (based on power system measured data arriving every 4 seconds or so) or phasor measurement data (at 30 samples or higher per second).

In assessing the voltage stability margins under contingency conditions, it is important that all the available reactive power supply and control capabilities be taken into account properly. They include generator reactive power limits (both short-term and long-term limits should be considered), shunt capacitors, static var systems, and tap-changing transformers.

A more comprehensive approach to voltage stability margin calculation is to recognize that the load increase is not restricted by a certain power factor. The objective here is to find the smallest load increase that will push the system operating point to the voltage stability limit. Figure 3.23 shows a hypothetical voltage stability boundary, noted as the singular surface. The first task is to start from the current operating condition (P_o, Q_o) to search for the boundary, which can be achieved via the continuation power flow method or the AQ-bus method to determine the points on the boundary. Then the shortest distance from (P_o, Q_o) to the stability boundary will be the voltage stability margin. Instead of directly computing all the points on the boundary, an iterative approach

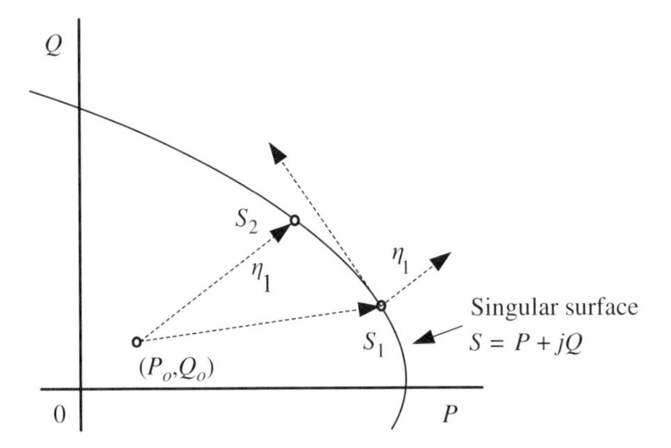

Figure 3.23 Voltage stability (P, Q) boundary.

is to use normal vectors (η_1) on the boundary surface to successively find the direction to the nearest voltage collapse point [6]. Figure 3.23 shows that the normal direction η_1 at the first estimate S_1 is used as the search for the next iterative value S_2.

Note that the voltage on the boundary surface is not necessarily a constant value. For a more complete treatment of loads that exhibit variable power factor, a complete *PQV* surface in three dimensions may be useful [26].

3.10.2 Voltage Sensitivities

Another group of voltage stability indicators is based on voltage sensitivities. Note that as the load is approaching the critical voltage point, the slope of a *PV* or *QV* curve will be quite steep. C. Taylor introduced the voltage collapse proximity index (VCPI) [25]. Referring to the stiff-source-to-load model in Figure 3.5, the reactive power equation can be expressed as

$$Q_S = Q + XI^2 = Q + X\frac{Q_S^2}{E^2} \tag{3.96}$$

where Q_S is the sending end reactive power, Q is the reactive power load, I is the line current magnitude, and XI^2 accounts for the reactive power losses in the line reactance. Rearrange (3.96) as a quadratic equation

$$Q_S^2 - \frac{E^2}{X}Q_S + \frac{E^2}{X}Q = 0 \tag{3.97}$$

to obtain the solutions

$$Q_S = \frac{1}{2}\left(\frac{E^2}{X} \pm E\sqrt{\frac{E^2}{X^2} - \frac{4Q}{X}}\right) \tag{3.98}$$

Then take the derivative to obtain the VCPI as

$$\left|\frac{dQ_S}{dQ}\right| = \frac{1}{\sqrt{1 - 4XQ/E^2}} = \frac{1}{\sqrt{1 - Q/Q_{\max}}} \tag{3.99}$$

As $Q \to Q_{\max}$, $|dQ_s/dQ| \to \infty$. That is, a small increase in Q will require a very large increase in Q_S.

Other sensitivities can also be used for determining whether the system operating condition is close to the voltage collapse point. For example, under the constant power-factor load assumption, the gradients $\partial V/\partial P$ and $\partial V/\partial Q$ can be computed. Near the voltage collapse point, these sensitivities will be large. These sensitivities are useful for establishing the proper voltage stability thresholds .

3.10.3 Singular Values and Eigenvalues of the Power Flow Jacobian Matrix

As discussed in Section 3.5, the full-order Jacobian matrix J (3.55) from power flow programs will become singular at the voltage collapse point. Thus by monitoring the smallest singular value σ_1 or smallest eigenvalue λ_1 of J when the load is increased, one can get a sense of how close the operating point is to the voltage collapse point. A threshold can be set up so that if either σ_1 or λ_1 is less than the threshold, the system would be in some critical voltage condition and immediate action would be needed to move the system operating condition away from the voltage collapse point. In addition, the singular vector corresponding to σ_1 or the eigenvector corresponding to λ_1 will provide information on the voltage collapse region [34].

3.11 Summary and Notes

This chapter discusses the fundamental aspects of steady-state voltage stability analysis. It starts with the stiff-source-to-load system to develop the PV curve. Then the Jacobian matrix of this system is used to illustrate the singularity issue at the maximum power transfer condition. For larger systems, singular values of Jacobian matrices are used to indicate the closeness of the current operating condition to the voltage collapse point. The AQ-bus method illustrates how the power flow solution at the voltage collapse point can be obtained without convergence problems, thus establishing it as an alternative to the continuation method. The effects of shunt capacitor banks and tap-charging transformers are also presented and illustrated with examples. Finally, various means of computing voltage stability margins and indices are summarized.

Since the 1990s, there has emerged a large body of literature on voltage stability analysis. A reader can find descriptions of additional voltage collapse events in real systems from [25], including the geomagnetic storm incident in Hydro Quebec [43]. There is recent interest in computing Thèvenin equivalents using measured power system data [44–46], which can be used for the VIP method for voltage stability margin calculation.

The discrete dynamics of under-load tap changing transformers under low voltage conditions have also been analyzed, with some fundamental results developed in [47, 48]. Load characteristics are also very important in voltage stability analysis [26]. The impact of induction motor stalling and restarting on system voltage is analyzed in Chapter 11.

Voltage stability is also impacted by the dynamics of the power equipment and their control systems [27]. For example, the over-excitation limit in an excitation system may have a strong impact in prolonged low-voltage conditions (Chapter 9). Furthermore, [27] describes various bifurcation processes associated with voltage instability.

Finally, test systems for voltage stability can be found in [49].

Problems

3.1 (Reactive power sources) List five sources of reactive power supply and two main consumptions of reactive power in a power grid.

3.2 (Transmission line active and reactive power flow) In Section 3.3, show that the receiving end voltage $\tilde{V}_R$ and current $\tilde{I}_R$ can be expressed in terms of V_S and I_S as

$$\tilde{V}_R = \tilde{V}_S \cosh(\gamma\ell) - Z_o\tilde{I}_S \sinh(\gamma\ell), \quad \tilde{I}_R = \tilde{I}_S \cosh(\gamma\ell) - Y_o\tilde{V}_S \sinh(\gamma\ell) \quad (3.100)$$

3.3 (Transmission lines) The characteristic impedance of a transmission line is $Z_o =$ 0.3 pu (on a 100 MVA base), and the product of its propagation constant and line length is $\gamma\ell = j0.12$. If the sending-end ($\ell = 0$) voltage is $\tilde{V}_s = 1.0\angle 0°$ pu and the sending-end current is $\tilde{I}_s = 3.33\angle 0°$ pu (this current is flowing into the transmission line), find the voltage phasor $\tilde{V}$ and current phasor $\tilde{I}$ at 0.75ℓ. Explain the results.

3.4 (Surge impedance loading) For the surge impedance loading case in Example 3.1, show analytically that the voltage profile and current magnitude are constant throughout the transmission line. Also show that

$$\frac{\tilde{V}}{\tilde{I}} = \frac{\tilde{V}_R}{\tilde{I}_R} \quad (3.101)$$

that is, the voltage and current on the transmission line have the same relative phase.

3.5 (Transmission lines) Using the parameters in Example 3.1, find Z_R in terms of Z_0 such that the voltage magnitude at the receiving end is 0.95 pu. Note that there may not be an easy closed-form solution, in which case Z_R may be found by trial and error. Compute the active and reactive power flow on the transmission line, and plot the voltage and current profiles.

3.6 (Transmission lines) Derive the equivalent π-model parameters (3.13) from the long-line model (3.7).

3.7 (Cable) For the underground cable given in Table 3.2, find the receiving end voltage with the load disconnected. Use the equivalent π model for the computation and assume the sending end voltage magnitude is unity.

3.8 (Transmission line models) Short transmission lines can be modeled as an equivalent π model. Multiple sections can be used if the line is longer. Consider a 345 kV, 100-mile long transmission line with $R = 0.0041$ pu, $X = 0.0423$ pu, and $B = 0.6866$ pu. Suppose the voltage at the sending end is 1.03 pu and the sending end current is 4 pu, with a power factor 0.95. Find the voltage at the midpoint and the receiving end of the transmission line using (a) long-line voltage expression, (b) one single section of the π model (only for the receiving end computation), and (c) two identical sections of the π model. Comment on the results.

3.9 (Stiff-source-to-load system) For the stiff-source-to-load system in Figure 3.5, use the normalized equation

$$p = \sqrt{v^2 - (q + v^2)^2} \tag{3.102}$$

to plot

1) the PV curve for $\tan\phi = 0.5, 0, -0.5$
2) the VQ curve for $p = 0.25, 0.5, 1.0$.

3.10 (Stiff-source-to-load system) The normalized voltage and power relationship of a load supplied by a transmission line connected to a stiff voltage source is given by

$$p = \sqrt{v^2 - (q + v^2)^2} \tag{3.103}$$

where v is the voltage and $p + jq$ is the load power.

1) Find v when $p = 0.4$ and $q = -0.1$.
2) For each value of v, find $\partial p/\partial v$, assuming that q is fixed at $q = -0.1$. From $\partial p/\partial v$, determine whether the operating point is voltage stable.

3.11 (Stiff-source-to-load system) Consider the stiff-source-to-load system of Figure 3.5 in which the source voltage is fixed at $E = 1.03$ pu and $X = 0.15$ pu. The load is characterized by $Q_L = 0.1P_L$. Find the maximum power transfer possible for this system and the critical voltage.

3.12 (Stiff-source-to-load system) In the equivalent power system shown in Figure 3.8, the equivalent voltage source E is fixed at 1.0 pu and $Z_{\text{Thev}} = 0.005 + j0.05$ pu.

1) Find the maximum active power transfer to the load and the critical voltage if the power factor of the load is 0.9 lagging.
2) Plot the active and reactive power flow from the voltage source and the load bus voltage as the load power is increased to maximum, with the load power factor kept at 0.9 lagging.

3.13 (Voltage stability analysis) This is a contingency analysis of the system in Problem 3.12, where the load is kept at 0.9 pf lagging.

1) Suppose the load is consuming 4.5 pu active power. What is the voltage stability margin of the system without considering any contingencies? For this example, the voltage stability margin is defined as the amount of active power load that can be increased until the critical voltage is reached.
2) In contingency analysis for secure dispatch (like $N - 1$ contingencies), voltage stability margins will be computed for all credible contingencies. Then the voltage stability margin will be the one that is most constraining. For simplicity, we will only evaluate one credible contingency. In this contingency, part of the transmission system between the load and the source is disconnected, such that the post-contingency Thevenin equivalent impedance becomes $Z_{\text{Thev}} = 0.010 + j0.075$ pu. There is no change to E_{Thev}. Calculate the voltage stability margin under this contingency. Note that if the voltage stability margin is negative, then the system operating condition is insecure.

3.14 (Thévenin equivalent model estimation)

1) For the equivalent power system shown in Figure 3.8, suppose one can measure the load bus voltage magnitude V and the load active and reactive power

components $P + jQ$ as the load varies. How many different data points will be needed to find E_{Thev} and $Z_{\text{Thev}} = jX_{\text{Thev}}$ (resistance neglected)?

2) Suppose two data points are available

$$(P, Q, V) = (1.483, 0, 0.95),\ (1.962, 0, 0.9) \tag{3.104}$$

Use MATLAB® functions *fsolve* to find E_{Thev} and X_{Thev}.

3) Suppose $Z_{\text{Thev}} = R_{\text{Thev}} + jX_{\text{Thev}}$. How many data points will be needed to find E_{Thev} and Z_{Thev}?

3.15 (Power flow solution) Compute the upper part of the PV curve for the 2-area system in Figure 2.7, which you have used in investigating power flow solutions, under the following conditions:

1) Generator 1 is the swing bus.
2) The total active power load on Buses 4 and 14 is always 27.52 pu.
3) The reactive power load at those two buses will remain unchanged at $j1.0$.
4) The load on Bus 14 starts at $14.76 + j1.0$ pu and is increased until the power flow would no longer solve.

Use a small stepsize in incrementing the load at Bus 14 to get as much of the PV curve as possible. (Because PST does not have the continuation power flow algorithm, a small stepsize is required.) Also, update the loads on Buses 4 and 14 in the power flow solution variable *bus_sol* and put *bus_sol* back into the function *loadflow* to improve the probability of convergence. Plot the voltage magnitude of Bus 14 versus the active power load on Bus 14 to obtain the PV curve.

3.16 (Jacobian matrix) Compute and comment on the singular values of the Jacobian matrix J for the last converged power flow solution in Problem 3.15. Remove from J the row corresponding to the active power equation for Bus 14 and the column corresponding to the Bus 14 voltage angle to form the reduced Jacobian J_R. Compute the singular values of J_R and compare to those of J.

3.17 (Maximum power transfer) A source with a fixed voltage supplying a load via a variable tap transformer is shown in Figure 3.24. The total reactance X_T for the transmission line and the transformer is 0.5 pu and the load impedance is $Z_L = 0.5 + j0.2$ pu. If the source voltage is $\tilde{V}_1 = 1.04$ pu, compute the tap ratio a such that maximum active power is transferred to the load. Compute the maximum power transfer and the load bus voltage $|\tilde{V}_2|$ at maximum power transfer. Assume that the tap ratio is continuous.

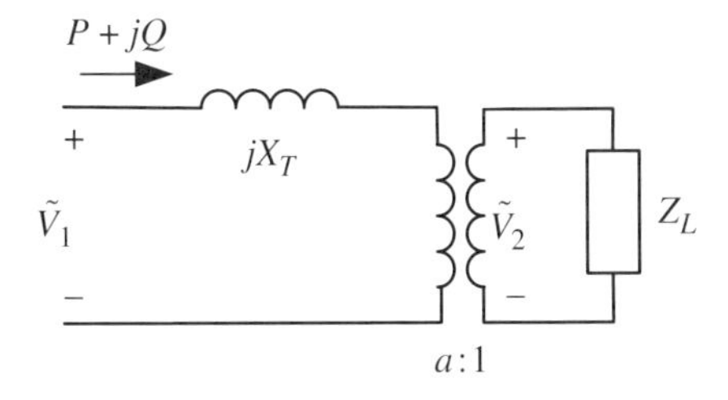

Figure 3.24 Circuit model for Problem 3.17.

4

Power System Dynamics and Simulation

4.1 Introduction

This chapter discusses power system dynamics and simulation using the electromechanical model for generators and the constant impedance model for loads. It introduces a reader to a modeling and simulation framework that remains valid for more detailed machine models with control equipment and for very large power systems. In some sense, it is a discussion on the modeling and simulation of dynamic networks [50], which are systems consisting of similar components interconnected via a network, such as mass-spring systems and heat-transfer systems.

Here the network is the AC power transmission network, which has been covered in Chapters 2 and 3. The dynamics of the AC power network itself, that is, electrical transients associated with transmission line inductances and capacitances, are assumed to be fast, and thus the power network is represented as a positive-sequence model with an instantaneous solution of the network equation $Y\tilde{V} = \tilde{I}$ (2.11). The purpose of this chapter is to show how generators and loads can be connected into the static power network to form a dynamic system.

First the model of a Single-Machine Infinite-Bus (SMIB) system is used to integrate the generator model into the power network. In this system the load is not important as it is represented by the infinite bus, which is modeled as the swing bus. Then multi-machine power system models are discussed, showing how the network admittance matrix Y can be extended to the generator internal nodes and to include constant-impedance loads.

Several integration methods are described for numerically computing the time response of such dynamic power systems to disturbances. A general framework for time-domain simulation consisting of a dynamic part modeled by differential equations (machines) and a static part modeled by algebraic equations (power network) is developed.

From the electromechanical models used in this chapter, power system stability concepts for multi-machine power systems can be introduced. Chapter 5 covers nonlinear power system models and discusses transient stability. Chapter 6 introduces linearization methods and analyzes linear power system models. After mastering these concepts and tools, the inclusion of detailed synchronous machine models and their control equipment into this framework will become more transparent. These topics are covered in subsequent chapters.

Power System Modeling, Computation, and Control, First Edition. Joe H. Chow and Juan J. Sanchez-Gasca.

Companion website: www.wiley.com/go/chow/power-system-modeling

4.2 Electromechanical Model of Synchronous Machines

Consider a synchronous generator on Bus i connected to a power network via a transformer with reactance X_T, as shown in Figure 4.1.a. The electromechanical model, also called the classical model, consists of two parts, an electrical model and a mechanical model.

The assumptions made for the electrical model include:

- the generator construction is of the round rotor type, so that saliency effects can be neglected
- the faster machine transients are neglected
- there is no saturation in the rotor magnetic field.

The electrical part of a classical model consists of a fixed voltage E' behind a transient reactance X'_d,[1] as shown in Figure 4.1.b. The generator internal voltage $\tilde{E}'$ consists of the fixed magnitude E', pointing in the direction of the rotor angle δ. The electrical model determines the electrical power output P_e supplied by the generator. Given the generator terminal bus voltage $\tilde{V}_T$, the generator output current is

$$\tilde{I}_G = \frac{1}{jX'_d}(\tilde{E}' - \tilde{V}_T) \tag{4.1}$$

and the electrical power supplied by the generator is

$$P_e = \text{Re}\{V_T I_G^*\} \tag{4.2}$$

The mechanical part of the classical model represents the motion of the generator rotor in response to system transient conditions. The angular acceleration of the rotating generator inertia is given by the so-called swing equation

$$\frac{2H}{\Omega}\frac{d^2\delta}{dt^2} = T_m - T_e \tag{4.3}$$

where the rotor angle δ is in radians,[2] the mechanical torque T_m applied by the turbine and the electrical torque T_e exerted on the rotor are given in pu (either on the system or machine base), the lumped inertia H of the turbine-generator is in sec (MW-sec/MVA), and $\Omega = 2\pi f_o$ rad/sec is the conversion factor from pu frequency to radians, where f_o is the system frequency in Hz.

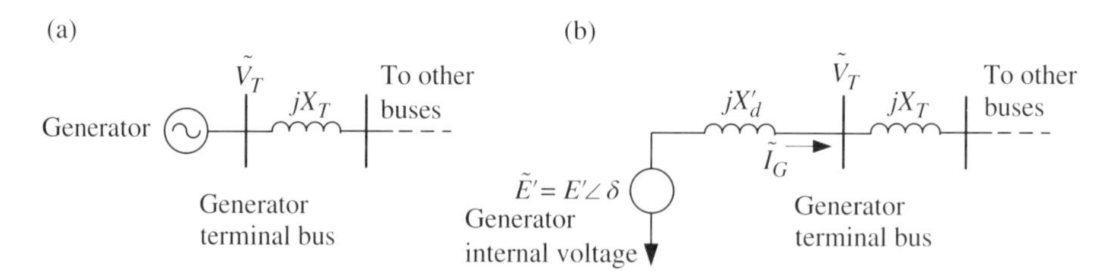

Figure 4.1 Classical model of synchronous generator: (a) one-line diagram and (b) electrical model.

1 X'_d is the direct-axis transient reactance.
2 The rotor angle δ points in the direction of the direct-axis of the rotor magnetic field, which is discussed in Chapter 7.

Note that power P is equal to the product of the torque T and the machine speed ω, that is, $P = T \cdot \omega$. However, with $\omega \simeq 1$ pu even in transient conditions, it is possible to use the approximation $T \simeq P$, resulting in the swing equation

$$\frac{2H}{\Omega}\frac{d^2\delta}{dt^2} = P_m - P_e - D(\omega - 1) \tag{4.4}$$

where damping effect intrinsic in the torque dynamics has been added via a damping coefficient D multiplying the speed deviation.[3]

The second-order swing equation (4.4) can also be written as two first-order differential equations

$$\frac{d\delta}{dt} = \Omega(\omega - 1), \quad 2H\frac{d\omega}{dt} = P_m - P_e - D(\omega - 1) \tag{4.5}$$

where ω is the rotor speed in pu on the nominal system frequency base. Defining the machine speed deviation as $\Delta\omega = \omega - 1$, (4.5) can be expressed as

$$\frac{d\delta}{dt} = \Omega\Delta\omega, \quad 2H\frac{d\Delta\omega}{dt} = P_m - P_e - D\Delta\omega \tag{4.6}$$

For the rest of the text, ω is mostly used to denote $\Delta\omega$. Also its usage should be clear from the context of how the variable is used in a dynamic equation.

In summary, the classical model of a synchronous generator is defined by the machine rating in MVA, and the parameters H, D, and X'_d, which, to be consistent, should be defined with respect to the machine MVA base. The advantage of this representation is that each of these parameters falls into a small range for practical equipment. For example, H is normally between 2.5 to 6 sec on the machine base, with combustion and hydraulic turbine-generators having lower inertias than fossil and nuclear units. The transient reactance X'_d is normally between 0.18 to 0.28 pu, and D, as mentioned earlier, nominally is either 0 or 1.

The fixed variables of mechanical input power P_m and internal voltage E' are computed based on the power flow solution. Let the electrical output power of the generator terminal bus be $P_e + jQ_e$ and the terminal voltage be $\tilde{V}_T = V_T\angle\theta$. Then the mechanical power is set to

$$P_m = P_e \tag{4.7}$$

and the internal voltage is computed as

$$\tilde{E}' = \tilde{V}_T + jX'_d\tilde{I}_G \tag{4.8}$$

where

$$\tilde{I}_G = ((P_e + jQ_e)/\tilde{V}_T)^* \tag{4.9}$$

is the generator current.

3 The coefficient D represents a composite of damping effect, such as governing and the dependence of loads on frequency [51]. For simulation software packages that use detailed machine models, D is typically set to zero. If (4.4) is used, a suitable value of D is the machine power output in pu, which often becomes $D = 1$.

Example 4.1: Generator parameter pu conversion and internal voltage computation

The parameters of a synchronous generator rated at 900 MVA are $H = 3.5$ sec and $X'_d = 0.24$ pu. The step-up transformer reactance is 15% on the machine base. The generator is supplying $S = P_e + jQ_e = 8.5 + j1.0$ pu power on the system base of 100 MVA. The generator terminal voltage obtained from a power flow solution is $\tilde{V}_T = 1.03\angle 30°$ pu. Express the system parameters and the transformer reactance on the system base of 100 MVA, and compute the steady-state values of P_m, E', and δ.

Solutions: First, the machine parameters on the machine base of 900 MVA are scaled to the system base of 100 MVA:

$$H = \frac{900}{100} \times 3.5 = 31.5 \text{ sec}, \quad X'_d = \frac{100}{900} \times 0.24 = 0.02667 \text{ pu} \tag{4.10}$$

The transformer reactance on 100 MVA base is

$$X_T = \frac{100}{900} \times 0.15 = 0.01667 \text{ pu} \tag{4.11}$$

which is modeled as branch data in the power flow data set. The mechanical power is $P_m = 8.5$ pu $= 850$ MW. The generator current (4.9) is computed as

$$\tilde{I}_G = \frac{8.5 - j1.0}{1.03\angle - 30°} = 7.6322 + j3.2854 \text{ pu} \tag{4.12}$$

From (4.8), the internal voltage is

$$\tilde{E}' = 1.03\angle 30° + j0.02667(7.6322 + j3.2854) = 1.0786\angle 41.773° \text{ pu} \tag{4.13}$$

Note that as the generator is supplying active power, $\delta = 41.773° > \theta = 30°$, and supplying reactive power, $E' = 1.0786$ pu $> V_T = 1.03$ pu. ∎

The following sections use the electromechanical model in a single-machine setting and then a multi-machine setting.

4.3 Single-Machine Infinite-Bus System

Consider a synchronous machine connected to an infinite bus via a transformer with a reactance X_T and a transmission line with a reactance X_L, as shown in Figure 4.2. The infinite bus is a fictitious entity representing an aggregate of a large number of generators. The reactances X_T and X_L are expressed on a common system base, such as 100 MVA. For a SMIB system, it is also quite common to use the machine base as the system base, in which case the transformer (leakage) reactance is about 0.15 pu, assuming that the transformer has the same rating as the synchronous machine. A SMIB system is often used to obtain the initial control parameter settings when a new generator is commissioned.

Before developing the dynamic model, the steady-state operating condition of the system is first established by specifying the voltages and the amount of active power P_e the generator is delivering to the infinite bus. The generator terminal voltage V_T is normally set to a few percent higher than 1 pu. The infinite-bus voltage magnitude V is also fixed, and normally set to 1 pu. The phase of the infinite bus voltage is set to 0° by modeling

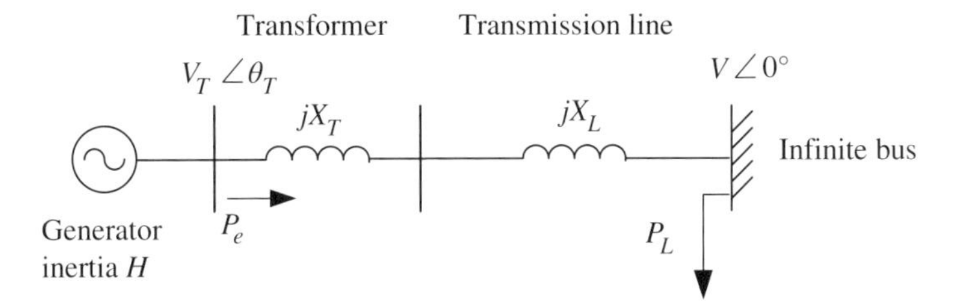

Figure 4.2 SMIB system.

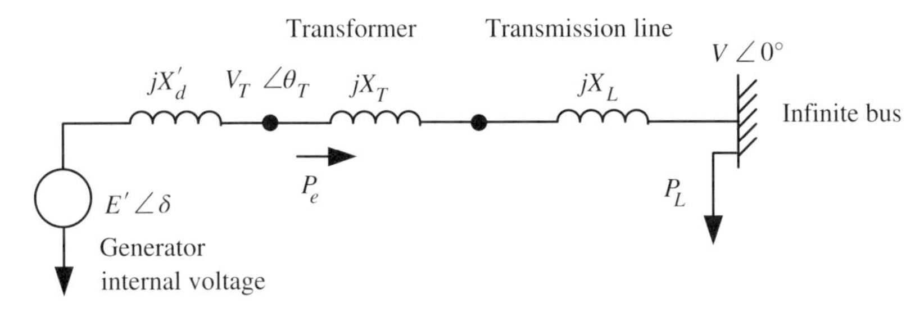

Figure 4.3 Electrical circuit representation of the SMIB system.

it as a swing bus in the power flow program with ample reactive power to support the bus voltage. The electrical power output P_e is normally set to 90–95% of the generator rating.

The generation of the infinite bus is effectively a load P_L which is equal to the generator electrical output power P_e. If there are no resistive losses,

$$P_L = P_e = \frac{V_T V \sin\theta_T}{X_T + X_L} \tag{4.14}$$

The terminal-bus voltage angle θ_T can be manually computed from (4.14) or obtained from a power flow solution.

Applying the electrical model of the electromechanical model to the SMIB system results in the electrical circuit diagram shown in Figure 4.3.

The next step in the nonlinear dynamic model development is to compute the machine internal voltage E' and the rotor angle δ, a process known as initialization, that is, finding the steady-state values of the states in a dynamic system. In Example 4.1, the power flow solution of V_T and θ_T at the terminal bus are used. Here this process is accomplished by starting from the infinite bus.

Using the voltage phasor at the infinite bus, the generator terminal voltage is

$$V_T\angle\theta_T = V\angle 0° + j(X_L + X_T)\,\tilde{I} \tag{4.15}$$

and the generator internal voltage is

$$E'\angle\delta = V\angle 0° + jX_{\text{eq}}\,\tilde{I}, \quad X_{\text{eq}} = X'_d + X_T + X_L \tag{4.16}$$

The electrical power equation (4.17) can be rewritten as

$$P_e = \frac{E' V \sin\delta}{X_{\text{eq}}} \tag{4.17}$$

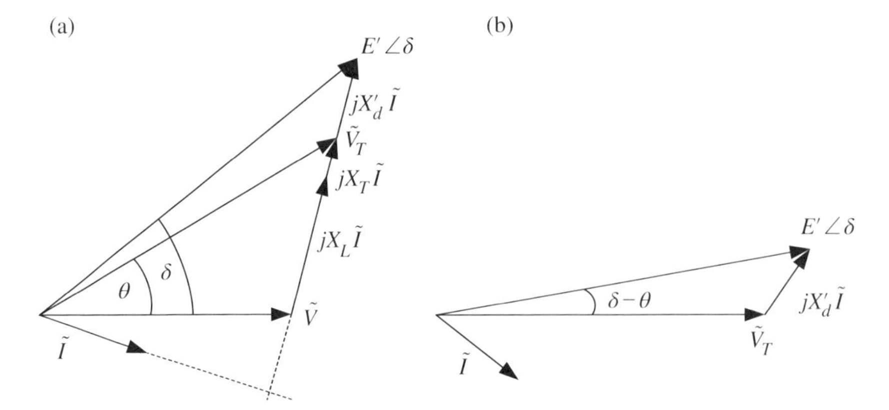

Figure 4.4 Voltage-current phasor diagrams: (a) infinite bus voltage used as reference and (b) generator terminal bus used as reference.

A common practice is to show a voltage-current phasor diagram to relate the voltages at various locations on the transmission system and inside a synchronous machine, as shown in Figure 4.4.[4] Figure 4.4.a shows the phasors relative to the infinite bus voltage phasor, which is pointing in the positive real direction. For multiple-machine systems without an infinite bus, it is more common to show the phasor diagram with the terminal bus voltage phasor pointing in the positive real direction for the synchronous machine of concern, as shown in Figure 4.4.b. In that reference frame, the rotor angle becomes $\delta' = \delta - \theta$.

The electrical power expression (4.17) can now be substituted into the swing equation (4.4) to form

$$\frac{2H}{\Omega}\frac{d^2\delta}{dt^2} = P_m - P_e = P_m - \frac{E'V}{X_{\text{eq}}}\sin\delta = P_m - P_{\max}\sin\delta \tag{4.18}$$

where the mechanical input power P_m is assumed to be constant, $X_{\text{eq}} = X'_d + X_T + X_L$, and $P_{\max} = E'V/X_{\text{eq}}$ is the maximum power transfer possible from the generator to the infinite bus. In equilibrium

$$\delta_{\text{ep}} = \sin^{-1}(P_m/P_{\max}), \text{and } P_m = P_{\max}\sin\delta_{\text{ep}} = P_e = P_L \tag{4.19}$$

The rotor angle δ is determined by the motion of the rotor. There are two torques acting on the rotor: the mechanical torque T_m in the same direction as the rotor rotation and the electrical torque T_e in the opposing direction, as shown in Figure 4.5.a. Note that the infinite bus is represented by a generator with an infinite inertia, and as a result its angle does not move. Thus it is shown as a fixed point. The reactance X_{eq} represents a nonlinear "spring" which provides the restoring force. This restoring force is nonlinear because it depends on $\sin\delta$ (4.18). An equivalent mass-spring system exhibiting a similar behavior with a linear spring is shown in Figure 4.5.b, where K is the spring constant. The fixed wall is analogous to the infinite bus.

4 Phasor diagrams for detailed machine models are more complex as additional reactances are included. They are shown in Chapter 7.

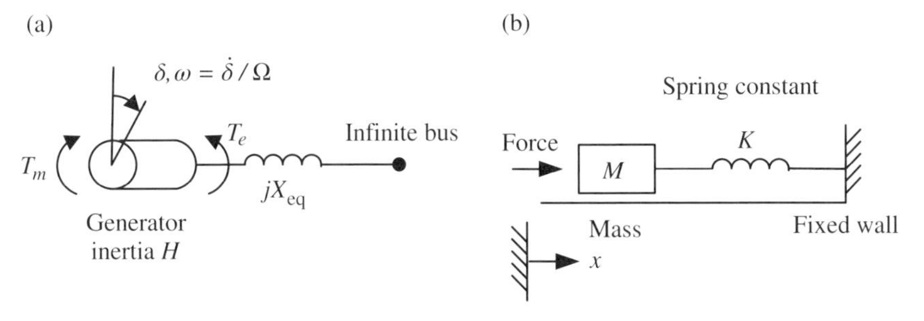

Figure 4.5 Free-body diagrams: (a) a lumped turbine-generator inertia with a reactance linked to an infinite bus and (b) a mass-spring system analog.

Example 4.2: Swing equation

For the SMIB system shown in Figure 4.3, the parameters on the machine base of 900 MVA are given as

$$V = 1.00 \text{ pu}, \quad V_T = 1.03 \text{ pu}, \quad X_L = 0.10 \text{ pu}, \quad X_T = 0.15 \text{ pu}, \quad X'_d = 0.24 \text{ pu}$$
$$H = 3.5 \text{ sec}, \quad D = 1.0 \text{ pu} \tag{4.20}$$

Suppose that the generator is supplying a load of 850 MW at the infinite bus. Calculate the steady-state values of E' and δ and write the swing equation. The nominal system frequency is $f_o = 60$ Hz.

Solutions: Note that the reactive power consumption is not specified in this example because it will be determined by the generator terminal voltage $\tilde{V}_T$. Here the line charging part of the transmission line is not represented. As $V_T > V$, reactive power will flow from the generator to the infinite bus. The computation in this example will take the machine MVA base as the system MVA. First, the angle separation between the generator terminal bus and the infinite bus is computed from

$$P_e = \frac{V_T V \sin\theta}{X_T + X_L} = \frac{1.03 \times 1.0 \sin\theta}{0.15 + 0.10} = \frac{850}{900} = 0.9444 \text{ pu} \tag{4.21}$$

so that

$$\theta_{ep} = \sin^{-1}\frac{0.25 \times 0.9444}{1.03} = 13.252° \tag{4.22}$$

The current flowing from the generator to the infinite bus is

$$\tilde{I} = \frac{V_T\angle 13.252° - 1.0}{j(X_T + X_L)} = 0.9445\angle -0.624° \text{ pu} \tag{4.23}$$

Hence

$$\tilde{E}' = V_T\angle 13.252° + j\tilde{I}X'_d = 1.1065\angle 24.724° \text{ pu} \tag{4.24}$$

The swing equation with $X_{eq} = X'_d + X_T + X_L = 0.49$ pu is given by

$$\frac{d\delta}{dt} = 2\pi f_o \omega, \quad 2 \times 3.5\frac{d\omega}{dt} = 0.9444 - \frac{1.1065 \times 1.0}{0.49}\sin\delta - 1.0\omega \tag{4.25}$$

that is,

$$\frac{d\delta}{dt} = 377\omega, \quad 7\frac{d\omega}{dt} = 0.9444 - 2.2581 \sin \delta - 1.0\omega \tag{4.26}$$

■

4.4 Power System Disturbances

To ensure the secure operation of a power system, the protection system should be set up so that a synchronous machine can maintain its stability when the power system is subjected to an array of disturbances, such as short-circuit faults due to lightning strikes. The impact of faults on power system stability is studied by simulating the resulting power system dynamics. A typical power system simulation consists of three periods:

1) Pre-fault period: this is the steady-state power flow condition as discussed in Chapter 2.
2) Fault-on period: the fault remains on until it is cleared by circuit breaker actions. With short-circuit faults, the generator rotors would in general accelerate as the electrical power transfer in the power network is disrupted.
3) Post-fault period: the fault has been cleared, mostly likely accompanied with a line trip, which results in a weaker power system. The purpose of dynamic simulation is to investigate whether the power system will converge to a post-fault equilibrium state.

This section discusses fault-on and post-fault electrical power computation for the SMIB system. A general framework using admittance matrices is discussed in Section 4.6.

4.4.1 Fault-On Analysis

One of the most severe disturbances is a 3-phase short-circuit (SC) fault close to a generator. Consider the system in Figure 4.6 in which the transmission line with reactance X_L is replaced by two parallel transmission lines, each with a reactance of $2X_L$. The fault is applied to one of the two transmissions lines near the high side of the transformer, as indicated in Figure 4.6.

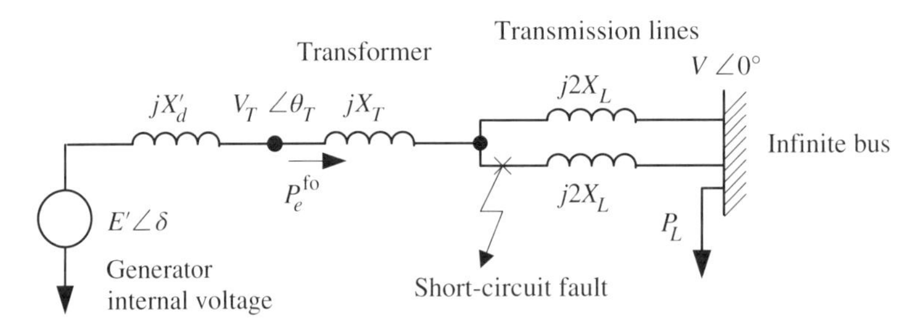

Figure 4.6 Circuit diagram showing short-circuit fault on one of the parallel lines.

With the SC fault on the high side of the generator transformer, the voltage V_T will be lowered, but the machine internal voltage E' is assumed to be fixed. Thus the generator current during the SC condition is

$$\tilde{I}_{gsc} = \frac{E'\angle\delta}{j(X'_d + X_T)} \tag{4.27}$$

The fault-on electrical power P_e^{fo} delivered by the generator to the system is zero because the current $\tilde{I}_{gsc}$ is delivered to a location with zero voltage through a pure reactance. Thus the rotor acceleration equation becomes

$$\frac{d^2\delta}{dt^2} = \frac{\Omega}{2H}P_m \tag{4.28}$$

If the fault is left on, the machine angle δ will become unbounded and the generator will lose synchronism (out-of-step), that is, it will be *transiently unstable.*

Example 4.3: Fault-on electrical power computation

This example illustrates the calculation of the electrical output power for a short-circuit fault away from the machine terminal bus. Consider a 3-phase fault at the mid-point of a transmission line, as shown in Figure 4.7. Find the fault-on electrical power provided by the generator, using the voltage and reactance parameters shown in the figure.

Solutions: The equivalent fault-on transmission system is shown in Figure 4.8. There are a number of ways to calculate the fault-on electrical power output P_e^{fo} from the generator. One approach is to develop a Thevènin equivalent. The transformer high-side voltage is chosen as the output port of the circuit. The two voltage sources $E'\angle\delta$ and $V\angle 0^\circ$ need to be combined into a single voltage source, and the transmission line and

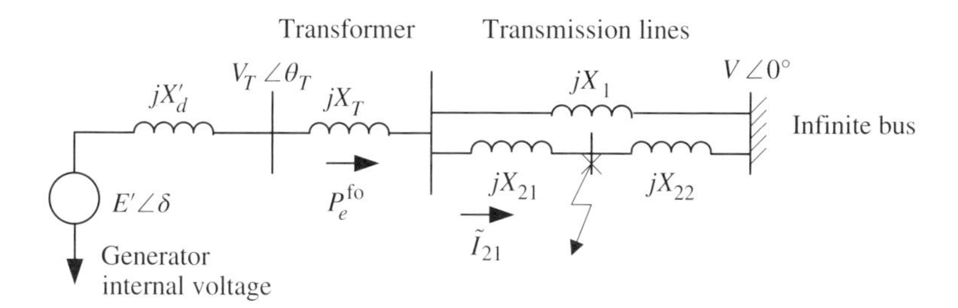

Figure 4.7 Short-circuit fault on transmission line.

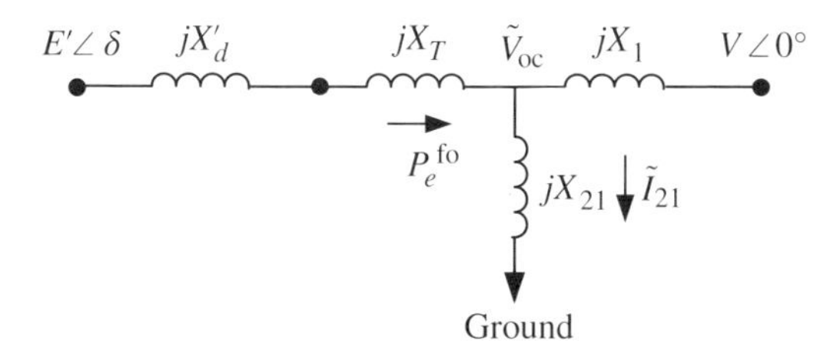

Figure 4.8 Equivalent transmission system of Figure 4.7.

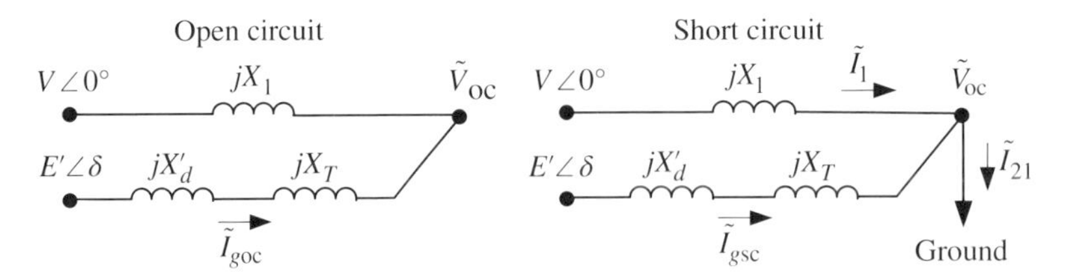

Figure 4.9 Open- and short-circuit representation.

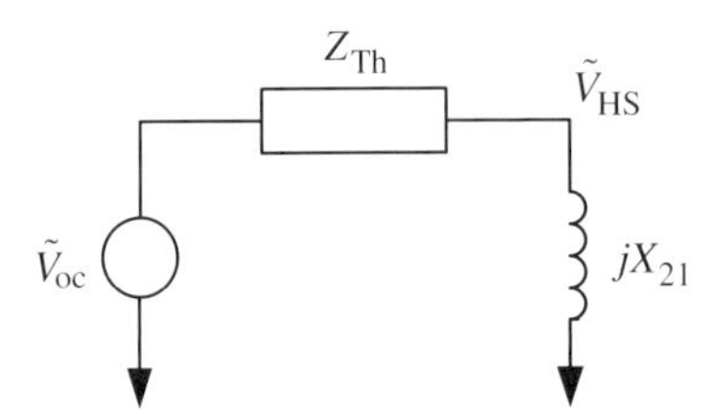

Figure 4.10 Thevènin equivalent circuit.

transformer reactances need to be combined into a single series reactance. Note that the reactance X_{22} is not needed for this calculation. Then the open-circuit voltage $\tilde{V}_{oc}$ and the short-circuit current $\tilde{I}_{21}$ in the transmission line with impedance jX_{21} shown in the two circuits of Figure 4.9 are computed as

1) Open-circuit condition:

$$\tilde{I}_{goc} = \frac{E'\angle\delta - V\angle 0°}{j(X'_d + X_T + X_1)}, \qquad \tilde{V}_{oc} = E'\angle\delta - j(X'_d + X_T)\tilde{I}_{goc} \tag{4.29}$$

2) Short-circuit condition:

$$\tilde{I}_{21} = \tilde{I}_1 + \tilde{I}_{gsc} = \frac{V\angle 0°}{jX_1} + \frac{E'\angle\delta}{j(X'_d + X_T)} \tag{4.30}$$

Thus the Thevènin equivalent voltage is $\tilde{V}_{oc}$ and the impedance is $Z_{Th} = \tilde{V}_{oc}/\tilde{I}_{21}$. The voltage on the high-side bus is given by

$$\tilde{V}_{HS} = \frac{jX_{21}}{Z_{Th} + jX_{21}}\tilde{V}_{oc} \tag{4.31}$$

as shown in Figure 4.10.

Note that $\tilde{V}_{HS}$ is a function of δ and thus varies with time. The output current and power of the generator are

$$\tilde{I}_g = \frac{E'\angle\delta - \tilde{V}_{HS}}{j(X'_d + X_T)}, \qquad P_e = \text{Re}(E'\angle\delta \cdot \tilde{I}_g^*) \tag{4.32}$$

■

4.4.2 Post-Fault Analysis

A short-circuit fault is typically cleared in 4–6 cycles (67–100 msec on a 60 Hz system) by protective relaying.[5] If the fault is on the transmission line, the circuit breaker at the

5 The clearing time consists of a fault detection time and a fault interruption (breaker opening) time.

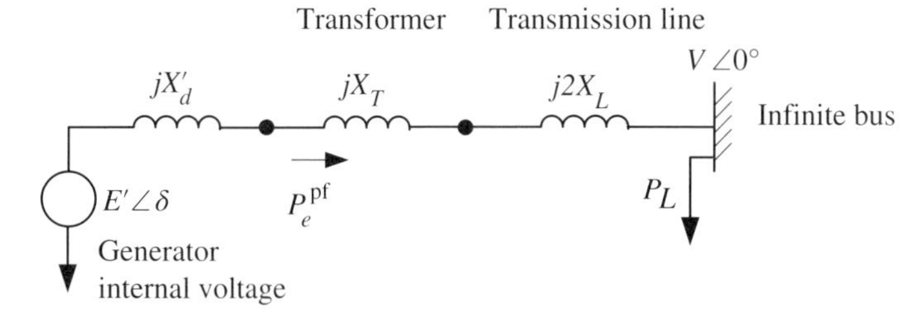

Figure 4.11 One-line diagram showing post-fault system.

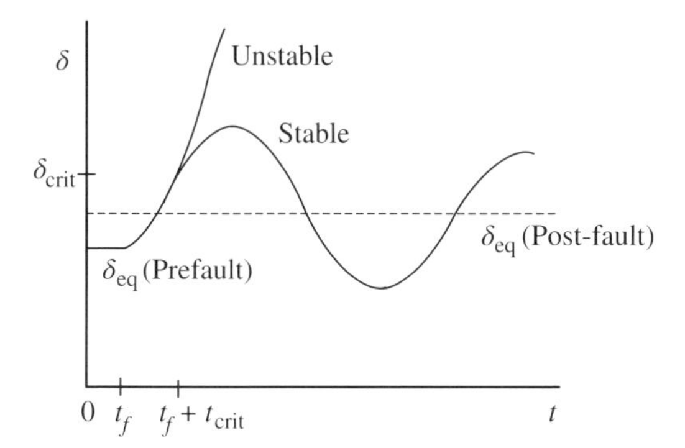

Figure 4.12 Plots showing stable and unstable generator angle responses.

end of the transmission line near the fault will open first. The circuit breaker at the other end will also open shortly afterwards.

For the SMIB system, after the faulted line is cleared, the post-fault system is shown in Figure 4.11. Whenever a major transmission line is tripped, the impedances in some of the transfer paths will be larger, such that the angular separation from the generators to the loads will also be larger. The system is now weaker and more highly stressed. It has less synchronizing capability to restore system stability compared to the situation in which the fault somehow manages to extinguish itself and the faulted line is not tripped.

For the post-fault system in Figure 4.11, the electrical power provided by the generator is

$$P_e^{pf} = \frac{E'V\sin\delta}{X'_d + X_T + 2X_L} \tag{4.33}$$

In the post-fault period, with the generator rotor having accelerated, system stability depends on whether the post-fault network is strong enough to hold the generator in synchronism. The longer a fault is left on, the higher the values of the generator rotor angle and speed when the fault is cleared such that the generator would lose synchronism (i.e., be unstable) with the power network. The *critical clearing time* t_{crit} is the longest period that a fault can stay on without causing the generator to be unstable. Figure 4.12 illustrates this situation showing a stable generator response when the fault clearing time $t_c = t_{crit}$. The generator is unstable if $t_c > t_{crit}$.

4.4.3 Other Types of Faults

In addition to 3-phase short-circuit faults, other common faults to be considered include single-phase-to-ground faults, line-to-line faults, and two-lines-to-ground faults. The fault current computation would require developing the positive-, negative-, and zero-sequences of the transmission circuits [52]. In a balanced 3-phase positive-sequence power system simulation program, the simulation of such faults can be modeled as a short-circuit fault through an appropriate impedance to ground at the fault location. These faults are normally less severe than 3-phase faults. Thus if a power system remains stable after a 3-phase fault, it is not necessary to simulate a single-phase fault at the same location. On the other hand, these less severe faults may be used in multiple-contingency studies. For example, a 3-phase fault is applied on a transmission line, which is cleared normally. Then a few seconds later, the compromised system is subject to a single-line-to-ground fault at a nearby transmission line and cleared by removing that phase.[6] Systems that can withstand such multiple contingencies are obviously more resilient.

4.5 Simulation Methods

Consider the second-order nonlinear swing equation of the SMIB system with damping included

$$\frac{d^2\delta}{dt^2} = \frac{\Omega}{2H}\left(P_m - P_e - D\omega\right) = \frac{\Omega}{2H}\left(P_m - \frac{E'V\sin\delta}{X_{\text{eq}}} - D\omega\right) \tag{4.34}$$

where the machine rotor angle δ is in radians, the machine rotor speed deviation ω is in pu, and $\Omega = 2\pi f_o$ ($\Omega = 377$ rad/sec for a 60 Hz system and 314 rad/sec for a 50 Hz system). The stability of a nonlinear system subject to a disturbance can be determined using step-by-step numerical integration.

General purpose integration routines are developed for a nonlinear dynamic system expressed as a system of first-order ordinary differential equations (ODEs). For example, the second-order swing equation can be written as a system of two first-order ODEs:

$$\frac{d\delta}{dt} = \Omega\omega, \quad \frac{d\omega}{dt} = \frac{1}{2H}\left(P_m - \frac{E'V\sin\delta}{X_{\text{eq}}} - D\omega\right) \tag{4.35}$$

In general, a power system consisting of an electrical network, generators, and controls (such as voltage regulators) can be modeled as a system of first-order nonlinear ODEs with t_o the initial time and x_o the initial condition

$$\dot{x} = f(x, u, t), \quad x(t_0) = x_0, \quad x, f \in R^n, u \in R^m \tag{4.36}$$

where x is the state vector of dimension n, u is the input vector of dimension m, and f is the vector of the nonlinear dynamics governing the system.

Let the numerical solution of (4.36) be

$$x(t_0) = x_0, x(t_1) = x_1, \ldots, x(t_k) = x_k, x(t_{k+1}) = x_{k+1}, \cdots \tag{4.37}$$

6 Many system operators will open the other two phases once the system is stabilized, to avoid operating an unbalanced system.

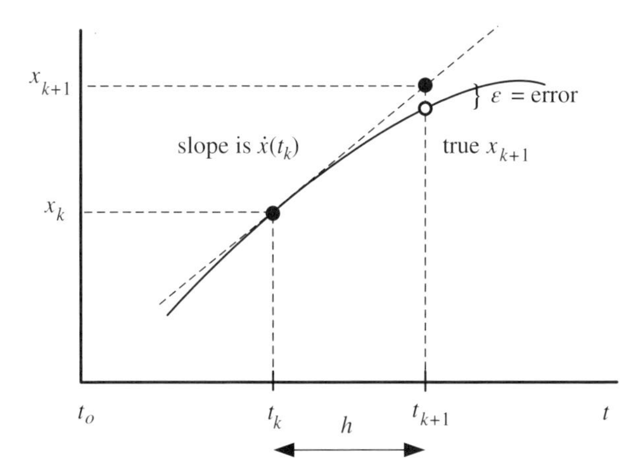

Figure 4.13 Euler method.

where t_k is the time at the kth time step and the time increment h is uniform, that is, $t_{k+1} = t_k + h$. The parameter h is also known as the integration stepsize.

The Taylor Series expansion of the solution to (4.36) up to the $(p+1)$st order is

$$x_{k+1} = x_k + h\dot{x}_k + \frac{h^2}{2}\ddot{x}_k + \cdots + \frac{h^p}{p!}x_k^{(p)} + \frac{h^{p+1}}{(p+1)!}x_k^{(p+1)}(\xi) \tag{4.38}$$

where $t_k \leq \xi \leq t_{k+1}$ and the superscript p in $x_k^{(p)}$ denotes the pth derivative of x_k.[7]

If $\dot{x} = f(x, u, t)$, then $x_k^{(p)}$ can be obtained by taking repeated partial derivatives with respect to the state x and the input u. For example, the second derivative is

$$\ddot{x} = \frac{\partial f}{\partial t} + \frac{\partial f}{\partial x}\dot{x} + \frac{\partial f}{\partial u}\dot{u} \tag{4.39}$$

If one only takes the first p terms in (4.38), then the $(p+1)$st term becomes the local truncation error ε. In the sequel, several integration algorithms will be developed using (4.38). The objective is to use methods that do not require partial derivatives and are sufficiently accurate without an excessive amount of evaluations of f.

4.5.1 Modified Euler Methods

In the Euler method, a nonlinear ODE is solved for a fixed stepsize h from $t = t_o$ to $t = t_f = t_o + N_h h$, using the formula

$$x(t_{k+1}) = x(t_k) + h \cdot \dot{x}(t_k) = x_k + h \cdot \dot{x}_k, \quad k = 0, 1, ..., N_h - 1 \tag{4.40}$$

Graphically, the integration is illustrated in Figure 4.13.

The Euler method is a first-order Runge-Kutta (RK) method.[8] The local truncation error is [54]

$$\varepsilon = \frac{h^2}{2}\ddot{x}(\xi) \simeq \frac{h^2}{2}\ddot{x}_k \simeq \frac{h^2}{2}\frac{(\dot{x}_{k+1} - \dot{x}_k)}{h} = \frac{h}{2}(\dot{x}_{k+1} - \dot{x}_k) \tag{4.41}$$

7 Recall the midpoint theorem from an advanced calculus text [53]. This is a higher order extension of the midpoint theorem.

8 The Euler method is also the first-order form of many other integration methods such as the Adam-Bashforth (AB) method. The second-order AB method is discussed in Section 4.5.2.

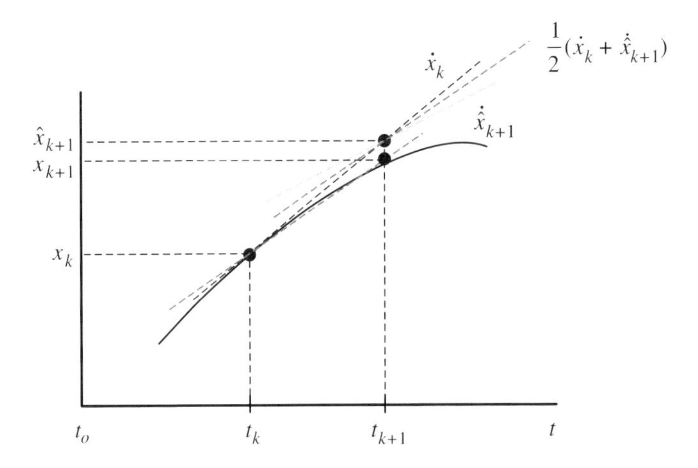

Figure 4.14 Euler method with full-step modification (see color plate section).

The Euler method requires one evaluation of f at each integration step, and is straightforward to implement.[9] In power system dynamic simulations, the Euler method is not normally used as it would require a very small stepsize to capture the lightly damped oscillations accurately.

4.5.1.1 Euler Full-Step Modification Method

In the Euler method with full-step modification, the integration is performed in two stages, as illustrated in Figure 4.14. In the predictor step, the value of $\hat{x}_{k+1}$ is obtained as

$$\hat{x}_{k+1} = x_k + h \cdot \dot{x}_k = x_k + h \cdot f(x_k, u_k, t_k) \tag{4.42}$$

where the slope of $\dot{x}_k$ is shown as the black dashed line. Then in the corrector step, the time-derivative of $\hat{x}_{k+1}$, shown as the green dashed line, is computed as

$$\dot{\hat{x}}_{k+1} = f(\hat{x}_{k+1}, u_{k+1}, t_{k+1}) \tag{4.43}$$

such that the predicted value x_{k+1} can be obtained using the average value of $\dot{x}_k$ and $\dot{\hat{x}}_{k+1}$, shown by the red dashed line, as

$$x_{k+1} = x_k + \frac{h}{2}(\dot{x}_k + \dot{\hat{x}}_{k+1}) \tag{4.44}$$

The corrector step can be used repeatedly to further improve the accuracy.

This predictor-corrector method is a second-order RK method with a local truncation error $\varepsilon \simeq (h^3/6)x_k^{(3)}$.[10] This improved accuracy comes at the expense of requiring two evaluations of f at each integration step.

9 Note that each evaluation of f in a power system simulation program involves sweeping through the dynamic and static equations of all the synchronous machines and their control equipment, and solving a network solution (sometimes iteratively), all of which are quite involved.

10 The Euler full-step modification method is used in the Power System Toolbox for the simulation of power system disturbances.

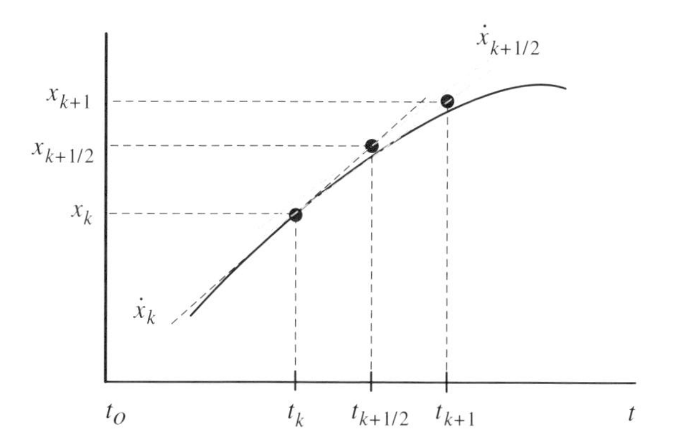

Figure 4.15 Euler method with half-step modification (see color plate section).

4.5.1.2 Euler Half-Step Modification Method

A variation of the Euler full-step modification is the half-step method, as illustrated graphically in Figure 4.15. The half-step method is also computed in two stages. First, the value of x at the midpoint between t_k and t_{k+1} is computed as

$$x_{k+1/2} = x_k + \frac{h}{2} \cdot \dot{x}_k \tag{4.45}$$

where the slope of $\dot{x}_k$ is indicated as the black dashed line. Then the time derivative of x at $t_{k+1/2}$ is computed using

$$\dot{x}_{k+1/2} = f(x_{k+1/2}, u_{k+1/2}, t_k + h/2) \tag{4.46}$$

the slope of which is indicated as the green dashed line passing through $x_{k+1/2}$. The value of x_{k+1} is computed using the time derivative at the midpoint

$$x_{k+1} = x_k + h \cdot \dot{x}_{k+1/2} \tag{4.47}$$

The Euler half-step modification method is also a second-order RK method, but the local truncation error $\varepsilon \simeq (h^3/24)x_k^{(3)}$ is smaller than that of the full-step modification method.

4.5.2 Adams-Bashforth Second-Order Method

The modified Euler methods require two evaluations of f per time step to achieve an error of $O(h^3)$.[11] The second f evaluation can be circumvented if the time derivative value at the previous time step is utilized. In the Adams-Bashforth second-order (AB2) method, the integration of (4.36) is performed as

$$x_{k+1} = x_k + h[1.5f(x_k) - 0.5f(x_{k-1})] \tag{4.48}$$

For initialization of AB2, the Euler method is used to integrate from x_0 to x_1.

11 A function $\varepsilon(h)$ is of order h^q, denoted by $O(h^q)$, where q is an integer, if for $h \to 0$, $\varepsilon(h)$ decays as fast as h^q.

Such a method is known as an explicit multi-step method because (1) there is no corrector part and (2) the new value depends on more than just the current value. The AB2 method is used in some commercial power system simulation packages.[12]

There are a number of ways to show the choice of coefficients 1.5 and −0.5 for the AB2 formula. Ref. [54] has the following derivation.

Let the general formula be

$$x_{k+1} = \alpha x_k + h[\beta_1 f(x_k) - \beta_2 f(x_{k-1})] \tag{4.49}$$

Performing the Taylor series expansion backwards from $k+1$ to k and $k-1$, this expression can be expanded as

$$\begin{aligned} x_{k+1} = \alpha[&x_{k+1} - hf(x_{k+1}) + (h^2/2)\dot{f}(x_{k+1}) + O(h^3)] \\ &+ \beta_1 h[f(x_{k+1}) - h\dot{f}(x_{k+1}) + O(h^2)] \\ &+ \beta_2 h[f(x_{k+1}) - 2h\dot{f}(x_{k+1}) + O(h^2)] \end{aligned} \tag{4.50}$$

Equating the like terms of h^0, h^1, and h^2, the resulting equations in matrix form

$$\begin{bmatrix} 1 & 0 & 0 \\ -1 & 1 & 1 \\ 0.5 & -1 & -2 \end{bmatrix} \begin{bmatrix} \alpha \\ \beta_1 \\ \beta_2 \end{bmatrix} = \begin{bmatrix} 1 \\ 0 \\ 0 \end{bmatrix} \tag{4.51}$$

yield the solution

$$\alpha = 1, \quad \beta_1 = 1.5, \quad \beta_2 = -0.5 \tag{4.52}$$

From the expression (4.50), it can readily be seen that the truncation error of AB2 is of the order of h^3. This is achieved with only one evaluation of $f(x_k)$ in the formula because $f(x_{k-1})$ has already been computed in the previous integration step.

4.5.3 Selecting Integration Stepsize

In addition to the truncation error, another important aspect of choosing a particular integration method is numerical stability. The electromechanical oscillatory modes in a power system are lightly damped. If the Euler method with a large stepsize is used, an oscillatory mode with a small positive damping may appear as an unstable oscillation in the dynamic simulation. The region of numerical stability as a function of the damping of the oscillatory mode and the integration stepsize h can be found in many textbooks on numerical integration [54]. Figure 4.16 shows the stability boundary of the RK1, RK2, and AB2 methods with respect to the real and oscillatory modes (eigenvalues) of linear continuous-time systems and the integration stepsize. Note that the axes are scaled by h, that is, the horizontal axis is $h\sigma$ and the vertical axis is $h\omega$, where σ is the damping and ω is the frequency of an oscillatory mode. The RK2 (modified Euler) methods have a larger stability region than the RK1 (Euler) method. The stability boundary of the AB2 methods is smaller than that of the RK2 method. Eigenvalues of linearized power systems are discussed in Chapter 6.

If the synchronous machines are modeled with classical models only, then the system dynamics consist of oscillatory modes between 0.25 to 3 Hz (i.e., about 1.6 to 20 rad/sec). An integration stepsize of 10 msec would result in $h\omega = 0.01 \times 20 = 0.2$, which

12 The Siemens-PTI PSS/E program and the GE PSLF program.

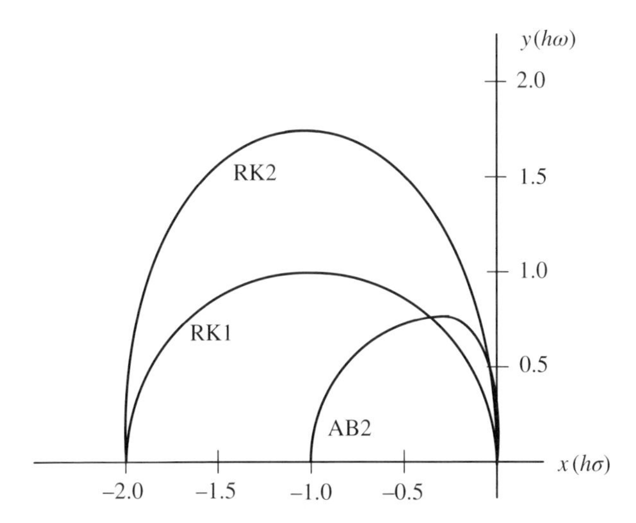

Figure 4.16 Numerical stability boundaries of the RK1, RK2, and AB2 integration methods showing only for positive values of ω (adopted from [55] and [54]).

will lie inside the RK2 and AB2 stability boundaries if the swing mode has a small positive damping. During the fault-on duration, a smaller stepsize like 5 msec may be used to improve accuracy.

If detailed models are used for the synchronous machines equipped with excitation systems and voltage regulators, the smallest time constants may be of the order of 10 msec. (These detailed models are described in later chapters.) For systems with high voltage regulator gains, the effective closed-loop time constants can be even smaller. Suppose the smallest time constant of the power system is $T = 10$ msec. Then $h\sigma = -h/0.01 = -100h$, such that h has to be less than 20 msec to be in the RK1 and RK2 stability regions according to Figure 4.16, and h has to be less than 10 msec to be in the AB2 stability region. Thus for RK1 and RK2 methods, h should be smaller than the smallest time constant, and for the AB2 method, h should be smaller than one-half of the smallest time constant. For systems with detailed machine models, an integration stepsize of 5 or 2.5 msec should be used, which is similar to the recommendation of an integration stepsize of a 1/4 cycle (240 points per sec for a 60-Hz system). Also during the fault-on period, the stepsize should be even smaller.

For transient stability analysis, a power system subject to a disturbance is typically simulated to 10 to 20 sec, until the system either settles to a steady state or becomes unstable. It is also good practice to simulate a small initial period in steady state before a fault is applied. The purpose is to show that the system starts from an equilibrium point and the simulation program initializes properly in steady-state condition. A proper initialization leads to all the variables that describe the system to remain constant during the pre-disturbance period; an incorrect initialization causes the system variables to drift away from their initial values and it is often due to erroneous system parameters.

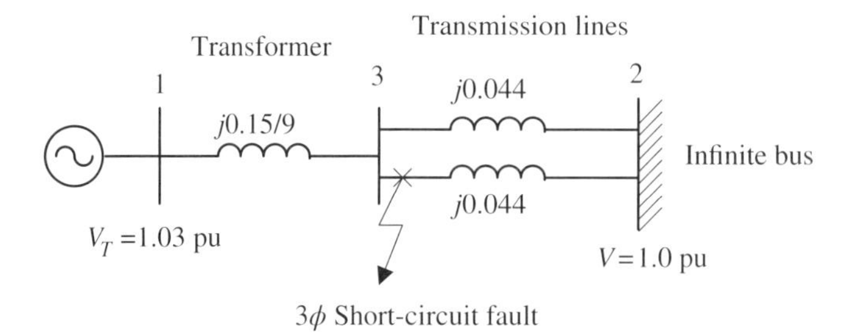

Figure 4.17 SMIB system for Example 4.4.

Example 4.4: Power system disturbance simulation

A SMIB system is shown in Figure 4.17. The electromechanical model parameters of Generator 1 on Bus 1 are base MVA = 900, H = 3.50 sec, D = 1 pu, and X'_d = 0.24 pu. The voltage at the infinite bus is 1.0 pu and the voltage at the generator bus is 1.03 pu. The transformer reactance is 0.15/9 = 0.0167 pu and the reactances of the two parallel transmission lines are each 0.044 pu, all on the system base of 100 MVA.

1) Solve the power flow for the system with P_e = 850 MW, using the infinite bus as the swing bus.
2) Simulate the system response up to 5 sec using the following script: no-fault simulation for 0 to 0.1 sec, a 3-phase short-circuit fault on one of the lines connecting Buses 2 and 3 on the Bus 3 side at 0.1 sec, fault cleared by opening the faulted line at 0.2 sec (6-cycle fault) at Bus 3 and 0.01 sec later at Bus 2. Plot $\delta(t)$ (in deg) and $\omega(t)$ (in pu) versus t. Use integration stepsizes of 5 msec for the pre-fault period, 1 msec from fault application to 0.5 sec, and 10 msec after 0.5 sec.

Solutions: The power flow solution shows that $V_1 = 1.03\angle 18.61°$, $V_2 = 1.0\angle 0°$, and $V_3 = 1.004\angle 10.73°$. The disturbance responses of Generator 1 machine angle and speed are shown in Figure 4.18. The system is stable for a fault clearing time of 6 cycles (60 Hz system). With the damping coefficient $D > 0$, these responses are damped, although the damping is small. If $D = 0$, the oscillations would be undamped. ■

4.5.4 Implicit Integration Methods

Explicit integration methods, such as the modified Euler methods and the AB2 method, require using a stepsize smaller than the smallest time constants in the system. In applications such as extended time simulation in which the simulation is performed over minutes and the fast dynamics are of less interest, using a bigger stepsize will speed up the simulation. Implicit integration methods can provide such a capability [56].

Consider a general nonlinear dynamic model governed by both differential and algebraic equations (DAEs)

$$\dot{x} = f(x, y, t), \quad y = \overline{g}(x, t), \quad x \in R^n, y \in R^p \tag{4.53}$$

where x is the state vector, y is a vector of algebraic variables, and f and $\overline{g}$ are nonlinear functions of x, y, and t. For a power system, the states in x represent the machine and

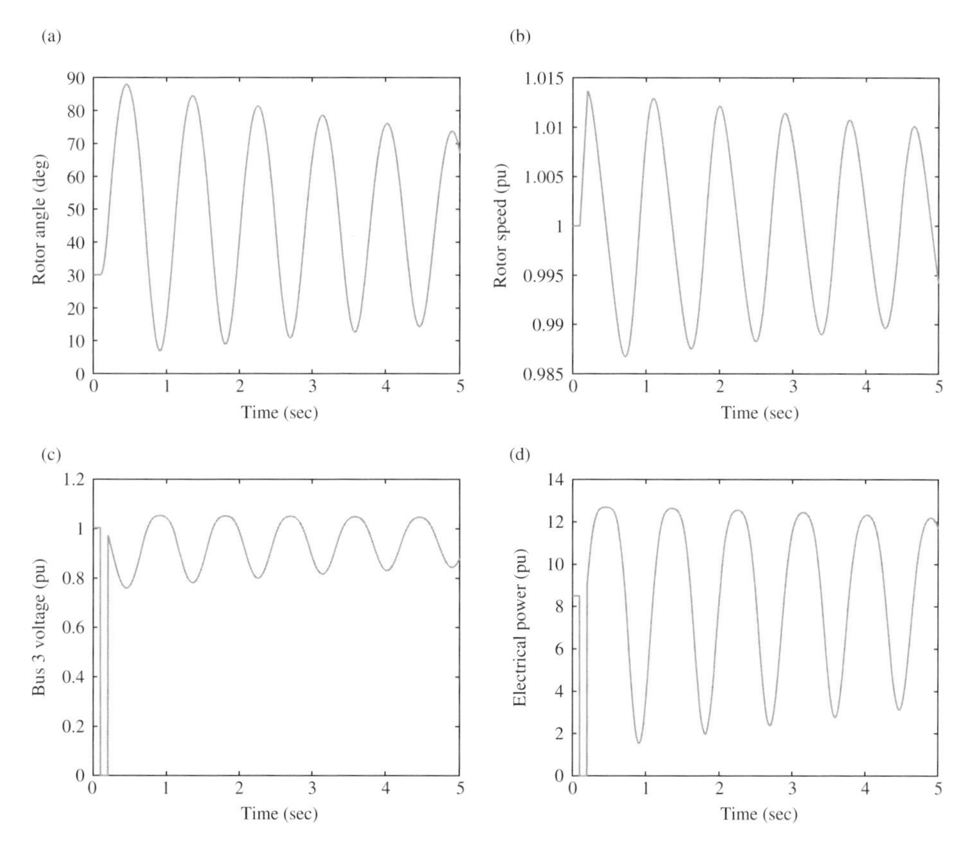

Figure 4.18 Generator 1 disturbance responses: (a) rotor angle, (b) rotor speed, (c) Bus 3 voltage magnitude, and (d) electrical power (see color plate section).

control equipment dynamics, and the algebraic variables in y represent the solution of the power network, such as bus voltages.

In the case in which the algebraic variable y does not have a closed form solution in terms of x and t, (4.53) is modeled as

$$\dot{x} = f(x, y, t), \quad 0 = g(x, y, t) \tag{4.54}$$

4.5.4.1 Integration of DAEs

Using the trapezoidal rule, for integration beyond $t = t_k$ with the state variable solution x_k, the state x_{k+1} is expressed as

$$x_{k+1} = x_k + \frac{h}{2}(f(x_k, y_k, t_k) + f(x_{k+1}, y_{k+1}, t_k + h)) \tag{4.55}$$

Including the algebraic equation (4.54), (4.55) can be rewritten as

$$x_{k+1} - x_k - \frac{h}{2}(f(x_k, y_k, t_k) + f(x_{k+1}, y_{k+1}, t_{k+1})) = 0 \tag{4.56}$$

$$g(x_{k+1}, y_{k+1}, t_{k+1}) = 0 \tag{4.57}$$

The unknowns x_{k+1} and y_{k+1} from this set of equations can be solved using a nonlinear equation solver, such as the Newton-Raphson method.

Implicit methods can be useful in the simulation of longer term power system dynamics, such as voltage collapse studies over minutes, taking into account over-excited current limiters in generators, and load-frequency control. In a multiple time-scale simulation method [57] which provides both transient and an extended-time simulation capability, the transient simulation is performed using the AB2 method. Once the transient has settled, the longer term simulation uses an implicit method. This method is an improvement over an ealier long-term power system dynamics simulation (LOTDYS) program [58].

4.6 Dynamic Models of Multi-Machine Power Systems

For a single machine connected via an impedance to an infinite bus, the swing equations given in (4.35) involve only one machine rotor angle δ and one machine rotor speed ω as the state variables. The dynamic behavior of a multi-machine system can be similarly developed by modeling the acceleration on each machine as the difference between the mechanical input torque and the electrical output torque as a system of swing equations

$$\frac{d\delta_i}{dt} = \Omega\omega_i, \quad \frac{d\omega_i}{dt} = \frac{1}{2H_i}(P_{mi} - P_{ei} - D_i\omega_i), \quad i = 1, 2, \dots, N_G \tag{4.58}$$

where N_G is the number of machines, δ_i is the machine angle, ω_i the machine speed deviation from the nominal speed, H_i the machine inertia, D_i the damping coefficient, P_{mi} the mechanical input power, and P_{ei} the electrical output power of machine i. Note that power is used in place of torque in (4.58) as the speed deviation from the nominal value is assumed to be small.

With the generators modeled by the classical model, E'_i and P_{mi} are assumed to be constant. As a result, there are only N_G machine rotor angle states and N_G machine speed states, for a total of $2N_G$ state variables.

The machines in (4.58) interact with each other through the electrical network, which determines the electrical output power of each machine. In a positive-sequence analysis program, the dynamics of the line inductances and capacitances, known as the network or electrical transients, are assumed to be fast and are neglected. Thus the electric output power P_e provided by the machines can be determined from an algebraic voltage-current relationship involving the network admittance matrix Y_N as

$$Y_N \tilde{V} = \tilde{I} \tag{4.59}$$

where $\tilde{I}$ is a vector of current injections at the buses, and the positive direction of $\tilde{I}$ means the current is flowing from a bus into the network.

Separating the generation and load buses, (4.59) can be rewritten as

$$\begin{bmatrix} Y_{11} & Y_{12} \\ Y_{21} & Y_{22} \end{bmatrix} \begin{bmatrix} \tilde{V}_G \\ \tilde{V}_L \end{bmatrix} = \begin{bmatrix} \tilde{I}_G \\ \tilde{I}_L \end{bmatrix} \tag{4.60}$$

where $\tilde{V}_G$ and $\tilde{V}_L$ are the voltages and $\tilde{I}_G$ and $\tilde{I}_L$ are the current injections of the generator and load buses, respectively. The admittance matrix Y_N includes line resistances, reactances and charging, shunt capacitors and reactors, and transformer taps (which are fixed during dynamic simulation).

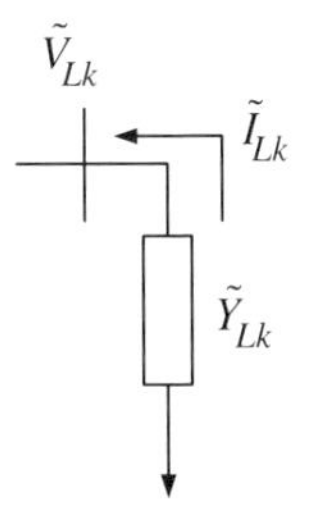

Figure 4.19 Constant-impedance load model for the *k*th load.

In (4.60), $\tilde{I}_G$ is the current supplied by the generators and $-\tilde{I}_L$ is the current consumed by the load (Figure 4.19). Note that (4.60) is initialized by the pre-fault power flow solution. For dynamic analysis of disturbances, the solution to (4.60) depends on the generator and load models.

4.6.1 Constant-Impedance Loads

A simple load model to analyze is the *constant-impedance load,* in which the load is represented by a constant impedance Z_L or a constant admittance Y_L (Figure 4.19).[13]

To determine the load admittance Y_{Lk} at Bus k, two pieces of information from the power flow solution are needed: the power consumption $P_{Lk} + jQ_{Lk}$ (positive values meaning consuming) and the voltage magnitude V_{Lk} at Bus k. From the power consumption equation

$$P_{Lk} + jQ_{Lk} = \tilde{V}_{Lk}(-\tilde{I}_{Lk})^* \tag{4.61}$$

the load current can be computed as

$$\begin{aligned} \tilde{I}_{Lk} &= -\left(\frac{P_{Lk} + jQ_{Lk}}{\tilde{V}_{Lk}}\right)^* = -\frac{P_{Lk} - jQ_{Lk}}{\tilde{V}^*_{Lk}} \cdot \frac{\tilde{V}_{Lk}}{\tilde{V}_{Lk}} \\ &= -\left(\frac{P_{Lk}}{|\tilde{V}_{Lk}|^2} - j\frac{Q_{Lk}}{|\tilde{V}_{Lk}|^2}\right)\tilde{V}_{Lk} = -(G_{Lk} + jB_{Lk})\tilde{V}_{Lk} = -Y_{Lk}\tilde{V}_{Lk} \end{aligned} \tag{4.62}$$

where $G_{Lk} = P_{Lk}/V_{Lk}^2$ and $B_{Lk} = -Q_{Lk}/V_{Lk}^2$ are the constant-impedance load components of the load at Bus k. These values, computed from the prefault steady-state condition are used throughout a dynamic simulation. Note that $B_{Lk} > 0$ if $Q_{Lk} < 0$, that is, the load is capacitive and generating reactive power.

The load components can be combined in matrix form as

$$\tilde{I}_L = \begin{bmatrix} -Y_{L1} & 0 & 0 \\ 0 & -Y_{L2} & 0 \\ 0 & 0 & \ddots \end{bmatrix} \tilde{V}_L = -Y_L\tilde{V}_L \tag{4.63}$$

Note that if there is no load on Bus k, like a connection bus, $Y_{Lk} = 0$.

---

13 The constant-impedance load model by itself is not realistic. A more realistic load model would also have *constant-current* and *constant-power* components. In addition, dynamics from induction motors can also be incorporated into load models. See Chapter 11 for more details.

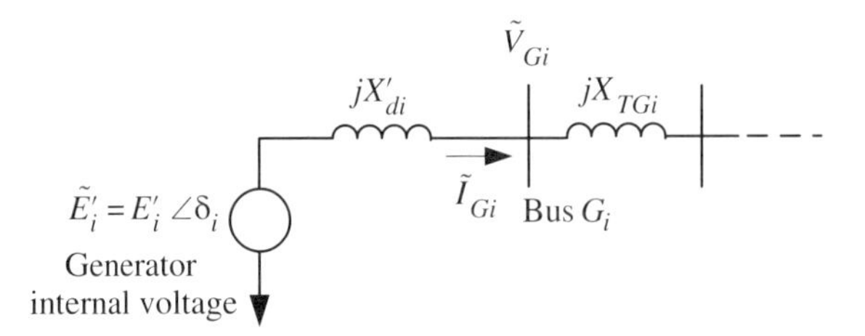

Figure 4.20 Classical generator electrical model for Generator *i*.

4.6.2 Generator Current Injections

The next step in the computation is to find the generator current injection $\tilde{I}_{Gi}$ for Generator i on Bus G_i. The electrical model of a classical model consists of a fixed voltage E'_i behind the d-axis transient reactance X'_{di}, as shown in Figure 4.20. To compute $\tilde{I}_{Gi}$, the generator rotor angle δ_i is used to form the voltage phasor $\tilde{E}'_i$. Knowing the generator terminal voltage phasor $\tilde{V}_{Gi}$,[14] the current injection from the generator into the generator terminal bus is

$$\tilde{I}_{Gi} = \frac{1}{jX'_{di}}(\tilde{E}'_i - \tilde{V}_{Gi}) \tag{4.64}$$

The generator current injections (4.64) can be combined and written in matrix form

$$\begin{bmatrix} \frac{1}{jX'_{d1}} & 0 & 0 \\ 0 & \frac{1}{jX'_{d2}} & 0 \\ 0 & 0 & \ddots \end{bmatrix} \tilde{E}' - \begin{bmatrix} \frac{1}{jX'_{d1}} & 0 & 0 \\ 0 & \frac{1}{jX'_{d2}} & 0 \\ 0 & 0 & \ddots \end{bmatrix} \tilde{V}_G = Y_d(\tilde{E}' - \tilde{V}_G) = \tilde{I}_G \tag{4.65}$$

4.6.3 Network Equation Extended to the Machine Internal Node

The load model (4.63) and the generator model (4.65) are now incorporated into the network equation (4.60) to form the network model extended to the generator internal node

$$\begin{bmatrix} Y_d & -Y_d & 0 \\ 0 & Y_{11} & Y_{12} \\ 0 & Y_{21} & Y_{22} \end{bmatrix} \begin{bmatrix} \tilde{E}' \\ \tilde{V}_G \\ \tilde{V}_L \end{bmatrix} = \begin{bmatrix} \tilde{I}_G \\ Y_d\tilde{E}' - Y_d\tilde{V}_G \\ -Y_L\tilde{V}_L \end{bmatrix} \tag{4.66}$$

which can be rewritten as

$$\begin{bmatrix} Y_d & -Y_d & 0 \\ -Y_d & Y'_{11} & Y_{12} \\ 0 & Y_{21} & Y'_{22} \end{bmatrix} \begin{bmatrix} \tilde{E}' \\ \tilde{V}_G \\ \tilde{V}_L \end{bmatrix} = Y_{\text{ex}} \begin{bmatrix} \tilde{E}' \\ \tilde{V}_G \\ \tilde{V}_L \end{bmatrix} = \begin{bmatrix} \tilde{I}_G \\ 0 \\ 0 \end{bmatrix} \tag{4.67}$$

where $Y'_{11} = Y_{11} + Y_d$ and $Y'_{22} = Y_{22} + Y_L$.

As shown in Chapter 8, this formulation of extending the network model into the generator internal node holds for detailed machine models as well, with the subtransient reactance used instead of the transient reactance.

---

14 The angles of the phasors $\tilde{E}'_i$ and $\tilde{V}_{Gi}$ must be on the same reference frame.

In a dynamic simulation, the rotor angles δ_i of a classical model will change with time, but the generator internal voltage E'_i will remain constant. At each integration step for a particular set of rotor angles, $\tilde{V}_G$, $\tilde{V}_L$, and $\tilde{I}_G$ can be solved in two stages.

- Stage 1: Extract $\tilde{V}_G$ and $\tilde{V}_L$ from (4.67) to form

$$\begin{bmatrix} Y'_{11} & Y_{12} \\ Y_{21} & Y'_{22} \end{bmatrix} \begin{bmatrix} \tilde{V}_G \\ \tilde{V}_L \end{bmatrix} = Y_{GL} \begin{bmatrix} \tilde{V}_G \\ \tilde{V}_L \end{bmatrix} = \begin{bmatrix} Y_d \\ 0 \end{bmatrix} \tilde{E}' \tag{4.68}$$

  and use LU decomposition to solve for $\tilde{V}_G$ and $\tilde{V}_L$, assuming that the phasor $\tilde{E}'$ is known. Note that the admittance matrix Y_{GL} is fixed for a given network configuration, so only one factorization is needed.
- Stage 2: Compute

$$\tilde{I}_G = Y_d(\tilde{E}' - \tilde{V}_G) \tag{4.69}$$

  From the solution of (4.69), the electrical power $P_e = \mathrm{Re}(\tilde{E}' \cdot \tilde{I}_G^*)$ can be computed. Note that Q_e is not needed, unless its limit needs to be checked.

4.6.4 Reduced Admittance Matrix Approach

Alternatively, from (4.68), $\tilde{V}_G$ and $\tilde{V}_L$ can be expressed in terms of $\tilde{E}'$ as

$$\begin{bmatrix} \tilde{V}_G \\ \tilde{V}_L \end{bmatrix} = \begin{bmatrix} Y'_{11} & Y_{12} \\ Y_{21} & Y'_{22} \end{bmatrix}^{-1} \begin{bmatrix} Y_d \\ 0 \end{bmatrix} \tilde{E}' = H_{\mathrm{VR}} \tilde{E}' \tag{4.70}$$

The matrix H_{VR} is known as the (complex) voltage reconstruction matrix. Note that the terms of this matrix are dimensionless. Note also that the network admittance Y_N is not invertible, but the admittance matrix Y_{GL} is, because of the transient reactance terms in the diagonal entries of Y'_{11}.

It follows that the generator currents can be directly computed from

$$\tilde{I}_G = \left(Y_d + \begin{bmatrix} -Y_d & 0 \end{bmatrix} H_{\mathrm{VR}}\right) \tilde{E}' = Y_{\mathrm{red}} \tilde{E}' \tag{4.71}$$

The matrix Y_{red} is known as the admittance matrix reduced to the machine internal node or the reduced admittance matrix. It is a full matrix because the voltage reconstruction bus is a full matrix. Thus even though Y_{red} is smaller in size than Y_N, sparse factorization cannot be applied to Y_{red}, and hence Y_{red} is not used in simulation programs for large power systems.

4.6.5 Method for Dynamic Simulation

A multi-machine simulation will make use of the swing equation (4.58), in which the electrical power P_{ei} is obtained from the network equations (4.68) and (4.69). Figure 4.21 shows the flowchart of a step-by-step integration method for dynamic simulation. The steps of the flowchart are described in more detail as follows.

1) *Initialization*: Using the generator terminal bus voltages $\tilde{V}_{Gi}$ and the generator electrical output power $P_{ei} + jQ_{ei}$ from the power flow solution, compute the generator current injection

$$\tilde{I}_{Gi} = \left(\frac{P_{ei} + jQ_{ei}}{\tilde{V}_{Gi}}\right)^*, \quad i = 1, \ldots, N_G \tag{4.72}$$

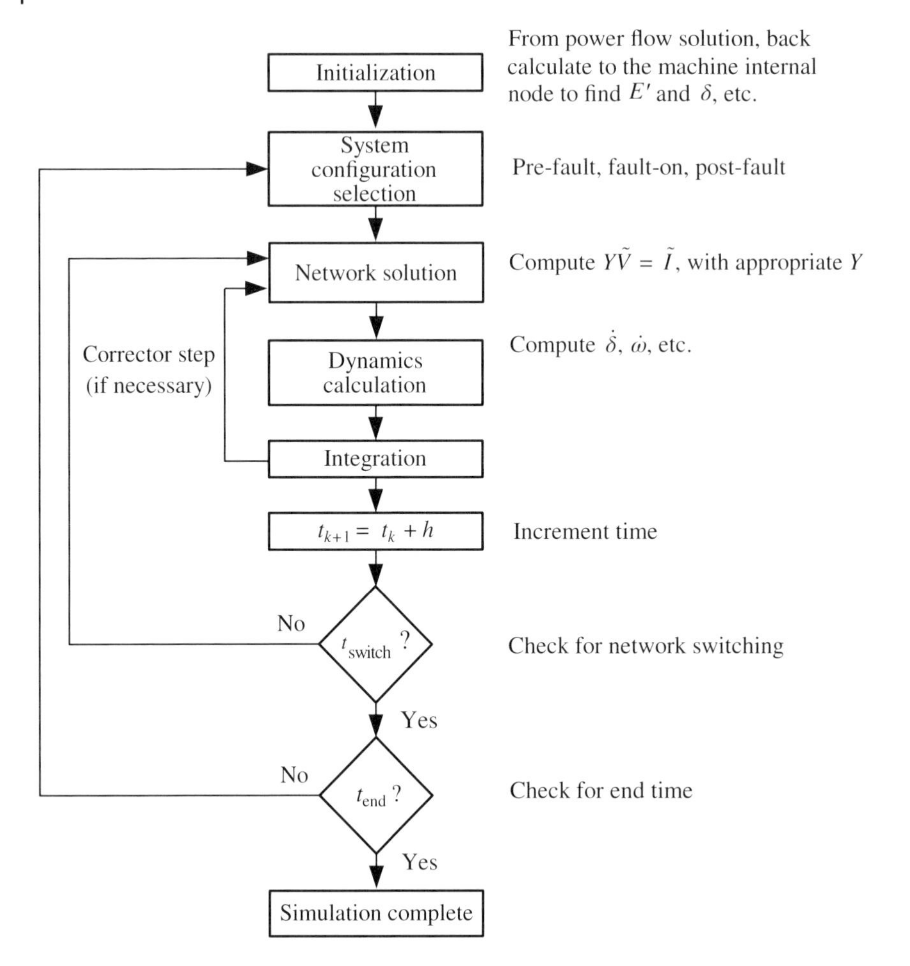

Figure 4.21 Power system dynamic simulation flowchart [3, 20].

Then the internal voltage phasors of the generators are given by

$$\tilde{E}'_i = E'_i \angle \delta_i = \tilde{V}_{Gi} + jX'_{di}\tilde{I}_{Gi}, \quad i = 1, \ldots, N_G \tag{4.73}$$

2) *System configuration selection*: There are typically three configurations for each disturbance simulation: (1) the steady-state or pre-fault condition, (2) the fault-on condition in which a short circuit is applied to a bus or a line, and (3) the post-fault condition in which the fault is cleared by removing a line. For each configuration, an admittance matrix extended to the generator internal nodes (4.67) with constant-impedance loads (4.62) is developed. Note that to approximate a short circuit on any Bus k, set the load admittance Y_{Lk} to a very large number, like 9999 pu, because computer programs will not accept $Y_{Lk} = \infty$.
3) *Network solution*: At each time step in the simulation, the network variables are solved from (4.67) using the generator rotor angles from the previous time step. First, the bus voltage phasors $\tilde{V}_G$ and $\tilde{V}_L$ are solved from (4.68), and then $\tilde{I}_G$ is solved from (4.69).

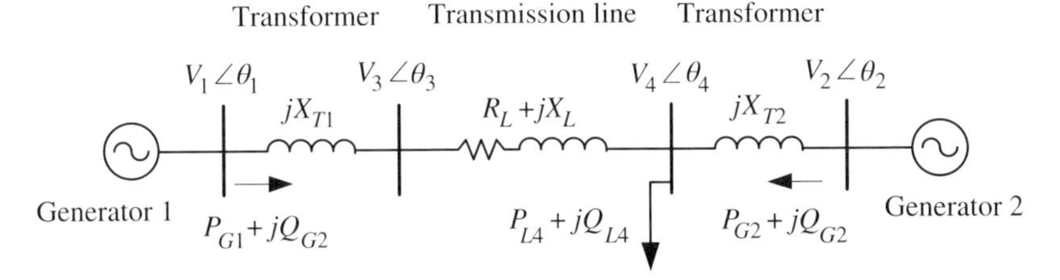

Figure 4.22 Power system for Example 4.5.

4) *Dynamics calculation*: The generator current injection $\tilde{I}_G$ is used to calculate the electric output power $P_e = \text{Re}(\tilde{E}' \cdot \tilde{I}_G^*)$, which is then used to calculate $\dot{\delta}$ and $\dot{\omega}$ from the swing equation (4.58).
5) *Integration*: The time derivatives $\dot{\delta}$ and $\dot{\omega}$ and the stepsize h are used to update the new values of δ and ω. Depending on the integration algorithm used (Section 4.5), provision is made for looping back to the network solution to compute additional correction steps in the integration process.
6) *Time increment and checking for network switching and simulation end time*: At the end of each complete integration cycle, the run time of the simulation is checked to see if a fault has been applied or cleared, necessitating switching to a different network admittance matrix, or if the simulation end time has been reached.

The initialization and construction of the various network matrices are illustrated in the next example. Dynamic simulation will be illustrated in the next section.

Example 4.5: Admittance matrix computation

The SMIB system in Figure 4.2 is expanded to form the 2-machine system in Figure 4.22, where the infinite bus is represented by Generator 2. The transmission line impedance is $R_L + jX_L = 0.005 + j0.05$ pu on the system base of 100 MVA. The base for Generator 1 is 900 MVA and for Generator 2 is 15,000 MVA. The transient reactances for these two generators and the reactances of their step-up transformers on their own base are the same: $X_T = 0.15$ pu and $X'_d = 0.24$ pu. The generator bus voltages are $V_1 = 1.05$ pu and $V_2 = 1.0$ pu, and $P_{L4} + jQ_{L4} = 10.0 + j2.0$ pu on 100 MVA base. Compute the power flow solution of this system and find (1) the Y matrix extended to the generator internal nodes, (2) the voltage reconstruction matrix H_{VR}, and (3) the reduced Y matrix Y_{red}.

Solutions: The power flow solution is given in Table 4.1.

The one-line diagram extended to the machine internal node is shown in Figure 4.23. Note that the generator step-up transformer reactance X_T and transient reactance X'_d have been converted to 100 MVA base. From the generator power output, the Generator 1 current injection and internal voltage are

$$\tilde{I}_{G1} = \left(\frac{P_{G1} + jQ_{G1}}{\tilde{V}_1}\right)^* = 8.1802 + j1.8582 \text{ pu} \tag{4.74}$$

$$\tilde{E}'_1 = \tilde{V}_1 + j0.024\tilde{I}_{G1} = 1.1366\angle 42.816^\circ \text{ pu} \tag{4.75}$$

Table 4.1 Power flow solution on 100 MVA base.

Bus	V (pu)	θ (deg)	P_g (pu)	Q_g (pu)	P_L (pu)	Q_L (pu)
1	1.0500	31.87	8.5000	2.5957	0	0
2	1.0000	0	1.8582	4.2017	0	0
3	1.0178	24.25	0	0	0	0
4	0.9958	−0.11	0	0	10.0	2.0

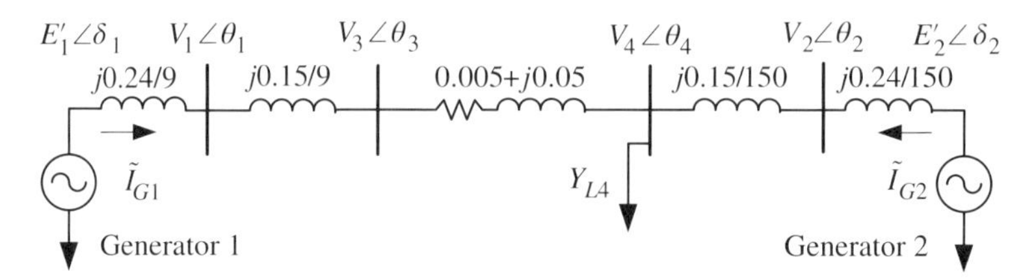

Figure 4.23 System showing machine internal nodes.

Similarly, for Generator 2

$$\tilde{I}_{G2} = 1.8582 - j4.2017 \text{ pu}, \quad \tilde{E}_2' = 1.0003 \text{ pu} \tag{4.76}$$

The load at Bus 4 is converted to a constant admittance as

$$Y_{L4} = \frac{P_{L4} - jQ_{L4}}{V_4^2} = 10.085 + j2.0169 \text{ pu} \tag{4.77}$$

Following the procedure provided earlier, the network equation extended to the machine internal node (4.67) is

$$Y_{\text{ex}} \begin{bmatrix} \tilde{E}_1' \\ \tilde{E}_2' \\ \tilde{V}_1 \\ \tilde{V}_2 \\ \tilde{V}_3 \\ \tilde{V}_4 \end{bmatrix} = \begin{bmatrix} \tilde{I}_{G1} \\ \tilde{I}_{G2} \\ 0 \\ 0 \\ 0 \\ 0 \end{bmatrix} \tag{4.78}$$

where the nonzero entries of Y_{ex} are given in Table 4.2. For example, the (1,3) entry of Y_{ex} is $j9/0.24 = j37.5$, which is the admittance between the internal node of Generator 1 and its terminal bus (Bus 1).

The 4×2 voltage reconstruction matrix is given by

$$H_{\text{VR}} = \begin{bmatrix} 0.7183 - j0.0150 & 0.2812 + j0.0119 \\ 0.0007 + j0.0000 & 0.9992 - j0.0007 \\ 0.5422 - j0.0243 & 0.4570 + j0.0193 \\ 0.0112 + j0.0005 & 0.9865 - j0.0111 \end{bmatrix} \tag{4.79}$$

Table 4.2 Nonzero entries of the extended Y matrix for Example 4.5.

i,j	$(Y_{ex})_{ij}$	i,j	$(Y_{ex})_{ij}$
1,1	$-j37.5$	2,2	$-j15000$
1,3	$j37.5$	2,4	$j15000$
3,1	$j37.5$	4,2	$j15000$
3,3	$-j97.5$	5,3	$j60$
4,4	$-j16000$	6,4	$j1000$
3,5	$j60$	5,5	$1.9802 + j79.802$
6,5	$-1.9802 + j19.802$	4,6	$j1000$
5,6	$-1.9802 + j19.802$	6,6	$12.065 - j1021.8$

Table 4.3 Reduced Y matrices used in disturbance simulation.

Y	(1,1)	(1,2)	(2,2)
Pre-fault	$-j15.306$	$j15.306$	$-j15.306$
Fault-on	$-j23.077$	$j0.0001$	$-j45.453$
Fault removed	$-j11.450$	$j11.450$	$-j34.178$
Line removed	$-j11.450$	$j11.450$	$-j11.450$

and the 2×2 reduced admittance matrix is

$$Y_{red} = \begin{bmatrix} 0.5608 - j10.5637 & -0.4464 + j10.5458 \\ -0.4464 + j10.5458 & 10.3726 - j12.6471 \end{bmatrix} \tag{4.80}$$

■

Example 4.6: Fault-on and post-fault admittance matrix computation

For the SMIB system disturbance simulation in Example 4.4, show the admittance matrix reduced to the generator internal node for:

1) the pre-fault condition
2) the fault-on condition
3) the post-fault condition.

Solutions: The reduced admittance matrices from PST are shown in Table 4.3.[15] Note that because $Y(2,1) = Y(1,2)$, it is not shown separately.

In the prefault condition, the $Y(1,2)$ is the equivalent admittance between the two generator internal nodes. In the fault-on period, $Y(1,2)$ is 0 because the two machines are separated and not synchronized. Here $Y(1,2)$ is not quite zero as in a computer

15 See the example file on how to get the reduced admittance matrices.

program, the admittance representing the short-circuit fault is a large but finite value. Also $Y(1,1)$ and $Y(2,2)$ are equivalent reactances from the generator internal nodes to the fault point. When the fault is cleared by opening the circuit breaker on Line 3-2 (second circuit) near Bus 3, $Y(1,2)$ recovers to a nonzero value. Note that $Y(2,2) \neq Y(1,1)$ as the generator representing the infinite bus still sees the faulted line from the remote end. When the faulted line is completely removed, then $Y(2,2) = Y(1,1)$. Of importance is that the post-fault equivalent susceptance 11.450 pu between the generators is less that the pre-fault equivalent susceptance 15.306 pu, implying that the restoring force between two generators would now be smaller. ■

4.7 Multi-Machine Power System Stability

When a multi-machine power system is subject to a short-circuit fault, the bus voltages near the fault will be depressed. As a result, the power consumption of the loads will be reduced, that is, the electrical power supplied by the generators will also be lowered. With the mechanical input power exceeding the electrical output power, a synchronous generator will accelerate. In general, the generators closest to the fault will accelerate the most. A simple definition of power system transient stability is that after the disturbance is cleared, the relative angle $\delta_i - \delta_j$ between any generators i and j remains finite.

Figure 4.24 contains the time response of three generators with respect to a short-circuit close to Generator 2, which was cleared in a few cycles. During the fault, the machine speeds will be above 1 pu, and thus the absolute values of their rotor angles will increase right after the fault is applied. In Figure 4.24.a, the difference between all the rotor angles are bounded. Thus the three generators remain in synchronism and the system is stable. On the other hand, with a longer clearing time, Figure 4.24.b shows that δ_2 is pulling away from the other two rotor angles. As a result, Generator 2 is transiently unstable.

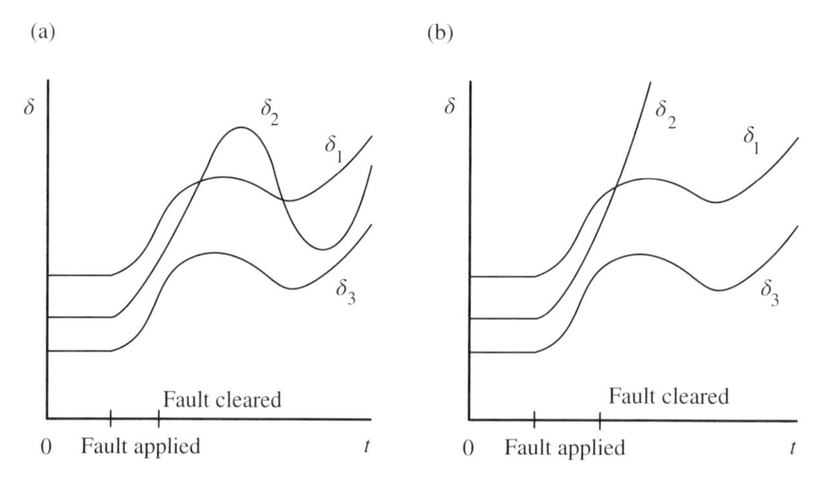

Figure 4.24 Simulation showing Generator 2: (a) remains stable with the other two generators and (b) is unstable with respect to the other two generators.

4.7.1 Reference Frames for Machine Angles

In a simulation involving a short circuit, because the generator rotors are accelerating initially, the rotor angles will be increasing. Thus instead of having all the angles moving, it is useful to select a reference generator, say N_{ref}, and examine the individual generator angles relative to the reference generator angle. Typically a generator with a large inertia is selected as the reference generator because it tends to move the least. The relative machine angle for Generator i with respect to the reference machine angle δ_{ref} is defined as

$$\hat{\delta}_i = \delta_i - \delta_{\text{ref}} \tag{4.81}$$

Instead of using a reference generator, an alternative is to define a center-of-inertia reference frame in which the center-of-inertia angle and speed are computed as

$$\delta_o = \frac{1}{H_T}\sum_{i=1}^{N_G} H_i\delta_i, \quad \omega_o = \frac{1}{H_T}\sum_{i=1}^{N_G} H_i\omega_i \tag{4.82}$$

where

$$H_T = \sum_{i=1}^{N_G} H_i \tag{4.83}$$

is the sum of all the machine inertias (expressed on the same base). Then the machine angles and speeds are calculated as

$$\overline{\delta}_i = \delta_i - \delta_o, \quad \overline{\omega}_i = \omega_i - \omega_o \tag{4.84}$$

Reference angles are illustrated in the next example.

Example 4.7: Disturbance simulation for a 16-machine system

The 16-machine NPCC power system is shown in Figure 4.25. Simulate the system response, using only the classical model for all the generators and constant impedance for all the loads, for a 3-phase short-circuit fault near Bus 29 on Line 29-28, cleared by tripping Line 29-28. Bus 29 is the high-side of the step-up transformer for Generator 9 (Maine Yankee). Perform the simulation for a 3-cycle fault and then for a 4-cycle fault.

Solutions: The fault is applied close to Generator 9. In steady state, Bus 29 is connected to Bus 26 via two parallel paths. By tripping Line 29-28, one of paths is taken out. Thus Generator 9 would be the most likely candidate to be unstable if the fault stays on long enough. The time responses for the 3-cycle fault and 4-cycle fault, using the bus, branch, and generator data provided in the data file *data16em.m*, are shown in Figures 4.26 and 4.27.

For the 3-cycle fault, the system is stable, meaning that all the machines remain in synchronism. Figure 4.26.a shows that all the generator rotor angles are increasing initially, a result due to the short-circuit reducing the electrical power supplied from the generators. When the generator angles are referred back to Generator 16, Figure 4.26.b shows that the oscillations are bounded.

For the 4-cycle fault, Generator 9 accelerates all by itself, whereas the other generator angles are clustered together. In this case, Generator 9 has lost synchronism with the

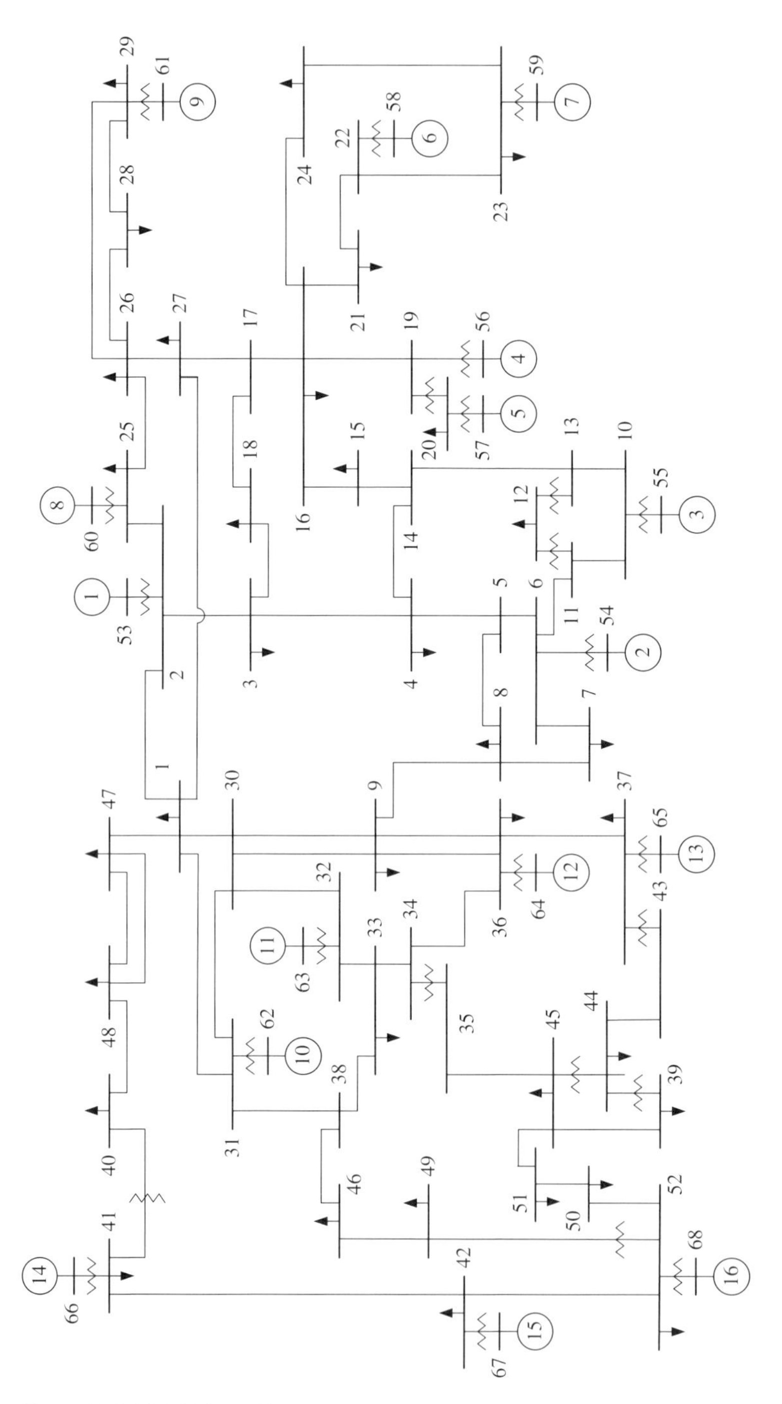

Figure 4.25 NPCC 16-machine system.

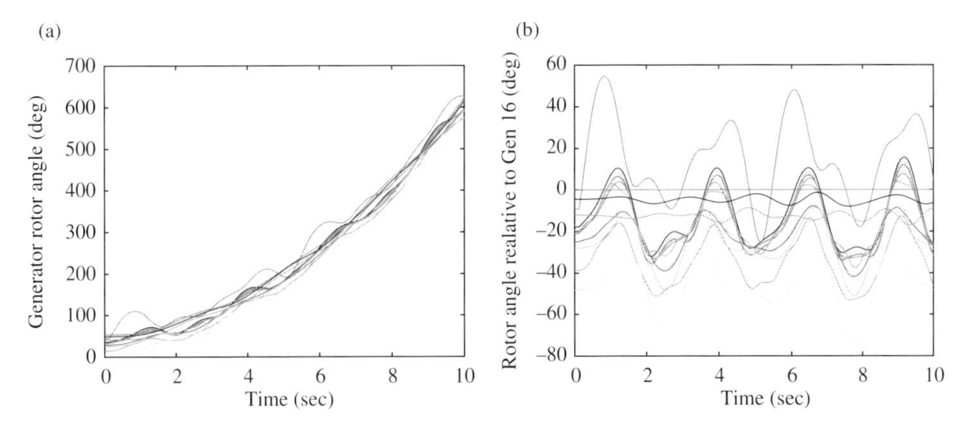

Figure 4.26 Generator disturbance responses for a 3-cycle fault: (a) absolute rotor angles and (b) rotor angles relative to Generator 16. Generator 9 has the highest angle variations (see color plate section).

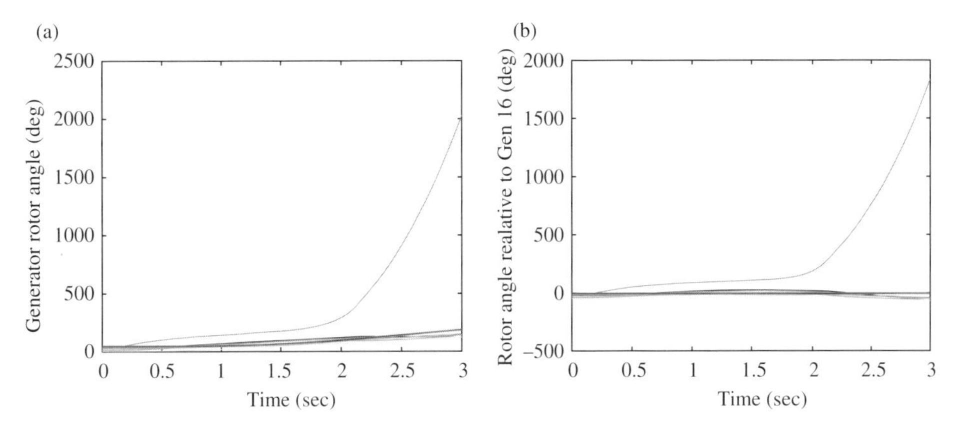

Figure 4.27 Generator disturbance responses for a 4-cycle fault: (a) absolute rotor angles and (b) rotor angles relative to Generator 16. Note that Generator 9 is unstable (see color plate section).

rest of the system, that is, it is unstable. This instability is visible from both the absolute angle plot in Figure 4.27.a or the relative angle plot in Figure 4.27.b. ■

4.8 Power System Toolbox

General purpose positive-sequence power system simulation programs are capable of simulating the disturbance response of a large power network with many generators and their control equipment. Some popular commercial software packages include PSLF® by General Electric Company, PSS®E by Siemens-PTI, and a suite of stability analysis packages from Power Tech. MATLAB®-based educational power system simulation software includes the Power System Toolbox (PST) [13] and Power System Analysis Toolbox (PSAT) [59], which has a Simulink® interface. PST scripts for the examples in this text are available, and a reader can modify them to solve most of the problems at

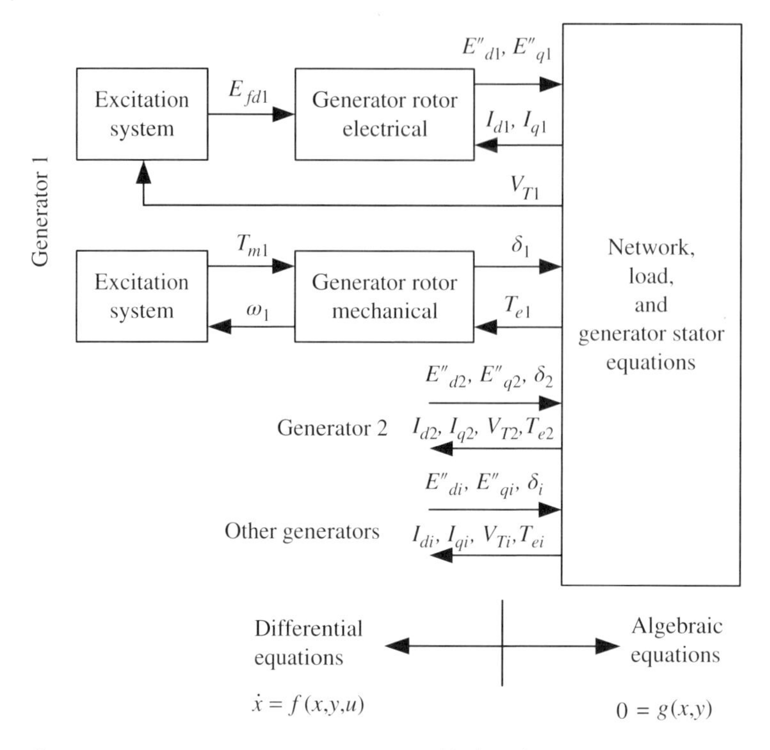

Figure 4.28 Power system computation variable flow diagram (adapted from [20]).

the end of each chapter. More information for using PST can be found from [3] and the PST manual [60].

The structure of a stability simulation package is shown in Figure 4.28, which depicts the computational components of the flowchart in Figure 4.21. Note that detailed machine models, excitation systems, prime movers, and turbine governors are discussed in later chapters. A reader can follow the various MATLAB® functions in the PST code, which is open source, to see how the network and dynamics computations are structured.

The input data for dynamic simulation consist of two sets of parameters: the power flow data consisting of the bus, line, and transformer data, which has been discussed in Section 2.2, and the dynamic model data with the inertias, impedances, and time constants of the generators as well as the parameters of the control equipment. It is customary to input the actual MVA base for the generator in the machine data, so that all the other parameters can be given in per unit on the machine base. In this way, an experienced power engineer can readily spot any anomaly in the machine parameters, as the parameters for practical generators are known to be in certain ranges, as determined by equipment geometry and physical laws.

In commercial power system simulation programs, a user can either upload the data from ASCII or binary files or interactively edit the data in a spreadsheet enabled by a user interface.

The flowchart in Figure 4.21 and the data flow diagram in Figure 4.28 are designed for a serial computer, which processes one computation at any given time. For a large power system, in this serial mode of computation, one can normally expect from experience that the network solution part and the dynamics calculation part would take about the same amount of time. For constant impedance loads, the network solution (4.68) is linear and can be computed directly. For more complex loads, such as constant-current and constant-power loads (see Chapter 11), the network solution has to be solved iteratively, requiring more computing time. Power system operations and planning typically involve analyzing the impact of many disturbances or contingencies. With the availability of multi-processor or massively parallel computers, these contingencies can be investigated with parallel computing, with one processor analyzing one contingency. Thus the total computing time can be reduced by a factor equal to the number of processors available [61].

4.9 Summary and Notes

This chapter uses the electromechanical model of synchronous machines to illustrate the simulation of power systems subject to disturbance conditions. For the SMIB system, a disturbance is studied using a standalone second-order nonlinear dynamic model which captures the power swing oscillation. For multi-machine systems, the power network is retained as an algebraic entity and its interface to the generator circuits is delineated. A framework for simulating the machine dynamics with interactions to other machines through the power network is shown in Figure 4.21. This framework is still applicable when detailed models are used for synchronous machines (4.28). Initialization of the electromechanical models using the power flow solution is also shown.

Two second-order integration methods are described, including the Adam-Bashforth-2 method used in commercial simulation programs. A reader interested in learning the large varieties of integration methods can consult numerical analysis textbooks such as [54].

A dynamic simulation is much more time consuming than a power flow solution. Thus an interesting question is whether the computation in Figures 4.21 and 4.28 can be parallelized so that the simulation of a single disturbance can be completed faster. A power system is a network with injection dynamics (such as generators) at certain nodes, which interact with each other through the network. Thus the dynamics part in Figures 4.21 and 4.28 can be computed in parallel, that is, the time derivatives of each generator, including its control equipment, can be computed independent of the time derivatives of all the other generators. Thus parallel computing, with minimal changes in computer code, can be used to greatly reduce the computing time of the dynamics. The network solution (4.68) solved by LU decomposition, however, is more difficult to parallelize and would require iterative solutions such as the Krylov's method [18]. This is still an active field of research [61].

Problems

4.1 Develop the electromechanical model of the 60-Hz power system shown in Figure 4.29, where the parameters on the machine base of 900 MVA are $H = 6.5$ sec, $X'_d = 0.30$ pu, and $X_T = 0.15$ pu, and the parameters on the system base of 100 MVA are $R_L = 0.0055$ pu, $X_L = 0.055$ pu, $P_L = 8.1$ pu, and $Q_L = 0.81$ pu. The

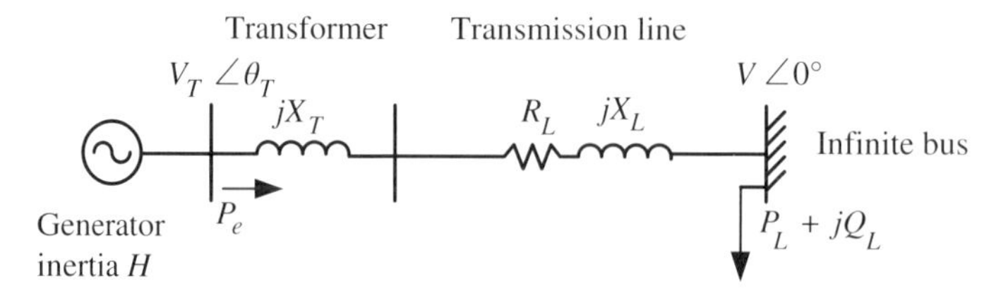

Figure 4.29 Figure for Problem 4.1.

load is located at the infinite bus which has a voltage of $V = 1.0$ pu. Use 900 MVA as the base for the electromechanical model.

Initialize the generator states and voltages and show the voltage and current vector diagram used for the initialization. Note that the voltage drop across the transmission line has two components: the resistive part in phase with the line current phasor and the reactive part perpendicular to the line current phasor. (Note that in this problem the voltage at the generator terminal bus is not specified. It can be calculated based on the reactive power consumption at the infinite bus.)

4.2 A fault is applied to the high-side transformer bus of the power system in Problem 4.1, as shown in Figure 4.30.

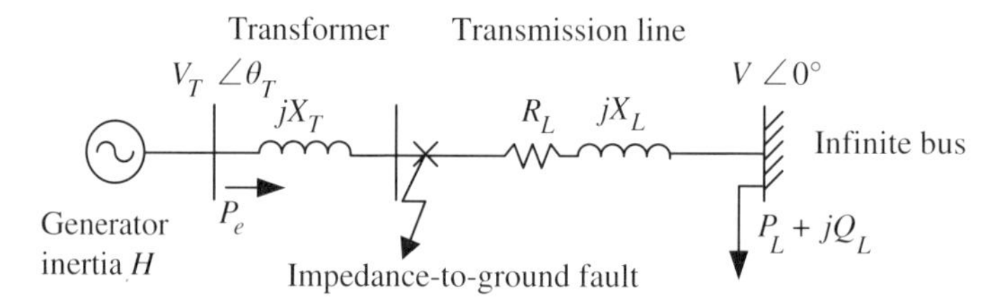

Figure 4.30 Figure for Problem 4.2.

Find the initial machine rotor acceleration (that is, $d\omega/dt$) at the instant of the fault, for:

1) a 3-phase short-circuit fault, and
2) an impedance-to-ground fault with a fault impedance of $j0.2$ pu (to model a less severe fault).

In addition, for each case, suppose the fault is cleared after 5 cycles. Find the machine speed and angle at the clearing time. For Part 2, assume that the acceleration during fault is constant.

4.3 Repeat Example 4.4 with $D = 0$. The time response of δ and θ should be purely oscillatory.

4.4 Use the bisection method to find the critical clearing time accurate to 0.01 sec for the system in Problem 4.3. Simulate and show the time response if the fault is cleared at (a) the critical clearing time and (b) 0.01 sec larger than the critical clearing time.

4.5 Repeat the simulation in Example 4.4 with a larger post-fault integration stepsize. For simulation time from 0 to 0.5 sec, use a stepsize of 0.001 sec. From 0.5 to 2 sec, use a stepsize of 0.005 sec. For 2 to 10 sec, use a stepsize of 0.07 sec. Comment on the simulation results and compare to those from Example 4.4.

4.6 For the NPCC 16-machine system, apply a 3-phase short-circuit fault on Line 32-30, near Bus 32. The fault is cleared by tripping Line 32-30. Find the critical clearing time of Generator 11, accurate to 0.01 sec. Plot the time response at the critical clearing time and at 0.01 sec larger than the critical clearing time.

5

Direct Transient Stability Analysis

5.1 Introduction

Chapter 4 uses the simulation of dynamic models of power systems to determine the stability of a power system subject to disturbances. In this and the next chapter, additional techniques are introduced to determine the stability of power systems. The objective of this chapter is to provide an analytical framework for determining the transient stability of a power system when disturbances perturb its states from their stable equilibrium values. The approach discussed in this chapter is based on Lyapunov or energy functions [62, 63] and the stability boundaries of stable equilibrium points [64, 65].

As discussed in Chapter 4, power system transient stability analysis typically consists of applying a short-circuit fault near a generator. The fault is cleared after a short period of time (4–6 cylces) by opening the circuit breakers on the faulted transmission line. During the fault, the generator terminal voltage is low, and thus the electrical power extracted from the generator will be reduced significantly. As the mechanical input torque applied to the turbine remains constant, it may greatly exceed the output electrical torque. Thus the machine rotor speed will accelerate and the machine will accumulate kinetic energy, until the fault is cleared. As the fault is cleared, the system topology is changed and the power system evolves as an autonomous system, that is, a system not subject to any more external forces. Whether the post-fault system is stable or not depends on the restoring or synchronizing torque available in the post-fault power network. The stronger the synchronizing torque, the larger is the region of stability of the post-fault equilibrium point.

Beside providing insights into the transient stability problem, an advantage of the energy function approach is that it can be an efficient method for determining the transient stability of a power system as it requires time simulation only up to the fault clearing time.

This chapter starts with the equal-area criterion for a Single-Machine Infinite-Bus (SMIB) system in Section 5.2. This method is generalized to energy functions in Section 5.3, and is illustrated with measured data of a disturbance event in a real power system in Section 5.4. Section 5.5 examines power system response as phase trajectories and Section 5.6 describes an algorithm for determining transient stability using energy functions. The extension of energy functions to multi-machine power systems is described in Section 5.7. Section 5.8 provides a brief description of the $N-1$ contingency criterion for dynamic security assessment.

Power System Modeling, Computation, and Control, First Edition. Joe H. Chow and Juan J. Sanchez-Gasca.

Companion website: www.wiley.com/go/chow/power-system-modeling

5.2 Equal-Area Analysis of a Single-Machine Infinite-Bus System

5.2.1 Power-Angle Curve

Consider the SMIB system in Figure 4.2, with the dynamic model given by

$$\frac{d\delta}{dt} = \omega, \quad 2H\frac{d\omega}{dt} = P_m - P_e \tag{5.1}$$

where δ is the rotor angle in rad (with the voltage angle at the infinite bus fixed at 0 rad), ω the machine rotor speed deviation from the nominal value in rad/sec, H the machine rotor inertia in sec, P_m the mechanical input power in pu, and P_e the electrical output power in pu.[1] The damping effect is neglected in (5.1). Furthermore the electrical output power is given by

$$P_e = \frac{E'V}{X_{eq}} \sin\delta = P_{max}\sin\delta \tag{5.2}$$

where E' is the generator internal voltage, V is the infinite-bus voltage, and the equivalent reactance between the generator and infinite bus is

$$X_{eq} = X'_d + X_T + X_L \tag{5.3}$$

in which X'_d is the d-axis transient reactance, X_T the transformer reactance, and X_L the transmission line reactance, all expressed on a common MVA base.

In steady state, for a given electrical output power P_{eo}, the rotor angle and mechanical power output of the generator are given by

$$\delta_{ep} = \sin^{-1}\left(\frac{P_{eo}}{P_{max}}\right), \quad P_m = \frac{E'V}{X_{eq}} \sin\delta_{ep} \tag{5.4}$$

Figure 5.1.a shows the variation of P_e versus δ according to (5.2). The equilibrium values of P_m and δ_{ep} are also indicated on the plot. This power-angle (P-δ) curve can be used to provide additional insight into transient stability analysis [12].

Suppose the swing equation is perturbed by moving the rotor angle away from its equilibrium value δ_{ep}. As a result, $d^2\delta/dt^2 \neq 0$ and thus $\delta(t)$ will start to vary as a function of time t. One can track the motion of δ on the P-δ curve in Figure 5.1.a. At Point a, $P_m > P_e$, resulting in $d^2\delta/dt^2 > 0$, that is, the generator rotor is accelerating. On the other hand, at Point b, $P_m < P_e$, resulting in $d^2\delta/dt^2 < 0$, that is, the generator is decelerating.

To study the rotor motion in more detail, suppose $\delta = \delta_a$ and $d\delta/dt = 0$ at $t = t_o$. Then for $t > t_o$, δ will increase because $d\delta/dt > 0$. When $\delta = \delta_{ep}$, $d^2\delta/dt^2 = 0$ as $P_e = P_m$. But $d\delta/dt$ is still greater than 0, so δ will continue to increase. However, $d^2\delta/dt^2 < 0$, implying that $d\delta/dt = 0$ at some value of δ, say, $\delta = \delta_b$. As t increases further, δ will start to decrease. For the oscillation to be stable, the trajectory δ should vary between δ_a and δ_b. In other words, if $\delta \neq \delta_{ep}$, there is a restoring force for δ to move back to δ_{ep}. Thus δ_{ep} is a stable equilibrium point, and will be denoted by δ_{sep}.

1 Note that the formulation (5.1) is different from the swing equation shown in (4.6), in which ω is given in pu frequency. The form (5.1) is used to conform with the energy function derived in texts such as [3, 62–63]. Otherwise the rad/sec-to-pu-frequency conversion factor Ω will appear in the kinetic energy. Also note that the inertia constant H for synchronous machines used in dynamic simulation programs is for ω in pu frequency. Thus in (5.1), the numerical value of H is to be scaled to H/Ω. More explanations are given later.

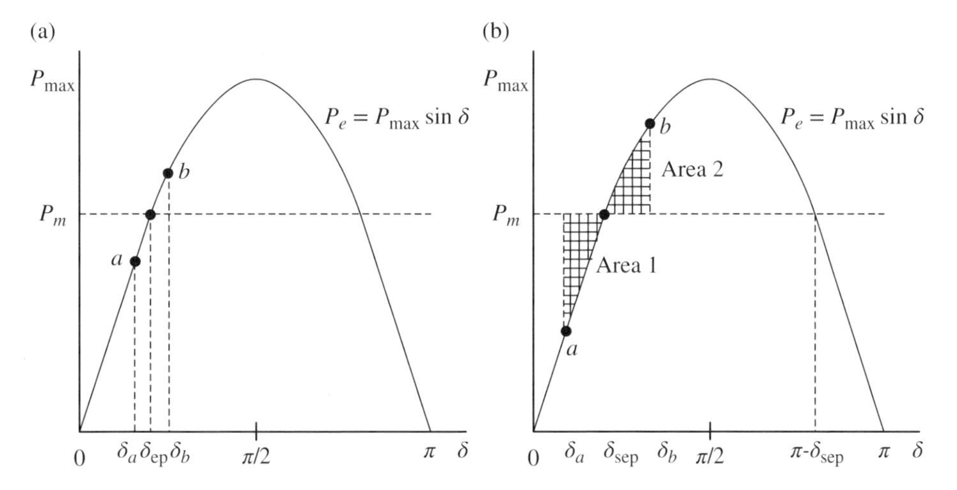

Figure 5.1 Power-angle curve: (a) equilibrium solution and (b) acceleration and deceleration areas.

Multiply both sides of (5.1) by $d\delta/dt$ to obtain

$$\frac{d^2\delta}{dt^2}\frac{d\delta}{dt} = \frac{1}{2}\frac{d}{dt}\left(\frac{d\delta}{dt}\right)^2 = \frac{1}{2H}(P_m - P_e)\frac{d\delta}{dt} \tag{5.5}$$

which can be expressed as

$$\frac{1}{2}d\left(\frac{d\delta}{dt}\right)^2 = \frac{1}{2H}(P_m - P_e)\, d\delta \tag{5.6}$$

Integrate δ from δ_a to δ_b to form

$$\frac{1}{2}\int_{\delta_a}^{\delta_b} d\left(\frac{d\delta}{dt}\right)^2 = \frac{1}{2H}\int_{\delta_a}^{\delta_b} (P_m - P_e)\, d\delta \tag{5.7}$$

resulting in

$$\frac{1}{2}\left(\frac{d\delta}{dt}\right)^2\Bigg|_{\delta_a}^{\delta_b} = \frac{1}{2H}\int_{\delta_a}^{\delta_b} (P_m - P_e)\, d\delta \tag{5.8}$$

The left-hand side (5.8) is zero because $d\delta/dt = 0$ at δ_a and δ_b. As a result, the right-hand side (5.8) can be decomposed into two terms, where the $1/(2H)$ factor is not needed:

$$\underbrace{\int_{\delta_a}^{\delta_{sep}} (P_m - P_e)d\delta}_{\text{Area 1}} + \underbrace{\int_{\delta_{sep}}^{\delta_b} (P_m - P_e)d\delta}_{-\text{Area 2}} = 0 \tag{5.9}$$

Eqn. (5.9) is the well-known *equal-area criterion* implying that the kinetic energy gained during acceleration, represented by Area 1 in Figure 5.1.b, is the same as the energy dissipated during deceleration, represented by Area 2 in Figure 5.1.b. In fact, stability is maintained if Area 1 is less than the largest possible Area 2 between P_e and P_m.

In this SMIB system scenario, P_m and X_{eq} are fixed regardless of the perturbation, that is, the equilibrium δ_{sep} remains the same throughout the perturbation. In transient stability studies, it is common to apply the most severe faults such that the largest amount of kinetic energy is accumulated by the machine rotor. If the system is stable for the most severe contingencies, then it should also be stable for other less severe disturbances because the system has sufficient restoring or synchronizing capability.

5.2.2 Fault-On and Post-Fault Analysis

Consider the 3-phase short-circuit fault applied to the SMIB system as shown in Figure 4.6. The electrical output power of the generator drops to zero, and as a result, the rotor acceleration is given by

$$\frac{d^2\delta}{dt^2} = \frac{d\omega}{dt} = \frac{1}{2H}P_m \tag{5.10}$$

where ω is the machine rotor speed deviation from the nominal value.

Let t_o be the time when the fault is applied and t_c the time when the fault is cleared. During this period, the kinetic energy accumulated by the machine rotor inertia is

$$V_{\text{KE}} = \frac{1}{2}(2H)\omega^2 = H\left(\int_{\delta_a}^{\delta_c} d\left(\frac{d\delta}{dt}\right)^2\right) \tag{5.11}$$

Using expression (5.6), this expression can be simplified to

$$V_{\text{KE}} = H\int_{\delta_a}^{\delta_c} \frac{1}{H}P_m\, d\delta = P_m \int_{\delta_a}^{\delta_c} d\delta = P_m(\delta_c - \delta_a) \tag{5.12}$$

which is the area A_1 shown in Figure 5.2.

After the fault is cleared by removing the faulted line, as shown in Figure 4.11, the equivalent reactance between the generator internal node and the infinite bus is

$$X_{\text{eq}}^{\text{pf}} = X'_d + X_T + 2X_L \tag{5.13}$$

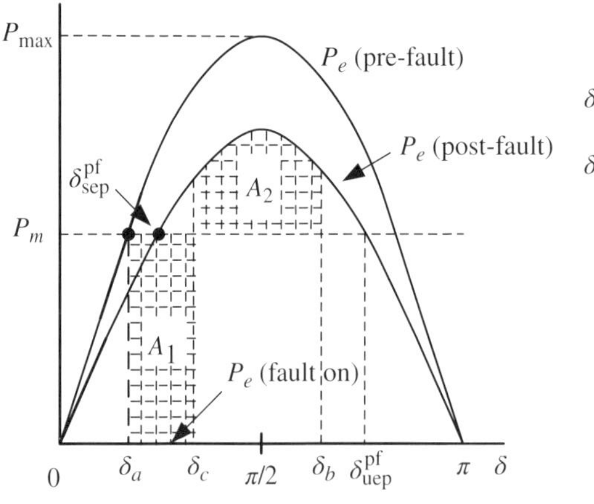

Figure 5.2 Pre-fault and post-fault power-angle plots.

such that the post-fault electrical power output is

$$P_e^{\rm pf} = \frac{E'V \sin\delta}{X_{\rm eq}^{\rm pf}} \tag{5.14}$$

The post-fault stable equilibrium point becomes

$$\delta_{\rm sep}^{\rm pf} = \sin^{-1}\left(\frac{1}{P_e^{\rm pf}}\frac{E'V}{X_{\rm eq}^{\rm pf}}\right) \tag{5.15}$$

Both $P_e^{\rm pf}$ and $\delta_{\rm sep}^{\rm pf}$ are shown in Figure 5.2, as well as the post-fault unstable equilibrium point

$$\delta_{\rm uep}^{\rm pf} = \pi - \delta_{\rm sep}^{\rm pf} \tag{5.16}$$

Transient stability can be analyzed using the power-angle curve in Figure 5.2. Let δ_b be the peak post-fault angle swing. Then the area A_2 between the post-fault electrical output power $P_e^{\rm pf}$ and the mechanical input power P_m represents the ability of the power network to absorb the energy accumulated in the generator rotor. From the equal-area criterion, δ will reach δ_b where $A_2 = A_1$. Figure 5.2 shows that the post-fault system is stable because $\delta_b < \delta_{\rm uep}^{\rm pf}$.

If the area $A_2(\delta_c, \delta_{\rm uep}^{\rm pf})$ between the post-fault $P_e^{\rm pf}$ and P_m from δ_c to $\delta_{\rm uep}^{\rm pf}$ is less than A_1, then the generator will be unstable. If the area $A_2(\delta_c, \delta_{\rm uep}^{\rm pf}) = A_1(\delta_a, \delta_c)$, then this fault clearing time is called the *critical clearing time* $t_{\rm crit}$, and $\delta_c = \delta_{\rm crit}$.

Example 5.1: Power-angle curve for Example 4.4

Plot the generator electrical output power versus the generator rotor angle in Example 4.4 for the pre-fault, fault-on, and post-fault periods.

Solutions: Using the simulation results of Example 4.4, the fault-on and post-fault trajectories of the electric power P_e versus δ are also shown as the purple curves in Figure 5.3. The arrows indicate the directions of the trajectories, with the numbers indicating the trajectory sequence. The post-fault P-δ curve is shown as a dashed line so it can be distinguished from the post-fault time response. Note that $\delta_b < \delta_{\rm uep}^{\rm pf}$, and thus the post-fault system is stable. If a damping term $-D\omega$ with $D > 0$ is included on the right-hand side of the swing equation (5.10), the post-fault trajectory will converge to $\delta_{\rm sep}^{\rm pf}$. ■

5.3 Transient Energy Functions

Lyapunov functions have been commonly used by control scientists to establish stability properties of nonlinear systems [66]. In many physical systems, Lyapunov functions have an energy function interpretation, which is the case with the electromechanical model of a power system.

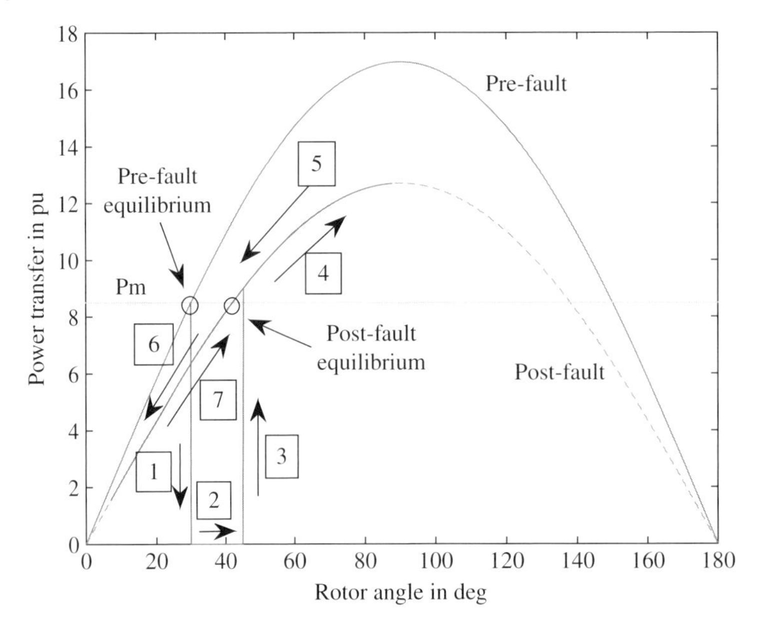

Figure 5.3 P-δ curve for Example 5.1 (see color plate section).

5.3.1 Lyapunov Functions

Consider a nonlinear autonomous and time-invariant dynamical system

$$\dot{x} = f(x) \tag{5.17}$$

where x is the state vector of dimension n and the n-dimensional column vector f represents the nonlinear dynamics without external inputs and is not dependent on t. Let x_{ep} be an equilibrium point, that is, $f(x_{ep}) = 0$.

Lyapunov Theorem: Let $V(x)$ be a positive definite scalar function of x (that is, $V(x) > 0$) in the neighborhood $\mathcal{D}$ of x_{ep} and $V(x_{ep}) = 0$. If

$$\dot{V}(x) = \frac{\partial V}{\partial x} \cdot \frac{dx}{dt} = \frac{dV}{dx} f(x) < 0 \tag{5.18}$$

where the derivative dV/dx is a row vector, for all $x \in \mathcal{D} - \{x_{ep}\}$, then x_{ep} is an asymptotically stable equilibrium, that is, $x \to x_{ep}$ as $t \to \infty$.

Without going into a detailed explanation, the Lyapunov Theorem can be interpreted as follows. As x evolves, the Lyapunov function $V(x)$ decreases. Thus $V(x)$ gradually approaches zero, its lowest value, and x approaches the equilibrium point x_{ep}. Additional results on Lyapunov stability are contained in [66].

5.3.2 Energy Function for Single-Machine Infinite-Bus Electromechanical Model

For physical systems such as mass-spring systems and power systems, one can use the energy function as the Lyapunov function. A common example is a ball in a bowl with uneven edges, shown in Figure 5.4. The ball of mass M at point A is initially at rest, at a height h above the lowest point. Its potential energy with respect to the lowest point is

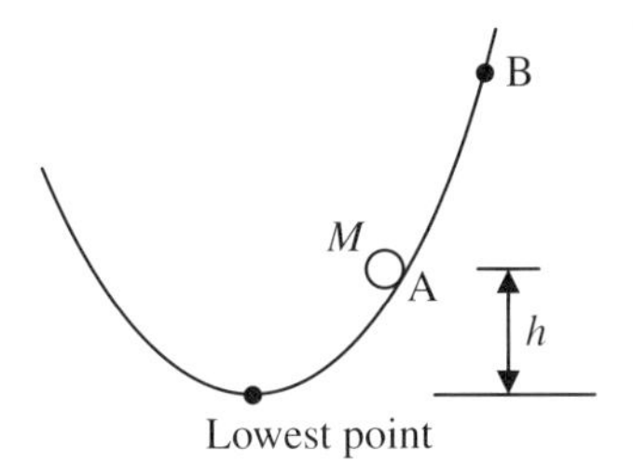

Figure 5.4 A ball in a bowl to illustrate energy functions.

Mgh, where g is the gravity constant, and its kinetic energy is zero. If the ball is released from A, it will accelerate and reach maximum kinetic energy at the lowest point in the bowl, when the potential energy is zero. As the ball goes up on the other side of the bowl, the maximum height it can reach would be level with A (assuming that there is no energy loss due to friction), when all the kinetic energy is converted back to potential energy. With friction inside the bowl, the ball will eventually settle to the lowest point on the bowl surface. However, if the ball is released at rest from B, it will go over the edge on the other side, resulting in an unstable trajectory.

Thus the stability of this ball-in-a-bowl system can be analyzed using the energy function

$$V_E = V_{\mathrm{KE}} + V_{\mathrm{PE}} = \frac{1}{2}Mv^2 + Mgx \tag{5.19}$$

where x is the ball location above the lowest point, v the ball velocity, V_{KE} the kinetic energy, and V_{PE} the potential energy.

To develop the energy function for the SMIB nonlinear power system model, start with the swing equation represented as two first-order differential equations

$$\dot{\delta} = \omega, \quad 2H\dot{\omega} = P_m - \frac{E'V}{X_{\mathrm{eq}}}\sin\delta - D\omega \tag{5.20}$$

where E' is the machine internal voltage, V the voltage at the infinite bus, X_{eq} the total reactance between the machine internal node and the infinite bus, and D represents the machine damping.

At the equilibrium point, the states are

$$\omega = 0, \quad \delta = \delta_{\mathrm{ep}} = \sin^{-1}\frac{P_m X_{\mathrm{eq}}}{E'V} \tag{5.21}$$

In the sequel, $(\delta_{\mathrm{ep}}, 0)$ will be taken to be the stable equilibrium point $(\delta_{\mathrm{sep}}, 0)$.

From (5.20), the restoring torque about the equilibrium point is, assuming that the rotor speed is close to 1 pu,

$$T_r = \frac{E'V}{X_{\mathrm{eq}}}\sin\delta - P_m \tag{5.22}$$

Define the potential energy as the integral of the restoring torque

$$V_{\mathrm{PE}}(\delta) = \int_{\delta_{\mathrm{sep}}}^{\delta}\left(-P_m + \frac{E'V}{X_{\mathrm{eq}}}\sin\phi\right)d\phi \tag{5.23}$$

Integrate (5.23) to obtain

$$V_{\mathrm{PE}}(\delta) = -P_m(\delta - \delta_{\mathrm{sep}}) - \frac{E'V}{X_{\mathrm{eq}}}\cos\delta + \frac{E'V}{X_{\mathrm{eq}}}\cos\delta_{\mathrm{sep}} \tag{5.24}$$

Note that $V_{\text{PE}}(\delta_{\text{sep}}) = 0$, and $V_{\text{PE}}(\delta) > 0$ about $\delta = \delta_{\text{sep}}$.

Define the kinetic energy of the rotating inertia as[2]

$$V_{\text{KE}}(\omega) = \frac{1}{2}(2H)\omega^2 = H\omega^2 \tag{5.25}$$

The total transient energy is then

$$V_E(\delta, \omega) = V_{\text{PE}}(\delta) + V_{\text{KE}}(\omega) > 0 \tag{5.26}$$

Take the time derivative to obtain

$$\begin{aligned}
\dot{V}_E = \frac{dV_E}{dt} &= \frac{dV_{\text{PE}}(\delta)}{d\delta}\frac{d\delta}{dt} + \frac{dV_{\text{KE}}(\omega)}{d\omega}\frac{d\omega}{dt} \\
&= \left(-P_m + \frac{E'V}{X_{\text{eq}}}\sin\delta\right)\omega + 2H\omega\frac{d\omega}{dt} \\
&= \omega\left(2H\frac{d\omega}{dt} - P_m + \frac{E'V}{X_{\text{eq}}}\sin\delta\right) = -D\omega^2 < 0
\end{aligned} \tag{5.27}$$

Thus from the Lyapunov Theorem, the equilibrium $(\delta, \omega) = (\delta_{\text{sep}}, 0)$ is asymptotically stable. Note that $\dot{V}_E = 0$ if $D = 0$, in which case the system is *marginally stable*.

Note that the energy functions V_{PE}, V_{KE}, and V_E contain the energy associated with the electromechanical mode of oscillations. Furthermore, the decay of V_E is not uniform as $\dot{V}_E = 0$ when ω crosses 0. An extension of the Lyapunov Theorem [67] still guarantees the asymptotic stability of the equilibrium point when $D > 0$, as $\omega(t)$ does not stay at 0 for an extended period of time t. If so, the trajectory has already converged to the equilibrium point.

Example 5.2: Energy function plots for Example 4.4

Using the simulated response for Example 4.4, compute and plot the kinetic, potential and total energy functions of the post-fault SMIB system.

Solutions: The post-fault potential energy function (5.24) requires the post-fault equivalent reactance

$$X_{\text{eq}}^{\text{pf}} = 0.24/9 + 0.15/9 + 0.044 = 0.08733 \text{ pu} \tag{5.28}$$

on 100 MVA base and the post-fault stable equilibrium point

$$\delta_{\text{sep}}^{\text{pf}} = \sin^{-1}\left(\frac{P_m X_{\text{eq}}^{\text{pf}}}{E'V}\right) = \sin^{-1}\left(\frac{8.5 \times 0.08733}{1.1088 \times 1.00}\right) = 0.7335 \text{ rad} = 42.03^\circ \tag{5.29}$$

The inertia H is $3.5 \times 900/100$ sec on 100 MVA base. The plots of V_{KE}, V_{PE}, and V_E are shown in Figure 5.5. Note that the simulation results are based on ω in pu frequency. Thus the kinetic energy is computed as $V_{\text{KE}}(\omega) = H\Omega\omega^2$. Also note that the energies are given in pu, which in absolute value is 100 MW-sec. In addition, note that due to a non-zero damping coefficient for the generator, the total energy V_E shows a trend of exponential decay. However, the actual decay rate varies within each oscillation period. When the speed (or V_{KE}) is near zero, the decay of V_E is almost zero. The decay is faster for high values of the rotor speed deviation. ■

2 With ω given in pu frequency, the kinetic energy expression will become $V_{\text{KE}}(\omega) = H\Omega\omega^2$.

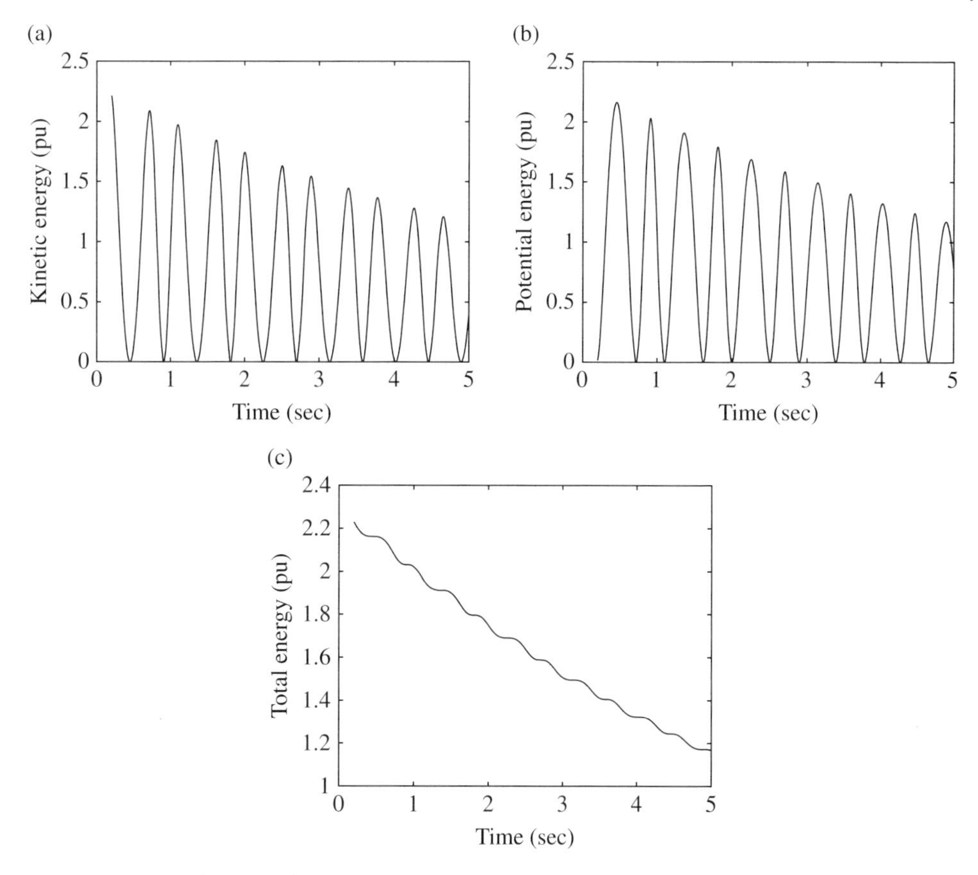

Figure 5.5 Energy functions for Example 5.2: (a) kinetic energy V_{KE}, (b) potential energy V_{PE}, and (c) total energy V_E.

5.4 Energy Function Analysis of a Disturbance Event

The concept of energy functions is now illustrated with high-sampling rate measurements captured during a power system disturbance [68].[3] A simplified schematic of the system is shown in Figure 5.6.a. Four generators connected to a common generator bus supply a large amount of power to a remote load center a few hundred miles away through two high-voltage transmission lines. Phasor measurement units are placed at Buses 1 and 2 to measure the bus voltage phasors and frequency, as well as the current phasor leaving Bus 1.

The measured voltage angle and frequency difference between Buses 1 and 2 are shown in Figures 5.7 and 5.8, respectively. The plots show a sudden increase, at about 60 seconds, of the transmission line impedance between the generators and the load center, resulting in a higher angle difference. Following the disturbance, a lightly damped oscillation at a frequency of about 0.578 Hz ensued for over 3 minutes.

3 The data were obtained using phasor measurement units [3, 8].

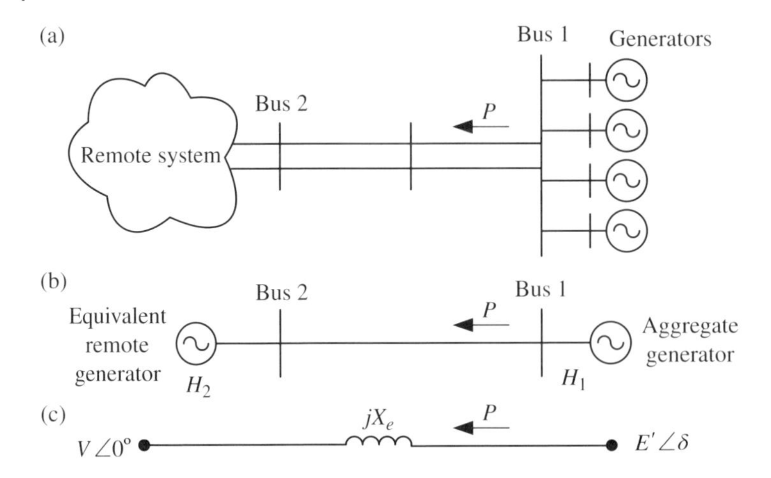

Figure 5.6 (a) Simplified schematic of the four-generator system, (b) 2-machine equivalent system, and (c) electrical equivalent circuit.

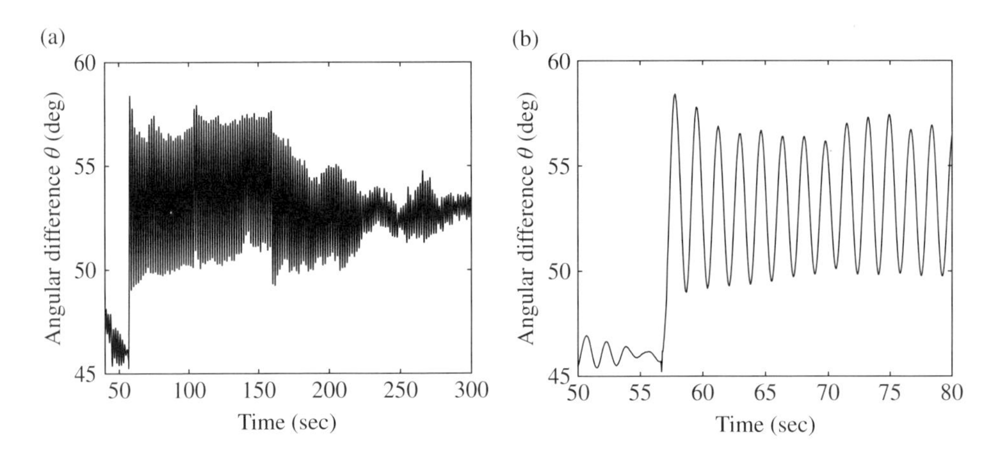

Figure 5.7 Angular difference between the generator substation and the remote system: (a) entire event and (b) enlarged plot at start of event.

One can approximately model this system by aggregating the four generators into a single generator and using a large inertia to model the remote system as shown in Figure 5.6.b. The electrical network model is shown in Figure 5.6.c. The generator and line parameters were not given. Using the oscillation amplitudes in the voltage and frequency measurements, the interarea model estimation (IME) algorithm [69, 70] was used to compute the pre-fault and post-fault equivalent reactance X_{eq}, and the machine inertias. For example, the post-fault equivalent reactance was found to be $X_{eq}^{pf} = 0.0903$ pu. The internal voltages of the equivalent machines were found to be $E' = 1.15$ pu and $V = 1.09$ pu. The aggregate inertias of the four generators and the load center were found to be $H_1 = 134.2$ sec and $H_2 = 1050$ sec, respectively. All these values are expressed on 100 MVA base. The active power flow from the equivalent generator to the load center also exhibited oscillations of the same frequency, and is shown in Figure 5.9.

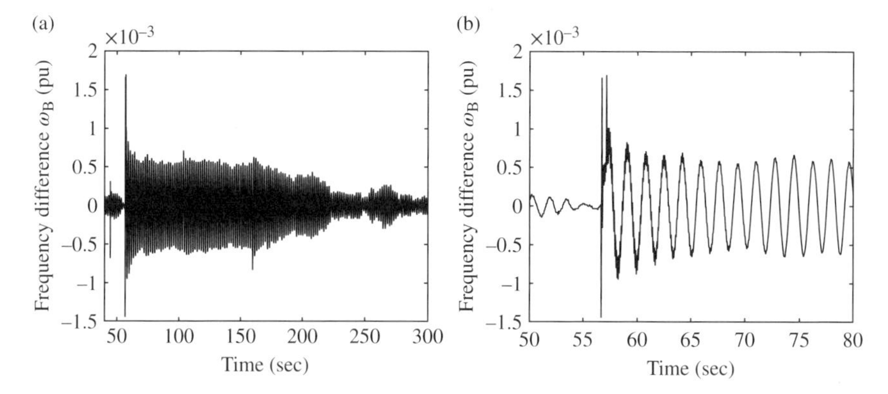

Figure 5.8 Frequency difference between the generator substation and the remote system: (a) entire event and (b) enlarged plot at start of event.

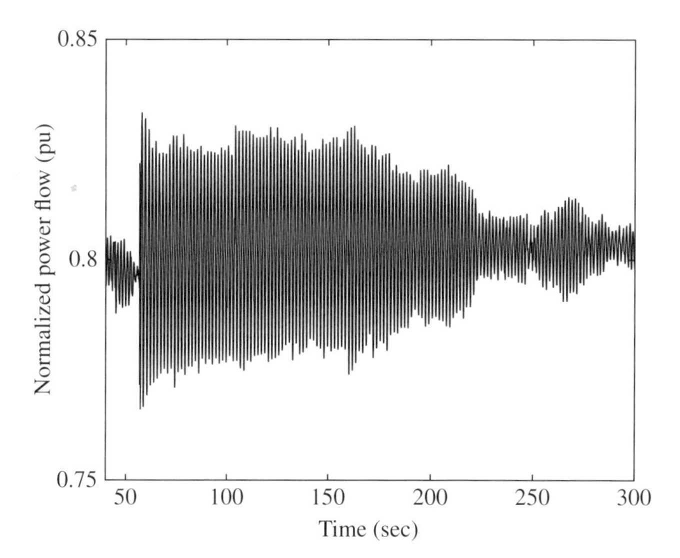

Figure 5.9 Active power flow from Bus 1 to Bus 2.

The estimated reactances are used to plot the power-angle curves before and after the disturbance (Figure 5.10). The measured PMU data are also plotted in Figure 5.10, showing close agreement.

The computation of the potential energy function requires the post-fault equilibrium angle difference between the two equivalent generators. Alternatively, this can be accomplished by removing the steady-state values of the angle difference through low-pass filtering of the angle-difference signal variable shown in Figure 5.7 to obtain the quasi-steady-state (QSS) component and then subtracting out the QSS component from the angle difference to obtain the oscillatory component. The individual angle difference signals are shown in Figure 5.11.a for the quasi-steady-state component and in Figure 5.11.b for the oscillatory component. Using the measured angle values and

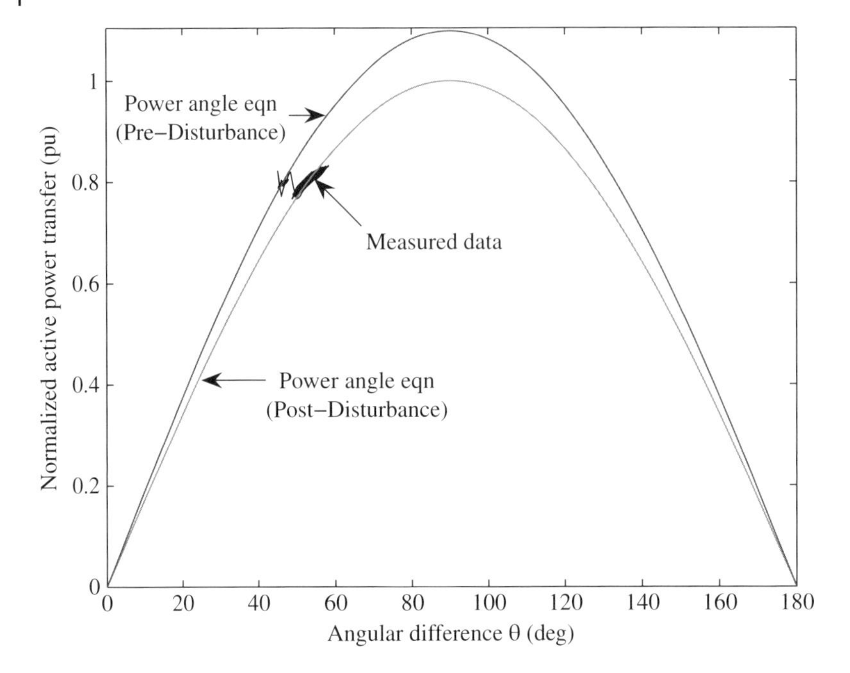

Figure 5.10 Power-angle curves (see color plate section).

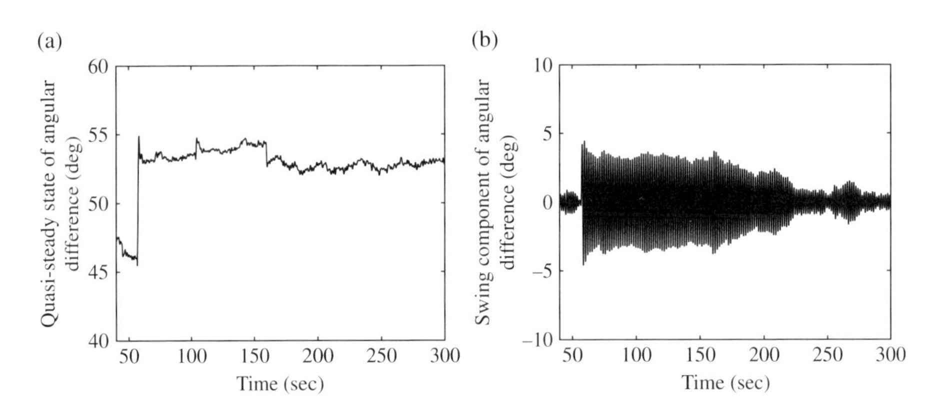

Figure 5.11 Angle difference: (a) quasi-steady-state and (b) oscillatory component.

the estimated X_{eq}^{pf} and inertias, the kinetic and potential energies can be computed and shown in Figures 5.12.a and 5.12.b, respectively. Note that the base unit of energy is 100 MW-sec.

Figure 5.12 shows the total energy function V_E obtained by adding the kinetic and potential energies. Note that electromechanical oscillations are barely visible in V_E. In addition, V_E clearly shows when the damping was negative during periods of increasing V_E and when the system was subject to some further external stimulus, indicated by jumps in V_E.

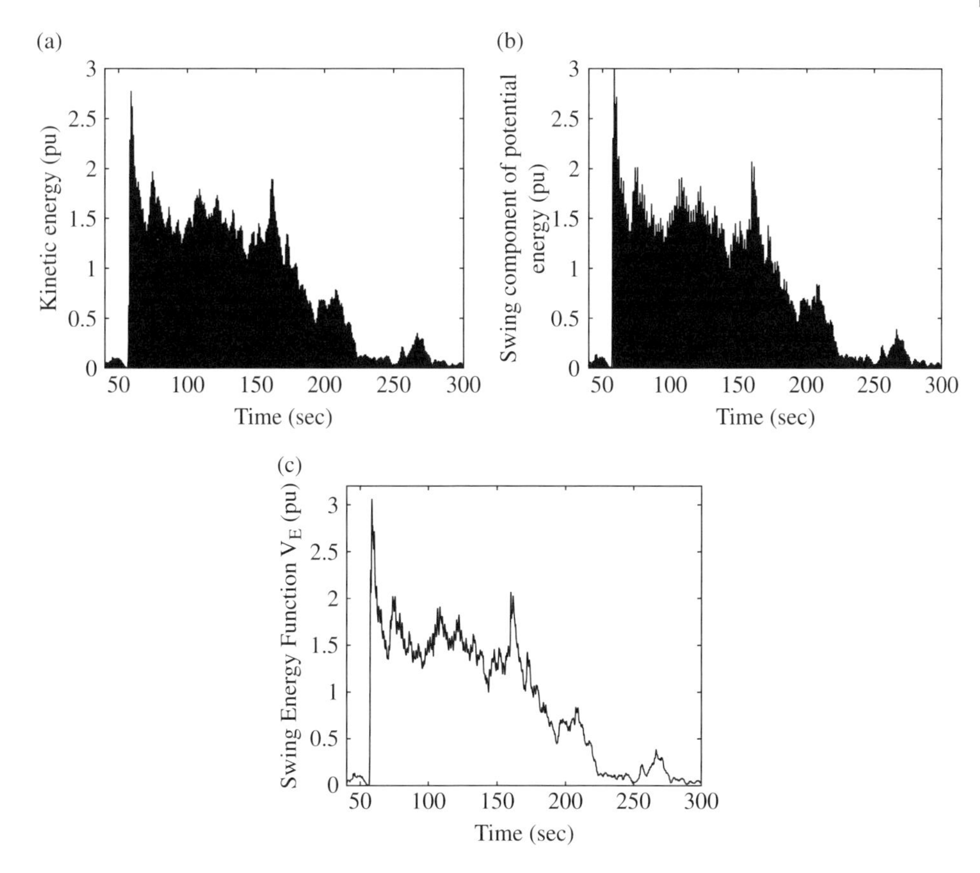

Figure 5.12 Energy functions: (a) kinetic energy, (b) potential energy, and (c) total energy.

5.5 Single-Machine Infinite-Bus Model Phase Portrait and Region of Stability

The SMIB system (5.20) is nonlinear because of the presence of the $\sin\delta$ term. If the system is perturbed from its equilibrium point, there is no closed-form solution of the time response to predict its stability against the disturbance. However, for such a second-order system, it is possible to depict the time response of (5.20) in a phase portrait, which is a common technique discussed in advanced textbooks on differential equations [71, 72].

First consider the SMIB system with the classical model representation (5.20)

$$\dot{\delta} = \Omega\omega, \quad 2H\dot{\omega} = P_m - \frac{E'V}{X_{\text{eq}}}\sin\delta - D\omega \tag{5.30}$$

which is a time-invariant system, that is, the right-hand side of (5.30) is not a function of time. Note that the unit for ω is pu frequency, as the study in this section does not use energy functions. For such systems, it is possible to determine the phase trajectories of the system based on the current states of the system.

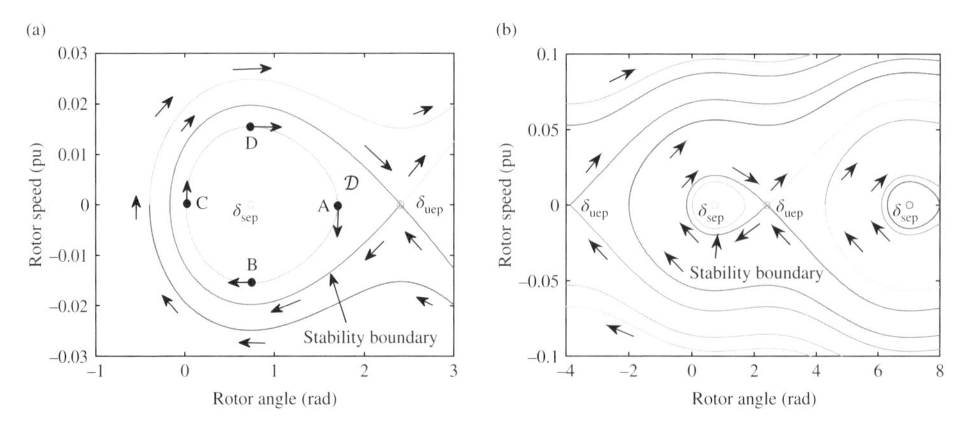

Figure 5.13 (a) Phase portraits and region of stability and (b) multiple equilibrium points (see color plate section).

To illustrate, consider the plot of phase trajectories on the δ-ω plane in Figure 5.13, in which the damping D is set to zero. Let $(\delta, \omega) = (\delta_{sep}, 0)$ be the stable equilibrium point, where

$$\delta_{sep} = \sin^{-1}\left(\frac{P_m X_{eq}}{E'V}\right) \tag{5.31}$$

and $(\delta, \omega) = (\delta_{uep}, 0)$ be the unstable equilibrium point, where

$$\delta_{uep} = \pi - \delta_{sep} \tag{5.32}$$

In the plot of rotor speed versus rotor angle in Figure 5.13, *phase flow* can be used to determine how the system states change, and as a result, the direction of the motion changes, which can be readily illustrated with a two-dimensional plot. Consider Point A in Figure 5.13.a, which is a small perturbation from the stable equilibrium point $(\delta_{sep}, 0)$. With $\omega = 0$ and $\delta > \delta_{sep}$, $\dot{\delta} = 0$ and $\dot{\omega} < 0$. Thus the direction of motion points downwards. For Point B with $\delta = \delta_{sep}$ and $\omega < 0$, $\dot{\delta} < 0$ and $\dot{\omega} = 0$. Thus the direction of motion points directly to the left. For Point C with $\omega = 0$ and $\delta < \delta_{sep}$, $\dot{\delta} = 0$ and $\dot{\omega} > 0$. Thus the direction of motion points upwards. Finally for Point D with $\delta = \delta_{sep}$ and $\omega > 0$, $\dot{\delta} > 0$ and $\dot{\omega} = 0$. Thus the direction of motion points directly to the right.

In fact, Points A, B, C, and D are on a stable orbit $((\delta(t), \omega(t))$ around the stable equilibrium $(\delta_{sep}, 0)$ in a clockwise direction. An easy way to remember this fact is: if $\omega > 0$, δ increases and hence moves to the right, and conversely, if $\omega < 0$, δ decreases and hence moves to the left. Note that although there is no closed-form solution to (5.20), it is a conservative system (without damping) and when the rotor angle or speed is plotted with respect to time, the response is periodic.

For a nonlinear system with a stable equilibrium point, there is a *Region of Stability* $\mathcal{D}$ defined as the region where if an initial condition is in $\mathcal{D}$, then the trajectory will stay in $\mathcal{D}$ as $t \to \infty$. For the system (5.30), the boundary of the stability region passes

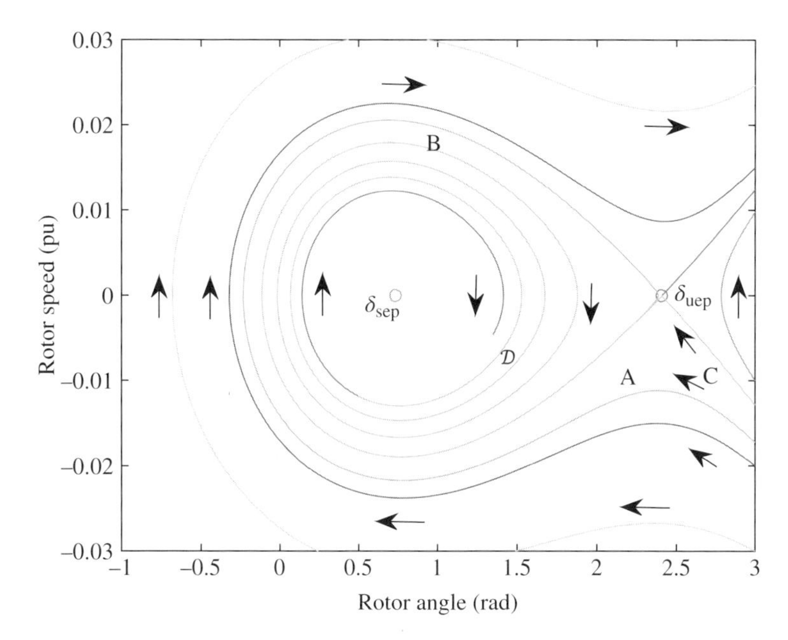

Figure 5.14 Phase portrait and region of stability with nonzero damping term D (see color plate section).

through the unstable equilibrium point $(\delta_{uep}, 0)$, as indicated in Figure 5.13.a. With zero damping, the stability boundary is a closed surface.

Figure 5.13.b shows several stable and unstable equilibrium points of (5.20). Note that each stable equilibrium point is displaced by 2π from its nearest stable equilibrium point, and the same holds for the unstable equilibrium points.

In a conservative system (that is, $D = 0$), the stability boundary of the region $\mathcal{D}$ is the largest area that can be enclosed by a constant energy value. In the SMIB system with an electromechanical model for the generator, this energy value is $V_{crit} = V_{PE}(\delta_{uep})$, that is, the value of the potential energy at the unstable equilibrium point, as the kinetic energy at this point is zero. Thus the stability boundary passes through the unstable equilibrium point and is a closed surface.

If $D > 0$, then $\mathcal{D}$ is no longer a closed region. In Figure 5.14, this stability region is bounded by the surface A-B-C. Note that the unstable equilibrium point $(\delta_{uep}, 0)$ is also on this boundary. If the initial condition is within the region $\mathcal{D}$, then the trajectory will stay within $\mathcal{D}$ and converge to δ_{eq}. The damping parameter is relatively small, and thus the stability region is only unbounded with $\omega < 0$. For higher values of D, the stability region would be unbounded with both $\omega < 0$ and $\omega > 0$. For power system transient stability analysis, the initial value would be in the region $\omega > 0$.

Note that Figures 5.13 and 5.14 are constructed using the parameter values from Example 4.4 in the post-fault condition and a general purpose simulation program. For the nonzero damping case, the damping is set to $D = 2$ pu on the generator base of 900 MVA. A general purpose simulation program is used as it allows for arbitrary initial conditions. This Simulink® program is provided as a course resource.

5.6 Direct Stability Analysis using Energy Functions

The phase trajectory and region of stability can be used for determining the stability of a power system: if the initial condition caused by a disturbance is in the stability region $\mathcal{D}$ of the post-fault system, then the system is stable with respect to that disturbance. For a conservative SMIB system, the total energy V_E at the stability boundary is called the critical energy V_{crit}. This critical energy is also equal to the potential energy at the unstable equilibrium point δ_{uep}

$$V_{\text{crit}} = V_{\text{PE}}(\delta_{\text{uep}}) \tag{5.33}$$

If at the fault clearing time, the accumulated total energy is larger than V_{crit}, then the system is unstable. An algorithm to determine system stability using energy functions is formulated as follows.

Algorithm for determining transient stability of a single-machine infinite-bus system using energy function

1) Find V_{crit} (5.33) by calculating δ_{sep} (5.31) and δ_{uep} (5.32) for the post-fault system.
2) Starting from the pre-fault system equilibrium, monitor $V_E = V_{\text{PE}} + V_{\text{KE}}$ for the fault-on system up to the time instant t_c when the fault is cleared, after which the system is in the post-fault configuration.
3) If $V_E(t_c) < V_{\text{crit}}$ when the fault is cleared, then the system is stable. Otherwise, if $V_E(t_c) > V_{\text{crit}}$, the system is unstable. The critical clearing time $t_c = t_{\text{crit}}$ is when $V_E(t_{\text{crit}}) = V_{\text{crit}}$.

An immediate advantage of this method is that it is no longer necessary to simulate the post-fault system to determine the system stability. The following example shows the phase portrait and the use of energy function to determine the critical clearing time.

Example 5.3: Energy function to determine critical clearing time

For the system in Examples 5.2, show the phase portraits for fault clearing times of 4, 6, and 9 cycles. Also plot the fault-on total energy function and the critical energy to determine the critical clearing time.

Solutions: The phase trajectories of the three clearing times are shown in Figure 5.15.a. For 4- and 6-cycle fault clearing, the fault-on trajectories remain in $\mathcal{D}$ and form closed orbits. The system is stable. For 9-cycle fault clearing, the fault-on trajectory leaves the stability region, resulting in an unstable response. The intersection of the fault-on trajectory and the stability boundary is known as the *exit point*. Plots of the fault-on total energy function and the critical energy are shown in Figure 5.15.b, where the critical energy is the potential energy at the unstable equilibrium point δ_{uep} and $\omega = 0$ pu. With $X_{\text{eq}}^{\text{pf}} = 0.0873$ pu, $E' = 1.109$ pu, $V = 1.0$ pu, $P_m = 8.5$ pu, $\delta_{\text{sep}} = 42.02°$, and $\delta_{\text{uep}} = 180° - 42.03° = 137.97°$, from (5.24)

$$V_{\text{crit}} = V_{\text{PE}}(\delta_{\text{uep}}) = 4.6283 \text{ pu} \tag{5.34}$$

The interaction of these two curves yields the critical clearing time of $t_{\text{crit}} = 0.1379$, slightly over 8 cycles. ■

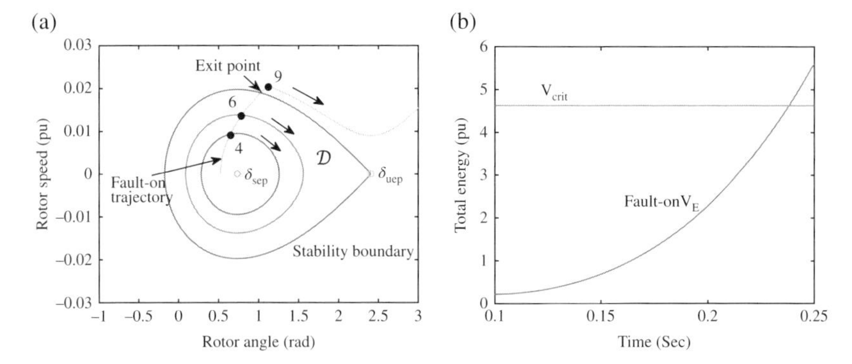

Figure 5.15 (a) Phase portraits for various fault clearing times (the numbers indicate the clearing times in cycles) and (b) potential energy at the unstable equilibrium point δ_{uep} and fault-on total energy (see color plate section).

For a SMIB system, there is only one unstable equilibrium point (except for the displacement of multiples of 2π in the rotor angle δ). For a multi-machine system, there is one stable equilibrium point, but many unstable equilibrium points, resulting in different potential energy levels. In such systems, the accuracy of the direct stability analysis method depends on finding the controlling or relevant unstable equilibrium point, which may not be easy to compute. This is the topic of the next section.

5.7 Energy Functions for Multi-Machine Power Systems

This section follows the discussion in [62, 63, 73, 74] to develop the energy function for multi-machine power systems. Consider the electromechanical model of a power system with N_G generators. The swing equations are given by, for $i = 1, 2, \ldots, N_G$,

$$\frac{d\delta_i}{dt} = \omega_i, \quad \frac{d\omega_i}{dt} = \frac{1}{2H_i}(P_{mi} - P_{ei} - D_i\omega_i) \tag{5.35}$$

where δ_i is the rotor angle in rad, ω_i the rotor speed in rad/sec, H_i the rotor inertia in seconds, D_i the pu damping coefficient, P_{mi} the pu mechanical input power, and P_{ei} the pu electrical output power of machine i.

The network voltage-current equation reduced to the machine internal node (Section 4.6) is

$$Y_{red}\tilde{E}' = \tilde{I}_G \tag{5.36}$$

Let Y_{ij} denote the equivalent admittance between Buses i and j

$$Y_{ij} = |Y_{ij}|\angle\varphi_{ij} = |Y_{ij}|(\cos\varphi_{ij} + j\sin\varphi_{ij}) = G_{ij} + jB_{ij}, \ i, j = 1, \ldots, N, \quad i \neq j \tag{5.37}$$

which is also the negative value of the (i, j) entry of Y_{red}.

The self-admittance terms (diagonal terms) of Y_{red} are

$$Y_{ii} = Y_i + \sum_{j=1, j\neq i}^{N} Y_{ij}, \quad i = 1, \ldots, N \tag{5.38}$$

where Y_i is the shunt admittance on Bus i.

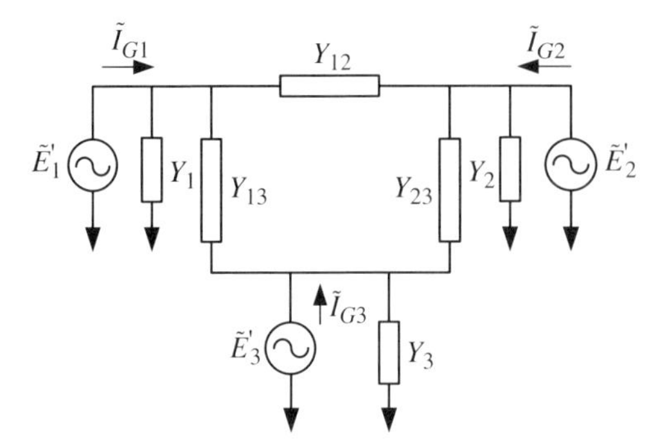

Figure 5.16 Three-machine system reduced to the machine internal nodes.

For a system with a symmetric Y matrix (no phase-shifting transformers), $Y_{ij} = Y_{ji}$. The network admittance representation of a 3-machine system is shown in Figure 5.16.

The electrical output power of machine i is

$$P_{ei} = \mathrm{Re}\{\tilde{E}'_i\tilde{I}^*_{Gi}\} = (E'_i)^2|Y_{ii}|\cos\varphi_{ii} + \sum_{j=1,j\neq i}^{N_G} E'_iE'_j|Y_{ij}|\cos(\varphi_{ij} - (\delta_i - \delta_j))$$
$$= (E'_i)^2G_{ii} + \sum_{j=1,j\neq i}^{N_G} E'_iE'_j(B_{ij}\sin\delta_{ij} + G_{ij}\cos\delta_{ij}) \tag{5.39}$$

where $\delta_{ij} = \delta_i - \delta_j$.

Denoting

$$C_{ij} = E'_iE'_jB_{ij},\quad D_{ij} = E'_iE'_jG_{ij},\quad P_i = P_{mi} - (E'_i)^2G_{ii} = P_{mi} - D_{ii} \tag{5.40}$$

the swing equation for machine i becomes

$$\frac{d\delta_i}{dt} = \omega_i,\quad \frac{d\omega_i}{dt} = \frac{1}{2H_i}\left(P_i - \sum_{j=1,j\neq i}^{N_G}(C_{ij}\sin\delta_{ij} + D_{ij}\cos\delta_{ij}) - D_i\omega_i\right) \tag{5.41}$$

The stable equilibrium point $(\delta_{\mathrm{sep}}, \omega_{\mathrm{sep}})$ is given by the individual machine equilibrium states as

$$\omega_{\mathrm{sep}i} = 0,\quad P_{mi} - D_{ii} = \sum_{j=1,j\neq i}^{N_G}(C_{ij}\sin\delta_{\mathrm{sep}ij} + D_{ij}\cos\delta_{\mathrm{sep}ij}) \tag{5.42}$$

The energy function can be developed using either the relative rotor angle formulation or the center-of-inertia (COI) (δ_o) formulation

$$\delta_o = \sum_{i=1}^{N_G} H_i\delta_i / \sum_{i=1}^{N_G} H_i,\quad \omega_o = \sum_{i=1}^{N_G} H_i\omega_i / \sum_{i=1}^{N_G} H_i \tag{5.43}$$

Here the COI reference frame is used, as fewer changes to the swing equations would be required.

Define the angles and speeds relative to the COI as

$$\hat{\delta}_i = \delta_i - \delta_o,\quad \hat{\omega}_i = \omega_i - \omega_o \tag{5.44}$$

It follows that

$$\delta_{ij} = \delta_i - \delta_j = (\delta_i - \delta_o) - (\delta_j - \delta_o) = \hat{\delta}_{ij} \tag{5.45}$$

The swing equations in the COI reference frame are, with $D_i = 0$, for $i = 1, 2, .., N_G$,

$$2H_i \frac{d^2\hat{\delta}_i}{dt^2} = P_i - \sum_{j=1, j\neq i}^{N_G} (C_{ij} \sin \hat{\delta}_{ij} + D_{ij} \cos \hat{\delta}_{ij}) - \frac{H_i}{H_T} P_{\text{COI}} \tag{5.46}$$

where

$$P_{\text{COI}} = \sum_{i=1}^{N_G} P_i - 2 \sum_{i=1}^{N_G-1} \sum_{j=i+1}^{N_G} D_{ij} \cos \hat{\delta}_{ij} \tag{5.47}$$

The term P_{COI} does not contain any sine terms because $C_{ij} = C_{ji}$ and sine is an odd function.

The energy functions for the individual machines are, for $i = 1, 2, .., N_G$

$$V_{Ei}(\hat{\delta}, \hat{\omega}) = V_{\text{KE}i}(\hat{\omega}_i) + V_{\text{PE}i}(\hat{\delta}) \tag{5.48}$$

where with $\delta_{\text{ep}i}$ being the post-fault equilibrium point, the kinetic and potential energies are, respectively,

$$V_{\text{KE}i}(\hat{\omega}_i) = \frac{1}{2} 2H_i \hat{\omega}_i^2 \tag{5.49}$$

$$V_{\text{PE}i}(\hat{\delta}) = \int_{\hat{\delta}_{\text{ep}i}}^{\hat{\delta}_i} \left(-P_i + \sum_{i=1, j\neq i}^{N_G} (C_{ij} \sin \hat{\phi}_{ij} + D_{ij} \cos \hat{\phi}_{ij}) + \frac{H_i}{H_T} P_{\text{COI}} \right) d\hat{\phi}_i \tag{5.50}$$

The total energy accounting for all the machines is

$$V_E(\hat{\delta}, \hat{\omega}) = V_{\text{KE}}(\hat{\omega}) + V_{\text{PE}}(\hat{\delta}) \tag{5.51}$$

where the total kinetic energy is

$$V_{\text{KE}}(\hat{\omega}) = \frac{1}{2} \sum_{i=1}^{N_G} 2H_i \hat{\omega}_i^2 \tag{5.52}$$

and the total potential energy, assuming $C_{ij} = C_{ji}$ and $D_{ij} = D_{ji}$, is

$$\begin{aligned} V_{\text{PE}}(\hat{\delta}) = \sum_{i=1}^{N_G} V_{\text{PE}i}(\hat{\delta}) &= - \sum_{i=1}^{N_G} P_i(\hat{\delta}_i - \delta_{\text{sep}i}) \\ &- \sum_{i=1}^{N_G-1} \sum_{j=i+1}^{N_G} \left[C_{ij}(\cos \hat{\delta}_{ij} - \cos \hat{\delta}_{\text{sep}ij}) - \int_{\hat{\delta}_{\text{sep}i} + \hat{\delta}_{\text{sep}j}}^{\hat{\delta}_i + \hat{\delta}_j} D_{ij} \cos \hat{\phi}_{ij} \, d(\hat{\phi}_i + \hat{\phi}_j) \right] \end{aligned} \tag{5.53}$$

Note that

$$\sum_{i=1}^{N_G} \frac{H_i}{H_T} \int_{\hat{\delta}_{\text{sep}i}}^{\hat{\delta}_i} P_{\text{COI}} d\hat{\phi}_i = 0 \tag{5.54}$$

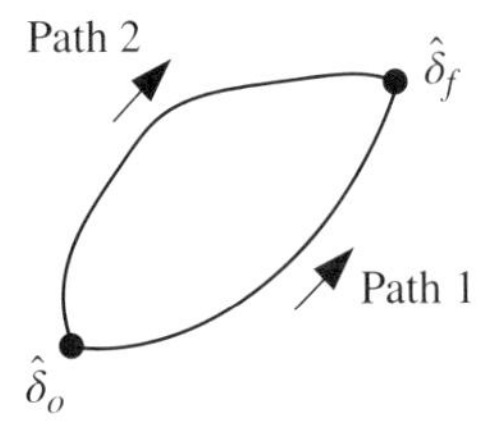

Figure 5.17 Path-dependent integrals.

The C_{ij} terms come out of the integral in (5.50), resulting in a closed-form solution. This part is conservative because the C_{ij} term represents the magnetic energy storage in the branch related to either releasing or storing energy in the line reactances.

The coefficient D_{ij} is the transfer conductance between machines i and j. It represents the losses or energy dissipation in the branch. Thus the integral involving the D_{ij} terms is a path-dependent term, that is, its value depends on the actual machine angle trajectory. It not only depends on the initial values of the machine angles, but also on the path it takes to get to the final values of the angles. Figure 5.17 shows two possible paths from $\hat{\delta}_o$ to $\hat{\delta}_f$. The integration of the system trajectories over these two paths will result in different energy functions. It is not known beforehand which path to use unless the system response is actually simulated. If D_{ij} is neglected, then V_{PE} is conservative and has a closed-form solution. Additional discussion on the impact of transfer conductance terms on energy function calculations can be found in [3].

The critical energy associated with an unstable equilibrium point in a multi-machine system is, if transfer conductances are neglected,

$$V_{\text{crit}} = V_{\text{PE}}(\delta_{\text{uep}}) = -\sum_{i=1}^{N_G} P_i(\hat{\delta}_{\text{uep}i} - \hat{\delta}_{\text{sep}i}) - \sum_{i=1}^{N_G-1}\sum_{j=i+1}^{N_G} C_{ij}(\cos\hat{\delta}_{\text{uep}ij} - \cos\hat{\delta}_{\text{sep}ij}) \tag{5.55}$$

which can be used for estimating the transient stability of a multi-machine system.

5.7.1 Direct Stability Analysis for Multi-Machine Systems

The counterpart of the energy function algorithm outlined in Section 5.6 for multi-machine systems is given as follows.

Multi-machine direct stability analysis algorithm

1) Determine V_{crit} (5.55) using a specific unstable equilibrium point $(\delta_{\text{uep}}, 0)$ or some other method.
2) Let the fault stay on until

$$V_E(\hat{\delta}, \hat{\omega}) = V_{\text{crit}} \tag{5.56}$$

to get the critical clearing time t_{crit}.

Although the steps in this algorithm look rather simple, Step 1, the computation of V_{crit} is an extremely difficult step, motivating the development of many theoretical results and practical algorithms.

5.7.2 Computation of Critical Energy

Let $(\delta_{\text{sep}}, 0)$ be the stable equilibrium point (SEP) of a power system with N_G machines. This stable equilibrium is unique, except for angle displacement of multiples of 2π. However, the stable equilibrium point will be surrounding by $2^{N_G} - 1$ unstable equilibrium points (UEPs), each corresponding to a specific mode of instability. For example, one mode of instability could be machine 1 becoming unstable versus the other $N_G - 1$ machines. A second mode of instability could be machine 2 becoming unstable versus the other $N_G - 1$ machines. Based on an unstable equilibrium point, the critical energy can be computed, which would be suitable for that particular mode of instability.

Thus if a multi-machine power systems is going to be unstable, it is important to know which machine or machines will actually be unstable. In the following, several methods proposed for choosing the appropriate unstable equilibrium point or the critical energy are discussed.

1) The closest unstable equilibrium point method [75]: Among all the UEPs, this method finds the one that is closest to the SEP. This UEP will also yield the smallest V_{crit} of all the unstable equilibrium points. The result is often conservative because the actual transient instability may not involve the same set of unstable machines as the closest UEP.
2) The potential energy boundary surface (PEBS) method [76, 77]: In this method, the fault-on system is simulated until the potential energy V_{PE} (with respect to the post-fault equilibrium point) reaches a maximum. This maximum value of V_{PE} is taken to be the critical energy, with the interpretation that it provides a local approximation of the stability boundary. Then the critical clearing time is found from equating the total energy V_E of the fault-on trajectory with this critical energy value.
3) The controlling UEP methods: The main idea in these methods is to find the appropriate UEP so that the correct critical energy is used to determine the critical clearing time [63, 78, 79]. Normally the machines closest to the fault in an electrical sense (admittance) or the machine with the largest accelerations (called the mode of disturbance (MOD)) are used to determine the UEP.
4) The **B**oundary-of-stability-region-based **C**ontrolling **U**nstable equilibrium point (BCU) method [73, 74]: This is also a controlling UEP method. It consists of the following major steps:
   a) Simulate the fault-on trajectory until a local maximum of the potential energy is reached (this is similar to the PEBS strategy).
   b) Clear the fault and continue the simulation of the post-fault trajectory until the minimum gradient point (MGP) is reached (this is explained in Example 5.4).
   c) Use the MGP to solve for an equilibrium of the post-fault system. It is presumed that this MGP will be close to the controlling UEP such that a Newton-type nonlinear equation solver will arrive at the desired controlling UEP.

   In addition, a number of calculations in the BCU method can be based on a reduced-order model to improve the computation speed.

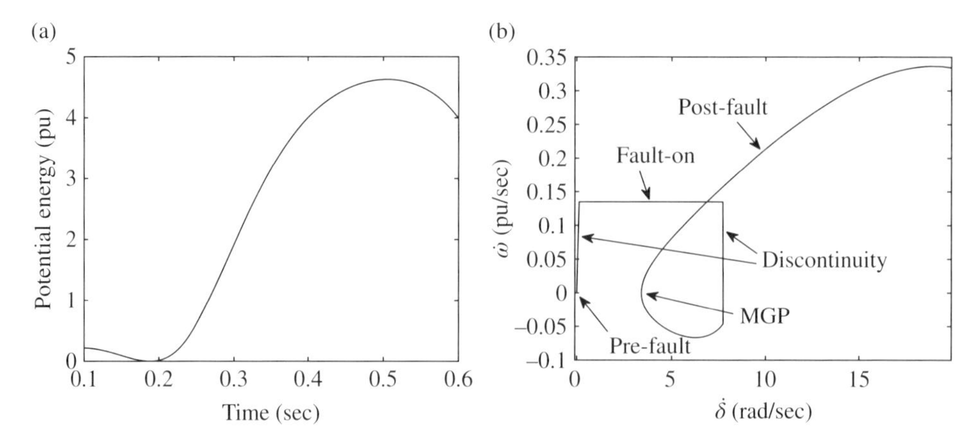

Figure 5.18 (a) Fault-on trajectory potential energy function and (b) plot of $\dot{\delta}$ versus $\dot{\omega}$ for fault-on and post-fault trajectories.

In the next two examples, the SMIB system and the NPCC 16-machine system are used to show the essential features of the PEBS and BCU methods.

Example 5.4: PEBS and BCU methods applied to SMIB system

Use the PEBS and BCU methods to determine the critical clearing time of the disturbance on the SMIB system in Examples 4.4, 5.2, and 5.3.

Solutions: First, for the PEBS method, the fault is applied until the potential energy reaches a local maximum. Figure 5.18 shows the change of potential energy along the fault-on trajectory. The peak of $V_{\text{PE}} = 4.6282$ pu is found, which is very close to the value $V_{\text{PE}}(\delta_{\text{uep}}) = 4.6285$ pu determined in Example 5.3. Thus the PEBS method will find an accurate critical clearing time for this disturbance.

Second, for the BCU method, consider the case where the fault is cleared at 9 cycles, when the system is already unstable. The post-fault trajectory is shown in Figure 5.15.a. The MGP point for the post-fault trajectory can be obtained by finding the minimum deviation from the equilibrium condition, which is defined as

$$d = ||\dot{\delta}||_2 + ||\dot{\omega}||_2 \tag{5.57}$$

Figure 5.18.b shows a plot of $\dot{\omega}$ versus $\dot{\delta}$, from which the first local minimum value of d is found at $\dot{\omega} = 0$ and $\dot{\delta} = 3.3847$, which occurs at $\omega = 0.009$ pu and $\delta = 2.4083$ rad. This is the point directly above δ_{uep} on the 9-cycle post-fault trajectory and thus is close to this unstable equilibrium point, as shown in Figure 5.15.a. Then the BCU method will start from this MGP point and solve the steady-state solution of the swing equation to get to the nearby UEP $(\delta_{\text{uep}}, 0)$, allowing it also to obtain an accurate critical clearing time. ■

Example 5.5: PEBS method applied to NPCC system

Use the disturbance simulation for the NPCC 16-machine system in Example 4.7 to illustrate the PEBS method for finding critical clearing time. Recall the disturbance is a

3-phase short-circuit fault near Bus 29 on Line 29-28, cleared by tripping Line 29-28. Generator 9 is the closest machine to Bus 29.

Solutions: There are a number of steps required to find the critical energy and clearing time:

1) Compute the stable equilibrium point: this can be accomplished by simulating the line trip without a fault and adding significant damping to the synchronous generators for fast decay of the oscillatory modes. Save the steady-state value of the machine rotor angles δ and internal voltages E' (also known as E'_q). Compute the center-of-inertia angles $\hat{\delta}$.
2) Obtain the post-fault reduced Y matrix from the simulation program. For this system, the first four rows and four columns of the reduced Y matrix are

$$\begin{bmatrix} 0.9978 - j15.2331 & 0.3314 + j0.7296 & 0.4106 + j0.9424 & 0.4205 + j0.8822 \\ 0.3314 + j0.7296 & 0.3271 - j9.1125 & 0.3574 + j1.4540 & 0.2404 + j0.4598 \\ 0.4106 + j0.9424 & 0.3574 + j1.4540 & 0.5024 - j10.9112 & 0.3177 + j0.6776 \\ 0.4205 + j0.8822 & 0.2404 + j0.4598 & 0.3177 + j0.6776 & 1.0340 - j12.4550 \end{bmatrix} \quad (5.58)$$

   Note that the transfer conductance terms (the real parts of the off-diagonal terms) are not small and cannot be neglected. From the reduced Y matrix and the generator internal voltages, construct the C_{ij} and D_{ij} terms.
3) Simulate a sufficiently long fault and calculate the potential energy V_{PE} for the fault-on trajectory using (5.53). For this disturbance, the variation of V_{PE} versus time is shown in Figure 5.19.a. The first maximum of V_{PE} is 0.69 pu occurring at 0.255 sec after the onset of the disturbance, which is taken as V_{crit}.
4) Compute the total energy V_E (5.51) and find the time when V_E equals V_{crit}. This step is shown in Figure 5.19.b, showing that the critical time is 0.053 sec. This is consistent with the simulation results in Example 4.7, that the critical clearing time is greater than 3 cycles, but less than 4 cycles.

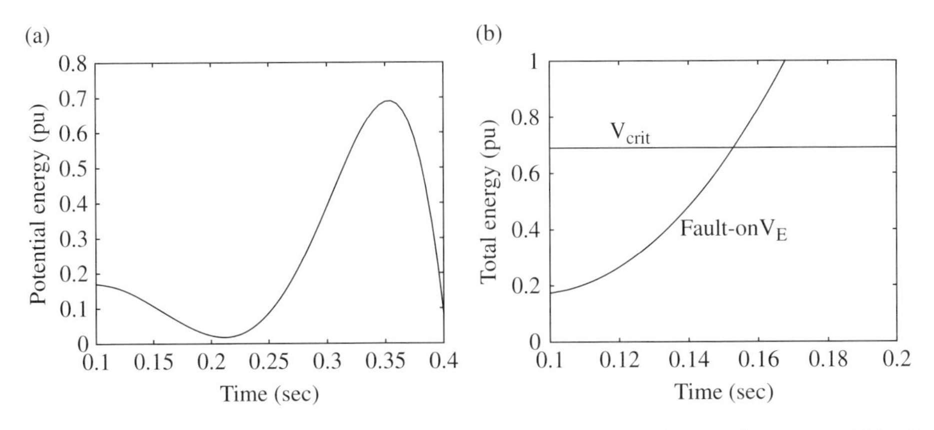

Figure 5.19 NPCC 16-machine system: (a) fault-on trajectory potential energy function and (b) critical energy from the PEBS method and fault-on total energy.

If the transfer conductance terms are not included in the calculation of the potential energy, this method would not have worked. In fact, the potential energy function would not have a local maximum.

The result in this example could be used to initiate the BCU method. To this end, clear the fault at a time slightly higher than the critical clearing time and integrate the post-fault system until the MGP is found. Then the MGP can be used at the starting point to solve for the controlling UEP, if such a nonlinear equation solver is available. ■

5.8 Dynamic Security Assessment

To ensure reliable power system operation, it is recognized that a power system has to survive through credible contingencies [80], such as the failure of a single transmission line or generator, double-circuit failure on the same tower, and delayed fault clearing. As a power system cannot be economically built to withstand all contingencies or sequences of contingencies, a common criterion used by many power grid operators is that a power grid should withstand all severe single contingencies in the high-voltage grid without disruption to the power delivery service. This is the so-called the $N-1$ contingency criterion (read as "all N components intact except for one"). For a system operator overseeing a large power grid, the $N-1$ contingency lists may have hundreds (or even thousands) of credible events. For each event, an operation engineer needs to simulate the disturbance and check the stability of the post-fault system, power transfer limits, and bus voltages. If the system is unstable, a line flow exceeds the steady-state thermal loading limits, or the voltages at critical buses are below a certain threshold, then the operating condition is deemed insecure and the system has to be re-dispatched, raising the cost of dispatch.

In real-time contingency analysis, because the amount of time to evaluate dynamic stability is limited, line flow limits and voltage constraints obtained from off-line detailed contingency analysis are often used to impose on the current operating conditions. If these limits are violated, then one of the $N-1$ contingencies will result in an undesirable operating condition. The flow limits and voltage constraints may vary from hour to hour due to the changing loading condition. When a system dispatch hits these limits, congestion pricing will result in some power importing areas paying higher energy prices.

The use of direct stability methods with reduced amount of simulation need will speed up the contingency analysis process. When applied in real time, these methods will provide more realistic operating limits and may be helpful in reducing the amount of redispatch needed.

5.9 Summary and Notes

This chapter develops the use of energy functions, which capture the kinetic and potential energy associated with the machine swing oscillations, for direct transient stability analysis of multi-machine systems. Phase trajectories of the machine rotor angles and speeds are provided to illustrate the importance of the stable and unstable equilibrium points for determining transient stability. Measurements from a real power system disturbance event illustrate the relevance of the energy function.

The energy function approach can also accommodate power systems with detailed machine models, excitation systems, and composite load models. Here the energy function would need to account for the impact of the control equipment and load models on the potential energy function, and the direct methods would compute the relevant unstable equilibrium point based on trajectories simulated using detailed machine models. Such direct transient stability analysis packages have been under consideration in power system control centers for real-time contingency analysis.

Problems

5.1 (Potential energy function) The potential energy function of a SMIB system is given by

$$V_{\rm PE} = -P_m(\delta - \delta_{\rm sep}) - \frac{P_m}{\sin(\delta_{\rm sep})}\cos\delta + \frac{P_m}{\sin(\delta_{\rm sep})}\cos(\delta_{\rm sep})$$

This function is positive for δ in a region about $\delta_{\rm sep}$.
1) Verify this property by plotting $V_{\rm PE}$ versus δ from $-60°$ to $120°$, using the parameters $P_m = 0.9$ pu and $\delta_{\rm sep} = 18.24°$.
2) Use asymptotic expansion to show that this is true analytically.

5.2 (Phase portrait for nonzero damping) Repeat Example 5.3 for the SMIB system by adding the damping term $D = 2$ pu (machine base) to the synchronous generator. Plot the phase trajectories and determine the critical clearing time using the critical energy.

5.3 Repeat Example 5.3 for the SMIB system by setting the reactance of the two parallel transmission lines from 0.044 pu to 0.066 pu. Explain why the critical energy is smaller.

5.4 Show the derivation of multi-machine potential energy function (5.53).

5.5 (NPCC system, Generator 9) Verify the PEBS results in Example 5.5 using the MATLAB© code provided in the example folder. Note that the simulated generator rotor angle and speed trajectories should be stored so that they can be used to compute the energy functions. Additional instructions are found in the example folder.

5.6 (NPCC system, Generator 11) Repeat Example 5.5, but the 3-phase short-circuit fault is applied on Line 32-30 close to Bus 32, which is cleared by tripping Line 32-30. Find the critical clearing time of Generator 11 using the PEBS method. Compare this result to the critical clearing time found in Problem 4.6.

5.7 The classical model swing equation for a SMIB system is given by

$$\dot{\delta} = \Omega\omega, \quad 2H\dot{\omega} = P_m - \frac{E'V}{X}\sin\delta \tag{5.59}$$

where $H = 3.0$ sec, $V = 1.0$ pu, $E' = 1.1$ pu, $P_m = 0.9$ pu, and $X = 0.5$ pu is the (post-fault) equivalent reactance from the infinite bus to the generator internal node. Note that ω is the machine speed deviation from the nominal speed of 1 pu.

1) Find the stable and unstable equilibrium operating conditions of the post-fault system.
2) Determine the values of the potential energy function at the stable and unstable equilibrium points.
3) Suppose after the fault is applied and cleared, the angle and speed of the synchronous machine are found to be $\delta = 36^\circ$ and $\omega = 0.04$ pu. Use the energy function to determine whether the system is stable.

6

Linear Analysis and Small-Signal Stability

6.1 Introduction

Two types of power system stability phenomena have been discussed: Chapter 3 discusses voltage stability analysis and techniques for computing voltage stability margins, and Chapters 4 and 5 discuss transient stability analysis for a power system subject to large disturbances using nonlinear system simulation and energy functions. This chapter presents the small-signal stability of power systems operating in "steady state," that is, the system stability of an equilibrium point. In practice, a power system is subject to constant perturbations in its operating condition, and thus is never truly in an equilibrium. For example, the morning load rise requires increasing the generator power setpoints, as determined by an economic dispatch. The ability of the power system to track the load changes requires the power system to be small-signal stable, that is, a power system can stably move from one operating condition to a slightly different operating condition.

The stability of a power system undergoing such incremental setpoint adjustments can be verified by performing a nonlinear simulation. However, such time simulation may not provide much more information other than an answer of "stable" or "unstable." In contrast, a linear analysis can provide additional information such as modes of instability and stability margins. In addition, control design to improve the damping of oscillatory modes would require linear models.

The main concern of small-signal stability is the study of the electromechanical modes, which represent the oscillatory interaction of the mechanical parts of the synchronous machines, that is, the machine inertias, against each other through the interconnected electrical power network, that is, the network impedances. The frequencies of these oscillatory modes normally range from 0.2 to 3 Hz. The damping on these modes is often quite small, with a damping ratio less than 5–10%. When the damping ratio is sufficiently small, a power system will exhibit sustained oscillations. As the system is constantly subject to small load perturbations, before the transients due to an earlier perturbation have decayed sufficiently, the power system is subjected to another disturbance resulting in additional oscillations, which are then superimposed on the earlier oscillations.

This chapter discusses electromechanical modes in power systems and techniques to compute and identify the frequency and damping of these modes. The approach used is to develop linearized system models that are valid in a region about the equilibrium operating point. Analytical and numerical linearization techniques are introduced and applied to the nonlinear power systems, followed by several examples. Although the

Power System Modeling, Computation, and Control, First Edition. Joe H. Chow and Juan J. Sanchez-Gasca.

Companion website: www.wiley.com/go/chow/power-system-modeling

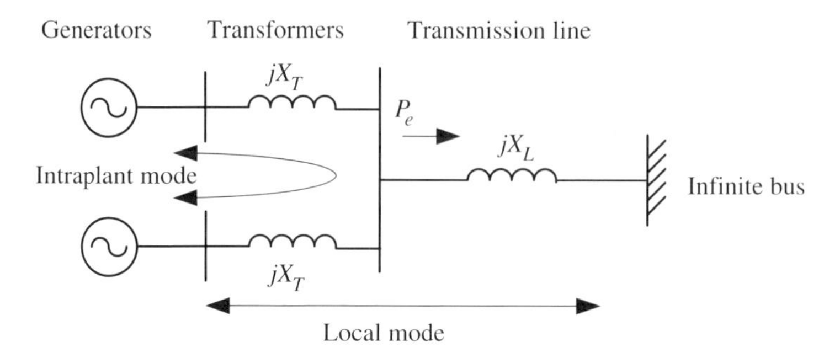

Figure 6.1 A power system showing a local mode and an intraplant mode.

classical model is used for the synchronous machines in this chapter, the techniques discussed here can be used for detailed machine models equipped with excitation systems and power system stabilizers.

6.2 Electromechanical Modes

A power system consists of many rotating inertias interchanging energy via interconnected transmission lines, and as a result exhibits oscillations of inertias against each other, similar to mass-spring systems used to illustrate Newtonian mechanics [81]. Electromechanical modes in a practical power system can be divided into three types:

1) Intraplant modes: these modes are due to the rotor oscillations between individual units within the same substation (Figure 6.1). It is quite common that hydraulic-turbine facilities consist of multiple units interconnected via step-up transformers to the same power plant. The effective impedances between the generators in the same power plant are small, and the generator rotor angles are about equal. In addition, the inertias of these machines are also lighter than those in fossil power plants. As a result, these modes have the highest oscillatory frequencies, typically between 2 and 3 Hz.
2) Local modes: these modes represent the oscillation of a synchronous machine against another machine not in the same substation or against a group of other machines. Figure 6.1 shows the local mode oscillation of the two generators in the same substation against a large group of generators, modeled as an infinite bus. Figure 6.2 shows the local mode oscillations of Generator 1 versus Generator 2 and Generator 11 versus Generator 12. The impedance of the generator to the infinite bus or the other machine is higher and the angle difference is larger. As a result, the frequency range of the local mode is between 1 to 2 Hz.
3) Interarea modes: these modes involve a group of generators oscillating against other groups of generators. Figure 6.2 shows an interarea mode of Generators 1 and 2 oscillating against Generators 11 and 12.[1] These interarea modes are lowest in frequency as both the combined inertias of the coherent machines and the impedances connecting these machine groups are higher. A more detailed treatment of interarea modes can be found in Chapter 16 on coherency and also in [50, 83].

1 The four-generator system shown in Figure 6.2 was introduced in [82] as a small system suitable for the study of electromechanical oscillations and damping controllers.

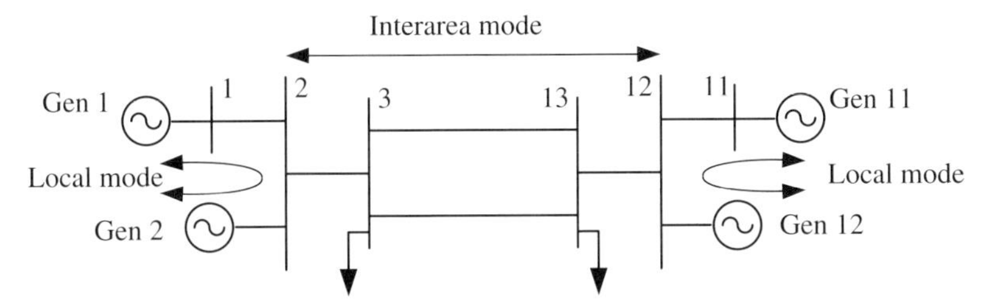

Figure 6.2 A 2-area, 4-machine system showing interarea and local modes.

The next section provides an overview of the development of linear models for nonlinear systems. Then the linearization process is applied to power systems for studying electromechanical oscillations.

6.3 Linearization

6.3.1 State-Space Models

Consider a nonlinear time-invariant system

$$\dot{x} = f(x, u), \quad x \in R^n,\ u \in R^r,\ f \in R^n \tag{6.1}$$

$$y = g(x, u), \quad y \in R^m,\ g \in R^m \tag{6.2}$$

where x, u, and y are the n-dimensional state vector, the r-dimensional input vector, and the m-dimensional output vector, respectively. The system dynamics and output are governed by the n- and m-dimensional nonlinear functions f and g, respectively, of the states and inputs.

Let x_o, u_o be an equilibrium point of (6.1) satisfying

$$f(x_o, u_o) = 0, \quad y_o = g(x_o, u_o) \tag{6.3}$$

To study the behavior of the system about this equilibrium point, perturb the system states and inputs from x_o, u_o by letting

$$x = x_o + \Delta x, \quad u = u_o + \Delta u \tag{6.4}$$

where the Δ terms denote small deviations from the equilibrium point.

Substituting these perturbation expressions into the nonlinear dynamic system (6.1) and (6.2), a Taylor series expansion yields

$$\begin{aligned} \dot{x}_o + \Delta\dot{x} &= f(x_o + \Delta x, u_o + \Delta u) \\ &= f(x_o, u_o) + \left.\frac{\partial f}{\partial x}\right|_o \Delta x + \left.\frac{\partial f}{\partial u}\right|_o \Delta u + \text{higher order terms} \end{aligned} \tag{6.5}$$

$$\begin{aligned} y_o + \Delta y &= g(x_o + \Delta x, u_o + \Delta u) \\ &= g(x_o, u_o) + \left.\frac{\partial g}{\partial x}\right|_o \Delta x + \left.\frac{\partial g}{\partial u}\right|_o \Delta u + \text{higher order terms} \end{aligned} \tag{6.6}$$

where the subscript "o" denotes function evaluation at the equilibrium point.

Neglecting the higher order terms results in the linearized model

$$\Delta\dot{x} = A\Delta x + B\Delta u, \quad \Delta y = C\Delta x + D\Delta u \tag{6.7}$$

where

$$A = \left.\frac{\partial f}{\partial x}\right|_o = \begin{bmatrix} \partial f_1/\partial x_1 & \cdots & \partial f_1/\partial x_n \\ \vdots & \ddots & \vdots \\ \partial f_n/\partial x_1 & \cdots & \partial f_n/\partial x_n \end{bmatrix}_o, \quad B = \left.\frac{\partial f}{\partial u}\right|_o = \begin{bmatrix} \partial f_1/\partial u_1 & \cdots & \partial f_1/\partial u_r \\ \vdots & \ddots & \vdots \\ \partial f_n/\partial u_1 & \cdots & \partial f_n/\partial u_r \end{bmatrix}_o$$
$$C = \left.\frac{\partial g}{\partial x}\right|_o = \begin{bmatrix} \partial g_1/\partial x_1 & \cdots & \partial g_1/\partial x_n \\ \vdots & \ddots & \vdots \\ \partial g_m/\partial x_1 & \cdots & \partial g_m/\partial x_n \end{bmatrix}_o, \quad D = \left.\frac{\partial g}{\partial u}\right|_o = \begin{bmatrix} \partial g_1/\partial u_1 & \cdots & \partial g_1/\partial u_r \\ \vdots & \ddots & \vdots \\ \partial g_m/\partial u_1 & \cdots & \partial g_m/\partial u_r \end{bmatrix}_o \tag{6.8}$$

In (6.7), A is known as the state matrix, B the input matrix, C the output matrix, and D the throughput (feedforward) matrix. Note that D denotes the direct influence of the input on the output, without any transition time through the states.

6.3.2 Input-Output Models

Applying the Laplace transform to the linear model (6.7) results in

$$sI\Delta x(s) = A\Delta x(s) + B\Delta u(s), \quad \Delta y(s) = C\Delta x(s) + D\Delta u(s) \tag{6.9}$$

where s is the Laplace operator, denoting d/dt, and I is an identity matrix with the same dimension as A.

Putting all the Δx terms of the state equation on the left-hand side yields

$$(sI - A)\Delta x(s) = B\Delta u(s) \tag{6.10}$$

which can be used to solve for $\Delta x(s)$ as

$$\Delta x(s) = (sI - A)^{-1}B\Delta u(s) \tag{6.11}$$

Substituting this expression into the output equation of (6.7) results in the transfer function model

$$\Delta y(s) = (C(sI - A)^{-1}B + D)\Delta u(s) = T(s)\Delta u(s) \tag{6.12}$$

If $r = m = 1$, then $T(s)$ is a scalar rational function. Otherwise, $T(s)$ is an $m \times r$ matrix transfer function.

6.3.3 Modal Analysis and Time-Domain Solutions

If x is a scalar ($n = 1$), the state equation (6.7) simplifies to

$$\Delta\dot{x} = a\Delta x + b\Delta u, \quad \Delta x(t_o) = \Delta x_o \tag{6.13}$$

where Δx_o is the perturbation of the initial condition at time t_o. The time-domain solution (response) to this differential equation is

$$\Delta x(t) = e^{a(t-t_o)}\Delta x_o + \int_{t_o}^{t} e^{a(t-\tau)}bu(\tau)d\tau \tag{6.14}$$

If x is a vector, then modal analysis can be applied to decompose the state equation into a system of n decoupled first-order systems.

Assuming all the eigenvalues of a matrix A are distinct, the eigenvalues λ_i, $i = 1, 2, \ldots, n$, and their corresponding nonzero (right) eigenvectors v_i are given by

$$Av_i = \lambda_i v_i, \quad i = 1, 2, \ldots, n \tag{6.15}$$

Rewrite (6.15) as

$$(\lambda_i I - A)v_i = 0 \tag{6.16}$$

which implies

$$\det(sI - A) = 0 \tag{6.17}$$

that is, v_i is in the one-dimensional null space of $(\lambda_i I - A)$. Thus v_i is unique in its direction, but not its magnitude. A common practice is to normalize the magnitude of v_i to unity, that is, the 2-norm $||v_i||_2 = 1$.

Assemble all the eigenvectors to form

$$A\underbrace{\begin{bmatrix} v_1 & v_2 & \cdots & v_n \end{bmatrix}}_{M} = \begin{bmatrix} v_1 & v_2 & \cdots & v_n \end{bmatrix}\underbrace{\begin{bmatrix} \lambda_1 & 0 & \cdots & 0 \\ 0 & \lambda_2 & \cdots & 0 \\ \vdots & \vdots & \ddots & \vdots \\ 0 & 0 & \cdots & \lambda_n \end{bmatrix}}_{\Lambda} \tag{6.18}$$

and compact into the matrix form

$$AM = M\Lambda \tag{6.19}$$

where Λ is the modal matrix containing the eigenvalues, and M is the modal transformation matrix of the right eigenvectors. Pre-multiplying (6.19) by M^{-1}, the modal decomposition is obtained as

$$M^{-1}AM = \Lambda \tag{6.20}$$

Let

$$M^{-1} = \begin{bmatrix} w_1^T \\ w_2^T \\ \vdots \\ w_n^T \end{bmatrix} \tag{6.21}$$

Then the column vectors w_i, $i = 1, 2, \ldots, n$, are the left eigenvectors satisfying

$$w_i^T A = \lambda_i w_i^T \quad \text{or} \quad A^T w_i = \lambda_i w_i \tag{6.22}$$

and the orthogonality condition

$$w_j^T v_i = \begin{cases} 1, & j = i \\ 0, & j \neq i \end{cases} \tag{6.23}$$

Consider the modal transformation using $\Delta q = M^{-1}\Delta x$ such that

$$\Delta \dot{q} = M^{-1}\Delta \dot{x} = M^{-1}AM\Delta q + M^{-1}B\Delta u = \Lambda \Delta q + B_M \Delta u \tag{6.24}$$

$$\Delta y = CM\Delta q + D\Delta u = C_M \Delta q + D\Delta u \tag{6.25}$$

This is the modal form in the n-dimensional modal variable Δq. Note that the throughput matrix D is independent of the state variable reference frame.

6.3.4 Time Response of Linear Systems

The time response of a linear system (6.7) with $n > 1$ can be expressed as

$$\Delta x(t) = e^{A(t-t_o)}\Delta x(t_o) + \int_{t_o}^{t} e^{A(t-\tau)}B\Delta u(\tau)d\tau \tag{6.26}$$

which consists of two components: one due to initial conditions $\Delta x(t_o) = \Delta x_o$ and the other to the input $\Delta u(t)$ [84]. In power system studies, linear analysis is concerned mostly with non-zero initial conditions, with $\Delta u(t) = 0$, such that

$$\Delta x(t) = e^{A(t-t_o)}\Delta x(t_o) = e^{At}\Delta x_o, \quad t_o = 0 \tag{6.27}$$

where e^{At} is known as the state transition matrix and is of dimension $n \times n$, the same as A. A Taylor series expansion of an exponential function holds also for a matrix exponent

$$e^{At} = I + At + \frac{A^2}{2!}t^2 + \cdots + \frac{A^n}{n!}t^n + \cdots \tag{6.28}$$

where I is the identify matrix of dimensions $n \times n$. This formula looks simple, but is not a good way to compute e^{At} numerically for a given A and t [85].

A robust method to compute e^{At} is to use eigenvalues and eigenvectors, which can be computed reliably using QR algorithms [19]. Using the Taylor series expansion formula, it can readily be shown that

$$e^{At} = Me^{\Lambda t}M^{-1}, \quad e^{\Lambda t} = \begin{bmatrix} e^{\lambda_1 t} & 0 & \cdots & 0 \\ 0 & e^{\lambda_2 t} & \cdots & 0 \\ \vdots & \vdots & \ddots & \vdots \\ 0 & 0 & \cdots & e^{\lambda_n t} \end{bmatrix} \tag{6.29}$$

Thus the time response due to nonzero initial conditions only can be computed as

$$\Delta x(t) = \begin{bmatrix} v_1 & v_2 & \cdots & v_n \end{bmatrix} \begin{bmatrix} e^{\lambda_1 t} & 0 & \cdots & 0 \\ 0 & e^{\lambda_2 t} & \cdots & 0 \\ \vdots & \vdots & \ddots & \vdots \\ 0 & 0 & \cdots & e^{\lambda_n t} \end{bmatrix} \begin{bmatrix} w_1^T \\ w_2^T \\ \vdots \\ w_n^T \end{bmatrix} \Delta x_o \tag{6.30}$$

Combining Δx_o with w_i, (6.30) becomes

$$\Delta x(t) = \begin{bmatrix} v_1 & v_2 & \cdots & v_n \end{bmatrix} \begin{bmatrix} e^{\lambda_1 t} & 0 & \cdots & 0 \\ 0 & e^{\lambda_2 t} & \cdots & 0 \\ \vdots & \vdots & \ddots & \vdots \\ 0 & 0 & \cdots & e^{\lambda_n t} \end{bmatrix} \begin{bmatrix} w_1^T \Delta x_o \\ w_2^T \Delta x_o \\ \vdots \\ w_n^T \Delta x_o \end{bmatrix}$$

$$= (w_1^T \Delta x_o)v_1 e^{\lambda_1 t} + \cdots + (w_n^T \Delta x_o)v_n e^{\lambda_n t} = \sum_{i=1}^{n}(w_i^T \Delta x_o)v_i e^{\lambda_i t} \tag{6.31}$$

where $w_i^T \Delta x_o$ is the projection of the initial condition Δx_o on mode i, and v_i is a vector denoting the mode shape of mode i, that is, how each mode i participates in each state. Thus $\Delta x(t)$ is a linear combination of the modal time response $e^{\lambda_i t}$.

As shown in Figure 6.3, there are two types of eigenvalues for a linear system: real and complex. A system is said to be (exponentially) stable if all its eigenvalues are in the open left-half plane,[2] that is, $\mathrm{Re}(\lambda_i) < 0$, for all i. It is unstable if any of the eigenvalues are in the

2 Open left-half plane is the left-half plane excluding the $j\omega$-axis.

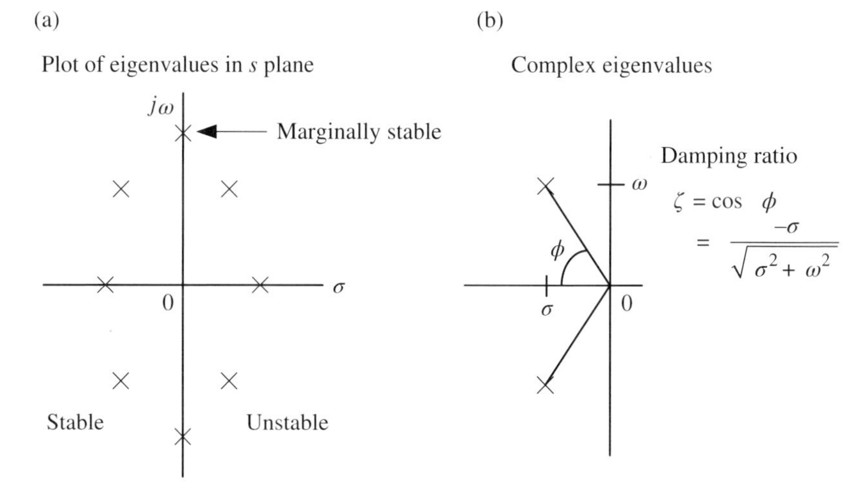

Figure 6.3 (a) Eigenvalues plotted in the complex s plane as "x" and (b) complex eigenvalues showing damping ratio.

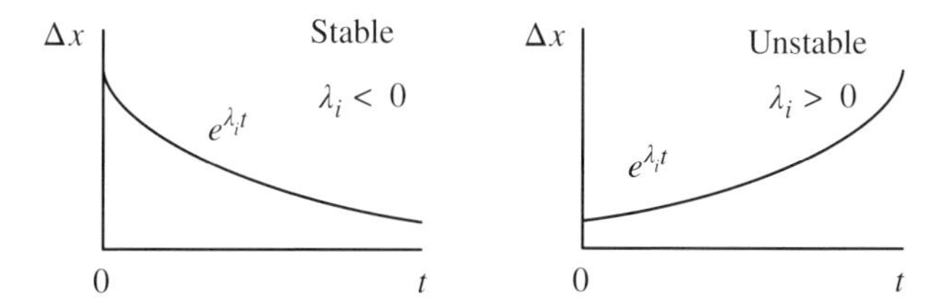

Figure 6.4 Stable and unstable time responses due to a real eigenvalue λ_i.

open right-half plane. If a system has eigenvalues on the $j\omega$-axis with multiplicity 1, then any non-zero initial condition on that eigenvalue will result in undamped oscillations, that is, the system is marginally stable.

Figure 6.4 shows the stable and unstable responses of real eigenvalues. For multivariable systems, a stable equilibrium is referred to as a node, and an unstable equilibrium is called a saddle. Figure 6.5 illustrates the phase trajectories for a two-dimensional system. In (a) with two stable real eigenvalues, the point A will converge to a node. In (b), one stable eigenvalue and one unstable eigenvalue will result in the point B moving away from the saddle point.[3]

Figure 6.6 shows the time responses due to a pair of stable and unstable complex eigenvalues. The time response in real form is given by $e^{\sigma_i t}\cos(\omega_i t + \phi_i)$, where ϕ_i is a phase shift determining the zero-crossing time. The response envelop is given by $e^{\sigma_i t}$, whereas the oscillatory part is due to $\cos(\omega_i t + \phi_i)$.

Figure 6.7 shows the phase trajectories for a two-dimensional system. In (a) with stable complex eigenvalues, the point A will spiral inward to the origin, also called a focus. In (b), the unstable eigenvalues will result in the phase trajectory starting from the point B and spiraling outward away from the unstable focus.

3 Note that the unstable equilibrium point in Figure 5.14 is a saddle point.

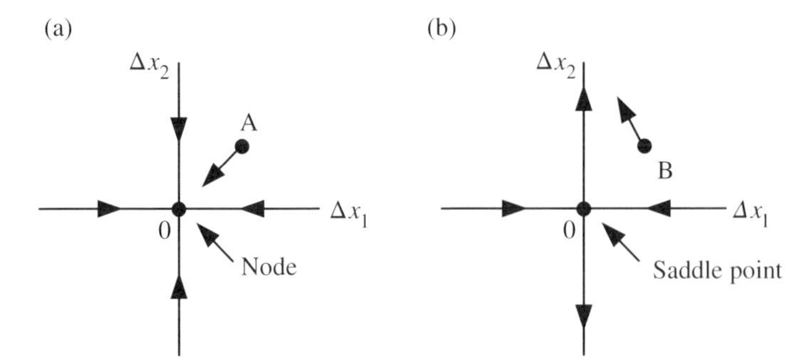

Figure 6.5 Phase plane trajectories showing (a) a node and (b) a saddle point.

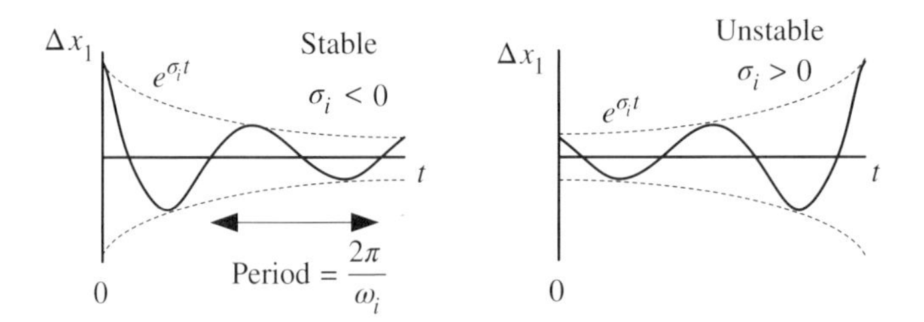

Figure 6.6 Stable and unstable time response due to a pair of complex eigenvalues $\lambda_i = \sigma_i + j\omega_i$ and $\lambda_i^* = \sigma_i - j\omega_i$.

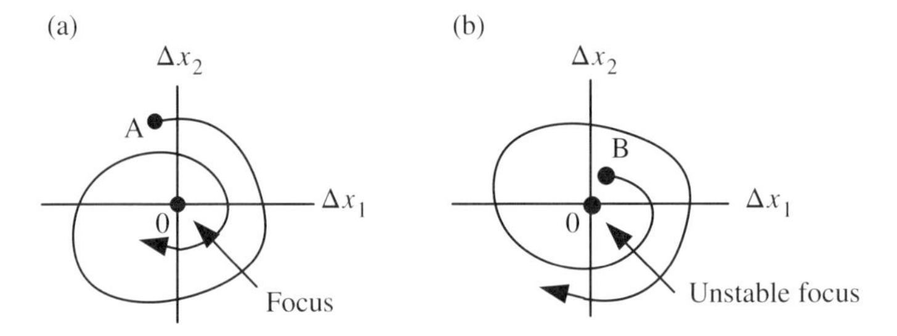

Figure 6.7 Phase plane trajectories showing (a) a focus and (b) an unstable focus.

6.3.5 Participation Factors

The eigenvectors can also be used to find the sensitivity of an eigenvalue to a change of a diagonal element of the state matrix A. Rewriting (6.20) as

$$\Lambda = \begin{bmatrix} w_1^T \\ w_2^T \\ \vdots \\ w_n^T \end{bmatrix} \begin{bmatrix} a_{11} & a_{12} & \cdots & a_{1n} \\ a_{21} & a_{22} & \cdots & a_{2n} \\ \vdots & \vdots & \ddots & \vdots \\ a_{n1} & a_{n2} & \cdots & a_{nn} \end{bmatrix} \begin{bmatrix} v_1 & v_2 & \cdots & v_n \end{bmatrix} \tag{6.32}$$

the sensitivity

$$\frac{\partial \lambda_i}{\partial a_{kk}} = w_{ik} v_{ik} \tag{6.33}$$

where w_{ik} and v_{ik} are the kth component of the left and right eigenvectors w_i and v_i, respectively, is defined as the participation factor of the kth state in the ith mode [86, 87]. All participation factors can be collected into a participation factor matrix as

$$P = \begin{bmatrix} w_{11}v_{11} & w_{12}v_{12} & \cdots & w_{1n}v_{1n} \\ w_{21}v_{21} & w_{22}v_{22} & \cdots & w_{2n}v_{2n} \\ \vdots & \vdots & \ddots & \vdots \\ w_{n1}v_{n1} & w_{n2}v_{n2} & \cdots & w_{nn}v_{nn} \end{bmatrix} \tag{6.34}$$

The participation factor is a good indicator of the importance of the kth state to the ith mode. In power systems, it is useful to assess whether a particular state variable can be useful as an input signal in a control design. This information can be useful for making decisions on the placement of controllers, such as a static var compensator, in a power system.

6.4 Linearized Models of Single-Machine Infinite-Bus Systems

Consider the Single-Machine Infinite-Bus (SMIB) system in Figure 6.8, in which the synchronous machine is delivering electrical power P_e to the infinite bus, which represents a bulk power system. The swing equation in state variable form can be written as

$$\dot{\delta} = \Omega\omega \tag{6.35}$$

$$\begin{aligned} 2H\dot{\omega} &= T_m - T_e - D\omega = T_m - P_e - D\omega \\ &= T_m - \frac{E'V\sin\delta}{X_{\text{eq}}} - D\omega \end{aligned} \tag{6.36}$$

where θ in rad is the generator rotor angle with respect to the infinite bus, ω in pu is the machine speed deviation, $\Omega = 2\pi f$ is the conversion factor from pu speed to rad, and $X_{\text{eq}} = X'_d + X_T + X_L$.

In steady state, the equilibrium satisfies

$$\omega = 0, \quad \frac{E'V\sin\delta_{\text{ep}}}{X_{\text{eq}}} = P_e = T_m \tag{6.37}$$

Following (6.5) and (6.6), the linearization of (6.35) and (6.36) at the equilibrium $\delta = \delta_{\text{ep}}$ and $\omega = 0$ results in

$$\Delta\dot{\delta} = \Omega\Delta\omega \tag{6.38}$$

$$\begin{aligned} 2H\Delta\dot{\omega} &= -\frac{\partial}{\partial\delta}\left(\frac{E'V\sin\delta}{X_{\text{eq}}}\right)\bigg|_o \Delta\delta - \frac{\partial}{\partial\omega}(D\omega)\bigg|_o \Delta\omega + \frac{\partial}{\partial T_m}(T_m)\bigg|_o \Delta T_m \\ &= -K_s\Delta\delta - D\Delta\omega + \Delta T_m \end{aligned} \tag{6.39}$$

where

$$K_s = \frac{E'V\cos\delta_{\text{ep}}}{X_{\text{eq}}} \tag{6.40}$$

is the synchronizing torque coefficient which is positive when $|\delta_{\text{ep}}| < \pi/2$. Note that K_s signifies the ability of the generator to stay synchronized to the infinite bus. A higher

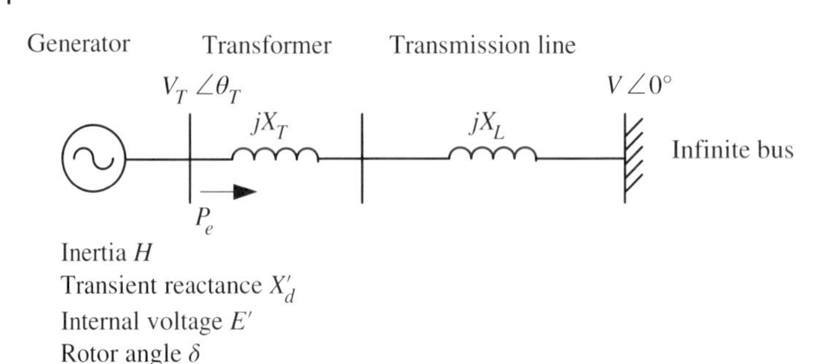

Figure 6.8 SMIB system.

value of K_s implies a stronger connection of the machine to the infinite bus. This term is also a part of the potential energy V_{PE} discussed in Chapter 5.

The linearized swing equation in matrix form is

$$\begin{bmatrix} \Delta\dot{\delta} \\ \Delta\dot{\omega} \end{bmatrix} = \begin{bmatrix} 0 & \Omega \\ -K_s/2H & -D/2H \end{bmatrix} \begin{bmatrix} \Delta\delta \\ \Delta\omega \end{bmatrix} + \begin{bmatrix} 0 \\ 1/2H \end{bmatrix} \Delta T_m \tag{6.41}$$

To find the eigenvalues, the determinant

$$\det\left(\begin{bmatrix} \lambda & 0 \\ 0 & \lambda \end{bmatrix} - \begin{bmatrix} 0 & \Omega \\ -K_s/2H & -D/2H \end{bmatrix} \right) = \det \begin{bmatrix} \lambda & -\Omega \\ K_s/2H & \lambda + D/2H \end{bmatrix} = 0 \tag{6.42}$$

is used to compute characteristic equation, yielding

$$\lambda\left(\lambda + \frac{D}{2H}\right) + \frac{K_s\Omega}{2H} = \lambda^2 + \frac{D}{2H}\lambda + \frac{K_s\Omega}{2H} = 0 \tag{6.43}$$

The eigenvalues are

$$\lambda = \frac{-(D/2H) \pm \sqrt{(D/2H)^2 - 4(K_s\Omega/(2H))}}{2} \tag{6.44}$$

Normally D is small, so that λ is complex and given by

$$\lambda = \frac{1}{4H}\left(-D \pm j\sqrt{8HK_s\Omega - D^2}\right) = \sigma \pm j\omega_s, \quad \sigma = -\frac{D}{4H}, \quad \omega_s \simeq \sqrt{\frac{K_s\Omega}{2H}} \tag{6.45}$$

where the unit of σ is 1/sec, and that of ω_s is rad/sec.

Example 6.1: Eigenvalue analysis of a SMIB system

The SMIB system shown in Figure 6.9 operates at 60 Hz. The classical model is used for the synchronous machine. The machine is rated at 900 MVA with an inertia $H = 3.5$ sec, damping $D = 1$ pu, and transient reactance $X'_d = 0.24$ pu. The step-up transformer reactance is $X_T = 0.15$ pu on the machine base. The high-side of the transformer is connected to an infinite bus with a transmission line reactance $X_L = 0.06$ pu on the system base of 100 MVA. From the power flow solution, the generator terminal bus voltage is $\tilde{V}_T = 1.03\angle 39.25°$ pu and the power supplied by the generator is ($P_e = 8.5$, $Q_e = 3.4338$) pu on the system base. The infinite bus voltage $\tilde{V}$ is $1.0\angle 0°$ pu. The load $P_L + jQ_L$

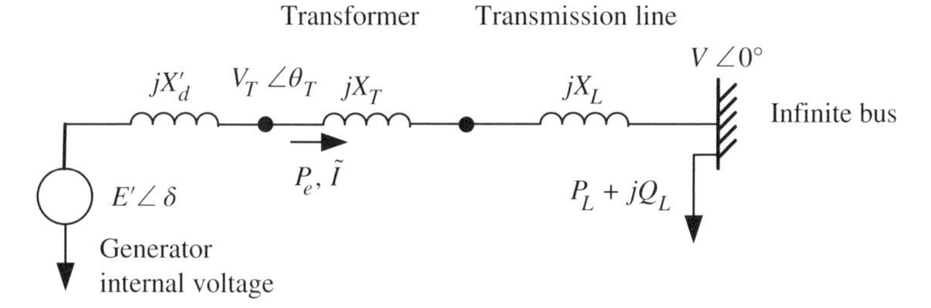

Figure 6.9 SMIB system for Example 6.1.

is located at the infinite bus. Find the linearized model of this system and its electromechanical mode.

Solutions: The computation here will be performed in pu on the system base of 100 MVA. From the generator output power at the terminal bus, the generator current can be computed as

$$\tilde{I} = \frac{(P_e + jQ_e)^*}{V_T} = 8.5 + j2.6395 \text{ pu} \tag{6.46}$$

Thus the machine internal voltage is

$$\begin{aligned} E'\angle\delta = \tilde{V}_T + jX'_d\tilde{I} &= 1.03\angle 39.25° + j(0.24/9)(8.5 + j2.6395) \\ &= 0.7273 + j0.8783 = 1.1140\angle 50.37° \text{ pu} \end{aligned} \tag{6.47}$$

The reactance from the infinite bus to the generator internal node on the system base is

$$X_{\text{eq}} = 0.06 + (0.15 + 0.24) \times 100/900 = 0.1033 \text{ pu} \tag{6.48}$$

Thus the synchronizing torque coefficient is computed as

$$K_s = \frac{1.1403 \times 1.0}{0.1033} \cos 50.37° = 7.0379 \tag{6.49}$$

resulting in the linearized dynamic system

$$\begin{bmatrix} \Delta\dot{\delta} \\ \Delta\dot{\omega} \end{bmatrix} = \begin{bmatrix} 0 & 377 \\ -0.1117 & -0.1429 \end{bmatrix} \begin{bmatrix} \Delta\delta \\ \Delta\omega \end{bmatrix} + \begin{bmatrix} 0 \\ 0.1429 \end{bmatrix} \Delta T_m \tag{6.50}$$

where $\Omega = 2\pi \times 60 = 377$ rad/sec. Note that in (6.50), ΔT_m is in pu on the machine base.

The characteristic equation is

$$\lambda^2 + 0.1429\lambda + 42.115 = 0 \tag{6.51}$$

yielding the complex eigenvalues

$$\lambda_{1,2} = -0.0714 \pm j6.4892 = \sigma \pm j\omega_s \tag{6.52}$$

The oscillatory mode frequency in Hz is $\omega_s/(2\pi) = 1.033$ Hz. This oscillatory frequency of the machine versus the infinite bus is known as the "local" mode. A local mode's frequency typically varies from 1 to 2 Hz.

There are some useful observations on the local-mode frequency ω_s:

1) The frequency ω_s is inversely proportional to the square root of H and X_{eq}.
2) As P_e (loading or power transfer) increases, ω_s decreases.
3) The synchronizing torque K_s is inversely proportional to X_{eq} and decreases as P_e (or δ) increases. Thus as K_s decreases, ω_s also decreases.

The normalized eigenvectors are

$$v_{1,2} = \begin{bmatrix} 0.9999 \\ -0.0002 \pm j0.0172 \end{bmatrix} \simeq \begin{bmatrix} 0.9999 \\ \pm j0.0172 \end{bmatrix} = \begin{bmatrix} v_\delta \\ \pm v_\omega \end{bmatrix} \tag{6.53}$$

where the eigenvector

$$\begin{bmatrix} 0.9999 \\ j0.0172 \end{bmatrix} \tag{6.54}$$

corresponds to the eigenvalue $-0.0714 + j6.4892$.

Performing Laplace transform of (6.38) to obtain

$$s\Delta\delta = \Omega\Delta\omega \tag{6.55}$$

the eigenvector relationship (6.54) can be seen from (6.55) with $s = j\omega_s$

$$v_\omega = \frac{j\omega_s v_\delta}{\Omega} = j\frac{6.4892}{377} \times 0.9999 = j0.0172 \tag{6.56}$$

■

6.5 Linearized Models of Multi-Machine Systems

For a power system with N_G generators, there will be $N_G - 1$ electromechanical modes. The reason that the number of electromechanical modes is one less than the number of generators is because one of the generators serves as the reference. In the SMIB system, the generator representing the infinite bus is the reference.

Following the development in Chapter 5, the swing equation of an N_G-machine power system using the Y matrix reduced to the machine internal node is

$$\dot{\delta}_i = \Omega\omega_i \tag{6.57}$$

$$\dot{\omega}_i = \frac{1}{2H_i}\left(P_{mi} - (E_i')^2 G_{ii} - D_i\omega_i - \sum_{j=1, j\neq i}^{N_G} (C_{ij}\sin(\delta_i - \delta_j) + D_{ij}\cos(\delta_i - \delta_j))\right) \tag{6.58}$$

for $i = 1, 2, \ldots, N_G$.

Linearizing around the equilibrium $\delta_i = \delta_{\text{ep}i}$, $\omega_i = 0$, yields

$$\Delta\dot{\delta}_i = \Omega\Delta\omega_i \tag{6.59}$$

$$\begin{aligned}
\Delta\dot{\omega}_i &= \frac{1}{2H_i}\left(\Delta T_{mi} - D_i\Delta\omega_i - \sum_{j=1,j\neq i}^{N_G} \frac{\partial}{\partial\delta_i}(C_{ij}\sin(\delta_i-\delta_j) + D_{ij}\cos(\delta_i-\delta_j))|_o\Delta\delta_i \right.\\
&\left. - \sum_{j=1,j\neq i}^{N_G} \frac{\partial}{\partial\delta_j}(C_{ij}\sin(\delta_i-\delta_j) + D_{ij}\cos(\delta_i-\delta_j))|_o\Delta\delta_j\right)\\
&= \frac{1}{2H_i}\left(\Delta T_{mi} - D_i\Delta\omega_i - K_{sii}\Delta\delta_i + \sum_{j=1,j\neq i}^{N_G} K_{sij}\Delta\delta_j\right)
\end{aligned} \tag{6.60}$$

where

$$K_{sii} = \sum_{j=1,j\neq i}^{N_G} (C_{ij}\cos(\delta_i-\delta_j) - D_{ij}\sin(\delta_i-\delta_j))|_o \tag{6.61}$$

$$K_{sij} = (C_{ij}\cos(\delta_i-\delta_j) - D_{ij}\sin(\delta_i-\delta_j))|_o \tag{6.62}$$

The state-space form of the linearized multi-machine power system, using the mechanical torque T_m as the input variable, is

$$\begin{bmatrix} \Delta\dot{\delta} \\ \Delta\dot{\omega} \end{bmatrix} = A \begin{bmatrix} \Delta\delta \\ \Delta\omega \end{bmatrix} + B\Delta T_m \tag{6.63}$$

where

$$\Delta\delta = \begin{bmatrix} \Delta\delta_1 \\ \Delta\delta_2 \\ \vdots \\ \Delta\delta_{N_G} \end{bmatrix}, \quad \Delta\omega = \begin{bmatrix} \Delta\omega_1 \\ \Delta\omega_2 \\ \vdots \\ \Delta\omega_{N_G} \end{bmatrix}, \quad \Delta T_m = \begin{bmatrix} \Delta T_{m1} \\ \Delta T_{m2} \\ \vdots \\ \Delta T_{mN_G} \end{bmatrix} \tag{6.64}$$

$$A = \begin{bmatrix} 0 & \Omega I \\ (2H)^{-1}K_s & -(2H)^{-1}D \end{bmatrix}, \quad B = \begin{bmatrix} 0 \\ (2H)^{-1} \end{bmatrix} \tag{6.65}$$

$$H = \begin{bmatrix} H_1 & 0 & \cdots & 0 \\ 0 & H_2 & \cdots & 0 \\ \vdots & \vdots & \ddots & \vdots \\ 0 & 0 & \cdots & H_{N_G} \end{bmatrix} = \mathrm{diag}(H_i) \tag{6.66}$$

$$D = \begin{bmatrix} D_1 & 0 & \cdots & 0 \\ 0 & D_2 & \cdots & 0 \\ \vdots & \vdots & \ddots & \vdots \\ 0 & 0 & \cdots & D_{N_G} \end{bmatrix} = \mathrm{diag}(D_i) \tag{6.67}$$

$$K_s = \begin{bmatrix} -K_{s11} & K_{s12} & \cdots & K_{s1N_G} \\ K_{s21} & -K_{s22} & \cdots & K_{s2N_G} \\ \vdots & \vdots & \ddots & \vdots \\ K_{sN_G1} & K_{sN_G2} & \cdots & -K_{sN_GN_G} \end{bmatrix} \tag{6.68}$$

Note that normally the diagonal entries of K_s are negative and the off-diagonal entries are positive. Sometimes the off-diagonal entries can be a small negative number.

6.5.1 Synchronizing Torque Matrix and Eigenvalue Properties

The synchronizing torque matrix K_s and the eigenvalues of A have a number of interesting properties:

- Each row of K_s sums up to zero, i.e,

$$K_s \begin{bmatrix} 1 \\ 1 \\ \vdots \\ 1 \end{bmatrix} = 0 \tag{6.69}$$

  Thus K_s is singular, meaning that one of the eigenvalues of K_s is zero.
- K_s is also negative semidefinite, that is, given any nonzero vector z of the same dimension as K_s, $z^T K_s z \leq 0$. Thus all the nonzero eigenvalues of K_s are negative.
- Let the eigenvalues of $(2H)^{-1}K_s$ be $0, \lambda_2, \ldots, \lambda_{N_G}$. Then the eigenvalues of A are approximated by

$$0, -2\sigma_1, -\sigma_2 \pm j\sqrt{-\Omega\lambda_2}, \ldots, -\sigma_{N_G} \pm j\sqrt{-\Omega\lambda_{N_G}} \tag{6.70}$$

  where σ_i are small.
- If the damping is uniform, that is, $(2H)^{-1}D = \sigma I$, then

$$\sigma_1 = \sigma_2 = \cdots = \sigma_{N_G} = \sigma \tag{6.71}$$

- The pair of eigenvalues 0 and $-2\sigma_1$ represent the system mode. The zero eigenvalue has the eigenvector

$$v_1 = \begin{bmatrix} \mathbf{1} \\ \mathbf{0} \end{bmatrix}, \quad \text{such that} \quad Av_1 = 0 \tag{6.72}$$

  where $\mathbf{1}$ is a vector of 1s and $\mathbf{0}$ is a vector of 0s. This is the motion of the machine rotor angle moving together as a group. This "drift" can be removed by setting one of the rotor angles as the reference.
- The second eigenvalue of $-2\sigma_1$ has the eigenvector

$$v_2 \simeq \begin{bmatrix} \mathbf{1} \\ -\frac{2\sigma_1}{\Omega}\mathbf{1} \end{bmatrix} \tag{6.73}$$

  This is the system frequency (speed) decay mode due to the overall system generation and load balance.
- The other $(N_G - 1)$ pairs of complex eigenvalues represent the oscillatory modes between the N_G machines. As discussed in Section 6.2, these modes can be one of three types: interarea modes in the range of 0.2 to 1 Hz, local modes of 1 to 2 Hz, and intraplant modes of 2 to 3 Hz.

6.5.2 Modeshapes and Participation Factors

There are additional insights that can be obtained from a modal analysis. In addition to eigenvalues, the eigenvectors can also be used to find out the modal contents in the states. How the right eigenvectors, also known as modeshapes, relate the state variables and the system eigenvalues can be seen by writing (6.31) as follows:

$$\begin{bmatrix} \Delta x_1(t) \\ \Delta x_2(t) \\ \vdots \\ \Delta x_n(t) \end{bmatrix} = c_1 \begin{bmatrix} v_{11}(t) \\ v_{21}(t) \\ \vdots \\ v_{n1}(t) \end{bmatrix} e^{\lambda_1 t} + c_2 \begin{bmatrix} v_{12}(t) \\ v_{22}(t) \\ \vdots \\ v_{n2}(t) \end{bmatrix} e^{\lambda_2 t} + \cdots + c_n \begin{bmatrix} v_{1n}(t) \\ v_{2n}(t) \\ \vdots \\ v_{nn}(t) \end{bmatrix} e^{\lambda_n t} \tag{6.74}$$

where $c_i = w_i^T \Delta x_o$, $i = 1, 2, \ldots, n$.

This equation shows how the relative magnitudes of the terms in each eigenvector provide a measure of the content of a given mode in the state variables. For instance, if the terms v_{ki}, v_{pi}, and v_{qi} of the ith eigenvector are significantly larger than the other terms in the eigenvector, then the ith eigenvalue would be mostly related to states k, p, and q. Conversely, if these terms were zero, then the dynamics associated with the ith eigenvalue would not be present in states k, p, and q. In other words, the $e^{\lambda_i t}$ term from (6.31) would not exist in the solution for states k, p, and q. The following example shows the application of these concepts.

Example 6.2: 2-area 4-machine system

The 2-area, 4-machine system in Example 2.3 is used to illustrate the concept of interarea and local modes. For this example, the inertias and power output of Generator 11 are decreased by 25% and those of Generator 12 are increased by 25%.[4] The resulting A matrix of the system is given by

$$A = \begin{bmatrix} 0 & \Omega I \\ (2H)^{-1}K_s & -(2H)^{-1}D \end{bmatrix} = \begin{bmatrix} A_{11} & A_{12} \\ A_{21} & A_{22} \end{bmatrix} \tag{6.75}$$

in which the ordering of the states is δ_1, δ_2, δ_{11}, δ_{12}, ω_1, ω_2, ω_{11}, ω_{12}, and $\Omega = 2\pi \times 60$. The matrices A_{ij} are all of dimension of 4×4. The matrices A_{21} and A_{22} computed from PST are

$$A_{21} = \begin{bmatrix} -7.3742 & 5.8736 & 0.5366 & 0.9641 \\ 6.4539 & -9.1792 & 0.9559 & 1.7704 \\ 0.8671 & 1.1622 & -9.7299 & 7.7006 \\ 1.0959 & 1.4943 & 4.8515 & -7.4417 \end{bmatrix} \times 10^{-2} \tag{6.76}$$

$$A_{22} = \text{diag}(-0.1539, -0.1539, -0.1539, -0.1539) \tag{6.77}$$

Set up the matrix A in (6.76) and compute the electromechanical modes and their mode-shapes.

Solutions: Using MATLAB®, the eigenvalues are found to be 0.0013, −0.1552, $-0.0769 \pm j4.0635$, $-0.0769 \pm j7.3957$, and $-0.0769 \pm j7.4778$. The eigenvectors for the two real eigenvalues 0.0013 and −0.1552 are,[5] respectively,

4 The data can be found in Appendix 2C. These changes are made so that the two areas are not replicates of each other, which will give rise to some resonance conditions in the local modes.

5 Note that the first eigenvalue is not exactly zero because a numerical linearization method is used to obtain the linearized model (see Section 6.6.2).

$$v_1 = \begin{bmatrix} 0.5 \\ 0.5 \\ 0.5 \\ 0.5 \\ 0 \\ 0 \\ 0 \\ 0 \end{bmatrix}, \quad v_2 = \begin{bmatrix} -0.5000 \\ -0.5000 \\ -0.5000 \\ -0.5000 \\ 0.0002 \\ 0.0002 \\ 0.0002 \\ 0.0002 \end{bmatrix} \tag{6.78}$$

Note that MATLAB® normalizes the eigenvalue to unit amplitude. For v_1, only the machine angles are involved, and their responses are in unison. In v_2, the machine speeds are also involved. Their amplitude is much smaller because of the pu frequency (or speed) to rad conversion factor $\Omega = 2\pi f$.

The electromechanical mode frequencies are 4.0635 rad/sec = 0.6467 Hz, 7.3957 rad/sec = 1.1771 Hz, and 7.4778 rad/sec = 1.1901 Hz. The complex eigenvector of the interarea mode 0.6467 Hz is

$$v_{3,4} = \begin{bmatrix} -0.5187 \pm j0.0000 \\ -0.3998 \pm j0.0000 \\ 0.5603 \\ 0.5078 \pm j0.0000 \\ 0.0001 \mp j0.0056 \\ 0.0001 \mp j0.0043 \\ -0.0001 \pm j0.0060 \\ -0.0001 \pm j0.0054 \end{bmatrix} \tag{6.79}$$

From the sign of the real components, machine angles δ_1 and δ_2 are swinging together against δ_{11} and δ_{12}. Note that these eigenvectors are also normalized with the largest entry being a real entry. This eigenvector is plotted in Figure 6.10, showing the components of the individual rotor angle direction of swings.

The eignevectors of the local modes 1.1771 Hz and 1.1901 Hz are, respectively,

$$v_{5,6} = \begin{bmatrix} -0.6004 \pm j0.0000 \\ 0.7335 \\ 0.2876 \pm j0.0000 \\ -0.1354 \mp j0.0000 \\ -0.0001 \pm j0.0118 \\ 0.0001 \mp j0.0144 \\ -0.0001 \pm j0.0056 \\ 0.0000 \mp j0.0027 \end{bmatrix}, \quad v_{7,8} = \begin{bmatrix} 0.1481 \mp j0.0000 \\ -0.2044 \pm j0.0000 \\ -0.7998 \\ 0.4427 \pm j0.0000 \\ -0.0000 \pm j0.0029 \\ -0.0002 \mp j0.0041 \\ 0.0002 \mp j0.0159 \\ -0.0001 \pm j0.0109 \end{bmatrix} \tag{6.80}$$

Note that $v_{5,6}$ is mainly the oscillation of δ_1 against δ_2, and $v_{7,8}$ is mainly the oscillation of δ_{11} against δ_{12}. These modeshapes are plotted in Figure 6.11. ■

6.6 Developing Linearized Models of Large Power Systems

There are two methods commonly used for generating linearized models: (1) using analytical expressions such as (6.38) and (6.39), and (2) using a numerical perturbation method. They are discussed in the following sections.

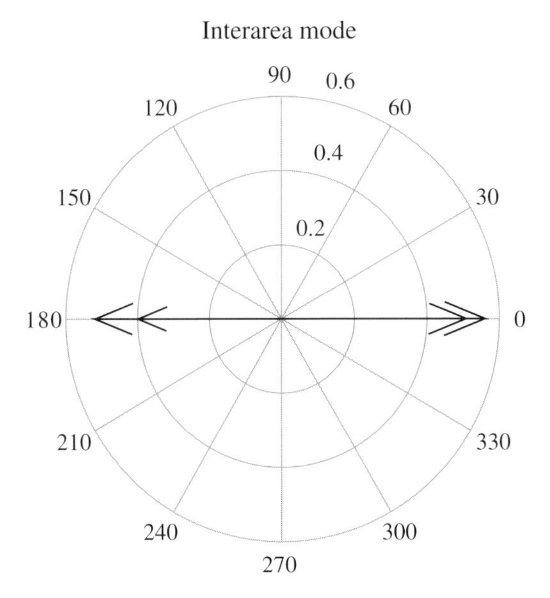

Figure 6.10 Interarea mode for Example 6.2.

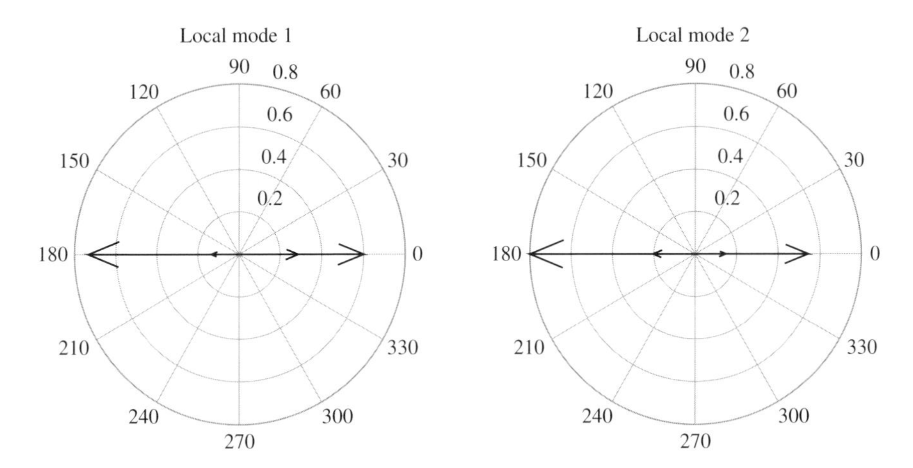

Figure 6.11 Local modes for Example 6.2.

6.6.1 Analytical Partial Derivatives

In this approach, analytical partial derivatives of the nonlinear dynamic equations are used to populate the (A, B, C, D) matrices of the linearized model (6.7) [21]. This is perhaps the most efficient approach in developing the linear model of a power system. Commercial software packages that use this technique include the Small Signal Analysis Tool from Powertech, and the PacDyn Software from CEPEL [88].

A nonlinear dynamic model can be described by the following three equations

$$\dot{x} = f(x, z, u), \quad 0 = h(x, z, u), \quad y = g(x, z, u) \tag{6.81}$$

where $f \in R^n, h \in R^q, g \in R^m$, the vector $x \in R^n$ consists of the state variables, the vector $z \in R^q$ consists of the algebraic network variables, the vector $u \in R^r$ consists of the input variables, the vector $y \in R^m$ consists of the output variables, and (f, g, h) are nonlinear functions of appropriate dimensions. In this framework, the network representation is kept.

For power systems, the states in x include machine rotor angles and speeds, and the function f includes the dynamic equations. The variables in z are the algebraic quantities such as the voltage magnitude and angle at a load bus, and the function h is essentially the network equations. For example, if the loads are assumed to be of constant impedance, then h reduces to the linear network equation $Y\tilde{V} = \tilde{I}$.[6] The input u consists of input points to actuators such as turbine control valves, voltage regulators, and capacitance settings for static var systems. The actuator states would be part of x and their dynamics would be part of f. The output variables can be a network variable, such as the current magnitude and the bus voltage magnitude.

At the steady-state operating point, the equilibrium point $(x_{ep}, z_{ep}, u_{ep}, y_{ep})$ of (x, z, u, y) satisfies

$$0 = f(x_{ep}, z_{ep}, u_{ep}), \quad 0 = h(x_{ep}, z_{ep}, u_{ep}), \quad y_{ep} = g(x_{ep}, z_{ep}, u_{ep}) \tag{6.82}$$

Computing the partial derivatives of the system at the equilibrium point, the linearized model of (6.81) becomes

$$\begin{bmatrix} \Delta\dot{x} \\ 0 \\ \Delta y \end{bmatrix} = \begin{bmatrix} A_{11} & A_{12} & B_1 \\ A_{21} & A_{22} & B_2 \\ C_1 & C_2 & D \end{bmatrix} \begin{bmatrix} \Delta x \\ \Delta z \\ \Delta u \end{bmatrix} \tag{6.83}$$

where

$$A_{11} = \frac{\partial f}{\partial x}, \quad A_{12} = \frac{\partial f}{\partial z}, \quad A_{21} = \frac{\partial h}{\partial x}, \quad A_{22} = \frac{\partial h}{\partial z} \tag{6.84}$$

$$B_1 = \frac{\partial f}{\partial u}, \quad B_2 = \frac{\partial h}{\partial u}, \quad C_1 = \frac{\partial g}{\partial x}, \quad C_2 = \frac{\partial g}{\partial z}, \quad D = \frac{\partial g}{\partial u} \tag{6.85}$$

There are some special structures for the state-space model (6.83):

- The matrix A_{11} is block-diagonal. For example, for a 3-machine system, A_{11} has the structure

$$A_{11} = \begin{bmatrix} A_{11}^{(1)} & 0 & 0 \\ 0 & A_{11}^{(2)} & 0 \\ 0 & 0 & A_{11}^{(3)} \end{bmatrix} \tag{6.86}$$

  in which one block is developed for each machine. These machines will interact with the rest of the system via the entries in the other A_{ij} matrices.
- The A_{ij} matrices are sparse, reflecting the characteristics of a power network.

The linearized model contains both differential and algebraic variables. There are efficient eigenvalue methods that can compute the eigenvalues of such differential-algebraic equations (DAE), taking advantage of sparsity. For large systems, one may only need to

6 This would not be the case for nonconforming loads (constant current and constant power), which will be discussed in Chapter 11.

compute the dominant eigenvalues of the power system, using techniques such as the Arnoldi method [89] and the Lanczo's method [90].

Alternatively one can eliminate the Δz variables by expressing it as

$$\Delta z = -A_{22}^{-1}(A_{21}\Delta x + B_2\Delta u) \tag{6.87}$$

such that the system model with only the states Δx and the output variables Δy becomes

$$\Delta \dot{x} = (A_{11} - A_{12}A_{22}^{-1}A_{21})\Delta x + (B_1 - A_{12}A_{22}^{-1}B_2)\Delta u = A\Delta x + B\Delta u \tag{6.88}$$

$$\Delta y = (C_1 - C_2A_{22}^{-1}A_{21})\Delta x + (D - C_2A_{22}^{-1}B_2)\Delta u = C\Delta x + \overline{D}\Delta u \tag{6.89}$$

Note that A is now a dense matrix.

This methodology is illustrated in the next example using a SMIB system.

Example 6.3

Use the analytical partial derivative approach to develop the linearized model of the SMIB system in Example 6.1, with the terminal bus voltage magnitude V_T as the output variable.

Solutions: In the example, the voltage magnitude at the infinite bus is denoted by V_{inf}. With an assumption of an infinite inertia, there is no need to include the swing equation for the infinite bus. The swing equations for the synchronous machine are written as

$$\dot{\delta} = \Omega\omega, \quad 2H\dot{\omega} = T_m - P_e - D\omega \tag{6.90}$$

where

$$P_e = \text{Re}\{(E'\angle\delta)(I\angle\phi)^*\} = E'_{\text{re}}I_{\text{re}} + E'_{\text{im}}I_{\text{im}} \tag{6.91}$$

and the real and imaginary parts of the generator voltage and current are $E'_{\text{re}} = E'\cos\delta$, $E'_{\text{im}} = E'\sin\delta$, $I_{\text{re}} = I\cos\phi$, and $I_{\text{im}} = I\sin\phi$, where the angle at the infinite bus is used as the reference. The state variables are $\Delta x_1 = \Delta\delta$ and $\Delta x_2 = \Delta\omega$ and the algebraic variables are $\Delta z_1 = \Delta I_{\text{re}}$ and $\Delta z_2 = \Delta I_{\text{im}}$. Hence the dynamic equations are

$$f_1 = \Omega\omega, \quad f_2 = \frac{1}{2H}\left(T_m - D\omega - I_{\text{re}}E'\cos\delta - I_{\text{im}}E'\sin\delta\right) \tag{6.92}$$

Next, the network equation $Y_{\text{red}}\tilde{E}_G = \tilde{I}_G$ is used to compute the generator current, where Y_{red} is the admittance matrix reduced to the generator internal nodes, $\tilde{E}_G$ is the generator internal voltage, and $\tilde{I}_G$ is the generator current. For the synchronous generator, the output current is given by

$$\frac{1}{jX_{\text{eq}}}((E'\angle\delta) - (V_{\text{inf}}\angle 0^\circ)) = \tilde{I}_G = I\angle\phi \tag{6.93}$$

which can be expressed by two real equations

$$h_1 = I_{\text{re}} - \frac{E'_{\text{im}}}{X_{\text{eq}}} = I_{\text{re}} - \frac{E'\sin\delta}{X_{\text{eq}}}, \quad h_2 = I_{\text{im}} + \frac{E'_{\text{re}} - V_{\text{inf}}}{X_{\text{eq}}} = I_{\text{im}} + \frac{E'\cos\delta - V_{\text{inf}}}{X_{\text{eq}}} \tag{6.94}$$

The terminal bus voltage magnitude can be computed in a number of ways. Here the following relationship is used:

$$\tilde{V}_T = (E' \angle \delta) - jX'_d I \angle \phi \tag{6.95}$$

from which the voltage magnitude can be obtained as

$$V_T = g = \sqrt{(E' \cos \delta + X'_d I_{\text{im}})^2 + (E' \sin \delta - X'_d I_{\text{re}})^2} \tag{6.96}$$

Taking the partial derivatives of these expressions, not including zero terms, the terms in the A_{11} matrix are

$$\frac{\partial f_1}{\partial \omega} = \Omega, \quad \frac{\partial f_2}{\partial \delta} = \frac{I_{\text{re}} E' \sin \delta - I_{\text{im}} E' \cos \delta}{2H}, \quad \frac{\partial f_2}{\partial \omega} = -\frac{D}{2H} \tag{6.97}$$

the terms in the A_{12} matrix are

$$\frac{\partial f_2}{\partial I_{\text{re}}} = -\frac{E' \cos \delta}{2H}, \quad \frac{\partial f_2}{\partial I_{\text{im}}} = -\frac{E' \sin \delta}{2H} \tag{6.98}$$

the terms in the A_{21} matrix are

$$\frac{\partial h_1}{\partial \delta} = -\frac{E' \cos \delta}{X_{eq}}, \quad \frac{\partial h_2}{\partial \delta} = -\frac{E' \sin \delta}{X_{eq}} \tag{6.99}$$

the terms in the A_{22} matrix are

$$\frac{\partial h_1}{\partial I_{\text{re}}} = 1, \quad \frac{\partial h_2}{\partial I_{\text{im}}} = 1 \tag{6.100}$$

and the term in the B_1 matrix is

$$\frac{\partial f_2}{\partial T_m} = \frac{1}{2H} \tag{6.101}$$

The output variable in the linearized model is ΔV_T. The term in the C_1 matrix is

$$\frac{\partial g}{\partial \delta} = -\frac{X'_d (I_{\text{re}} E' \cos \delta + I_{\text{im}} E' \sin \delta)}{V_T} \tag{6.102}$$

and the terms in the C_2 matrix are

$$\frac{\partial g}{\partial I_{\text{re}}} = -\frac{(E' \sin \delta - X'_d I_{\text{re}}) X'_d}{V_T}, \quad \frac{\partial g}{\partial I_{\text{im}}} = \frac{(E' \cos \delta + X'_d I_{\text{im}}) X'_d}{V_T} \tag{6.103}$$

Substituting the equilibrium values into these expressions and eliminating I_{re} and I_{im}, one can obtain the linearized model in Example 6.1. This is left as an exercise. ■

Note that in Example 6.3, it is also possible to use the polar coordinate form of I and ϕ as the algebraic variables, instead of the rectangular coordinate form I_{re} and I_{im}. However, this approach would require additional manipulation of the network power flow equations. With rectangular coordinates, the Y matrix can be used as is. Note also that when there are many buses and generators, (6.94) will be in matrix form.

6.6.2 Numerical Linearization

An alternative to the analytical partial derivative method is to evaluate the partial derivatives numerically by perturbing the state and input variables from their equilibrium values. The discussion in Section 6.6.1 can be readily applied to numerical partial derivative evaluations. For example, the partial derivative $\partial f/\partial x$ (6.84) for x_i would be approximated by

$$\begin{aligned}\frac{\partial f}{\partial x_i} &\simeq \frac{f(x_{\rm ep} + \Delta x^{(i)}, z_{\rm ep}, u_{\rm ep}) - f(x_{\rm ep}, z_{\rm ep}, u_{\rm ep})}{\Delta x_i} \\ &= \frac{f(x_{\rm ep} + \Delta x^{(i)}, z_{\rm ep}, u_{\rm ep})}{\Delta x_i} = i\text{th column of } A_{11}\end{aligned} \tag{6.104}$$

where $\Delta x^{(i)}$ is a zero vector except for the ith component, which is the small perturbation Δx_i. Typically a 1% perturbation of the equilibrium value can be considered small. If the equilibrium value is zero, then a value of 0.01 could be used. For improved accuracy, a small negative perturbation $-\Delta x_i$ can also be performed. Then the average of the positive and negative perturbations can be used to obtain a more accurate value of the partial derivative.

The complete set of partial derivatives in (6.84) and (6.85) can be obtained by perturbing all the components in x, z, and u sequentially, one component at a time.

Another method of numerical linearization is to take advantage of the structure of a dynamic simulation program to obtain the model (6.88)–(6.89) directly. The idea is to use the network solution computer code and compute the impact on the network variables due to perturbations in the state x and the input u. Starting from the $0 = h(x, z, u)$ equation in (6.81), perturb the ith component of x by a small positive quantity $\Delta x^{(i)}$ in the ith entry of x, so that $x = x_{\rm ep} + \Delta x^{(i)}$. As a result, z is also perturbed to $z = z_{\rm ep} + \Delta z^{(i)}$, satisfying

$$0 = h(x_{\rm ep} + \Delta x^{(i)}, z_{\rm ep} + \Delta z^{(i)}, u_{\rm ep}) \tag{6.105}$$

Note that although $\Delta x^{(i)}$ denotes a perturbation only in the ith entry of x, $\Delta z^{(i)}$ is a result of that perturbation with changes possibly in all entries of z. For a power network, this part is the network solution for the perturbed variables. It is linear if the constant-impedance load model is used, and is thus straightforward to compute.

Then the ith column of the state matrix A can be computed as

$$\begin{aligned}\frac{\partial f}{\partial x_i} &\simeq \frac{f(x_{\rm ep} + \Delta x^{(i)}, z_{\rm ep} + \Delta z^{(i)}, u_{\rm ep}) - f(x_{\rm ep}, z_{\rm ep}, u_{\rm ep})}{\Delta x_i} \\ &= \frac{f(x_{\rm ep} + \Delta x^{(i)}, z_{\rm ep} + \Delta z^{(i)}, u_{\rm ep})}{\Delta x_i} = i\text{th column of } A\end{aligned} \tag{6.106}$$

In addition, suppose that the output vector y is a function of $\Delta x^{(i)}$ and $\Delta z^{(i)}$. Then the ith column of the linearized output matrix C is

$$\begin{aligned}\frac{\partial g}{\partial x_i} &\simeq \frac{g(x_{\rm ep} + \Delta x^{(i)}, z_{\rm ep} + \Delta z^{(i)}, u_{\rm ep}) - g(x_{\rm ep}, z_{\rm ep}, u_{\rm ep})}{\Delta x_i} \\ &= \frac{g(x_{\rm ep} + \Delta x^{(i)}, z_{\rm ep} + \Delta z^{(i)}, u_{\rm ep})}{\Delta x_i} = i\text{th column of } C\end{aligned} \tag{6.107}$$

The same process can be applied to the input u to obtain the input B matrix and, if applicable, the throughput D matrix, using the following expressions for the jth input

$$\frac{\partial f}{\partial u_j} \simeq \frac{f(x_{ep}, z_{ep} + \Delta z^{(i)}, u_{ep} + \Delta u^{(j)}) - f(x_{ep}, z_{ep}, u_{ep})}{\Delta u_j}$$
$$= \frac{f(x_{eq}, z_{ep} + \Delta z^{(i)}, u_{ep} + \Delta u^{(j)})}{\Delta u_j} = j\text{th column of } B \tag{6.108}$$

$$\frac{\partial f}{\partial u_j} \simeq \frac{g(x_{ep}, z_{ep} + \Delta z^{(i)}, u_{ep} + \Delta u^{(j)}) - g(x_{ep}, z_{ep}, u_{ep})}{\Delta u_j}$$
$$= \frac{g(x_{ep}, z_{ep} + \Delta z^{(i)}, u_{ep} + \Delta u^{(j)})}{\Delta u_j} = j\text{th column of } D \tag{6.109}$$

Note that this method produces the model (6.88)–(6.89) directly, as Δz is automatically taken into account and eliminated.

Procedure for Numerical Linearization

A procedure for computing linearized models using the perturbation method is formulated with the following steps:

1) Solve for the power flow solution and initialize the states and the other necessary variables.
2) Perturb sequentially each of the states and then each of the inputs.
3) Compute $\Delta y^{(i)}$ from the network solution code. These values can be used to compute the C and D matrices if the output variables are from the power network.
4) Use the state variable derivative code to compute columns of the A and B matrices.
5) Go back to Step 2 to perturb the next state or input variable.

The advantage of this numerical linearization method is that the initialization code (Step 1), the network solution code (Step 3), and the derivative code (Step 4) are already in a power system simulation program. The additional code needed is to develop a for-loop for Step 2 and Step 5 to sequence the perturbations. This is the method used in the Power System Toolbox.

Although we have only used classical models for synchronous machines, both the analytical and numerical linearization methods can be applied to detailed machine models equipped with excitation systems and turbine-governors.

Example 6.4

Use the numerical linearization method to compute the state matrix A and the input matrix B of the SMIB model of Example 6.1.

Solutions: The equilibrium values of the states are $\delta_{ep} = 0.8792$ rad and $\omega_{ep} = 0$ pu. Thus the perturbation values are set at $\Delta\delta = 0.008792$ rad and $\Delta\omega = 0.01$ pu. Using numerical derivatives, the entries of the A matrix are

$$a_{11} = \frac{\Omega\omega_{ep}}{\Delta\delta} = 0, \quad a_{12} = \frac{\Omega\Delta\delta}{\Delta\delta} = \Omega \tag{6.110}$$

$$a_{21} = \frac{1}{2 \times 31.5} \frac{8.5 - \frac{1.1403 \times 1}{0.1033} \sin(0.8792 + \Delta\delta) - 9\omega_{ep}}{\Delta\delta} = -0.1111 \tag{6.111}$$

$$a_{22} = \frac{1}{2 \times 31.5} \frac{-9\Delta\omega}{\Delta\omega} = -0.1429 \tag{6.112}$$

which match the entries of the A matrix in Example 6.1, except for a small error in the a_{21} entry (which is found to be -0.1117 in Example 6.1) because of the sine nonlinearity.

The B matrix from numerical linearization is identical to the one in Example 6.1 because the mechanical power appears linearly in the swing equation (6.36). ■

6.7 Summary and Notes

This chapter develops linear models of power systems to facilitate the analysis of the stability of the electromechanical modes. Concepts such as small-signal stability, local and interarea modes, synchronizing torque, modeshapes, and participation factors are introduced. Linearized models are used in later chapters for designing damping controllers.

A numerical linearization method is described in this chapter. Analytical linearization using partial derivatives and network equations can be found in [21, 3], and [91]. The numerical linearization method is also valid when detailed machine models and control systems are included. This aspect is demonstrated in later chapters, with linear models having higher dimensional state variables to accommodate additional machine dynamics and control systems.

Modeshapes can be used for a variety of system dynamic analysis. For example, [92] describes the application of modeshapes to analyze modal interactions in the US Western power grid. Furthermore, modeshapes are used in Chapter 16 for determining slow coherent machines in large power systems.

Problems

6.1 (Analytical linearization) Use the analytical expressions in Example 6.3 to compute the linearized system matrix for Example 6.1. Compare the results to (6.53) and comment on the accuracy of the numerical linearization method.

6.2 (Numerical linearization) Obtain the system and input matrices of the linearized model for Example 6.1 using the numerical linearization method. Use $\Delta\delta = 0.01$ rad and $\Delta\omega = 0.01$ pu as the perturbations.

6.3 (Local mode) A SMIB model shown in Figure 6.12 with $X_L = 0.02$ pu on 100 MVA base. The base for the generator is 600 MVA. The transformer reactance is $X_T = 0.15$ pu, the generator transient reactance is $X'_d = 0.24$ pu, the machine inertia is $H = 3.5$ sec, and the damping coefficient is $D = 1.0$ pu, all on the generator base. The system frequency is 60 Hz. The load is $P_L = 550$ MW and $Q_L = 70$ MVar. The infinite-bus voltage is 1.0 pu. Develop the nonlinear electromechanical model on the system base of 100 MVA.

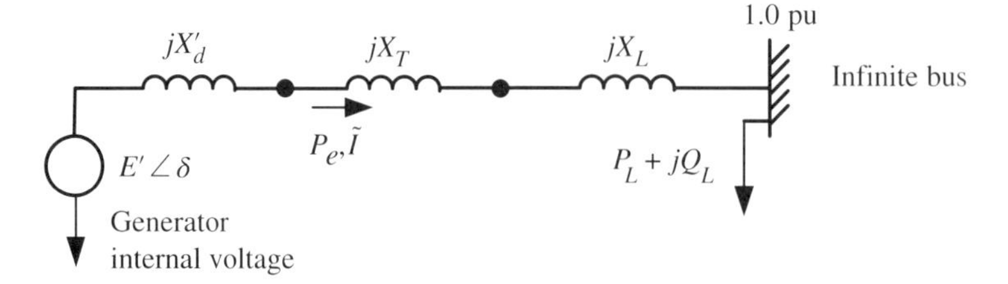

Figure 6.12 SMIB system for Problem 6.2.

1) Develop the linear model and calculate the local mode.
2) Repeat Part 1 for $P_L = 400$ MW.
3) Repeat Part 1 for $H = 5.0$ sec.
4) Comment on the variation of the local mode frequency.

6.4 (Intra-plant mode) The example power system in Problem 6.2 is expanded into the system shown in Figure 6.13, in which the 600 MVA machine is split into two identical 300 MVA machines with equal amount of generation. Use the linearization program *svm_mgen* from PST to develop the linearized model and find the oscillatory modes in the system. Note that you have to make up the power flow and dynamic data in PST format for this system. Extract the synchronizing torque coefficient matrix and find its eigenvalues and eigenvectors. What are the mode shapes for the oscillatory modes? Use the MATLAB© function *compass* to plot the modeshapes. The mode in which the two machines oscillate against each other is the intra-plant mode, and the one in which they oscillate together against the infinite bus is the local mode in Problem 6.2. Note that you need to use the generator terminal bus voltage from Problem 6.2 as the desired voltages on the two generator buses in the expanded system in the bus data. Compare the local mode of this expanded system to that of the original system.

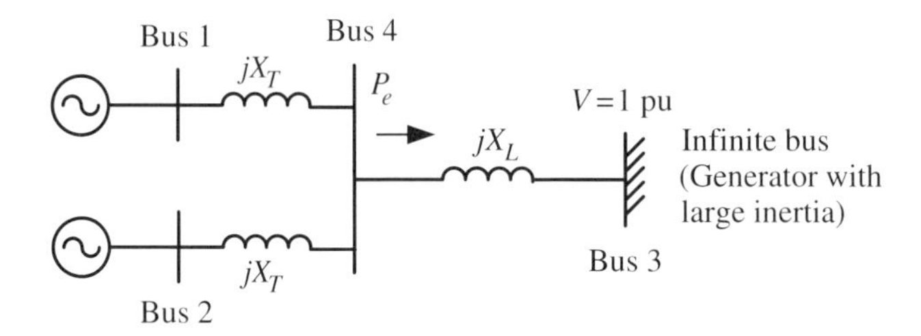

Figure 6.13 System with intra-plant mode for Problem 6.4.

6.5 (Synchronizing torque matrix) Find the synchronized torque matrix K_s for Example 6.2 using the A_{21} matrix. Show that the row sums of K_s are zero.

6.6 (Interarea mode)

1) Obtain the linearized model of the 2-area, 4-machine system in the data file *data2aemh.m*, using only the electromechanical model data and the

constant-impedance load model. Note that in this data set, the inertias of the machines in Area 2 are changed so that the machine and network parameters are not duplications of those in Area 1. In this way a resonance condition in which the local modes of these two areas interact is avoided. Use the PST function *svm_mgen* to obtain the linearized model and compute the oscillatory modes. Obtain the synchronizing torque coefficient matrix and find its eigenvalues and eigenvectors. Show the mode shapes for the oscillatory modes.

2) Increase the power transfer from Area 1 to Area 2 by changing the load at Bus 4 to $11.76 + j1.0$ pu and the load at Bus 14 to $15.67 + j1.0$ pu. Calculate the electromechanical modes and compare them with those in Part 1.

6.7 The swing equations of a system with two machines are given by

$$\dot{\delta}_1 = \Omega\omega_1, \quad \dot{\omega}_1 = \frac{1}{2H_1}(P_{m1} - 4\sin(\delta_1 - \delta_2) - D_1\omega_1)$$
$$\dot{\delta}_2 = \Omega\omega_2, \quad \dot{\omega}_2 = \frac{1}{2H_2}(P_{m2} - 4\sin(\delta_2 - \delta_1) - D_2\omega_2) \quad (6.113)$$

where $\Omega = 2\pi \times 60$, $H_1 = 3$ sec, $H_2 = 6$ sec, $D_1 = 3$ pu, and $D_2 = 6$ pu. In steady state, the machine angles are given by $\delta_1 = 0°$ and $\delta_2 = 60°$.

1) Develop the linearized model of the swing equations.
2) Compute the modes and mode shapes of the linearize model. What is the impact of the inertias on the mode shape?

Part II

Synchronous Machine Models and their Control Systems

7

Steady-State Models and Operation of Synchronous Machines

7.1 Introduction

In the discussion on single-machine and multi-machine power systems in Chapters 4 to 6, the electromechanical model is used for synchronous machines to provide a relatively simple treatment of the transient stability and linearization concepts through analysis and simulation. In Chapters 7 through 12, detailed models of the synchronous machines and their control equipment, and load models are introduced. With this approach, each control system can be added to a power system model and studied accordingly, using linear analysis and nonlinear simulation.

The synchronous machine models are discussed in this and the next chapter. This chapter describes steady-state modeling and operation of synchronous machines. The next chapter describes dynamic models for stability analysis.

The materials covered in this chapter include:

- synchronous machine concepts based on Faraday's Law with the DC field current on a rotor interacting with the stator sinusoidal AC voltages and currents
- two-axis representation using the $dq0$ (Park) transformation, allowing the sinusoidal stator voltages and currents to become fixed variables on a reference frame that rotates at the same speed as the rotor
- equal, reciprocal per-unit system useful for developing simplified equivalent circuits
- equivalent circuits to represent flux linkages and voltages in the dq-axes
- steady-state operation
- saturation effect on the field voltage.

There are a number of excellent texts on synchronous machines [6, 93]. The discourse in this chapter follows the materials and organization in [94].

7.2 Physical Description

A synchronous machine consists of a rotor generating a constant-speed rotating magnetic field inducing alternating currents in the windings on the stator. The magnetic field on the rotor is generated by the field winding controlled by the field voltage. The schematic of a rotor with two poles and a stator with three stator windings, separated equally by 120°, is shown in Figure 7.1, with the rotor turning in the counter-clockwise direction at an angular velocity of ω_r.

Power System Modeling, Computation, and Control, First Edition. Joe H. Chow and Juan J. Sanchez-Gasca.

Companion website: www.wiley.com/go/chow/power-system-modeling

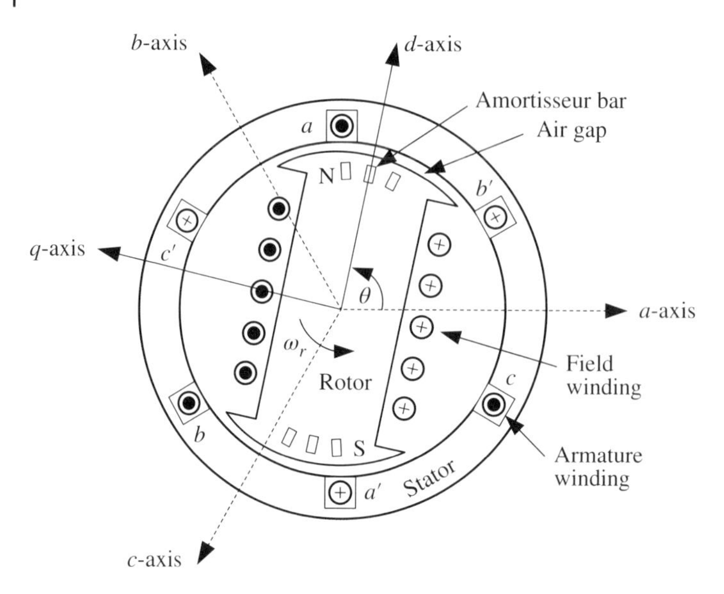

Figure 7.1 Schematic of a 3-phase synchronous machine.

The direction of the rotor magnetic field defines the direct- or d-axis. The quadrature- or q-axis is perpendicular to the d-axis and rotate in the direction of the rotor. The three windings of stator circuits are called phase a, phase b, and phase c. The axes of the phases are defined with respect to the d-axis, as indicated in Figure 7.1. The relative positions of the three stator windings are such that a 120° counterclockwise rotation after the d-axis is aligned with phase a, the d-axis will be aligned with phase b, and then another 120° counterclockwise rotation later, the d-axis will be aligned with phase c.

Synchronous machines are mainly of two types depending on the construction of the rotor. Round-rotor machines have a solid iron rotor with a uniform air gap between the rotor and the stator (Figure 7.2). These machines normally have two poles, and are primarily intended for high-speed synchronous machines (3600 or 3000 rev/min) such as steam turbines. Salient-pole machines have a non-uniform air gap between the rotor and stator, and normally have the number of poles $p > 2$, where p is an even number (Figure 7.3). These machines are primarily used with lower speed hydraulic turbines. Note that for machines with more than two poles, the rotor will always have alternating north and south polefaces.

A power system can readily operate with synchronous machines consisting of different numbers of poles. A common approach for accommodating the calculation of field fluxes is to define an electrical angle θ. The mechanical angle θ_m is the physical angle of the d-axis of a pole pair made with an arbitrary fixed reference axis and goes through 360° every time the d-axis of that pole-pair rotates through that reference axis. For a two-pole machine rotating at 3600 rev/min for a 60-Hz system, $\theta = \theta_m$. For a p-pole machine with the rotor rotating at $3600 \times 2/p$ rev/min, with p a multiple of 2, the relationship between the electrical and mechanical angles is

$$\theta = \frac{p}{2}\theta_m \tag{7.1}$$

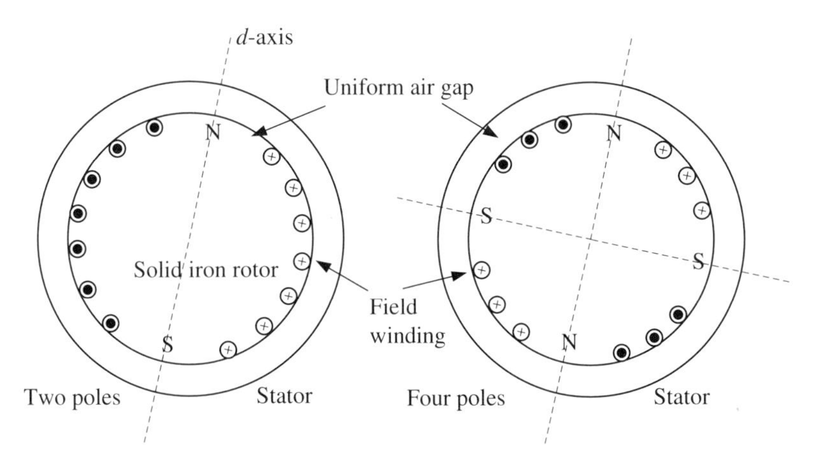

Figure 7.2 Two- and four-pole round-rotor synchronous machines.

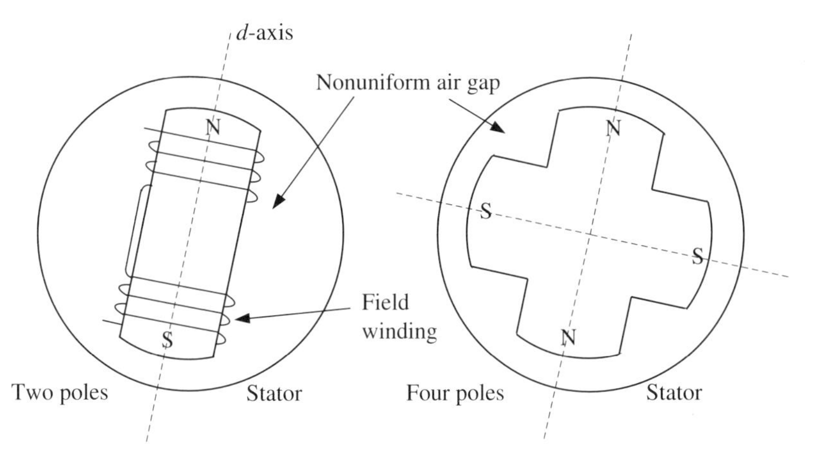

Figure 7.3 Two- and four-pole salient-pole synchronous machines.

In this definition, the electrical angle between two adjacent north and south poles will always be 180°. With this definition of the electrical angle, the derivation of a mathematical model of a synchronous machine can be based on a two-pole configuration.

7.2.1 Amortisseur Bars

Amortisseur bars or damper windings are normally added to the rotor surface to damp harmonic oscillations [93, 95]. They are typically embedded in the pole face, as shown in Figure 7.1. The bars are connected at the end by rings to form a short circuit. Their effect is captured in detailed synchronous machine models.

7.3 Synchronous Machine Model

Following the schematic of the synchronous machine in Figure 7.1, Figure 7.4 shows the dq-circuits on the rotor and the individual phase circuits on the stator. On the rotor,

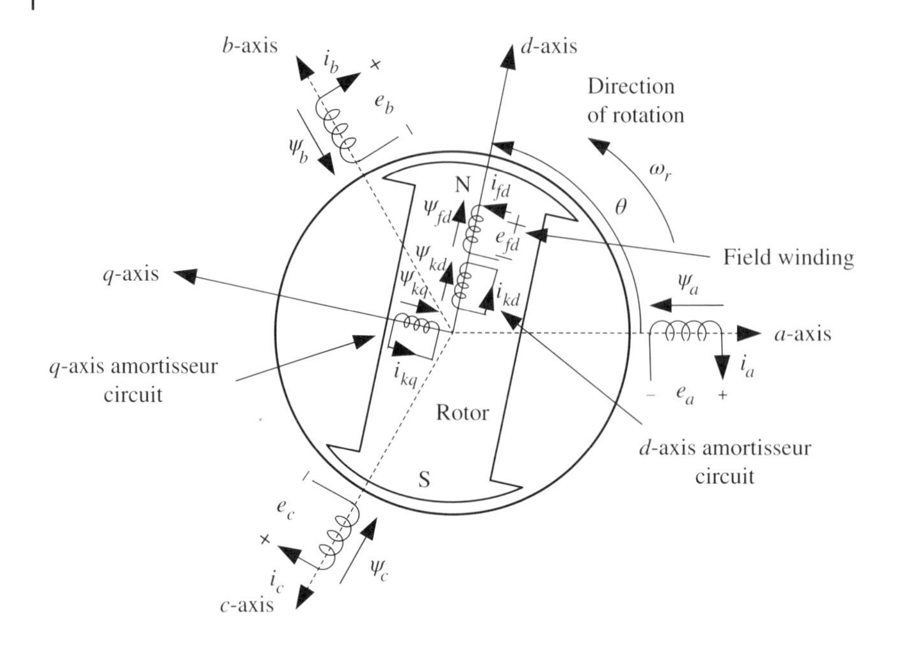

Notations:
a,b,c : stator phase windings
fd : field winding
kd : d-axis amortisseur circuit
kq : q-axis amortisseur circuit
$k = 1,2,\ldots,n$, n = number of amortisseur circuits
θ = d-axis phase lead relative to
center line of phase a winding
ω_r = rotor angular velocity

Figure 7.4 Stator and rotor circuits of a synchronous machine.

the field winding is aligned with the d-axis. The amortisseur circuits are included on both the d- and q-axes, and are assumed to be decoupled. To avoid clutter, only one d-axis amortisseur circuit and one q-axis amortisseur circuit are shown in Figure 7.4. If more than one amortisseur circuit is modeled, it may be denoted by $k = 1, 2, \ldots$, for as many circuits as needed (although in practice k is seldom greater than 2). The angle between the d-axis of the rotor and the magnetic axis of phase a is denoted by θ. The rotor speed is denoted by ω_r (in the counterclockwise direction) and the frequency of the stator voltage and current by ω_s. In this chapter on steady-state analysis, six circuit equations are developed: three for the stator (phases a, b, and c) and three for the rotor (one for the field and two for the amortisseurs). In the next chapter two amortisseur circuits on the q-axis are used to derive a six-state round-rotor model including subtransient effects (also known as the 2-axis model), which is the most common detailed synchronous machine model used in dynamic simulations.

The stator magnetomotive force (mmf) rotates because of the sinusoidal nature of the AC currents, and the rotor mmf also rotates because of the motion of the rotor. They

Table 7.1 Nomenclature used for stator and rotor winding equations.

ψ_a, e_a, i_a	stator phase a field flux linkage, voltage, current
ψ_b, e_b, i_b	stator phase b field flux linkage, voltage, current
ψ_c, e_c, i_c	stator phase c field flux linkage, voltage, current
$\psi_{fd}, e_{fd}, i_{fd}$	rotor field flux linkage, voltage, current
$\psi_{kd}, e_{kd}, i_{kd}$	rotor d-axis damping winding flux linkage, voltage, current
$\psi_{kq}, e_{kq}, i_{kq}$	rotor q-axis damping winding flux linkage, voltage, current
$\ell_{aa}, \ell_{bb}, \ell_{cc}$	stator self inductances
$\ell_{ab}, \ell_{ac}, \ell_{bc}$	stator mutual inductances
$\ell_{afd}, \ell_{bfd}, \ell_{cfd}$	stator mutual inductances with the field circuit
$\ell_{akd}, \ell_{bkd}, \ell_{ckd}$	stator mutual inductances with d-axis damper winding
$\ell_{akq}, \ell_{bkq}, \ell_{ckq}$	stator mutual inductances with q-axis damper winding
$\ell_{ffd}, \ell_{kkd}, \ell_{kkq}$	rotor self inductances
ℓ_{fkd}, ℓ_{kfd}	rotor mutual inductances
R_a, R_b, R_c	stator circuit resistances
R_{fd}, R_{kd}, R_{kq}	rotor circuit resistances

are both functions of t or, more specifically, functions of the stator frequency ω_s and the rotor speed ω_r. These time varying functions are complex to analyze. However, in steady state, the stator and rotor mmfs are stationary with respect to each other. Thus if the dq-axes of the rotor are taken as the reference frame, the rotor mmf will remain stationary in this rotating reference frame. In steady state, if $\omega_s = \omega_r$, then the stator mmf will also be stationary with respect to the dq reference frame.

7.3.1 Flux Linkage and Voltage Equations

From electromagnetic induction and circuit laws, the flux linkages and voltages in a synchronous machine can be modeled by four sets of equations. The stator equations are in terms of the (a, b, c) phases and the rotor equations are on of the d- and q-axes, using the nomenclature in Table 7.1. Note that the mutual inductances are only listed once in Table 7.1. Due to reciprocity, the mutual inductances between two different windings are the same, for example $\ell_{ab} = \ell_{ba}$ and $\ell_{afd} = \ell_{fad}$.

Stator flux-linkage equations:

$$\psi_a = \ell_{aa}i_a - \ell_{ab}i_b - \ell_{ac}i_c - \ell_{afd}i_{fd} - \ell_{akd}i_{kd} - \ell_{akq}i_{kq} \tag{7.2}$$

$$\psi_b = -\ell_{ab}i_a + \ell_{bb}i_b - \ell_{bc}i_c - \ell_{bfd}i_{fd} - \ell_{bkd}i_{kd} - \ell_{bkq}i_{kq} \tag{7.3}$$

$$\psi_c = -\ell_{ac}i_a - \ell_{bc}i_b + \ell_{cc}i_c - \ell_{cfd}i_{fd} - \ell_{ckd}i_{kd} - \ell_{ckq}i_{kq} \tag{7.4}$$

Rotor flux-linkage equations:

$$\psi_{fd} = -\ell_{fad}i_a - \ell_{fbd}i_b - \ell_{fcd}i_c + \ell_{ffd}i_{fd} + \ell_{fkd}i_{kd} \tag{7.5}$$

$$\psi_{kd} = -\ell_{kad}i_a - \ell_{kbd}i_b - \ell_{kcd}i_c + \ell_{kfd}i_{fd} + \ell_{kkd}i_{kd} \tag{7.6}$$

$$\psi_{kq} = -\ell_{kaq}i_a - \ell_{kbq}i_b - \ell_{kcq}i_c + \ell_{kkq}i_{kq} \tag{7.7}$$

Stator voltage equations:

$$e_a = -\frac{d\psi_a}{dt} - R_a i_a \tag{7.8}$$

$$e_b = -\frac{d\psi_b}{dt} - R_b i_b \tag{7.9}$$

$$e_c = -\frac{d\psi_c}{dt} - R_c i_c \tag{7.10}$$

Rotor voltage equations:

$$e_{fd} = \frac{d\psi_{fd}}{dt} + R_{fd} i_{fd} \tag{7.11}$$

$$e_{kd} = \frac{d\psi_{kd}}{dt} + R_{kd} i_{kd} \tag{7.12}$$

$$e_{kq} = \frac{d\psi_{kq}}{dt} + R_{kq} i_{kq} \tag{7.13}$$

Define the following vector and matrix notations:

$$\psi_s = \begin{bmatrix} \psi_a \\ \psi_b \\ \psi_c \end{bmatrix}, \quad \psi_r = \begin{bmatrix} \psi_{fd} \\ \psi_{kd} \\ \psi_{kq} \end{bmatrix}, \quad i_s = \begin{bmatrix} i_a \\ i_b \\ i_c \end{bmatrix}, \quad i_r = \begin{bmatrix} i_{fd} \\ i_{kd} \\ i_{kq} \end{bmatrix}$$
$$e_s = \begin{bmatrix} e_a \\ e_b \\ e_c \end{bmatrix}, \quad e_r = \begin{bmatrix} e_{fd} \\ e_{kd} \\ e_{kq} \end{bmatrix} \tag{7.14}$$

$$\ell_s = \begin{bmatrix} \ell_{aa} & -\ell_{ab} & -\ell_{ac} \\ -\ell_{ba} & \ell_{bb} & -\ell_{bc} \\ -\ell_{ca} & -\ell_{cb} & \ell_{cc} \end{bmatrix}, \quad \ell_{sr} = \begin{bmatrix} \ell_{afd} & \ell_{akd} & \ell_{akq} \\ \ell_{bfd} & \ell_{bkd} & \ell_{bkq} \\ \ell_{cfd} & \ell_{ckd} & \ell_{ckq} \end{bmatrix}$$
$$\ell_{rs} = \begin{bmatrix} \ell_{fad} & \ell_{fbd} & \ell_{fcd} \\ \ell_{kad} & \ell_{kbd} & \ell_{kcd} \\ \ell_{kaq} & \ell_{kbq} & \ell_{kcq} \end{bmatrix}, \quad \ell_r = \begin{bmatrix} \ell_{ffd} & \ell_{fkd} & 0 \\ \ell_{kfd} & \ell_{kkd} & 0 \\ 0 & 0 & \ell_{kkq} \end{bmatrix} \tag{7.15}$$

$$R_s = \begin{bmatrix} R_a & 0 & 0 \\ 0 & R_b & 0 \\ 0 & 0 & R_c \end{bmatrix}, \quad R_r = \begin{bmatrix} R_{fd} & 0 & 0 \\ 0 & R_{kd} & 0 \\ 0 & 0 & R_{kq} \end{bmatrix} \tag{7.16}$$

Equations (7.2) to (7.13) can be organized in matrix form as

$$\psi_s = -\ell_s i_s + \ell_{sr} i_r, \quad \psi_r = -\ell_{rs} i_s + \ell_r i_r \tag{7.17}$$

$$e_s = \frac{d\psi_s}{dt} - R_s i_s, \quad e_r = \frac{d\psi_r}{dt} - R_r i_r \tag{7.18}$$

Note that the negative sign associated with the stator flux linkages ψ_s is generally omitted in (7.17) and (7.18) for notational convenience,

7.3.2 Stator (Armature) Self and Mutual Inductances

For a round-rotor machine with a uniform air gap, the self inductances are constant and independent of the rotor position:

$$\ell_{aa} = \ell_{bb} = \ell_{cc} = L_{ag} + L_{a\ell} = L_{aa0} \tag{7.19}$$

where L_{ag} is due to the air-gap flux and $L_{a\ell}$ is the leakage inductance. For a salient-pole machine with non-uniform air gap, the self inductance will depend on the rotor position given by

$$\ell_{aa} = L_{aa0} + L_{aa2}\cos(2\theta) \tag{7.20}$$

which contains a second harmonic variation. The inductance ℓ_{aa} is maximum when $\theta =$ 0 or 180° (when the rotor aligns with the phase-a axis), and is minimum when $\theta = 90°$ or 270° (when the rotor is perpendicular to the phase-a axis).

The self inductances of phases b and c can be derived in a similar manner. These self inductances are shifted by $\pm 120°$:

$$\ell_{bb} = L_{aa0} + L_{aa2}\cos\left(2\left(\theta - \frac{2\pi}{3}\right)\right) \tag{7.21}$$

$$\ell_{cc} = L_{aa0} + L_{aa2}\cos\left(2\left(\theta + \frac{2\pi}{3}\right)\right) \tag{7.22}$$

For a round-rotor machine, the mutual inductances between the phases are constant and equal

$$\ell_{ab} = \ell_{bc} = \ell_{ca} = L_{ag} + L_{ab\ell} = L_{ab0} \tag{7.23}$$

where $L_{ab\ell}$ is the leakage reactance.

To find the mutual inductance between phases a and b for a salient-pole machine, the phase-a flux is projected at $(\theta - \pi/6)$ and the phase-b flux is projected at $((\theta - 2\pi/3) - \pi/6)$, resulting in the mutual inductance

$$\ell_{ab} = L_{ab0} + L_{aa2}\cos\left(2\left(\theta + \frac{\pi}{6}\right)\right) \tag{7.24}$$

The mutual inductances also have a second harmonic variation. For ℓ_{ab}, the magnitude peaks at $\theta = 5\pi/6 = 150°$ and $-\pi/6 = -30°$.

Similarly, the mutual inductances between phases b and c and between phases c and a are, respectively,

$$\ell_{bc} = \ell_{cb} = L_{ab0} + L_{aa2}\cos\left(2\left(\theta - \frac{\pi}{2}\right)\right) \tag{7.25}$$

$$\ell_{ca} = \ell_{ac} = L_{ab0} + L_{aa2}\cos\left(2\left(\theta + \frac{5\pi}{6}\right)\right) \tag{7.26}$$

7.3.3 Mutual Inductances between Stator and Rotor

For both the round-rotor and salient-pole machines, the variation of the d-axis mutual inductances due to the relative position of the stator phase a is

$$\ell_{afd} = L_{afd}\cos\theta \tag{7.27}$$

Similarly, for phases b and c

$$\ell_{bfd} = L_{afd}\cos\left(\theta - \frac{2\pi}{3}\right), \quad \ell_{cfd} = L_{afd}\cos\left(\theta + \frac{2\pi}{3}\right) \tag{7.28}$$

For the d-axis amortisseur circuit, the mutual inductances are

$$\ell_{akd} = L_{akd} \cos\theta, \quad \ell_{bkd} = L_{akd} \cos\left(\theta - \frac{2\pi}{3}\right), \quad \ell_{ckd} = L_{akd} \cos\left(\theta + \frac{2\pi}{3}\right) \tag{7.29}$$

For the q-axis, the mutual inductance with phase a is

$$\ell_{akq} = L_{akq} \cos\left(\theta + \frac{\pi}{2}\right) = -L_{akq} \sin\theta \tag{7.30}$$

The mutual inductances with the other two phases are

$$\ell_{bkq} = -L_{akq} \sin\left(\theta - \frac{2\pi}{3}\right), \quad \ell_{ckq} = -L_{akq} \sin\left(\theta + \frac{2\pi}{3}\right) \tag{7.31}$$

Note that for phase a, ℓ_{afd} and ℓ_{akd} achieve maximum amplitude at $\theta = 0°$ and $180°$, and ℓ_{akq} achieves maximum amplitude at $\theta = 90°$ and $270°$.

7.3.4 Rotor Self and Mutual Inductances

For both round-rotor and salient-pole machines, the rotor self inductances are constant

$$\ell_{ffd} = L_{ffd}, \quad \ell_{kkd} = L_{kkd}, \quad \ell_{kkq} = L_{kkq} \tag{7.32}$$

The mutual inductance between the field winding and the d-axis amortisseur winding is constant

$$\ell_{fkd} = L_{fkd} \tag{7.33}$$

as it is independent of the rotor position. The mutual reactances between the d- and q-axis are zero

$$\ell_{fkq} = \ell_{kdkq} = 0 \tag{7.34}$$

which as a result are not shown in the rotor inductance matrix ℓ_r (7.15).

Consolidating the dependence of self and mutual inductances on the rotor angle with respect to the magnetic a-axis, the expressions for the self and inductance matrices can be expressed as follows:

Stator self-inductance matrix - round-rotor machines:

$$\ell_s = \begin{bmatrix} L_{aa0} & -L_{ab0} & -L_{ab0} \\ -L_{ab0} & L_{aa0} & -L_{ab0} \\ -L_{ab0} & -L_{ab0} & L_{aa0} \end{bmatrix} \tag{7.35}$$

Stator self-inductance matrix - salient-pole machines:

$$\ell_s = \begin{bmatrix} L_{aa0} + L_{aa2}\cos 2\theta & -\left(L_{ab0} + L_{aa2}\cos 2\left(\theta + \frac{\pi}{6}\right)\right) & -\left(L_{ab0} + L_{aa2}\cos 2\left(\theta + \frac{5\pi}{6}\right)\right) \\ -\left(L_{ab0} + L_{aa2}\cos 2\left(\theta + \frac{\pi}{6}\right)\right) & L_{aa0} + L_{aa2}\cos 2\left(\theta - \frac{2\pi}{3}\right) & -\left(L_{ab0} + L_{aa2}\cos 2\left(\theta - \frac{\pi}{2}\right)\right) \\ -\left(L_{ab0} + L_{aa2}\cos 2\left(\theta + \frac{5\pi}{6}\right)\right) & -\left(L_{ab0} + L_{aa2}\cos 2\left(\theta - \frac{\pi}{2}\right)\right) & L_{aa0} + L_{aa2}\cos 2\left(\theta + \frac{2\pi}{3}\right) \end{bmatrix} \tag{7.36}$$

Rotor self-inductance matrix:

$$\ell_r = \begin{bmatrix} L_{ffd} & L_{fkd} & 0 \\ L_{fkd} & L_{kkd} & 0 \\ 0 & 0 & L_{kkq} \end{bmatrix} \tag{7.37}$$

Stator and rotor mutual-inductance matrix:

$$\ell_{sr} = \begin{bmatrix} L_{afd}\cos\theta & L_{akd}\cos\theta & -L_{akq}\sin\theta \\ L_{afd}\cos\left(\theta - \frac{2\pi}{3}\right) & L_{akd}\cos\left(\theta - \frac{2\pi}{3}\right) & -L_{akq}\sin\left(\theta - \frac{2\pi}{3}\right) \\ L_{afd}\cos\left(\theta + \frac{2\pi}{3}\right) & L_{akd}\cos\left(\theta + \frac{2\pi}{3}\right) & -L_{akq}\sin\left(\theta + \frac{2\pi}{3}\right) \end{bmatrix}, \quad \ell_{rs} = \ell_{sr}^T \tag{7.38}$$

7.4 Park Transformation

The stator and rotor self and mutual inductances (7.36)–(7.38) are functions of θ, which varies with time. Such time-varying equations are difficult to analyze. However, it is possible to simplify these flux-linkage expressions by transforming the stator currents into the dq reference frame, which is a reference frame that rotates at the same speed as the rotor and is known as the Park transformation [96].

Writing out the rotor winding flux linkages reveals the dependence on θ as

$$\psi_{fd} = -L_{afd}\left(i_a\cos\theta + i_b\cos\left(\theta - \frac{2\pi}{3}\right) + i_c\cos\left(\theta + \frac{2\pi}{3}\right)\right) + L_{ffd}i_{fd} + L_{fkd}i_{kd} \tag{7.39}$$

$$\psi_{kd} = -L_{akd}\left(i_a\cos\theta + i_b\cos\left(\theta - \frac{2\pi}{3}\right) + i_c\cos\left(\theta + \frac{2\pi}{3}\right)\right) + L_{fkd}i_{fd} + L_{kkd}i_{kd} \tag{7.40}$$

$$\psi_{kq} = -L_{akq}\left(-i_a\sin\theta - i_b\sin\left(\theta - \frac{2\pi}{3}\right) - i_c\sin\left(\theta + \frac{2\pi}{3}\right)\right) + L_{kkq}i_{kq} \tag{7.41}$$

Collecting the θ dependent terms on the right-hand sides of d-axis flux linkage equations (7.39) and (7.40), the d-axis current is defined as

$$i_d = K_d\left(i_a\cos\theta + i_b\cos\left(\theta - \frac{2\pi}{3}\right) + i_c\cos\left(\theta + \frac{2\pi}{3}\right)\right) \tag{7.42}$$

where K_d is a scaling factor to be selected so that i_d has a meaningful relationship with the phase currents.

For balanced operation in which the currents are varying at the synchronous frequency ω_s,

$$i_a = I_m\cos(\omega_s t + \phi), \quad i_b = I_m\cos\left(\omega_s t - \frac{2\pi}{3} + \phi\right), \quad i_c = I_m\cos\left(\omega_s t + \frac{2\pi}{3} + \phi\right) \tag{7.43}$$

where ϕ is the phase of i_a at $t = 0$. Substituting (7.43) into (7.42), it can be shown after some simple manipulations that

$$i_d = \frac{3}{2}K_d I_m\cos(\omega_s t - \theta + \phi) \tag{7.44}$$

Selecting $K_d = 2/3$ results in

$$i_d = I_m\cos(\omega_s t - \theta + \phi) \tag{7.45}$$

such that i_d and the phase currents have the same peak amplitude.

Similarly, collecting the θ dependent terms on the right-hand side of the q-axis flux linkage equation (7.39), the q-axis current is defined as

$$i_q = -K_q\left(i_a \sin\theta + i_b \sin\left(\theta - \frac{2\pi}{3}\right) + i_c \sin\left(\theta + \frac{2\pi}{3}\right)\right) \tag{7.46}$$

Substituting (7.43) into (7.46) yields

$$i_q = \frac{3}{2}K_q I_m \sin(\omega_s t - \theta + \phi) \tag{7.47}$$

Thus if K_q is also set to 2/3, then

$$i_q = I_m \sin(\omega_s t - \theta + \phi) \tag{7.48}$$

such that i_q and the phase currents also have the same peak amplitude.

Under synchronous conditions, θ rotates at the same speed of ω_s and hence $(\theta - \omega_s t)$ is constant and is equal to the value θ_o of θ at $t = 0$. Thus

$$i_d = I_m \cos(\phi - \theta_o), \quad i_q = I_m \sin(\phi - \theta_o) \tag{7.49}$$

are constant valued, time-invariant functions.

The current i_d may be interpreted as the instantaneous current in a fictitious armature winding located on the d-axis of the rotor and rotating at the same speed as the rotor. The magnitude of i_d is such that it gives the same mmf on the d-axis as do the actual three instantaneous armature phase currents. The current i_q may be interpreted in a similar manner except that it acts on the q-axis of the rotor.

To completely specify the $dq0$ transformation, a third quantity – the zero-sequence current – is needed

$$i_0 = \frac{1}{3}(i_a + i_b + i_c) \tag{7.50}$$

which is zero in balanced operation.

To summarize, the Park transformation T_P in matrix form is

$$\begin{bmatrix} i_d \\ i_q \\ i_0 \end{bmatrix} = \frac{2}{3}\begin{bmatrix} \cos\theta & \cos\left(\theta - \frac{2\pi}{3}\right) & \cos\left(\theta + \frac{2\pi}{3}\right) \\ -\sin\theta & -\sin\left(\theta - \frac{2\pi}{3}\right) & -\sin\left(\theta + \frac{2\pi}{3}\right) \\ \frac{1}{2} & \frac{1}{2} & \frac{1}{2} \end{bmatrix}\begin{bmatrix} i_a \\ i_b \\ i_c \end{bmatrix} = T_P\begin{bmatrix} i_a \\ i_b \\ i_c \end{bmatrix} \tag{7.51}$$

This transformation can also be used to obtain ψ_d, ψ_q, ψ_0 from ψ_a, ψ_b, ψ_c, and to obtain e_d, e_q, e_0 from e_a, e_b, e_c.

The inverse Park transformation matrix T_P^{-1} is given by

$$\begin{bmatrix} i_a \\ i_b \\ i_c \end{bmatrix} = T_P^{-1}\begin{bmatrix} i_d \\ i_q \\ i_0 \end{bmatrix} = \begin{bmatrix} \cos\theta & -\sin\theta & 1 \\ \cos\left(\theta - \frac{2\pi}{3}\right) & -\sin\left(\theta - \frac{2\pi}{3}\right) & 1 \\ \cos\left(\theta + \frac{2\pi}{3}\right) & -\sin\left(\theta + \frac{2\pi}{3}\right) & 1 \end{bmatrix}\begin{bmatrix} i_d \\ i_q \\ i_0 \end{bmatrix} \tag{7.52}$$

Using both the $dq0$ transformation and its inverse transform, the stator flux-linkage equations become

$$\psi_d = -L_d i_d + L_{afd} i_{fd} + L_{akd} i_{kd}, \quad L_d = L_{aa0} + L_{ab0} + \frac{3}{2}L_{aa2} \tag{7.53}$$

$$\psi_q = -L_q i_q + L_{akq} i_{kq}, \quad L_q = L_{aa0} + L_{ab0} - \frac{3}{2}L_{aa2} \tag{7.54}$$

$$\psi_0 = -L_0 i_0, \quad L_0 = L_{aa0} - 2L_{ab0} \tag{7.55}$$

The rotor flux-linkage equations simplify to

$$\psi_{fd} = L_{ffd} i_{fd} + L_{fkd} i_{kd} - \frac{3}{2} L_{afd} i_d \tag{7.56}$$

$$\psi_{kd} = L_{fkd} i_{fd} + L_{kkd} i_{kd} - \frac{3}{2} L_{akd} i_d \tag{7.57}$$

$$\psi_{kq} = L_{kkq} i_{kq} - \frac{3}{2} L_{akq} i_q \tag{7.58}$$

Note that all inductance terms in the stator and rotor flux-linkage expressions are now constant and independent of θ and t. Also note that some of the mutual inductances are not reciprocal. For example, the mutual inductance of ψ_d with respect to the field current i_{fd} is L_{afd}, whereas the mutual inductance of ψ_{fd} with respect to the d-axis current i_d is $(3/2)L_{afd}$. This discrepancy will create problems in developing a flux-linkage circuit. It is resolved in a later section by choosing a proper basis for the field quantities.

Applying the inverse Park transformation to the phase-a voltage (7.8)[1]

$$e_a = \frac{d\psi_a}{dt} - R_a i_a \tag{7.59}$$

results in an equation involving e_d, e_q, and e_0:

$$\begin{aligned} e_a &= e_d \cos\theta - e_q \sin\theta + e_0 \\ &= \frac{d}{dt}(\psi_d \cos\theta - \psi_q \sin\theta + \psi_0) - R_a(i_d \cos\theta - i_q \sin\theta + i_0) \\ &= -\psi_d \sin\theta \frac{d\theta}{dt} + \frac{d\psi_d}{dt} \cos\theta - \psi_q \cos\theta \frac{d\theta}{dt} - \frac{d\psi_q}{dt} \sin\theta + \frac{d\psi_0}{dt} \\ &\quad - R_a(i_d \cos\theta - i_q \sin\theta + i_0) \end{aligned} \tag{7.60}$$

Equating the like terms of θ on both sides of the equation, the stator voltages in the $dq0$ reference frame are given by

$$e_d = \frac{d\psi_d}{dt} - \omega_r \psi_q - R_a i_d \tag{7.61}$$

$$e_q = \frac{d\psi_q}{dt} + \omega_r \psi_d - R_a i_q \tag{7.62}$$

$$e_0 = \frac{d\psi_0}{dt} - R_a i_0 \tag{7.63}$$

where $\omega_r \psi_d$ and $\omega_r \psi_q$ are called the speed-voltage terms and ω_r is the rotor speed.

Note that the rotor voltage equations (7.11)–(7.13) do not depend on the armature voltages and currents, and hence remain the same in the $dq0$ reference frame.

1 Note that the negative sign on the flux variable has been dropped.

7.4.1 Electrical Power in $dq0$ Variables

The electrical power can be computed as the sum of the power delivered by all three phases

$$P_e = e_a i_a + e_b i_b + e_c i_c = \begin{bmatrix} e_a & e_b & e_c \end{bmatrix} \begin{bmatrix} i_a \\ i_b \\ i_c \end{bmatrix}$$
$$= \begin{bmatrix} e_d & e_q & e_0 \end{bmatrix} (T_P^{-1})^T T_P^{-1} \begin{bmatrix} i_d \\ i_q \\ i_0 \end{bmatrix} = \begin{bmatrix} e_d & e_q & e_0 \end{bmatrix} \begin{bmatrix} \frac{3}{2} & 0 & 0 \\ 0 & \frac{3}{2} & 0 \\ 0 & 0 & 3 \end{bmatrix} \begin{bmatrix} i_d \\ i_q \\ i_0 \end{bmatrix}$$
$$= \frac{3}{2}(e_d i_d + e_q i_q + 2e_0 i_0) \tag{7.64}$$

Note that the zero-sequence term in the expression is zero in steady state. Thus it is normally not included in simulation code.

Substituting the $dq0$ flux linkage expressions into (7.64), the electrical power expression becomes

$$P_e = \frac{3}{2}\left[\left(i_d \frac{d\psi_d}{dt} + i_q \frac{d\psi_q}{dt} + 2i_0 \frac{d\psi_0}{dt}\right) + (\psi_d i_q - \psi_q i_d)\omega_r - (i_d^2 + i_q^2 + 2i_0^2)R_a\right] \tag{7.65}$$

where the three $d\psi/dt$ terms are the rate of change of the stator magnetic energy, the two ω_r terms are the electrical power across the air gap, and the three R_a terms represent the armature losses.

From (7.65), the air-gap torque can be defined as

$$T_e = \frac{3}{2}(\psi_d i_q - \psi_q i_d) \tag{7.66}$$

Example 7.1: Park transformation for negative sequence currents

The Park transformation can be applied to various periodic phase voltages and currents. Consider a 3-phase synchronous machine operating at synchronous speed ω_s and in steady-state condition. Negative-sequence currents

$$i_a = I_m \cos(\omega_s t + \phi), \quad i_b = I_m \cos(\omega_s t + 2\pi/3 + \phi)$$
$$i_c = I_m \cos(\omega_s t - 2\pi/3 + \phi) \tag{7.67}$$

are injected into the stator phases. Use the Park transformation to calculate i_d, i_q, and i_o.

Solutions: Applying the $dq0$ transformation to the negative sequence currents, with much of the intermediate manipulation omitted, the required currents are obtained as

$$i_d = \frac{2}{3}\left[i_a \cos(\omega_s t) + i_b \cos\left(\omega_s t - \frac{2\pi}{3}\right) + i_c \cos\left(\omega_s t + \frac{2\pi}{3}\right)\right]$$
$$= \frac{2I_m}{3}\left[\cos(\omega_s t + \phi)\cos(\omega_s t) + \cos\left(\omega_s t + \frac{2\pi}{3} + \phi\right)\cos\left(\omega_s t - \frac{2\pi}{3}\right)\right.$$
$$\left. + \cos\left(\omega_s t - \frac{2\pi}{3} + \phi\right)\cos\left(\omega_s t + \frac{2\pi}{3}\right)\right]$$

$$= \frac{2I_m}{3}\left[\frac{3}{2}\cos(\omega_s t + \phi)\cos(\omega_s t) - \frac{3}{2}\sin(\omega_s t + \phi)\sin(\omega_s t)\right]$$
$$= I_m \cos(2\omega_s t + \phi) \tag{7.68}$$

$$i_q = -\frac{2}{3}\left[i_a \sin(\omega_s t) + i_b \sin\left(\omega_s t - \frac{2\pi}{3}\right) + i_c \sin\left(\omega_s t + \frac{2\pi}{3}\right)\right]$$
$$= -I_m \sin(2\omega_s t + \phi) \tag{7.69}$$

$$i_0 = \frac{1}{3}(i_a + i_b + i_c) = 0 \tag{7.70}$$

Note that i_d and i_q are still sinusoidal signals, except that the signals are rotating at twice the synchronous speed. This fact is not surprising as a forward 3-phase rotating synchronous frequency signal (positive-sequence signal) is transformed to a DC quantity; a backward 3-phase rotating synchronous frequency signal would now be rotating at twice the synchronous speed. ■

7.5 Reciprocal, Equal L_{ad} Per-Unit System

A per-unit system is useful to define the base values of various components of a synchronous machine, including the stator, rotor, and amortisseur, so that the machine parameters are relatively invariant with respect to the machine sizes. A careful definition of the per-unit values can also allow for the development of equivalent circuits, thus permitting more effective modeling and computation.

7.5.1 Stator Base Values

The stator voltage, current, and frequency bases of a synchronous machine are normally based on the machine ratings:

- base power $(3\phi\text{VA})_{\text{base}}$ = rated total VA of all 3 phases
- base voltage E_{sbase} = rated line-to-line voltage in rms value
- base frequency f_{base} = rated frequency in Hz.

From the rms values of MVA and kV ratings, the voltage and current bases can be defined as

$$E_{\text{sbase}(\ell-n)} = \frac{(\text{Rated kV})\cdot 1000}{\sqrt{3}}\text{V}, \quad I_{\text{sbase}} = \frac{(\text{Rated MVA})\cdot 10^6}{\sqrt{3}(\text{Rated kV})\cdot 1000}\text{ A} \tag{7.71}$$

In terms of the peak values of the stator phase voltage e_s and current i_s, the voltage and current bases are

$$e_{\text{sbase}} = \sqrt{2}E_{\text{sbase}}, \quad i_{\text{sbase}} = \sqrt{2}I_{\text{sbase}} \tag{7.72}$$

The peak values of the base quantities are used in synchronous machine equations because

- the original equations are written in terms of the instantaneous values of the variables, and

- the transformed stator variables are related to the peak values of the phase quantities under balanced load conditions.

The dependent stator bases become

$$Z_{sbase} = \frac{e_{sbase}}{i_{sbase}} = \frac{E_{sbase}}{I_{sbase}} \text{ Ohms} \tag{7.73}$$

$$\omega_{base} = 2\pi f_{base} \text{ (electrical) rad/sec} \tag{7.74}$$

$$L_{sbase} = \frac{Z_{sbase}}{\omega_{base}} \text{ Henries} \tag{7.75}$$

$$\psi_{sbase} = L_{sbase} i_{sbase} = \frac{e_{sbase}}{\omega_{base}} \text{ Weber-turns} \tag{7.76}$$

$$(3\phi\text{VA})_{sbase} = 3E_{sbase} I_{sbase} = \frac{3}{2} e_{sbase} i_{sbase} \text{ Volt-Amperes} \tag{7.77}$$

$$\text{Torque base} = \frac{(3\phi\text{VA})_{sbase}}{\omega_{base}} = \frac{3}{2} \psi_{sbase} i_{sbase} \text{ Newton-meters} \tag{7.78}$$

$$t_{rad} = \omega_{base} t_{sec} \text{ rad} \tag{7.79}$$

7.5.2 Stator Voltage Equations

From (7.70)–(7.70), the stator base voltage can be expressed in several forms

$$e_{sbase} = i_{sbase} Z_{sbase} = i_{sbase} \omega_{base} L_{sbase} = \omega_{base} \psi_{sbase} \tag{7.80}$$

Divide each of the terms in the d-axis stator voltage equation

$$e_d = \frac{d\psi_d}{dt} - \psi_q \frac{d\theta}{dt} - R_a i_d \tag{7.81}$$

where θ is in radians and t in seconds, by e_{sbase} to obtain

$$\underbrace{\frac{e_d}{e_{sbase}}}_{\bar{e}_d} = \frac{d}{dt_{sec}} \left(\frac{1}{\omega_{base}} \underbrace{\frac{\psi_d}{\psi_{sbase}}}_{\bar{\psi}_d} \right) - \frac{1}{\omega_{base}} \underbrace{\frac{\psi_q}{\psi_{sbase}}}_{\bar{\psi}_q} \frac{d\theta}{dt_{sec}} - \underbrace{\frac{R_a}{Z_{sbase}}}_{\bar{R}_a} \underbrace{\frac{i_d}{i_{sbase}}}_{\bar{i}_d} \tag{7.82}$$

Thus the d-axis stator voltage equation in per unit becomes

$$\bar{e}_d = \frac{d\bar{\psi}_d}{dt_{rad}} - \bar{\psi}_q \frac{d\theta}{dt_{rad}} - \bar{R}_a \bar{i}_d \tag{7.83}$$

where the bar above a variable denotes its per-unitized value. Similarly, the q-axis stator voltage equation in per unit is

$$\bar{e}_q = \frac{d\bar{\psi}_q}{dt_{rad}} + \bar{\psi}_d \frac{d\theta}{dt_{rad}} - \bar{R}_a \bar{i}_q \tag{7.84}$$

and the zero-sequence stator voltage equation in per unit is

$$\bar{e}_0 = \frac{d\bar{\psi}_0}{dt_{rad}} - \bar{R}_a \bar{i}_0 \tag{7.85}$$

7.5.3 Rotor Base Values

For now, arbitrary base values for the voltage and current for each rotor circuit will be used. Subsequently, in Section 7.5.7, they will be selected to achieve the desired "reciprocal, equal L_{ad}" requirement.

For the field circuit, let $e_{fd\text{base}}$ and $i_{fd\text{base}}$ be the base values. Then

$$Z_{fd\text{base}} = \frac{e_{fd\text{base}}}{i_{fd\text{base}}}, \quad \omega_{\text{base}} = 2\pi f_{\text{base}}, \quad L_{fd\text{base}} = \frac{Z_{fd\text{base}}}{\omega_{\text{base}}}$$
$$\psi_{fd\text{base}} = L_{fd\text{base}} i_{fd\text{base}}, \quad (\text{VA})_{fd\text{base}} = e_{fd\text{base}} i_{fd\text{base}} \tag{7.86}$$

For the d-axis amortisseur circuit, let $e_{kd\text{base}}$ and $i_{kd\text{base}}$ be the base values. Then

$$Z_{kd\text{base}} = \frac{e_{kd\text{base}}}{i_{kd\text{base}}}, \quad L_{kd\text{base}} = \frac{Z_{kd\text{base}}}{\omega_{\text{base}}}$$
$$\psi_{kd\text{base}} = L_{kd\text{base}} i_{kd\text{base}}, \quad (\text{VA})_{kd\text{base}} = e_{kd\text{base}} i_{kd\text{base}} \tag{7.87}$$

For the q-axis amortisseur circuit, let $e_{kq\text{base}}$ and $i_{kq\text{base}}$ be the base values. Then

$$Z_{kq\text{base}} = \frac{e_{kq\text{base}}}{i_{kq\text{base}}}, \quad L_{kq\text{base}} = \frac{Z_{kq\text{base}}}{\omega_{\text{base}}}$$
$$\psi_{kq\text{base}} = L_{kq\text{base}} i_{kq\text{base}}, \quad (\text{VA})_{kq\text{base}} = e_{kq\text{base}} i_{kq\text{base}} \tag{7.88}$$

7.5.4 Rotor Voltage Equations

From (7.83), the rotor base voltage can be expressed in several forms:

$$e_{fd\text{base}} = i_{fd\text{base}} Z_{fd\text{base}} = i_{fd\text{base}} \omega_{\text{base}} L_{fd\text{base}} = \omega_{\text{base}} \psi_{fd\text{base}} \tag{7.89}$$

Divide each of the terms in the rotor field voltage equation

$$e_{fd} = \frac{d\psi_{fd}}{dt} + R_{fd} i_{fd} \tag{7.90}$$

by $e_{fd\text{base}}$ with the appropriate base-value combinations to obtain

$$\frac{e_{fd}}{e_{fd\text{base}}} = \frac{d}{dt} \frac{\psi_{fd}}{\omega_{\text{base}} \psi_{fd\text{base}}} + \frac{R_{fd} i_{fd}}{Z_{fd\text{base}} i_{fd\text{base}}} \tag{7.91}$$

Thus the field circuit equation in per-unit becomes

$$\bar{e}_{fd} = \frac{d\bar{\psi}_{fd}}{dt_{\text{rad}}} + \bar{R}_{fd} \bar{i}_{fd} \tag{7.92}$$

Similarly, the per-unitized amortisseur circuits are

$$\bar{e}_{kd} = \frac{d\bar{\psi}_{kd}}{dt_{\text{rad}}} + \bar{R}_{kd} \bar{i}_{kd} = 0, \quad \bar{e}_{kq} = \frac{d\bar{\psi}_{kq}}{dt_{\text{rad}}} + \bar{R}_{kq} \bar{i}_{kq} = 0 \tag{7.93}$$

These equations are set to zero because the amortisseur circuits do not have any applied voltages.

7.5.5 Stator Flux-Linkage Equations

Per-unitizing the d-axis flux linkage equation

$$\psi_d = -L_d i_d + L_{afd} i_{fd} + L_{akd} i_{kd} \tag{7.94}$$

with respect to ψ_{sbase} results in

$$\underbrace{\frac{\psi_d}{\psi_{\text{sbase}}}}_{\overline{\psi}_d} = -\underbrace{\frac{L_d}{L_{\text{sbase}}}}_{\overline{L}_d}\underbrace{\frac{i_d}{i_{\text{sbase}}}}_{\overline{i}_d} + \underbrace{\frac{L_{afd} i_{fd\text{base}}}{L_{\text{sbase}} i_{\text{sbase}}}}_{\overline{L}_{afd}}\underbrace{\frac{i_{fd}}{i_{fd\text{base}}}}_{\overline{i}_{fd}} + \underbrace{\frac{L_{akd} i_{kd\text{base}}}{L_{\text{sbase}} i_{\text{sbase}}}}_{\overline{L}_{akd}}\underbrace{\frac{i_{kd}}{i_{kd\text{base}}}}_{\overline{i}_{kd}} \tag{7.95}$$

that is,

$$\overline{\psi}_d = -\overline{L}_d \overline{i}_d + \overline{L}_{afd} \overline{i}_{fd} + \overline{L}_{akd} \overline{i}_{kd} \tag{7.96}$$

Similarly, the q-axis and zero sequence flux linkage equations are given by

$$\overline{\psi}_q = -\overline{L}_q \overline{i}_q + \overline{L}_{akq} \overline{i}_{kq}, \quad \overline{L}_{akq} = \frac{L_{akq} i_{kq\text{base}}}{L_{\text{sbase}} i_{\text{sbase}}} \tag{7.97}$$

$$\overline{\psi}_0 = -\overline{L}_0 \overline{i}_0 \tag{7.98}$$

7.5.6 Rotor Flux-Linkage Equations

Divide the field flux linkage equation

$$\psi_{fd} = -\frac{3}{2} L_{afd} i_d + L_{ffd} i_{fd} + L_{fkd} i_{kd} \tag{7.99}$$

by $\psi_{fd\text{base}}$ to obtain the expression in per unit

$$\underbrace{\frac{\psi_{fd}}{\psi_{fd\text{base}}}}_{\overline{\psi}_{fd}} = -\underbrace{\frac{(3/2) L_{afd} i_{\text{sbase}}}{L_{fd\text{base}} i_{fd\text{base}}}}_{\overline{L}_{fad}}\underbrace{\frac{i_d}{i_{\text{sbase}}}}_{\overline{i}_d} + \underbrace{\frac{L_{ffd}}{L_{fd\text{base}}}}_{\overline{L}_{ffd}}\underbrace{\frac{i_{fd}}{i_{fd\text{base}}}}_{\overline{i}_{fd}} + \underbrace{\frac{L_{fkd} i_{kd\text{base}}}{L_{fd\text{base}} i_{fd\text{base}}}}_{\overline{L}_{fkd}}\underbrace{\frac{i_{kd}}{i_{kd\text{base}}}}_{\overline{i}_{kd}} \tag{7.100}$$

that is,

$$\overline{\psi}_{fd} = -\overline{L}_{fad} \overline{i}_d + \overline{L}_{ffd} \overline{i}_{fd} + \overline{L}_{fkd} \overline{i}_{kd} \tag{7.101}$$

7.5.7 Equal Mutual Inductance

The freedom available in choosing the rotor bases will now be used to ensure that mutual inductances in pu values for the stator and rotor circuits are the same (that is, reciprocal) so that the stator and rotor circuits can be coupled. For the per-unit mutual inductances between the stator d-axis and the field winding to be reciprocal requires

$$\overline{L}_{afd} = \overline{L}_{fad} \tag{7.102}$$

that is

$$\frac{L_{afd} i_{fd\text{base}}}{L_{\text{sbase}} i_{\text{sbase}}} = \frac{(3/2) L_{afd} i_{\text{sbase}}}{L_{fd\text{base}} i_{fd\text{base}}} \tag{7.103}$$

implying

$$L_{fd\text{base}} i_{fd\text{base}}^2 = \frac{3}{2} L_{s\text{base}} i_{s\text{base}}^2 \tag{7.104}$$

Multiplying on both sides by ω_s results in

$$Z_{fd\text{base}} i_{fd\text{base}}^2 = \frac{3}{2} Z_{s\text{base}} i_{s\text{base}}^2 \tag{7.105}$$

which can be rewritten as

$$e_{fd\text{base}} i_{fd\text{base}} = \frac{3}{2} e_{s\text{base}} i_{s\text{base}} \tag{7.106}$$

or

$$(\text{VA})_{fd\text{base}} = (3\phi\text{VA})_{s\text{base}} \tag{7.107}$$

In a similar manner, all per-unit mutual or magnetizing inductances would be reciprocal if $(\text{VA})_{\text{base}}$ is the same in all windings, that is,

$$(3\phi\text{VA})_{s\text{base}} = (\text{VA})_{fd\text{base}} = (\text{VA})_{kd\text{base}} = (\text{VA})_{kq\text{base}} \tag{7.108}$$

Note that in (7.96), the VA base for the stator is the sum of the three individual phase VA. Also note that in (7.100), only the product of $e_{fd\text{base}}$ and $i_{fd\text{base}}$ is specified. This freedom will be used for setting additional mutual inductances to be the same.

For multiple-coil transformers, it would be convenient if the base value used for each coil is selected such that the per-unit values of mutual inductances between the coils are equal. Such a scheme will simplify the circuit representation. Figure 7.5 shows the circuit representation of a three-winding transformer when the mutual inductances between all three windings are the same. Using the same strategy for the mutual inductances between the synchronous machine stator and rotor circuits will simplify the equivalent-circuit representation.

The stator mutual flux linkages can be separated into two parts:

1) the flux linking only the stator windings.
2) the flux linking a rotor circuit.

For the stator dq-axes, the synchronous inductances are

$$L_d = L_\ell + L_{ad}, \quad L_q = L_\ell + L_{aq} \tag{7.109}$$

In per unit, the synchronous inductances are expressed as

$$\overline{L}_d = \overline{L}_\ell + \overline{L}_{ad}, \quad \overline{L}_q = \overline{L}_\ell + \overline{L}_{aq} \tag{7.110}$$

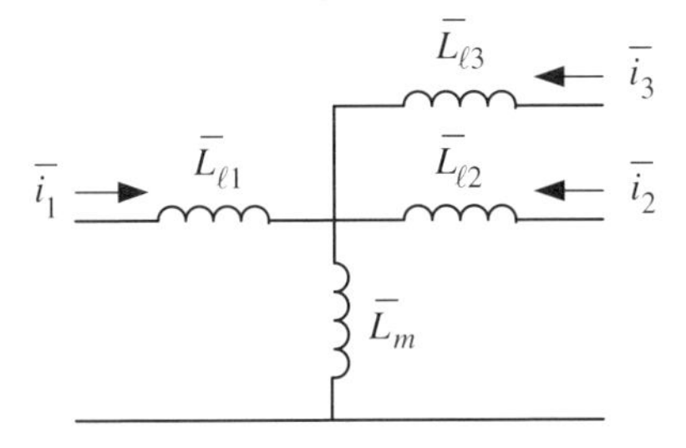

Figure 7.5 Coupling of three transformer windings, with the subscript m denoting mutual inductance and the subscript ℓ denoting leakage inductance.

Equal per-unit mutual inductances of the d-axis stator, the field winding, and the d-axis amortisseur winding require that

$$\overline{L}_{ad} = \overline{L}_{afd} = \overline{L}_{akd}, \quad \overline{L}_{aq} = \overline{L}_{akq} \tag{7.111}$$

The per-unitized d-axis mutual or magnetizing inductances are

$$\overline{L}_{ad} = \frac{L_{ad}}{L_{sbase}}, \quad \overline{L}_{afd} = \frac{L_{afd}}{L_{sbase}} \frac{i_{fdbase}}{i_{sbase}}, \quad \overline{L}_{akd} = \frac{L_{akd}}{L_{sbase}} \frac{i_{kdbase}}{i_{sbase}} \tag{7.112}$$

Equating the various terms yields

$$i_{fdbase} = \frac{L_{ad}}{L_{afd}} i_{sbase}, \quad i_{kdbase} = \frac{L_{ad}}{L_{akd}} i_{sbase} \tag{7.113}$$

Similarly, equal q-axis mutual or magnetizing inductances require

$$i_{kqbase} = \frac{L_{aq}}{L_{akq}} i_{sbase} \tag{7.114}$$

Normally i_{kdbase} and i_{kqbase} are not required. The field current base i_{fdbase} can be calculated by performing machine testing to measure L_{ad} and L_{afd}.

The rotor inductances can also be partitioned into a leakage component and a mutual component as

$$\overline{L}_{ffd} = \overline{L}_{\ell fd} + \overline{L}_{fkd}, \quad \overline{L}_{kkd} = \overline{L}_{\ell kd} + \overline{L}_{akd}, \quad \overline{L}_{kkq} = \overline{L}_{\ell kq} + \overline{L}_{akq} \tag{7.115}$$

A good approximation commonly used is

$$\overline{L}_{fkd} = \overline{L}_{ad} \tag{7.116}$$

which further simplifies the equivalent circuit representation.

The following example is from Chapter 2 of [97]. It illustrates the various aspect in the per-unitization of inductances in a synchronous machine.

Example 7.2: Per-unit value calculation

A 60-cycle, 3-phase, 4-pole synchronous generator is rated at 20 MVA and 12.1 kV. It has the following resistance and inductance values:

$R_a = 0.0147$ ohms, $L_\ell = 0.00214$ H, $L_{ad} = 0.02145$ H, $L_{aq} = 0.0191$ H
$L_{afd} = 0.045$ H, $L_{akd} = 0.017$ H, $L_{akq} = 0.007$ H
$L_{ffd} = 0.1538$ H, $R_{fd} = 0.0208$ ohms, $L_{fkd} = 0.0541$ H
$L_{kkd} = 0.0226$ H, $R_{kkd} = 0.1315$ ohms
$L_{kkq} = 0.00446$ H, $R_{kkq} = 0.0611$ ohms

1) Calculate the reciprocal equal L_{ad} per-unit system base quantities for the stator, rotor, and amortisseur windings.
2) Calculate the per-unit inductance and resistance parameters for the stator, rotor, and amortisseur windings.

Solutions: The stator ratings are

$$(3\phi\text{VA})_{base} = 20 \text{ MVA} = 20 \times 10^6 \text{ VA} \tag{7.117}$$

$$\text{base kV} = 12.1 \text{ kV}, \quad f_{\text{base}} = 60 \text{ Hz} \tag{7.118}$$

Thus the derived stator base values are

$$E_{s\text{base}} = \frac{\text{base kV}}{\sqrt{3}} = 6.986 \text{ kV (rms)} \tag{7.119}$$

$$I_{s\text{base}} = \frac{\text{base VA}}{3E_{s\text{base}}} = 954.3 \text{ A (rms)} \tag{7.120}$$

$$i_{s\text{base}} = \sqrt{2}I_{s\text{base}} = 1350 \text{ A (peak)} \tag{7.121}$$

$$Z_{s\text{base}} = \frac{E_{s\text{base}}}{I_{s\text{base}}} = 7.32 \text{ ohms}, \quad L_{s\text{base}} = \frac{Z_{s\text{base}}}{2\pi f_b} = 0.0195 \text{ H} \tag{7.122}$$

The rotor base is taken to be 20 MVA, the same as the stator base. From the reciprocal, equal L_{ad} system

$$i_{fd\text{base}} = \frac{L_{ad}}{L_{afd}} i_{s\text{base}} = 643.3 \text{ A}, \quad e_{fd\text{base}} = \frac{\text{base VA}}{i_{fd\text{base}}} = 31.09 \text{ kV} \tag{7.123}$$

$$Z_{fd\text{base}} = \frac{e_{fd\text{base}}}{i_{fd\text{base}}} = 48.33 \text{ ohms}, \quad L_{fd\text{base}} = \frac{Z_{fd\text{base}}}{2\pi f_b} = 0.1282 \text{ H} \tag{7.124}$$

The amortisseur base is also taken to be 20 MVA. To achieve equal L_{ad} and L_{aq} per-unit values

$$i_{kd\text{base}} = \frac{L_{ad}}{L_{akd}} i_{s\text{base}} = 1703 \text{ A}, \; i_{kq\text{base}} = \frac{L_{aq}}{L_{akq}} i_{s\text{base}} = 3682 \text{ A} \tag{7.125}$$

$$Z_{kkd\text{base}} = \frac{\text{base VA}}{I^2_{kd\text{base}}} = 6.897 \text{ ohms}, \; Z_{kkq\text{base}} = \frac{\text{base VA}}{I^2_{kq\text{base}}} = 1.475 \text{ ohms} \tag{7.126}$$

$$L_{kkd\text{base}} = \frac{Z_{kkd\text{base}}}{2\pi f_b} = 0.0183 \text{ H}, \; L_{kkq\text{base}} = \frac{Z_{kkq\text{base}}}{2\pi f_b} = 0.0039 \text{ H} \tag{7.127}$$

$$L_{kfd\text{base}} = \frac{i_{fd\text{base}}}{i_{kd\text{base}}} L_{fd\text{base}} = 0.0484 \text{ H} \tag{7.128}$$

Using these base values, the per-unit parameter values can be calculated. For the stator

$$\overline{R}_a = \frac{R_a}{Z_{s\text{base}}} = 0.0020 \text{ pu}, \quad \overline{L}_\ell = \frac{L_\ell}{L_{s\text{base}}} = 0.1102 \text{ pu} \tag{7.129}$$

$$\overline{L}_{ad} = \frac{L_{ad}}{L_{s\text{base}}} = 1.1046 \text{ pu}, \quad \overline{L}_{aq} = \frac{L_{aq}}{L_{s\text{base}}} = 0.9836 \text{ pu} \tag{7.130}$$

$$\overline{L}_d = \overline{L}_\ell + \overline{L}_{ad} = 1.2148 \text{ pu}, \quad \overline{L}_q = \overline{L}_\ell + \overline{L}_{aq} = 1.0938 \text{ pu} \tag{7.131}$$

The mutual reactances from the base formulation are

$$\overline{L}_{afd} = \overline{L}_{akd} = \overline{L}_{ad} = 1.1046 \text{ pu} \tag{7.132}$$

$$\overline{L}_{akq} = \overline{L}_{aq} = 0.9836 \text{ pu} \tag{7.133}$$

For the field winding, the parameters are

$$\overline{R}_{fd} = \frac{R_{fd}}{Z_{fd\text{base}}} = 4.304 \times 10^{-4} \text{ pu}, \quad \overline{L}_{ffd} = \frac{L_{ffd}}{L_{fd\text{base}}} = 1.1997 \text{ pu} \tag{7.134}$$

$$\overline{L}_{\ell fd} = \overline{L}_{ffd} - \overline{L}_{afd} = 0.0951 \text{ pu}, \quad \overline{L}_{fkd} = \frac{L_{fkd}}{L_{kfd\text{base}}} = 1.1171 \text{ pu} \tag{7.135}$$

For the amortisseur circuits, the parameters are

$$\overline{R}_{kd} = \frac{R_{kkd}}{Z_{kkd\text{base}}} = 0.0191 \text{ pu}, \quad \overline{R}_{kq} = \frac{R_{kkq}}{Z_{kkq\text{base}}} = 0.04143 \text{ pu} \tag{7.136}$$

$$\overline{L}_{kkd} = \frac{L_{kkd}}{L_{kkd\text{base}}} = 1.2353 \text{ pu}, \quad \overline{L}_{kkq} = \frac{L_{kkq}}{L_{kkq\text{base}}} = 1.1400 \text{ pu} \tag{7.137}$$

$$\overline{L}_{\ell kd} = \overline{L}_{kkd} - \overline{L}_{akd} = 0.1306 \text{ pu}, \quad \overline{L}_{\ell kq} = \overline{L}_{kkq} - \overline{L}_{aq} = 0.1564 \text{ pu} \tag{7.138}$$

■

7.6 Equivalent Circuits

7.6.1 Flux-Linkage Circuits

The three per-unitized reciprocal d-axis flux-linkage equations are summarized as

$$\overline{\psi}_d = -(\overline{L}_\ell + \overline{L}_{ad})\overline{i}_d + \overline{L}_{ad}\overline{i}_{fd} + \overline{L}_{ad}\overline{i}_{kd} \tag{7.139}$$

$$\overline{\psi}_{fd} = -\overline{L}_{ad}\overline{i}_d + \overline{L}_{ffd}\overline{i}_{fd} + \overline{L}_{fkd}\overline{i}_{kd} \tag{7.140}$$

$$\overline{\psi}_{kd} = -\overline{L}_{ad}\overline{i}_d + \overline{L}_{fkd}\overline{i}_{fd} + \overline{L}_{kkd}\overline{i}_{kd} \tag{7.141}$$

These three equations can be combined into a single d-axis equivalent flux-linkage circuit as shown in Figure 7.6. Note that in Figure 7.6, $\overline{L}_{\ell fd} = \overline{L}_{ffd} - \overline{L}_{fkd}$ and $\overline{L}_{\ell kd} = \overline{L}_{kkd} - \overline{L}_{fkd}$ are leakage inductances. Note also that only one d-axis amortisseur circuit is shown in Figure 7.6. If more amortisseur circuits are included, the $\overline{L}_{\ell kd}$ branch is replaced by an appropriate number of parallel branches of inductances $\overline{L}_{\ell id}$ driven by fluxes $\overline{\psi}_{id}$, $i = 1, 2, ..., n_d$, where n_d is the number of individual d-axis circuits.

The two per-unitized reciprocal q-axis flux-linkage equations are summarized as

$$\overline{\psi}_q = -(\overline{L}_\ell + \overline{L}_{aq})\overline{i}_q + \overline{L}_{aq}\overline{i}_{kq} \tag{7.142}$$

$$\overline{\psi}_{kq} = -\overline{L}_{aq}\overline{i}_q + \overline{L}_{kkq}\overline{i}_{kq} \tag{7.143}$$

These two equations can be combined into a single q-axis equivalent flux-linkage circuit as shown in Figure 7.7, where $\overline{L}_{\ell kq} = \overline{L}_{kkq} - \overline{L}_{aq}$ is the leakage inductance. Again note

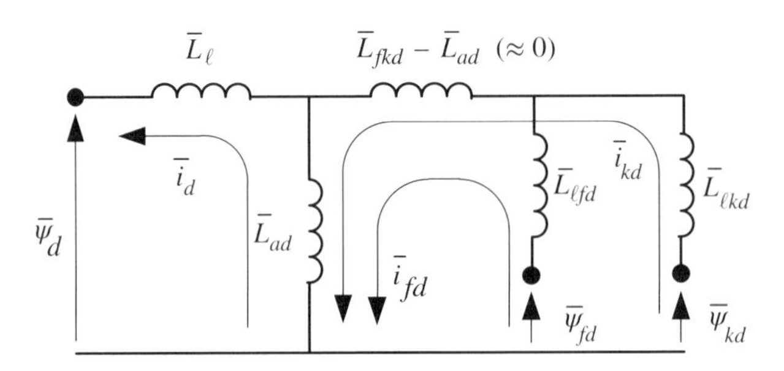

Figure 7.6 d-axis magnetic circuit.

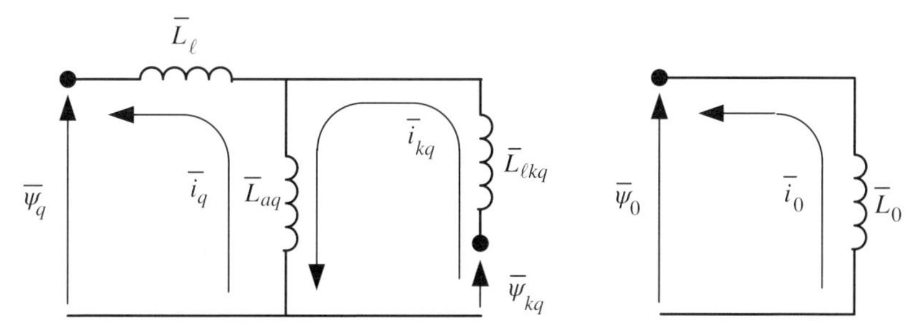

Figure 7.7 q-axis and 0-sequence magnetic circuits.

that only one q-axis amortisseur circuit is shown in Figure 7.7. In the subtransient synchronous machine model, the q-axis amortisseur winding is represented by two parallel circuits.

Finally, the zero-sequence flux-linkage equation is

$$\overline{\psi}_0 = -\overline{L}_0\overline{i}_0 \tag{7.144}$$

which is in general not modeled.

7.6.2 Voltage Equivalent Circuits

The flux-linkage equations involve instantaneous algebraic quantities. The voltage equations, which involve time derivatives of the flux linkages, represent dynamic equations used for time simulations. The three d-axis voltage equations in reciprocal per-unit bases are

$$\begin{aligned}\overline{e}_d &= \frac{d\overline{\psi}_d}{dt} - \overline{\omega}\,\overline{\psi}_q - \overline{R}_a\overline{i}_d \\ &= -(\overline{L}_\ell + \overline{L}_{ad})\frac{d\overline{i}_d}{dt} + \overline{L}_{ad}\frac{d\overline{i}_{fd}}{dt} + \overline{L}_{ad}\frac{d\overline{i}_{kd}}{dt} - \overline{\omega}\,\overline{\psi}_q - \overline{R}_a\overline{i}_d\end{aligned} \tag{7.145}$$

$$\begin{aligned}\overline{e}_{fd} &= \frac{d\overline{\psi}_{fd}}{dt} + \overline{R}_{fd}\overline{i}_{fd} \\ &= -\overline{L}_{ad}\frac{d\overline{i}_d}{dt} + (\overline{L}_{\ell fd} + \overline{L}_{ad})\frac{d\overline{i}_{fd}}{dt} + \overline{L}_{ad}\frac{d\overline{i}_{kd}}{dt} + \overline{R}_{fd}\overline{i}_{fd}\end{aligned} \tag{7.146}$$

$$\begin{aligned}\overline{e}_{kd} &= \frac{d\overline{\psi}_{kd}}{dt} + \overline{R}_{kd}\overline{i}_{kd} \\ &= -\overline{L}_{ad}\frac{d\overline{i}_d}{dt} + \overline{L}_{ad}\frac{d\overline{i}_{fd}}{dt} + (\overline{L}_{\ell kd} + \overline{L}_{ad})\frac{d\overline{i}_{kd}}{dt} + \overline{R}_{kd}\overline{i}_{kd} = 0\end{aligned} \tag{7.147}$$

Like the flux-linkage equations, these three per-unit equations can also be combined into a single d-axis equivalent voltage circuit as shown in Figure 7.8.

The per-unitized reciprocal q-axis voltage equations are

$$\overline{e}_q = \frac{d\overline{\psi}_q}{dt} + \overline{\omega}\,\overline{\psi}_d - \overline{R}_a\overline{i}_q \tag{7.148}$$

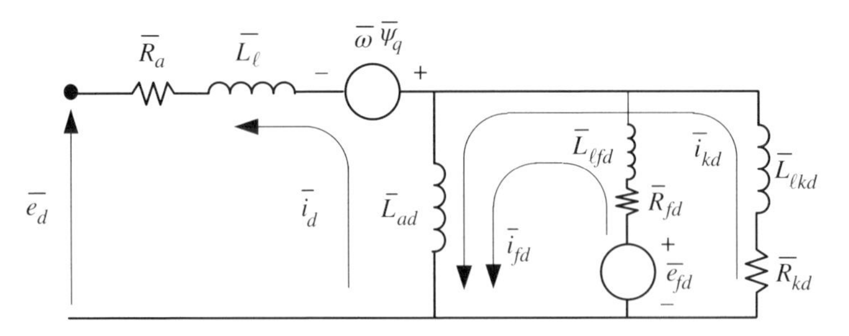

Figure 7.8 d-axis electrical circuit.

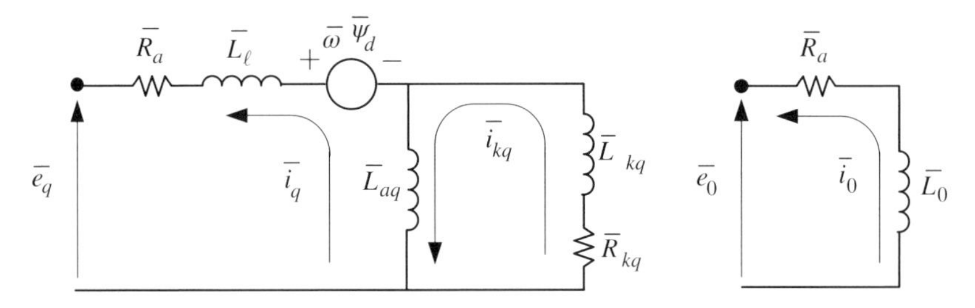

Figure 7.9 q-axis and 0-sequence electrical circuits.

$$\overline{e}_{kq} = \frac{d\overline{\psi}_{kq}}{dt} + \overline{R}_{kq}\overline{i}_{kq} = 0 \tag{7.149}$$

which can be combined into the q-axis circuit shown in Figure 7.9.

The 0-axis voltage equation is given by

$$\overline{e}_0 = \frac{d\overline{\psi}_0}{dt} - \overline{R}_a\overline{i}_0 \tag{7.150}$$

which can be modeled by a separate circuit, also shown in Figure 7.9.

The stator flux and voltage equations in the reciprocal, equal mutual-reactance per-unit system are summarized below, with t in radians

$$\overline{\psi}_d = -\overline{L}_d\overline{i}_d + \overline{L}_{afd}\overline{i}_{fd} + \overline{L}_{akd}\overline{i}_{kd} \tag{7.151}$$

$$\overline{\psi}_q = -\overline{L}_q\overline{i}_q \qquad\qquad + \overline{L}_{akq}\overline{i}_{kq} \tag{7.152}$$

$$\overline{\psi}_0 = -\overline{L}_0\overline{i}_0 \tag{7.153}$$

$$\overline{e}_d = \frac{d\overline{\psi}_d}{dt} - \overline{\omega}\,\overline{\psi}_q - \overline{R}_a\overline{i}_d \tag{7.154}$$

$$\overline{e}_q = \frac{d\overline{\psi}_q}{dt} + \overline{\omega}\,\overline{\psi}_d - \overline{R}_a\overline{i}_q \tag{7.155}$$

$$\overline{e}_0 = \frac{d\overline{\psi}_0}{dt} \qquad\qquad - \overline{R}_a\overline{i}_0 \tag{7.156}$$

The field and amortisseur winding flux and voltage equations in the reciprocal, equal mutual-reactance per-unit system are summarized below

$$\overline{\psi}_{fd} = -\overline{L}_{afd}\overline{i}_d + \overline{L}_{ffd}\overline{i}_{fd} + \overline{L}_{fkd}\overline{i}_{kd} \tag{7.157}$$

$$\overline{\psi}_{kd} = -\overline{L}_{ad}\overline{i}_d + \overline{L}_{fkd}\overline{i}_{fd} + \overline{L}_{kkd}\overline{i}_{kd} \tag{7.158}$$

$$\overline{\psi}_{kq} = -\overline{L}_{aq}\overline{i}_q + \overline{L}_{kkq}\overline{i}_{kq} \tag{7.159}$$

$$\overline{e}_{fd} = \frac{d\overline{\psi}_{fd}}{dt} + \overline{R}_{fd}\overline{i}_{fd} \tag{7.160}$$

$$\overline{e}_{kd} = \frac{d\overline{\psi}_{kd}}{dt} + \overline{R}_{kd}\overline{i}_{kd} \tag{7.161}$$

$$\overline{e}_{kq} = \frac{d\overline{\psi}_{kq}}{dt} + \overline{R}_{kq}\overline{i}_{kq} \tag{7.162}$$

7.7 Steady-State Analysis

In steady-state balanced-load synchronous-speed operation, $\overline{i}_{kd} = 0$, $\overline{i}_{kq} = 0$, $\overline{i}_0 = 0$, $d\psi/dt = 0$, and $\overline{\omega} = 1$. Then the stator equations in steady state simplify to

$$\begin{aligned} &\overline{\psi}_d = -\overline{L}_d\overline{i}_d + \overline{L}_{afd}\overline{i}_{fd}, \quad \overline{\psi}_q = -\overline{L}_q\overline{i}_q, \quad \overline{\psi}_0 = 0 \\ &\overline{e}_d = -\overline{\psi}_q - \overline{R}_a\overline{i}_d, \quad \overline{e}_q = \overline{\psi}_d - \overline{R}_a\overline{i}_q, \quad \overline{e}_0 = 0 \end{aligned} \tag{7.163}$$

and the field and amortisseur equations simplify to

$$\overline{\psi}_{fd} = -\overline{L}_{afd}\overline{i}_d + \overline{L}_{ffd}\overline{i}_{fd}, \quad \overline{\psi}_{kd} = -\overline{L}_{ad}\overline{i}_d + \overline{L}_{fkd}\overline{i}_{fd}, \quad \overline{\psi}_{kq} = -\overline{L}_{aq}\overline{i}_q, \quad \overline{e}_{fd} = \overline{R}_{fd}\overline{i}_{fd} \tag{7.164}$$

These equations are now used to study the operation of synchronous machines.

7.7.1 Open-Circuit Condition

For a synchronous machine operating at synchronous speed, under the open-circuit condition with $i_a = i_b = i_c = 0$, $\overline{i}_d = 0$ and $\overline{i}_q = 0$, the steady-state equations (7.163) and (7.164) reduce to

$$\overline{\psi}_d = \overline{L}_{afd}\overline{i}_{fd}, \quad \overline{\psi}_q = 0, \quad \overline{e}_d = 0, \quad \overline{e}_q = \overline{\psi}_d = \overline{L}_{afd}\overline{i}_{fd} \tag{7.165}$$

From the terminal voltage

$$\overline{e}_t = \sqrt{\overline{e}_d^2 + \overline{e}_q^2} = \overline{e}_q \tag{7.166}$$

define $\overline{E}_I$ to be a voltage proportional to $\overline{i}_{fd}$ and equal to $\overline{e}_t$

$$\overline{E}_I = \overline{L}_{ad}\overline{i}_{fd} \tag{7.167}$$

Combining the field voltage

$$\overline{e}_{fd} = \overline{R}_{fd}\overline{i}_{fd} \tag{7.168}$$

with the $\overline{e}_q$ equation in (7.165) results in

$$\overline{e}_q = \overline{L}_{afd}\frac{\overline{e}_{fd}}{\overline{R}_{fd}} \tag{7.169}$$

Then define $\overline{E}_{fd}$ to be a voltage proportional to $\overline{e}_{fd}$ and equal to $\overline{e}_t$ to obtain

$$\overline{E}_{fd} = \frac{\overline{L}_{ad}}{\overline{R}_{fd}} \overline{e}_{fd} \tag{7.170}$$

Combining the field flux linkage

$$\overline{\psi}_{fd} = \overline{L}_{ffd} \overline{i}_{fd} \tag{7.171}$$

with the $\overline{e}_q$ expression in (7.165) results in

$$\overline{e}_q = \frac{\overline{L}_{ad}}{\overline{L}_{ffd}} \overline{\psi}_{fd} \tag{7.172}$$

Define $\overline{e}'_q$ to be a voltage proportional to $\overline{\psi}_{fd}$ and equal to $\overline{e}_q = \overline{e}_t$

$$\overline{e}'_q = \frac{\overline{L}_{ad}}{\overline{L}_{ffd}} \overline{\psi}_{fd} \tag{7.173}$$

Applying these new variables to the stator equations (7.163) results in

$$\overline{\psi}_d = -\overline{L}_d \overline{i}_d + \overline{E}_I, \quad \overline{\psi}_q = -\overline{L}_q \overline{i}_q, \quad \overline{e}_d = -\overline{\psi}_q - \overline{R}_a \overline{i}_d, \quad \overline{e}_q = \overline{\psi}_d - \overline{R}_a \overline{i}_q \tag{7.174}$$

and to the rotor equations (7.164) results in

$$\overline{\psi}_{fd} = \overline{L}_{ffd} \overline{i}_{fd} - \overline{L}_{ad} \overline{i}_d, \quad \overline{e}_{fd} = \overline{R}_{fd} \overline{i}_{fd} \tag{7.175}$$

Two important voltages are $\overline{e}'_q$ and E_I. From (7.173) and (7.175), the expression

$$\overline{e}'_q \frac{\overline{L}_{ffd}}{\overline{L}_{ad}} = L_{ffd} \frac{\overline{E}_I}{\overline{L}_{ad}} - \overline{L}_{ad} \overline{i}_d \tag{7.176}$$

simplifies to

$$\overline{e}'_q = \overline{E}_I - \frac{\overline{L}_{ad}^2}{\overline{L}_{ffd}} \overline{i}_d \tag{7.177}$$

Define the d-axis transient inductance as

$$\overline{L}'_d = \overline{L}_\ell + \frac{1}{\frac{1}{\overline{L}_{ad}} + \frac{1}{\overline{L}_{\ell fd}}} = \overline{L}_\ell + \frac{\overline{L}_{ad} \overline{L}_{\ell fd}}{\overline{L}_{ad} + \overline{L}_{\ell fd}} \tag{7.178}$$

Then

$$\overline{L}_d - \overline{L}'_d = \overline{L}_{ad} + \overline{L}_\ell - \overline{L}_\ell - \frac{\overline{L}_{ad} \overline{L}_{\ell fd}}{\overline{L}_{ad} + \overline{L}_{\ell fd}} = \frac{\overline{L}_{ad}^2}{\overline{L}_{ffd}} \tag{7.179}$$

That is,

$$\overline{e}'_q = \overline{E}_I - (\overline{L}_d - \overline{L}'_d) \overline{i}_d \tag{7.180}$$

Combining (7.170), (7.168), and (7.167) yields the equation

$$\overline{E}_{fd}\frac{\overline{R}_{fd}}{\overline{L}_{ad}} = \overline{R}_{fd}\frac{\overline{E}_I}{\overline{L}_{ad}} \tag{7.181}$$

which simplifies to

$$\overline{E}_{fd} = \overline{E}_I \tag{7.182}$$

From the open-circuit operating condition $\overline{e}_d = \overline{e}_0 = 0$ and $\overline{e}_q = \overline{E}_I = \overline{e}_t$, the individual phase voltages can be obtained from the Park transformation as

$$\overline{e}_a = -\overline{E}_I \sin\theta, \quad \overline{e}_b = -\overline{E}_I \sin(\theta - 2\pi/3), \quad \overline{e}_c = -\overline{E}_I \sin(\theta + 2\pi/3) \tag{7.183}$$

7.7.2 Loaded Condition

Consider a synchronous machine connected to a load as shown in Figure 7.10. The generator is represented by the field voltage $\overline{E}_I$ and the generator terminal voltage magnitude in per-unit rms value is denoted by $\overline{E}_t$.

Assume that $\overline{E}_t$ lags $\overline{E}_I$ by an angle δ, which defines the q-axis relative to $\overline{E}_t$. The stator voltages in steady state are

$$\overline{e}_a = -\overline{E}_t \sin(\theta - \delta) = \overline{E}_t \sin(\delta - \theta) \tag{7.184}$$

$$\overline{e}_b = -\overline{E}_t \sin(\theta - 2\pi/3 - \delta) = \overline{E}_t \sin(\delta - \theta + 2\pi/3) \tag{7.185}$$

$$\overline{e}_c = -\overline{E}_t \sin(\theta + 2\pi/3 - \delta) = \overline{E}_t \sin(\delta - \theta - 2\pi/3) \tag{7.186}$$

Using the Park transformation, the $dq0$-axes voltages are

$$\overline{e}_d = \overline{E}_t \sin\delta, \quad \overline{e}_q = \overline{E}_t \cos\delta, \quad \overline{e}_0 = 0 \tag{7.187}$$

as shown in Figure 7.11.

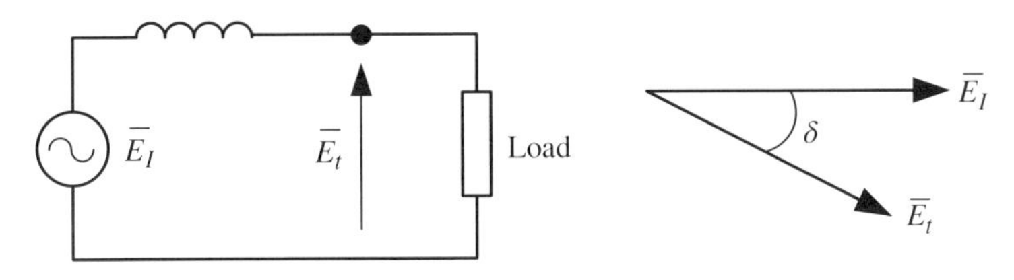

Figure 7.10 Generator supplying a load.

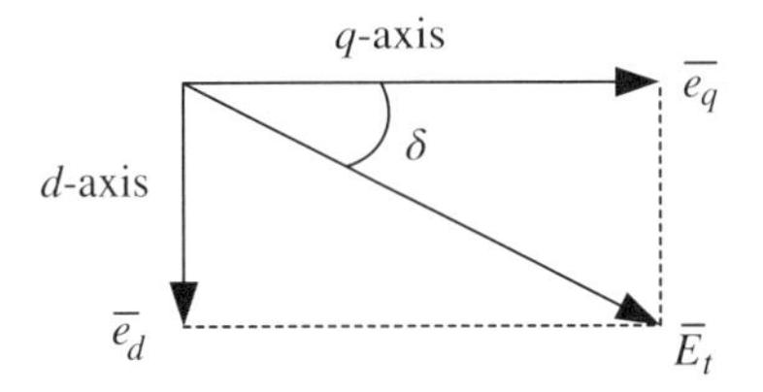

Figure 7.11 Decomposition of $\overline{E}_t$ onto the dq-axes.

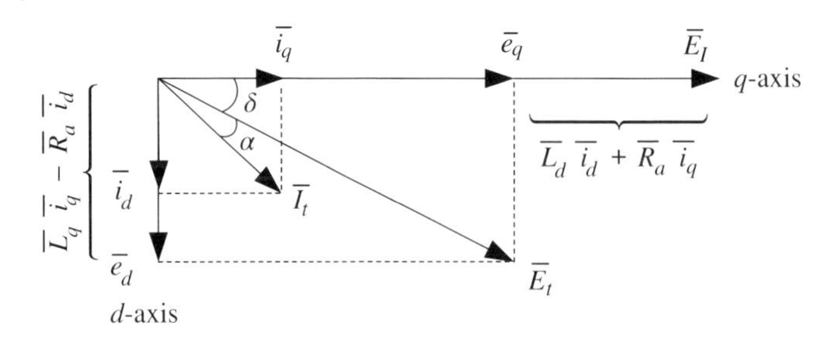

Figure 7.12 Voltage-current phasor diagram.

Substituting the $\overline{\psi}_d$ and $\overline{\psi}_q$ terms into the stator d- and q-axis voltage equations (7.163) yields

$$\overline{e}_q + \overline{R}_a\overline{i}_q = -\overline{L}_d\overline{i}_d + \overline{E}_I, \quad \overline{e}_d + \overline{R}_a\overline{i}_d = \overline{L}_q\overline{i}_q \tag{7.188}$$

which can be rewritten as

$$\overline{E}_I = \overline{e}_q + \overline{R}_a\overline{i}_q + \overline{L}_d\overline{i}_d, \quad \overline{e}_d = \overline{L}_q\overline{i}_q - \overline{R}_a\overline{i}_d \tag{7.189}$$

Equation (7.189) involves the decomposition of the generator current $\overline{I}_t$ into $\overline{i}_d$ and $\overline{i}_q$. The current components can be added to the diagram in Figure 7.11 to form the voltage-current phasor diagram Figure 7.12. If the current $\overline{I}_t$ lags the terminal voltage $\overline{E}_t$ by an angle of α, which is known as the power angle, then

$$\overline{i}_d = \overline{I}_t \sin(\delta + \alpha), \quad \overline{i}_q = \overline{I}_t \cos(\delta + \alpha) \tag{7.190}$$

Given $\overline{e}_d$, $\overline{e}_q$, and $\overline{E}_I$, (7.188) can be rearranged as

$$\begin{bmatrix} \overline{L}_d & \overline{R}_a \\ -\overline{R}_a & \overline{L}_q \end{bmatrix} \begin{bmatrix} \overline{i}_d \\ \overline{i}_q \end{bmatrix} = \begin{bmatrix} \overline{E}_I - \overline{e}_q \\ \overline{e}_d \end{bmatrix} \tag{7.191}$$

from which $\overline{i}_d$ and $\overline{i}_q$ can be calculated as

$$\begin{aligned} \begin{bmatrix} \overline{i}_d \\ \overline{i}_q \end{bmatrix} &= \frac{1}{\overline{L}_d\overline{L}_q + \overline{R}_a^2} \begin{bmatrix} \overline{L}_q & -\overline{R}_a \\ \overline{R}_a & \overline{L}_d \end{bmatrix} \begin{bmatrix} \overline{E}_I - \overline{e}_q \\ \overline{e}_d \end{bmatrix} \\ &= \frac{1}{\overline{L}_d\overline{L}_q + \overline{R}_a^2} \begin{bmatrix} \overline{L}_q(E_I - \overline{e}_q) - \overline{R}_a\overline{e}_d \\ \overline{L}_d\overline{e}_d + \overline{R}_a(\overline{E}_I - \overline{e}_q) \end{bmatrix} \end{aligned} \tag{7.192}$$

7.7.3 Drawing Voltage-Current Phasor Diagrams

The voltage-current phasor diagram allows the computation of the dq-axes voltages and currents in a synchronous machine in under-load conditions. It is also useful in obtaining the initial conditions of a synchronous machine before the application of a disturbance. In such calculations, the per-unitized generator terminal voltage phasor $\tilde{\overline{E}}_t$ and the per-unitized generator terminal current phasor $\tilde{\overline{I}}_t$ are given, as determined from a power flow solution. The key step is to systematically locate the dq-axes, so that the relevant voltages can be computed. Because reactances are used in simulation programs, the notation will be switched to reactance $X = \omega_s L$ in pu, with ω_s at 1 pu.

Procedure to draw a voltage-current phasor diagram

1) Determine the per-unit phasors $\tilde{\bar{E}}_t$ and $\tilde{\bar{I}}_t$ using the power factor angle α. Normally $\tilde{\bar{E}}_t$ is placed on the positive real axis.
2) Calculate $\tilde{\bar{E}}_q = (\bar{R}_a + j\bar{X}_q)\tilde{\bar{I}}_t$ to locate the q-axis and the d-axis is a 90° clockwise rotation of the q-axis.
3) Project $\tilde{\bar{E}}_t$ and $\tilde{\bar{I}}_t$ on the d- and q-axes to find $\tilde{\bar{e}}_d$, $\tilde{\bar{e}}_q$, $\tilde{\bar{i}}_d$, and $\tilde{\bar{i}}_q$.
4) Calculate the voltage behind the transient reactance $\tilde{\bar{e}}' = \tilde{\bar{E}}_t + (\bar{R}_a + j\bar{X}'_d)\tilde{\bar{I}}_t$.
5) Use $\tilde{\bar{i}}_d$ to find the field current $\tilde{\bar{E}}_I = \tilde{\bar{E}}_q + j(\bar{X}_d - \bar{X}_q)\tilde{\bar{i}}_d$.
6) From $\tilde{\bar{E}}_I$, find the field flux linkage $\tilde{\bar{e}}'_q = \tilde{\bar{E}}_I - j(\bar{X}_d - \bar{X}'_d)\tilde{\bar{i}}_d$. Alternatively, project $\tilde{\bar{e}}'$ onto the q-axis to find $\tilde{\bar{e}}'_q$.

The quantities computed in this procedure are shown in Figure 7.13.

Example 7.3: *V-I* phasor diagram

A generator rated at 20 MVA and 4 kV has the following parameters in pu on the machine base:

$$R_a = 0.03, \quad X_q = 1.65, \quad X_d = 1.76, \quad X'_d = 0.285 \tag{7.193}$$

The machine is operating under steady-state synchronous-speed condition at rated voltage ($E_t = 1.0$ pu) and delivering 18.0 MW at 0.9 power factor. The load is balanced. Find $\tilde{E}_I$ and $\tilde{e}'_q$.

Solutions: Convert the electrical power P_e into per unit to obtain

$$P_e = (18.0 \times 10^6)/(20 \times 10^6) = 0.90 \text{ pu} \tag{7.194}$$

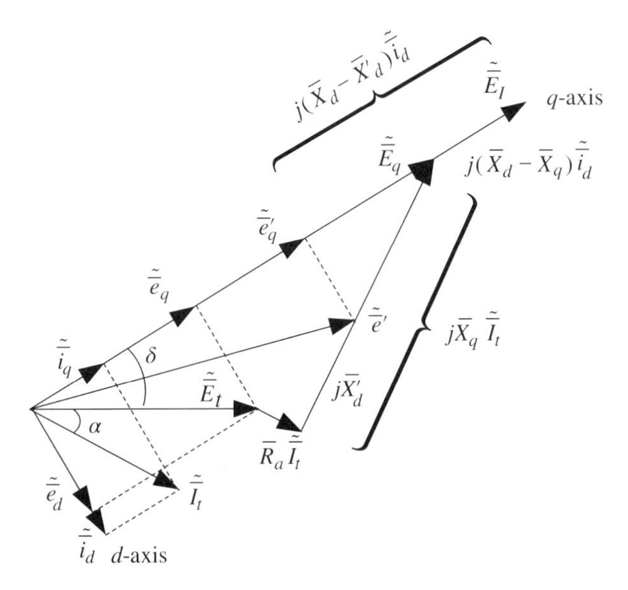

Figure 7.13 Voltage-current phasor diagram.

The power angle and the magnitude of the apparent power output of the synchronous machine are

$$\alpha = \cos^{-1}(0.9) = 25.84^\circ, \quad |S| = P_e/0.9 = 1.0 \text{ pu} \tag{7.195}$$

The apparent power is given by

$$S = P_e + jQ_e = 0.90 + j|S|\sin(\alpha) = 0.90 + j0.4359 \text{ pu} \tag{7.196}$$

Setting the angle of the terminal voltage to be zero, that is, $\tilde{E}_t = 1$ pu, the terminal current is given by

$$\tilde{I}_t = S^*/\tilde{E}_t = 0.90 - j0.4359 = 1.0e^{-j25.84^\circ} \text{ pu} \tag{7.197}$$

Then $\tilde{E}_q$ is computed to locate the q-axis:

$$\tilde{E}_q = \tilde{E}_t + (\overline{R}_a + j\overline{X}_q)\tilde{I}_t = 1.7462 + j1.4719 = 2.2838e^{j40.13^\circ} \text{ pu} \tag{7.198}$$

The angle $\delta = 40.13^\circ$ is the rotor angle relative to the terminal bus phasor $\tilde{E}_t$ and defines the q-axis. The voltage behind the transient reactance is

$$\tilde{e}' = \tilde{E}_t + (\overline{R}_a + j\overline{X}'_d)\tilde{I}_t = 1.1512 + j0.2434 = 1.1767e^{j11.94^\circ} \text{ pu} \tag{7.199}$$

The terminal voltage is decomposed into the q-axis component

$$\tilde{e}_q = |\tilde{E}_t| \cos\delta\varepsilon^{j\delta} = 0.5846 + j0.4928 = 0.7646e^{j\delta} \text{ pu} \tag{7.200}$$

and the d-axis component

$$\tilde{e}_d = |\tilde{E}_t| \sin\delta\varepsilon^{j(\delta-90^\circ)} = 0.4154 - j0.4928 = 0.6445e^{j(\delta-90^\circ)} \text{ pu} \tag{7.201}$$

Similarly, the current $\tilde{i}_t$ can be decomposed into

$$\tilde{i}_q = |\tilde{I}_t| \cos(\delta + \alpha)e^{j\delta} = 0.3114 + j0.2624 = 0.4072e^{j\delta} \text{ pu} \tag{7.202}$$

$$\tilde{i}_d = |\tilde{I}_t| \sin(\delta + \alpha)e^{j(\delta-90^\circ)} = 0.5886 - j0.6983 = 0.9133e^{j(\delta-90^\circ)} \text{ pu} \tag{7.203}$$

The field voltage can then be computed as

$$\tilde{E}_I = \tilde{E}_q + j(\overline{X}_d - \overline{X}_q)\tilde{i}_d = 1.8230 + j1.5367 = 2.3843e^{j\delta} \text{ pu} \tag{7.204}$$

and the field flux linkage

$$\tilde{e}'_q = \tilde{E}_I - j(\overline{X}_d - \overline{X}'_d)\tilde{i}_d = 0.7930 + j0.6684 = 1.0371e^{j\delta} \text{ pu} \tag{7.205}$$

It should be noted that the generator is providing reactive power to the system. Thus in terms of pu voltage values, the machine internal voltages, particularly the field voltage, will be higher than the terminal voltage. ∎

7.8 Saturation Effects

Thus far in this chapter the magnetic circuits in a synchronous machine are assumed to be linear, so that the inductances are independent of the field current. However, a synchronous machine is normally designed so that magnetic saturation will occur if it is operated at its rated load condition. Magnetic saturation is illustrated using the

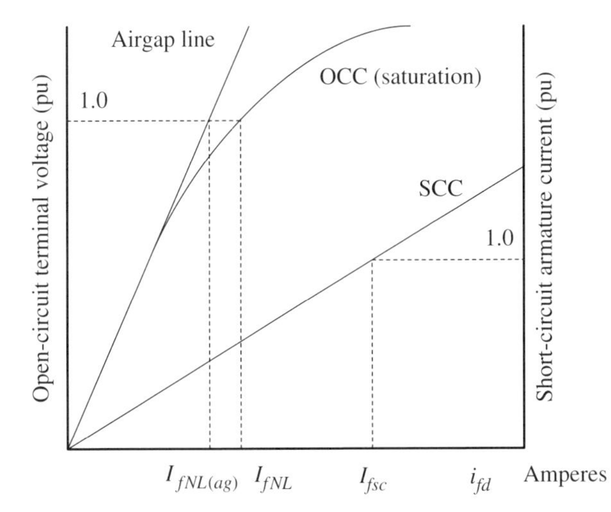

Figure 7.14 Synchronous machine open-circuit voltage and short-circuit current curves.

open-circuit characteristics (OCC) between the field current i_{fd} and the open-circuit terminal voltage, as shown in the Figure 7.14.

For a synchronous machine operating at rated speed and no-load (open circuit), the fluxes, voltages, and currents are

$$\bar{i}_d = \bar{i}_q = \bar{\psi}_q = \bar{e}_d = 0, \quad \bar{E}_t = \bar{e}_q = \bar{\psi}_d = \bar{L}_{ad}\bar{i}_{fd} \tag{7.206}$$

The airgap line in Figure 7.14 assumes no saturation effect. The slope of the airgap line is the unsaturated $\bar{L}_{ad}$. The OCC curve shows the effect of saturation, requiring a higher field current to achieve rated voltage operation of the synchronous machine.

Under short-circuit condition, the terminal voltage of the synchronous machine is $\bar{E}_t = 0$. The loading on the machine is negligible and there is no saturation. Thus the current $\bar{i}_d$ is linear with respect to $\bar{i}_{fd}$. At 1.0 pu armature current, the internal voltage is proportional to the short-circuit field current $\bar{I}_{fsc}$, that is,

$$K_{\text{SCC}} \cdot \bar{I}_{fsc} = 1.0 \cdot \bar{X}_{s(\text{unsat})} \tag{7.207}$$

where K_{SCC} is a constant, and $\bar{X}_s$ is the synchronous reactance (from the synchronous reactance $\bar{L}_{ad}$). However, the same constant K_{SCC} also applies to OCC:

$$K_{\text{SCC}} \cdot \bar{I}_{fNL(ag)} = 1.0 \tag{7.208}$$

Thus the reactances and the short-circuit ratio (SCR) are

$$\bar{X}_{s(\text{unsat})} = \frac{\bar{I}_{fsc}}{\bar{I}_{fNL(ag)}}, \quad \bar{X}_{s(\text{sat})} = \frac{\bar{I}_{fsc}}{\bar{I}_{fNL}} < \bar{X}_{s(\text{unsat})}, \quad \text{SCR} = \frac{1}{\bar{X}_{s(\text{sat})}} \tag{7.209}$$

7.8.1 Representations of Magnetic Saturation

There are in general two ways of representing magnetic saturation in power system dynamic simulation programs. The saturation may be represented separately or jointly

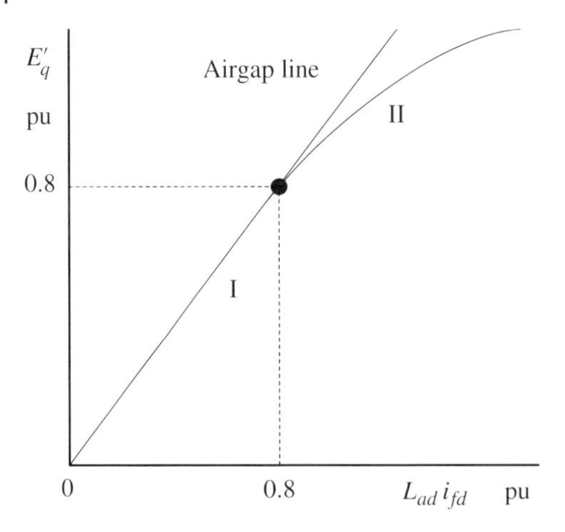

Figure 7.15 Exponential saturation representation.

for the d- and q-axes. The representations typically divide the saturation curve into two parts or segments: Segment I is linear (typically up to 0.8 pu voltage) and Segment II is nonlinear.

1. Exponential function

The exponential saturation representation is shown in Figure 7.15, in which Segment II is determined from

$$\overline{L}_{ad}\overline{i}_{fd} = A_{\text{sat}} e^{B_{\text{sat}}(\overline{E}_q' - 0.8)} \tag{7.210}$$

from which A_{sat} and B_{sat} are estimated from measured data obtained in open-circuit generator testing.

2. Quadratic polynomial fit

In the second scheme, Segment II is represented by a quadratic polynomial as

$$\overline{L}_{ad}\overline{i}_{fd} = \alpha_0 + \alpha_1 \overline{E}_q' + \alpha_2 (\overline{E}_q')^2 \tag{7.211}$$

Figure 7.16 facilitates the calculation of the coefficients α_0, α_1, and α_2. To compute these coefficients, three points (a, b, c) on the saturation curve at $\overline{E}_q' = 0.8, 1.0, 1.2$ pu are specified. The values of $L_{ad}i_{fd} = s_2$ at $E_q' = 1.0$ pu, and $L_{ad}i_{fd} = s_4$ at $E_q' = 1.2$ pu are obtained.

These three data points, when substituted into (7.205), yield the matrix equation

$$\begin{bmatrix} \alpha_0 & \alpha_1 & \alpha_2 \end{bmatrix} \begin{bmatrix} 1 & 1 & 1 \\ 0.8 & 1 & 1.2 \\ 0.8^2 & 1 & 1.2^2 \end{bmatrix} = \begin{bmatrix} 0.8 & s_2 & s_4 \end{bmatrix} \tag{7.212}$$

For polynomial saturation curves, the saturation data are normally specified as

$$S_{E1.0} = \frac{s_2}{s_1} - 1 = s_2 - 1, \quad S_{E1.2} = \frac{s_4}{s_3} - 1 = \frac{s_4}{1.2} - 1 \tag{7.213}$$

The three points (a, b, c) can also be used to determine A_{sat} and B_{sat} of exponential saturation curves.

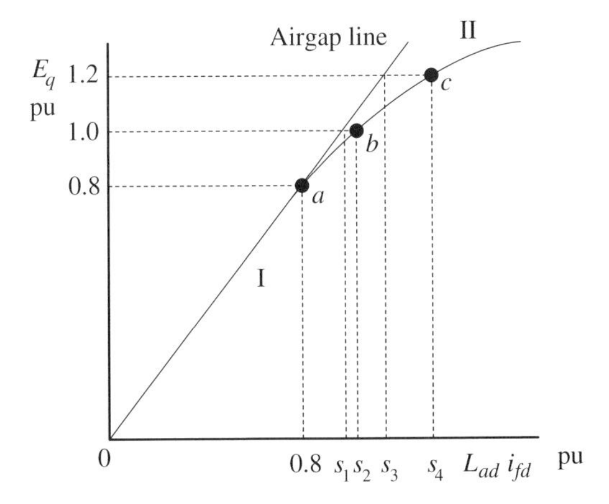

Figure 7.16 Polynomial saturation representation.

Example 7.4: Saturation curve parameters

In an open-circuit test of a synchronous machine, it is found that $i_{fd} = 378.4$ A when $E'_q = 0.8$ pu, $i_{fd} = 524$ A when $E'_q = 1.0$ pu, and $i_{fd} = 800$ A when $E'_q = 1.2$ pu. Find the magnetic saturation values $S_E(1.0)$ and $S_E(1.2)$, given that the base of the field current is 473 A. Also find the coefficients for a quadratic fit of the saturation function, and use it to find i_{fd} for $E_t = 1.1$ pu.

Solutions: The coefficients s_i are computed as

$$s_0 = \frac{378.4}{473} = 0.8, \quad s_2 = \frac{524}{473} = 1.1078, \quad s_4 = \frac{800}{473} = 1.6913 \tag{7.214}$$

such that

$$S_{E1.0} = 1.1078 - 1 = 0.1078, \quad S_{E1.2} = \frac{1.6913}{1.2} - 1 = 0.4094 \tag{7.215}$$

From (7.212), the magnetic saturation beyond $\bar{i}_{fd} = 0.8$ pu is approximated by

$$\begin{bmatrix} \alpha_0 & \alpha_1 & \alpha_2 \end{bmatrix} = \begin{bmatrix} 2.3256 & -4.6638 & 3.4461 \end{bmatrix} \tag{7.216}$$

The saturation curve determined by these parameters is shown in Figure 7.17. At $E_t =$ 1.1 pu,

$$i_{fd} = 2.3256 - 4.6638 \times 1.1 + 3.4461 \times 1.1^2 = 1.365 \text{ pu} = 645.7 \text{ A}. \tag{7.217}$$

■

7.9 Generator Capability Curves

In steady-state operation, the maximum active and reactive power output is limited by several thermodynamic and stability constraints. Figure 7.18 shows the typical steady-state operating capability of a generator as a function of the P and Q output,

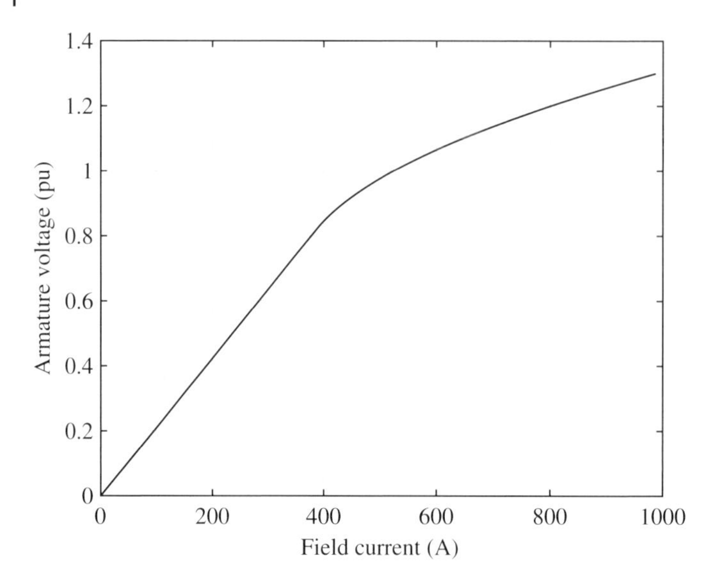

Figure 7.17 Saturation characteristic.

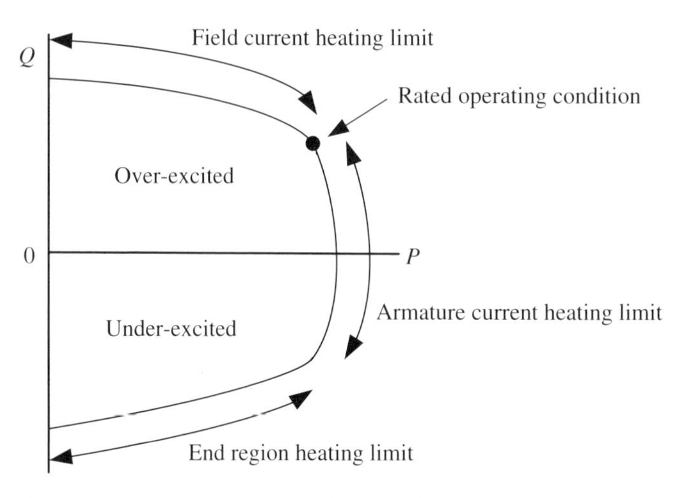

Figure 7.18 Generator capability shown in the *PQ* plane [94].

where positive values of P and Q imply that the generator is operating in the over-excited region and is supplying the power system with active and reactive power.

The generator limits are determined by three physical phenomena. The maximum active power output is limited primarily by the armature winding current heating limit. In over-excited operation, the reactive power supply is limited by the rotor field current heating limit. This is known as the over-excitation limit, denoted by OEL or OXL. A generator is normally operated in the over-excited region as it provides reactive power to support the transfer of active power from the generator. In the under-excited region, the reactive power is limited by the armature winding end-region heating limit, which is denoted by UEL (under-excitation limit). More detailed descriptions of synchronous generator protection can be found in [98].

These OEL and UEL limits are used by excitation systems to protect the operation of a generator, as discussed in Chapter 9 on excitation systems. Each of the limits is a function of time, allowing a generator to operate with overloaded currents for a short period of time. Such capabilities provide momentary voltage support, buying some time for operators to schedule additional reactive power support from switched shunt capacitors.

7.10 Summary and Notes

This chapter is the first part of developing analytical models of synchronous machines for use in dynamic simulation programs. The derivations include establishing the Park transformation from 3-phase circuits to the dq-axes, thus making the equations time-independent. The reciprocal, equal-mutual-inductance per-unit system is used so that equivalent flux-linkage and voltage circuits can be established. The equivalent circuits can be used to compute the d- and q-axis voltages and currents in the generator. These circuit models are also used in the next chapter to derive dynamic models for synchronous machines.

The discussion in this chapter follows the lecture notes [94] used in the General Electric Company's Power System Engineering Course in the 1970s. The authors are grateful to Dale Swann for permission to use his notes. Some of the problems are also adapted from [94] and [97] and are labelled as such.

A reader interested in a more in-depth discussion on topics such as synchronous machines under fault conditions should consult texts such as Concordia [93]. A more modern discussion of synchronous machine models is provided by Kundur [6], which also influences the development of this and the next chapter.

Problems

7.1 (D. Swann) Figure 7.19 indicates the relative relationships between the windings for a hypothetical machine having two stator windings a and b, which are identical, and a single pair of poles on the rotor, representing by the field winding f. The angular velocity ω of the rotor is constant.

1) The self inductances are L_{aa}, L_{bb}, L_{ff}, and the mutual inductances are L_{ab}, L_{ba}, L_{af}, L_{bf}. Find the relationships between these inductances (for example, some of them may be equal or zero, and some others maybe constant). Note that the mutual inductances between the stator and the rotor are a function of the angle $\theta = \omega t + \theta_o$ between the a-axis and the f-axis (the direct axis of the field winding).
2) Assume that the load is balanced and that the phase currents are given by

$$i_a = \sqrt{2} I_a \sin(\omega t + \theta_o + \phi), \quad i_b = -\sqrt{2} I_a \cos(\omega t + \theta_o + \phi) \tag{7.218}$$

where ϕ is the phase angle between e_a and i_a. Derive an expression for the field voltage e_{fd}.

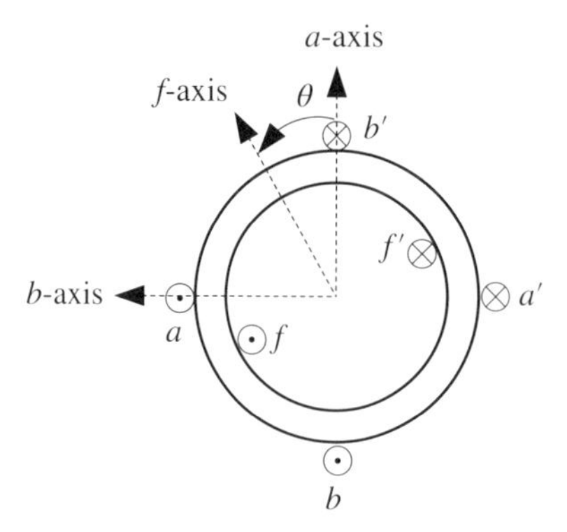

Figure 7.19 A 2-phase synchronous machine.

7.2 Show the derivation of Eqns. (7.45) and (7.48).

7.3 A machine is operating at a synchronous speed ω, with phases b and c open circuit such that

$$i_a = I_m \sin(\omega t - \phi), \quad i_b = i_c = 0 \tag{7.219}$$

where ϕ is an arbitrary, constant phase angle and I_m is the maximum value of the current. Find the instantaneous $dq0$ components of the current. Assume $\theta = 0$. What frequencies are these components?

7.4 When a 3-phase, terminal fault is applied to a generator, the dc components of the phase currents will increase. Suppose these components are

$$i_a = I_m/2, \quad i_b = I_m/2, \quad i_c = -I_m \tag{7.220}$$

where I_m is a constant (DC) current. Find the $dq0$ components due to these phase currents. Assume $\theta_o = 0$. What frequencies are these components?

7.5 For a 3-phase synchronous machine operating at steady-state synchronous speed ω_s, zero-sequence currents are injected into the stator

$$i_a = i_b = i_c = I_m \cos(\omega_s t + \phi) \tag{7.221}$$

Calculate i_d, i_q, and i_o.

7.6 (D. Swann) For a synchronous machine operating in steady state at synchronous speed and supplying a balanced load, indicate which of the following quantities are either 0 (zero), DC (constant), a function of ω_o (60 Hz), or a function of $2\omega_o$ (120 Hz):

i_a e_a ψ_a
i_d e_d ψ_d $d\psi_d/dt$
i_q e_q ψ_q $d\psi_q/dt$
i_{fd} e_{fd} ψ_{fd}
i_{kd} ψ_{kd}
i_{kq} ψ_{kq}

7.7 Use the expressions for ψ_a, ψ_b, and ψ_c (7.2)–(7.4) to derive the expressions for ψ_d, ψ_q, and ψ_0 (7.53)–(7.53), the flux linkages in the $dq0$ reference frame.

7.8 ([63], Example 4.1) A 3-phase, 60 Hz synchronous generator is rated at 160 MVA and 15 kV (line-to-line). It has the following parameters:

$$L_\ell = 0.0005595\text{ H}, \quad L_d = 0.006341\text{ H}, \quad L_q = 0.001423\text{ H}$$
$$L_{afd} = 0.005782\text{ H}, \quad L_{ffd} = 2.189\text{ H}, \quad R_{fd} = 0.00185\text{ ohms}$$

1) Calculate the following base quantities: $e_{s\text{base}}$, $i_{s\text{base}}$, $Z_{s\text{base}}$, $L_{s\text{base}}$, $i_{fd s\text{base}}$, $e_{fd\text{base}}$, and $L_{fd\text{base}}$.
2) Calculate the following per-unit parameters: $\overline{L}_d$, $\overline{L}_q$, $\overline{L}_{ad}$, $\overline{L}_{aq}$, $\overline{L}_{afd}$, $\overline{L}_{ffd}$, and $\overline{R}_{fd}$.

7.9 A generator rated at 426 MVA and 22.0 kV has the following parameters in pu:

$$\overline{R}_a = 0.002, \quad \overline{X}_d = 1.21, \quad \overline{X}_q = 1.09, \quad \overline{X}'_d = 0.193 \tag{7.222}$$

The machine is operating under steady-state synchronous speed conditions at rated voltage with an output of 380 MW and a leading 0.95 power factor (that is, the generator is over-excited). The load is balanced. Draw the synchronous machine voltage-current phasor diagram and find $\overline{E}_I$ and $\overline{E}'_q$.

7.10 A generator rated at 20 MVA and 4.0 kV has the following parameters in pu:

$$\overline{R}_a = 0.03, \quad \overline{X}_d = 1.40, \quad \overline{X}_q = 0.90, \quad \overline{X}'_d = 0.25$$

The machine is operating under steady-state synchronous speed conditions at rated voltage and 11.2 MW at 0.9 power factor lagging. The load is balanced. Calculate all the relevant d- and q-axes voltages and currents, as well as $\tilde{e}'$.

7.11 (Saturation function) Repeat Example 7.4 to find the coefficient A_{sat} and B_{sat} so that the saturation is represented by the exponential function

$$\overline{L}_{ad}\overline{i}_{fd} = A_{\text{sat}} e^{B_{\text{sat}}(\overline{E}'_q - 0.8)} \tag{7.223}$$

Note that this function is valid only for $\overline{E}'_q \geq 0.8$.

8

Dynamic Models of Synchronous Machines

8.1 Introduction

Following the discussion on the steady-state operation of synchronous machines in Chapter 7, this chapter develops positive-sequence synchronous machine models suitable for dynamic simulation of power system disturbances. A synchronous machine subject to a 3-phase fault exhibits a variety of time responses in different time scales, namely, the transient and subtransient effects, as it settles to a new steady state after the fault is cleared. These dynamics can be captured in detailed synchronous machine models based on the d- and q-axis circuits already used in Chapter 7 for steady-state analysis. Circuit models, however, are not convenient to use in a dynamic simulation program. Instead, the components of the d- and q-axis circuits are used to develop transient and subtransient time constants and reactances, such that a transfer-function-based model can be derived to represent a synchronous generator. This leads to a 4-electrical-state and 2-mechanical-state subtransient model, also known as the round-rotor model. The model includes two stator circuits and two rotor circuits and is commonly used for detailed synchronous machine representation for large turbine generators in dynamic simulation studies. From the detailed machine model, simplified models, including the 2-state classical model, can be readily obtained. For completeness, this chapter also discusses how to interface the detailed synchronous machine model to the power network, and how to develop a linearized synchronous machine model.

8.2 Machine Dynamic Response During Fault

To observe the dynamic phenomena present in a synchronous machine, consider a balanced 3-phase short-circuit fault applied to the power system near the machine terminals. Figure 8.1 shows a typical cycle-by-cycle response of the a-phase current and the field current. Note that the oscillations in these plots are due to the 3-phase nominal system frequency components. The dynamics of the response can be observed from the envelopes of these oscillations.

There are usually four distinct components of dynamic response (Figure 8.2):

1) DC offset and decay – the immediate time response following the short-circuit fault, when the phase currents are responding individually. The dynamics of the DC offset and decay are readily seen in the field current i_{fd} in Figure 8.1.

Power System Modeling, Computation, and Control, First Edition. Joe H. Chow and Juan J. Sanchez-Gasca.

Companion website: www.wiley.com/go/chow/power-system-modeling

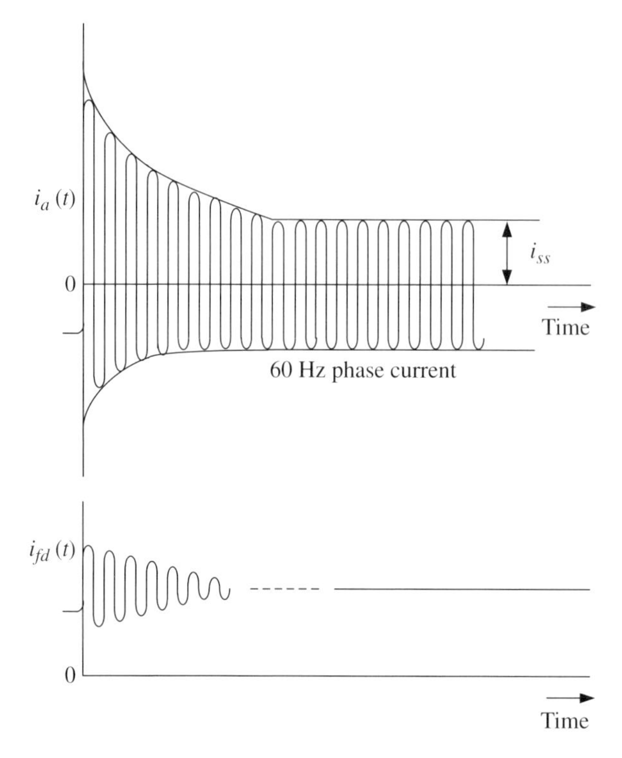

Figure 8.1 Oscillogram of cycle-by-cycle current response.

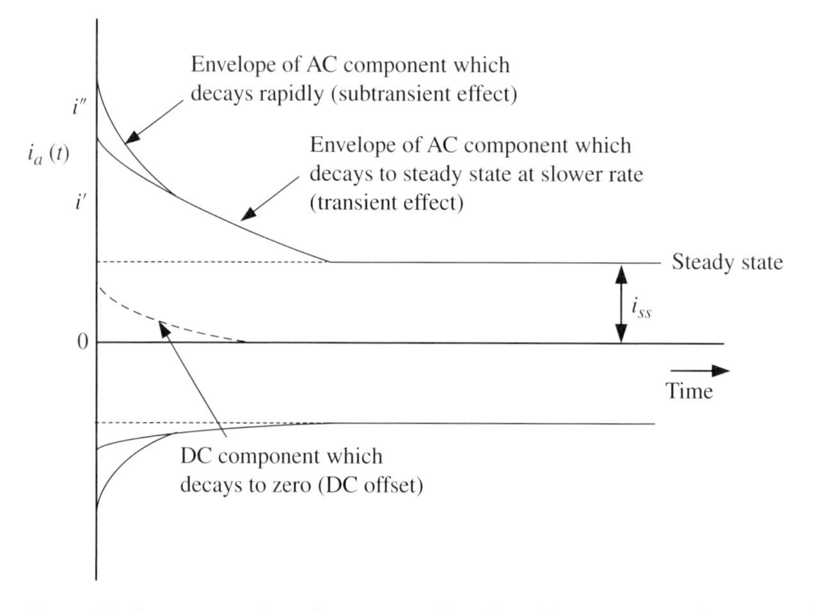

Figure 8.2 Components of synchronous-machine dynamic response to a short-circuit fault.

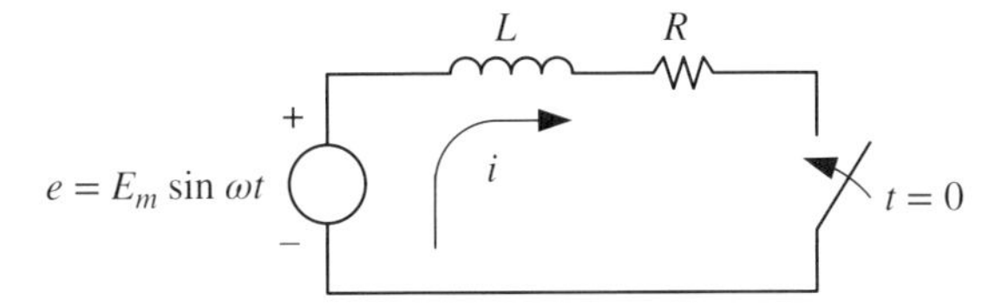

Figure 8.3 *RL* circuit illustrating DC offset.

2) Subtransient effect – this fast initial response is in the range of several cycles.
3) Transient effect – this response lasts for seconds and represents the dynamic response decaying to a steady-state value.
4) Steady-state condition – this is the operating condition after all the transients due to the fault have decayed. This operating condition may be different from the initial operating condition if the network configuration has changed.

These phenomena are discussed in more detail in the next two sections.

8.2.1 DC Offset and Stator Transients

The DC offset phenomenon arises due to the fact that all the currents start out at different values at the onset of the three-fault short-circuit fault. In a dynamic system, due to the inherent energy storage in certain components, some quantities do not change instantaneously, such as the current in an inductor and the voltage across a capacitor. Consider the *RL* circuit driven by a sinusoidal voltage source $e = E_m \sin \omega t$ in Figure 8.3. Initially for $t < 0$, the current i is zero and the switch is closed at $t = 0$.

The differential equation describing this system is

$$L\frac{di}{dt} + Ri = e(t) = E_m \sin(\omega t) \tag{8.1}$$

The solution to this differential equation, that is, the current response, after the switch is closed is given by

$$i(t) = Ae^{-(R/L)t} + \frac{E_m}{Z}\sin(\omega t - \phi), \quad Z = \sqrt{R^2 + \omega^2 L^2}, \quad \phi = \tan^{-1}\left(\frac{\omega L}{R}\right) \tag{8.2}$$

At $t = 0$, $i(0) = 0$, as the current in an inductor does not change instantaneously. Thus the amplitude A is determined from

$$0 = A - \frac{E_m}{Z}\sin\phi \quad \Rightarrow \quad A = \frac{E_m}{Z}\sin\phi \tag{8.3}$$

The term $Ae^{-(R/L)t}$ is known as the DC offset, which will decay to zero.

In the stability analysis of a synchronous machine, the DC offset is, in general, of no consequence, and is normally removed from the model. Consider the stator voltage equations (7.154) and (7.155)

$$\bar{e}_d = \frac{d\bar{\psi}_d}{dt} - \bar{\omega}\,\bar{\psi}_q - \bar{R}_a\bar{i}_d, \quad \bar{e}_q = \frac{d\bar{\psi}_q}{dt} + \bar{\omega}\,\bar{\psi}_d - \bar{R}_a\bar{i}_q \tag{8.4}$$

If the $d\bar{\psi}_d/dt$ and $d\bar{\psi}_q/dt$ terms are neglected, then the currents $\bar{i}_d$ and $\bar{i}_q$ do not have to be continuous, and thus the DC-offset phenomenon is not represented. Because $\bar{\psi}_d$

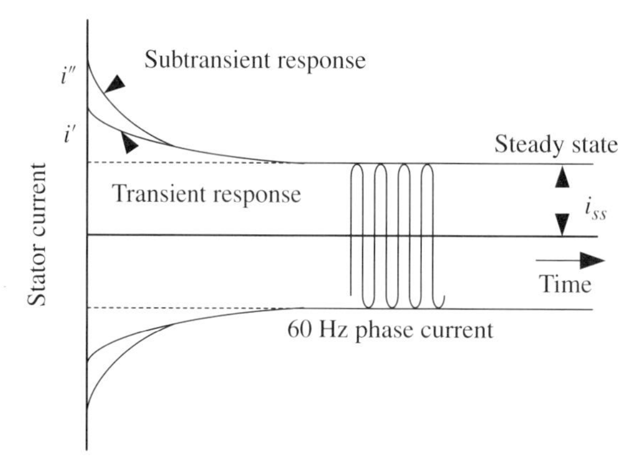

Figure 8.4 Envelope of stator current response.

and $\overline{\psi}_q$ are flux linkages in the stator circuits, neglecting their derivatives is referred to as neglecting stator transients.

In [3], the $d\overline{\psi}_d/dt$ and $d\overline{\psi}_q/dt$ terms are taken to be the fast dynamics, decaying rapidly to zero, and thus can be ignored. An analytical derivation of this approximation based on the singular perturbations method is also provided in [3].

Without the DC offset, the stator current response envelope of the fundamental frequency sinusoid would resemble that shown in the Figure 8.4. Note that i'' is the subtransient component, i' is the transient component, and i_{ss} is the steady-state component of the current.

8.3 Transient and Subtransient Reactances and Time Constants

To facilitate the modeling of the transient and subtransient phenomena, the dq-axes flux linkage circuits will be used. In particular, the flux linkage and current variables in the circuits are perturbed from their equilibrium values as

$$\begin{aligned}
&\overline{\psi}_d = \overline{\psi}_{do} + \Delta\overline{\psi}_d, \quad \overline{\psi}_q = \overline{\psi}_{qo} + \Delta\overline{\psi}_q \\
&\overline{\psi}_{fd} = \overline{\psi}_{fdo} + \Delta\overline{\psi}_{fd}, \quad \overline{\psi}_{kd} = \overline{\psi}_{kdo} + \Delta\overline{\psi}_{kd}, \quad \overline{\psi}_{kq} = \overline{\psi}_{kqo} + \Delta\overline{\psi}_{kq} \\
&\overline{i}_d = \overline{i}_{do} + \Delta\overline{i}_d, \quad \overline{i}_q = \overline{i}_{qo} + \Delta\overline{i}_q \\
&\overline{i}_{fd} = \overline{i}_{fdo} + \Delta\overline{i}_{fd}, \quad \overline{i}_{kd} = \overline{i}_{kdo} + \Delta\overline{i}_{kd}, \\
&\overline{i}_{1q} = \overline{i}_{1qo} + \Delta\overline{i}_{1q}, \quad \overline{i}_{kq} = \overline{i}_{kqo} + \Delta\overline{i}_{kq}
\end{aligned} \tag{8.5}$$

where the subscript "o" denotes an equilibrium value and Δ denotes a small perturbation.

Figure 8.5 shows the small-perturbation flux-linkage circuits, derived from Figures 7.6 and 7.7. Note that in Figure 8.5, the q-axis amortisseur representation has been expanded to two windings: the "$1q$" inductance is the dominant q-axis amortisseur

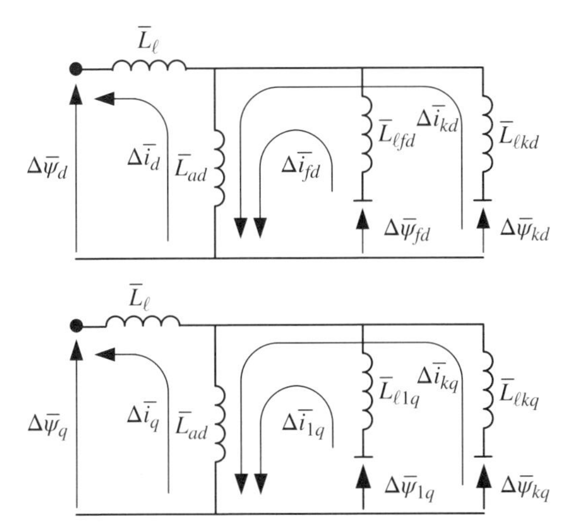

Figure 8.5 Small perturbation model of the *d*-axis flux-linkage circuit (top) and the *q*-axis flux-linkage circuit (bottom).

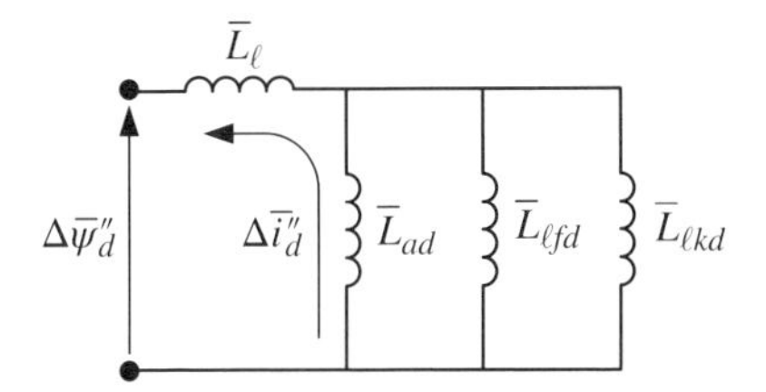

Figure 8.6 *d*-axis subtransient flux-linkage circuit.

winding representation and the "*kq*" inductance represents the net impact of all the other amortisseur windings.[1]

To derive the subtransient model, the subtransient variables in Figure 8.5 will be denoted by the superscript ″. At $t = 0^+$, $\Delta\overline{\psi}''_{fd} = \Delta\overline{\psi}''_{kd} = \Delta\overline{\psi}''_{1q} = \Delta\overline{\psi}''_{kq} = 0$. Thus equivalent circuits with two rotor windings per axis can be obtained. The *d*-axis circuit is shown in Figure 8.6.

The *d*-axis subtransient inductance can be obtained by looking into the circuit from the terminals, and is given by

$$\begin{aligned}\overline{L}''_d &= \overline{L}_\ell + \frac{1}{(1/\overline{L}_{ad}) + (1/\overline{L}_{\ell fd}) + (1/\overline{L}_{\ell kd})}\\ &= \overline{L}_\ell + \frac{\overline{L}_{ad}\overline{L}_{\ell fd}\overline{L}_{\ell kd}}{\overline{L}_{ad}\overline{L}_{\ell fd} + \overline{L}_{ad}\overline{L}_{\ell kd} + \overline{L}_{\ell fd}\overline{L}_{\ell kd}}\end{aligned} \tag{8.6}$$

which is the series connection of the leakage reactance with the parallel connection of the mutual inductance and two rotor-winding leakage inductances.

1 It is also common to use "1*d*" for "*kd*" and "2*q*" for "*kq*" [3, 6, 95]. Here ψ_{kd} and ψ_{kq} are used to be consistent with the notation in [99].

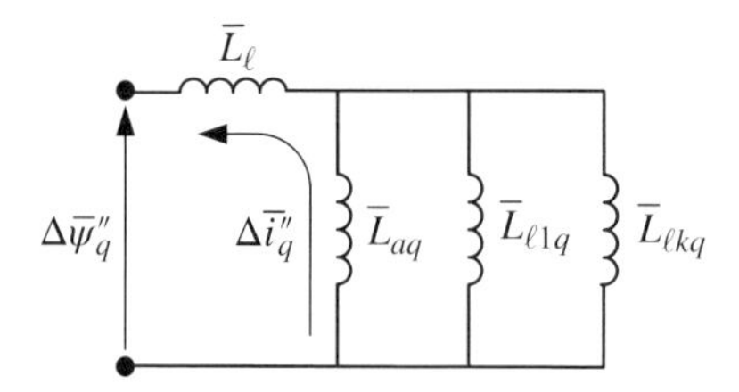

Figure 8.7 *q*-axis subtransient flux-linkage circuit.

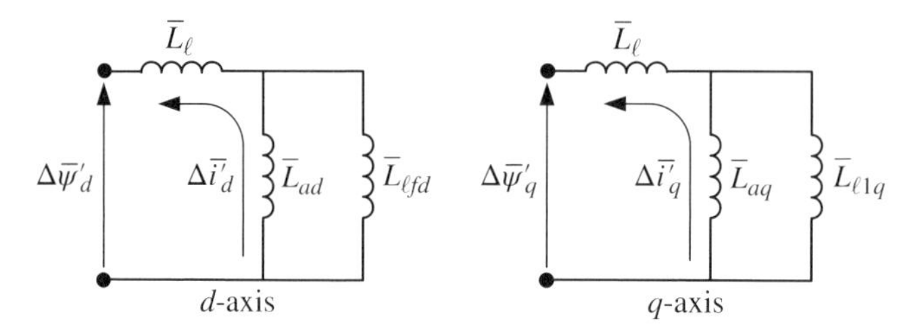

Figure 8.8 *dq*-axes transient flux-linkage circuits.

Similarly, the q-axis subtransient circuit is shown in Figure 8.7. The q-axis subtransient inductance is given by

$$\begin{aligned}\overline{L}''_q &= \overline{L}_\ell + \frac{1}{(1/\overline{L}_{aq}) + (1/\overline{L}_{\ell 1q}) + (1/\overline{L}_{\ell kq})} \\ &= \overline{L}_\ell + \frac{\overline{L}_{aq}\overline{L}_{\ell 1q}\overline{L}_{\ell kq}}{\overline{L}_{aq}\overline{L}_{\ell 1q} + \overline{L}_{aq}\overline{L}_{\ell kq} + \overline{L}_{\ell 1q}\overline{L}_{\ell kq}}\end{aligned} \tag{8.7}$$

After the subtransient currents $\Delta\overline{i}''_d$ and $\Delta\overline{i}''_q$ have decayed to zero, the $\overline{L}_{\ell kd}$ and $\overline{L}_{\ell kq}$ rotor windings can be neglected to obtain the dq-axes transient flux-linkage circuits, shown in Figure 8.8 with one rotor winding per axis. The variables associated with the transient phenomenon are denoted by the superscript $'$.

The d-axis transient inductance looking into the circuit is given by

$$\overline{L}'_d = \overline{L}_\ell + \frac{1}{(1/\overline{L}_{ad}) + (1/\overline{L}_{\ell fd})} = \overline{L}_\ell + \frac{\overline{L}_{ad}\overline{L}_{\ell fd}}{\overline{L}_{ffd}} \tag{8.8}$$

where

$$\overline{L}_{ffd} = \overline{L}_{ad} + \overline{L}_{\ell fd} \tag{8.9}$$

Similarly, the q-axis transient inductance is given by

$$\overline{L}'_q = \overline{L}_\ell + \frac{1}{(1/\overline{L}_{aq}) + (1/\overline{L}_{\ell 1q})} = \overline{L}_\ell + \frac{\overline{L}_{aq}\overline{L}_{\ell 1q}}{\overline{L}_{11q}} \tag{8.10}$$

where

$$\overline{L}_{11q} = \overline{L}_{aq} + \overline{L}_{\ell 1q} \tag{8.11}$$

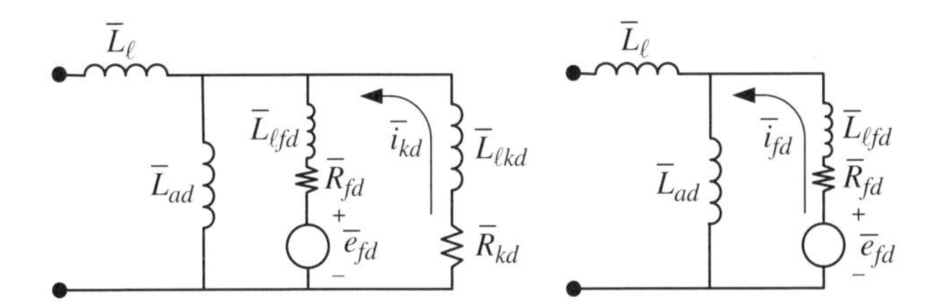

Figure 8.9 *d*-axis flux-linkage circuits with resistances included: (left) subtransient circuit and (right) transient circuit.

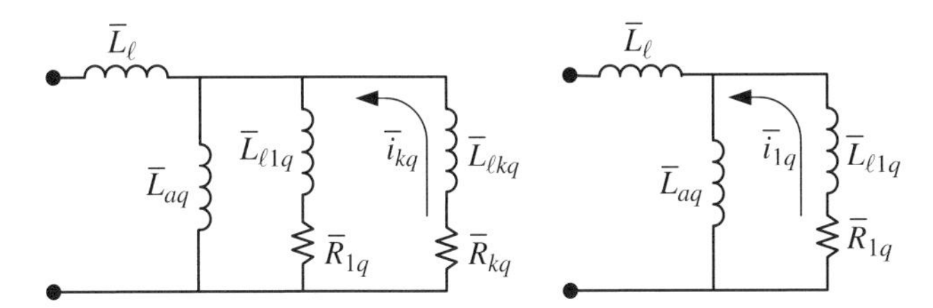

Figure 8.10 *q*-axis flux-linkage circuits with resistances included: (left) subtransient circuit and (right) transient circuit.

In steady state, the $L_{\ell fd}$ and $L_{\ell 1q}$ windings can be neglected such that the synchronous reactances for the d- and q-axes are given by, respectively,

$$\overline{L}_d = \overline{L}_\ell + \overline{L}_{ad}, \quad \overline{L}_q = \overline{L}_\ell + \overline{L}_{aq} \tag{8.12}$$

In addition to the subtransient and transient inductances, the subtransient and transient time constants associated with the decay of the subtransient and transient responses can also be determined from open-circuit conditions. For this purpose, the resistances in the rotor circuits are included. Figure 8.9 shows the subtransient and transient flux-linkage circuits including the field winding resistance $\overline{R}_{fd}$ and the damper winding resistance $\overline{R}_{kd}$.

The decay of $\overline{i}_{kd}$ is governed by the direct-axis subtransient open-circuit time constant T''_{do}. With $\overline{R}_{fd}$ neglected, T''_{do} can be computed by treating the d-axis flux linkage model as an RL circuit:

$$T''_{do} = \frac{\overline{L}_{\ell kd} + \overline{L}_{ad} || \overline{L}_{\ell fd}}{\overline{R}_{kd}} = \frac{1}{\overline{R}_{kd}} \left(\overline{L}_{\ell kd} + \frac{\overline{L}_{ad}\overline{L}_{\ell fd}}{\overline{L}_{ffd}} \right) \tag{8.13}$$

Assuming that $\overline{i}_{kd}$ has decayed, and thus neglecting the damper winding, the decay in $\overline{i}_{fd}$ due to a change in $\overline{e}_{fd}$ is governed by the direct-axis transient open-circuit time constant given by

$$T'_{do} = \frac{\overline{L}_{\ell fd} + \overline{L}_{ad}}{\overline{R}_{fd}} = \frac{\overline{L}_{ffd}}{\overline{R}_{fd}} \tag{8.14}$$

For the q-axis circuits including resistances, Figure 8.10 shows that the decay of $\overline{i}_{kq}$ is governed by the quadrature-axis subtransient open-circuit time constant with $\overline{R}_{1q}$ neglected

$$T''_{qo} = \frac{\overline{L}_{\ell kq} + \overline{L}_{aq}||\overline{L}_{\ell 1q}}{\overline{R}_{kq}} = \frac{1}{\overline{R}_{kq}}\left(\overline{L}_{\ell kq} + \frac{\overline{L}_{aq}\overline{L}_{\ell 1q}}{\overline{L}_{11q}}\right) \tag{8.15}$$

and the decay of $\overline{i}_{1q}$ is governed by the quadrature-axis transient open-circuit time constant

$$T'_{qo} = \frac{\overline{L}_{\ell 1q} + \overline{L}_{aq}}{\overline{R}_{1q}} = \frac{\overline{L}_{11q}}{\overline{R}_{1q}} \tag{8.16}$$

At this point in the chapter, two notational changes are introduced as the focus is switched to notations more commonly used in simulation models. First, instead of using inductance values, reactance values $X = \omega_o L$ are used, where ω_o is the nominal system frequency. Second, as reactances are normally given in pu values on some machine or system base, the $\overline{\ }$ symbol on the voltage and current variables and the reactance parameters will be dropped. Also note at synchronous speed, X and L are equal in pu values.

Example 8.1: Computation of reactances and time constants

The pu reactances and resistances of the dq-axis voltage-current circuits of a synchronous machine on its own base are given as

$$\begin{aligned}
&X_\ell = 0.155, \quad X_{ad} = 1.605, \quad X_{\ell fd} = 0.1414, \quad X_{\ell kd} = 0.0812\\
&X_{aq} = 1.495, \quad X_{\ell 1q} = 0.4317, \quad X_{\ell kq} = 0.0588\\
&R_{fd} = 0.000551, \quad R_{kd} = 0.01556, \quad R_{1q} = 0.01161, \quad R_{kq} = 0.01514
\end{aligned} \tag{8.17}$$

Find the synchronous, transient, and subtransient reactances and the transient and subtransient time constants for this synchronous machine. The nominal frequency f_o of the system is 60 Hz.

Solutions: Here the inductances used in the earlier formulas will be changed into reactances. For the d-axis, the synchronous reactance is

$$X_d = X_\ell + X_{ad} = 1.76 \text{ pu} \tag{8.18}$$

the transient reactance is

$$X'_d = X_\ell + \frac{X_{ad}X_{\ell fd}}{X_{ffd}} = 0.285 \text{ pu} \tag{8.19}$$

and the subtransient reactance is

$$X''_d = X_\ell + \frac{X_{ad}X_{\ell fd}X_{\ell kd}}{X_{ad}X_{\ell fd} + X_{ad}X_{\ell kd} + X_{\ell fd}X_{\ell kd}} = 0.205 \text{ pu} \tag{8.20}$$

The d-axis transient open-circuit time constant is

$$T'_{do} = \frac{X_{\ell fd} + X_{ad}}{2\pi f_o R_{fd}} = 8.40 \text{ sec} \tag{8.21}$$

and the subtransient time constant is

$$T''_{do} = \frac{1}{2\pi f_o R_{kd}}\left(X_{\ell kd} + \frac{X_{ad}X_{\ell fd}}{X_{ad} + X_{\ell fd}}\right) = 0.036 \text{ sec} \tag{8.22}$$

Note that the time in the per-unit system is in radians, and thus the conversion factor $2\pi f_o$ is needed to convert the time constants to seconds.

Similarly, for the q-axis, the synchronous reactance is

$$X_q = X_\ell + X_{aq} = 1.65 \text{ pu} \tag{8.23}$$

the transient reactance is

$$X'_q = X_\ell + \frac{X_{aq}X_{\ell 1q}}{X_{11q}} = 0.490 \text{ pu} \tag{8.24}$$

and the subtransient reactance is

$$X''_q = X_\ell + \frac{X_{aq}X_{\ell 1q}X_{\ell kq}}{X_{aq}X_{\ell 1q} + X_{aq}X_{\ell kq} + X_{\ell 1q}X_{\ell kq}} = 0.205 \text{ pu} \tag{8.25}$$

The q-axis transient time constant is

$$T'_{qo} = \frac{X_{\ell 1q} + X_{aq}}{2\pi f_o R_{1q}} = 0.44 \text{ sec} \tag{8.26}$$

and the subtransient time constant is

$$T''_{qo} = \frac{1}{2\pi f_o R_{kq}}\left(X_{\ell kq} + \frac{X_{aq}X_{\ell 1q}}{L_{aq} + L_{\ell 1q}}\right) = 0.069 \text{ sec} \tag{8.27}$$

Note that the parameters of this machine have been used in Example 7.3 to compute the steady-state voltages and currents. ■

Although the parameters used in Example 8.1 belong to a particular synchronous machine, the reactance pu values on the machine base and the time constant values are typical of large generators. For example, X_d ranges from 1.5 to 2.0 pu, X'_d from 0.2 to 0.3 pu, and T'_{do} from 5 to 10 sec. If incorrect values are used in the synchronous machine model, problems will arise in the initialization of the machine states and the simulation of machine dynamics following a disturbance.

8.4 Subtransient Synchronous Machine Model

Using the d- and q-axis flux linkage-current and voltage-current equations, this section derives the detailed synchronous machine dynamic model in [99]. This synchronous generator model is included in the IEEE Standard on Synchronous Machine Modeling [95]. The model is available in commercial simulation programs and is known as the round-rotor or 2-axis model. Simplified synchronous machine models can readily be derived from this detailed synchronous machine model.

The d- and q-axis flux linkage-current equations derived in Chapter 7 are given by, using reactances instead of inductances

$$\psi_d = -X_d i_d + X_{ad} i_{fd} + X_{ad} i_{kd} \tag{8.28}$$

$$\psi_{fd} = -X_{ad} i_d + X_{ffd} i_{fd} + X_{ad} i_{kd} \tag{8.29}$$

$$\psi_{kd} = -X_{ad} i_d + X_{ad} i_{fd} + X_{kkd} i_{kd} \tag{8.30}$$

$$\psi_q = -X_q i_q + X_{aq} i_{1q} + X_{aq} i_{kq} \tag{8.31}$$

$$\psi_{1q} = -X_{aq}i_q + X_{11q}i_{1q} + X_{aq}i_{kq} \tag{8.32}$$

$$\psi_{kq} = -X_{aq}i_q + X_{aq}i_{1q} + X_{kkq}i_{kq} \tag{8.33}$$

Similarly, the d- and q-axis voltage-current equations are given by

$$e_d = -\omega\psi_q - R_a i_d \tag{8.34}$$

$$e_{fd} = \frac{d\psi_{fd}}{dt} + R_{fd}i_{fd} \tag{8.35}$$

$$e_{kd} = \frac{d\psi_{kd}}{dt} + R_{kd}i_{kd} = 0 \tag{8.36}$$

$$e_q = \omega\psi_d - R_a i_q \tag{8.37}$$

$$e_{1q} = \frac{d\psi_{1q}}{dt} + R_{1q}i_{1q} = 0 \tag{8.38}$$

$$e_{kq} = \frac{d\psi_{kq}}{dt} + R_{kq}i_{kq} = 0 \tag{8.39}$$

Note that the $d\psi_d/dt$ term has been neglected in the e_d equation (8.34) and the $d\psi_q/dt$ term neglected in the e_q equation (8.37). This approximation has been discussed earlier in Section 8.2.

The objective is to manipulate these equations to develop four dynamic equations for the states

$$E_q' = \frac{X_{ad}}{X_{ffd}}\psi_{fd}, \quad E_d' = \frac{X_{aq}}{X_{11q}}\psi_{1q}, \quad \psi_{kd}, \quad \psi_{kq} \tag{8.40}$$

where E_q' is a voltage proportional to the field flux and E_d' is a voltage proportional to the flux linkage of the first amortisseur circuit on the q-axis of the rotor. The variables (8.40) represent the internal voltages and flux linkages in the synchronous machine.

The d-axis models are developed first. Solve i_{fd} and i_{kd} from (8.29) and (8.30) as

$$\begin{bmatrix} X_{ffd} & X_{ad} \\ X_{ad} & X_{kkd} \end{bmatrix} \begin{bmatrix} i_{fd} \\ i_{kd} \end{bmatrix} = \begin{bmatrix} \psi_{fd} + X_{ad}i_d \\ \psi_{kd} + X_{ad}i_d \end{bmatrix} \tag{8.41}$$

such that

$$\begin{bmatrix} i_{fd} \\ i_{kd} \end{bmatrix} = \frac{1}{\Delta_d} \begin{bmatrix} X_{kkd} & -X_{ad} \\ -X_{ad} & X_{ffd} \end{bmatrix} \begin{bmatrix} \psi_{fd} + X_{ad}i_d \\ \psi_{kd} + X_{ad}i_d \end{bmatrix} \tag{8.42}$$

where the determinant $\Delta_d = X_{ffd}X_{kkd} - X_{ad}^2 > 0$, as the mutual reactance X_{ad} is part of the reactances X_{ffd} and X_{kkd}. In term of E_q', these currents can be expressed as

$$i_{fd} = \frac{X_{kkd}(X_{ffd}/X_{ad})E_q' - X_{ad}\psi_{kd} + X_{\ell kd}X_{ad}i_d}{\Delta_d} \tag{8.43}$$

$$i_{kd} = \frac{X_{ffd}(-E_q' + \psi_{kd}) + X_{\ell fd}X_{ad}i_d}{\Delta_d} \tag{8.44}$$

Note that these d-axis current quantities are now functions of the rotor d-axis flux linkages and the stator d-axis current i_d.

Eliminate i_{fd} and i_{kd} in the ψ_d equation (8.28) to obtain

$$\begin{aligned}
\psi_d &= -X_d i_d + X_{ad}(i_{fd} + i_{kd}) \\
&= \frac{X_{ad}}{\Delta_d}\left(X_{kkd}\frac{X_{ffd}}{X_{ad}} - X_{ffd}\right)E'_q - \frac{X_{ad}}{\Delta_d}(X_{ad} - X_{ffd})\psi_{kd} \\
&\quad -\left(X_d - \frac{X_{ad}^2 X_{\ell kd}}{\Delta_d} - \frac{X_{ad}^2 X_{\ell fd}}{\Delta_d}\right) i_d \\
&= \frac{X_{ffd}}{\Delta_d}(X_{kkd} - X_{ad})E'_q + \frac{X_{ad}X_{\ell fd}}{\Delta_d}\psi_{kd} - \left(X_\ell + \frac{X_{ad}X_{\ell fd}X_{\ell kd}}{\Delta_d}\right) i_d
\end{aligned} \tag{8.45}$$

which can be further simplified into

$$\psi_d = \frac{X''_d - X_\ell}{X'_d - X_\ell}E'_q + \frac{X'_d - X''_d}{X'_d - X_\ell}\psi_{kd} - X''_d i_d \tag{8.46}$$

The e_{kd} voltage equation (8.36) is rewritten as

$$\frac{d\psi_{kd}}{dt} = -R_{kd} i_{kd} \tag{8.47}$$

Multiply both sides by T''_{do} to obtain

$$\begin{aligned}
T''_{do}\frac{d\psi_{kd}}{dt} &= -\frac{X_{\ell kd} + X_{ad}X_{\ell fd}/X_{ffd}}{R_{kd}} R_{kd} i_{kd} = -\left(X_{\ell kd} + \frac{X_{ad}X_{\ell fd}}{X_{ffd}}\right) i_{kd} \\
&= -\frac{\Delta_d}{X_{ffd}}\frac{X_{ffd}(-E'_q + \psi_{kd}) + X_{\ell fd}X_{ad} i_d}{\Delta_d}
\end{aligned} \tag{8.48}$$

Further simplification of (8.48) results in the d-axis subtransient flux decay equation

$$T''_{do}\frac{d\psi_{kd}}{dt} = E'_q - \psi_{kd} - (X'_d - X_\ell) i_d \tag{8.49}$$

Now express the field voltage equation (8.35) as

$$\frac{d\psi_{fd}}{dt} = e_{fd} - R_{fd} i_{fd} \tag{8.50}$$

From the definition of E'_q, (8.50) can be used to derive

$$\begin{aligned}
\frac{dE'_q}{dt} &= \frac{X_{ad}}{X_{ffd}}\frac{d\psi_{fd}}{dt} = \frac{X_{ad}}{X_{ffd}}(e_{fd} - R_{fd} i_{fd}) \\
&= \frac{R_{fd}}{X_{ffd}}\left(\frac{X_{ad}}{R_{fd}} e_{fd} - X_{ad} i_{fd}\right) = \frac{1}{T'_{do}}(E_{fd} - X_{ad} i_{fd})
\end{aligned} \tag{8.51}$$

Thus the d-axis transient flux decay equation is

$$T'_{do}\frac{dE'_q}{dt} = E_{fd} - X_{ad} i_{fd} \tag{8.52}$$

The $X_{ad}i_{fd}$ term in (8.52) can be expressed as

$$\begin{aligned} X_{ad}i_{fd} &= \frac{X_{kkd}X_{ffd}E'_q - X_{ad}^2\psi_{kd} + X_{\ell kd}X_{ad}^2 i_d}{\Delta_d} \\ &= \frac{X_{ad}^2}{\Delta_d}(E'_q - \psi_{kd} + X_{\ell kd}X_{ad}^2 i_d) + \frac{X_{kkd}X_{ffd} - X_{ad}^2}{\Delta_d}E'_q \\ &= \frac{(X'_d - X''_d)(X_d - X'_d)}{(X'_d - X_\ell)^2}\left(E'_q - \psi_{kd} + \frac{(X'_d - X_\ell)(X''_d - X_\ell)}{(X'_d - X''_d)}i_d\right) + F_d(E'_q) \end{aligned} \tag{8.53}$$

where $F_d(E'_q)$ is the d-axis saturation function discussed earlier in Section 7.8.

Similarly, for the q-axis, start by eliminating i_{1q} and i_{kq} to obtain

$$\psi_q = \frac{X''_q - X_\ell}{X'_q - X_\ell}E'_d + \frac{X'_q - X''_q}{X'_q - X_\ell}\psi_{kq} - X''_q i_q \tag{8.54}$$

The q-axis subtransient flux decay equation is

$$T''_{qo}\frac{d\psi_{kq}}{dt} = E'_d - \psi_{kq} - (X'_q - X_\ell)i_q \tag{8.55}$$

and the q-axis transient flux decay equation is

$$T'_{qo}\frac{dE'_d}{dt} = -X_{aq}i_{1q} \tag{8.56}$$

where

$$X_{aq}i_{1q} = \frac{(X'_q - X''_q)(X_q - X'_q)}{(X'_q - X_\ell)^2}\left(E'_d - \psi_{kq} + \frac{(X'_q - X_\ell)(X''_q - X_\ell)}{(X'_q - X''_q)}i_q\right) + F_q(E'_d) \tag{8.57}$$

where $F_q(E'_d)$ represents the q-axis saturation function.

This model with four electrical states E'_q (8.52), E'_d (8.56), ψ_{kd} (8.49), and ψ_{kq} (8.55) is known as the subtransient model (or the detailed model). It is also known as the synchronous machine model with one d-axis damper circuit and two q-axis damper circuits or the 2.2 model (2 windings on the d-axis and 2 windings on the q-axis) [95].

Finally, the generator terminal voltages are developed using the states from the synchronous machine dynamic model. From the d- and q-axis voltage equations, with ω close to 1 pu, (8.34) and (8.37) become

$$e_d = -\omega\psi_q - R_a i_d = -\psi_q - R_a i_d \tag{8.58}$$

$$e_q = \omega\psi_d - R_a i_q = \psi_d - R_a i_q \tag{8.59}$$

Define the voltages

$$E''_d = -\psi''_q = -\left(\frac{X''_q - X_\ell}{X'_q - X_\ell}E'_d + \frac{X'_q - X''_q}{X'_q - X_\ell}\psi_{kq}\right) \tag{8.60}$$

$$E''_q = \psi''_d = \frac{X''_d - X_\ell}{X'_d - X_\ell}E'_q + \frac{X'_d - X''_d}{X'_d - X_\ell}\psi_{kd} \tag{8.61}$$

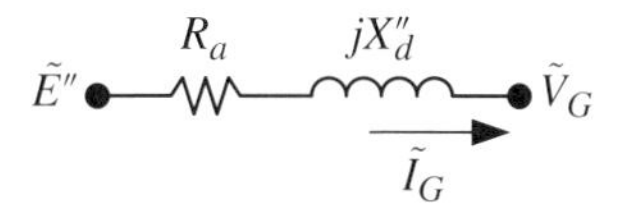

Figure 8.11 Generator subtransient circuit model with saliency neglected.

which are functions of the d- and q-axis voltages. Using the ψ_d equation (8.46) and the ψ_q equation (8.54), (8.60) and (8.61) can be expressed as

$$E''_d - e_d = R_a i_d - X''_q i_q, \quad E''_q - e_q = R_a i_q + X''_d i_d \tag{8.62}$$

Note that e_d and e_q comprise the generator terminal voltage. Thus E''_d and E''_q are the voltages behind the subtransient reactances. If subtransient saliency is neglected, that is, $X''_d = X''_q$, then the two equations in (8.62) can be combined into a single equation

$$\tilde{E}'' = \tilde{V}_G + (R_a + jX''_d)\tilde{I}_G \tag{8.63}$$

where $\tilde{E}'' = E''_d + jE''_q$ is the generator internal voltage, $\tilde{V}_G = e_d + je_q$ is the generator terminal voltage, and $\tilde{I}_G = i_d + ji_q$ is the generator output current. Equation (8.63) can be represented by the circuit shown in Figure 8.11.

A block diagram representation of the subtransient synchronous machine model is shown in Figure 8.12 [99]. Note that this block diagram only accounts for saturation in the d-axis. A more inclusive model also includes saturation effects in the q-axis [95].

Example 8.2

As mentioned earlier, the parameters used in Example 7.3 are from Example 8.1. Now use the steady-state voltages and currents computed in Example 7.3 and the additional dq-axes parameters computed in Example 8.1 to find the initial conditions of the four machine electrical states in (8.40).

Solutions: The steady-state values of the two d-axis states E'_q and ψ_{kd} and the two q-axis states E'_d and ψ_{kq} are computed by setting the derivatives of these state variables with respect to time to zero. The d-axis transient flux decay equations (8.52) and (8.53) yield, using the given reactance values

$$E_{fdo} - 6.982(E'_{qo} - \psi_{kdo} + 0.08125 i_{do}) - E'_{qo} = 0 \tag{8.64}$$

where the subscript "o" denotes equilibrium values. The d-axis subtransient flux decay equation (8.49) yields

$$E'_{qo} - \psi_{kdo} - 0.13 i_{do} = 0 \tag{8.65}$$

The q-axis transient flux decay equations (8.56) and (8.57) yield

$$2.946(E'_{do} - \psi_{kqo} + 0.05877 i_{qo}) + E'_{do} = 0 \tag{8.66}$$

The q-axis subtransient flux decay equation (8.55) yields

$$E'_{do} - \psi_{kqo} - 0.335 i_{do} = 0 \tag{8.67}$$

From the voltage-current diagram of Example 7.3,

$$i_{do} = 0.9133 \text{ pu}, \quad i_{qo} = 0.4072 \text{ pu}, \quad E_{fdo} = E_I = 2.3843 \text{ pu} \tag{8.68}$$

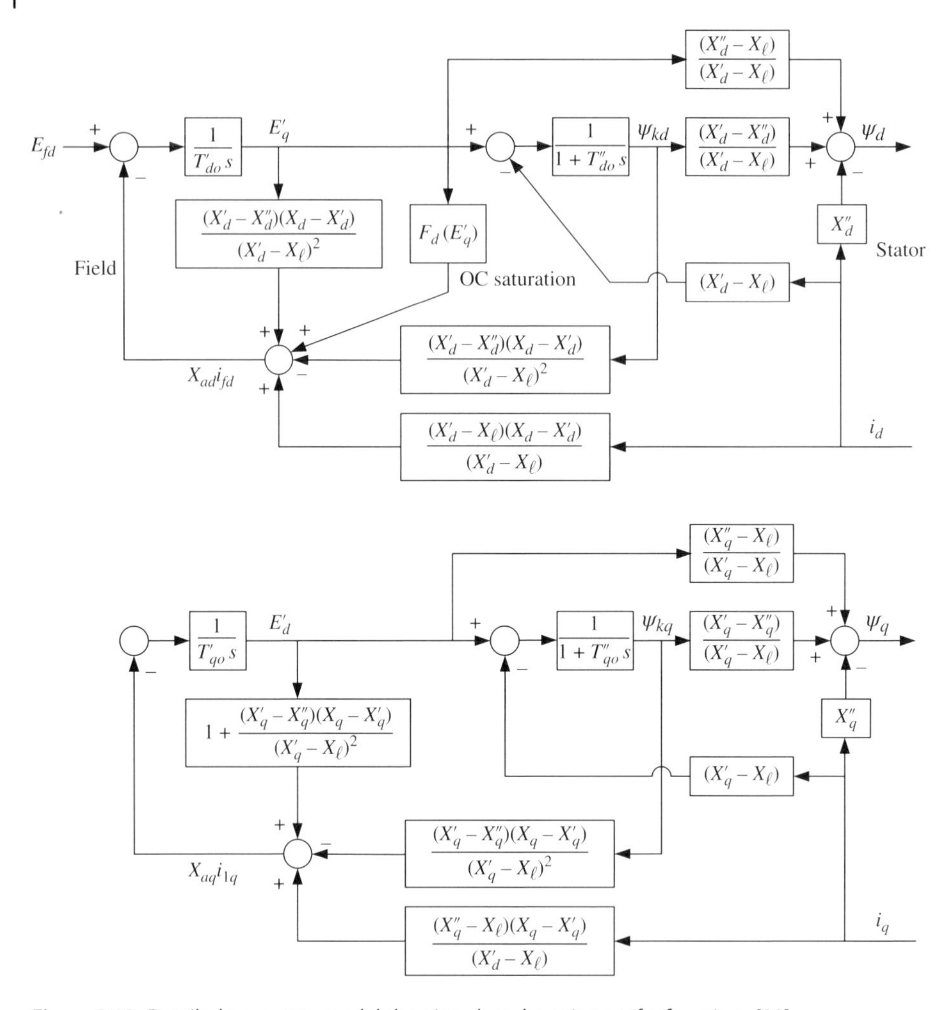

Figure 8.12 Detailed generator model showing *d*- and *q*-axis transfer functions [99].

Thus substituting (8.65) into (8.64) results in

$$E'_{qo} = 2.3843 - 6.982(0.1187 + 0.0742) = 1.0372 \text{ pu} \tag{8.69}$$

which is the value of e'_q computed in Example 7.3. From (8.69)

$$\psi_{kdo} = 1.0372 - 0.1187 = 0.9185 \text{ pu} \tag{8.70}$$

Similarly, substituting (8.67) into (8.66) yields

$$E'_{do} = -2.946(0.1364 + 0.0239) = -0.4724 \text{ pu} \tag{8.71}$$

such that

$$\psi_{kqo} = -0.4724 - 0.1364 = -0.6088 \text{ pu} \tag{8.72}$$

These equilibrium values are used for the machine states to initiate a disturbance simulation. ■

8.5 Other Synchronous Machine Models

As mentioned earlier, additional amortisseur circuits can be added in parallel to the dq-axes circuits shown in Figures 8.6 and 8.7 to more accurately model the very fast dynamic behavior of synchronous machines. The IEEE Guide to Synchronous Machine Modeling [95] provides a matrix of $n_d.n_q$ machine models, in which n_d and n_q denote the number of windings in the d- and q-axes, respectively. For example, the model in Figure 8.12 is a 2.2 model representing a round-rotor synchronous machine. The 2.1 model is used to represent a salient-pole synchronous machine suitable for a hydraulic generating unit. The laminated construction of a salient-pole machine reduces the circulation of the eddy currents in the rotor, and as a result, only the subtransient effect of the q-axis is represented. Ref. [95] emphasizes that there is seldom a need to go beyond a 3.3 model.

Synchronous machine model dynamic parameters are supplied by the manufacturers. Two sets of reactance values are typically provided, with the subscript "u" denoting the unsaturated condition values and the subscript "sat" denoting the saturated condition values. The unsaturated values are used for synchronous machine models in stability studies, as the saturation is accounted for separately. The subscript "u" for the reactance values is nominally omitted.

The 2.2 synchronous machine model can be used to derive two common reduced-order models, as shown in the following subsections.

8.5.1 Flux-Decay Model

The flux-decay model, also known as the transient model, can be derived by assuming that the subtransient dynamics are much faster than the transient dynamics, that is, the subtransient time constants T''_{do} and T''_{qo} are much smaller than the transient time constants T'_{do} and T'_{qo}. Based on the theory of singular perturbations [100], after the fast dynamics associated with ψ_{kd} and ψ_{kq} have decayed in the dynamic equations (8.49) and (8.55), the quasi-steady states of ψ_{kd} and ψ_{kq} satisfy[2]

$$E'_q - \psi_{kd} - (X'_d - X_\ell)i_d = 0 \quad \Rightarrow \quad \psi_{kd} = E'_q - (X'_d - X_\ell)i_d \tag{8.73}$$

$$E'_d - \psi_{kq} - (X'_q - X_\ell)i_q = 0 \quad \Rightarrow \quad \psi_{kq} = E'_d - (X'_q - X_\ell)i_q \tag{8.74}$$

Substituting ψ_{kd} into (8.52) and simplifying, the d-axis equation of the transient model simplifies to the flux-decay model

$$e_q = \psi_d = E'_q - X'_d i_d, \quad T'_{do}\frac{dE'_q}{dt} = E_{fd} - X_{ad}i_{fd} = E_{fd} - (X_d - X'_d)i_d - F_d(E'_q) \tag{8.75}$$

where $F_d(E'_q)$ is the saturation function (Figure 8.13). If saturation is neglected, then $F_d(E'_q) = E'_q$.

Similarly, substituting ψ_{kq} into (8.56) and simplifying, the q-axis model simplifies to

$$e_d = -\psi_q = -E'_d + X'_q i_q, \quad T'_{qo}\frac{dE'_d}{dt} = -(X_q - X'_q)i_q - F_q(E'_d) \tag{8.76}$$

where $F_q(E'_d)$ is the saturation function. If saturation is neglected, then $F_q(E'_d) = E'_d$.

---

2 These equations also describe the so-called slow manifold [3].

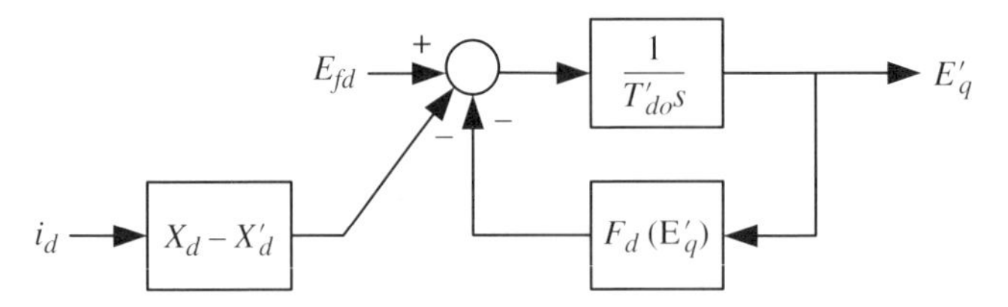

Figure 8.13 Generator flux-decay model.

The electrical power can still be computed as

$$P_e = e_d i_d + e_q i_q \tag{8.77}$$

This particular model is a 2-electrical-state flux decay model. Further simplification is possible if q-axis transient time constant T'_{qo} is much smaller than the d-axis transient time constant T'_{do}. Under this assumption, (8.76) reduces to the quasi-steady-state equation

$$-(X_q - X'_q)i_q - E'_d = 0 \quad \Rightarrow \quad E'_d = -(X_q - X'_q)i_q \tag{8.78}$$

such that

$$e_d = X_q i_q \tag{8.79}$$

Equations (8.75) and (8.79) form the first-order flux-decay model.

Note that in both the first- and second-order flux decay models, an excitation system can be used to control the field voltage E_{fd} which is the input to the E'_q dynamic equation (8.75), as shown in the block diagram in Figure 8.13.

The flux-decay model is suitable for use in studies in which the subtransient effect of a synchronous machine is not important, such as longer term dynamic voltage stability studies.

8.5.2 Classical Model

The open-circuit direct-axis time constant T'_{do} is in the range of 5–10 seconds. Thus E'_q is a slow variable and thus can be assumed to be constant. Furthermore, it is assumed that there is no transient saliency, so that $X'_q = X'_d$.

Replacing the subtransient quantities with transient quantities, (8.62) becomes

$$E'_d - e_d = R_a i_d - X'_d i_q, \quad E'_q - e_q = R_a i_q + X'_d i_d \tag{8.80}$$

which can be combined into

$$\tilde{E}' = \tilde{V}_G + j\tilde{I}_G(R_a + jX'_d) \tag{8.81}$$

where $\tilde{E}' = E'_d + jE'_q$ is the voltage behind the transient reactance, $\tilde{V}_G = e_d + je_q$ is the generator terminal voltage, and $\tilde{I}_G = i_d + ji_q$ is the generator output current. The circuit diagram of (8.81) is shown in Figure 8.14.

This is the constant-voltage (or constant flux-linkage) behind transient reactance model, also known as the classical model or the electromechanical model. This model has been used in earlier chapters.

The classical model has no electrical states, and hence, has only 2 mechanical states (δ and ω), as compared to 6 states in the subtransient model (4 electrical states and 2

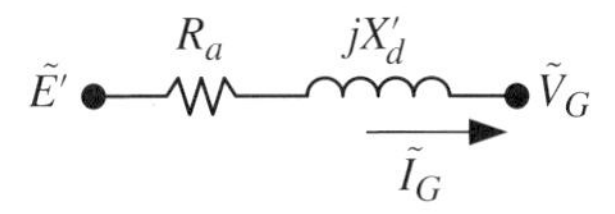

Figure 8.14 Generator transient circuit model.

mechanical states), and 3 states in the flux-decay model (1 electrical state and 2 mechanical states). Note that although the voltage magnitude of $\tilde{E}'$ is constant, the phase of $\tilde{E}'$ is determined by the rotor angle δ.

8.6 *dq*-axes Rotation Between a Generator and the System

The detailed synchronous machine model consists of the flux-linkage model for ψ_{kd} (8.49), ψ_{kq} (8.55), E'_q (8.52)–(8.53), and E'_d (8.56)–(8.57), and the electromechanical model

$$\frac{d\delta}{dt} = \Omega\omega, \quad 2H\frac{d\omega}{dt} = P_m - P_e - D\omega \tag{8.82}$$

for δ, the machine rotor angle, and ω, the machine speed deviation from the nominal frequency. Note that the rotor angle δ defines the q-axis of a particular machine. Additionally, in the development of the detailed synchronous machine models in Chapter 7 and to this point in this chapter, the rotor angle δ is referenced to the generator terminal voltage $\tilde{V}_G$ (the notation $\tilde{E}_t$ has also been used earlier to denote the generator terminal voltage).

In a power system with multiple synchronous machines, each of the machines will have its own d- and q-axes, that is, the voltages and currents E''_{di}, E''_{qi}, e_{di}, e_{qi}, i_{di}, and i_{qi} in (8.62) of machine i are on that machine's dq-axes. As discussed in Chapter 4 on multi-machine power system simulation, these machines need to be connected to each other via the power network. The power network itself has a common system DQ-axes coordinate. As a result, each machine's dq-axes will be projected onto the system DQ-axes, so that all the machine voltage and current variables will be consistent with the network DQ-axes, as shown in Figure 8.15.

From Figure 8.15, the projection of E''_{di} and E''_{qi} of machine i on the system DQ-axes is given by

$$\begin{bmatrix} E''_{Di} \\ E''_{Qi} \end{bmatrix} = \begin{bmatrix} \cos\delta_i & -\sin\delta_i \\ \sin\delta_i & \cos\delta_i \end{bmatrix} \begin{bmatrix} E''_{di} \\ E''_{qi} \end{bmatrix} = T \begin{bmatrix} E''_{di} \\ E''_{qi} \end{bmatrix} \tag{8.83}$$

where T is the rotation matrix.

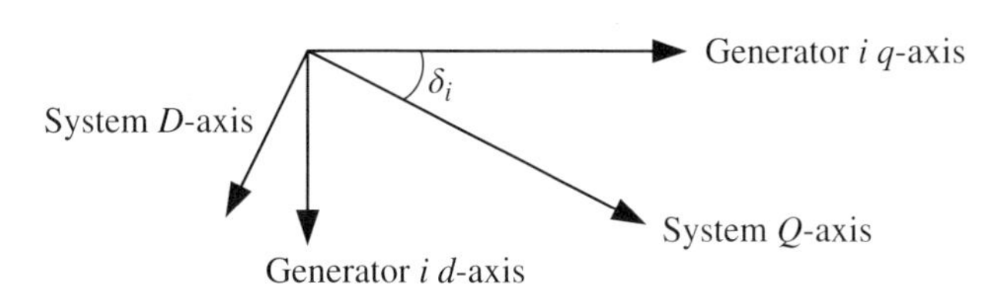

Figure 8.15 The generator i dq-axes and the system DQ-axes, in which δ_i is the machine rotor angle.

Furthermore i_{di} and i_{qi} can be expressed as

$$\begin{bmatrix} i_{di} \\ i_{qi} \end{bmatrix} = \begin{bmatrix} \cos\delta_i & \sin\delta_i \\ -\sin\delta_i & \cos\delta_i \end{bmatrix} \begin{bmatrix} i_{Di} \\ i_{Qi} \end{bmatrix} = T^{-1} \begin{bmatrix} i_{Di} \\ i_{Qi} \end{bmatrix} \tag{8.84}$$

Thus the network equation for the generator (8.62) can be expressed as

$$\begin{bmatrix} E''_{Di} - e_{Di} \\ E''_{Qi} - e_{Qi} \end{bmatrix} = \begin{bmatrix} R_{ai} & -X''_{qi} \\ X''_{di} & R_{ai} \end{bmatrix} \begin{bmatrix} i_{Di} \\ i_{Qi} \end{bmatrix} \tag{8.85}$$

or

$$\begin{bmatrix} i_{Di} \\ i_{Qi} \end{bmatrix} = \begin{bmatrix} R_{ai} & -X''_{qi} \\ X''_{di} & R_{ai} \end{bmatrix}^{-1} \begin{bmatrix} E''_{Di} - e_{Di} \\ E''_{Qi} - e_{Qi} \end{bmatrix} = Y_{Gi} \begin{bmatrix} E''_{Di} - e_{Di} \\ E''_{Qi} - e_{Qi} \end{bmatrix} \tag{8.86}$$

Note that the only known quantities in (8.86) are the generator internal voltages (E''_{Di}, E''_{Qi}). To calculate the other quantities in (8.86) as well as the overall network variables, the internal generator voltages (E''_{Di}, E''_{Qi}) and the generator terminal bus voltages, recast as $e_{Di} = V_{GDi}$ and $e_{Qi} = V_{GQi}$, are combined with the load buses to form a set of network voltage and current injection equations, when expressed in matrix form, as

$$\begin{bmatrix} Y_G & -Y_G & 0 \\ -Y_G & Y'_{11} + Y_G & Y_{12} \\ 0 & Y_{21} & Y'_{22} \end{bmatrix} \begin{bmatrix} E' \\ V_G \\ V_L \end{bmatrix} = \begin{bmatrix} I_G \\ 0 \\ 0 \end{bmatrix} \tag{8.87}$$

where

$$E' = \begin{bmatrix} E''_{D1} + jE''_{Q1} \\ \vdots \\ E''_{DN_G} + jE''_{QN_G} \end{bmatrix}, \quad V_G = \begin{bmatrix} V_{GD1} + jV_{GQ1} \\ \vdots \\ V_{GDN_G} + jV_{GQN_G} \end{bmatrix}$$

$$V_L = \begin{bmatrix} V_{LD(N_G+1)} + jV_{LQ(N_G+1)} \\ \vdots \\ V_{LDN_B} + jV_{LQN_B} \end{bmatrix}, \quad I_G = \begin{bmatrix} i_{D1} + ji_{Q1} \\ \vdots \\ i_{DN_G} + ji_{QN_G} \end{bmatrix} \tag{8.88}$$

N_G is the number of generators, N_B is the total number of buses, and Y_{ij} are the network admittance matrices in complex form. Note that Y'_{11} and Y'_{22} contain the bus load admittance terms.[3]

Equation (8.87) can be used to solve for the unknown variables i_D, i_Q, V_{GD}, V_{GQ}, V_{LD}, and V_{LQ}.

Note that this formulation is similar to that in Section 4.6 for a classical generator model. Here the subtransient flux variables E'' and reactances X''_d and X''_q are used in place of the fixed machine internal flux E' and the transient reactance X'_d.

8.7 Power System Simulation using Detailed Machine Models

As this point, the steps of simulating the dynamic response of a synchronous machine with subtransient dynamics can be formulated. In this discussion, the field voltage E_{fd} is assumed to be fixed. In practice, it is determined by the voltage regulator and excitation system, which are discussed in the next chapter.

3 Here the load is assumed to be of constant impedance. Loads that have constant current and constant power components are known as non-conforming loads and are discussed Chapter 3.

8.7.1 Power System Simulation Algorithm

1) Initialization: use the power flow solution to develop the machine voltage-current phasor diagram and compute the initial values of the machine rotor angle δ and the machine voltages and fluxes E'_q, E'_d, ψ_{kd}, and ψ_{kq} for each machine as shown in Example 8.2. Note that the machine speed deviation ω in steady state is always equal to zero.
2) Network equation solution: compute E''_q and E''_d and use them to solve for the bus voltages in the system DQ-axis frame from

$$\begin{bmatrix} Y'_{11} + Y_G & Y_{12} \\ Y_{21} & Y'_{22} \end{bmatrix} \begin{bmatrix} V_G \\ V_L \end{bmatrix} = Y_G \begin{bmatrix} E''_D \\ E''_Q \end{bmatrix} \tag{8.89}$$

such that all the generator and line currents can be computed.
3) Dynamics computation: compute the derivatives $\dot{\delta}$ and $\dot{\omega}$ of the machine mechanical states, and the derivatives $\dot{E}'_q$, $\dot{E}'_d$, $\dot{\psi}_{kd}$, and $\dot{\psi}_{kq}$, of the generator electrical states, and integrate to obtain the new values for the state variables.

Example 8.3

At some point in time after a disturbance, the electrical states of a synchronous machine are calculated to be

$$E'_d = -0.6325 \text{ pu}, \quad E'_q = 0.6762 \text{ pu}, \quad \psi_{kd} = 0.6204 \text{ pu}, \quad \psi_{kq} = -0.7524 \text{ pu} \tag{8.90}$$

These variables are used to calculate the generator current injected into the network. This would be a fairly cumbersome manual calculation. To simplify the problem, the machine terminal voltage is given in the machine dq-axes components as

$$e_d = 0.8232 \text{ pu}, \quad e_q = 0.5263 \text{ pu} \tag{8.91}$$

Calculate the generator current, given the machine parameters

$$X_\ell = 0.15 \text{ pu}, \quad X'_d = 0.245 \text{ pu}, \quad X''_d = 0.2 \text{ pu}, \quad X'_q = 0.42 \text{ pu}, \quad X''_q = 0.2 \text{ pu} \tag{8.92}$$

Solution: First the machine internal voltage is calculated from (8.60) and (8.61) as

$$E''_d = -\psi''_q = -\left(\frac{X''_q - X_\ell}{X'_q - X_\ell} E'_d + \frac{X'_q - X''_q}{X'_q - X_\ell} \psi_{kq} \right) = 0.7302 \text{ pu} \tag{8.93}$$

$$E''_q = \psi''_d = \frac{X''_d - X_\ell}{X'_d - X_\ell} E'_q + \frac{X'_d - X''_d}{X'_d - X_\ell} \psi_{kd} = 0.6498 \text{ pu} \tag{8.94}$$

Then based on the voltage difference between the generator internal voltage and terminal voltage, the current components can be found using the subtransient reactances as

$$i_d = \frac{E''_q - e_q}{X''_d} = 0.6171 \text{ pu} \tag{8.95}$$

$$i_q = -\frac{E''_d - e_d}{X''_q} = 0.4696 \text{ pu} \tag{8.96}$$

■

The disturbance performance of detailed synchronous machines are postponed to the next chapter, as such a detailed machine model requires the use of an excitation system to provide a proper response for the field voltage in low network voltage conditions such as short-circuit faults.

8.8 Linearized Models

Linearized models of power systems using classical models for generators have been discussed in Chapter 6. The linearization of power systems with subtransient generator models will increase the dimensions of the state variables and the system matrix. The linearized model is illustrated with Example 8.4. A reader interested in deriving the linear model using analytical expressions can consult [3].

Example 8.4: Linearization of detailed generator model

The single-machine infinite-bus (SMIB) system shown in Figure 8.16 operates at 60 Hz. The machine is rated at 991 MVA with an inertia $H = 2.88$ sec and damping $D = 0$ pu. The generator reactances and time constants are given by

$$X_d = 2.00 \text{ pu}, \quad X_q = 1.91 \text{ pu}, \quad X_\ell = 0.15 \text{ pu}, \quad R_a = 0 \text{ pu}$$
$$X'_d = 0.245 \text{ pu}, \quad X'_q = 0.42 \text{ pu}, \quad X''_d = 0.200 \text{ pu}, \quad X''_q = 0.200 \text{ pu}$$
$$T'_{d0} = 5.00 \text{ sec}, \quad T'_{q0} = 0.66 \text{ sec}, \quad T''_{d0} = 0.031 \text{ sec}, \quad T''_{q0} = 0.061 \text{ sec}$$

The infinite bus is modeled with a synchronous generator with a rated MVA base of 100,000, and classical model parameters of $X'_d = 0.01$ pu, $H = 3.00$ sec, and $D = 2.00$ pu on the machine base. The 991 MVA generator is connected to the infinite bus via a step-up transformer of 0.15 pu on the machine base, and two transmission lines, each having a reactance of 0.04 pu on the system base of 100 MVA. The voltage at the machine terminal is specified to be 1.05 pu and the voltage at the infinite bus is 1.0 pu. The load at the infinite bus is consuming 950 MW. Solve the power flow for this system loading condition using the infinite bus as the swing bus with an angle of 0°. Then obtain the linearized model of the system, compute the eigenvalues of the system matrix, and identify the modes.

Solution: The power flow solution is given in Table 8.1. Using the linearization function in PST, the initial conditions of the states of the generator are

$$\delta_{1o} = 65.84°, \quad \omega_{1o} = 0 \text{ pu}, \quad E'_{q1o} = 0.9250 \text{ pu}$$

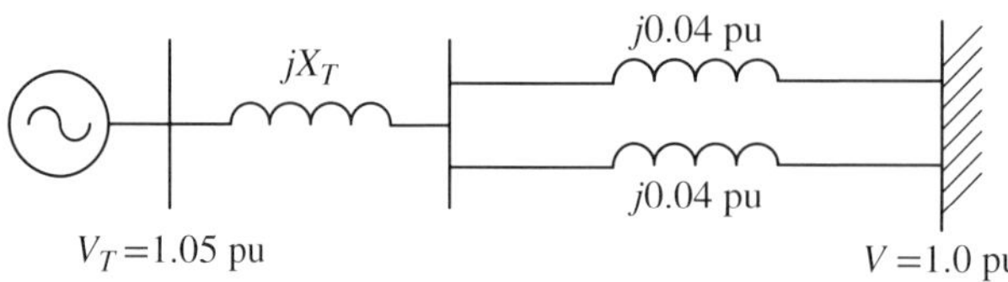

Figure 8.16 SMIB system for Example 8.4 with reactances on 100 MVA base.

Table 8.1 Power flow solution on 100 MVA base.

Bus	V (pu)	θ (deg)	P_g (pu)	Q_g (pu)	P_L (pu)	Q_L (pu)
1	1.0500	18.54	9.5000	3.0444	0	0
2	1.0000	0	0	0.1272	9.5000	0
3	1.0154	10.79	0	0	0	0

$$\psi_{kd1o} = 0.8424 \text{ pu}, \quad E'_{d1o} = -0.6020 \text{ pu}, \quad \psi_{kq1o} = -0.7111 \text{ pu}$$

and the states of the infinite-bus machine are

$$\delta_{2o} = -0.0054^\circ \simeq 0^\circ, \quad \omega_{2o} = 0 \text{ pu}$$

The states of the linearized system models are in the order

$$\Delta\delta_1,\ \Delta\omega_1,\ \Delta E'_{q1},\ \Delta\psi_{kd1},\ \Delta E'_{d1},\ \Delta\psi_{kq1},\ \Delta\delta_2,\ \Delta\omega_2$$

The linearized system model is

$$A = \begin{bmatrix} 0 & 377 & 0 & 0 & 0 & 0 & 0 & 0 \\ -0.3144 & 0 & -0.1521 & -0.1369 & -0.0240 & -0.1056 & 0.3144 & 0 \\ -0.3074 & 0 & -2.1275 & 1.5905 & 0 & 0 & 0.3074 & 0 \\ -5.0998 & 0 & 29.3164 & -34.9055 & 0 & 0 & 5.0998 & 0 \\ -0.3120 & 0 & 0 & 0 & -8.4693 & 6.1917 & 0.3120 & 0 \\ -3.3031 & 0 & 0 & 0 & 14.8985 & -22.9712 & 3.3032 & 0 \\ 0 & 0 & 0 & 0 & 0 & 0 & 0 & 377 \\ 0.0030 & 0 & 0.0014 & 0.0013 & 0.0002 & 0.0010 & -0.0030 & -0.3333 \end{bmatrix} \tag{8.97}$$

Note that the entries in the last row of A are quite small due to the large inertia of the infinite bus.

The eigenvalues of A are

$$\lambda_1 = 0,\ \lambda_2 = -0.3302,\ \lambda_{3,4} = -0.5676 \pm j10.482$$
$$\lambda_5 = -0.2093,\ \lambda_6 = -36.111,\ \lambda_7 = -3.3724,\ \lambda_8 = -27.648$$

The eigenvectors can be used to associate the states to these modes. The first four modes are the electromechanical modes, with λ_1 the system angle mode, λ_2 the system damping mode, and $\lambda_{3,4}$ the oscillation of Generator 1 vs the infinite bus. Their corresponding eigenvectors are

$$v_1 = \begin{bmatrix} \mathbf{0.7071} \\ 0 \\ 0 \\ 0 \\ 0 \\ 0 \\ \mathbf{0.7071} \\ 0 \end{bmatrix}, \quad v_2 = \begin{bmatrix} \mathbf{0.7085} \\ -0.0006 \\ -0.0034 \\ -0.0033 \\ -0.0008 \\ -0.0009 \\ \mathbf{0.7057} \\ -0.0006 \end{bmatrix}, \quad v_{3,4} = \begin{bmatrix} \mathbf{-0.9743} + j0.0000 \\ 0.0015 \mp j0.0271 \\ -0.0046 \mp j0.0475 \\ 0.1187 \mp j0.0768 \\ 0.0034 \mp j0.0852 \\ 0.0991 \mp j0.1030 \\ \mathbf{0.0093} \pm j0.0003 \\ -0.0000 \pm j0.0003 \end{bmatrix}$$

The other four modes are the flux linkage modes, with λ_5 the d-axis decay mode of E'_{q1}, λ_6 the decay mode of ψ_{kd1}, λ_7 the q-axis mode of E'_{d1}, and λ_8 the decay mode of ψ_{kq1}. The corresponding eigenvectors are

$$v_5 = \begin{bmatrix} 0.5863 \\ -0.0003 \\ \mathbf{-0.5437} \\ -0.5442 \\ -0.1660 \\ -0.1923 \\ 0.0094 \\ -0.0000 \end{bmatrix}, \quad v_6 = \begin{bmatrix} -0.0340 \\ 0.0033 \\ -0.0470 \\ \mathbf{0.9982} \\ 0.0021 \\ -0.0110 \\ 0.0003 \\ -0.0000 \end{bmatrix}, \quad v_7 = \begin{bmatrix} -0.2160 \\ 0.0019 \\ -0.0452 \\ -0.0068 \\ \mathbf{0.7584} \\ 0.6133 \\ 0.0023 \\ -0.0000 \end{bmatrix}, \quad v_8 = \begin{bmatrix} -0.0408 \\ 0.0030 \\ -0.0018 \\ 0.0215 \\ -0.3075 \\ \mathbf{0.9504} \\ 0.0004 \\ -0.0000 \end{bmatrix}$$

Note that the eigenvalues show that the decay of the ψ_{kd1} and ψ_{kq1} modes to be very fast. The decay of E'_{d1} is not as fast as those of ψ_{kd1} and ψ_{kq1} but is still quite fast. On the other hand, the decay of E'_{q1} is slower, due to a large T'_{do}. Such eigenvalue assessment justifies the use of flux-decay models in longer-term dynamics simulation. ■

8.9 Summary and Notes

This chapter is the second part of the development of analytical models of synchronous machines for use in dynamic simulation programs. Starting from the equivalent circuits in the last chapter, transient and subtransient reactances and time constants are developed. Then the subtransient machine model (also called the 2-axis model or the round rotor model) is derived, which can be represented by the block diagram shown in Figure 8.12, and is found in the manuals of many commercial simulation programs. In addition, this chapter shows the systematic reduction of the 2-axis model to simpler models, including the flux-decay model and the electromechanical model.

It should be noted that Sections 8.2 and 8.3 follow closely the lecture notes [94].

An important topic that is not treated in the two chapters on synchronous machines is on methods for obtaining the synchronous generator parameters, so that an accurate representation is used in a dynamic simulation program. There are a variety of tests, such as open-circuit, short-circuit, and standstill-frequency-response (SSFR) tests, that can be used to obtain specific parameters [101]. For example, from the open-circuit test shown in Figure 7.14,the inductance L_{ad} and saturation function can be obtained. In SSFR tests, currents consisting of frequencies from 0.001 Hz to 100 Hz are injected into the stator or rotor windings with the rotor aligned such that the phase-a axis is either in parallel or perpendicular to the d-axis. Then the measured frequency response is fitted with a transfer function, from which reactances and time constants can be extracted. Chapter 4 in [6] has a comprehensive coverage of this generator parameter topic.

It should also be noted that finite-element methods have been proposed to compute the machine parameters [102].

In addition, there is a body of work using time-domain data obtained from phasor measurement units [8] to tune or validate the generator parameters, such as the results reported in [103] using Kalman filtering techniques and in [104] using trajectory sensitivities.

Problems

8.1 Derive the q-axis flux and voltage equations given by (8.54)–(8.57).

8.2 Show that for steady-state calculation such as initialization, the electrical states of a synchronous machine can be computed from

$$E'_{qo} = E_{fdo} - (X_d - X'_d)i_{do}, \quad \psi_{kdo} = E'_{qo} - (X'_d - X_\ell)i_{do}$$
$$E'_{do} = -(X_q - X'_q)i_{qo}, \quad \psi_{kqo} = -(X_q - X_\ell)i_{qo} \tag{8.98}$$

Use these expressions to verify the initial values obtained in Example 8.2.

8.3 Show that the subtransient model (8.49), (8.55), (8.52)–(8.53), and (8.56)–(8.57) simplifies to the flux-decay model when $X''_d = X'_d$ and $X''_q = X'_q$.

8.4 The parameters of a 60 Hz synchronous machine in the form employed by most manufacturers are as follows:

$$X_d = 1.21 \text{ pu}, \quad X_q = 1.09 \text{ pu}, \quad X_\ell = 0.110 \text{ pu}, \quad R_a = 0.002 \text{ pu}$$
$$X'_d = 0.193 \text{ pu}, \quad X'_q = 0.270 \text{ pu}, \quad X''_d = 0.161 \text{ pu}, \quad X''_q = 0.240 \text{ pu}$$
$$T'_{d0} = 7.39 \text{ sec}, \quad T'_{q0} = 1.50 \text{ sec}, \quad T''_{d0} = 0.030 \text{ sec}, \quad T''_{q0} = 0.035 \text{ sec}$$

The equivalent circuit of a synchronous machine requires the following parameters:

$$X_\ell, X_{ad}, X_{\ell fd}, X_{\ell kd}, X_{aq}, X_{\ell 1q}, X_{\ell kq}, R_{fd}, R_{kd}, R_{1q}, R_{kq}, R_a.$$

Calculate the necessary equivalent circuit parameters and sketch the direct-axis and quadrature-axis voltage equivalent circuits. Note that the unit for the time constants given here is seconds.

8.5 Use the machine parameters in Problem 8.4 to perform the necessary calculations to draw the steady-state voltage-current phasor diagram when the machine is operating at rated terminal voltage ($E_t = 1$ pu) and rated terminal current ($I_t = 1$ pu) at 0.9 leading power factor. Assume no saturation.

8.6 Use the machine parameters and the machine loading in Problems 8.4 and 8.5 to initialize the synchronous machine model in Figure 8.12, that is, find the steady-state values for E'_d, E'_q, ψ_{kd}, and ψ_{kq}.

8.7 Repeat Problems 8.5 and 8.6, for the case when the machine is operating at rated terminal voltage ($E_t = 1$ pu) and rated terminal current ($I_t = 1$ pu) at unity power factor. Assume no saturation.

8.8 Use the machine parameters and the power flow solution (Bus 1 active and reactive power output and voltage magnitude and angle) to verify the initial conditions of the states in Example 8.4. Use the machine base of 991 MVA for the calculation.

8.9 Develop analytical expressions for the entries $A(i,j)$ for $(i,j) = (3,3), (3,4), (4,3), (4,4), (5,5), (5,6), (6,5)$, and (6,6) in Example 8.4, and use them to calculate the A matrix values.[4] The block diagram in Figure 8.12 can be used. Note that you will also need the sensitivities of Δi_d and Δi_q with respect to the state variables, which are given in Table 8.2.[5] It is more convenient to do the calculations on the machine base of 991 MVA. Thus the coefficients in Table 8.2 need to be scaled accordingly. No saturation model is used for the synchronous machine.

Table 8.2 Sensitivities of Δi_d and Δi_q given on the system base of 100 MVA.

	$\Delta E'_q$	$\Delta \psi_{kd}$	$\Delta E'_d$	$\Delta \psi_{kq}$
Δi_d	9.5127	8.5614	0	0
Δi_q	0	0	3.3471	14.7270

8.10 Repeat the linearization process in Example 8.4 by reducing the load at the infinite bus to 900 MW. Use PST to solve the power flow and obtain the linear model. Compare the results to those of Example 8.4, especially the entries of the A matrix.

8.11 Use PST to linearize the 2-area power system in Figure 2.7 using detailed machine models for the base case operating condition stated in Example 2.3, given in the file *data2a_problem_08_11.m*. Identify the states associated with the system A matrix. Compute the eigenvalues and associate the modes with the states.

4 Analytical expressions for entries of the A matrix that involve network variables, such as $A(3,1)$, can be quite complex and are best left to computer calculations.
5 A user of PST will find these sensitivities in the arrays *c_curd* and *c_curq*.

9

Excitation Systems

9.1 Introduction

An excitation system supplies the field voltage to the rotor circuit of a synchronous generator, allowing the conversion of mechanical power to electrical power. With appropriate feedback loops, an excitation system also provides control functionalities to make the synchronous machine, and, by extension, an interconnected power system, operate more reliably. The performance enhancement is accomplished by adjusting the field voltage to regulate the generator terminal voltage. As discussed in the voltage stability chapter, an excitation system can also participate in hierarchical voltage control.

The block diagram of a generic excitation system is shown in Figure 9.1. The excitation system monitors the terminal bus voltage V_T and compares it to a desired reference voltage V_{ref}. The error, which is the difference between the desired and actual signals, is used by the voltage regulator (VR) to drive several control circuits. The output signal from the voltage regulator is then sent to an exciter, which can be a rotating alternator or power-electronic rectifier, to provide the field voltage to the rotor circuit of a synchronous generator.

There are two main functions of an excitation system:

1) In steady state, the voltage regulator controls the generator bus terminal voltage to a desired value.
2) In disturbance conditions (such as short-circuit faults), the excitation control system supplies additional reactive power to the post-fault system to maintain the generator terminal voltage, thus increasing the synchronizing torque and allowing the generator to maintain synchronism to improve the transient stability of the interconnected system.

High-response ratio excitation systems, which are mostly power-electronics-based exciters with high feedback gains, provide more benefits in maintaining synchronism, but will reduce damping on the local and interarea modes. The voltage regulator control input, however, allows the implementation of a supplementary signal to add damping to the oscillatory modes via a power system stabilizer (PSS). PSS design is the topic of the next chapter.

The remaining sections of this chapter are organized as follows. Section 9.2 describes the three main types of excitation systems. Then each excitation system type is described in a separate section (Sections 9.3–9.5) and illustrated with examples showing linearized

Power System Modeling, Computation, and Control, First Edition. Joe H. Chow and Juan J. Sanchez-Gasca.

Companion website: www.wiley.com/go/chow/power-system-modeling

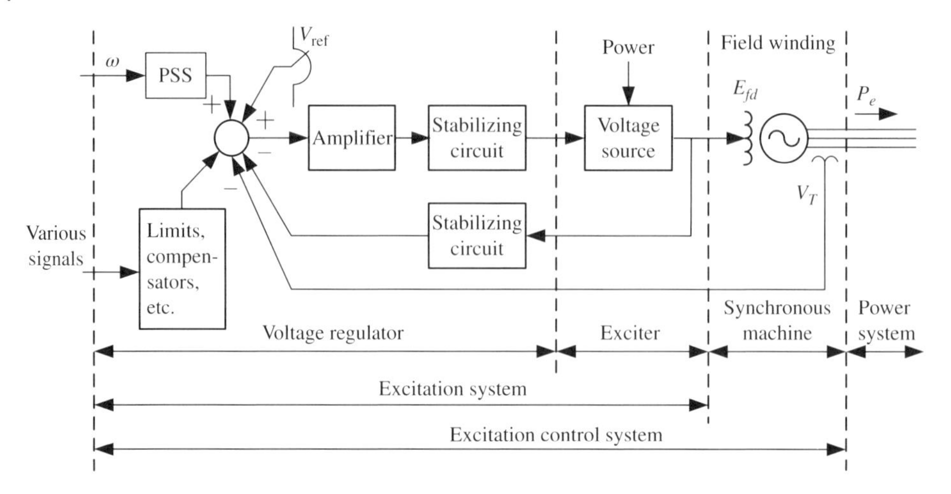

Figure 9.1 Excitation system block diagram (adopted from M. L. Crenshaw [20]).

models and performance during disturbances. Section 9.6 describes some of the protection functions provided by excitation systems.

9.2 Excitation System Models

Excitation systems can be organized into three types based on the mechanism used to generate the field voltage E_{fd} for a synchronous generator. The older types of excitation systems are based on a rotating alternator driven by the generator shaft to provide the field voltage. Modern excitation systems use power electronics to generate a DC field voltage, with the control functions in Figure 9.1 and the protection functions implemented on a printed circuit board, or more recently coded in a microprocessor, so that functionalities can be readily reprogrammed. The purpose of this discussion is to point out some of the most important features of these excitation systems. More detailed descriptions can be found in [5, 6, 105, 106].

An excitation system has two main components: (1) an exciter acting as the power supply to the generator rotor field circuit and (2) control circuits or functions for performing voltage regulation.

1) The exciter provides the DC field voltage and current to the generator rotor circuit. Exciters are classified based on how the DC field voltage is generated.
   - Direct current commutator exciters (Type DC) use a commutator to generate a direct current.
   - Alternator supplied rectifier excitation systems (Type AC) use an alternator with a rectifier to produce direct current.
   - Static excitation systems (Type ST) supply excitation voltage through transformers or auxiliary generator windings and rectifiers. In contrast to Types DC and AC, there is no rotating equipment. Newer excitation systems are of this type.
2) The control circuits amplify the voltage error (the difference between the reference voltage value and the actual terminal voltage), allowing the field voltage to control the generator terminal voltage. Frequently stabilizing circuits (transient-gain

reduction or low-pass circuits) are added to stabilize the control loop, particularly when the amplifier gain for voltage regulation is high.

There are many variations of Type AC, DC and ST excitation system models currently in use in power systems, as synchronous machines are normally designed to operate for 40 or more years. Some older units have been refitted with more modern excitation systems to improve their performance. In this chapter, only selected models of these three types of excitation systems are discussed. It should be noted that the control block diagrams in this section are mostly based on those found in the 1992 IEEE excitation system standard document [105].

9.3 Type DC Exciters

There are several DC excitation system models in [105]. The Type DC1A model,[1] based on a DC commutator exciter and shown in Figure 9.2, is discussed in this section. In the DC1A model, the transfer function blocks from V_R to the E_{fd} comprise the exciter model, and the balance of the transfer functions represents the voltage regulator and protection functions. The DC2A model (Figure 9.2) is similar to the DC1A model, except that the voltage regulator limits are now proportional to the terminal voltage V_T, as the field-controlled DC commutator exciters are supplied by the generator bus. The variables, parameters, and functions for the DC1A and DC2A models in Figure 9.2 are listed in Table 9.1.

9.3.1 Separately Excited DC exciter

In a separately excited DC exciter, a DC power source is supplied to the field winding of the exciter, which after commutation provides a DC voltage to the synchronous generator rotor field winding. If the exciter is supplied by the generator terminal voltage, the DC2A model is used to represent the excitation system. A schematic of the exciter is shown in Figure 9.3.a.

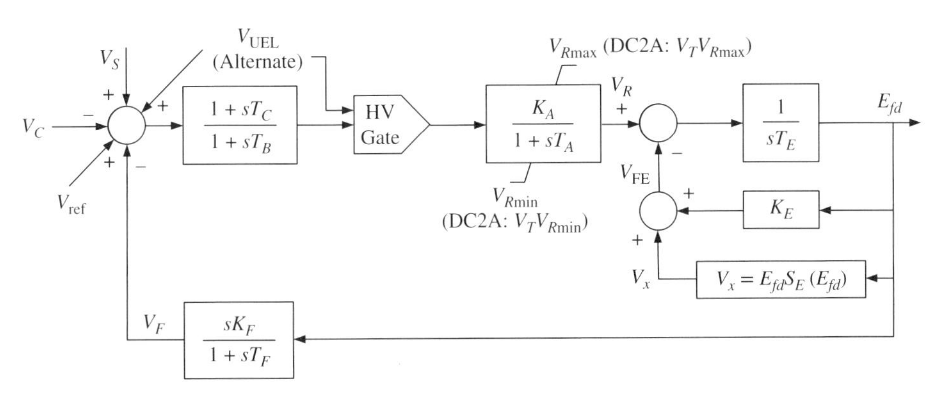

Figure 9.2 Type DC1A excitation system model. For Type DC2A models, the limits V_{Rmax} and V_{Rmin} are replaced by $V_T V_{Rmax}$ and $V_T V_{Rmin}$, respectively.

1 The "A" designation refers to models in [105] and represents updates of the models in [107].

Table 9.1 DC1A excitation system model variables, parameters, and functions [105].

Variables
V_{ref}: reference voltage value
V_C: measured voltage to be controlled
V_s: supplemental input voltage (typically from power system stabilizer)
V_F: excitation system stabilizer output
V_{UEL}: under-excitation voltage limit
V_R: input signal to exciter
$V_{R\max}, V_{R\min}$: upper and lower limits of V_R
V_T: generator terminal bus voltage
V_{FE}: signal proportional to exciter field current
V_x: signal proportional to exciter saturation
E_{fd}: field voltage applied to generator rotor field winding
Parameters
K_A: voltage regulator gain
T_A, T_B, T_C: voltage regulator time constants
T_F: excitation control system stabilizer time constant
K_E: exciter constant related to field voltage
T_E: exciter time constant (integration rate associated with exciter control)
Functions
HV Gate: high-value gate selecting high value of inputs
$S_E[E_{fd}]$: exciter saturation function value at corresponding exciter voltage E_{fd}

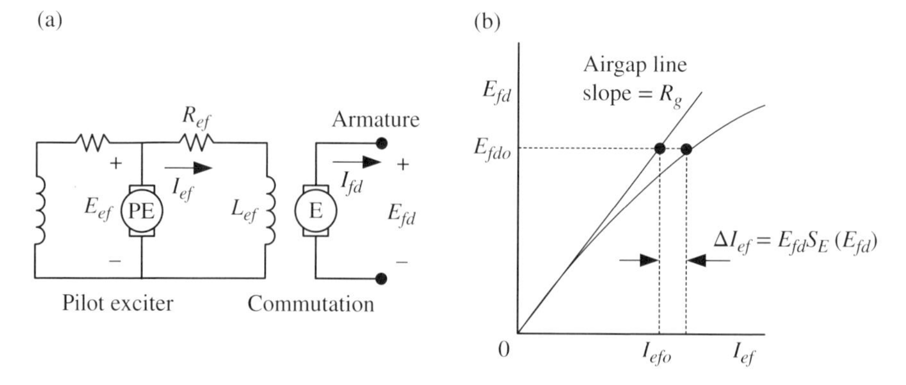

Figure 9.3 Separately excited DC exciter: (a) schematic (adopted from [5]) and (b) saturation function.

The field circuit of the exciter is modeled by (the subscripts "*ef*" denote the exciter field)

$$E_{ef} = R_{ef} I_{ef} + \frac{d\psi_{ef}}{dt}, \quad \psi_{ef} = L_{ef} I_{ef} \tag{9.1}$$

where E_{ef} is the applied voltage, I_{ef} is the field current, ψ_{ef} is the flux in the inductor, and R_{ef} and L_{ef} are the field circuit resistance and inductance, respectively. The output voltage of the exciter is

$$E_{fd} = K_x \psi_{ef} \tag{9.2}$$

where K_x is a constant.

The saturation in the magnetic circuit is modeled by

$$I_{ef} = \frac{E_{fd}}{R_g} + \Delta I_{ef}, \quad \Delta I_{ef} = E_{fd} S_E(E_{fd}) \tag{9.3}$$

where R_g is the slope of the airgap line and ΔI_{ef} is the saturated component of the field current (Figure 9.3.b). Thus (9.1) can be rewritten as

$$E_{ef} = R_{ef} \left(\frac{E_{fd}}{R_g} + E_{fd} S_E(E_{fd}) \right) + \frac{1}{K_x} \frac{dE_{fd}}{dt} \tag{9.4}$$

The exciter quantities can be per-unitized using the generator rotor field circuit base, namely,

$$E_{ef\text{base}} = E_{fd\text{base}}, \quad I_{ef\text{base}} = \frac{E_{fd\text{base}}}{R_g} \tag{9.5}$$

Divide the E_{ef} equation (9.1) by $E_{ef\text{base}}$ to obtain

$$\frac{E_{ef}}{E_{ef\text{base}}} = \frac{R_{ef}}{R_g} \frac{E_{fd}}{E_{ef\text{base}}} + R_{ef} S_E(E_{fd}) \frac{E_{fd}}{E_{ef\text{base}}} + \frac{1}{K_x} \frac{d}{dt} \frac{E_{fd}}{E_{ef\text{base}}} \tag{9.6}$$

which when written in per-unit values becomes

$$\overline{E}_{ef} = \frac{R_{ef}}{R_g} \overline{E}_{fd} + \frac{R_{ef}}{R_g} \overline{S}_E(\overline{E}_{fd}) \overline{E}_{fd} + \frac{1}{K_x} \frac{d\overline{E}_{fd}}{dt} \tag{9.7}$$

where the exciter saturation in per unit is

$$\overline{S}_E(\overline{E}_{fd}) = R_g S_E(E_{fd}) = R_g \frac{\Delta I_{ef}}{E_{fd}} \frac{E_{fd\text{base}}}{E_{fd\text{base}}} = \frac{\Delta I_{ef}}{I_{ef\text{base}}} \frac{1}{\overline{E}_{fd}} = \frac{\Delta \overline{I}_{ef}}{\overline{E}_{fd}} \tag{9.8}$$

The exciter saturation curve in per unit is shown in Figure 9.4.

In power system dynamic simulation programs, the saturation function is parametrized as

$$\overline{S}_E(\overline{E}_{fd}) = \frac{A - B}{B} \tag{9.9}$$

where A is on the saturation curve and B is on the airgap line, as determined by the operating value of E_{fd}. Typically this saturation is defined for $\overline{E}_{fd}$ at the exciter ceiling voltage, and at 75% of the exciter ceiling voltage.

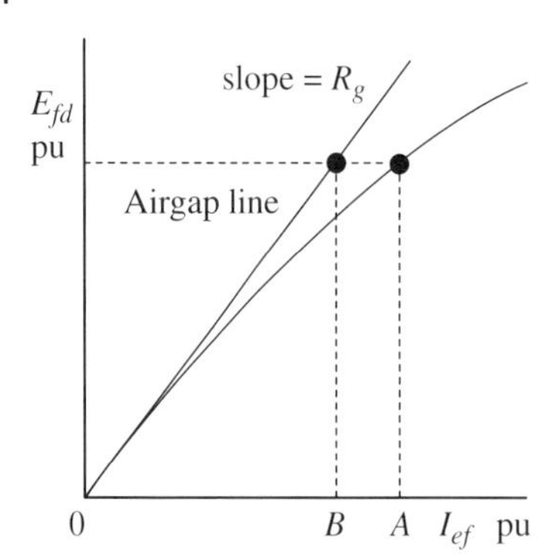

Figure 9.4 Exciter saturation function in per unit.

Continuing the derivation for the other parameters, from (9.2),

$$K_x = \frac{E_{fd}}{\psi_{ef}} = \frac{E_{fd}}{L_{ef}I_{ef}}\frac{E_{fd\text{base}}}{E_{fd\text{base}}} = \frac{R_g}{L_{ef}}\frac{\overline{E}_{fd}}{\overline{I}_{ef}} \tag{9.10}$$

At any operating point (I_{efo}, E_{xo}), define

$$L_{fu} = L_{ef}\frac{\overline{I}_{efo}}{\overline{E}_{fdo}} \quad \Rightarrow \quad K_x = \frac{R_g}{L_{fu}} \tag{9.11}$$

where the subscript "u" indicates an unsaturated variable or parameter. Thus (9.7) can be rewritten as

$$\overline{E}_{ef} = K_E\overline{E}_{fd} + S_E(\overline{E}_{fd})\overline{E}_{fd} + T_E\frac{d\overline{E}_{fd}}{dt} \tag{9.12}$$

where

$$K_E = \frac{R_{ef}}{R_g}, \quad T_E = \frac{L_{fu}}{R_g}, \quad S_E(\overline{E}_{fd}) = \overline{S}_E(\overline{E}_{fd})\frac{R_{ef}}{R_g} \tag{9.13}$$

The exciter gain constant K_E is normally equal to 1. The exciter time constant T_E is in general between 0.5 and 1.5 seconds.

Rearrange (9.12) to obtain

$$T_E\frac{d\overline{E}_{fd}}{dt} = \overline{E}_{ef} - K_E\overline{E}_{fd} - V_x(\overline{E}_{fd}) \tag{9.14}$$

where $V_x(\overline{E}_{fd})$ is commonly approximated by

$$V_x(\overline{E}_{fd}) = \overline{E}_{fd}S_E(\overline{E}_{fd}) \simeq A_{\text{EX}}e^{B_{\text{EX}}\overline{E}_{fd}} \tag{9.15}$$

Taking Laplace transform of (9.15) yields

$$\overline{E}_{fd} = \frac{1}{sT_E}(\overline{E}_{ef} - K_E\overline{E}_{fd} - V_x(\overline{E}_{fd})) \tag{9.16}$$

where the input voltage signal is $\overline{E}_{ef} = V_R$ when referred to the block diagram in Figure 9.2.

Example 9.1: Saturation model parameters

A set of saturation parameters (all in pu) of a DC1A exciter model is given by

$$E_{fd1} = 2.3, \quad S_E(E_{fd1}) = 0.10; \quad E_{fd2} = 3.1, \quad S_E(E_{fd2}) = 0.33 \tag{9.17}$$

Compute the parameters A_{EX} and B_{EX} for the saturation model (9.15), and then compute V_x for $E_{fd} = 2.8$ pu.

Solutions: The values in (9.17) are used to set the numerical values of (9.15) as

$$2.3 \times 0.10 = A_{\text{EX}} e^{2.3B_{\text{EX}}}, \quad 3.1 \times 0.33 = A_{\text{EX}} e^{3.1B_{\text{EX}}} \tag{9.18}$$

which can be put into a matrix form by taking the logarithm on both sides

$$\begin{bmatrix} 1 & 2.3 \\ 1 & 3.1 \end{bmatrix} \begin{bmatrix} \ln(A_{\text{EX}}) \\ B_{\text{EX}} \end{bmatrix} = \begin{bmatrix} \ln(0.23) \\ \ln(1.023) \end{bmatrix} \tag{9.19}$$

The solutions to (9.19) are $A_{\text{EX}} = 0.00315$ and $B_{\text{EX}} = 1.8655$. Thus for $E_{fd} = 2.8$ pu, $V_x = 0.5845$ pu. ■

9.3.2 Self-Excited DC Exciter

The exciter voltage needed to generate the flux linkage of a DC exciter can also be obtained from the alternator output voltage, as shown in Figure 9.5. This arrangement is called a self-excited DC exciter and can also be represented by the DC1A model. The resistance on the exciter field side can be fixed as shown in Figure 9.5.a or can be adjusted to trim the input voltage V_R, often called the "buck-boost" mode, as shown in Figure 9.5.b.

With the sign of V_R as given in Figure 9.5, the effective exciter field voltage in pu is

$$\overline{E}_{ef} = \overline{E}_{fd} + \overline{V}_R \tag{9.20}$$

Following the derivations for the separately excited DC exciter, the self-excited exciter model equation in per unit is

$$\overline{E}_{fd} + \overline{V}_R = \frac{R_{ef}}{R_g}\overline{E}_{fd} + \frac{R_{ef}}{R_g}\overline{S}_E(\overline{E}_{fd})\overline{E}_{fd} + \frac{1}{K_x}\frac{d\overline{E}_{fd}}{dt} \tag{9.21}$$

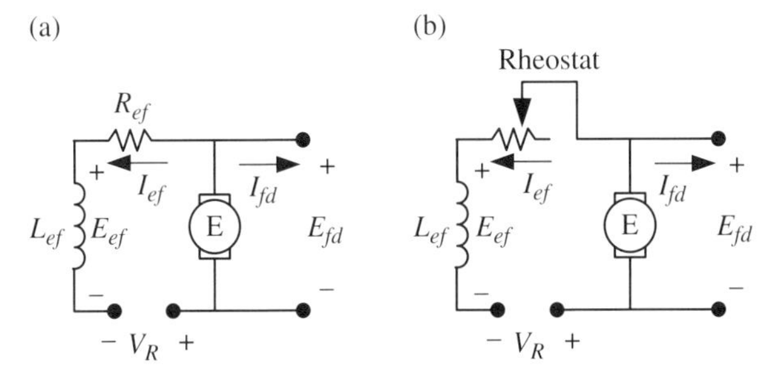

Figure 9.5 Self-excited DC exciter: (a) fixed resistance and (b) variable resistance as controlled by a rheostat.

that is,

$$\overline{V}_R = K_E \overline{E}_{fd} + S_E(\overline{E}_{fd})\overline{E}_{fd} + T_E \frac{d\overline{E}_{fd}}{dt} \tag{9.22}$$

where

$$K_E = \left(\frac{R_{ef}}{R_g} - 1\right), \quad T_E = \frac{L_{fu}}{R_g}, \quad S_E(\overline{E}_{fd}) = \overline{S}_E(\overline{E}_{fd})\frac{R_{ef}}{R_g} \tag{9.23}$$

Note that K_E is not a constant, as R_{ef} varies with the setting of the rheostat.

In the input data of a self-excited DC exciter, if K_E is given, then it is kept. If K_E is zero or not given, then in the initialization stage of a simulation program, K_E is selected such that $V_R = 0$.[2] As a result K_E can be computed from

$$0 = K_E \overline{E}_{fd} + S_E(\overline{E}_{fd})\overline{E}_{fd} \quad \Rightarrow \quad K_E = -S_E(\overline{E}_{fd}) \tag{9.24}$$

As a result, a negative K_E would indicate that the exciter is of the self-excited type.

9.3.3 Voltage Regulator

The main transfer function of the voltage regulator is an amplifier circuit modeled as

$$\frac{K_A}{1 + sT_A} \tag{9.25}$$

where K_A is the voltage regulator gain to bring the measured controlled voltage V_C, typically the generator terminal voltage, close to the reference voltage V_{ref}. The output of this circuit is the control signal V_R driving the exciter. The signal V_R is bounded by the upper limit $V_{R\max}$ and the lower limit $V_{R\min}$, which are functions of the equipment properties. To properly represent the physical equipment, the transfer function is implemented with anti-windup limits. A discussion on anti-windup limit implementation is given in Appendix 9.A.

The amplifier circuit (9.25) is a first-order transfer function with a DC gain of K_A and a bandwidth of $1/T_A$ rad/sec. A high value of K_A allows the voltage regulator to tightly control the desired voltage, for example the generator terminal bus voltage. If T_A is small, the bandwidth of the voltage regulator is high. Thus the gain K_A can impact other modes in the $1/T_A$ rad/sec bandwidth, such as the electromechanical modes. As a counter measure, a bandwidth reduction circuit, commonly known as a stabilizing circuit or transient gain reduction, is implemented in many voltage regulators.

There are two ways to implement a stabilizing circuit in the voltage regulator. In the forward path, a lowpass filter

$$\frac{1 + sT_C}{1 + sT_B} \tag{9.26}$$

where $T_C < T_B$, can be used. Alternatively, in the feedback path, a derivative block

$$\frac{sK_F}{1 + sT_F} \tag{9.27}$$

2 This selection will allow V_{ref} to be same as V_C, the controlled voltage.

having a steady-state gain of zero, can be used. Both of these functions can reduce the regulator gain at high frequency, without affecting steady-state regulation. The effect of implementing (9.27) to achieve transient-gain reduction of the regulation function is shown in Example 9.3.

9.3.4 Initialization of DC Type Exciters

In steady state, $d\overline{E}_{fd}/dt$ is zero, and thus the exciter equation (9.14) reduces to

$$V_R - K_E\overline{E}_{fd} - V_x(\overline{E}_{fd}) = 0 \quad \Rightarrow \quad V_R = K_E\overline{E}_{fd} + V_x(\overline{E}_{fd}) \tag{9.28}$$

For a self-exciter DC exciter, V_R can be set to 0. For a separately excited DC exciter, solve for V_R using K_E (which is normally equal to 1) and E_{fd}, which is computed in the initialization of the machine variables. Note that V_R is the output of the voltage regulator, given by the Laplace transform

$$V_R(s) = \frac{K_A}{1+sT_A}(V_{\text{ref}} - V_C(s)) \tag{9.29}$$

where V_C is the controlled voltage. In steady state, there is no need to account for the gain reduction (lowpass) block in the forward loop or the rate circuit in the feedback path as $s = 0$, resulting in

$$V_R = K_A(V_{\text{ref}} - V_C) \tag{9.30}$$

The voltage V_C is obtained from the power flow solution, allowing the determination of the reference voltage as

$$V_{\text{ref}} = V_C + \frac{V_R}{K_A} \tag{9.31}$$

For a self-excited DC exciter, $V_R = 0$, so $V_{\text{ref}} = V_C$, implying that there is no steady-state error in voltage regulation.

Example 9.2: Excitation system steady-state values

In the initialization of a synchronous generator, the field voltage is found to be $E_{fdo} = 2.391$ pu and the voltage at the generator terminal bus, $V_T = 1.0$ pu, is used as V_C. The field voltage is controlled by a Type DC1A separately excited excitation system with the parameters

$$K_A = 46,\ T_A = 0.06 \text{ sec},\ K_E = 1.0,\ T_E = 0.46 \text{ sec},\ K_F = 0.1,\ T_F = 1.0 \text{ sec} \tag{9.32}$$

and the saturation function obtained in Example 9.1. Initialize the states and find the reference voltage V_{ref} of the excitation system. Repeat the computation for a self-excited excitation system with the same parameters and loading condition, except that the value of K_E is adjusted such that $V_{Ro} = 0$.

Solutions: There are three states in the excitation system, E_{fd}, V_R, and V_F, where V_F is the integrator state for the transient-gain reduction block in the feedback path. For

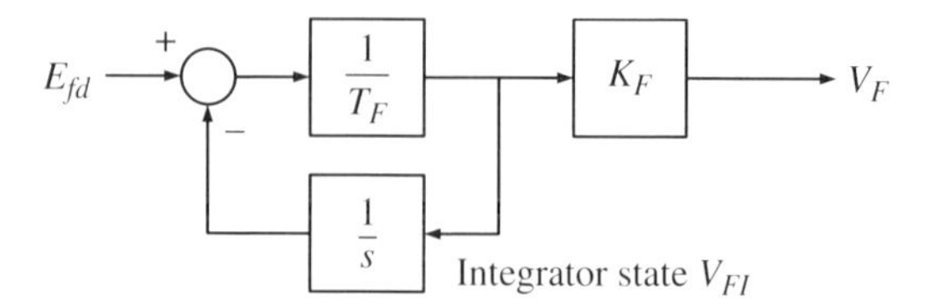

Figure 9.6 Block diagram representation of the derivative block $sK_F/(1 + sT_F)$.

initialization in steady state of the separately excited DC1A model, the inputs to all the integrators have to be zero. Thus at the input to the E_{fd} integrator, from (9.28)

$$V_{Ro} = 1.0 \times 2.391 + 2.391 \times 0.0032e^{1.8655\times2.391} = 2.391 + 0.6517 = 3.0427 \text{ pu} \quad (9.33)$$

At the input to the voltage regulator integrator

$$K_A(V_{\text{ref}} - V_T) - V_{Ro} = 0 \quad \Rightarrow \quad V_{\text{ref}} = 1.0 + \frac{3.0427}{46} = 1.0661 \text{ pu} \quad (9.34)$$

The derivative function is redrawn as a block diagram shown in Figure 9.6. The zero at the origin of the transfer function is implemented as an integrator in the feedback path. Setting the input to the integrator to zero results in

$$E_{fdo} - V_{FIo} = 0 \quad \Rightarrow \quad V_{FIo} = 2.391 \text{ pu} \quad (9.35)$$

Note that the time constants are not used in the initialization process.

For the self-excited DC1A model,

$$K_E = -S_E = 0.0032e^{1.8655\times2.391} = -0.2725 \quad (9.36)$$

Then the initial values are given as $V_{Ro} = 0$ pu, $V_{FIo} = 2.391$ pu, and $V_{\text{ref}} = V_T = 1.0$ pu. ■

9.3.5 Transfer Function Analysis

To obtain a better understanding of the performance of the voltage regulator, linearized models can be used for a frequency response analysis of the excitation system. In progressing to dynamic analysis, it is assumed that all exciter variables are in pu and thus the "bar" denoting pu values will be dropped.

The linearized or small-signal model of the exciter model (9.12) is, with all the values in pu,

$$T_E \frac{d\Delta E_{fd}}{dt} = -K_E' \Delta E_{fd} + \Delta V_R \quad (9.37)$$

where

$$K_E' = K_E + S_E(E_{fdo}) + E_{fdo} \left. \frac{\partial S_E}{\partial E_{fd}} \right|_{E_{fdo}} \quad (9.38)$$

Then take the Laplace transform to obtain

$$T_E s \Delta E_{fd}(s) = -K_E' \Delta E_{fd}(s) + \Delta V_R(s) \quad \Rightarrow \quad \Delta E_x(s) = \frac{1}{sT_E + K_E'} \Delta V_R(s) \quad (9.39)$$

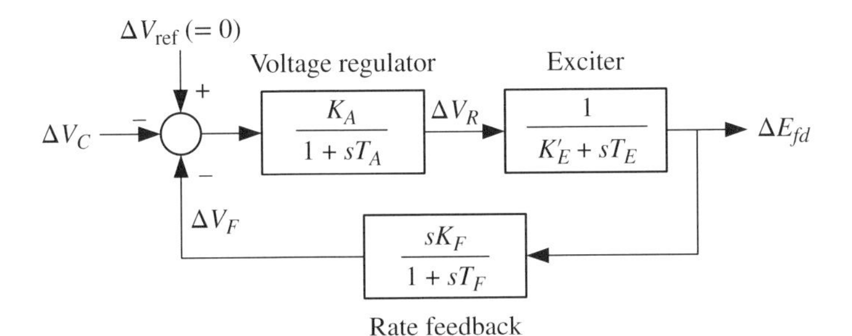

Figure 9.7 Linearized Type DC1A excitation system block diagram without input lowpass filter.

This transfer function can be inserted into the exciter model of Figure 9.2 to obtain a linearized model with ΔV_C as the input and ΔE_{fd} as the output, as shown in Figure 9.7. Note that the stabilizing circuit $(1 + sT_C)/(1 + sT_B)$ and the nonlinear elements are not included in the diagram.

From the linearized block diagram in Figure 9.7, the forward-path transfer function from ΔV_{ref} to ΔE_{fd} without the rate feedback path is

$$\Delta E_{fd} = \frac{1}{sT_E + K'_E}\frac{K_A}{1+sT_A}(\Delta V_{ref}(s) - \Delta V_C(s)) = T_{ngr}(s)(\Delta V_{ref}(s) - \Delta V_C(s)) \quad (9.40)$$

Note that the sign of the transfer function from ΔV_C to ΔE_{fd} is negative; that is, if V_C (also ΔV_C) increases, then E_{fd} (also ΔE_{fd}) will decrease.

With the derivative block in place, the closed-loop transfer function becomes

$$\frac{\Delta E_{fd}(s)}{\Delta V_{ref}(s) - \Delta V_C(s)} = \frac{\text{Forward path gain}}{1 + \text{Feedback loop gain}} = \frac{\frac{1}{sT_E+K'_E}\frac{K_A}{1+sT_A}}{1 + \frac{1}{sT_E+K'_E}\frac{K_A}{1+sT_A}\frac{sK_f}{1+sT_f}} = T_{tgr}(s) \quad (9.41)$$

which becomes, after simplifying the denominator terms

$$T_{tgr}(s) = \frac{K_A(1+sT_f)}{(sT_E + K'_E)(1+sT_A)(1+sT_f) + sK_AK_f} \quad (9.42)$$

The following example illustrates the difference between the transfer functions $T_{ngr}(s)$ and $T_{tgr}(s)$.

Example 9.3: Transient-gain reduction

Given the Type DC2A excitation system parameters $K_A = 300$, $T_A = 0.01$ sec, $K_E = 1.0$, $T_E = 1.33$ sec, $K_F = 0.1$, and $T_F = 0.675$ sec [105], find and plot the transfer functions $T_{ngr}(s)$ and $T_{tgr}(s)$.

Solutions: From (9.40) and (9.41), these two transfer functions are computed as

$$T_{ngr}(s) = \frac{300}{1 + 0.47s + 0.0046s^2}, \quad \text{DC gain} = 300 \quad (9.43)$$

$$T_{tgr}(s) = \frac{300(1+0.675s)}{1 + 31.15s + 0.3219s^2 + 0.003105s^3}, \quad \text{DC gain} = 300 \quad (9.44)$$

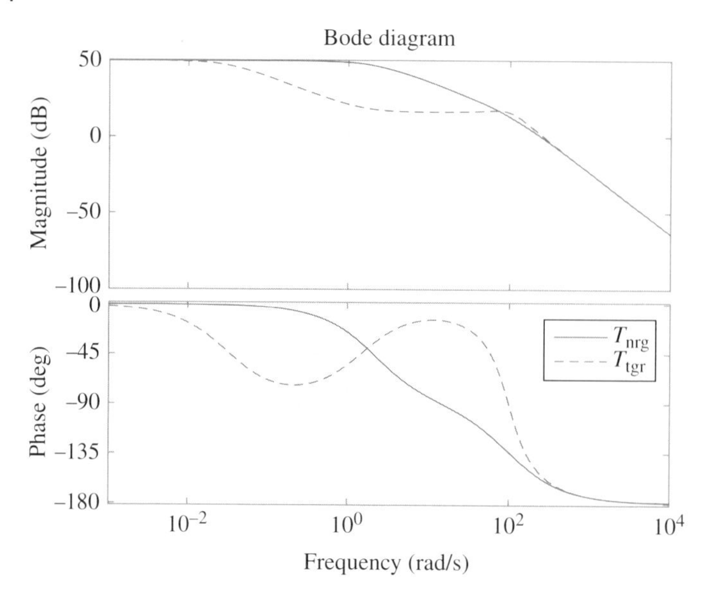

Figure 9.8 Frequency response plot of $T_{ngr}(s)$ and $T_{tgr}(s)$ (see color plate section).

The DC gain of both transfer function is $20\ \log_{10}300 = 49.54$ dB. Thus the transient-gain reduction does not change the steady-state regulation.

The frequency response plots of these two functions are shown in Figure 9.8. Note that the gain of $T_{tgr}(s)$ at 1 rad/sec is about 25 dB lower than that of $T_{ngr}(s)$. ■

Transient-gain reduction decreases the destabilizing effect of a high-gain voltage regulator on the power system swing modes, at the expense of reducing the speed of response of the voltage regulator and thus the synchronizing torque. Transient-gain reduction in excitation systems is, in general, not used in power systems with long transmission lines, like the US western system (WECC), because high-gain exciters are needed to maintain transient stability after disturbances. Without gain reduction, the local and interarea mode damping will be reduced, thus requiring power system stabilizers to restore the damping of the swing modes. More details are provided in the next chapter on power system stabilizer design.

9.3.6 Generator and Exciter Closed-Loop System

The schematic of an excitation system controlling the field voltage of a generator connected to an infinite bus is shown in Figure 9.9. The terminal voltage V_T of the generator is used as the controlled variable V_C. The analytical model of this closed-loop system can be developed by combining the generator dynamic equations in Sections 8.4 and 8.5 with the exciter model, like the DC1A model in this section. Note that as the machine voltages and currents are subject to dq-axes rotation, the excitation system deals with only the magnitude of the terminal voltage and provides a DC value of the field voltage to the d-axis generator rotor circuit. Hence no quantities with real and imaginary parts are present in excitation system control loops, although some bus-fed compound-source excitation systems would require some computation with real and imaginary parts.

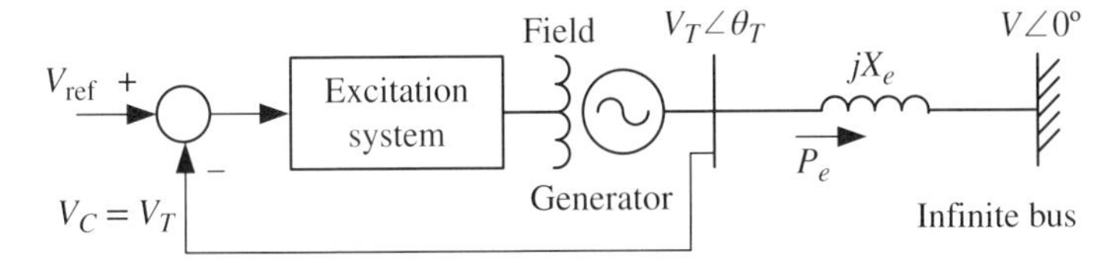

Figure 9.9 Closed-loop excitation system control.

The states in the closed-loop model include (1) a synchronous machine with 2 mechanical states and up to 4 electrical states, and (2) an excitation system with 2 to 3 states for the exciter, amplifier, and transient-gain reduction. Thus the model of one synchronous generator with an excitation system would have up to 9 states.

The complexity of generator-exciter closed-loop systems would be cumbersome to analyze manually, and thus such systems are normally analyzed using computer software. The next two examples illustrate a linear model and a disturbance response for a Single-Machine Infinite-Bus (SMIB) system.

Example 9.4: Linearized closed-loop system

For the SMIB system in Example 8.4 and Figure 8.6 where the synchronous machine is modeled in detailed and controlled by a Type DC2A excitation system with the parameters in Example 9.3, generate the linearized model and compute the eigenvalues. Compare the eigenvalues to those obtained in Example 8.4.

Solutions: The linearization function from PST yields the state-space model

$$\Delta\dot{x} = A\Delta x \tag{9.45}$$

where the states in x are

$$x = \begin{bmatrix} \delta & \omega & E'_q & \psi_{kd} & E'_d & \psi_{kq} & V_R & E_{fd} & V_F \end{bmatrix}^T \tag{9.46}$$

$$A = \begin{bmatrix} 0 & 377 & 0 & 0 & 0 & 0 & 0 & 0 & 0 \\ -0.315 & 0 & -0.152 & -0.137 & -0.0240 & -0.106 & 0 & 0 & 0 \\ -0.307 & 0 & -2.13 & 1.59 & 0 & 0 & 0 & 0.2 & 0 \\ -5.100 & 0 & 29.3 & -34.9 & 0 & 0 & 0 & 0 & 0 \\ -0.312 & 0 & 0 & 0 & -8.47 & 6.19 & 0 & 0 & 0 \\ -3.303 & 0 & 0 & 0 & 14.9 & -23.0 & 0 & 0 & 0 \\ 3480 & 0 & -6801 & -6121 & 2594 & 11413 & -100 & -4444 & 4444 \\ 0 & 0 & 0 & 0 & 0 & 0 & 0.752 & -1.283 & 0 \\ 0 & 0 & 0 & 0 & 0 & 0 & 0 & 1.48 & -1.48 \end{bmatrix} \tag{9.47}$$

Note that the rotor angle and speed of the generator modeling the infinite bus are not included.

The eigenvalues of this exciter controller system are listed in Table 9.2, together with the eigenvalues from Example 8.4 without the excitation system. The impact of the excitation system is clearly visible from this eigenvalue comparison. While the modes associated with δ, ω, E'_d, ψ_{kd}, and ψ_{kq} remain about the same, the E'_q state now interacts

Table 9.2 SMIB system eigenvalues with and without excitation system.

With excitation system		Without excitation system	
States	**Eigenvalues**	**States**	**Eigenvalues**
δ, ω	$-0.5132 \pm j10.47$	δ, ω	-0.5676 ± 10.482
ψ_{kd}	-35.49	ψ_{kd}	-36.111
E'_d	-3.473	E'_d	-3.3724
ψ_{kq}	-27.67	ψ_{kd}	-27.648
E'_q, E_{fd}, V_F	$-0.3789 \pm j0.7424$	E'_q	-0.2093
and V_R	$-51.41 \pm j31.01$		

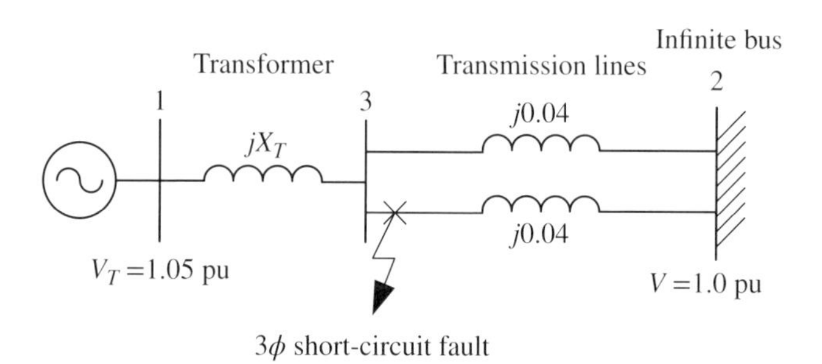

Figure 9.10 Closed-loop excitation system control.

strongly with the exciter states. Thus the -0.2093 mode of E'_q associated with $T'_{do} = 5$ sec would now respond faster as the $-0.3789 \pm j0.7424$ mode. As will be discussed later, this E'_q and exciter mode is also important in power system stabilizer design. ■

Example 9.5: Disturbance response of closed-loop system

Continuing from Example 9.4, apply a fault on one of the transmission lines near Bus 3, which is cleared in 5 cycles (0.083 sec) by removing the faulted line (Figure 9.10). Simulate the system disturbance response and plot the variations of the Bus 3 voltage, machine rotor angle δ, field voltage E_{fd}, and voltage regulator output V_R versus time.

Solutions: Performing a 10-second disturbance response simulation, with the fault applied at 0.1 sec and cleared at 0.183 sec, the time responses are shown in Figure 9.11. The voltage and E_{fd} time responses are the most informative about the excitation system performance. After the fault has been cleared, the generator terminal (Bus 1) voltage is only at 0.85 pu and drops further to 0.7 pu. The excitation system responds by increasing E_{fd}. During the first 2 sec of the time response, E_{fd} increases from 2.4 pu to over 3 pu. In so doing, the terminal voltage is able to recover and system stability is preserved. Once the terminal voltage recovers to 1 pu, the field voltage also drops back to its pre-fault value. The electromechanical oscillation at $10.44/(2\pi) = 1.66$ Hz is visible in all the variables plotted in Figure 9.11. The oscillation is lightly damped, as pre-fault system damping ratio is $\zeta = \cos(\cot^{-1}(0.5132/10.47)) = 0.0490$.

Note that the generator is unstable for a 6-cycle fault. ■

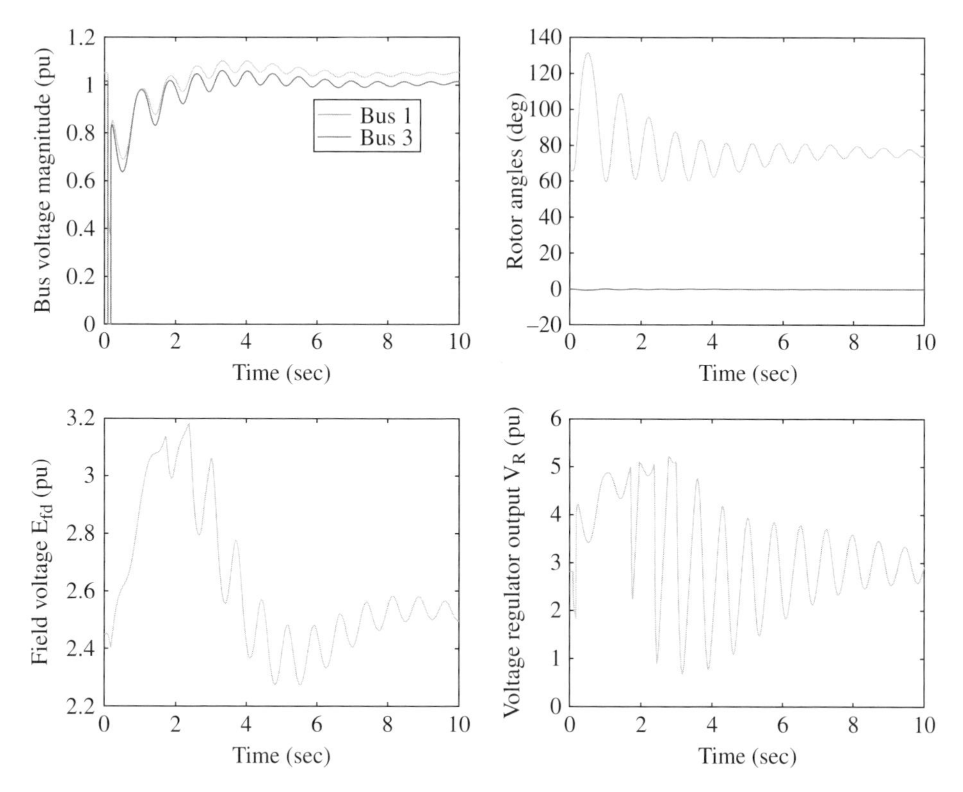

Figure 9.11 Five-cycle fault disturbance response for Example 9.5 (see color plate section).

9.3.7 Excitation System Response Ratios

Example 9.5 shows the benefit of the excitation system boosting the field voltage to a higher value to provide a stronger synchronizing torque in the post-fault condition. Thus the speed of response of an excitation system is an important performance measure, which is known as the *excitation system nominal response* [106]. Figure 9.12 [106] shows the definition of the nominal response ratio. From the rated exciter voltage operating condition, a step signal is applied to the voltage regulator to drive the exciter voltage to its ceiling. Figure 9.12 shows the exciter voltage starting at Point a and increasing to Point b. The time period of interest is 0.5 sec (the length of oe), as during large disturbances, the rotor angle swing typically peaks between 0.4 and 0.75 sec [6]. Let the area enclosed by the triangle acd be equal to the area enclosed by abd. Then

$$\text{nominal response} = \frac{ce - ao}{ao}\frac{1}{oe} \tag{9.48}$$

Thus a higher response ratio implies a more responsive excitation system. Note that the nominal response ratio is a function of the time and gain constants as well as the maximum exciter voltage limit of the excitation system. A higher maximum exciter voltage limit will result in a higher response ratio.

Because it would be difficult to simulate this step response in a nonlinear power system simulation program, the next example is an illustration of (9.48) using a linearized model.

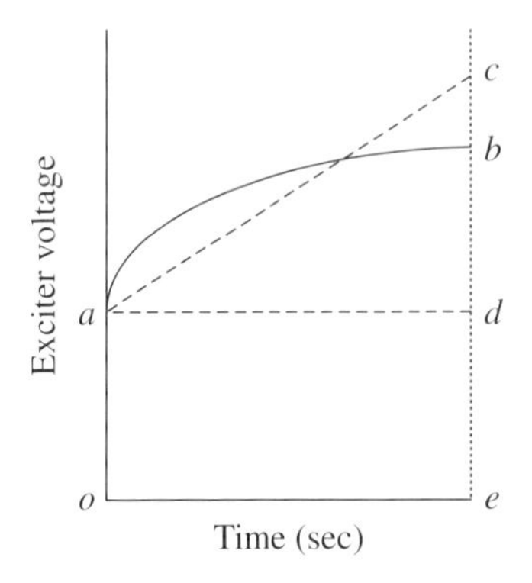

Figure 9.12 Excitation system nominal response.

Example 9.6: Exciter nominal response computation

From Example 9.4, obtain a linearized system model from V_{ref} to E_{fd}. In steady-state, the field voltage is 2.45 pu and the maximum field voltage is 4.95 pu. Apply a step input of appropriate magnitude to compute (9.48) for this system.

Solutions: In the linearized model

$$\Delta\dot{x} = A\Delta x + B\Delta u, \quad \Delta y = C\Delta x \tag{9.49}$$

with x the states in Example 9.4, $u = V_{\mathrm{ref}}$ and $y = E_{fd}$, the A matrix is the same as the one given in the solutions to Example 9.4, the B matrix is

$$B^T = \begin{bmatrix} 0 & 0 & 0 & 0 & 0 & 0 & 3000 & 0 & 0 \end{bmatrix}$$

and the C matrix is

$$C = \begin{bmatrix} 0 & 0 & 0 & 0 & 0 & 0 & 0 & 1 & 0 \end{bmatrix}$$

A unit step response of the system (9.49) reaches a maximum exciter voltage value of 8.94 pu. As the voltage in the nonlinear exciter can reach only 4.95 pu starting from 2.45 pu, the step response needs to be scaled down by a factor of $(4.95 - 2.45)/8.94 = 0.28$. The scaled step response plotted on top of the steady-state field voltage is shown in Figure 9.13 from 0 to 0.5 sec. The area enclosed by the step response and the steady-state field voltage is found to be 0.9806, implying that $cd = 3.92$ pu. Thus the nominal response is 3.20. ■

9.4 Type AC Exciters

The second type of exciters is the AC excitation systems which use an alternator-supplied rectifier to provide the field voltage to the generator. In this type of exciters, the direct current for the generator field is supplied by stationary rectifiers from a DC regulator, or by rotating rectifiers from an AC regulator fed by the generator terminal currents and voltages.

With rectifier control, the response of the excitation system is faster. However, in self-excited Type AC excitation systems, the rectifiers divert some current from the exciter output to supply the alternator field, thus reducing the field current I_{fd} supplied

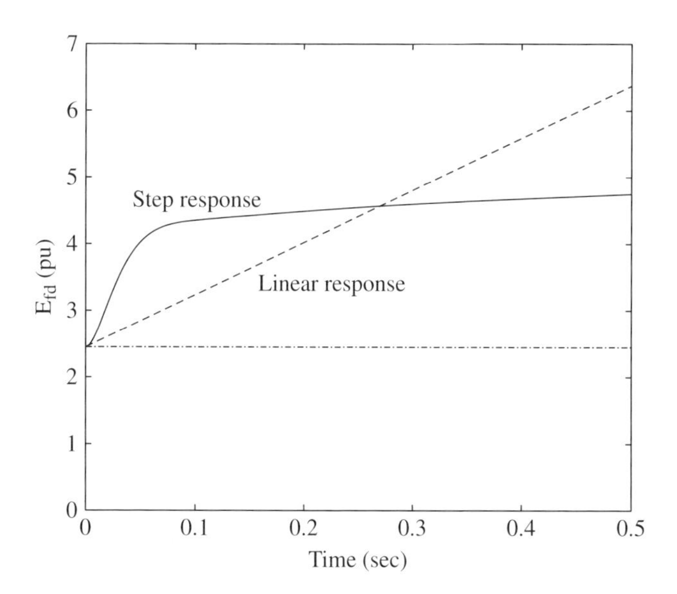

Figure 9.13 Excitation system nominal response illustration of Example 9.6.

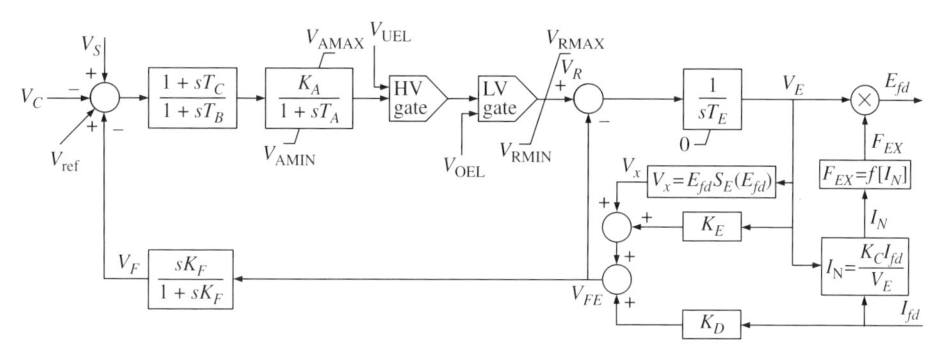

Figure 9.14 Type AC1A - alternator-rectifier excitation system with noncontrolled rectifiers and feedback from exciter field current.

to the generator. This loading effect may be significant and is captured in the Type AC excitation system models as a reduction of the exciter voltage output. To alleviate this situation, the field current for the AC exciter can be separately supplied.

Figure 9.14 shows the block diagram of the Type AC1A model, which is similar to the DC1A model except that the exciter voltage V_E is no longer the field voltage E_{fd}. The parameters of the AC1A model that are not already in Table 9.1 are given in Table 9.3. Note that the exciter voltage V_E is always positive because of the rectification.

In the AC1A model, the voltage signal that is proportional to the exciter field current is now given by

$$V_{FE} = (K_E + S_E(V_E))V_E + K_D I_{fd} \tag{9.50}$$

Table 9.3 AC1A excitation system model variables, parameters, and functions [105].

Variables
F_{EX}: rectifier loading factor, a function of I_N I_N: normalized exciter load current V_E: exciter voltage back of commutating reactance
Parameters
K_C: rectifier loading factor proportional to commutating reactance K_D: demagnetizing factor, a function of exciter alternator reactances
Functions
LV Gate: low-value gate selecting low value of inputs

Table 9.4 Rectifier regulation equations for $f(I_N)$.

I_N	$F_{EX} = f(I_N)$
$0 < I_N < 0.433$	$1 - 0.5774I_N$
$0.433 < I_N < 0.75$	$\sqrt{0.75 - (I_N)^2}$
$0.75 < I_N < 1$	$1.732(1 - I_N)$
$I_N > 1$	0

which has a demagnetizing term $K_D I_{fd}$ proportional to the field current I_{fd}. The field voltage is

$$E_{fd} = F_{EX} V_E = f(I_N) V_E \tag{9.51}$$

where the normalized exciter load current is

$$I_N = \frac{K_C I_{fd}}{V_E} \tag{9.52}$$

as determined by the rectifier loading factor K_C. The function f is a nonlinear rectifier regulation function depending on I_N. An approximate representation of the function is given in Table 9.4.

Other Type AC excitation systems vary in details from the Type AC1A model in the modeling of the exciters and the exciter loading effects, and are described in [105].

9.5 Type ST Excitation Systems

In newer Type ST excitation systems, the field voltage is supplied directly from a rectifier. The dynamics of the rectifier is assumed to be fast and thus neglected in the models. There are two groups of such exciters: the potential-source exciters using only the terminal bus voltage to drive the rectifiers, and the compound-source exciters using both

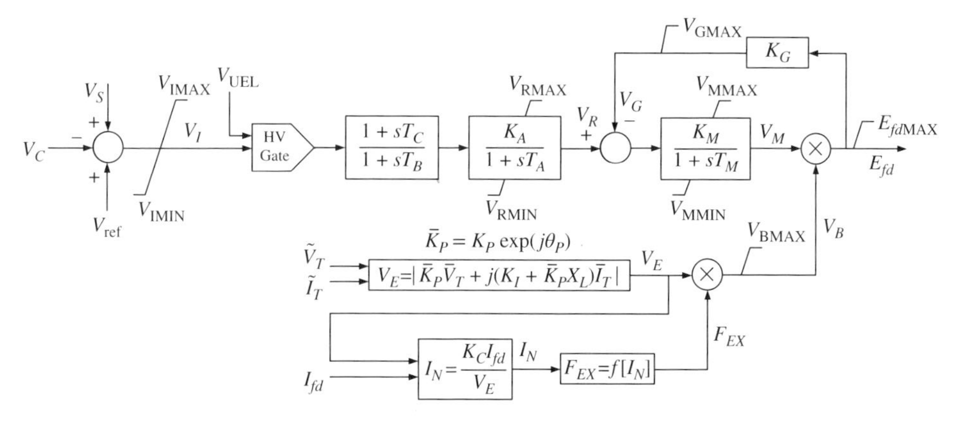

Figure 9.15 Type ST3A potential or compound-source controlled-rectifier exciter with field voltage control loop.

the generator terminal bus voltage and current to drive the rectifiers. The latter implementation would have better performance for close-by faults, as the reduction of the bus voltage is compensated by the increase in the generator output current.

These exciters typically can provide high output voltages during fault. For example, the upper limit $V_{R\max}$ of V_R can be as high as 7 pu. With a higher voltage ceiling and smaller time constants, these excitation systems can achieve a higher response ratio.

Figure 9.15 shows the block-diagram representation of the Type ST3A model, which uses both voltage and current to drive the rectifiers (shown in the lower part of the block diagram). Note that the loading effect of the field current I_{fd} is accounted for in the available exciter voltage V_B. To deal with the nonlinearity associated with the source voltage, a regulator is used to control the field voltage E_{fd}. In Figure 9.15, the regulator is represented by the feedback loop from V_R to E_{fd}. With K_G set to 1, E_{fd} will follow V_R if K_M is a relatively high gain.

The nomenclature of the variables and parameters in the ST3A model not included in the previous tables is shown in Table 9.5.

At this point, it would be useful to compare the performance of the ST3A excitation system with the the DC1A excitation system.

Example 9.7: Linearized closed-loop system with ST3A model

Repeat Example 9.4 to obtain a linearized SMIB model using a ST3A excitation system model with the parameters:

$$\begin{aligned}
K_A &= 200,\ T_A = 0\,\text{sec},\ T_B = 6.67\,\text{sec},\ T_C = 1.0\,\text{sec}\\
V_{I\max} &= 0.2\,\text{pu},\ V_{I\min} = -0.2\,\text{pu},\ V_{R\max} = 7.57\,\text{pu},\ V_{R\min} = 0\,\text{pu}\\
K_M &= 7.04,\ T_M = 0.4\,\text{sec},\ V_{M\max} = 7.57\,\text{pu},\ V_{M\max} = 0\,\text{pu}\\
E_{fd\max} &= 6.53\,\text{pu},\ K_G = 1,\ V_{G\max} = 6.53\,\text{pu}\\
K_P &= 4.365,\ \theta_P = 20^\circ,\ K_I = 4.83,\ X_L = 0.091\,\text{pu},\ K_C = 1.096
\end{aligned}$$

Compare the eigenvalues to those obtained in Example 9.4.

Table 9.5 ST3A excitation system model variables and parameters [105].

Variables
$\tilde{I}_T$: synchronous machine terminal current phasor
V_B: available exciter voltage
V_G: inner-loop voltage feedback
V_I: internal signal within voltage regulator
V_M: output factor of converter bridge corresponding to firing angle command to thyristors
V_T, $\tilde{V}_T$: synchronous machine terminal voltage magnitude and phasor
θ_P: potential circuit phase angle (in degrees)
Parameters
K_G: feedback gain constant of the inner-loop field regulator
K_I: potential circuit gain coefficient
K_M: forward gain constant of the inner-loop field regulator
$\overline{K}_P$: complex potential circuit gain coefficient
T_M: forward time constant of the inner-loop field regulator
X_L: reactance associated with potential source

Solutions: The linearization function from PST yields the state-space model

$$\Delta \dot{x} = A \Delta x \tag{9.53}$$

where there are 8 states in x

$$x = \begin{bmatrix} \delta & \omega & E'_q & \psi_{kd} & E'_d & \psi_{kq} & V_R & V_M \end{bmatrix}^T \tag{9.54}$$

and the system matrix is

$$A = \begin{bmatrix} 0 & 377 & 0 & 0 & 0 & 0 & 0 & 0 \\ -0.315 & 0 & -0.152 & -0.137 & -0.0240 & -0.106 & 0 & 0 \\ 0.147 & 0 & -2.15 & 2.35 & -0.0134 & -0.588 & 0 & 1.40 \\ -5.10 & 0 & 29.3 & -34.9 & 0 & 0 & 0 & 0 \\ -0.312 & 0 & 0 & 0 & -8.47 & 6.19 & 0 & 0 \\ -3.30 & 0 & 0 & 0 & 14.9 & -23.0 & 0 & 0 \\ 3.48 & 0 & -6.80 & -6.12 & 2.59 & 11.4 & -0.150 & 0 \\ 21.2 & 0 & -117 & -174 & 46.8 & 206 & 15.0 & -126 \end{bmatrix} \tag{9.55}$$

Note that the rotor angle and speed of the generator modeling the infinite bus are not included. Also, the $A(3,1)$ entry is now positive, which will have a destabilizing effect on the swing mode. This aspect is discussed in more detail in the next chapter.

The eigenvalues of the ST3A exciter controller system are listed in Table 9.6, together with the eigenvalues from Example 9.4 with the DC1A excitation system. The E'_d, ψ_{kd}, and ψ_{kq} modes are similar for both excitation systems. Just like the DC1A model, E'_q is coupled with the exciter state V_R. Perhaps the most important observation is that the real part of the local mode for the ST3A model is only half that of the DC1A model. This local mode becomes unstable when one of the two parallel lines connecting Bus 3 to the infinite bus is tripped, as will be shown in the next example. ■

Table 9.6 SMIB system eigenvalues with and without ST3A excitation system.

With DC1A excitation system		With ST3A excitation system	
States	**Eigenvalues**	**States**	**Eigenvalues**
δ, ω	$-0.5132 \pm j10.47$	δ, ω	-0.2938 ± 10.53
ψ_{kd}	-35.49	ψ_{kd}	-34.37
E'_d	-3.473	E'_d	-4.355
ψ_{kq}	-27.67	ψ_{kd}	-27.73
E'_q, E_{fd}, V_F,	$-0.3789 \pm j0.7424$	E'_q, V_M,	$-1.150 \pm j0.6893$
and V_R	$-51.41 \pm j31.01$	V_R	-124.9

Example 9.8: Disturbance response of closed-loop system with ST3A model

Repeat the disturbance simulation in Example 9.5 using the ST3A model with parameters used in Example 9.7, but the fault is cleared in 6 cycles (0.1 sec). Plot the variations of the Bus 3 voltage, machine rotor angle δ, field voltage E_{fd}, and voltage regulator output V_R versus time.

Solutions: Performing a 10-second disturbance response simulation, with the fault applied at 0.1 sec and cleared at 0.2 sec, the time responses are shown in Figure 9.16. The system is transient stable for a fault clearing time of 0.1 sec. Note that during the fault and immediately after the fault is cleared, E_{fd} reaches its maximum almost instantaneously, such that the synchronous generator can supply the highest amount of electrical power to slow down the initial acceleration of the generator rotor. The problem with this ST3A model shows up after the first swing in the post-fault system, as the local mode exhibits sustained oscillation and is unstable. Such behavior is clearly undesirable. The next chapter discusses the design of power system stabilizers to damp power swing oscillations. ■

9.6 Load Compensation Control

In the previous sections, it has been assumed that the measured voltage signal V_C is the generator terminal voltage, which is the most common application. However, V_C can also be any reasonable voltage in the power network to obtain a more appropriate voltage control response from an excitation system. Although a voltage can be measured anywhere in the network and the signal then communicated to the excitation system, such a control scheme may be expensive to implement and maintain. An alternative is to use the terminal voltage $\tilde{V}_T$ and generator current $\tilde{I}_T$ to emulate a voltage using a suitable resistance value R_C and reactance value X_C in the form

$$V_C = |\tilde{V}_T + (R_C + jX_C)\tilde{I}_T| \tag{9.56}$$

In case $R_C = 0$ and $X_C = 0$, V_C is simply the terminal voltage magnitude.

The implementation of the load compensation scheme is shown in Figure 9.17, in which a transducer circuit with a time constant of T_R is added to represent the amount of time needed to obtain the processed signal V_C.

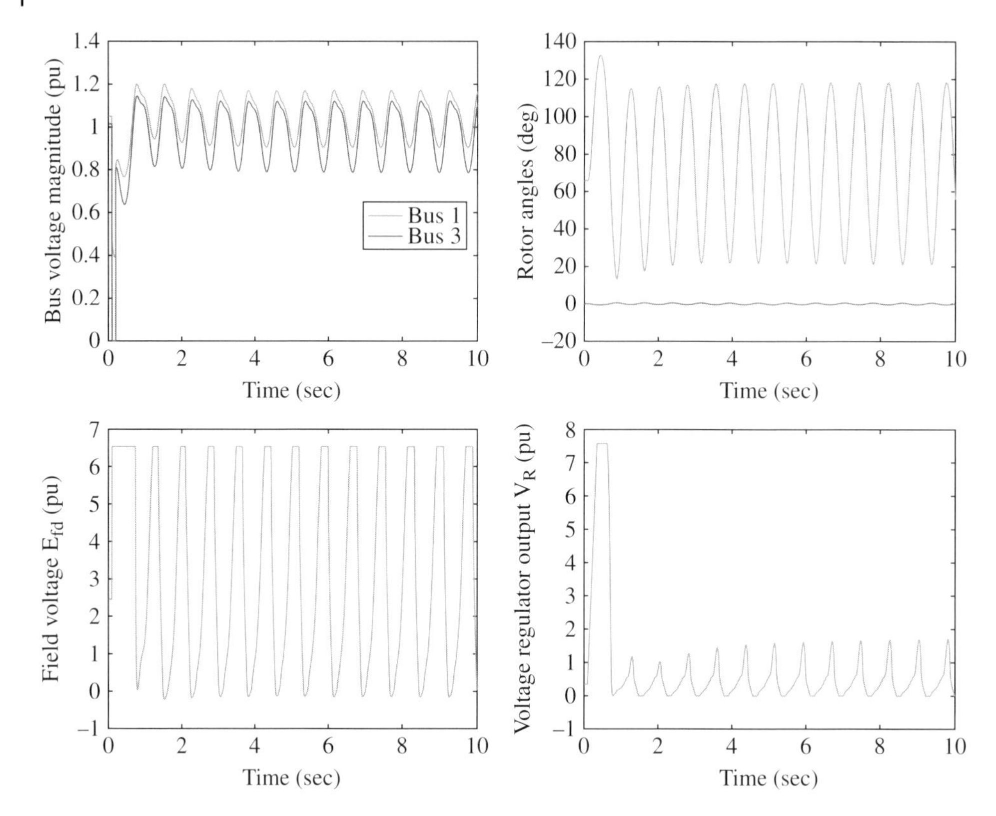

Figure 9.16 Disturbance response for Example 9.8 (see color plate section).

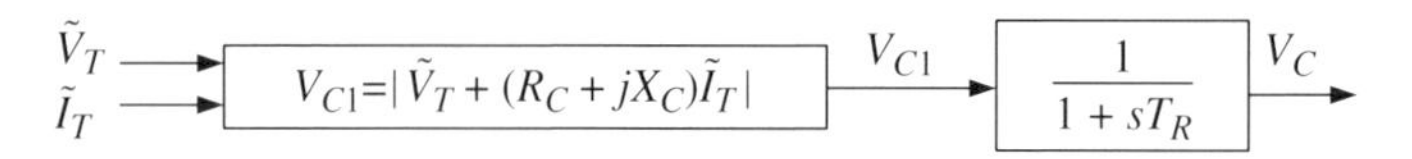

Figure 9.17 Load compensation scheme for voltage control [105].

There are two common types of compensations used:

1) Droop compensation: If multiple generators are connected to a common bus bar with no impedance between them (such as those of transformers), the compensator creates an artificial impedance for the units to provide reactive power appropriately, such that the unit with the highest internal voltage setpoint value does not have to provide a major share of the reactive power. Thus droop compensation allows for the regulation of voltage at a point internal to the synchronous generator circuits. In this application, R_C and X_C would have positive values.
2) Line-drop compensation: In the case when a tighter voltage control is desired, the excitation system may control the voltage at a point beyond the generator terminal, such as part way through the step-up transformer. In this application, R_C and X_C would have negative values. By controlling the voltage closer to the load, the voltage stability margin (as discussed in Chapter 3) may be increased.

Example 9.9: Voltage regulation at high-side of generator step-up transformer

For the generator in Example 9.2, the voltage at the high-side of the generator step-up transformer is 0.995 pu, which is lower than the voltage at the generator terminal (the low-side of the step-up transformer). Suppose line-drop compensation is used for the excitation system to regulate the high-side voltage at $V_C = 0.995$ pu. Calculate V_{ref}.

Solutions: The initial values of all the variables in the excitation system will remain the same except for V_{ref}, because V_{Ro} still has to be 3.0427 pu to maintain the terminal bus voltage. However, the V_{ref} equation (9.34) is modified to yield

$$K_A(V_{\text{ref}} - V_C) = V_{Ro} \quad \Rightarrow \quad V_{\text{ref}} = 0.995 + \frac{3.0427}{46} = 1.0611 \text{ pu} \tag{9.57}$$

Note that V_{ref} is lower now as $V_C < V_T$. ■

9.7 Protective Functions

An excitation system controls the generator field voltage, which determines the generator current and voltage output. As such, an excitation system needs to respect the operation limits of the generator that it controls. Thus it needs to modify the field voltage to steer the generator away from its limits. Figure 9.18 shows some of the protection functions normally implemented in an excitation system, in addition to the control and sensing functions.

Some of the limits have already been used in the examples, such as $E_{fd\max}$ and $V_{R\max}$. These limits are shown in the three excitation system model block diagrams discussed in this chapter. These limits are typically implemented with non-windup limits for the integrator that stores these values, reflecting how the physical equipment actually functions.

Two additional limits are incorporated into excitation system models: the under-excitation limiter (UEL) and the over-excitation limit (OEL) (see Section 7.9). Of particular interest in voltage stability studies is the OEL, which is coordinated by an over-current relay. This relay consists of an instantaneous over-current function and a time-over-current function having an inverse-time characteristic as shown in Figure 9.19. In [98], for a particular machine, the relay operating time is 7.0 sec at 218% of full-load current. Prescribed times for other overload levels of the full-load current can also be set accordingly.

9.8 Summary and Notes

This chapter discusses the main features of three types of excitation systems: DC, AC, and ST. The emphasis here is on the functionalities of the exciters and the transfer function blocks in the voltage regulator. Frequency responses are used to illustrate some of the transfer function blocks. Excitation system initialization and linearization are described. The destabilizing effect of high-gain excitation systems is illustrated with an example. In the next chapter, power system stabilizers are introduced to provide damping to counter such destabilizing effects. Concepts such as excitation

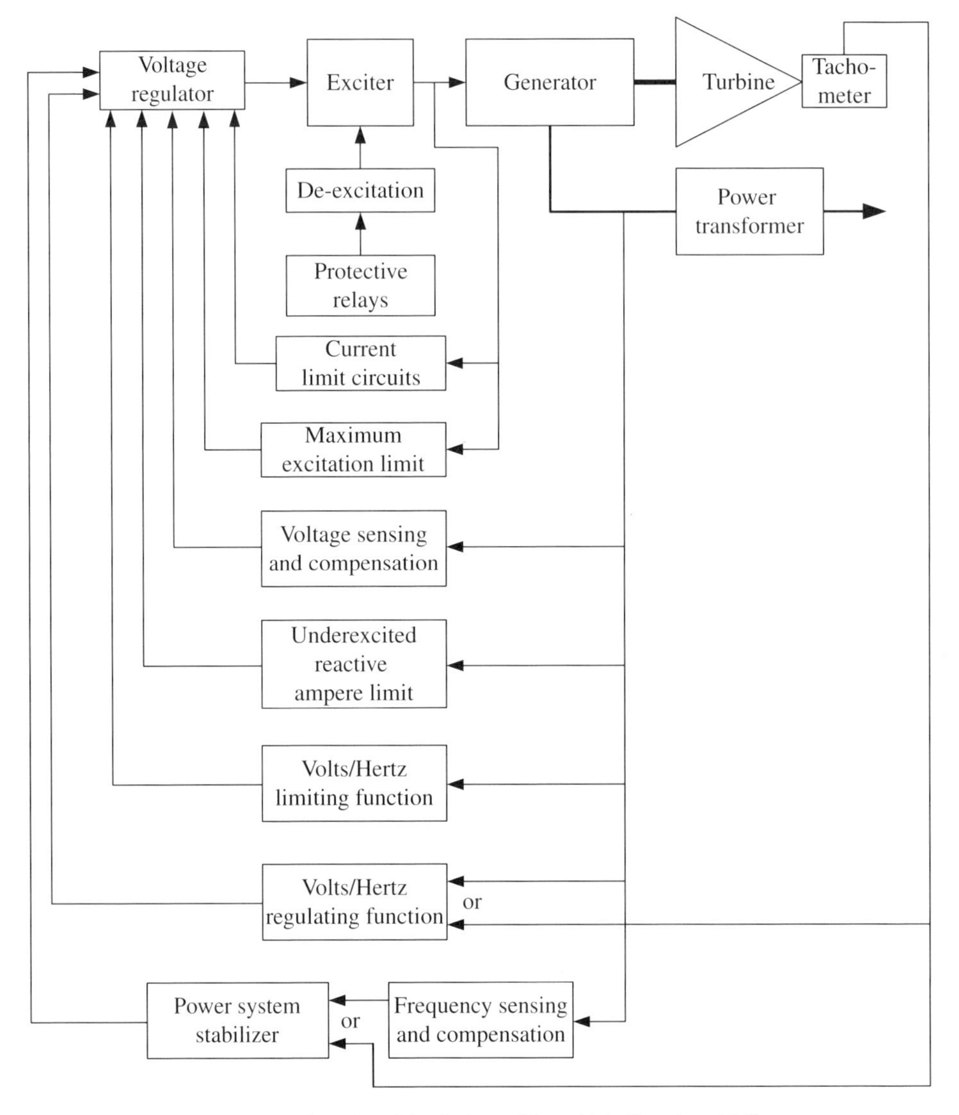

Figure 9.18 Excitation system functionalities (adopted from M. L. Crenshaw [20]).

system response ratio and load compensation control are discussed and illustrated with examples.

A reader interested in other models of excitation systems can consult [105–107].

Appendix 9.A Anti-Windup Limits

Control functions need to take into account equipment limitations, such as the voltage regulator output discussed in Section 9.3 and the capacitors used in FACTS controllers in Chapter 14. Figure 9.20.a shows the voltage regulation block from Figure 9.2, and

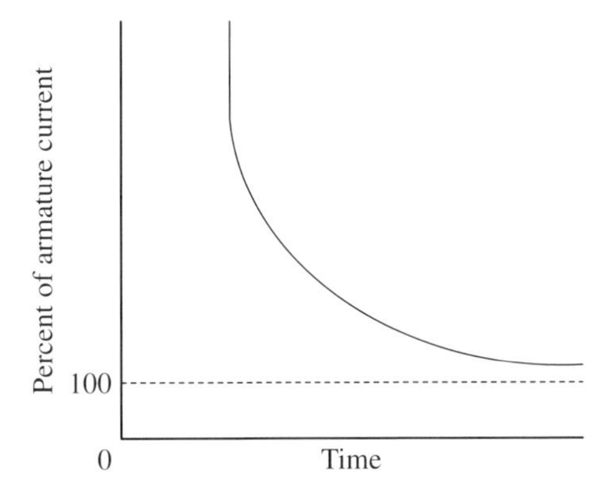

Figure 9.19 Generator short-time thermal capability.

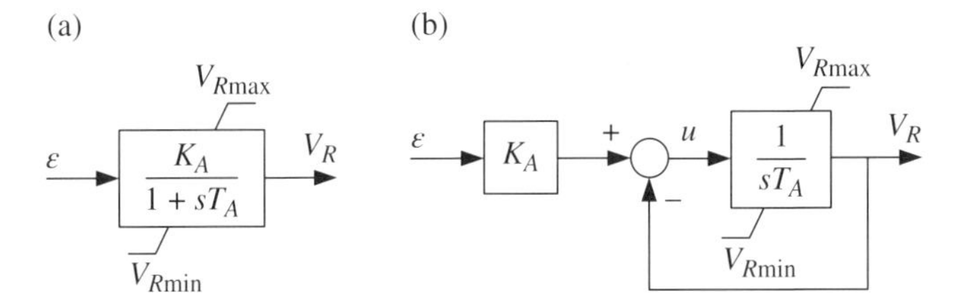

Figure 9.20 (a) Transfer function with anti-windup limits and (b) implementation showing integrator.

Figure 9.20.b shows the implementation with an integrator. The input ε is a voltage error signal and V_R is the state of the integrator. For severe faults, the value of V_R can be driven to a value higher than $V_{R\max}$. However, the real equipment can only respond with $V_R = V_{R\max}$. If the integrator state V_R is allowed to go higher than $V_{R\max}$, then when the input signal u turns around and becomes negative, it will take some time for V_R to reduce to $V_{R\max}$, such that the control signal can be useful again.

Anti-windup logic can prevent such windup situations by imposing the following rules when V_R is at its limits:

$$\text{if } V_R = V_{R\max} \text{ and } u > 0, \text{ then set } u = 0 \text{ and integrator state } V_R = V_{R\max} \tag{9.58}$$

$$\text{if } V_R = V_{R\min} \text{ and } u < 0, \text{ then set } u = 0 \text{ and integrator state } V_R = V_{R\min} \tag{9.59}$$

When anti-windup limits are not implemented properly in a power system simulation program, electromechanical swings will become larger in amplitude. Damping control signals may also be rendered ineffective during the period when the integrator needs to reduce its state value to below the output limit.

Problems

9.1 Example 9.3 illustrates transient-gain reduction in the rate-feedback path of Figure 9.7. For this problem, disconnect the rate-feedback path, and derive the transfer function from ΔV_{ref} to E_{fd} by including the transient-gain reduction block $(1 + sT_C)/(1 + sT_B)$. Use the parameters $K_A = 300$, $T_A = 0.01$ sec,

$K_E = 1.0$, $T_E = 1.33$ sec, $T_B = 10$ sec, and $T_C = 1$ sec to plot the transfer functions $T_{\text{ngr}}(s)$ and $T_{\text{tgr}}(s)$.

9.2 The paper [104] describes the performance testing of an excitation system. A simplified model of the generator (flux-decay model) and the excitation system is shown in Figure 9.21. Initialize the model using a value of $V_T = 1.00$ pu, using the parameters $K_A = 46$, $T_A = 60$ msec, $T_B = 1.6$ sec, $T_C = 0.8$ sec, $T_R = 6.5$ msec, and $T'_{d0} = 8.40$ s. Assume there is no saturation, that is, the saturation block is a linear function. Build a Simulink® diagram or use MATLAB® to represent the model. Introduce a step change of 5% to V_{ref} and simulate the terminal voltage V_T response of the simplified model.

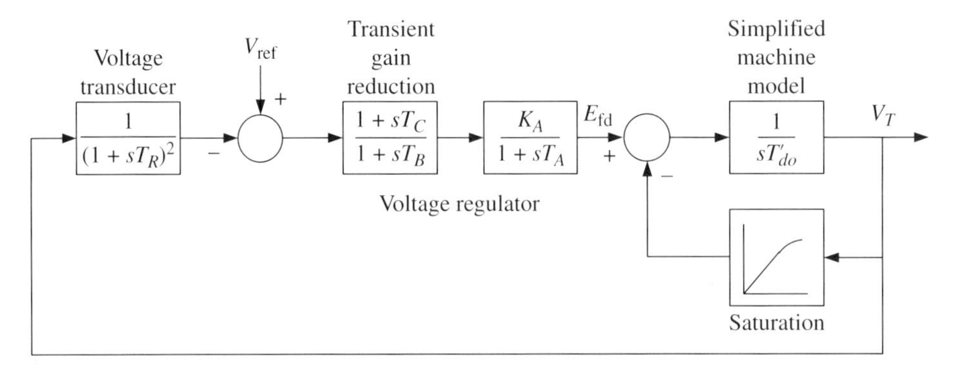

Figure 9.21 Excitation system control diagram for Problem 9.2.

9.3 The association of the states and modes in Table 9.2 was obtained by examining the eigenvectors. Compute the participation factors for the modes of the A matrix for Example 9.4 and use them to determine the significance of the states in each of the modes.

9.4 Repeat Example 9.5 with the DC2A excitation system subject to a 6-cycle fault. Show that the system is unstable.

9.5 Following Example 9.6, generate a closed-loop linear system from the input V_{ref} to the output V_T (generator terminal voltage) for the system in Example 9.4. Plot the closed-loop frequency response and the step response for $\Delta V_{\text{ref}} = 0.05$ pu.

9.6 Obtain the closed-loop linear model from the input V_{ref} to the output V_T for the ST3A excitation system in Example 9.7. Plot the step response for $\Delta V_{\text{ref}} = 0.05$ pu.

9.7 Calculate the nominal response ratio for the ST3A excitation system in Example 9.8. The initial value of E_{fd} is 2.45 pu and $E_{fd\max}$ is 6.53 pu. Assume that E_{fd} will immediately reach saturation after the fault is applied.

9.8 Repeat Example 9.8 for a 7-cycle fault. Is the system stable?

9.9 Example 9.2 shows the initialization of the excitation system states when $V_C = V_T$. Calculate the steady-state values of the exciter states and variables if load compensation is used:

1) Line-drop compensation for $R_C = 0$ and $X_C = -0.1$ pu
2) Droop compensation for $R_C = 0$ and $X_C = 0.05$ pu

Assume that the generator output current is 0.9 pu and is lagging the terminal voltage $\tilde{V}_T$ by 30°.

10

Power System Stabilizers

10.1 Introduction

As discussed in the last chapter, a high-gain excitation system improves the transient stability of a power system, but reduces the damping of the electromechanical modes. Using the transient gain reduction function in the voltage regulation loop lessens this destabilizing effect. However, many power systems with long transmission lines, such as the US western power system and some of the Canadian power systems, do not use transient gain reduction so as to preserve the transient stability enhancement benefit from the excitation systems. To restore the swing-mode damping, a common practice is to apply power system stabilizers (PSSs), which are quite inexpensive to install. In modern excitation systems, it is a matter of turning on the PSS function already provided on the circuit board or in the microprocessor.

The design objective of a PSS is summarized using the eigenvalue plot in Figure 10.1. The electromechanical modes in the s-plane can be used to depict the impact of changing the damping and synchronous torque on the synchronous machine. To illustrate, let the local mode without voltage regulator action be located at Point a in the s-plane. With the voltage regulator action, which adds synchronizing torque but reduces damping torque, the local mode will be shifted to Point b with a larger imaginary part but a smaller magnitude in the real part. Thus the role of the PSS is to move the local mode away from the $j\omega$-axis to improve the local-mode damping. A desirable location for the resulting local mode would be Point c, in which the damping is improved, without compromising on the synchronizing torque. This chapter will discuss the process of improving damping on the local mode by adding a supplementary input signal to the voltage regulator. In so doing, the damping enhancement can be implemented without requiring a new actuator.

To facilitate the PSS design, the chapter starts with a small-signal analysis approach to build the concepts of synchronizing and damping torques. It will serve as the motivation to develop the structure of the PSS, as well as a design process for tuning the gain and time constants of the PSS. A design example on a PSS using the generator speed as the input signal is provided. The use of other signals for PSS design, including a dual-input-signal design, is also discussed.

Power System Modeling, Computation, and Control, First Edition. Joe H. Chow and Juan J. Sanchez-Gasca.

Companion website: www.wiley.com/go/chow/power-system-modeling

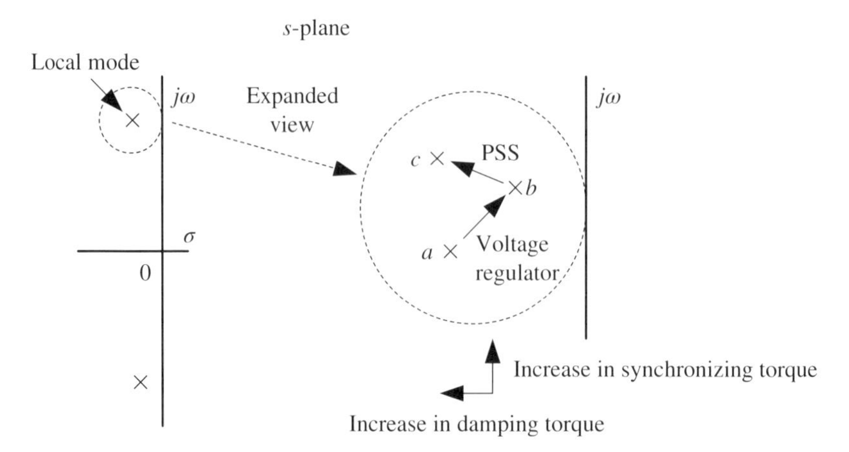

Figure 10.1 Design objective of power system stabilizer.

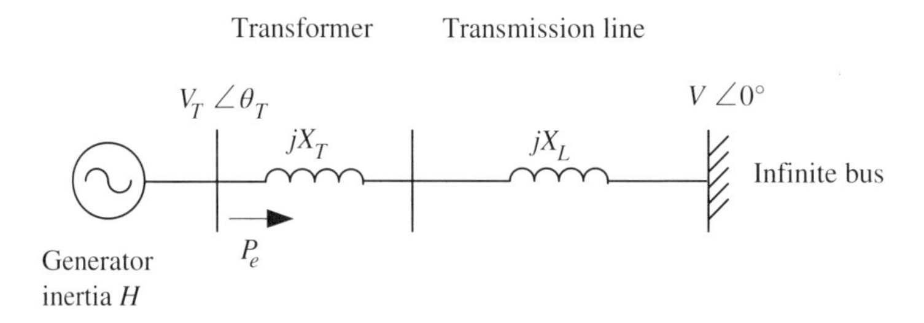

Figure 10.2 Single-machine infinite-bus model.

10.2 Single-Machine Infinite-Bus System Model

The SMIB system in Figure 10.2 is used to discuss the synchronizing and damping torque concept which is useful in understanding the impact of excitation systems on the swing-mode damping and in developing the intuitions for damping control design. The impedances from the machine terminal bus to the infinite bus are lumped into

$$Z_e = j(X_T + X_L) = jX_e \tag{10.1}$$

For the damping analysis of the swing modes, the subtransient phenomenon in a synchronous machine can be neglected. As a result, the first-order flux decay model described by (8.75), (8.77), and (8.79) will be used in the analysis, with E'_q the only electrical state for the generator model.

The electromechanical model of the SMIB model is given by

$$\dot{\delta} = \Omega\omega, \quad \dot{\omega} = \frac{1}{2H}(T_m - T_e - D\omega) \tag{10.2}$$

where δ in radians is the rotor angle relative to the infinite bus, ω in pu is the machine speed deviation from the nominal frequency, and the various machine damping effect is lumped into the damping coefficient D.

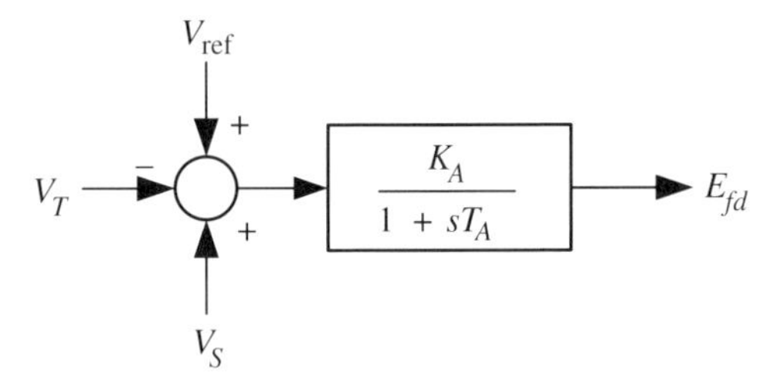

Figure 10.3 Block diagram of a simple voltage regulator.

The interface equations between the machine model and the network is represented by

$$V_T^2 = e_d^2 + e_q^2, \quad e_d = -\psi_q = X_q i_q, \quad e_q = \psi_d = E_q' - X_d' i_d, \quad T_e = E_q i_q \tag{10.3}$$

$$E_q = E_q' + (X_q - X_d')i_d, \quad i_d = \frac{E_q - V\cos\delta}{X_e + X_q}, \quad i_q = \frac{V\sin\delta}{X_e + X_q} \tag{10.4}$$

where (e_d, e_q) and (i_d, i_q) are d- and q-axis voltages and currents, respectively, at the machine terminal bus. These equations can be readily derived using expressions from Section 8.5.1, except that the expressions of i_d and i_q are derived in terms of the machine internal voltage, the infinite-bus voltage, and the transformer and transmission line reactances.

The flux-decay model is

$$T_{do}'\dot{E}_q' = E_{fd} - E_q' - (X_d - X_d')i_d \tag{10.5}$$

The field voltage E_{fd} is controlled by a fast-acting thyristor-type voltage regulator shown in Figure 10.3, represented by the transfer function

$$E_{fd} = \frac{K_A}{1 + sT_A}(V_{\text{ref}} - V_T + V_S) \quad \Rightarrow \quad T_A\, sE_{fd} = -E_{fd} + K_A(V_{\text{ref}} - V_T + V_S) \tag{10.6}$$

which in the time domain becomes

$$T_A\dot{E}_{fd} = -E_{fd} + K_A(V_{\text{ref}} - V_T + V_S) \tag{10.7}$$

The fourth-order system defined by (10.2), (10.5), and (10.7) constitutes the model for developing the synchronizing and damping torque concept. The following discussion is based on [108].

Denote the equilibrium values of the four states as $\delta_o, \omega_o = 0, E_{qo}'$, and E_{fdo}, the machine and network variables at equilibrium as $i_{do}, i_{qo}, E_{qo}, e_{do}, e_{qo}$, and V_{To}, and the steady-state input values as T_{mo} and V_{So}. Note that the infinite-bus voltage V does not change, that is, $V = V_o$, and the reference voltage in the excitation system V_{ref} is fixed. Also, nominally $V_{So} = 0$. The perturbations away from the equilibrium values are defined as

$$\begin{aligned} &\Delta\delta = \delta - \delta_o, \quad \Delta\omega = \omega - \omega_o, \quad \Delta E_q' = E_q' - E_{qo}' \\ &\Delta E_{fd} = E_{fd} - E_{fdo}, \quad \Delta T_m = T_m - T_{mo}, \quad \Delta V_S = V_S - V_{So} \end{aligned} \tag{10.8}$$

Thus the linearized or small-signal model is

$$
\begin{aligned}
\Delta\dot{\delta} &= \Omega\Delta\omega \\
\Delta\dot{\omega} &= \frac{1}{2H}(\Delta T_m - D\Delta\omega - \Delta T_e) \\
\Delta\dot{E}'_q &= \frac{1}{T'_{do}}\left(-\frac{1}{K_3}\Delta E'_q - K_4\Delta\delta + \Delta E_{fd}\right) \\
\Delta\dot{E}_{fd} &= \frac{1}{T_A}\left(-\Delta E_{fd} + K_A\Delta V_S - K_A\Delta V_T\right)
\end{aligned}
\tag{10.9}
$$

with the outputs

$$\Delta T_e = K_1\Delta\delta + K_2\Delta E'_q, \quad \Delta V_T = K_5\Delta\delta + K_6\Delta E'_q \tag{10.10}$$

The K_i coefficients are, using the notation in [108]

$$
\begin{aligned}
K_1 &= \text{the change in electrical torque for a change in } \delta \\
&= \frac{\partial T_e}{\partial \delta} = \frac{X_q - X'_d}{X_e + X'_d} i_{qo} V \sin\delta_o + \frac{E_{qo} V \cos\delta_o}{X_e + X_q}
\end{aligned}
\tag{10.11}
$$

$= $ the synchronizing torque coefficient

$$
\begin{aligned}
K_2 &= \text{the change in electrical torque for a change in } E'_q \\
&= \frac{\partial T_e}{\partial E'_q} = \frac{V \sin\delta_o}{X_e + X'_d}
\end{aligned}
\tag{10.12}
$$

$$
\begin{aligned}
K_3 &= \text{an impedance factor} \\
&= \frac{1}{1 + (X_d - X'_d)\frac{\partial i_d}{\partial E'_q}} = \frac{X'_d + X_e}{X_d + X_e}
\end{aligned}
\tag{10.13}
$$

$$
\begin{aligned}
K_4 &= \text{the demagnetizing effect for a change in } \delta \\
&= \frac{1}{K_3}\frac{\partial E'_q}{\partial \delta} = \frac{X_d - X'_d}{X_e + X'_d} V \sin\delta_o
\end{aligned}
\tag{10.14}
$$

$$
\begin{aligned}
K_5 &= \text{the terminal bus voltage sensitivity for a change in } \delta \\
&= \frac{\partial V_T}{\partial \delta} = \frac{X_q}{X_e + X_q}\frac{e_{do}}{V_{To}} V \cos\delta_o - \frac{X'_d}{X_e + X'_d}\frac{e_{qo}}{V_{To}} V \sin\delta_o
\end{aligned}
\tag{10.15}
$$

$$
\begin{aligned}
K_6 &= \text{the terminal bus voltage sensitivity for a change in } E'_q \\
&= \frac{\partial V_T}{\partial E'_q} = \frac{X_e}{X_e + X'_d}\frac{e_{qo}}{V_{To}}
\end{aligned}
\tag{10.16}
$$

The modifications to the coefficients $K_1, \ldots, K_6$ to include the effect of the transmission line resistance in Z_e can be found in [108].

Example 10.1: Sensitivity factors

Derive the expressions (10.11) for K_1 and (10.12) for K_2.

Solutions: From (10.4), at the equilibrium point denoted by the subscript o,

$$K_1 = \left.\frac{\partial T_e}{\partial \delta}\right|_o = i_{qo}\left.\frac{\partial E_q}{\partial \delta}\right|_o + E_{qo}\left.\frac{\partial i_q}{\partial \delta}\right|_o \tag{10.17}$$

Substituting i_d into E_q (10.4) results in

$$E_q = \frac{X_e + X_q}{X_e + X'_d}\left(E'_q - \frac{X_q - X'_d}{X_e + X_q} V\cos\delta\right) \tag{10.18}$$

Hence

$$\left.\frac{\partial E_q}{\partial \delta}\right|_o = \frac{X_q - X'_d}{X_e + X'_d} V \sin\delta_o \tag{10.19}$$

From (10.4),

$$\left.\frac{\partial i_q}{\partial \delta}\right|_o = \frac{V\cos\delta_o}{X_e + X_q} \tag{10.20}$$

Thus K_1 (10.11) follows from combining (10.17), (10.19), and (10.20). It also follows that

$$K_2 = \left.\frac{\partial T_e}{\partial E'_q}\right|_o = i_{qo}\left.\frac{\partial E_q}{\partial E'_q}\right|_o = \frac{V\sin\delta_o}{X_e + X'_d} \tag{10.21}$$

as i_q is not dependent on E'_q. ∎

The values of the coefficients $K_1, \ldots, K_6$ depend on the system parameters as well as the operating condition. For stable system operation, $0 < \delta_o < 90°$, the terms K_1–K_4 and K_6 are positive. For a lightly loaded system with small X_e and hence small δ_o, (10.15) implies that $K_5 > 0$. However, for a heavily loaded system with large X_e, δ_o would be sufficiently large to cause $K_5 < 0$. The coefficients $K_1, \ldots, K_6$ will be used to analyze synchronizing and damping torques. Plots of $K_1, \ldots, K_6$ as a function of X_e and T_e can be found in [109].

Applying Laplace transform, the linearized model (10.9) becomes

$$\Delta\delta(s) = \frac{1}{s}\Omega\Delta\omega(s) \tag{10.22}$$

$$\Delta\omega(s) = \frac{1}{2Hs}(\Delta T_m(s) - D\Delta\omega(s) - \Delta T_e(s)) \tag{10.23}$$

$$sK_3 T'_{do}\Delta E'_q(s) = -\Delta E'_q(s) + K_3(-K_4\Delta\delta(s) + \Delta E_{fd}(s)) \tag{10.24}$$

$$sT_A\Delta E_{fd}(s) = -\Delta E_{fd}(s) + K_A\Delta V_S(s) - K_A\Delta V_T(s) \tag{10.25}$$

Equations (10.24) and (10.25) can be expressed in transfer function form, respectively, as

$$\Delta E'_q(s) = \frac{K_3}{1 + sK_3T'_{do}}(-K_4\Delta\delta(s) + \Delta E_{fd}(s)) \tag{10.26}$$

$$\Delta E_{fd}(s) = \frac{K_A}{1 + sT_A}(\Delta V_S(s) - \Delta V_T(s)) \tag{10.27}$$

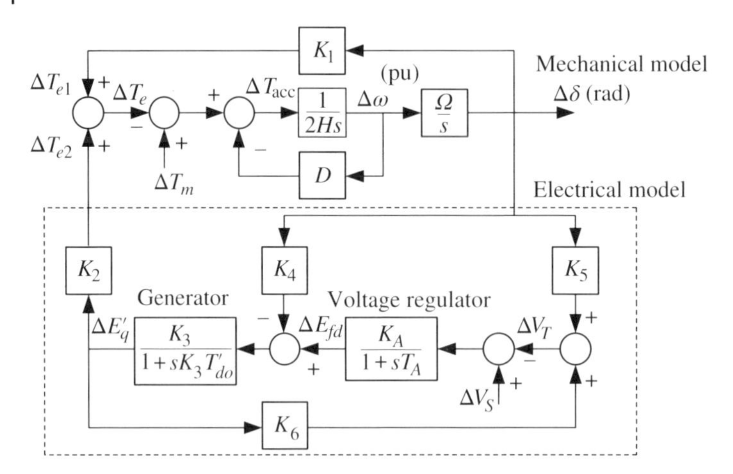

Figure 10.4 Modal decomposition of synchronous machine showing synchronizing and damping torques.

The block diagram in Figure 10.4 is constructed using these transfer function blocks [108, 110], and is also known as the Heffron-Phillips model or the modal decomposition.

Furthermore, the linearized model (10.9) and (10.10) can be put in state-space form as

$$\dot{x} = Ax + Bu, \quad y = Cx + Du \tag{10.28}$$

where

$$x = \begin{bmatrix} \Delta\delta \\ \Delta\omega \\ \Delta E'_q \\ \Delta E_{fd} \end{bmatrix}, \quad u = \begin{bmatrix} \Delta T_m \\ \Delta V_S \end{bmatrix}, \quad y = \begin{bmatrix} \Delta T_e \\ \Delta V_T \end{bmatrix} \tag{10.29}$$

$$A = \begin{bmatrix} 0 & \Omega & 0 & 0 \\ -K_1/(2H) & -D/(2H) & -K_2/(2H) & 0 \\ -K_4/T'_{d0} & 0 & -1/(K_3 T'_{d0}) & 1/T'_{d0} \\ -K_A K_5/T_A & 0 & -K_A K_6/T_A & -1/T_A \end{bmatrix}, \quad B = \begin{bmatrix} 0 & 0 \\ 1/(2H) & 0 \\ 0 & 0 \\ 0 & K_A/T_A \end{bmatrix}$$

$$C = \begin{bmatrix} K_1 & 0 & K_2 & 0 \\ K_5 & 0 & K_6 & 0 \end{bmatrix}, \quad D = \begin{bmatrix} 0 & 0 \\ 0 & 0 \end{bmatrix} \tag{10.30}$$

Thus if a computer program is available to generate the linearized model (10.28)–(10.30), the coefficients $K_1, \ldots, K_6$ can be readily obtained. The following example examines this process using a 2-state flux-decay synchronous machine model controlled by a 1-state excitation system.

Example 10.2

For the system in Example 8.4, the subtransient generator model is reduced to the flux decay model and saliency is neglected by setting $X'_q = X'_d$. The machine model reactance parameters in pu on the machine base (991 MVA) are $X_d = 2.0$, $X_q = 1,91$, and $X'_d = X'_q = 0.245$. The d-axis transient time constant is $T'_{do} = 5.0$ sec. The transformer reactance is 0.15 pu on the machine base and the transmission reactance is 0.02 pu on the system base (100 MVA). The bus voltages are $V_T = 1.05$ pu and $V = 1.0$ pu. The voltage regulator gain is $K_A = 200$ and the time constant is $T_A = 0.05$ sec. The steady-state values are computed as

$$\begin{aligned} \delta_o &= 1.1491 \text{ rad}, \ E'_{qo} = 0.925 \text{ pu}, \ E'_{do} = 0.6727 \text{ pu}, \ E_{fdo} = 2.4508 \text{ pu} \\ e_{do} &= 0.7717 \text{ pu}, \ e_{qo} = 0.7120 \text{ pu}, \ i_{do} = 8.6159 \text{ pu}, \ i_{qo} = 4.0041 \text{ pu} \end{aligned} \tag{10.31}$$

where the currents i_{do} and i_{qo} are given on the system base. The A and C matrices of the resulting linearized model are given by

$$A = \begin{bmatrix} 0 & 376.99 & 0 & 0 & 0 \\ -0.29082 & 0 & -0.2674 & 0.1199 & 0 \\ -0.53982 & 0 & -0.7916 & 0 & 0.2 \\ 1.7398 & 0 & 0 & -5.7672 & 0 \\ 525.25 & 0 & -1592.3 & -1725.9 & -20.0 \end{bmatrix} \tag{10.32}$$

$$C = \begin{bmatrix} 1.6726 & 0 & 1.5379 & -0.6897 & 0 \\ -0.1313 & 0 & 0.3981 & 0.4315 & 0 \end{bmatrix} \tag{10.33}$$

where the states and the outputs are

$$x = \begin{bmatrix} \Delta\delta & \Delta\omega & \Delta E'_q & \Delta E'_d & \Delta E_{fd} \end{bmatrix}^T, \quad y = \begin{bmatrix} \Delta T_e & \Delta V_T \end{bmatrix}^T \tag{10.34}$$

Note that the states and portions of the A and C matrices associated with the infinite bus are not shown. Find the coefficients $K_1, \ldots, K_6$.

Solutions: Using the form of the A and C matrix shown in (10.30), the K_i coefficients can either be obtained by inspection or with simple calculation as

$$\begin{aligned} K_1 &= C(1,1) = 1.6726 \text{ pu/rad}, \quad K_2 = C(1,3) = 1.5379 \text{ pu/pu} \\ K_3 &= 1/(T'_{d0} \times A(3,3)) = 0.25265, \quad K_4 = -T'_{d0} \times A(3,1) = 2.6691 \text{ pu/rad} \\ K_5 &= C(2,1) = -0.1313 \text{ pu/rad}, \quad K_6 = C(2,3) = 0.3981 \text{ pu/pu} \end{aligned} \tag{10.35}$$

At this loading level, $K_5 < 0$. Note that the expressions for K_2, K_3, K_4, and K_6 derived earlier for the 1-state flux decay model are still applicable to the 2-state flux decay model. For example,

$$K_2 = \frac{V \sin \delta_o}{X_e + X'_d} = \frac{1.0 \sin(1.1491)}{0.02 \times 9.91 + 0.15 + 0.245} = 1.5381 \tag{10.36}$$

on the machine base of 991 MVA. However, the expressions for K_1 and K_5 derived earlier for the 1-state flux decay model cannot be used for the 2-state flux decay model, as they do not take into account E'_d. New expressions for K_1 and K_5 for the 2-state flux decay model can be derived and verified by those values obtained here (Problem 10.2). ■

10.3 Synchronizing and Damping Torques

The block diagram in Figure 10.4 can be divided into two parts: the upper part with the state variables $\Delta\delta$ and $\Delta\omega$ represents the rotor mechanical motion, and the lower part with the state variables $\Delta E'_q$ and ΔE_{fd} represents the electrical flux dynamics in the generator circuits. The electrical part can have additional states if more detailed models including the subtransient effect are used.

The most important consideration here is the accelerating torque acting on the rotor of the turbine-generator, which from (10.23) is

$$\Delta T_{\rm acc}(s) = \Delta T_m(s) - D\Delta\omega(s) - \Delta T_e(s) \tag{10.37}$$

The mechanical torque ΔT_m comes from the action of the governor, which is considered to be inactive during periods of small speed deviation as determined by a deadband. The coefficient $D > 0$ contains various sources of damping, which generates a damping torque as it is in the direction of $-\Delta\omega$. The electrical torque ΔT_e has two components:

1) The component $\Delta T_{e1} = K_1\Delta\delta$ is the synchronizing torque from the transmission network (transmission lines, transformers, and shunt capacitors) that "holds" the synchronous machine together with the other machines, which in this case would only be the infinite bus. As K_1 is a real coefficient, T_{e1} is a purely synchronizing torque pointing in the direction of $\Delta\delta$ and hence is the main part of the synchronizing torque. On the other hand, if $\delta_o > 90°$, that is, the system is operating beyond its stability limit, $K_1 < 0$. Any random disturbance would result in the generator losing synchronism with the infinite bus. It is not possible for practical power systems to operate in such a manner. Based on ΔT_{e1}, the dynamic equation for the rotor mechanics of Figure 10.4 neglecting the small damping coefficient D is

$$2H\Delta\ddot{\delta} = -K_1\Omega\Delta\delta \tag{10.38}$$

The eigenvalues of this system, known as the local mode, are $\pm j\omega_s$, where $\omega_s = \sqrt{K_1\Omega/(2H)}$ rad/s.

2) The other component, ΔT_{e2}, is the electrical torque obtained by manipulating the field voltage via the excitation system. It consists of several transfer functions originating from $\Delta\delta$. Suppose a swing mode of frequency ω_s has low damping. Then a study of $T_{e2}(s)$ can be accomplished by assuming $s = j\omega_s$.[1] In this approach, the composite transfer function from $\Delta\delta$ to ΔT_{e2} can be represented by a complex quantity, allowing ΔT_{e2} to have both synchronizing and damping torque. This method will now be discussed in more detail.

10.3.1 ΔT_{e2} Under Constant Field Voltage

Accounting for the flux decay effect but without voltage regulation action, the transfer function from $\Delta\delta$ to ΔT_{e2} (Figure 10.4) at $s = j\omega_s$ is

$$\begin{aligned}\Delta T_{e2}(j\omega_s) &= -\left.\frac{K_2K_3K_4}{1 + sK_3T'_{do}}\right|_{s=j\omega_s}\Delta\delta(j\omega_s) = -K_2K_3K_4\frac{1 - j\omega_sK_3T'_{do}}{1 + (\omega_sK_3T'_{do})^2}\Delta\delta(j\omega_s)\\ &= -\frac{K_2K_3K_4}{1 + (\omega_sK_3T'_{do})^2}\Delta\delta(j\omega_s) + \frac{K_2K_3^2K_4T'_{do}\Omega}{1 + (\omega_sK_3T'_{do})^2}\Delta\omega(j\omega_s)\end{aligned} \tag{10.39}$$

using the relationship $j\omega_s\Delta\delta(j\omega_s) = \Omega\Delta\omega(j\omega_s)$ from (10.22).

---

1 A reader familiar with state-space model techniques can think of this approach as a modal analysis specialized to a local mode that is lightly damped, so that it can be approximated by just the imaginary part.

In (10.39), the first term proportional to $\Delta\delta$ is the synchronizing torque component with a negative coefficient. Thus it is a destabilizing torque, because the machine internal voltage is subject to decay. The second term proportional to $\Delta\omega$ is the damping torque component, which is positive, implying that the flux decay effect improves damping. The situation, however, will be quite different if the field voltage E_{fd} can be controlled.

10.3.2 ΔT_{e2} With Excitation System Control

With the voltage regulator function included, the transfer function from $\Delta\delta$ to $\Delta T_{e2}(s)$ of the block diagram inside the dashed lines in Figure 10.4 can be simplified to

$$\Delta T_{e2}(s) = -K_2 \frac{\frac{K_3}{1+sK_3T'_{do}}}{1+K_6\frac{K_A}{1+sT_A}\frac{K_3}{1+sK_3T'_{do}}}\left(K_4 + \frac{K_5K_A}{1+sT_A}\right)\Delta\delta(s) \tag{10.40}$$

which is shown in Figure 10.5.

Simplification of (10.40) yields

$$\Delta T_{e2}(s) = -\frac{K_2(K_4 + K_5K_A + sK_4T_A)}{(1/K_3 + K_6K_A) + s(T_A/K_3 + T'_{do}) + s^2T_AT'_{do}}\Delta\delta(s) \tag{10.41}$$

Substituting $s = j\omega_s$, (10.41) becomes

$$\begin{aligned}\Delta T_{e2}(j\omega_s) &= -\frac{K_2(K_4 + K_5K_A + j\omega_sK_4T_A)}{(1/K_3 + K_6K_A) + j\omega_s(T_A/K_3 + T'_{do}) - \omega_s^2T_AT'_{do}}\Delta\delta(j\omega_s) \\ &= -\frac{a+jb}{c+jd}\Delta\delta(j\omega_s) = -\frac{(ac+bd)+j(bc-ad)}{c^2+d^2}\Delta\delta(j\omega_s)\end{aligned} \tag{10.42}$$

where

$$\begin{aligned} &a = K_2(K_4 + K_5K_A), \quad b = K_2(\omega_sK_4T_A) \\ &c = 1/K_3 + K_6K_A - \omega_s^2T_AT'_{do}, \quad d = \omega_s(T_A/K_3 + T'_{do})\end{aligned} \tag{10.43}$$

The expression (10.42) is decomposed into a synchronizing torque and a damping torque as

$$\Delta T_{e2}(j\omega_s) = \Delta T_{s2}(j\omega_s) + \Delta T_{d2}(j\omega_s) \tag{10.44}$$

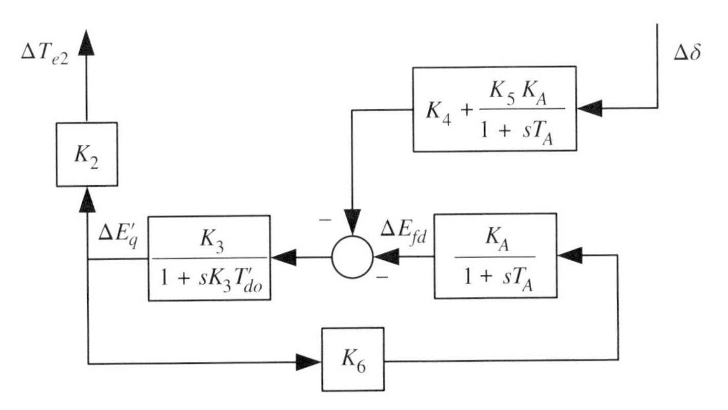

Figure 10.5 Simplification of the $\Delta\delta$ to ΔT_{e2} path.

The synchronizing torque component is

$$\Delta T_{s2}(j\omega_s) = -\frac{ac + bd}{c^2 + d^2}\Delta\delta(j\omega_s) \tag{10.45}$$

which can be approximated by, assuming K_A is large,

$$\Delta T_{s2}(j\omega_s) \simeq -\frac{K_2 K_5}{K_6}\Delta\delta(j\omega_s) \tag{10.46}$$

The damping torque component is

$$\Delta T_{d2}(j\omega_s) = -\frac{bc - ad}{c^2 + d^2} j\Delta\delta(j\omega_s) = -\frac{bc - ad}{c^2 + d^2}\frac{\Omega}{\omega_s}\Delta\omega(j\omega_s) \tag{10.47}$$

which can be approximated by, again assuming K_A is large,

$$\Delta T_{d2} = \frac{K_2 K_A K_5 (T_A/K_3 + T'_{do})\Omega}{(1/K_3 + K_6 K_A - \omega_s^2 T'_{do} T_A)^2 + (T_A/K_3 + T'_{do})^2 \omega_s^2}\Delta\omega = K_D \Delta\omega \tag{10.48}$$

As shown earlier, $K_5 < 0$ in a heavily loaded system. Equation (10.46) shows that ΔT_{e2} provides additional synchronizing torque, improving the transient stability of the synchronous machine subject to large disturbances. This is consistent with the action of the excitation system to provide large field currents to restore low system voltages. On the other hand, in (10.48), $K_5 < 0$ implies $K_D < 0$, that is, the voltage regulation action has a destabilizing effect, reducing the damping torque applied to the synchronous machine. This may lead to small-signal instability. This is not an uncommon situation faced by a control design engineer in which an actuator used to enhance certain system properties has a detrimental effect on some other system aspects. Fortunately, in this case, additional damping can be provided to the synchronous generator by implementing a power system stabilizer, which is the subject of the following sections.

Referring back to Example 10.2, the eigenvalues and their associated states of the A are given in Table 10.1. Note that the local mode $0.2133 \pm j11.462$ is unstable, which is due to a negative K_5 as a result of the action of the voltage regulator.

To summarize, the small-signal electrical torque is approximated as

$$\Delta T_e = \Delta T_s + \Delta T_d, \quad \Delta T_s = \left(K_1 - \frac{K_2 K_5}{K_6}\right)\Delta\delta, \quad \Delta T_d = K_D \Delta\omega \tag{10.49}$$

in which the total synchronizing torque ΔT_s is a function of K_1 due to network admittances, as well as the action of the voltage regulator.

The electrical torque ΔT_e (10.49) for various paths in Figure 10.4 is plotted in Figure 10.6. In Figure 10.6, the vertical axis is $\Delta\delta$ and the horizontal axis pointing to the

Table 10.1 Eigenvalues and modal states of Example 10.2.

Eigenvalues	States
$0.2133 \pm j11.462$	$\Delta\delta, \Delta\omega$
$-11.557 \pm j14.898$	E'_q, E_{fd}
-3.875	E'_d

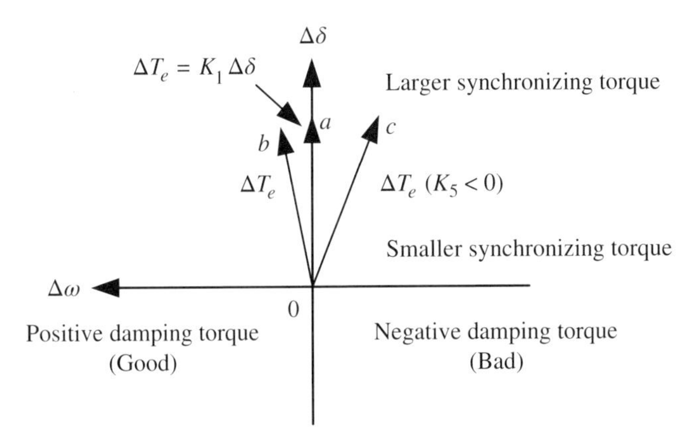

Figure 10.6 Plot of ΔT_e in the $\Delta\delta$–$\Delta\omega$ plane.

left is the positive $\Delta\omega$ direction, noting that $\Delta\omega$ leads $\Delta\delta$ by 90°. This placement of the axes matches up with the s-plane used for plotting eigenvalues, such as the one shown in Figure 10.1. The projection of ΔT_e on the $\Delta\delta$-axis is the synchronizing component, which determines mostly the oscillatory mode frequency: a larger synchronizing torque will increase the local mode frequency, and vice versa. The projection of ΔT_e on the $\Delta\omega$-axis is the damping component: a larger positive projection on the $\Delta\omega$-axis will increase the damping on the local model.

Taking into account only the network, $\Delta T_e = K_1 \Delta\delta$ lies on the vertical axis, as labelled by a. Including ΔT_{e2} from (10.39) for the flux decay generator model but no voltage regulator, $\Delta T_e = K_1 \Delta\delta + \Delta T_{e2}$ is labelled as b in Figure 10.6. Finally, including the excitation system, ΔT_{e2}, labeled as c, will have a component pointing in the negative $\Delta\omega$ direction.

10.4 Power System Stabilizer Design using Rotor Speed Signal

The purpose of a power system stabilizer (PSS) is to restore or add damping torque to the swing modes. There are three components to a control system: an input signal, a compensator, and an actuator. There is a range of options for these components. For actuator selection, a common practice is to use an existing actuator, thus saving the cost of adding a new actuator. Note that a voltage regulator makes use of the voltage error signal $V_{\text{err}} = V_{\text{ref}} - V_T$ to regulate the field voltage E_{fd}. Thus it is relatively straightforward to add a supplemental signal V_S to V_{err}, so that the new voltage error signal becomes

$$V_{\text{err}} = V_{\text{ref}} - V_T + V_S \tag{10.50}$$

allowing the damping function to act through an excitation system.

The input signal selection may be more varied and often depends on the design objectives and system configurations. This section discusses the use of the generator rotor speed deviation $\Delta\omega$ as the input signal. This choice is fundamental as all other signals have to be referenced back to $\Delta\omega$ in the control design. The use of other signals for PSS design is discussed in the next section.

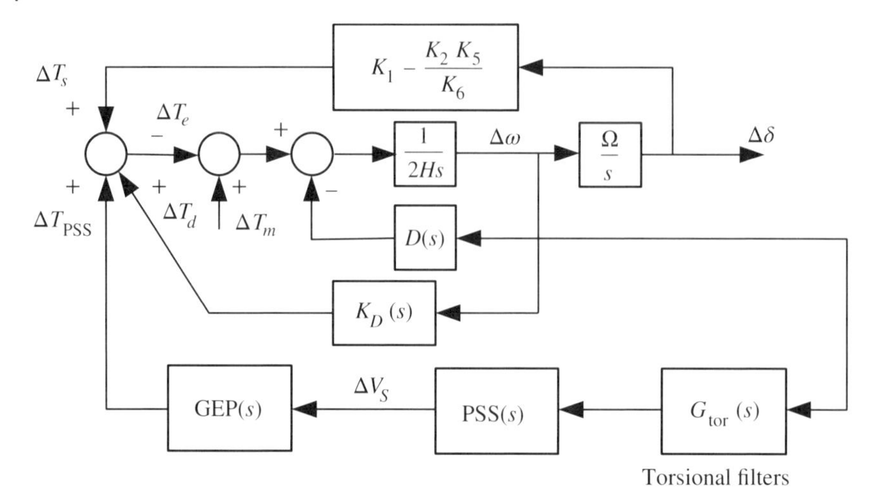

Figure 10.7 Block diagram showing the PSS implementation using the rotor speed as the input signal.

In a steam turbine, the rotor speed is typically measured on the end of the rotor shaft on the turbine side, which is known as the front standard speed. In older units with alternator-type exciters, the speed measurement at the generator end of the shaft, known as the rear standard speed, is actually the exciter alternator shaft speed. This rear standard speed contains more of the exciter alternator shaft torsional mode, and thus is a less desirable signal than the front standard speed. With modern static excitation systems without an alternator on the generator rotating shaft, the torsional oscillation content in the rear standard speed is less of a concern. A discussion on the torsional model of turbine-generators can be found in Chapter 12.

The PSS implementation is shown in Figure 10.7. The voltage regulator influence on the local mode from Figure 10.4 is aggregated into the $K_D(s)$ block, except for the term K_2K_5/K_6,[2] which is included in the K_1 block.

The transfer function of the excitation system is represented by

$$\text{GEP}(s) = \frac{K_2(K_A/(1+sT_A))(K_3/(1+sK_3T'_{do}))}{1+K_6(K_A/(1+sT_A))(K_3/(1+sK_3T'_{do}))} \tag{10.51}$$

The PSS control path also includes a transfer function $G_{\rm tor}(s)$ to represent the effect of torsional filters, which are used to reduce the torsional signal amplitude in the speed signal. For the basic PSS design, the torsional filters will be neglected. The impact of the torsional filters on the PSS design is discussed in a later section.

10.4.1 PSS Design Requirements

From Figure 10.7, the electrical torque $\Delta T_{\rm PSS}$ due to the PSS can be expressed as

$$\Delta T_{\rm PSS}(s) = \text{GEP}(s) \cdot \text{PSS}(s)\Delta\omega \tag{10.52}$$

2 The term K_2K_5/K_6 is the synchronizing torque coefficient obtained by letting $K_A \rightarrow \infty$ in the "electrical model" block in Figure 10.4.

There are several design requirements for the compensator PSS (s):

1) The PSS output should be zero in steady state, thus requiring a derivative circuit, also known as a washout filter

$$G_{wo}(s) = \frac{T_w s}{T_w s + 1} \tag{10.53}$$

where the time constant T_w is between 5 to 10 sec, so that the gain of $G_{wo}(s)$ is mostly flat (the high-pass frequency region) in the frequency range of the swing mode.

2) The phase of the cascade connection GEP(s) · PSS(s) at the local mode with frequency ω_s should ideally be

$$\angle(\text{GEP}(s) \cdot \text{PSS}(s)|_{s=j\omega_s} = 0 \tag{10.54}$$

so that ΔT_{PSS} is in phase with $\Delta\omega$. In some cases, it is also advisable to make the phase compensation slightly negative (like $-10°$) such that the PSS also provides some synchronizing torque. Because GEP (s) has lowpass characteristics, the PSS has to provide phase lead to accomplish (10.54). This phase lead is provided by lead-lag circuits.

3) The output signal of PSS (s) is usually limited to the range ± 0.10 pu. This precaution is needed as the PSS may initially output a very large signal at the onset of a severe disturbance, counteracting the normal function of a voltage regulator to restore voltage and provide synchronizing torque.

10.4.2 PSS Control Blocks

Following (10.53) and (10.54), the PSS transfer function consists of a cascade of a washout filter $G_{wo}(s)$ and a two-stage lead-lag filter $G_{\text{ldlg}}(s)$

$$\text{PSS}(s) = G_{\text{ldlg}}(s)G_{wo}(s) = \left(K_{\text{PSS}}\frac{1+sT_1}{1+sT_2}\frac{1+sT_3}{1+sT_4}\right)\left(\frac{T_w s}{T_w s+1}\right) \tag{10.55}$$

as shown in Figure 10.8, with the output V_S limited by $V_{S\max}$ and $V_{S\min}$. Note that only a two-stage lead-lag filter is needed as the required phase-lead compensation is normally within 120°. The number of lead-lag stages can be adjusted to suit particular design needs.

The design of $G_{\text{ldlg}}(s)$ is accomplished in two stages: first, the time constants $T_1, \ldots, T_4$ are adjusted to provide the appropriate phase compensation, and second, the gain K_{PSS} is tuned to achieve the desired damping response.

Denote the path from the speed signal $\Delta\omega$ to ΔT_{PSS} without $G_{\text{ldlg}}(s)$

$$G_{\text{uncomp}}(s) = \text{GEP}(s)\frac{T_w s}{T_w s + 1} \tag{10.56}$$

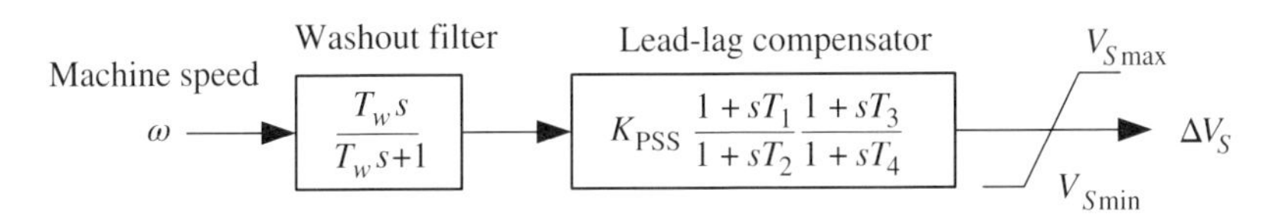

Figure 10.8 Block diagram of a power system stabilizer.

as the *uncompensated* transfer function. Similarly denote the same path including $G_{\mathrm{ldlg}}(s)$

$$G_{\mathrm{comp}}(s) = \mathrm{GEP}(s)G_{\mathrm{ldlg}}(s)\frac{T_w s}{T_w s + 1} \tag{10.57}$$

as the *compensated* transfer function.

Ideally the desired phase of $G_{\mathrm{comp}}(s)$ is about 0° at the local mode frequency $s = j\omega_s$; that is, $G_{\mathrm{comp}}(j\omega_s)$ is a mostly real quantity, so that the feedback path adds maximum damping torque. In practice, it is also desirable that the phase of $G_{\mathrm{comp}}(s)$ be about 0° for a range of frequencies about the local mode $s = j\omega_s$, because the local mode frequency may change according to the system loading and configuration. Such a design will achieve robustness.

The Bode or frequency response plot in Figure 10.9, which can also be found in [111, 112], shows the phase characteristics of several transfer functions. The phase of $G_{\mathrm{uncomp}}(s)$ ranges from -60° to -100° between 0.25 to 2 Hz. Thus phase lead of such magnitude needs to be added in this frequency range. It can be provided by a phase-lead compensator, which for a single stage, has the form [113]

$$G_{1s}(s) = K\frac{1 + sT_1}{1 + sT_2}, \quad T_1 > T_2, K > 0 \tag{10.58}$$

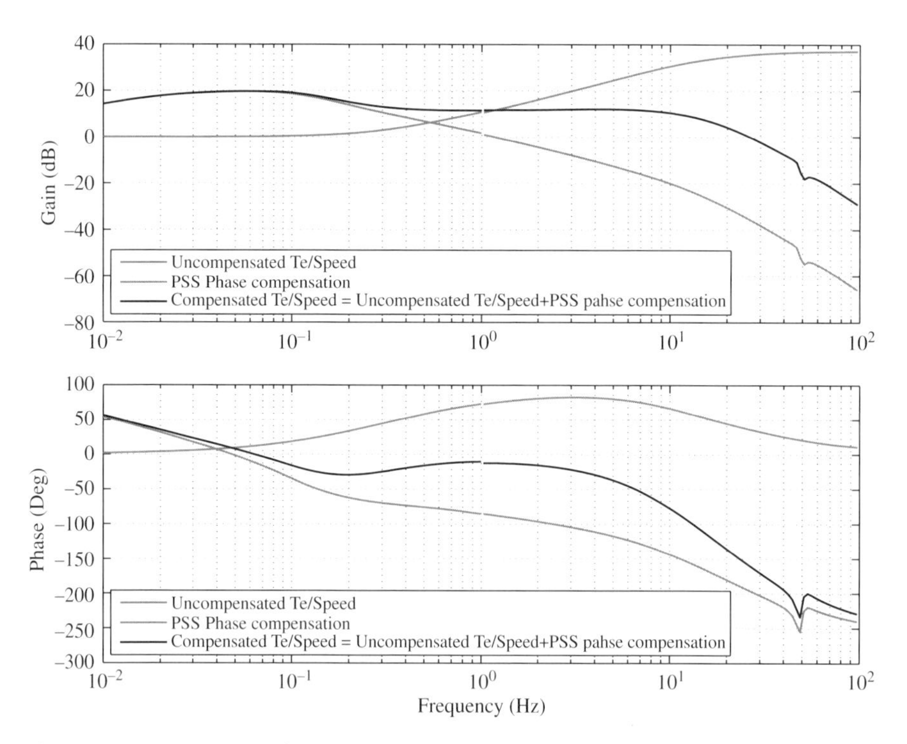

Figure 10.9 Uncompensated, compensated, and PSS phase plot (courtesy of GE Energy Consulting) (see color plate section).

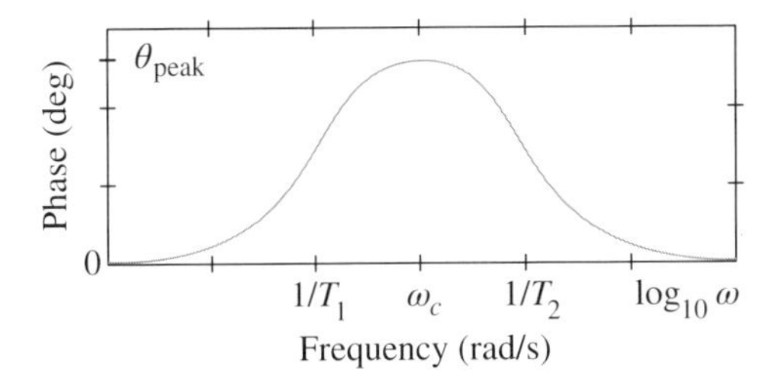

Figure 10.10 Phase characteristic of a phase-lead compensator versus frequency (see color plate section).

The phase characteristic of (10.58) is shown in Figure 10.10, where the zero of $G_{1s}(s)$ is $-1/T_1$ and the pole is $-1/T_2$.

Note that in Figure 10.10, the phase lead peaks at the

$$\theta_{peak} = \sin^{-1}\left(\frac{\alpha_{ld} - 1}{\alpha_{ld} + 1}\right) \tag{10.59}$$

where

$$\alpha_{ld} = (1/T_2)/(1/T_1) = T_1/T_2 \geq 1 \tag{10.60}$$

is the pole-zero ratio. The peak phase lead occurs at the center frequency

$$\omega_c = \sqrt{\frac{1}{T_1} \cdot \frac{1}{T_2}} \tag{10.61}$$

which is the geometric mean of $1/T_1$ and $1/T_2$. Note that for any given θ_{peak}, α_{ld} can be computed from

$$\alpha_{ld} = (1 + \sin\theta_{peak})/(1 - \sin\theta_{peak}) \tag{10.62}$$

Figure 10.9 also shows the phase of the compensated transfer function $G_{comp}(s)$, using the lead compensation shown in the top part of the same figure. Note that the compensated phase varies between $[-20°, -10°]$ over the frequency range of 0.5–2 Hz.

The need for phase compensation can also be visualized in Figure 10.11, using the same set of axes as in Figure 10.6. An uncompensated ΔT_{PSS} with over 90° in phase lag will point in the direction of positive synchronizing torque but negative damping torque. A phase-lead compensation will rotate the compensated ΔT_{PSS} counter-clockwise so that it will point in the direction of positive synchronizing and damping torques.

10.4.3 PSS Design Methods

There are many design methods for PSSs [112], two of which using linearized models are discussed here. These methods require setting up a linearized system model with the appropriate input and output relevant for the PSS design, and computing the local mode frequency ω_s. The system model could be a SMIB system, or include several generators close to the generator requiring the PSS tuning.

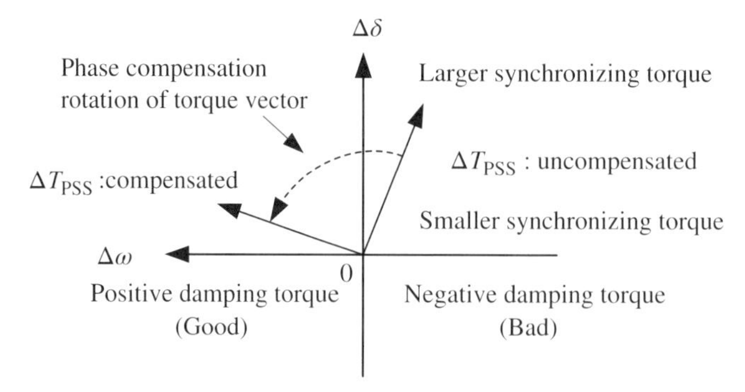

Figure 10.11 Rotation of ΔT_{PSS} through phase compensation (see color plate section).

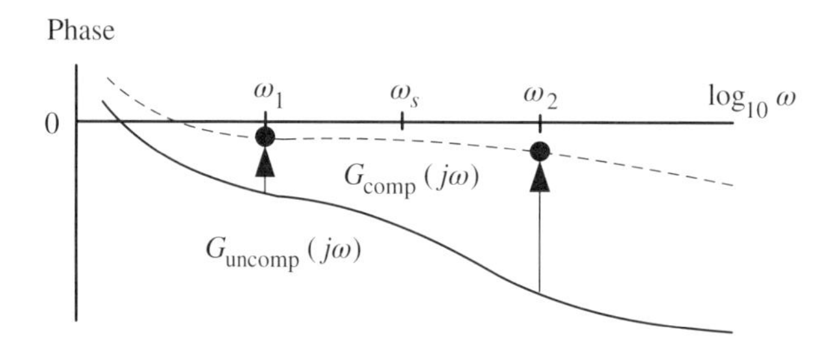

Figure 10.12 Phase characteristics of G_{uncomp} and G_{comp} versus frequency.

1. Phase Compensation Method

The phase compensation method follows readily from the earlier discussion. The design steps are listed as follows:

PC1. Make a Bode (frequency response) plot for $G_{\mathrm{uncomp}}(s)$, focusing only on the phase characteristic as shown in Figure 10.12.

PC2. Select a frequency $\omega_1 < \omega_s$ and a frequency $\omega_2 > \omega_s$, and set the compensated phase at these two frequency values to be between $-10°$ and $-20°$.

PC3. Select the time constants T_1, T_2, T_3, and T_4 to meet the desired compensated phase values. An optimization routine can be used to obtain these time constants, but a trial-and-error method can also be used. Note that the two lead-lag circuits do not have to be the same; that is, T_1 can be different than T_3 and T_2 can be different than T_4.

PC4. Select a value for $K_{\mathrm{PSS}} > 0$ such that the desired damping is achieved. The design can be verified by nonlinear time simulation.

2. Root-Locus Method

The phase compensation method does not directly work with the closed-loop system poles. The root-locus method, however, can be used to design PSS based on directly placing the close-loop poles. A reader not familiar with the root-locus method can refer to a control textbook, such as [113]. This method is more accessible as root-locus tools are readily available.

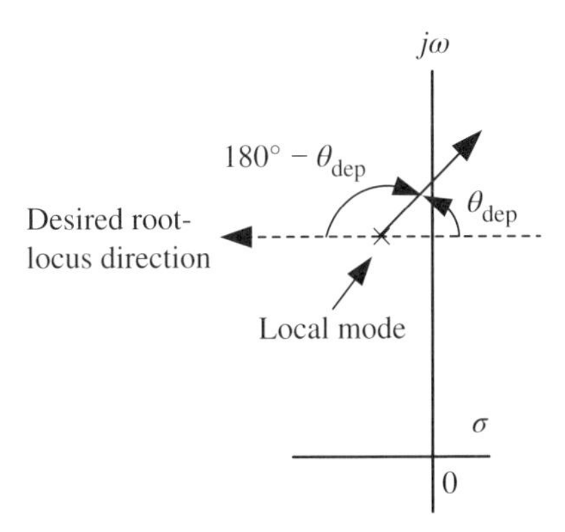

Figure 10.13 Root-locus plot of the uncompensated system $G(s)$.

The design steps are listed as follows for a two-stage lead-lag compensator:

RL1. Referring back to Figure 10.7, take out the PSS block and generate a linear state-space model $G(s)$ using ΔV_S as the input and $\Delta\omega$ as the output. Denote $G(s)$ as the uncompensated system.

RL2. Set $K_{PSS}(s) = K$, a proportional gain, and vary K from 0 to ∞ to generate a root-locus plot for the uncompensated system $-G_{tor}(s)G_{wo}(s)G(s)$, which at the local mode will resemble that shown in Figure 10.13. If no torsional filter is used, $G_{tor}(s) = 1$.

RL3. Find the angle of departure θ_{dep} of the root-locus branch leaving the local mode.

RL4. Set $\theta_{peak} = (180^\circ - \theta_{dep})/2$. A value somewhat less than this θ_{peak} would also be acceptable.

RL5. Find α_{ld} from the formula (10.62) or a chart of θ_{peak} versus α_{ld} (Figure 10.14).

RL6. Use the local mode frequency ω_s as the center frequency ω_c to find $T_1 = T_3$ and $T_2 = T_4$ according to (10.61).

RL7. Perform a root-locus analysis for the compensated system $-G_{ldlg}(s)G_{tor}(s)G_{wo}(s)G(s)$ to determine the gain K_{PSS} to satisfy the damping ratio requirement of the local mode. Verify the design with nonlinear time simulation.

This design method is illustrated in Example 10.3.

Example 10.3: PSS design for SMIB system

Example 9.8 shows that the generator in a SMIB system equipped with a ST3A excitation system, although improving the critical clearing time, results in a small-signal unstable system, with undamped oscillations. Use the root-locus design method to develop a two-stage lead-lag speed-input PSS to improve the damping ratio of the local mode to 0.1. Compute the closed-loop system eigenvalues. Implement the designed PSS in the system data file and simulate the time response for the same disturbance as in Example 9.8. In the disturbance simulation, limit the PSS output to ± 0.1 pu. Note that no torsional filter is used in this example.

Solutions: For Step RL1, the linearized model of the SMIB system is obtained as

$$G(s): \ \dot{x} = Ax + Bu, \quad y = Cx \tag{10.63}$$

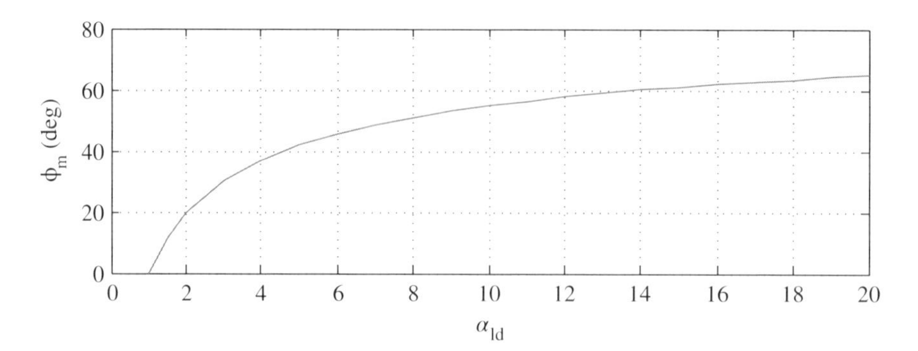

Figure 10.14 The maximum phase angle θ_{peak} versus α_{ld} for a lead compensator (see color plate section).

in which the states are

$$x = \begin{bmatrix} \Delta\delta & \Delta\omega & \Delta E'_q & \Delta\psi_{kq} & \Delta E'_d & \Delta\psi_{kq} & \Delta V_R & \Delta V_M \end{bmatrix}^T \tag{10.64}$$

the input is $u = \Delta V_s$, the output is $\Delta\omega$. The A matrix is given in (9.55), and the input and output matrices are given by

$$B = \begin{bmatrix} 0 & 0 & 0 & 0 & 0 & 0 & 29.99 & 527.7 \end{bmatrix}^T \tag{10.65}$$

$$C = \begin{bmatrix} 0 & 1 & 0 & 0 & 0 & 0 & 0 & 0 \end{bmatrix} \tag{10.66}$$

Then the washout filter with $T_w = 10$ sec is connected in series to $G(s)$ to form $G'(s)$.

For Step RL2, because the output of the PSS enters into the voltage regulator summing junction with a positive sign, a root-locus analysis is performed for $-G'(s)$ resulting in the root-locus plot shown in Figure 10.15.

In Step RL3, the angle of departure is found to be about 90°. Thus in Step RL4, the PSS design will be based on two identical single-stage lead-lag compensators, each with 40° in phase lead.

For Steps RL5 and RL6, the center frequency of these compensators will be set to 10 rad/sec, which is very close to the local mode frequency. Using (10.59)–(10.62), the lead-lag compensator will have the transfer function

$$G_{\text{ldlg}}(s) = K_{\text{PSS}} \left(\frac{1 + 0.2144s}{1 + 0.04663s} \right)^2 \tag{10.67}$$

where $T_1 = T_3 = 0.2144$ sec and $T_2 = T_4 = 0.04663$ sec.

In Step RL7, to find the gain K_{PSS}, a root-locus analysis of the compensated system $-G_{\text{ldlg}}(s)G'(s)$ is performed, with the resulting root-locus plot shown in Figure 10.16. Note that the root-locus branches leaving the local mode go directly left. At the gain $K_{\text{PSS}} = 2.19$, a damping ratio of about 0.1 for the local mode is achieved.

In the closed-loop system, three more states have been added, such that the new state vector is

$$x = \begin{bmatrix} \Delta\delta & \Delta\omega & \Delta E'_q & \Delta\psi_{kq} & \Delta E'_d & \Delta\psi_{kq} & \Delta V_R & \Delta V_M & x_{\text{wo}} & x_{\text{ldlg1}} & x_{\text{ldlg2}} \end{bmatrix}^T \tag{10.68}$$

The closed-loop system poles with the PSS compared to the system poles without the PSS are shown in Table 10.2.

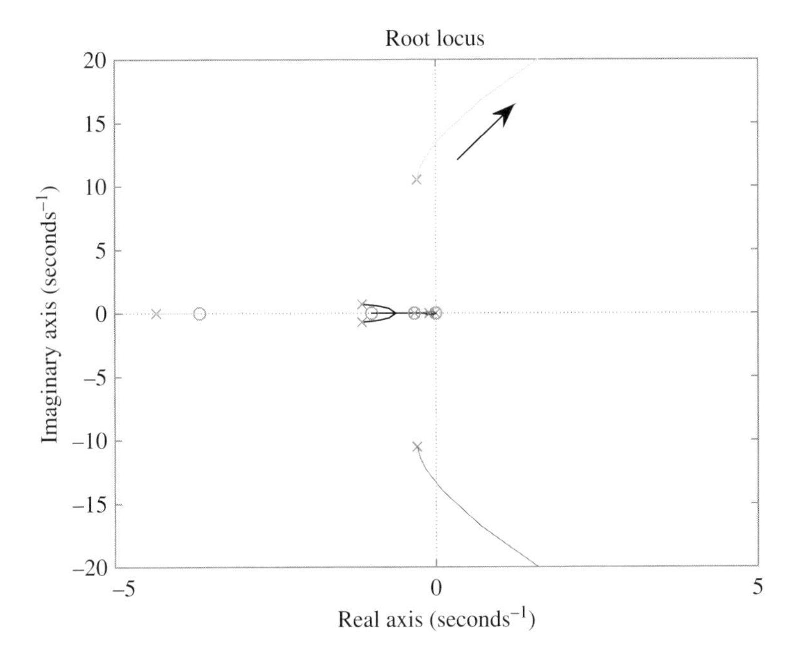

Figure 10.15 Root-locus plot of the uncompensated system (note that the scales of the horizontal and vertical axes are not the same) (see color plate section).

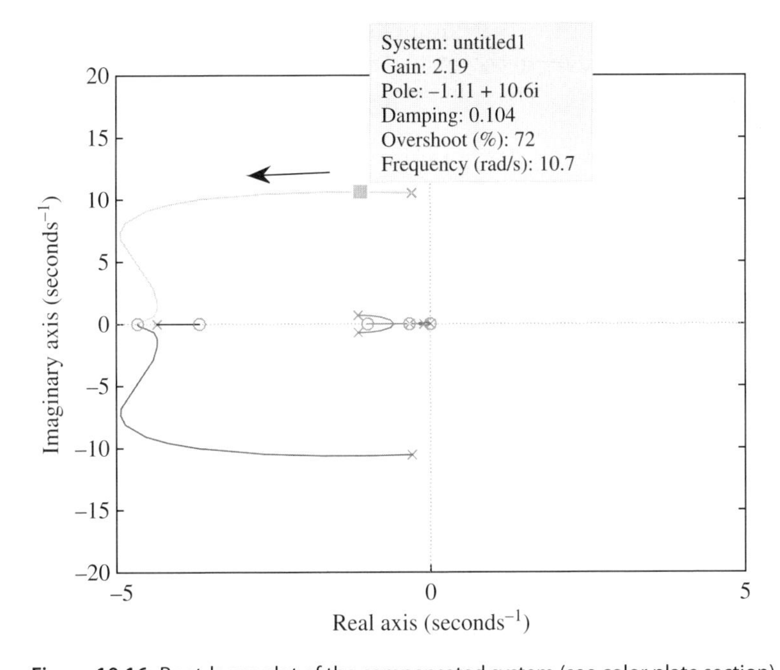

Figure 10.16 Root-locus plot of the compensated system (see color plate section).

Table 10.2 SMIB system eigenvalues with and without PSS.

With ST3A excitation system		With ST3A excitation system and PSS	
States	Eigenvalues	States	Eigenvalues
δ, ω	$-0.2938 \pm j10.53$	δ, ω	$-1.1304 \pm j10.620$
ψ_{kd}	-34.37	ψ_{kd}	-38.23
E'_d	-4.355	E'_d	-4.355
ψ_{kq}	-27.73	ψ_{kd}	-27.75
E'_q, V_M	$-1.150 \pm j0.6893$	E'_q, V_M	$-1.137 \pm j0.6901$
V_R	-124.9	V_R	-125.2
		$x_{\text{ldlg1}}, x_{\text{ldlg2}}$	-18.52 ± 7.990
		and x_{wo}	-0.1002

Note that the PSS introduces the pole at -0.1002 due to the washout filter, and the well-damped oscillatory mode $-18.52 \pm j7.990$ due to the lead-lag compensator interaction with the local mode. The real part of the local mode shifts from -0.2938 to -1.1304, whereas the imaginary part does not decrease, thus maintaining the synchronizing torque. The synchronous machine electrical modes and the excitation system modes remain largely unchanged. Note that the poles in Table 10.2 are for the system before the disturbance.

Applying the short-circuit fault to Bus 3 and opening the line connecting Buses 3 and 2, the performance of the designed PSS is shown in Figure 10.17. The output of the PSS signal is limited to ± 0.1 pu, which is reached during the initial part of the disturbance. The PSS reduces the system oscillation to acceptable levels. Also note that as the oscillation is damped, the output of the PSS also diminishes. In steady state, the output of the PSS is zero. ■

The effectiveness of the PSS can also be illustrated with a spectral analysis of the disturbance time response. The results of applying Fourier transform to the generator terminal bus (Bus 3) voltage magnitude response without the PSS (Figure 9.16) and to the Bus 3 voltage magnitude response with the PSS (Figure 10.17) are shown in Figure 10.18. Without the PSS, the local mode shows a sharp peak at 1.33 Hz, indicating the presence of a lightly damped mode.[3] With the application of the PSS, the peak value of the 1.33 Hz mode is significantly lower.

10.4.4 Torsional Filters

As mentioned earlier, it is important that the addition of new control functions does not excite the torsional modes in a turbine-generator. One way to reduce the torsional interaction is to include a cascade of notch or band-reject filters such as those described in [114] (Figure 10.19). The frequency of a particular torsional mode is not fixed. Rather, it varies within a small frequency range as a function of the mechanical loading. Each

3 The post-fault local mode frequency at 1.33 Hz is lower than the pre-fault local mode frequency of 1.68 Hz, for the transmission system is weaker with the loss of one of the parallel transmission lines. Also because of the nonlinearity in the time response, several harmonic components are also observable in the spectrum.

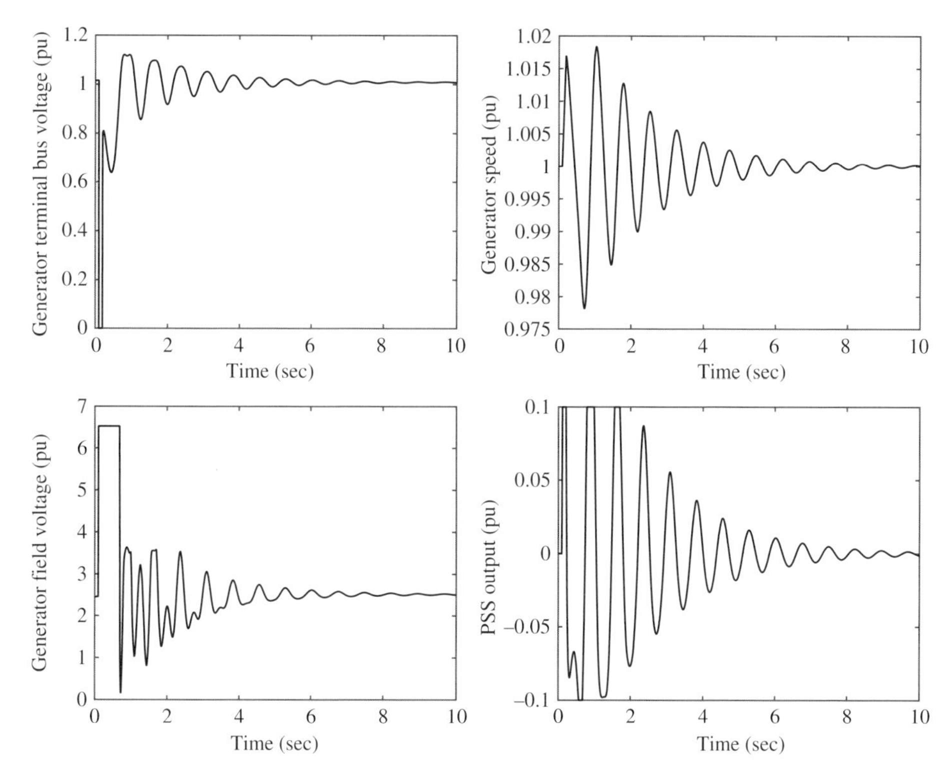

Figure 10.17 Time response of generator terminal bus voltage, generator rotor speed, generator field voltage, and the PSS output signal.

of the torsional filters will target a specific frequency in that range. If these filters are placed in cascade, the resulting characteristic will show an amplitude reduction over a desired frequency range, thus reducing the feedback gain on the torsional modes.

The lowest torsional modes are normally in the range of 12–15 Hz, which is far beyond the electromechanical mode frequency range of 0.3–3 Hz. Hence it is not necessary to represent these torsional filters in full detail. Instead, $G_{\text{tor}}(s)$ is an approximation of the lower frequency portion of the torsional filter that impacts the PSS design. As in many control tasks, the phase lag of $G_{\text{tor}}(j\omega)$ has a more profound impact on the control design than its magnitude decrease.

To illustrate, an approximation of a torsional filter is [112]

$$G_{\text{tor}}(s) = \frac{1}{1 + 0.061s + 0.0017s^2} \tag{10.69}$$

At 1 Hz, the frequency response is

$$G_{\text{tor}}(j2\pi) = 0.9915\angle - 22.34^\circ \tag{10.70}$$

which is very close to unity in magnitude. The angle, however, is at −22.34°. At 2 Hz, the frequency response becomes

$$G_{\text{tor}}(j4\pi) = 0.9438\angle - 46.34^\circ \tag{10.71}$$

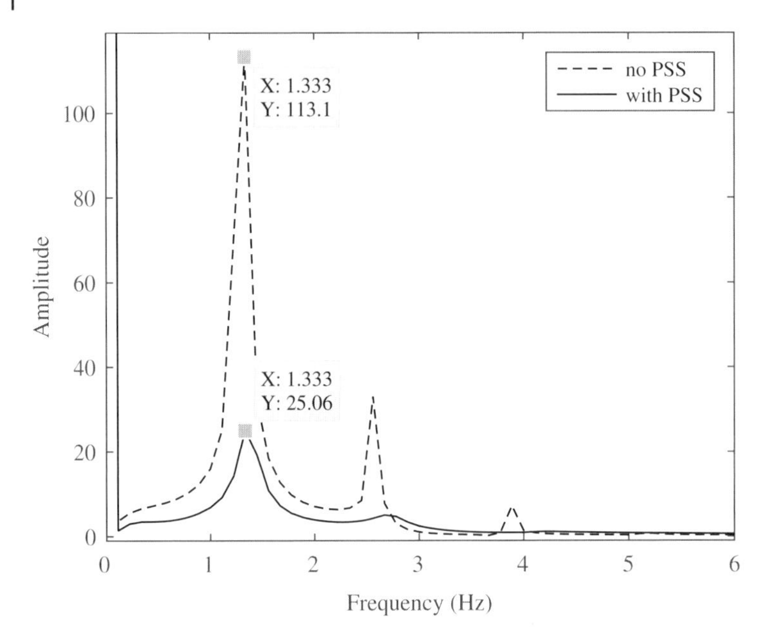

Figure 10.18 Spectral analysis of the post-fault generator terminal voltage response with and without power system stabilizer (see color plate section).

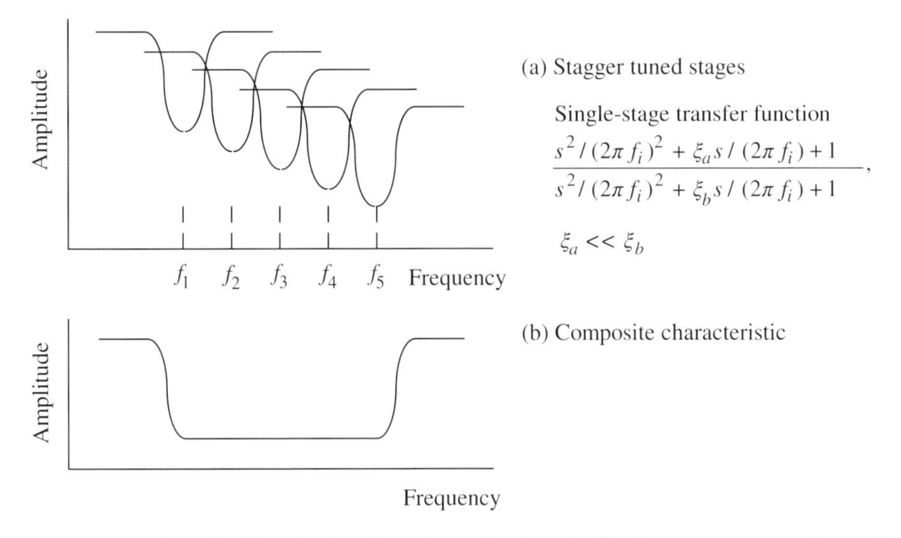

Figure 10.19 Cascade of band-reject filters for torsional mode filtering using staggered tuned biquad filter stages [114].

The magnitude is still only off by 5%, but the phase is now at $-46.34°$. Thus the net impact of the torsional filter is that PSS will need a higher phase-lead compensation.

When torsional filtering is included in the PSS design consideration, the uncompensated system will become $G_{\text{tor}}(s)G(s)$, which should be used in the phase compensation and root-locus design methods. Problem 10.5 addresses PSS design including a torsional filter $G_{tor}(s)$.

10.4.5 PSS Field Tuning

Typically a PSS is commissioned during the startup of a new power generation plant. The initial design is performed via a system simulation study, using a small system with a few generators and an infinite bus. During the commissioning, the gain of the PSS is adjusted to provide a sufficient gain margin. The following steps serve as a guideline.

FT1. Set up a simple power system model with the new generator and an infinite bus or several nearby machines to model the power network around the new generator.
FT2. Generate linearized models at several operating conditions.
FT3. Design a PSS valid for all the operating conditions using the phase compensation method or the root-locus method.
FT4. During commissioning, keep the time constants T_1, T_2, T_3, and T_4 unchanged (that is, the phase compensation is fixed from the off-line design result) and increase K_{PSS} until spontaneous oscillations in V_T are observed. Reduce K_{PSS} to either 1/2 (for a 6 dB gain margin) or 1/3 (for a 10 dB gain margin) of the oscillation gain.[4] Inject a small step command (0.01 pu) into V_{ref} input of the voltage regulator to verify that a satisfactory response of the terminal voltage is obtained.

10.4.6 Interarea Mode Damping

As discussed in Chapter 6, for large power systems with long-distance power transmission, groups of machines would oscillate against other groups of machines, the so-called interarea modes. These interarea modes are of lower frequencies, in the range of 0.25 to 1 Hz, compared to local mode oscillations at 1 to 2 Hz. These interarea modes are in general lightly damped, and would require supplemental damping control. Although a PSS is tuned for improving the damping of the local mode, it can also offer benefits to the interarea mode. Referring back to Figure 10.12, it can be seen that the phase lag of the uncompensated transfer function at the interarea mode frequency range is less than that in the local mode frequency range. Thus interarea mode damping improvement requires less phase lead, which is consistent with the phase characteristics of a lead compensator. It should be noted that (1) the PSSs on machines close to the "node" of the interarea oscillation would not be as effective in providing damping, and (2) interarea damping is generally contributed by PSSs on generators that participate in its oscillations. It is important for many generators to contribute to interarea mode damping. In case one of the generators providing damping is out of service, the rest of the generators can still contribute adequate damping.

For systems with several lightly damped modes, a multiple-band PSS can be designed [115]. In this approach, a PSS can be tuned to damp a local mode and an interarea mode. Interarea modes can also be damped using HVDC and FACTS, which are discussed in later chapters.

---

4 This is akin to a root-locus tuning using real equipment. When the root-locus branch is close to the $j\omega$-axis, the system is close to instability. Any random noise in the real system will generate a sustained oscillation.

10.5 Other PSS Input Signals

Although the generator rotor speed is the most commonly used input signal for a PSS, there are also other signals that are suitable. In this section, two alternative signals, the generator terminal bus frequency deviation Δf and the output power P_e are discussed. Sensors for these signals are readily available in a power plant. More detailed discussions on these signals can be found in [116].

10.5.1 Generator Terminal Bus Frequency

The bus frequency f is measured with a frequency transducer. In a dynamic simulation program, the bus frequency deviation from the nominal value can be computed from the bus voltage angle using a transfer function

$$\Delta f(s) = \frac{1}{\Omega} \frac{s}{1 + sT_r} \theta(s) \tag{10.72}$$

where T_r is the filter time constant to prevent step changes in Δf during disturbances and $\Omega = 2\pi f_o$ is the conversion factor from radians to pu speed, f_o being the nominal system frequency in Hz. Note that although generator rotor angles do not change instantaneously, bus voltage angles do because faults will change instantaneously the power flow in a system.

In general, the generator terminal bus Δf follows closely the rotor speed measurement of the generator. In an SMIB model, the local mode oscillation in Δf mimics that in the rotor speed $\Delta\omega$, perhaps with a small phase shift and a slightly lower amplitude.

The advantages of using Δf include:

- There is less torsional mode content in Δf.
- There are more swing modes observable in Δf. Thus it is useful for interarea mode damping.
- For a power plant with several generators, such as a hydraulic turbine-generator complex, the speed signal $\Delta\omega$ also contains intra-plant modes (Figure 10.20). The Δf signal measured at the high side of the step-up transformers, being at the node of the oscillation, would have minimal intra-plant mode content. Thus using this signal for the PSS would not excite the intra-plant mode.

The disadvantage of using the Δf signal is that the local mode content is not as strong, especially in a large system.

10.5.2 Electrical Power Output ΔP_e

The main advantages of using ΔP_e as the input signal are:

- There are less torsional mode contents in ΔP_e.
- From the swing equation (10.2)

$$2Hs\Delta\omega = \Delta T_m - \Delta P_e \tag{10.73}$$

With $s = j\omega_s$ and $\Delta T_m = 0$, this equation becomes

$$-\Delta P_e = 2Hj\omega_s\Delta\omega \tag{10.74}$$

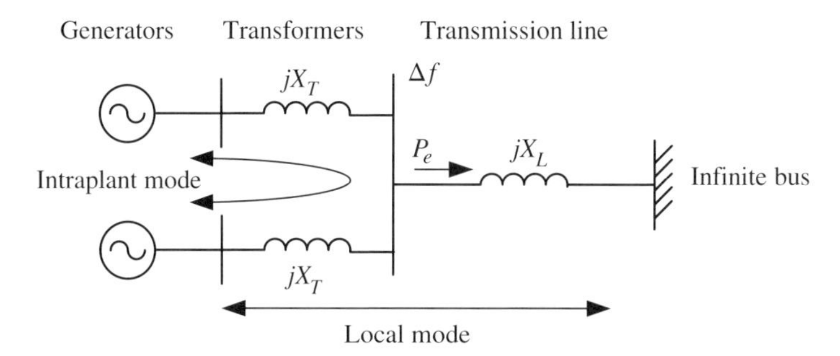

Figure 10.20 Two generators in a power plant showing intra-plant oscillations.

Thus the signal $-\Delta P_e$ is 90° ahead of $\Delta\omega$, requiring less phase compensation in the PSS.

The disadvantages of using ΔP_e include:

- It is not a good variable to use during the onset of large disturbances, as its magnitude may drop significantly.
- The signal is nonzero during load ramping, leading to unnecessary control actions.

10.6 Integral-of-Accelerating-Power or Dual-Input PSS

Although in practice most of the PSSs implemented are of the single-input type, there is also a dual-input PSS [117–119] that is now offered by manufacturers. The input signals to this PSS are the machine rotor speed and generator electrical output power. This PSS is known as a dual-input integral-of-accelerating-power PSS.

With a change from mechanical torque to power, the swing equation (10.73) becomes

$$2H\Delta\dot{\omega} = \Delta P_m - \Delta P_e = \Delta P_{\mathrm{acc}} \tag{10.75}$$

where P_{acc} is the accelerating power, such that the mechanical power can be obtained as

$$\Delta P_m(s) = 2Hs\Delta\omega(s) + \Delta P_e(s) \tag{10.76}$$

Figure 10.21 shows an implementation of (10.76) in which the lowpass filter $G(s)$ is used to allow only slower changes of the mechanical power signal $\Delta P'_m$ to go through.

The rotor speed can be estimated by

$$\Delta\omega = \frac{1}{2H}\int(\Delta P'_m - \Delta P_e)dt = \frac{1}{2H}\int \Delta P_{\mathrm{acc}}\, dt \tag{10.77}$$

Taking the Laplace transform of (10.77) gives

$$\mathcal{L}\left\{\frac{1}{2H}\int \Delta P_{\mathrm{acc}}\, dt\right\} = -\frac{1}{2Hs}\Delta P_e(s) + G(s)\left(\frac{\Delta P_e}{2Hs} + \Delta\omega(s)\right) \tag{10.78}$$

whose implementation is shown in Figure 10.22. This integral of accelerating power system is then passed into a phase-lead compensator to provide adequate damping to the local mode. This dual-input PSS is represented by the IEEE PSS2A model in

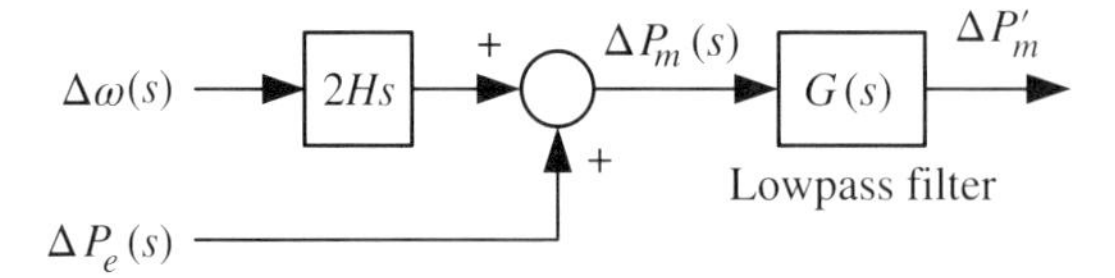

Figure 10.21 Block diagram showing mechanical power estimation.

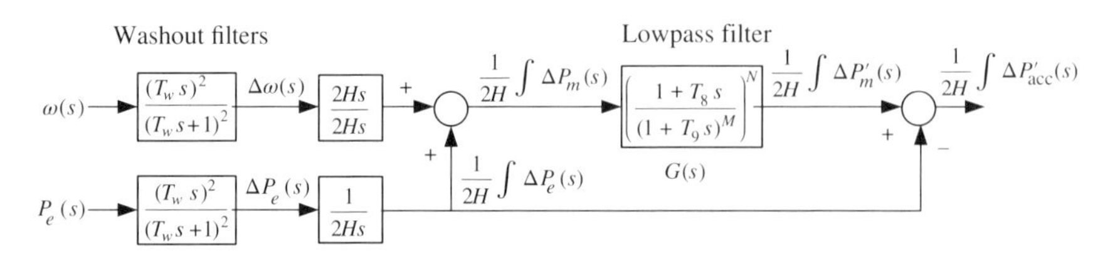

Figure 10.22 Estimation of the accelerating power signal.

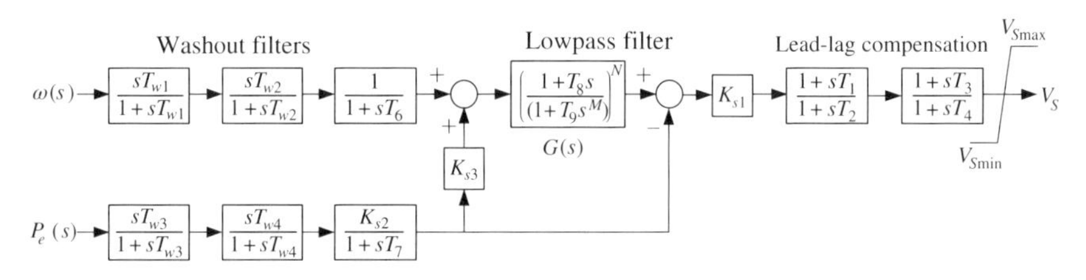

Figure 10.23 IEEE type PSS2A dual-input power system stabilizer.

Figure 10.23 [105]. The use of the PSS2A model for damping of the local mode is illustrated in Example 10.4.

The speed synthesized by this PSS, $\Delta\omega$, is not equal to the generator speed; rather, $\Delta\omega$ in (10.77) has lower levels of torsional frequencies and has dynamic characteristics of generator speed (at lower frequencies) and electrical power (at higher frequencies) [118].

Example 10.4: Dual-input PSS design for SMIB system

Repeat Example 10.3 using the dual-input PSS, using the parameters in [118] given as, where the unit of the time constants is in seconds

$$\begin{gathered}
T_{w1} = T_{w2} = T_{w3} = T_{w4} = 10, \quad K_{s2} = 1.136, \quad K_{s3} = 1.0, \quad T_6 = 0, \quad T_7 = 10 \\
T_8 = 0.5, \quad T_9 = 0.1, \quad N = 1, \quad M = 5 \\
K_{s1} = 20, \quad T_1 = 0.12, \quad T_2 = 0.035, \quad T_3 = 0.10, \quad T_4 = 0.02
\end{gathered} \tag{10.79}$$

The example is done in four parts:

1) Generate a linear model of the SMIB system with the input ΔV_S (the input into the voltage summing junction of the voltage regulator) and with the outputs $\Delta\omega$ and ΔP_e. Call this system $G_{21}(s)$.

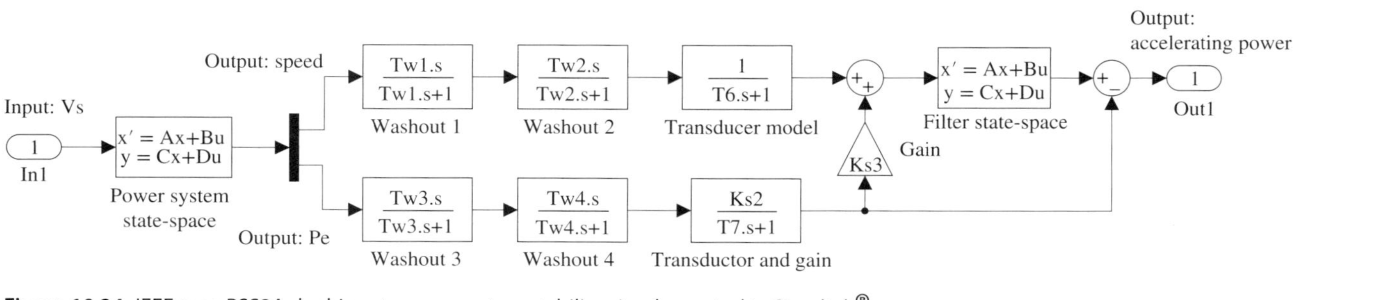

Figure 10.24 IEEE type PSS2A dual-input power system stabilizer implemented in Simulink®.

2) Build the transfer functions to obtain the integral-of-accelerating power signal starting from $\Delta\omega$ and ΔP_e. Call this system $G_{acc}(s)$.
3) Perform a root-locus analysis of $-G_{acc}(s)G_{21}(s)$ and determine the angle of departure from the local mode.
4) Let $G_{PSS}(s)$ be the lead-lag portion of the PSS. Perform a root-locus analysis of $-G_{PSS}(s)G_{acc}(s)G_{21}(s)$.

Solutions: Part (1) is computed using PST and Part (2) is developed using Simulink, which is shown in Figure 10.24. The uncompensated root-locus plot is shown in Figure 10.25. The angle of departure of the root-locus branch from the local mode is slightly less than 90°. Incorporating the lead-lag compensator, the compensated root-locus plot is shown in Figure 10.26. Note that the root-locus branches leaving the local mode mostly go toward to the left in the direction of improved damping. With

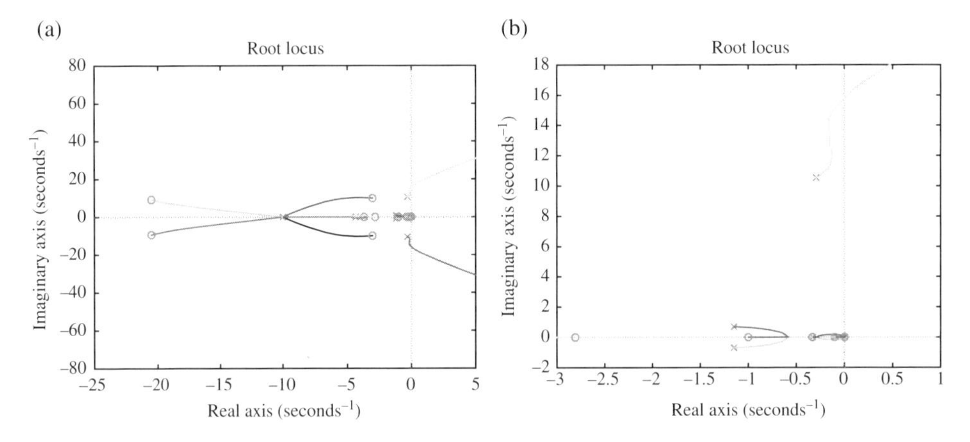

Figure 10.25 (a) Uncompensated root-locus plot of Example 10.4 and (b) zoomed-in plot (see color plate section).

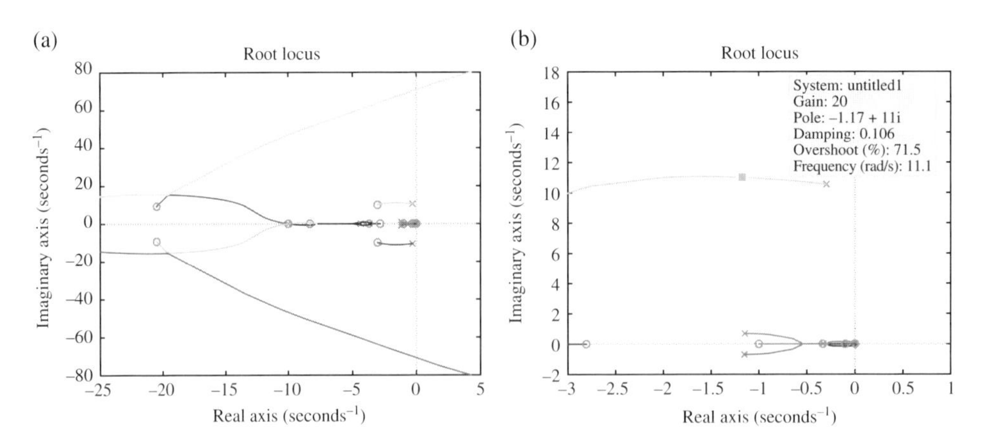

Figure 10.26 (a) Compensated root-locus plot of Example 10.4 and (b) zoomed-in plot (see color plate section).

the gain of $K_{s1} = 20$, the location of the local mode becomes $-1.17 \pm j11.01$, with a damping ratio of 0.106. Thus the damping improvement using the PSS2A is comparable to the speed-input PSS in Example 10.3. However, the torsional filter is not included in Example 10.3. With low-pass filters reducing the torsional mode in the machine speed signal, the PSS2A does not need any torsional filters. ■

10.7 Summary and Notes

This chapter starts by describing the Heffron-Phillips model to introduce the synchronizing and damping torques acting on a synchronous machine. This model can be used to explain the destabilizing effect of high-gain excitation systems. It also establishes the basis for designing a power system stabilizer to provide additional damping torque to improve the damping on the swing mode using phase-lead compensation.

The root-locus method and a phase-compensation method for PSS design are discussed. The root-locus method is illustrated with an example. State-space methods such as pole-placement and H_2 and H_∞ control have also been proposed for PSS design [21]. These space-state methods develop full-state feedback controllers, which would be difficult to implement directly on real systems. Controller model reduction would be needed, which may not be a trivial task [112].

Note that the root-locus methods are also used in later chapters for damping design of other control equipment, such as flexible AC transmission systems (FACTS).

Problems

10.1 Derive the expressions for the coefficients K_3 to K_6 in (10.13)–(10.16).

10.2 Derive K_1 to K_6 for the 2-state flux decay model in Example 10.2.

10.3 Compute and plot the frequency response of the lead-lag compensator

$$G_{\text{ldlg}}(s) = \frac{(1 + 0.18s)(1 + 0.2s)}{(1 + 0.035s)(1 + 0.04s)} \tag{10.80}$$

10.4 Generate the phase characteristics of the uncompensated $G_{\text{uncomp}}(s)$ and the compensated $G_{\text{comp}}(s)$ (similar to Figure 10.9) for the design in Example 10.3. To obtain GEP(s) of the system, which has a 2-axis synchronous generator model, the expression (10.51) for a flux-decay model is no longer applicable. Thus the procedure here is to first generate a state-space model with ΔV_S as the input and ΔP_e as the output. Then remove the states $\Delta\delta$ and $\Delta\omega$ from the model by deleting the first two rows of the A and B matrices and the first two columns of the A and C matrices. What remains is GEP(s).

10.5 Repeat the speed-input PSS design in Example 10.3, but this time include the effect of the torsional filter modeled as

$$G_{\text{tor}}(s) = \frac{1}{0.0027s^2 + 0.0762s + 1} \tag{10.81}$$

The design can be done using Simulink® after generating a linear model of the power system. In addition to the locus-locus analysis, perform a simulation of the compensated system by increasing V_{ref} by 0.1 pu.

10.6 Repeat Example 10.4. In addition to the root-locus analysis, perform a simulation of the compensated system by increasing V_{ref} by 0.1 pu, using the same PSS2A settings as given in Example 10.4.

11

Load and Induction Motor Models

11.1 Introduction

In power flow formulation, loads are assumed to be fixed and independent of the bus voltage. In practice, loads come in many forms and can be a complex function of bus voltages and frequencies when a system is subject to a disturbance. Incandescent and fluorescent lights and electronic equipment, whose power consumption is mostly a function of the voltage, are known as static loads, modeled with algebraic expressions. Induction motors, used in air conditioning, fans, and factory machinery, have a rotating inertia and are considered as dynamic loads modeled with differential equations.

The complexity of load models is further compounded by the fact that most loads are small. A 10 MW load fed by a distribution substation may consist of thousands of lights and induction motors, ranging from 10 W to several kW. Such diversity in loads and load composition would make accurate modeling of the loads supplied by a feeder a complicated if not impossible task. In addition, induction motors can have different slip-torque characteristics depending on their applications [120].

In a power system with central generating stations providing tens of thousands of MW, it is impractical to model individual load components. The industry practice is to lump loads of the same kind into aggregate models of different load types, which is known as the composite load modeling approach. Figure 11.1 shows a simplified diagram of a composite load model at a lower voltage bus [121]. A composite load model allows a user to enter four different types of induction motors, and other loads such as static loads and electronic loads. This model can also include small generating units such as diesel generators, rooftop solar photo-voltaic (PV) systems, and combined heat and power (CHP) systems, which are generally known as distributed energy resources (DER). DERs will not be considered in this text, and a reader is referred to textbooks such as [122].

Various load model data have been reported in [123–126]. To facilitate data gathering, load models have been classified into categories such as residential, commercial, industrial, and agricultural [121]. Each load category has a distinct load composition. For example, residential loads consist of a mix of air conditioners, refrigerators, heat pumps, pool pumps, water heaters, air compressors, electronics (constant power load), and incandescent and LED lighting ([121], Table 3.1). Industrial loads consist of mostly induction motors of various types. The actual composition is a function of the geographical location of the residential load or the machinery used in industrial production as well as the time of day. If a feeder supplies a mix of residential and industrial loads,

Power System Modeling, Computation, and Control, First Edition. Joe H. Chow and Juan J. Sanchez-Gasca.

Companion website: www.wiley.com/go/chow/power-system-modeling

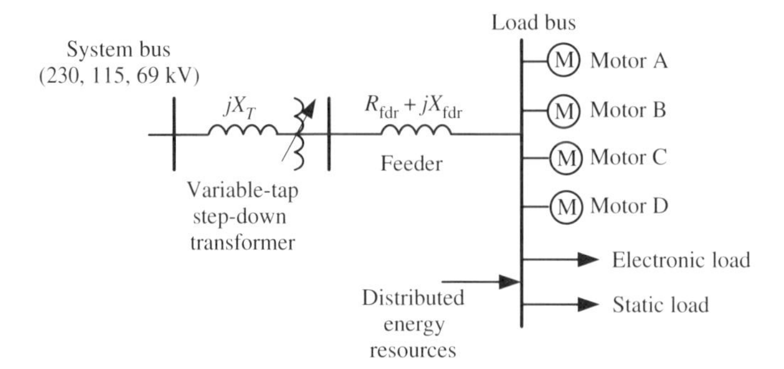

Figure 11.1 WECC composite load model [121].

then their components are aggregated to form the composite model in Figure 11.1; for example, motor A load is the sum of the individual Type A motors in the residential and industrial loads.

The main objective of this chapter is to discuss the impact of these static and dynamic loads on power system disturbance response and stability. Up to this point, a constant impedance load model has been used in disturbance simulation and stability analysis of power systems. Using different load models will affect the analysis results. Besides developing mathematical models for the various load models, the focus is to show the impact of load models on system voltage and reactive power demand, given the phenomenon of fault-induced delayed voltage recovery (FIDVR) [127, 128] in load centers with high air-conditioning loads.

This chapter starts with a description of static load models, and continues with a discussion on incorporating such loads into the power system network solution process. Here the network solution is no longer linear, such that it has to be solved iteratively. Then the chapter develops dynamic models of inductor motor models and discusses the motor initialization and simulation process.

11.2 Static Load Models

Two types of static load models are considered: the exponential load model and the polynomial load model, also known as the ZIP model. They are nonlinear models of the bus voltage magnitude, which are also known as non-conforming loads.

11.2.1 Exponential Load Model

Consider a load $P + jQ$, where P is the active power load and Q the reactive power load. The measured normalized load sensitivity of the active power load P with respect to the bus voltage can be rewritten as

$$\frac{\Delta P\,(\%)}{\Delta V\,(\%)} = \frac{\Delta P/P_o}{\Delta V/V_o} = \frac{V_o}{P_o}\frac{\Delta P}{\Delta V} \tag{11.1}$$

where P_o and V_o are the initial active power load and voltage values, respectively. Similarly, the normalized load sensitivity of the reactive power load Q with respect to the voltage is

Table 11.1 Typical exponential load parameters [26].

Load type	α	β
Incandescent lamps	1.54	0
Room air conditioner	0.50	2.5
Furnace fan	0.08	1.6
Battery charger	2.59	4.06
Conventional fluorescent	2.07	3.21

$$\frac{\Delta Q\,(\%)}{\Delta V\,(\%)} = \frac{V_o}{Q_o}\frac{\Delta Q}{\Delta V} \tag{11.2}$$

where Q_o is the initial reactive power load.

A load model that can be constructed based on the measured sensitivities $\partial P/\partial V$ and $\partial Q/\partial V$ is the exponential model

$$P = P_o \overline{V}^{\alpha}, \quad Q = Q_o \overline{V}^{\beta} \tag{11.3}$$

where the bus voltage magnitude V is normalized to the initial voltage V_o as $\overline{V} = V/V_o$, and α and β are load parameters. Note that the analytical normalized sensitivities of the active power load P at V_o in (11.1) is approximated by

$$\frac{V_o}{P_o}\frac{\partial P}{\partial V} = \frac{V_o}{P_o}\frac{\partial}{\partial V}\left(P_o\frac{V^{\alpha}}{V_o^{\alpha}}\right) = \frac{V_o}{P_o}\alpha P_o\frac{V^{\alpha-1}}{V_o^{\alpha}} = \alpha\left(\frac{V}{V_o}\right)^{\alpha-1} \tag{11.4}$$

which is equal to α at the nominal voltage $V = V_o$. Similarly, based on (11.2), the analytical normalized sensitivities of the reactive power load Q at V_o in (11.2) is approximated by

$$\frac{V_o}{Q_o}\frac{\partial Q}{\partial V} = \beta\left(\frac{V}{V_o}\right)^{\beta-1} \tag{11.5}$$

which is equal to β at the nominal voltage $V = V_o$.

Some typical values of α and β for various loads are given in Table 11.1.

11.2.2 Polynomial Load Model

A more common model of a static load that also includes the impact of frequency deviation Δf is the polynomial load model in the form

$$P = P_o\left(\overline{\alpha}_1\overline{V}^2 + \overline{\alpha}_2\overline{V} + \overline{\alpha}_3\right)\left(1 + K_{pf}\Delta f\right), \quad \overline{\alpha}_1 + \overline{\alpha}_2 + \overline{\alpha}_3 = 1 \tag{11.6}$$

$$Q = Q_o\left(\overline{\beta}_1\overline{V}^2 + \overline{\beta}_2\overline{V} + \overline{\beta}_3\right)\left(1 + K_{qf}\Delta f\right), \quad \overline{\beta}_1 + \overline{\beta}_2 + \overline{\beta}_3 = 1 \tag{11.7}$$

where $\overline{\alpha}_i$ and $\overline{\beta}_i$, $i = 1, 2, 3$, are the load coefficients, and

$$K_{pf} = \frac{\Delta P\,(\%)}{\Delta f\,(\%)}, \quad K_{qf} = \frac{\Delta Q\,(\%)}{\Delta f\,(\%)} \tag{11.8}$$

are the frequency-dependent coefficients. This model readily incorporates three load types:

1) A portion of the load $\overline{\alpha}_1 P_o + j\overline{\beta}_1 Q_o$ is proportional to the square of the bus voltage magnitude, which is called the constant-impedance (Z) load.
2) A portion of the load $\overline{\alpha}_2 P_o + j\overline{\beta}_2 Q_o$ is proportional to the bus voltage magnitude, which is called the constant-current (I) load.
3) A portion of the load $\overline{\alpha}_3 P_o + j\overline{\beta}_3 Q_o$ is independent of the bus voltage magnitude, which is called the constant-power (P) load.

As a result, this model is known as the ZIP load model, which is also the so-called non-conforming load model, in contrast to just a constant impedance load, which can be readily absorbed into the admittance matrix when solving the network equation during a dynamic simulation.

Note that frequency dependency has been added to the models (11.6) and (11.7). The same frequency variation can also be added to the exponential load model (11.3). The exponential and polynomial load models are interchangeable in some cases. For example, a constant current load can be represented as an exponential load model with $\alpha = 1$.

In dynamic simulation for stability analysis, it is a common industry practice to use constant current for active power loads ($\overline{\alpha}_2 = 1$, $\overline{\alpha}_1 = \overline{\alpha}_3 = 0$ in the ZIP model) and constant impedance for reactive power loads ($\overline{\beta}_1 = 1$, $\overline{\beta}_2 = \overline{\beta}_3 = 0$), which was the load model recommended in [130, p. 15]. This selection is within the range of observed α and β from Table 11.1. This load proportion can be applied to some or all of the loads, by specifying $\overline{\alpha}_i$ and $\overline{\beta}_i$, for $i = 2, 3$ (the constant-impedance part can be automatically computed). The proportions of ZIP load components used to model loads vary depending on the geographic region [131].

When a particular load is assigned as a constant-power or constant-current load, either for the active power or the reactive power or both, the network solution during the fault-on period may not have a feasible solution, if the bus is close to the fault. The explanation for this phenomenon could be that at very low voltages it is not feasible to deliver the required amount of power to a particular load. In practice, a constant power or current assumption is valid at normal voltage levels, but at low voltages the load may no longer demand such power consumption. In some simulation programs, a constant power or constant current load at a bus is converted to a constant impedance load when the bus voltage drops below a certain threshold, say, 0.5 pu. This temporary modification will most likely yield a feasible network flow solution and allow the simulation to continue.

11.3 Incorporating ZIP Load Models in Dynamic Simulation and Linear Analysis

This section discusses the incorporation of non-conforming load models (11.6) and (11.7) into the network solution computation. Recall that the network solution extended to the generator internal nodes is represented by the complex linear equation (8.87), repeated below as

$$\begin{bmatrix} Y_d & -Y_d & 0 \\ -Y_d & Y'_{11} & Y_{12} \\ 0 & Y_{21} & Y'_{22} \end{bmatrix} \begin{bmatrix} \tilde{E}' \\ \tilde{V}_G \\ \tilde{V}_L \end{bmatrix} = \begin{bmatrix} \tilde{I}_G \\ 0 \\ 0 \end{bmatrix} \tag{11.9}$$

where Y_{11}, Y_{12}, Y_{21}, and Y_{22} are admittances connecting the generator buses, denoted by the subscript G, and the load buses, denoted by the subscript L; $Y'_{11} = Y_{11} + Y_d$, where Y_d is the generator transient or subtransient admittances; and $Y'_{22} = Y_{22} + Y_L$, where Y_L is the admittances of all the loads represented by the constant impedance model. This network solution does not require any iterations. With the ZIP model, the constant Z component of the loads will be converted to Y_L and added to Y_{22}. The constant I and P components are now modeled as current injections and placed on the right-hand side of the equation. As the current injections will depend on the voltage solution, this new network solution will require an iterative process. For faster convergence, the current injection sensitivities can be added to the admittance matrix. These steps are described below. To be consistent with the earlier derivation, the subscript L is omitted from V_L.

Rewrite the non-conforming load models (11.6) and (11.7) as

$$P = P_o \left(\alpha_1 V^2 + \alpha_2 V + \alpha_3\right) \tag{11.10}$$

$$Q = Q_o \left(\beta_1 V^2 + \beta_2 V + \beta_3\right) \tag{11.11}$$

with the scalings

$$\alpha_1 = \overline{\alpha}_1 / V_o^2, \quad \alpha_2 = \overline{\alpha}_2 / V_o, \quad \alpha_3 = \overline{\alpha}_3 \tag{11.12}$$

$$\beta_1 = \overline{\beta}_1 / V_o^2, \quad \beta_2 = \overline{\beta}_2 / V_o, \quad \beta_3 = \overline{\beta}_3 \tag{11.13}$$

Note that the frequency dependence in (11.6) and (11.7) has been omitted for the time being. Their incorporation into the process is discussed at the end of this section.

The current injections can be obtained by dividing the P, Q load equations by the load bus voltage magnitude as

$$I_{LP}(V) = \frac{P}{V} = \alpha_1 P_o V + \alpha_2 P_o + \alpha_3 P_o \frac{1}{V} = G_o V + I_{LPo} + \alpha_3 P_o \frac{1}{V} \tag{11.14}$$

$$I_{LQ}(V) = \frac{Q}{V} = \beta_1 Q_o V + \beta_2 Q_o + \beta_3 Q_o \frac{1}{V} = B_o V + I_{LQo} + \beta_3 Q_o \frac{1}{V} \tag{11.15}$$

where $G_o = \alpha_1 P_o$, $B_o = \beta_1 Q_o$, $I_{LPo} = \alpha_2 P_o$, and $I_{LQo} = \beta_2 Q_o$. Note that the constant current portion $\left(I_{LPo} + jI_{LQo}\right)$ will remain fixed at the initial value during a dynamic simulation.

Equations (11.14) and (11.15) are developed assuming that the phase θ of the bus voltage is zero. Thus a phase rotation is needed so that the current injections are in the proper reference frame. Let the bus voltage angle be θ relative to the system reference frame. Then the real and imaginary parts of the current injections $\tilde{I} = I_{\text{re}}\left(\tilde{V}\right) + jI_{\text{im}}\left(\tilde{V}\right)$ in the system reference frame are, respectively,

$$\begin{aligned} I_{\text{re}}\left(\tilde{V}\right) &= I_{LP}(V)\cos\theta + I_{LQ}(V)\sin\theta \\ &= \left(G_o V + I_{LPo} + \alpha_3 P_o \frac{1}{V}\right)\frac{V\cos\theta}{V} + \left(B_o V + I_{LQo} + \beta_3 Q_o \frac{1}{V}\right)\frac{V\sin\theta}{V} \\ &= G_o V_{\text{re}} + B_o V_{\text{im}} + I'_{\text{re}}\left(\tilde{V}\right) \end{aligned} \tag{11.16}$$

where

$$I'_{\rm re}\left(\tilde{V}\right) = I_{LPo}\frac{V_{\rm re}}{V} + I_{LQo}\frac{V_{\rm im}}{V} + \alpha_3 P_o \frac{V_{\rm re}}{V^2} + \beta_3 Q_o \frac{V_{\rm im}}{V^2} \tag{11.17}$$

and

$$\begin{aligned} I_{\rm im}\left(\tilde{V}\right) &= I_{LP}(V)\sin\theta - I_{LQ}(V)\cos\theta \\ &= \left(G_o V + I_{LPo} + \alpha_3 P_o \frac{1}{V}\right)\frac{V\sin\theta}{V} - \left(B_o V + I_{LQo} + \beta_3 Q_o \frac{1}{V}\right)\frac{V\cos\theta}{V} \\ &= -B_o V_{\rm re} + G_o V_{\rm im} + I'_{\rm im}\left(\tilde{V}\right) \end{aligned} \tag{11.18}$$

where

$$I'_{\rm im}\left(\tilde{V}\right) = -I_{LQo}\frac{V_{\rm re}}{V} + I_{LPo}\frac{V_{\rm im}}{V} - \beta_3 Q_o \frac{V_{\rm re}}{V^2} + \alpha_3 P_o \frac{V_{\rm im}}{V^2} \tag{11.19}$$

$V^2 = V_{\rm re}^2 + V_{\rm im}^2$, $V_{\rm re} = V\cos\theta$, and $V_{\rm im} = V\sin\theta$.

From these expressions, the constant impedance part of the non-conforming load $G_o + jB_o$ can be added to form the Y'_{22} matrix in (11.9) and $I'_{\rm re}\left(\tilde{V}\right) + jI'_{\rm im}\left(\tilde{V}\right)$ can be added to the right-hand side so that the the third part of expression (11.9) becomes

$$Y_{21}\overline{V}_G + \left(Y'_{22} - \text{diag}\left(G_o + jB_o\right)\right)\overline{V} = I'_{\rm re}\left(\tilde{V}\right) + jI'_{\rm im}\left(\tilde{V}\right) \tag{11.20}$$

where $V_{\rm re} + jV_{\rm im}$ is part of the load bus voltage phasor vector $\overline{V}$ and the notation "diag(·)" means an appropriately sized diagonal matrix with the entries from its argument. There are no changes to the admittances and injections at constant impedance load buses in (11.20).

Note that the convention used in the network equation (11.9) is that current injections flow into the network from the buses. For loads, by taking $P_o < 0$ (load consuming active power), the terms in the active power load will be negative. A negative G_o will become a positive shunt conductance in the Y'_{22} matrix, which is consistent with constant impedance loads. The terms in the reactive power loads will depend on whether the load has a leading power factor (supplying var), resulting in $B_o > 0$, or a lagging power factor (consuming var), resulting in $B_o < 0$.

As $\overline{V}_G$ is known (and fixed at each time step of time simulation), the solution of (11.20) is obtained by using the new value of $\tilde{V}$ to calculate $I'_{\rm re}\left(\tilde{V}\right) + jI'_{\rm im}\left(\tilde{V}\right)$ and then solve for a new value of $\tilde{V}$. This process is repeated until the solution $I'_{\rm re}\left(\tilde{V}\right) + jI'_{\rm im}\left(\tilde{V}\right)$ converges to a value within a tolerance. This is akin to a Gauss-type of iterative methods. For faster convergence, a Newton-like method can be developed by incorporating the sensitivities of the current injections $I'_{\rm re}\left(\tilde{V}\right) + jI'_{\rm im}\left(\tilde{V}\right)$ with respect to the voltage in the admittance matrix.

The idea is to expand $I'_{\rm re}\left(\tilde{V}\right) + jI'_{\rm im}\left(\tilde{V}\right)$ in terms of $\Delta V_{\rm re}$ and $\Delta V_{\rm im}$ as

$$I'_{\rm re}\left(\tilde{V}\right) \simeq I'_{\rm reo}\left(\tilde{V}\right) + \frac{\partial I'_{\rm re}}{\partial V_{\rm re}}\Delta V_{\rm re} + \frac{\partial I'_{\rm re}}{\partial V_{\rm im}}\Delta V_{\rm im} \tag{11.21}$$

$$I'_{\rm im}\left(\tilde{V}\right) \simeq I'_{\rm imo}\left(\tilde{V}\right) + \frac{\partial I'_{\rm im}}{\partial V_{\rm re}}\Delta V_{\rm re} + \frac{\partial I'_{\rm im}}{\partial V_{\rm im}}\Delta V_{\rm im} \tag{11.22}$$

where $I'_{\rm reo}\left(\tilde{V}\right)$ and $I'_{\rm imo}\left(\tilde{V}\right)$ are the values of $I'_{\rm re}\left(\tilde{V}\right)$ and $I'_{\rm im}\left(\tilde{V}\right)$ at the current iteration.

After some manipulations, the sensitivities can be expressed as

$$\frac{\partial I'_{\rm re}}{\partial V_{\rm re}} = I_{\rm LPo}\frac{V_{\rm im}^2}{V^3} - I_{\rm LQo}\frac{V_{\rm re}V_{\rm im}}{V^3} - \alpha_3 P_o \frac{V_{\rm re}^2 - V_{\rm im}^2}{V^4} - 2\beta_3 Q_o \frac{V_{\rm re}V_{\rm im}}{V^4} \tag{11.23}$$

$$\frac{\partial I'_{\rm re}}{\partial V_{\rm im}} = -I_{{\rm LP}o}\frac{V_{\rm re}V_{\rm im}}{V^3} + I_{{\rm LQ}o}\frac{V^2_{\rm re}}{V^3} - 2\alpha_3 P_o \frac{V_{\rm re}V_{\rm im}}{V^4} + \beta_3 Q_o \frac{V^2_{\rm re} - V^2_{\rm im}}{V^4} \tag{11.24}$$

$$\frac{\partial I'_{\rm im}}{\partial V_{\rm re}} = -I_{{\rm LP}o}\frac{V_{\rm re}V_{\rm im}}{V^3} - I_{{\rm LQ}o}\frac{V^2_{\rm im}}{V^3} - 2\alpha_3 P_o \frac{V_{\rm re}V_{\rm im}}{V^4} + \beta_3 Q_o \frac{V^2_{\rm re} - V^2_{\rm im}}{V^4} \tag{11.25}$$

$$\frac{\partial I'_{\rm im}}{\partial V_{\rm im}} = I_{{\rm LP}o}\frac{V^2_{\rm re}}{V^3} + I_{{\rm LQ}o}\frac{V_{\rm re}V_{\rm im}}{V^3} + \alpha_3 P_o \frac{V^2_{\rm re} - V^2_{\rm im}}{V^4} + 2\beta_3 Q_o \frac{V_{\rm re}V_{\rm im}}{V^4} \tag{11.26}$$

With these sensitivity coefficients, the Newton iteration to solve (11.20) is, using the real form of the network admittance matrix,

$$\begin{bmatrix} {\rm Re}\left(Y'_{22}\right) + {\rm diag}\left(G_o - \frac{\partial I'_{\rm re}}{\partial V_{\rm re}}\right) & -{\rm Im}\left(Y'_{22}\right) - {\rm diag}\left(B_o + \frac{\partial I'_{\rm re}}{\partial V_{\rm im}}\right) \\ {\rm Im}\left(Y'_{22}\right) + {\rm diag}\left(B_o - \frac{\partial I'_{\rm im}}{\partial V_{\rm re}}\right) & {\rm Re}\left(Y'_{22}\right) + {\rm diag}\left(G_o - \frac{\partial I'_{\rm im}}{\partial V_{\rm im}}\right) \end{bmatrix} \begin{bmatrix} \Delta V_{\rm re} \\ \Delta V_{\rm im} \end{bmatrix} = \begin{bmatrix} {\rm Re}\left(I'_{\rm mis}\left(\tilde{V}\right)\right) \\ {\rm Im}\left(I'_{\rm mis}\left(\tilde{V}\right)\right) \end{bmatrix} \tag{11.27}$$

where the current mismatch is

$$\tilde{I}'_{\rm mis}\left(\tilde{V}\right) = I'_{\rm re}\left(\tilde{V}\right) + jI'_{\rm im}\left(\tilde{V}\right) - \left(Y_{21}\overline{V}_G + \left(Y'_{22} - {\rm diag}\left(G_o + jB_o\right)\right)\overline{V}\right) \tag{11.28}$$

In this method, $I'_{\rm re}\left(\tilde{V}\right)$, $V'_{\rm im}\left(\tilde{V}\right)$, and the entries (11.23)–(11.26) are updated at each iteration when the new values of

$$V^{\rm new}_{\rm re} = V^{\rm old}_{\rm re} + \Delta V_{\rm re}, \quad V^{\rm new}_{\rm im} = V^{\rm old}_{\rm im} + \Delta V_{\rm im} \tag{11.29}$$

are computed, where the superscript "new" denotes the latest iteration and the superscript "old" denotes the previous iteration. The solution converges when the error

$$\varepsilon = \sqrt{\Delta V^2_{\rm re} + \Delta V^2_{\rm im}} \tag{11.30}$$

is less than a pre-specified tolerance. Otherwise, use $V^{\rm new}_{\rm re}$ and $V^{\rm new}_{\rm im}$ to repeat the iteration.

The frequency-dependent ZIP loads can be modeled as

$$P = P_o\left(\alpha_1 V^2 + \alpha_2 V + \alpha_3\right)\left(1 + K_{pf}\Delta f\right) \tag{11.31}$$

$$Q = Q_o\left(\beta_1 V^2 + \beta_2 V + \beta_3\right)\left(1 + K_{qf}\Delta f\right) \tag{11.32}$$

where Δf is the frequency deviation in pu from the nominal frequency. Once Δf is known at each time step, (11.31) and (11.32) can readily be incorporated as non-conforming loads. This model requires frequency computation at each load bus, but bus frequency is not normally computed in a positive-sequence dynamic simulation program. One can apply a rate (high-pass) filter to the bus voltage angle. Instead, if it is desired to approximate the bus frequency deviation at every bus by the overall system frequency deviation, the computation would be much simpler, using the expression

$$\Delta f = \Delta\omega = \frac{1}{H_{\rm Total}}\sum_{i=1}^{N_G} H_i \Delta\omega_i, \quad H_{\rm Total} = \sum_{i=1}^{N_G} H_i \tag{11.33}$$

where H_i and $\Delta\omega_i$ are the inertia and speed for machine i, respectively, and N_G is the number of machines.

The next example illustrates the impact of load modeling on system stability.

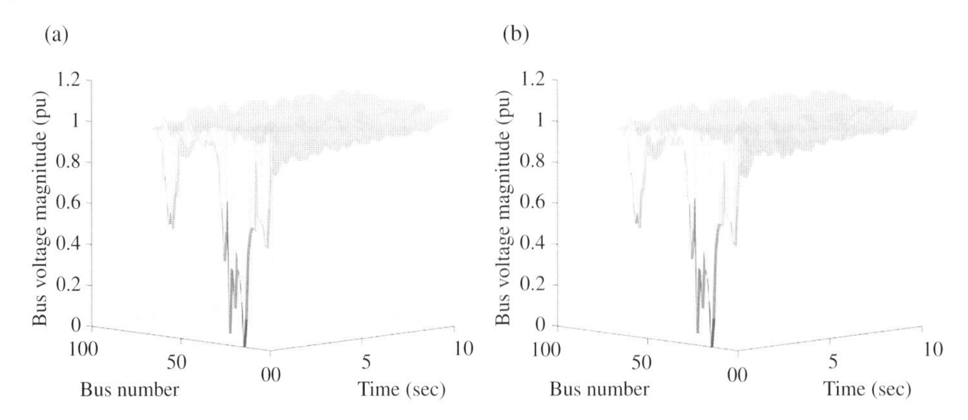

Figure 11.2 Eight-cycle fault voltage magnitude response for (a) constant-Z active power load model and (b) constant-I active power load model (see color plate section).

Example 11.1: Impact of static load models on transient stability

For the NPCC 16-machine system (with the electromechanical model used for all the generators) in Example 4.7, a 3-phase short-circuit fault is applied near Bus 16 for 8 cycles and for 9 cycles, which is cleared by opening the line connecting Buses 16 and 15. Find the time response for (a) constant-impedance load models and (b) constant-current for active power loads and constant-impedance for reactive power loads. The generator damping term D is set to half of the inertia $H/2$, for all machines.

Solutions: For an 8-cycle fault, both types of load models result in a stable post-fault system. The voltage magnitude profiles of all the buses are shown in Figure 11.2. Note that at the time of fault clearing, Bus 16 voltage $V_{16} = 1.0257$ pu and Bus 15 voltage $V_{15} = 0.9285$ pu for the constant-Z model. However, for the constant-I model, $V_{16} = 1.0190$ pu and $V_{15} = 0.9233$ pu. In general, the constant-I model is more "demanding" on the system, resulting in somewhat lower voltages over the whole system.

For a 9-cycle fault, the constant-Z load model still yields a stable post-fault response, as shown in Figure 11.3.a. The network solution of the constant-I active power load model simulation fails to converge when the fault is cleared, as shown in Figure 11.3.b. Evidently there is not enough reactive power support to achieve a feasible power flow solution. Thus changing the load models at low-voltage buses to constant-Z would be helpful to achieve a feasible solution. ■

In linear analysis, ZIP load models will impact the swing modes, as illustrated by the following example.

Example 11.2: Impact of static load models on electromechanical mode damping

For the NPCC 16-machine system in Example 11.1, compute the damping of the eigenvalues of the steady-state pre-fault linearized model for (a) constant-impedance load model and (b) constant-current model for active power loads and constant-impedance model for reactive power loads.

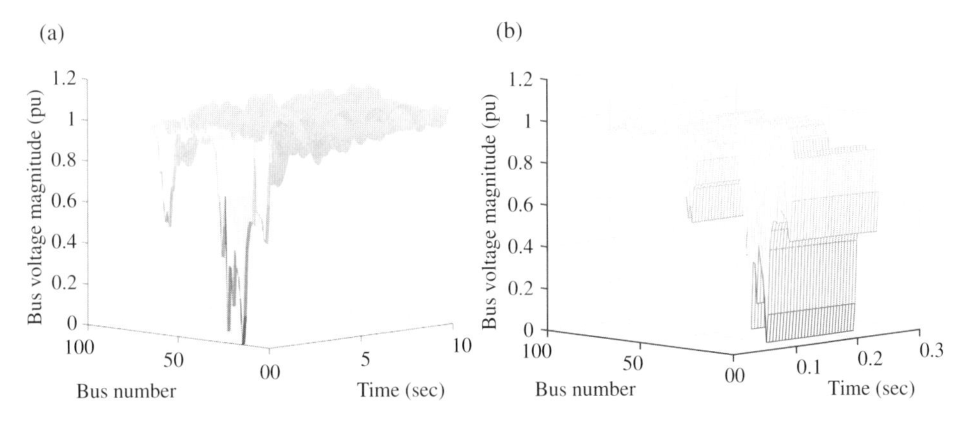

Figure 11.3 Nine-cycle fault voltage magnitude response for (a) constant-Z active power load model, and (b) constant-I active power load model (see color plate section).

Solutions: First note that with all the damping D set to half of the inertia $H/2$, the damping of all the swing modes will be equal. In fact, the real parts of all the swing modes of the linear model for both Part (a) and Part (b) are all equal to -0.1250. However, the imaginary parts (oscillatory frequency) are different because the ZIP load models will affect the synchronizing torque. For example, the mode $-0.1250 \pm j2.4505$ for Part (a) becomes the lower frequency mode $-0.1250 \pm j2.4118$ in Part (b). On the other hand, the mode $-0.1250 \pm j3.1449$ for Part (a) becomes a higher frequency mode $-0.1250 \pm j3.2352$ for Part (b). There are no general guidelines as to how the frequency will change with respect to the load models. ■

For generators equipped with excitation systems, the choice of ZIP load models will affect the swing mode damping. However, there are also no general rules on whether constant-I or constant-P load models will reduce damping as compared to the constant-Z model [82]. Thus if a constant-Z load model is used for an analysis or design, the constant-I model for active power loads should also be checked to ensure that there are no significant differences in the damping of dominant swing modes.

11.4 Induction Motors: Steady-State Models

Induction motors of all sizes are used in factories and households, and can be a significant part of the total load. As an induction motor is a rotating machine with an inertia and consumes both active and reactive power for its operation, its dynamics need to be captured in stability simulation programs. In this and the next section, it is assumed that the stator is in balanced 3-phase operation at synchronous speed, and generates an electromagnetic flux to induce a torque on the rotor, which runs slightly lower than synchronous speed. The steady-state model discussion in this section starts with the stator and rotor flux linkages, similar to the steady-state model treatment of synchronous machines in Chapter 7, and continues with equivalent circuits for inductor motors. The materials will follow closely the modeling approach in [6].

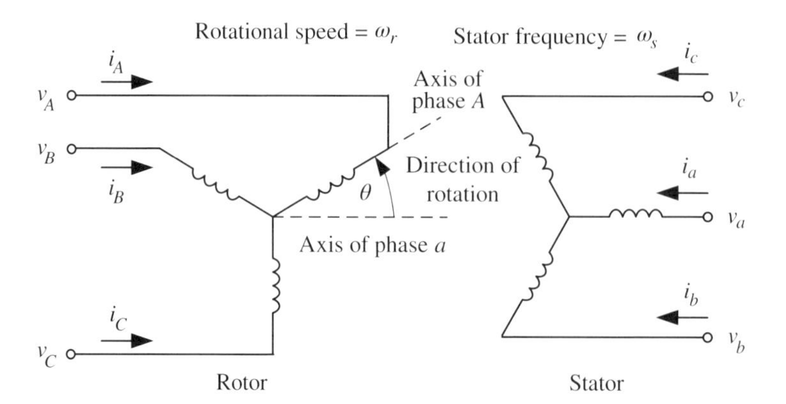

Figure 11.4 Three-phase stator and rotor circuit model of an induction machine.

11.4.1 Physical Description

An induction motor consists of a stator, which is connected to the grid power supply, and a rotor. The 3-phase stator and rotor circuits of an induction motor are shown in Figure 11.4. There are in general two types of rotor winding structures: (1) a wound rotor with three slip rings on the shaft and (2) a squirrel-cage rotor with uninsulated bars and short-circuited by end rings at both ends. It is also possible to connect the rotor winding to an external circuit as a means of control. The rotor windings do not have a voltage source to control its current. Instead, the rotor windings allow the stator currents to induce a torque on the rotor to drive a load, such as a fan or a compressor.

With balanced 3-phase operation at the nominal frequency f_s, the stator windings produce a magnetic field rotating at synchronous speed n_s given by

$$n_s = 2\pi f_s / p_f \text{ rad/sec} = 120\pi f_s / p_f \text{ rad/min} \tag{11.34}$$

where p_f is the number of poles. The frequency f_r of the induced rotor currents depends on the relative speeds of the stator field and the rotor. This relative speed or slip is defined as

$$s = \left(n_s - n_r\right) / n_s \text{ pu} \tag{11.35}$$

where

$$n_r = 2\pi f_r / p_f \text{ rad/sec} \tag{11.36}$$

In an induction motor, the rotor speed n_r is less than the synchronous speed n_s, and thus $s > 0$. Furthermore, the frequency of the induced rotor voltages and currents is sf_s.[1]

11.4.2 Mathematical Description

11.4.2.1 Modeling Equations

Following the synchronous machine modeling approach, denoting the stator phases with a, b, and c, and the rotor phases with A, B, and C, the voltage-flux-linkage equations for the stator and the rotor windings are expressed, respectively, as

1 If the rotor is driven by a prime mover at a speed $n_r > n_s$, then the machine operates as an induction generator (such as a wind turbine).

$$v_a = \frac{d\psi_a}{dt} + R_s i_a, \quad v_b = \frac{d\psi_b}{dt} + R_s i_b, \quad v_c = \frac{d\psi_c}{dt} + R_s i_c \tag{11.37}$$

and

$$v_A = \frac{d\psi_A}{dt} + R_r i_A, \quad v_B = \frac{d\psi_B}{dt} + R_r i_B, \quad v_C = \frac{d\psi_C}{dt} + R_r i_C \tag{11.38}$$

where ψ denotes the flux linkage, R_s is the stator winding resistance, and R_r is the rotor winding resistance.

Because of the symmetry in the rotor circuits in an induction machine, the self and mutual inductances on the rotor and stator are symmetric

$$L_{AA} = L_{BB} = L_{CC}, \quad L_{AB} = L_{BC} = L_{CA} \tag{11.39}$$

$$L_{aa} = L_{bb} = L_{cc}, \quad L_{ab} = L_{bc} = L_{ca} \tag{11.40}$$

The mutual inductances between the rotor and stator circuits are also symmetric, such that when the rotor circuits are aligned with the stator circuits

$$L_{aA} = L_{Aa} = L_{bA} = L_{Bb} = L_{cC} = L_{Cc} \tag{11.41}$$

The actual mutual inductances will depend on the rotor angle θ.

The flux linkage in phase a of the stator is given by

$$\psi_a = L_{aa} i_a + L_{ab}\left(i_b + i_c\right) + L_{aA}\left[i_A \cos\theta + i_B \cos\left(\theta + \frac{2\pi}{3}\right) + i_C \cos\left(\theta - \frac{2\pi}{3}\right)\right] \tag{11.42}$$

where θ is the angle of the phase A rotor winding leading the phase a stator winding (Figure 11.4). Similar expressions can be developed for phases b and c.

The flux linkage in phase A of the rotor is given by

$$\psi_A = L_{AA} i_A + L_{AB}\left(i_B + i_C\right) + L_{aA}\left[i_a \cos\theta + i_b \cos\left(\theta - \frac{2\pi}{3}\right) + i_c \cos\left(\theta + \frac{2\pi}{3}\right)\right] \tag{11.43}$$

Similar expressions can be developed for phases B and C.

In balanced operation and for ungrounded windings, the rotor and stator phase currents sum to zero

$$i_a + i_b + i_c = 0, \quad i_A + i_B + i_C = 0 \tag{11.44}$$

because there are no zero-sequence currents.

By defining the reactances in the stator and rotor windings as

$$L_{ss} = L_{aa} - L_{ab}, \quad L_{rr} = L_{AA} - L_{AB} \tag{11.45}$$

and using (11.44), the flux linkage equations (11.42) and (11.43) can be simplified, respectively, to

$$\psi_a = L_{ss} i_a + L_{aA}\left[i_A \cos\theta + i_B \cos\left(\theta + \frac{2\pi}{3}\right) + i_C \cos\left(\theta - \frac{2\pi}{3}\right)\right] \tag{11.46}$$

$$\psi_A = L_{rr} i_A + L_{aA}\left[i_a \cos\theta + i_b \cos\left(\theta - \frac{2\pi}{3}\right) + i_c \cos\left(\theta + \frac{2\pi}{3}\right)\right] \tag{11.47}$$

11.4.2.2 Reference Frame Transformation

Similar to a synchronous machine model, the 3-phase representation can be simplified to a dq reference frame, in which the q-axis leads the d-axis by 90°. In a synchronous machine, the d-axis rotates with the rotor field circuit. In an induction machine, the d-axis rotates at the synchronous speed ω_s referenced to the stator.

Following (7.42) and (7.46), the transformation of the stator phase currents i_a, i_b, and i_c into the d-axis current i_{ds} and the q-axis current i_{qs} yields

$$i_{ds} = \frac{2}{3}\left[i_a \cos\left(\omega_s t\right) + i_b \cos\left(\omega_s t - \frac{2\pi}{3}\right) + i_c \cos\left(\omega_s t + \frac{2\pi}{3}\right)\right] \tag{11.48}$$

$$i_{qs} = -\frac{2}{3}\left[i_a \sin\left(\omega_s t\right) + i_b \sin\left(\omega_s t - \frac{2\pi}{3}\right) + i_c \sin\left(\omega_s t + \frac{2\pi}{3}\right)\right] \tag{11.49}$$

The inverse transformation from the dq quantities back to the phase quantities consists of

$$i_a = i_{ds} \cos\left(\omega_s t\right) - i_{qs} \sin\left(\omega_s t\right) \tag{11.50}$$

$$i_b = i_{ds} \cos\left(\omega_s t - \frac{2\pi}{3}\right) - i_{qs} \sin\left(\omega_s t - \frac{2\pi}{3}\right) \tag{11.51}$$

$$i_c = i_{ds} \cos\left(\omega_s t + \frac{2\pi}{3}\right) - i_{qs} \sin\left(\omega_s t + \frac{2\pi}{3}\right) \tag{11.52}$$

The rotor speed of an inductor motor is less than the synchronous frequency, and thus the rotor d-axis is not stationary with respect to phase A winding, as is the case for a synchronous generator. Now define θ_r as the angle of the d-axis leading the A-axis of the rotor. Then given a slip of s, θ_r is advancing at the speed of

$$\frac{d\theta_r}{dt} = \omega_r = s\omega_s \tag{11.53}$$

Also, from Figure 11.4

$$\theta = \omega_r t = (1 - s)\,\omega_s t \tag{11.54}$$

such that

$$\theta_r = \omega_s t - \theta \tag{11.55}$$

Then the transformation of the rotor phase currents i_A, i_B, and i_C into the rotor d-axis current i_{dr} and q-axis current i_{qr} is given by

$$i_{dr} = \frac{2}{3}\left[i_A \cos\theta_r + i_B \cos\left(\theta_r - \frac{2\pi}{3}\right) + i_C \cos\left(\theta_r + \frac{2\pi}{3}\right)\right] \tag{11.56}$$

$$i_{qr} = -\frac{2}{3}\left[i_A \sin\theta_r + i_B \sin\left(\theta_r - \frac{2\pi}{3}\right) + i_C \sin\left(\theta_r + \frac{2\pi}{3}\right)\right] \tag{11.57}$$

which are functions of θ_r. The inverse transformation is

$$i_A = i_{dr} \cos\left(\theta_r\right) - i_{qr} \sin\left(\theta_r\right) \tag{11.58}$$

$$i_B = i_{dr} \cos\left(\theta_r - 2\pi/3\right) - i_{qr} \sin\left(\theta_r - \frac{2\pi}{3}\right) \tag{11.59}$$

$$i_C = i_{dr} \cos\left(\theta_r + 2\pi/3\right) - i_{qr} \sin\left(\theta_r + \frac{2\pi}{3}\right) \tag{11.60}$$

The same dq transformation applies to both the stator and rotor flux linkages and voltages, resulting in the stator flux linkage expressions

$$\psi_{ds} = L_{ss} i_{ds} + L_m i_{dr}, \quad \psi_{qs} = L_{ss} i_{qs} + L_m i_{qr} \tag{11.61}$$

the rotor flux linkage expressions

$$\psi_{dr} = L_{rr} i_{dr} + L_m i_{ds}, \quad \psi_{qr} = L_{rr} i_{qr} + L_m i_{qs} \tag{11.62}$$

where $L_m = (2/3) L_{aA}$, the stator voltage expressions

$$v_{ds} = R_s i_{ds} - \omega_s \psi_{qs} + \frac{\psi_{ds}}{dt}, \quad v_{qs} = R_s i_{qs} + \omega_s \psi_{ds} + \frac{\psi_{qs}}{dt} \tag{11.63}$$

and the rotor voltage expressions

$$v_{dr} = R_r i_{dr} - s\omega_s \psi_{qr} + \frac{\psi_{dr}}{dt} = 0, \quad v_{qr} = R_r i_{qr} + s\omega_s \psi_{dr} + \frac{\psi_{qr}}{dt} = 0 \tag{11.64}$$

as there are no voltage sources for the rotor windings. In (11.63) and (11.64), the $\omega_s \psi_{qs}$ and $\omega_s \psi_{ds}$ terms are voltages created in the stator windings by the synchronously rotating flux waves. The $s\omega_s \psi_{qr}$ and $s\omega_s \psi_{dr}$ terms are voltages created in the rotor windings that move at the slip frequency.

The instantaneous power provided to the stator from the grid is

$$P_s = v_a i_a + v_b i_b + v_c i_c = \frac{3}{2} \left(v_{ds} i_{ds} + v_{qs} i_{qs} \right) \tag{11.65}$$

The instantaneous power input delivered to the rotor is

$$P_r = \frac{3}{2} \left(v_{dr} i_{dr} + v_{qr} i_{qr} \right) \tag{11.66}$$

Substituting in the rotor voltage expressions (11.64) with $d\psi_{dr}/dt = 0$ and $d\psi_{qr}/dt = 0$ in steady state, P_r becomes

$$P_r = \frac{3}{2} i_r^2 R_r - s\omega_s T_e \tag{11.67}$$

where

$$i_r^2 = i_{dr}^2 + i_{qr}^2, \quad T_e = \frac{3}{2} \left(\psi_{qr} i_{dr} - \psi_{dr} i_{qr} \right) \tag{11.68}$$

The term T_e (11.68) is the electromagnetic torque.

In case of motor startup or a disturbance, the rotor position is determined from the electro-mechanical equation

$$J \frac{d\omega_r}{dt} = J \frac{d^2 \theta_r}{dt^2} = T_e - T_m \tag{11.69}$$

where T_e (11.68) is the electromagnetic input torque, T_m is the load mechanical torque, and J is the combined inertias of the induction motor rotor and the load. Note that the rotor speed ω_r is in pu based on the synchronous speed.

For a rotating machine, the mechanical load torque demanded by the load is often a function of the rotating speed ω_r. Thus it is common to express T_m as a function of ω_r raised to an exponent m:

$$T_m = T_{mo} \omega_r^m \tag{11.70}$$

A more detailed load model may include several different exponents of ω_r, such as a quadratic function of the rotor speed

$$T_m = T_{mo}\left(\alpha_2\omega_r^2 + \alpha_1\omega_r + \alpha_o\right) \tag{11.71}$$

where α_o, α_1, and α_2 are load coefficients.

11.4.3 Equivalent Circuits

To connect the electrical model of an induction motor to the power system model, the stator and rotor circuits will be coupled via flux linkages to form an equivalent circuit. In addition, the equivalent circuit also defines the parameters required for modeling the inductor motor.

The stator voltage and current when expressed as phasors become, respectively,

$$\tilde{I}_s = I_{ds} + jI_{qs}, \quad I_{ds} = \frac{i_{ds}}{\sqrt{2}}, \quad I_{qs} = \frac{i_{qs}}{\sqrt{2}} \tag{11.72}$$

$$\tilde{V}_s = V_{ds} + jV_{qs}, \quad V_{ds} = \frac{v_{ds}}{\sqrt{2}}, \quad V_{qs} = \frac{v_{qs}}{\sqrt{2}} \tag{11.73}$$

Similarly, the rotor voltage and current when expressed as phasors become, respectively,

$$\tilde{I}_r = I_{dr} + jI_{qr}, \quad I_{dr} = \frac{i_{dr}}{\sqrt{2}}, \quad I_{qr} = \frac{i_{qr}}{\sqrt{2}} \tag{11.74}$$

$$\tilde{V}_r = V_{dr} + jV_{qr}, \quad V_{dr} = \frac{v_{dr}}{\sqrt{2}}, \quad V_{qr} = \frac{v_{qr}}{\sqrt{2}} \tag{11.75}$$

In steady state, with $d\psi_{ds}/dt = d\psi_{qs}/dt = 0$, the dq-axes stator voltages can be expressed as

$$v_{ds} = R_s i_{ds} - \omega_s L_{ss} i_{qs} - \omega_s L_m i_{qr} \tag{11.76}$$

$$v_{qs} = R_s i_{qs} + \omega_s L_{ss} i_{ds} + \omega_s L_m i_{dr} \tag{11.77}$$

Thus the voltage and current relationship in the stator circuit is

$$\begin{aligned}
\tilde{V}_s &= R_s\left(\frac{i_{ds}}{\sqrt{2}} + j\frac{i_{qs}}{\sqrt{2}}\right) + \omega_s L_{ss}\left(j\frac{i_{ds}}{\sqrt{2}} - \frac{i_{qs}}{\sqrt{2}}\right) + \omega_s L_m\left(j\frac{i_{dr}}{\sqrt{2}} - \frac{i_{qr}}{\sqrt{2}}\right) \\
&= R_s\tilde{I}_s + j\omega_s\left(L_{ss}\tilde{I}_s + L_m\tilde{I}_r\right) \\
&= R_s\tilde{I}_s + j\omega_s\left(L_{ss} - L_m\right)\tilde{I}_s + j\omega_s L_m\left(\tilde{I}_s + \tilde{I}_r\right) \\
&= R_s\tilde{I}_s + jX_s\tilde{I}_s + jX_m\left(\tilde{I}_s + \tilde{I}_r\right)
\end{aligned} \tag{11.78}$$

where

$$X_s = \omega_s\left(L_{ss} - L_m\right) \tag{11.79}$$

is the stator leakage reactance and

$$X_m = \omega_s L_m \tag{11.80}$$

is the magnetizing reactance.

Using

$$v_{dr} = 0 = R_r i_{dr} - s\omega_s \left(L_{\mathrm{rr}} i_{qr} + L_m i_{qs} \right) \tag{11.81}$$

$$v_{qr} = 0 = R_r i_{qr} + s\omega_s \left(L_{\mathrm{rr}} i_{dr} + L_m i_{ds} \right) \tag{11.82}$$

the voltage and current relationship in the rotor circuit is

$$\tilde{V}_r = 0 = R_r \tilde{I}_r + js\omega_s \left(L_{\mathrm{rr}} \tilde{I}_r + L_m \tilde{I}_s \right) \tag{11.83}$$

that is,

$$0 = \frac{R_r}{s} \tilde{I}_r + jX_r \tilde{I}_r + jX_m \left(\tilde{I}_s + \tilde{I}_r \right) \tag{11.84}$$

where

$$X_r = \omega_s \left(L_{\mathrm{rr}} - L_m \right) \tag{11.85}$$

is the rotor leakage reactance.

Equations (11.78) and (11.84) can be combined to form an equivalent circuit shown in Figure 11.5. The left-hand side of the circuit is the stator model and the right-hand side is the rotor model. Note that the direction of the current $\tilde{I}_r$ in Figure 11.5 is opposite to the direction used in (11.78) and (11.84). It is common to direct $\tilde{I}_r$ towards the load to denote that active power is delivered from the stator to the rotor [135]. Given the stator voltage $\tilde{V}_s$, the slip s, and the various circuit elements, the currents $\tilde{I}_s$ and $\tilde{I}_r$ can be computed according to the flow directions indicated in Figure 11.5.

It is important to recognize that the inductor motor circuit representation is similar to those of the transformer and synchronous machines in that the stator circuit (primary winding in a transformer) and the rotor circuit (secondary winding in a transformer) are coupled via the *mutual* inductance, shown as a shunt branch, and the series components are the *leakage* reactances in the windings.

The active power transfer across the airgap to the rotor is

$$P_{\mathrm{ag}} = \frac{R_r}{s} I_r^2 \tag{11.86}$$

where I_r is the magnitude of $\tilde{I}_r$. Because the rotor resistive loss is

$$P_{\mathrm{rl}} = R_r I_r^2 \tag{11.87}$$

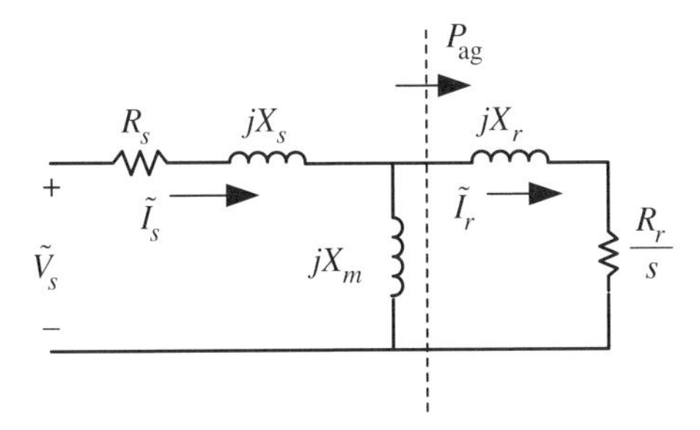

Figure 11.5 Induction motor equivalent circuit.

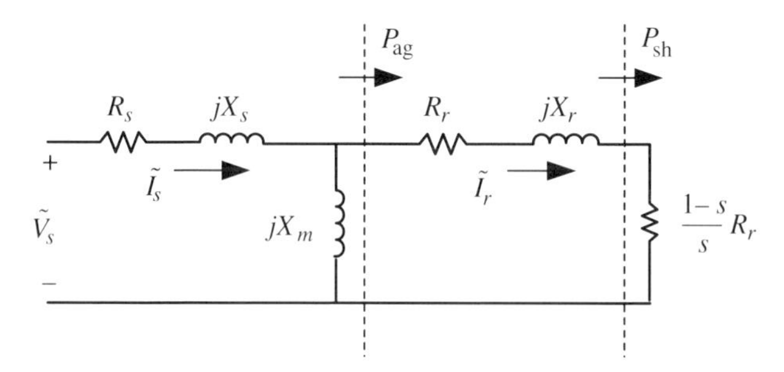

Figure 11.6 Induction motor equivalent circuit showing shaft power.

the mechanical power transferred to the rotor shaft is

$$P_{\text{sh}} = P_{\text{ag}} - P_{\text{rl}} = R_r \frac{1-s}{s} I_r^2 \tag{11.88}$$

This separation of resistive loss and power delivered by the equivalent circuit is shown in Figure 11.6.

For a p_f-pole 3-phase motor, the electromagnetic torque is $T_e = 3P_e/\omega_m$ where

$$\omega_m = \omega_r \frac{2}{p_f} = \omega_s (1-s) \frac{2}{p_f} \tag{11.89}$$

is the angular velocity of the rotor in mechanical rad/sec, such that

$$T_e = 3 \frac{p_f}{2} \frac{R_f}{s\omega_s} I_r^2 \tag{11.90}$$

11.4.4 Per-Unit Representation

To incorporate induction motor models into a dynamic simulation program, the model parameters need to be put into per-unit values. The per-unitization of the parameter values of an inductor motor is similar to the per-unitization of the parameters for synchronous machines and transformers.

The base values for the stator are:

v_{sbase} = peak value of rated phase voltage, V
i_{sbase} = peak value of rated phase current, A
f_{base} = rated frequency, Hz

Other base values can be derived from these definitions as:

$\omega_{\text{sbase}} = 2\pi f_{\text{base}}$, electrical rad/sec
$Z_{\text{sbase}} = v_{\text{sbase}}/i_{\text{sbase}}$, ohm
$L_{\text{sbase}} = Z_{\text{sbase}}/\omega_{\text{base}}$, H
$\psi_{\text{sbase}} = v_{\text{sbase}}/\omega_{\text{base}}$, Wb-Turns
3-ϕ $\text{VA}_{\text{base}} = (3/2)\left(v_{\text{sbase}} i_{\text{sbase}}\right)$, VA
Torque base $= (3/2)\left(p_f/2\right) \psi_{\text{sbase}} i_{\text{sbase}}$, N-m
$v_{\text{sbase}} = Z_{\text{sbase}} i_{\text{sbase}} = \omega_{\text{sbase}} \psi_{\text{sbase}}$

As an illustration, the per unitization of v_{ds} yields

$$\frac{v_{ds}}{v_{\text{sbase}}} = \frac{R_s}{Z_{\text{sbase}}}\frac{i_{ds}}{i_{\text{sbase}}} - \frac{\omega_s}{\omega_{\text{sbase}}}\frac{\psi_{qs}}{\psi_{\text{sbase}}} + \frac{d}{dt}\left(\frac{1}{\omega_{\text{sbase}}}\frac{\psi_{ds}}{\psi_{\text{sbase}}}\right) \tag{11.91}$$

which becomes

$$\overline{v}_{ds} = \overline{R}_s\overline{i}_{ds} - \overline{\omega}_s\overline{\psi}_{qs} + \frac{d\overline{\psi}_{ds}}{dt} \tag{11.92}$$

where the bar denotes a per-unitized parameter or variable.

The rotor base quantities are the same as stator base quantities. Thus all the reactances, resistances, and currents in the circuits in Figures 11.5 and 11.6 are on the same base.

In addition, the electromechanical model for the rotor in per unit is given by

$$\frac{d\overline{\omega}_r}{dt} = \frac{1}{2H\omega_{\text{base}}}\left(\overline{T}_e - \overline{T}_m\right) \tag{11.93}$$

where

$$H = \frac{1}{2}\frac{J\omega_{\text{base}}^2}{\text{VA}_{\text{base}}} \tag{11.94}$$

and J is the combined inertia of the motor and the load.

11.4.5 Torque-Slip Characteristics

To compute T_e given $\tilde{V}_s$ and the slip s, a Thévenin equivalent can be developed for the stator and magnetizing components in Figure 11.5 to obtain the equivalent circuit shown in Figure 11.7, where the Thévenin voltage

$$\tilde{V}_e = \frac{jX_m}{R_s + j\left(X_s + X_m\right)}\tilde{V}_s \tag{11.95}$$

is the voltage across the reactance X_m without the rotor circuit and the Thévenin impedance

$$R_e + jX_e = \frac{jX_m\left(R_s + jX_s\right)}{R_s + j\left(X_s + X_m\right)} \tag{11.96}$$

is the parallel combination of jX_m and $R_s + jX_s$. Note that the rotor $\tilde{I}_r$ is the same in both Figures 11.5 and 11.7, but the stator current $\tilde{I}_s$ is no longer an entity in Figure 11.7. Furthermore, neither $\tilde{V}_e$ nor $R_e + jX_e$ are functions of s. If the stator voltage $\tilde{V}_s$ is fixed assuming that the grid has infinite capacity, then $\tilde{V}_e$ is also a constant value.

The rotor current can be readily computed from the circuit in Figure 11.7 as

$$\tilde{I}_r = \frac{\tilde{V}_e}{\left(R_e + R_s/s\right) + j\left(X_e + X_r\right)} \tag{11.97}$$

The Thévenin impedance looking from the resistance R_r/s into the stator in Figure 11.7 is

$$Z_e' = R_e + j\left(X_e + X_r\right) \tag{11.98}$$

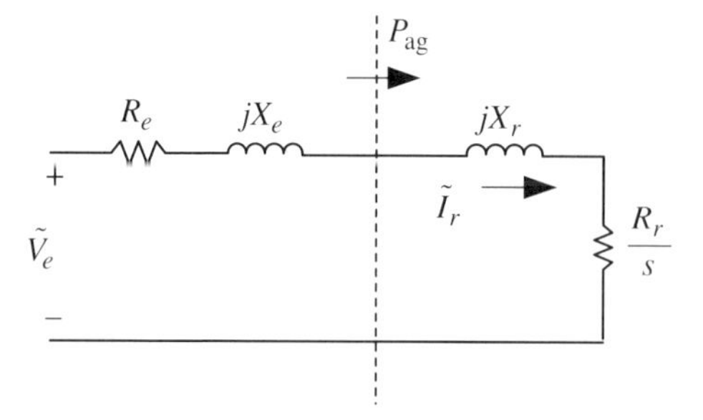

Figure 11.7 Equivalent circuit for torque-slip calculation.

The maximum power delivered to the rotor occurs when the rotor resistance R_r/s is equal to the magnitude of the sum of the Thévenin impedance and the rotor reactance,[2] that is,

$$R_r/s = |Z_e'| = \sqrt{R_e^2 + \left(X_e + X_r\right)^2} \tag{11.99}$$

Thus the maximum torque occurs when the slip is

$$s = \frac{R_r}{|Z_e'|} \tag{11.100}$$

such that

$$T_{emax} = 3\frac{p_f}{2\omega_s}\frac{0.5V_e^2}{R_e + |Z_e'|} = 3\frac{1}{\omega_m}\frac{0.5V_e^2}{R_e + |Z_e'|} \tag{11.101}$$

The variation of T_e versus slip or speed is shown in Figure 11.8 of the following example.

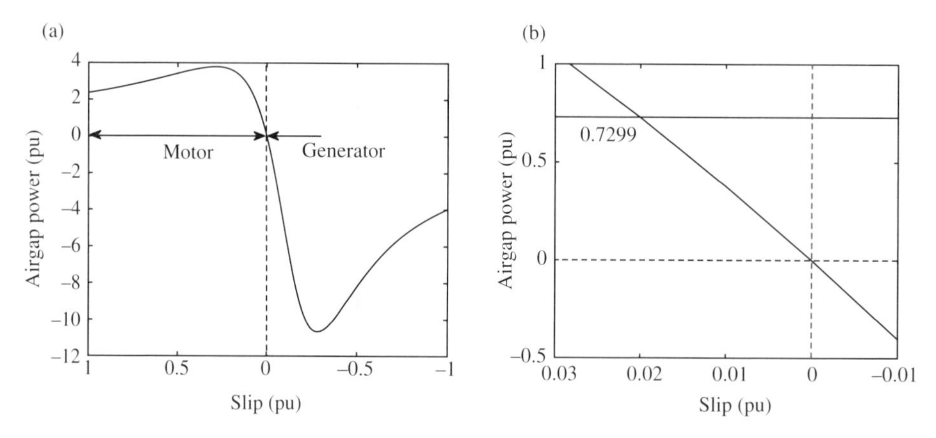

Figure 11.8 (a) Torque-slip characteristic of an induction machine and (b) enlarged plot.

2 Recall the condition for maximum power transfer based on the PV curve in voltage stability analysis (Chapter 3).

Example 11.3: Slip calculation

An induction motor is rated at 137 kVA, 400 V, and 198 A ([6, p. 305]). The equivalent circuit parameters are given in pu as

$$R_s = 0.0425,\ X_s = 0.0435,\ R_r = 0.0249,\ X_r = 0.0329,\ X_m = 2.9745 \tag{11.102}$$

Plot the airgap power (that is, the torque-slip curve) for $s = 1$ pu to $s = -1$ pu. Suppose that 100 kW is delivered to the rotor. Find the value of the slip s at that operating condition.

Solutions: Using (11.95)–(11.97), the plot of the airgap power P_{ag} (11.86) versus the slip s is given in Figure 11.8.a, with an enlarged plot for -0.01 pu $< s < 0.03$ pu shown in Figure 11.8.b.

The amount of airgap power delivered to the load is

$$P_{\text{ag}} = 100\text{ kW} = \frac{100}{137}\text{ pu} = 0.7299\text{ pu} \tag{11.103}$$

From Figure 11.8.b, the slip is estimated to be $s = 0.02$ pu. ■

The torque-slip curve of the induction motor in Figure 11.8 shows that an induction motor typically operates at a slip value of a few percent. The torque at $s = 1$ pu is the startup torque, which can be quite high, and needs to be provided by the power grid. The impact on system voltage during motor startup will be illustrated in Example 11.7.

11.4.6 Reactive Power Consumption

In normal operation, it is assumed that the bus voltage $\tilde{V}_s$ is a stiff voltage source, supported by sufficient reactive power supply. However, in weaker systems, such as subtransmission level, the bus voltage may be affected by the reactive power requirement of the induction motor.

Referring back to Figure 11.5, the stator current can be computed as

$$\tilde{I}_s = \frac{\tilde{V}_s}{R_s + jX_s + \left(jX_m || \left(R_r/s + jX_r\right)\right)} \tag{11.104}$$

which can be expressed as

$$\tilde{I}_s = \frac{\tilde{V}_s\left(R_r/s + j\left(X_m + X_r\right)\right)}{\left(R_s\left(R_r/s\right) - X_mX_r - X_s\left(X_m + X_r\right)\right) + j\left(\left(X_m + X_s\right)R_r/s + R_s\left(X_m + X_r\right)\right)} \tag{11.105}$$

The apparent power delivered from the stator to the induction motor is

$$S = \tilde{V}_s\tilde{I}_s^* = P_s + jQ_s \tag{11.106}$$

in which Q_s can be expressed as

$$Q_s = V_s^2\frac{\left(R_r/s\right)^2\left(X_m + X_s\right) + \left(X_m + X_r\right)\left(X_mX_r + X_mX_s + X_sX_r\right)}{\left(\left(R_sR_r/s\right) - X_mX_r - X_s\left(X_m + X_r\right)\right)^2 + \left(\left(X_m + X_s\right)R_r/s + R_s\left(X_m + X_r\right)\right)^2} \tag{11.107}$$

Note that $Q_s > 0$, that is, the induction machine always requires reactive power supply whether as a load ($s > 0$) or generator ($s < 0$) because there are only inductive circuits inside an induction machine.

At $s = 0$, that is, at synchronous speed, Q_s reduces to

$$Q_s = V_s^2 \frac{X_m + X_s}{R_s^2 + (X_s + X_m)^2} \tag{11.108}$$

From the circuit in Figure 11.5, this Q_s expression can be seen as the reactive power required by the stator circuit for magnetizing the rotor circuit.

Example 11.4: Reactive power consumption

For the induction motor in Example 11.3, compute and plot the reactive power Q_s for for $s = 1$ pu to $s = -1$ pu.

Solutions: The reactive power consumption is shown in Figure 11.9. Note that the reactive power requirement rises quite drastically as s moves away by a few percent from the nominal frequency. The lowest Q_s consumption point occurs at $s = 0$, which from (11.108) is $Q_s = 0.3313$ pu. Thus the ratio of the inductive power required by the motor during startup and in steady-state operation is about 10:1. ■

11.4.7 Motor Startup

As is evident from the torque-slip analysis, the rotor winding is designed to achieve efficient operation at a low slip value. However, when starting a motor from standstill ($s = 1$ pu), it would be desirable to have lower values of currents and higher values of torque.

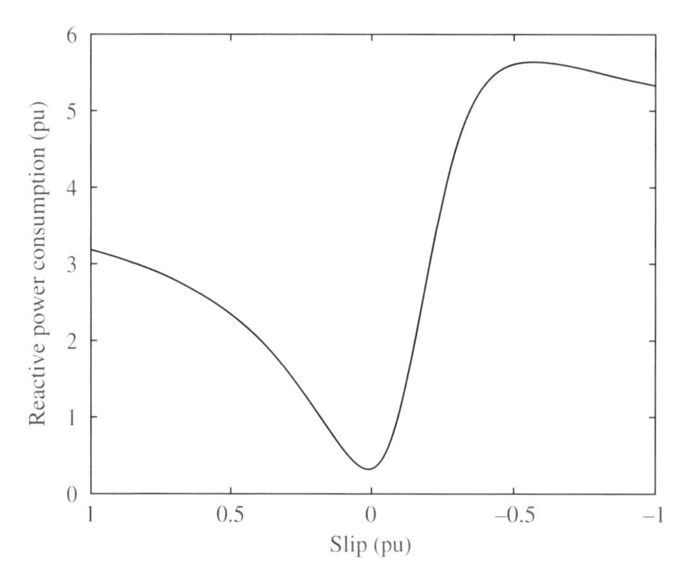

Figure 11.9 Reactive power consumption of an induction machine vs slip.

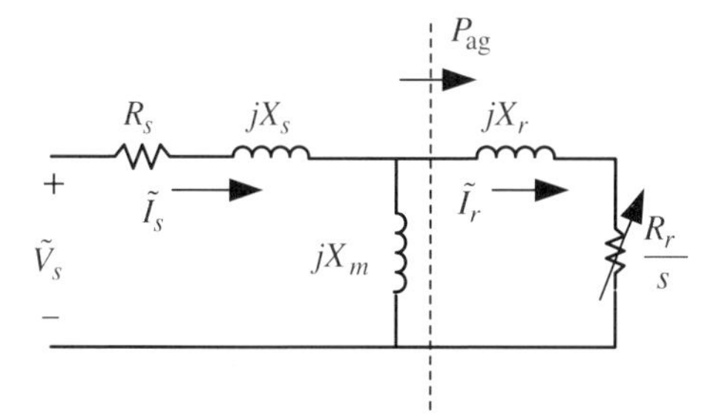

Figure 11.10 Equivalent circuit for wound-rotor induction motor with variable external resistance.

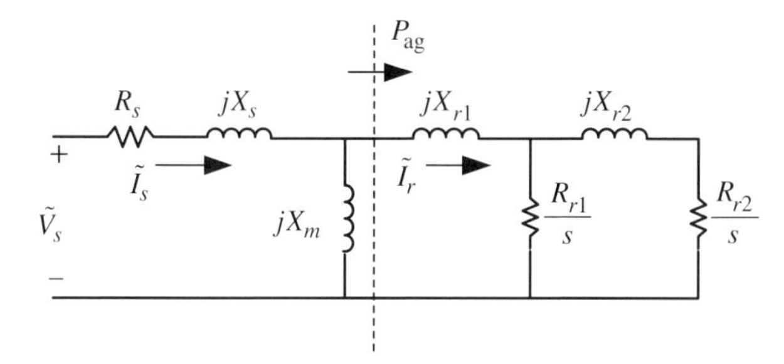

Figure 11.11 Equivalent circuit representation for double-cage induction motor.

For a wound-rotor type induction motor, the startup process can be made more efficient by connecting the rotor circuits via the slip rings to an adjustable external resistance. The equivalent circuit representation is shown in Figure 11.10, where R_r is adjustable. By adjusting R_r, the startup torque can be increased and the stator current can be decreased.

In addition, it is possible to design the rotor with a double-squirrel-cage configuration with two layers of conducting bars shorted as both ends. The bars closer to the stator have a higher resistance R_{r1} and provide a path of higher reluctance, resulting in a lower leakage inductance X_{r1}. The second set of bars is farther from the stator and has a lower resistance R_{r2} and a higher leakage resistance X_{r2}. The equivalent circuit of this induction motor is shown in Figure 11.11. At the startup stage with s close to 1 pu, the rotor current will mostly flow through the outer bars, that is, R_{r1}. At normal operation, the current will flow in both R_{r1} and R_{r2} as the value of slip is small.

In addition, it is possible to construct the rotor with deep bars resulting in variable leakage reactance versus s, to achieve the same effect as a double-cage motor. Additional discussion on motor startup can be found in [135].

11.5 Induction Motors: Dynamic Models

In this section, the induction motor circuits are used to develop a dynamic model with three states: two states to represent the variations of flux linkages and one state for the slip. These states are subject to perturbations as the induction motor is subjected to a disturbance.[3] The derivation will make use of the voltage and flux linkage equations (11.63) and (11.61) for the stator, and equations (11.64) and (11.62) for the rotor. The development is in two parts: the network representation used in solving the system network flows and the dynamic equations.

From (11.62), the rotor currents can be solved as

$$i_{dr} = \frac{\psi_{dr} - L_m i_{ds}}{L_{rr}}, \quad i_{qr} = \frac{\psi_{qr} - L_m i_{qs}}{L_{rr}} \tag{11.109}$$

Substitute the rotor currents into (11.61) to obtain

$$\psi_{ds} = L_{ss} i_{ds} + \frac{L_m}{L_{rr}} \left(\psi_{dr} - L_m i_{ds} \right) = \frac{L_m}{L_{rr}} \psi_{dr} + \left(L_{ss} - \frac{L_m^2}{L_{rr}} \right) i_{ds} \tag{11.110}$$

$$\psi_{qs} = L_{ss} i_{qs} + \frac{L_m}{L_{rr}} \left(\psi_{qr} - L_m i_{qs} \right) = \frac{L_m}{L_{rr}} \psi_{qr} + \left(L_{ss} - \frac{L_m^2}{L_{rr}} \right) i_{qs} \tag{11.111}$$

Define the voltages

$$V_d' = -\omega_s \frac{L_m}{L_{rr}} \psi_{qr}, \quad V_q' = \omega_s \frac{L_m}{L_{rr}} \psi_{dr} \tag{11.112}$$

and the transient reactance

$$X_s' = \omega_s \left(L_{ss} - \frac{L_m^2}{L_{rr}} \right) \tag{11.113}$$

Then the ψ_{ds} and ψ_{qs} expressions are substituted into the stator voltage equations with the stator transients neglected, that is, $d\psi_{ds}/dt = d\psi_{qs}/dt = 0$, to obtain

$$V_{ds} = R_s i_{ds} + V_d' - X_s' i_{qs}, \quad V_{qs} = R_s i_{qs} + V_q' + X_s' i_{ds} \tag{11.114}$$

The two voltage equations in (11.114) are combined to obtain a single phasor equation

$$V_{ds} + jV_{qs} = \left(R_s + jX_s' \right) \left(i_{ds} + j i_{qs} \right) + \left(V_d' + jV_q' \right) \tag{11.115}$$

that is,

$$\tilde{V}_s = \left(R_s + jX_s' \right) \tilde{I}_s + \tilde{V}' \tag{11.116}$$

which can be represented by the circuit in Figure 11.12. The voltage $\tilde{V}'$ is the induction motor internal voltage, and $R_s + jX_s'$ is the transient impedance of the motor. The transient impedance is a constant and does not depend on the system voltages or the motor slip s. This circuit links the inductor motor windings to the power network via the stator voltage $\tilde{V}_s$ and current $\tilde{I}_s$. The voltage $\tilde{V}'$ consists of two states, which are added to the network equation (11.9) in the phasor form.

With the induction motor circuits added, the network equation (11.9) expands to

$$\begin{bmatrix} Y_d & -Y_d & 0 & 0 & 0 \\ -Y_d & Y_{11}' & 0 & Y_{121} & Y_{122} \\ 0 & 0 & Y_{\text{im}} & -Y_{\text{im}} & 0 \\ 0 & Y_{211} & -Y_{\text{im}} & Y_{221}' & Y_{222}' \\ 0 & Y_{212} & 0 & Y_{223}' & Y_{224}' \end{bmatrix} \begin{bmatrix} \tilde{E}' \\ \tilde{V}_G \\ \tilde{V}' \\ \tilde{V}_s \\ \tilde{V}_L' \end{bmatrix} = \begin{bmatrix} \tilde{I}_G \\ 0 \\ -\tilde{I}_s \\ 0 \\ 0 \end{bmatrix} \tag{11.117}$$

3 The development here assumes there is only one set of rotor circuit parameters (R_r, X_r), which is valid when the motor is running at its optimal slip. For a double-cage rotor, it would be more accurate to include two additional states to model the dynamics due to the rotor parameters $(R_{r1}, X_{r1}, R_{r2}, X_{r2})$. In the latter case, the second circuit parameters are known as the subtransient effects.

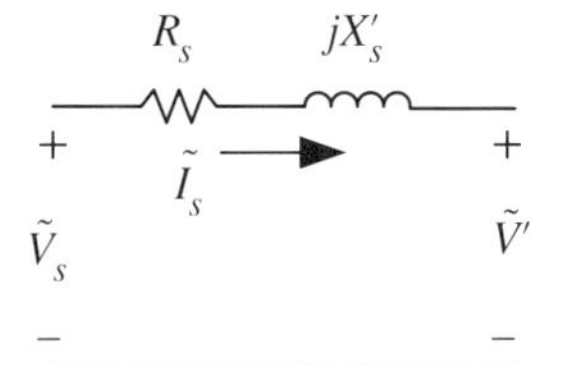

Figure 11.12 Equivalent circuit for dynamic model.

In (11.117), the internal voltage $\tilde{V}'$ of the induction motors is added to the network variables and the load bus voltages $\tilde{V}_L$ are separated into the voltage $\tilde{V}_s$ of the induction motor buses and the voltages $\tilde{V}'_L$ for all the other load buses. The admittance $Y_{\text{im}} = 1/\left(R_s + jX'_s\right)$ is the inverse of the transient impedance. The admittance matrices Y'_{12}, Y'_{21}, and Y'_{22} are also separated into submatrices of appropriate dimensions. In addition, Y'_{221} also contains the effect of Y_{im}. Note also that $\tilde{I}_s$ carries a negative sign as current injection is regarded as positive going into the network.

To develop the dynamic equations, start by eliminating the rotor currents in the rotor voltage equations (11.64) to obtain

$$0 = R_r \frac{\psi_{dr} - L_m i_{ds}}{L_{\text{rr}}} - s\omega_s \psi_{qr} + \frac{d\psi_{dr}}{dt} \tag{11.118}$$

$$0 = R_r \frac{\psi_{qr} - L_m i_{qs}}{L_{\text{rr}}} + s\omega_s \psi_{dr} + \frac{d\psi_{qr}}{dt} \tag{11.119}$$

Substituting ψ_{dr} and ψ_{qr} from (11.112) results in

$$0 = \frac{R_r}{L_{\text{rr}}} \left(\frac{L_{\text{rr}}}{\omega_s L_m} V'_q - L_m i_{ds} \right) + s\omega_s \frac{L_{\text{rr}}}{\omega_s L_m} V'_d + \frac{d}{dt} \left(\frac{L_{\text{rr}}}{\omega_s L_m} V'_q \right) \tag{11.120}$$

$$0 = \frac{R_r}{L_{\text{rr}}} \left(-\frac{L_{\text{rr}}}{\omega_s L_m} V'_d - L_m i_{qs} \right) + s\omega_s \frac{L_{\text{rr}}}{\omega_s L_m} V'_q - \frac{d}{dt} \left(\frac{L_{\text{rr}}}{\omega_s L_m} V'_d \right) \tag{11.121}$$

These two equations can be rearranged to form the equations

$$\frac{dV'_q}{dt} = -\frac{R_r}{L_{\text{rr}}} \left(V'_q - \frac{\omega_s L_m^2}{L_{\text{rr}}} i_{ds} \right) + s\omega_s V'_d \tag{11.122}$$

$$\frac{dV'_d}{dt} = -\frac{R_r}{L_{\text{rr}}} \left(V'_d + \frac{\omega_s L_m^2}{L_{\text{rr}}} i_{qs} \right) + s\omega_s V'_q \tag{11.123}$$

Defining

$$T'_o = \frac{L_{\text{rr}}}{R_r} = \frac{X_{\text{rr}}}{\omega_s R_r}, \quad \frac{\omega_s L_m^2}{L_{\text{rr}}} = \omega_s L_{\text{ss}} - \omega_s \left(L_{\text{ss}} - \frac{L_m^2}{L_{\text{rr}}} \right) = X_{\text{ss}} - X'_s \tag{11.124}$$

the dynamic equations for the voltage $\tilde{V}'$ are given by

$$\frac{dV'_q}{dt} = -\frac{1}{T'_o} \left(V'_q - \left(X_{\text{ss}} - X'_s \right) i_{ds} \right) + s\omega_s V'_d \tag{11.125}$$

$$\frac{dV'_d}{dt} = -\frac{1}{T'_o} \left(V'_d + \left(X_{\text{ss}} - X'_s \right) i_{qs} \right) + s\omega_s V'_q \tag{11.126}$$

Note that T_o' is known as the transient open-circuit time constant. Because the rotor circuits are symmetric, T_o' applies to both the d- and q-axes.

The electromagnetic torque in terms of the stator voltages and currents can be computed as

$$T_e = \psi_{qr} i_{dr} - \psi_{dr} i_{qr} = \frac{V_d' i_{ds} + V_q' i_{qs}}{\omega_s} \tag{11.127}$$

where the stator frequency ω_s is assumed to be constant at the nominal frequency $2\pi f_s$ rad/sec where $f_s = 50$ or 60 Hz. Thus the dynamic equation for the slip is

$$2H\frac{ds}{dt} = T_m - T_e \tag{11.128}$$

in which the slip of a motor below synchronous speed is taken as positive. The term T_m is the combined motor load and losses, which can be approximated as a constant or a function of ω_r.

Note that V_d' and V_q' form the voltage $\tilde{V}'$ required in the network equation (11.117). Thus the synchronous machine internal voltage $\tilde{E}'$ and the induction motor internal voltages $\tilde{V}'$ are fixed during the network solution iterations, as they are states, such that the other voltages $\tilde{V}_G$, $\tilde{V}_s$, and $\tilde{V}_L'$ can be solved at each integration step.

Example 11.5: Induction motor dynamic model parameters

Using the motor parameters in Example 11.3, compute the synchronous reactance X_{ss}, transient reactance X_s', and transient time constant T_o'.

Solutions: The synchronous reactance is

$$X_{ss} = X_s + X_m = 3.018 \text{ pu} \tag{11.129}$$

The transient reactance is

$$X_s' = X_{ss} - X_m^2/X_{rr} = 3.018 - 2.9745^2/(2.9745 + 0.0249) = 0.0760 \text{ pu} \tag{11.130}$$

The transient time constant is

$$T_o' = \frac{X_{rr}}{R_s \Omega} = \frac{2.9994}{0.0249 \times 377} = 0.3204 \text{ sec} \tag{11.131}$$

■

11.5.1 Initialization

The purpose of the initialization of an induction motor running in steady state is to compute the slip of the motor to coincide with the active power load specified in the power flow formulation, and determine the reactive power compensation needed at the induction motor bus. Knowing the slip, the voltages V_d' and V_q' can be computed. It is also possible to include the motor steady-state equations in the power flow solution [136, 137]. In this approach, no adjustment to the external reactive power compensation is needed.

From the circuit diagram in Figure 11.13, one can increase the slip s from zero to achieve the desired load P_{mo}. A faster iterative Newton method can be systematically developed.

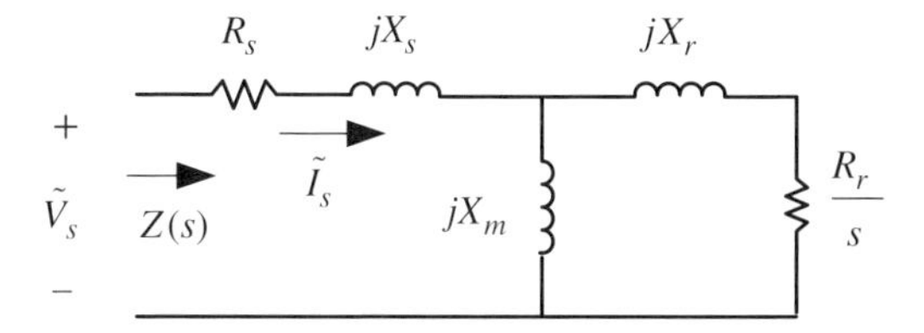

Figure 11.13 Equivalent circuit for initialization.

From Figure 11.13, the impedance looking into the induction motor from the stator winding as a function of the slip is denoted as $Z(s)$

$$Z(s) = R_s + jX_s + \frac{(R_r/s + jX_r)(jX_m)}{R_r/s + j(X_r + X_m)} \tag{11.132}$$

After some manipulation, (11.132) can be expressed as[4]

$$Z(s) = Z(y) = Z_r(y) + jZ_i(y) \tag{11.133}$$

where

$$Z_r(y) = R_s + y(s)\frac{X_{ss} - X_s'}{1 + y^2(s)}, \quad Z_i(y) = X_s' + \frac{X_{ss} - X_s'}{1 + y^2(s)}, \quad y(s) = T_o's \tag{11.134}$$

The active power consumed by the induction motor is

$$P_m(y) = \text{Re}(\tilde{V}_s\tilde{I}_s^*) = \text{Re}\left(\tilde{V}_s\frac{\tilde{V}_s^*}{Z(y)^*}\right) = |V_s|^2\frac{Z_r(y)}{Z_r^2(y) + Z_i^2(y)} \tag{11.135}$$

This nonlinear equation is used to find y (and then s). Let $y = y_k$ at iteration k. Expand (11.135) as a Taylor series about y_k with an increment of Δy_k and equate it to P_{mo}

$$P_m(y_k) + \left.\frac{\partial P_m}{\partial y}\right|_{y_k}\Delta y_k + \text{H.O.T.} = P_{mo} \tag{11.136}$$

where

$$\frac{\partial P_m}{\partial y} = |V_s|^2\left(\frac{dZ_r(y)/dy}{Z_r^2(y) + Z_i^2(y)} - 2Z_r(y)\frac{(dZ_r(y)/dy)Z_r + (dZ_i(y)/dy)Z_i}{(Z_r^2(y) + Z_i^2(y))^2}\right) \tag{11.137}$$

$$\frac{dZ_r(y)}{dy} = (X_{ss} - X_s')\frac{1 - y^2}{(1 + y^2)^2}, \quad \frac{dZ_i(y)}{dy} = -(X_{ss} - X_s')\frac{2}{(1 + y^2)^2} \tag{11.138}$$

The update equation at each iteration can be computed as

$$y_{k+1} = y_k + \Delta y_k \tag{11.139}$$

where

$$\Delta y_k = (P_{mo} - P_m(y_k)) / (\partial P_m/\partial y)|_{y_k} \tag{11.140}$$

4 The derivation here follows the PST code developed by Graham Rogers.

The iteration converges if

$$|y_{k+1} - y_k| \leq \varepsilon \tag{11.141}$$

where ε is a pre-specified tolerance. At convergence, the slip can be computed as

$$s = y_{k+1}/T_o' \tag{11.142}$$

Once s is known, $Z(s)$ can be used to compute the stator current phasor $\tilde{I}_s = \tilde{V}_s/Z(s)$ and hence the power supplied to the motor $S = \tilde{V}_s(\tilde{I}_s)^* = P_{mo} + jQ_{mo}$. In the initialization, only P_{mo} is used to compute s. However, the value of Q_{mo} may not be the same as the reactive power Q_L specified in the power flow solution. To handle this discrepancy, the power at the load bus is allocated as an induction motor power consumption of $P_{mo} + jQ_{mo}$ and a reactive power consumption of $Q_L - Q_{mo}$ is added as a shunt susceptance in the power flow bus data.

Example 11.6: Induction motor initialization

Using the motor parameters in Example 11.3, compute the slip s in steady state, and V'_{do} and V'_{qo}, given $P_{mo} = 100/137 = 0.7299$ pu.

Solutions: Starting from $s = 0.02$ pu, the Newton solution converges in three iterations, yielding $s = 0.0192$ pu. Note this value of s is less than that found in Example 11.3, which came from the airgap $P_{ag} = 0.7299$ pu. With resistive losses in the motor, the airgap power will be smaller, and hence the slip is also smaller.

At this value of $s = 0.0192$ pu, $Z(0.0192) = 1.1112 + j0.5363$ pu and $\tilde{I}_s = 0.7299 - j0.3523$ pu. As a result, $P_{mo} = 0.7299$ pu (which is expected) and $Q_{mo} = 0.3523$ pu, which will be used to adjust the reactive load on the bus, so that the power flow solution is still valid.

The motor internal voltage is computed by

$$\tilde{V}_o' = \tilde{V}_s' - (R_s + jX_s')\,\tilde{I}_s = 0.9422 - j0.0405 \text{ pu} \tag{11.143}$$

such that

$$V'_{do} = 0.9422 \text{ pu}, \quad V'_{qo} = -0.0405 \text{ pu} \tag{11.144}$$

■

11.5.2 Reactive Power Requirement during Motor Stalling

From Figure 11.9, it can be seen that the reactive power consumption by an induction motor is high when s is close to 1 pu, also known as standstill. It has been reported [127, 128] that in systems with a high amount of induction motor loads, such as the southern part of the USA on hot days with heavy air-conditioning loads, the system voltage after a short-circuit fault (especially single-phase faults) took a long time (sometimes 30 sec or more) to recover. This fault-induced delayed voltage recovery (FIDVR) phenomenon has been attributed to induction motor stalling due to low bus voltage during the fault-on period.

Figure 11.14 has been used to describe FIDVR [129]. The figure shows that, after the fault is cleared, the voltage takes about 20 sec to return to its pre-fault value. The

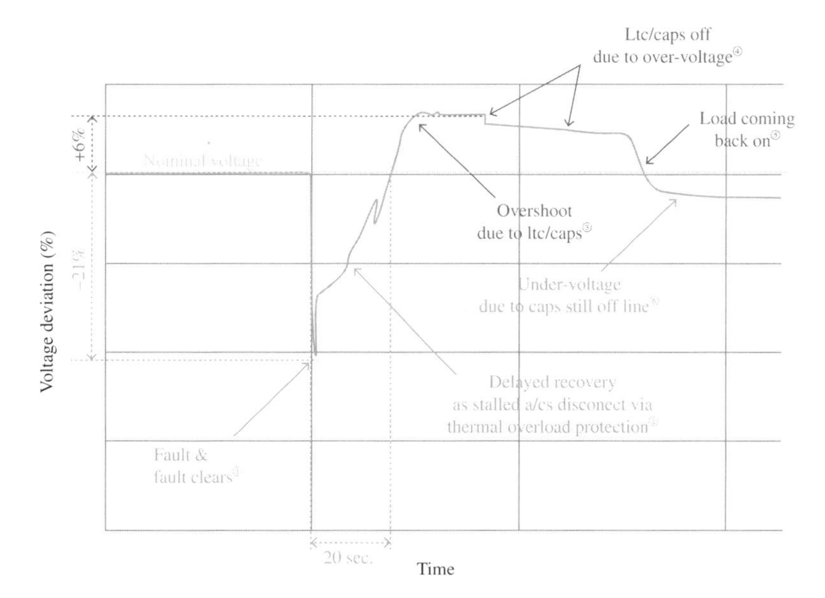

Figure 11.14 Fault-induced delayed voltage recovery [129] (see color plate section).

delayed recovery is caused by the stalled air-conditioner (A/C) single-phase motors. These motors would stall following the fault due to low voltages and remain on-line until their thermal protection trips them. When the motors are stalled, their reactive power consumption is much larger than under normal operating conditions; depending on the system conditions, this may cause the prolonged depressed voltage shown in Figure 11.14. While the motors are stalled, LTCs and shunt capacitors are used to increase the voltage; however, once the stalled motors are able to restart, the voltage will return to near normal levels.

Instead of trying to duplicate a FIDVR event with repeated induction motor stalling and restarting, the next example illustrates the system voltage behavior during a motor startup in a simple system, showing a significant voltage drag on the system.

Example 11.7: Motor startup

A power system consisting of a generator on Bus 1 supplying a load on Bus 3 is shown Figure 11.15. The system base is 100 MVA, the Bus 3 load is $5 + j0$ pu, and the generator is rated at 600 MVA. An induction motor rated at 15 MVA is located on Bus 4, with the following parameters: synchronous reactance $X_{ss} = 3.2$ pu, transient reactance $X'_s = 0.15$ pu, transient time constant $T'_o = 1$ sec, and rotor inertia $H = 0.4$ sec. At $t = 0$, the induction motor is switched on to synchronize with the grid. Once synchronized, the induction motor will consume 7.5 MW. The generator is modeled with subtransient dynamics, and is equipped with a type DC1A excitation system and a turbine-governor model. The power flow data and the parameters of the different dynamic models are given in the example data file. Simulate the time response of the motor startup. Note that the integration stepsize during startup has to be quite small, like 0.001 sec.

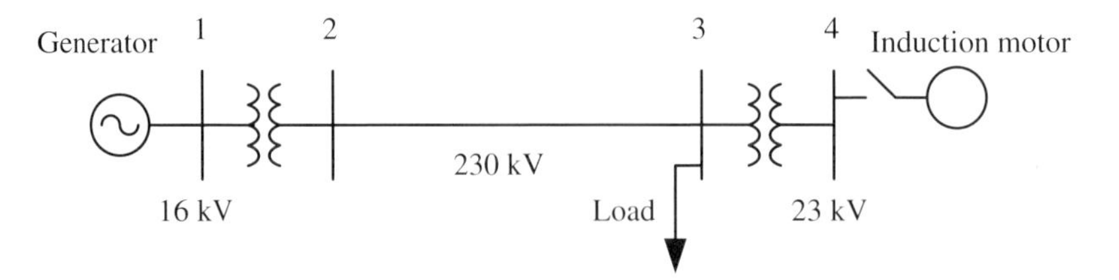

Figure 11.15 System for simulating motor startup.

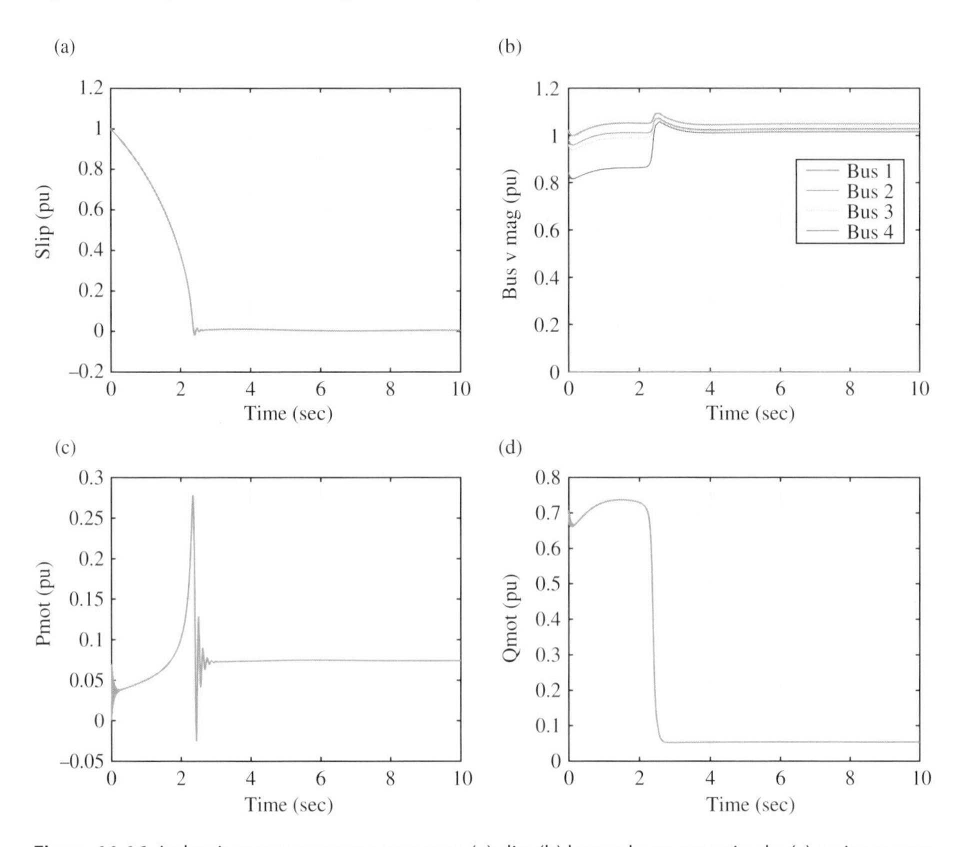

Figure 11.16 Induction motor startup response (a) slip, (b) bus voltage magnitude, (c) active power consumption, and (d) reactive power consumption (see color plate section).

Solutions: Note that some computer programs can directly make use of the parameters X_{ss}, X_s', and T_o'. PST only accepts the stator and rotor reactances and resistances. To translate the parameters, it is assumed that $R_s = 0$ and $X_s = X_r$. Under these assumptions, an equivalent set of parameters is

$$R_r = 0.0085 \text{ pu}, \quad X_s = X_r = 0.0759 \text{ pu}, \quad X_m = 3.1241 \text{ pu} \tag{11.145}$$

The startup time response is shown in Figure 11.16. At the start of the simulation, the reactive power consumption jumps to over 0.6 pu. The voltage on Bus 4 drops from 1.0269 pu in steady state to about 0.8 pu. Meanwhile the slip moves from 1 pu to a small value in the synchronizing process. During this period of slightly over 2 sec, the voltage on Bus 4 remains low. Once the motor is synchronized, the reactive power consumption drops significantly, allowing the bus voltages to recover to a safe level. ■

As a continuation of Example 11.6, sensitivity studies can be performed for this system by adjusting the induction motor rating and the reactance of Line 2-3. As the rating of the induction motor is increased, the voltage on Bus 4 will drop further, and the startup period will be longer. For example, if the induction motor rating is increased to 30 MVA with a power consumption of 15 MW, the voltage at Bus 4 will be depressed to 0.75 pu for about 4 sec. The same conclusions can be obtained by increasing the line reactance, making it more difficult to deliver power. For example, if the reactance between Line 2-3 is increased from 0.02 pu to 0.08 pu, the voltage at Bus 4 will be depressed to below 0.7 pu for over 5 sec. In both cases, the startup time will be longer as well. When the motor rating or the line reactance is sufficiently high, the low bus voltage would not recover, resulting in a "dynamic" voltage collapse.

Note that motor loads in simulation studies are aggregated. In the restart process after stalling, it is important to stagger the starting of induction motors (such as in a large factory with many motors). The next motor should be started only after a previous motor has been successfully started.

11.6 Summary and Notes

This chapter provides more comprehensive load models for power system dynamic simulation and stability analysis. Non-conforming loads such as constant power and constant current are introduced, together with procedures to interface them into the network equations. Then induction motor models are discussed, with the calculation of the torque versus slip and the reactive power requirement versus slip during startup. A motor-startup example illustrating low voltage condition during motor startup is useful for understanding the fault-induced delayed voltage recovery (FIDVR) phenomenon at large load centers on hot days.

There are a variety of induction motor applications, including pumps, compressors, and mechanical punches, requiring induction motors to have specific torque-slip characteristics. A reader interested to learn more about the different types of induction motors can search on the website of the National Electrical Manufacturers Association (NEMA).

Induction machines have also gained popularity with their application in variable-speed wind turbine-generators. Their versatility is improved by using power electronic converters to insert a voltage into the rotor circuit. Induction generators are covered in Chapter 15 on wind energy.

In the future, it is anticipated that detailed load models for large load centers will be needed to improve the fidelity of dynamic simulation programs. This process will require extensive testing and continuous measurement of load parameters, accomplished with probing signals or recorded disturbance data [126]. Although including detailed load models would increase the simulation time, such models would be necessary if a load center is close to a short-circuit fault.

The motor model in PST was developed by Graham Rogers. A reader can read [132–134] for a sample of his work on motors.

Problems

11.1 Example 4.7 shows that using classical machine models and constant impedance loads, Generator 9 in the NPCC 16-machine system remains stable for a 3-cycle 3-phase short-circuit fault on Bus 29, cleared by tripping Line 29-28. Generator 9 will be unstable if the fault is cleared in 4 cycles. Repeat the simulation using the constant current load to model the active power loads and the constant impedance load to model the reactive power loads. Determine the stability of Generator 9 for the fault cleared in 3, 4, and 5 cycles.

11.2 For the NPCC system used in Problem 5.6, a fault was applied on Line 32-30 near Bus 32, and cleared by removing the line. The critical clearing time was computed using the constant-impedance model for all the loads. In this problem, repeat the critical clearing time computation, accurate to the nearest 0.01 sec, using constant power for all the active power loads and constant impedance for all the reactive power loads. Take the system as unstable if the simulation terminates because the network solution does not converge,

11.3 For the two-area, four-machine system shown in Figure 6.2 and used for Problem 6.6, compute the damping and frequency of the three swing modes (two local modes and one interarea mode) using (a) constant-impedance load model for both P and Q loads, (b) constant-current load model for both P and Q loads, and (c) constant-power load model for both P and Q loads. The system data file is provided in the Problem 11.3 folder. Use only the Part 1 operating condition in Problem 6.6.

11.4 Derive the T_{emax} expression (11.101).

11.5 An induction motor is rated at 1.5 MVA, with the equivalent circuit parameters

$$R_s = 0.0120 \text{ pu}, \; X_s = 0.09 \text{ pu}, \; R_r = 0.009 \text{ pu}, \; X_r = 0.08 \text{ pu}, \; X_m = 3.0 \text{ pu} \tag{11.146}$$

Plot the airgap power (that is, the torque-slip curve) for $s = 1$ pu to $s = -1$ pu. Suppose 1 MW is delivered to the rotor. Find the value of the slip s at that operating condition. Assume that the stator voltage V_s is at 1 pu.

11.6 For the induction motor model in Example 11.3, plot the torque-slip curve by changing the value of R_r to $2R_r$, $3R_r$, $4R_r$, and $5R_r$ (via an external variable resistance). Why does the maximum torque for each of these resistance values remain the same?

11.7 Derive the Q_s expression (11.107).

11.8 Plot Q_s vs slip for the motor in Problem 11.5.

11.9 Derive the $Z(s)$ expressions (11.132) and (11.133).

11.10 Compute the dynamic model parameters X_{ss}, X_s', and T_o' for the motor in Problem 11.5.

11.11 Repeat Example 11.6 if the induction motor is consuming $P_{mo} = 90$ kW from the power grid.

11.12 Repeat the induction motor startup problem in Example 11.7 by increasing the rating of the induction motor to 30, 40, and 50 MVA. Can the motor start properly?

11.13 Repeat the induction motor startup problem in Example 11.7 by changing the reactance between Buses 2 and 3 from 0.02 pu to 0.08 pu. Keep the induction motor rating at 15 MVA. How long does it take for the Bus 4 voltage to reach 1 pu?

12

Turbine-Governor Models and Frequency Control

12.1 Introduction

This chapter discusses dynamic models of turbine-generators as energy supply and for frequency regulation. Depending on the prime mover drive mechanisms, turbine-governor control models are in general categorized into three types: (1) steam turbines,[1] (2) hydraulic turbines, and (3) gas turbines and combined-cycle co-generation plants. Renewable generation such as wind turbines and photovoltaic panels is also becoming an important source of energy. The rapidly increasing penetration of wind and solar energy resources into the power grid will present a significant challenge to engineers in planning and operating future power systems. This chapter describes the models of conventional prime movers. Renewable resources typically have a power-electronic interface and are discussed in Chapter 15, after power conversion with power electronics is covered in Chapter 14.

This chapter starts with a discussion on steam turbine models, which is followed by a discussion of hydraulic turbine models. Gas turbine models, which are more complex, are briefly discussed. The reminder of the chapter deals with system frequency response and regulation. Both primary frequency response and automatic generation control (AGC) are described.

In this chapter, torsional models for large turbine-generators are also developed and the interaction between the generator and electrical network is discussed. This presentation is relevant to the discussion in Chapter 10 on including torsional filters in the design of power system stabilizers to reduce the impact of swing-mode damping controller interaction with the torsional modes.

Several parts of this chapter are developed based on the Power System Operation and Control Course notes by Donald Ewart [138]. Some examples, problems, and block diagrams used in this chapter are also adapted from [138, 139].

1 The steam is produced by burning coal or oil in a boiler or by nuclear fission in a nuclear reactor.

Power System Modeling, Computation, and Control, First Edition. Joe H. Chow and Juan J. Sanchez-Gasca.

Companion website: www.wiley.com/go/chow/power-system-modeling

12.2 Steam Turbines

12.2.1 Turbine Configurations

The design of a steam turbine requires analyzing the steam conditions at various turbine stages to extract as much useful energy from the steam as possible ([140], Chapter 6). The steam turbine models used for stability and frequency regulation simulation are developed by neglecting most of the complex steam dynamics. Figure 12.1 shows two common steam turbine configurations with multiple turbine sections.

In Figure 12.1, each of the turbine sections (high-pressure (HP), intermediate pressure (IP), and low-pressure (LP)) consists of a set of concentric wheels with increasing diameters and increasing size of blades or buckets mounted on the wheels. As steam expands in a turbine section, it pushes against the blades, which are optimally curved according to aerodynamic design, to produce rotational mechanical power. Figure 12.2 shows the LP sections of a steam turbine.

In a non-reheat unit (Figure 12.1.a), when the steam exits the HP section, it is at a lower pressure and temperature. This steam is then passed through the crossover conduit into the LP sections. Each LP section has a set of larger concentric wheels mounted with longer blades. The exhaust steam is then cooled in the condenser to a liquid state, so that water (still relatively warm) can be pumped back into the boiler, forming a closed-loop system. Steam units are typically located near a river, a large lake, or an ocean so that an ample supply of cooling water is available. In a non-reheat unit, the HP section normally provides about 30% of the mechanical power output, whereas the LP sections provide about 70%.

In a reheat turbine (Figure 12.1.b), the steam exiting from the HP section is passed into a reheater in which the steam is heated up to a higher temperature and higher pressure. Then the steam is passed into the IP section, and then to the LP sections. The reheater is added so that the exit steam from the HP section can be more optimally used. Not shown

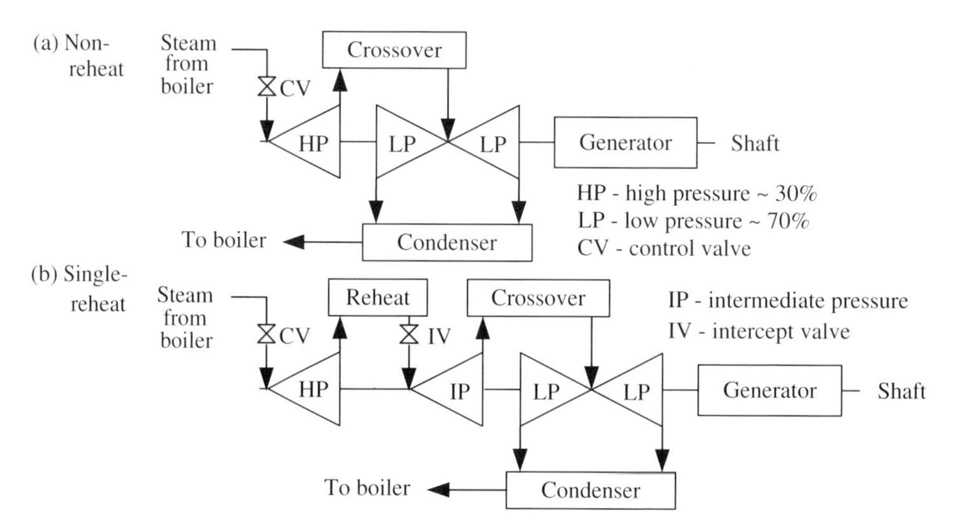

Figure 12.1 Steam turbine configurations: (a) a non-reheat unit and (b) a single-reheat unit.

Figure 12.2 Low-pressure sections of a steam turbine (Courtesy of General Electric Company) (see color plate section).

in Figure 12.1 are steam extractions at various turbine stages for additional optimization of the entire thermodynamical cycle [140].

Both turbine-generators shown in Figure 12.1 have all the rotating components on the same shaft. The inertia constant H used in transient stability analysis as part of the generator data is the combined inertias of the turbine sections and the generator. For a steam unit, the turbine sections are an important part of the total inertia. With multiple stages, a steam turbine is normally laid horizontally, with the shaft parallel to the ground.

It is also possible to use one steam supply to drive two separate turbine shafts. One turbine shaft typically consists of the HP and IP sections and a generator, and the other shaft consists of the LP sections and a separate generator. The steam supply is first passed into the HP section on the first shaft, and the reheated steam is provided to the IP section on the same shaft. Then the steam from the IP section is sent to the LP sections on the second shaft, which possibly operate at a lower rotational speed. Such a configuration is called a cross-compound unit.

The steam flow into the HP section is controlled by the control valve (CV). The control valve is typically constructed with four different compartments [11]. Each compartment is fully utilized to allow maximum steam flow before another compartment is open. This so-called valve-point loading allows for more economical operation. The steam flow through the control valve is initially linear with respect to the valve opening, also known as the valve stroke, but becomes nonlinear when the valve is nearly fully open (Figure 12.3). This nonlinearity can be compensated with control circuitry if needed. The control valve is operated by a governor for frequency regulation. The motion of a valve, however, has to be minimized as it is a mechanical device and subject to wear and tear.

Most steam turbines are equipped with an intercept valve (IV), which is mostly located after the HP exit steam. The IV protects a turbine from overspeeding if the unit has to be suddenly tripped during load rejection or when it is transiently unstable due to

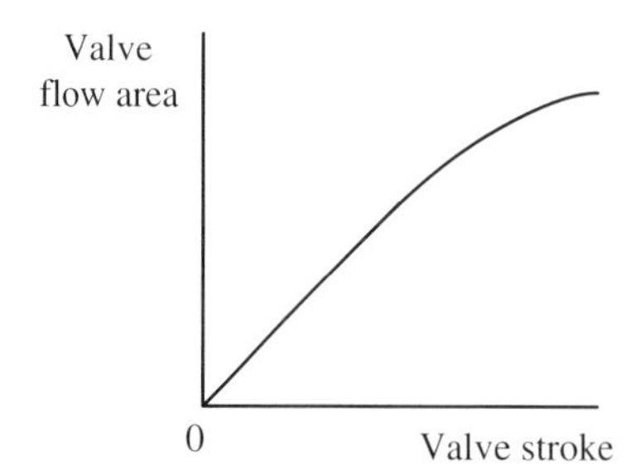

Figure 12.3 Control valve nonlinearity.

some disturbances or equipment malfunction. When it operates, the intercept valve will open fully to divert the steam from continuing on to the IP turbine, thus reducing the mechanical torque. If an early valving actuation (EVA) circuit can be implemented, the IV can be made responsive to an overspeed condition from a disturbance, to enhance the transient stability of the turbine-generator [141].

In a turbine model for dynamic stability and frequency regulation studies, the steam pressure and temperature are assumed to remain constant in each section. For example, the steam chest is an "unlimited" supply of steam, that is, opening the control valves can increase the flow of the steam without diminishing the steam quality. The steam is supplied either by a boiler using coal, oil, or natural gas as fuel, or from a nuclear reactor. Dual-fuel units can use more than one type of fuel, allowing them to take advantage of a cheaper fuel or meet emission constraints by switching to a less polluting fuel. The time constant for the boiler is of the order of 20 seconds or more, and thus is mostly unnecessary in transient stability studies when a 10-second time simulation of a disturbance is normally quite adequate.

The dynamic model for a reheat turbine is shown in Figure 12.4. It consists of a cascade of first-order steam transport transfer function blocks, modeling the flow path of the steam through various turbine sections. The fractions of the total power from the HP, IP, and LP turbines are denoted by F_{HP}, F_{IP}, and F_{LP}, respectively, which sum to unity:

$$F_{HP} + F_{IP} + F_{LP} = 1 \tag{12.1}$$

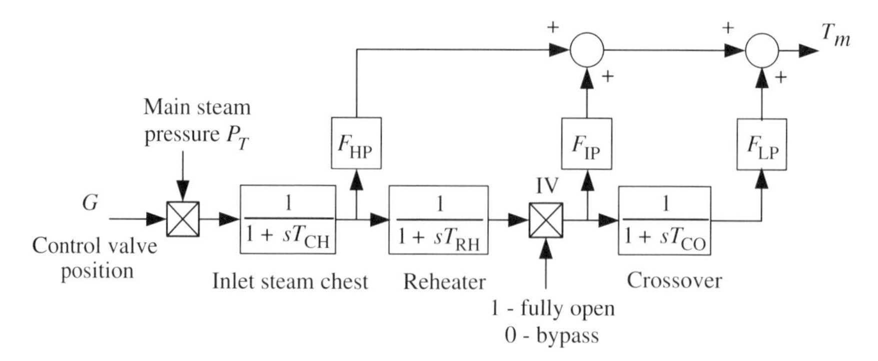

Figure 12.4 Block diagram of a reheat steam turbine.

A representative set of parameter values for a large steam turbine-generator is given by

$$T_{\text{CH}} = 0.3\,\text{sec},\quad T_{\text{RH}} = 5.0\ \text{sec},\quad T_{\text{CO}} = 0.5\ \text{sec},\quad F_{\text{HP}} = 0.3,\quad F_{\text{IP}} = 0.3,\quad F_{\text{LP}} = 0.4 \tag{12.2}$$

Because the steam chest and crossover time constants are smaller than the reheater time constant, the dynamics in the turbine sections can be neglected, that is, their block diagrams replaced with unity gains. The reduced transfer function of the incremental change in the gate position ΔG to the incremental change in the mechanical torque ΔT_m becomes

$$\frac{\Delta T_m}{\Delta G} = F_{\text{HP}} + (1 - F_{\text{HP}})\frac{1}{1 + sT_{RH}} = \frac{1 + F_{\text{HP}}T_{\text{RH}}s}{1 + T_{RH}s} \tag{12.3}$$

which has a pole at $s = -1/T_{\text{RH}}$ and a zero at $s = -1/(F_{\text{HP}}T_{\text{RH}})$.

12.2.2 Steam Turbine-Governors

The power output of a turbine-generator is normally dispatched to a setpoint, as determined from an optimal power flow or a real-time dispatch program. In a deregulated electricity market, the setpoint is issued by the independent system operator to the power plant operator. However, during disturbances resulting in significant generation and load imbalance, such as the loss of a large generator of 800 MW or more, the generators are required to temporarily provide additional power to reduce the power imbalance. This task is accomplished by the governing control on the control valve. This governor operates on the same principle as the flyball governor invented by James Watt in 1788 to control a steam engine.[2]

A turbine-generator governor senses the frequency deviation in the power system to adjust the control valve, as shown in the schematic in Figure 12.5, using either one of the deadband mechanisms shown in Figure 12.6. Although ideally a power grid is supposed to operate precisely at its nominal frequency f_o, its frequency moves about f_o due to random load variations. A governor is not supposed to balance out these random load variations because a control valve is a mechanical device that would be subjected to excessive wear and tear if it chases after small frequency deviations. Thus the frequency deviation $\Delta\omega$ is passed through a deadband function with a properly set threshold for activation. The deadband characteristic of Type 1 in Figure 12.6.a is more responsive than that of Type 2 in Figure 12.6.b.

The parameter R, with a unit of frequency/power, in Figure 12.5 is known as the droop. For units providing regulation, the droop R is set at 3–5% normalized to the generator

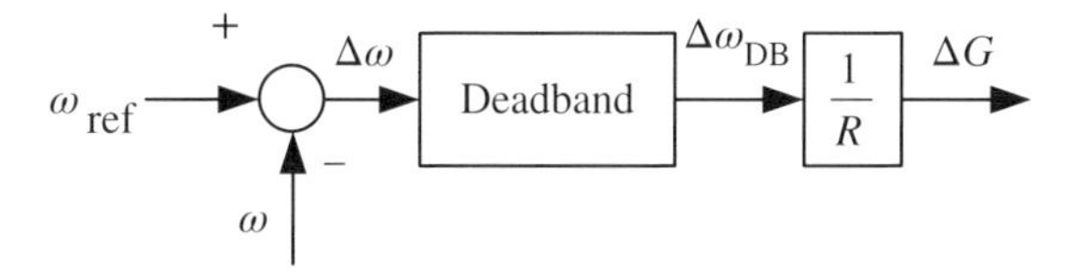

Figure 12.5 Block diagram of a governor.

2 A reader can readily find many illustrations of the flyball governor on the Internet.

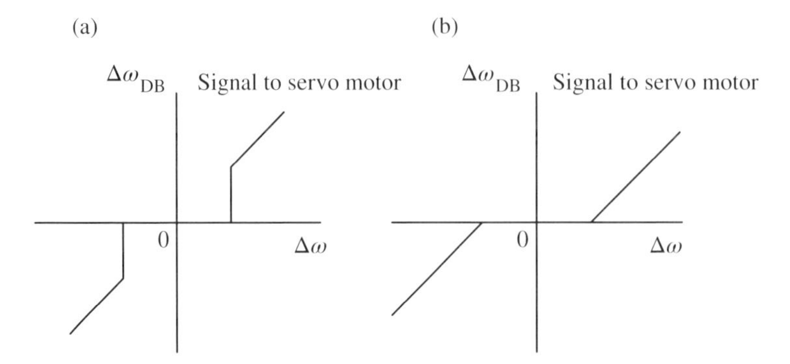

Figure 12.6 Two deadband functions: (a) Type 1 and (b) Type 2.

rating. For example, if $R = 0.05$ such that $1/R = 20$, then a 1% change in frequency will result in a 20% change in ΔG, assuming that the control valve position is roughly equal to the power output of the turbine. In the US eastern power grid, tripping a 1000 MW generator will result in a frequency drop of about 30–60 mHz, that is, 0.0005–0.0010 pu frequency. In addition, regulation service is provided by multiple generators, such that each unit may end up contributing a relatively small amount of additional power output.

The turbine-generator rotor equation of motion is

$$2H\frac{d\omega}{dt} = T_m - T_e = \frac{P_m}{\omega} - T_e \tag{12.4}$$

where T_m is the input mechanical torque and T_e the generator output electrical torque. Linearizing about the operating condition P_{mo}, T_{eo} and $\omega_o = 1$ pu, the small-signal model is

$$2H\frac{d\Delta\omega}{dt} = \frac{1}{\omega_o}\Delta P_m - \frac{P_{mo}}{\omega_o^2}\Delta\omega - \Delta T_e = \Delta P_m - D\Delta\omega - \Delta P_e \tag{12.5}$$

where $D = P_{mo}$ is known as the "steam damping" [140] and P_e is used in place of T_e because T_e is not computed for classical and flux-decay generator models.

The small-signal model in Laplace transform is

$$(2Hs + D)\Delta\omega = \Delta P_m - \Delta P_e \quad \Rightarrow \quad \Delta\omega = \frac{1}{2Hs + D}(\Delta P_m - \Delta P_e) \tag{12.6}$$

Note that in system frequency regulation analysis, the rotor angle does not have to be included. The closed-loop turbine-governor system for frequency regulation study using a simplified turbine control model is shown in Figure 12.7.

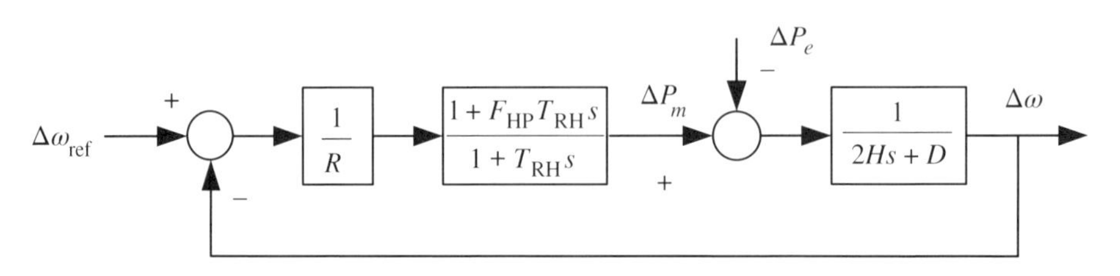

Figure 12.7 Closed-loop turbine-governor system.

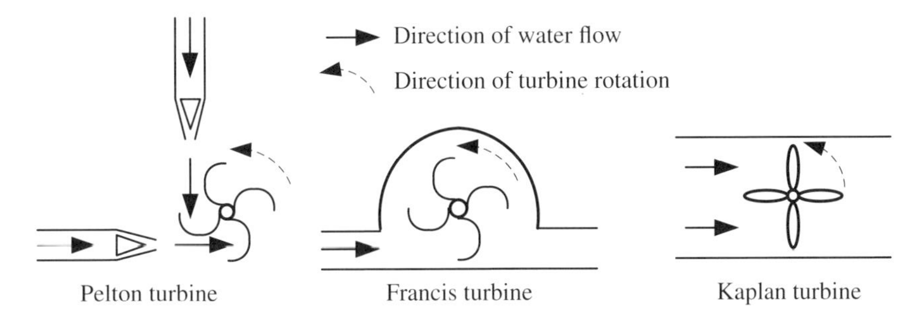

Figure 12.8 Types of hydraulic turbines.

12.3 Hydraulic Turbines

Hydraulic turbines are driven by large volumes of water flowing at high speed. Figure 12.8 shows three common types of hydraulic turbines, described as follows:

1) Impulse-type (Pelton) turbines [144]: The turbine gates create several high-pressure water jets that impact the turbine blades. Water is supplied from a reservoir with a high head. The turbine blades are not submerged in water.
2) Reaction-type (Francis) turbines: These turbines are most commonly used for large hydraulic units. The turbine blades are submerged in water. Water is supplied from a reservoir.
3) Kaplan turbines: These turbines have blades that look like propellers. They are most suited for low-head, run-of-river hydraulic units, typically with many units located along a river. There is no need for a reservoir, thus eliminating the need to construct a dam and create a reservoir.

In contrast to steam turbines, hydraulic turbines have only a single turbine stage with relatively simple construction. As such, a hydraulic turbine has a smaller inertia, such that the H value of a hydraulic turbine-generator is smaller than that of a steam turbine-generator. The hydraulic turbine drives a generator on the same vertical shaft, with the generator mounted on top of the turbine, allowing the water to turn the turbines most efficiently.

Large hydraulic turbines of the Francis and Pelton types use a penstock to supply water to the turbines. The flow dynamics in the penstock are factored into the hydraulic turbine model. Consider the schematic of a hydraulic turbine shown in Figure 12.9. It is assumed that the penstock is inelastic (that is, its concrete wall does not expand and contract) and the water is incompressible.

The turbine in Figure 12.9 can be represented by a nonlinear model with one dynamic equation and two algebraic equations[3]

$$\frac{d\mu}{dt} = \frac{g}{L}(H_o - H) \tag{12.7}$$

$$H = \mu^2 (\frac{1}{G})^2 \tag{12.8}$$

$$P_m = H \cdot \mu \cdot A \tag{12.9}$$

3 Note that the symbol H in Section 12.3 denotes a measure of pressure, not inertia as in the rest of book.

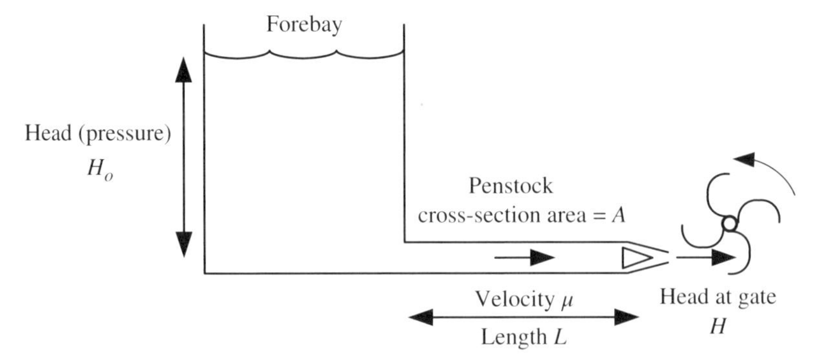

Figure 12.9 Hydro turbine and penstock.

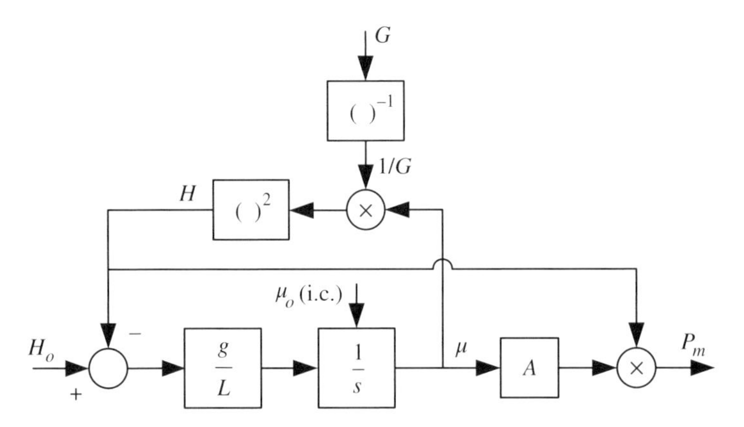

Figure 12.10 Block diagram of the nonlinear hydraulic turbine model.

where μ is the water velocity in the penstock, g the gravity constant, L the length of the penstock, H_o the head (pressure) of the reservoir, H the head at the gate, G the effective gate position, P_m the turbine mechanical power, and A the cross-section area of the penstock. These equations can be combined into the block diagram shown in Figure 12.10, where μ_o is the initial condition of the water velocity.

In steady state, $\mu = \mu_o$ is constant. Thus the equilibrium values of the various variables, denoted by the subscript "o," are

$$H = H_o, \quad G = \mu_o\sqrt{\frac{1}{H_o}} = G_o, \quad G_o^2 = \frac{\mu_o^2}{H_o}, \quad P_m = H_o\mu_o A = P_{mo} \tag{12.10}$$

Example 12.1

Given the normalized value of $G_o = 0.9$ pu, $H_o = 1$ pu, and $\mu_o = 0.9$ pu, with $L/g = 2$ sec and $A = 1$ pu, simulate the time response of the model (12.7)–(12.9) for a step increase in G of 0.05 pu. Describe the time response.

Solutions: The time responses due to a change in the gate position $G = G_o + \Delta G$ are plotted in Figure 12.11. The sequence of events from the simulation plots is described below:

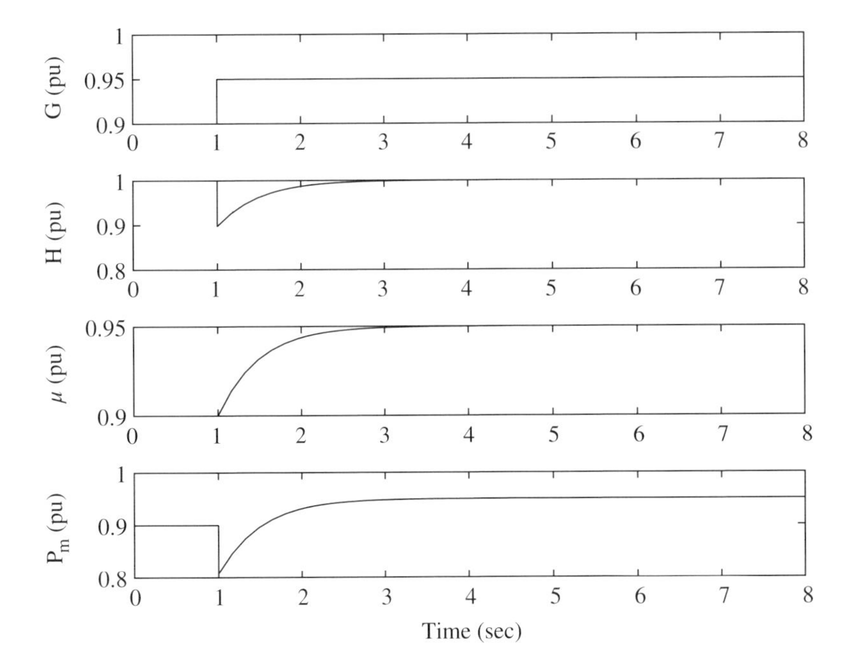

Figure 12.11 Step response of a hydraulic turbine.

1) The change in the gate position ΔG is a step command.
2) The pressure H at the gate abruptly drops because the gate is opened wider. In steady state, the pressure at the gate will eventually be equalized to the water pressure in the reservoir, which is assumed to be constant.
3) As the gate is opened wider, the water flow velocity μ in the penstock will increase. Note that the water velocity in the penstock is continuous, that is, it does not jump abruptly to a new value.
4) The mechanical power P_m drops immediately as the gate is opened, as the pressure is lowered (see (12.9)). Eventually, as more water is flowing at a higher velocity, the mechanical power will increase, achieving its new dispatch value.

This phenomenon of having a drop in mechanical power before it is eventually increased is quite common in physical systems that have storage or flow continuity constraints.[4] ■

To perform a more rigorous analysis of this phenomenon, perturbations are introduced to the variables in the nonlinear model (12.7)–(12.9) to obtain

$$\frac{d(\mu_o + \Delta\mu)}{dt} = \frac{g}{L}(H_o - (H_o + \Delta H)) \tag{12.11}$$

$$H_o + \Delta H = (\mu_o + \Delta\mu)^2 \left(\frac{1}{G_o + \Delta G} \right)^2 \tag{12.12}$$

$$P_{mo} + \Delta P_m = (H_o + \Delta H)(\mu_o + \Delta\mu)A \tag{12.13}$$

4 In economic systems, such a phenomenon of the economic situation getting worse before it gets better, like in currency devaluation, is known as a "J" curve [143].

Expanding and keeping only the first-order terms results in

$$\frac{d\Delta\mu}{dt} = -\frac{g}{L}\Delta H \tag{12.14}$$

$$\Delta H = \frac{2\mu_o}{G_o^2}\Delta\mu - \frac{2\mu_o^2}{G_o^3}\Delta G \tag{12.15}$$

$$\Delta P_m = H_o A\Delta\mu + \mu_o A\Delta H \tag{12.16}$$

Eliminating ΔH yields

$$\frac{d\Delta\mu}{dt} = -\frac{2g\mu_o}{LG_o^2}\Delta\mu + \frac{2g\mu_o^2}{LG_o^3}\Delta G = -\frac{2gH_o}{\mu_o L}\Delta\mu + \frac{2gH_o}{\mu_o L}\frac{\mu_o}{G_o}\Delta G \tag{12.17}$$

$$\Delta P_m = H_o A\Delta\mu + \frac{2\mu_o^2 A}{G_o^2}\Delta\mu - \frac{2\mu_o^3 A}{G_o^3}\Delta G \tag{12.18}$$

Dividing these two equations by μ_o and P_{mo}, respectively, results in

$$\frac{d}{dt}\frac{\Delta\mu}{\mu_o} = -\frac{2gH_o}{\mu_o L}\frac{\Delta\mu}{\mu_o} + \frac{2gH_o}{\mu_o L}\frac{\Delta G}{G_o} \tag{12.19}$$

$$\frac{\Delta P_m}{P_{mo}} = \frac{H_o A\Delta\mu}{P_{mo}} + \frac{2\mu_o^2 A\Delta\mu}{P_{mo}G_o^2} - \frac{2\mu_o^3 A}{P_{mo}G_o^2}\frac{\Delta G}{G_o} \tag{12.20}$$

The second equation (12.20) becomes, using the expression $P_{mo} = H_o\mu_o A$,

$$\frac{\Delta P_m}{P_{mo}} = 3\frac{\Delta\mu}{\mu_o} - 2\frac{\Delta G}{G_o} \tag{12.21}$$

Define the normalized variables

$$\Delta\overline{\mu} = \frac{\Delta\mu}{\mu_o}, \quad \Delta\overline{G} = \frac{\Delta G}{G_o}, \quad \Delta\overline{P}_m = \frac{\Delta P_m}{P_{mo}} \tag{12.22}$$

Equations (12.19) and (12.21) in the normalized variables are

$$\frac{d\Delta\overline{\mu}}{dt} = -\frac{2}{T_w}\Delta\overline{\mu} + \frac{2}{T_w}\Delta\overline{G} \quad \Rightarrow \quad \frac{T_w}{2}\frac{d\Delta\overline{\mu}}{dt} = -\Delta\overline{\mu} + \Delta\overline{G} \tag{12.23}$$

$$\Delta\overline{P}_m = 3\Delta\overline{\mu} - 2\Delta\overline{G} \tag{12.24}$$

where $T_w = (\mu_o L)/(gH_o)$ is known as the water starting time constant.

Apply Laplace transform to (12.23) to obtain

$$\frac{T_w}{2}s\Delta\overline{\mu}(s) = -\Delta\overline{\mu}(s) + \Delta\overline{G}(s) \quad \Rightarrow \quad \left(\frac{T_w}{2}s + 1\right)\Delta\overline{\mu}(s) = \Delta\overline{G}(s) \tag{12.25}$$

that is,

$$\Delta\overline{\mu}(s) = \frac{1}{T_w s/2 + 1}\Delta\overline{G}(s) \tag{12.26}$$

such that the transfer function from $\Delta\overline{G}$ to $\Delta\overline{P}_m$ becomes

$$\Delta\overline{P}_m = \left(3\frac{1}{T_w s/2 + 1} - 2\right)\Delta\overline{G} = \left(\frac{1 - T_w s}{1 + T_w s/2}\right)\Delta\overline{G} \tag{12.27}$$

In (12.27), the water starting time constant T_w provides an indication of the time required for the water velocity to change from an initial speed to a final speed. In practical hydraulic units, T_w is normally in the range of 0.5–4.0 seconds, which is dependent on the pressure arising from the reservoir and the length of the penstock. A typical value of T_w is 2 seconds.

The pole of the transfer function (12.27) is $-2/T_w$ (in the left half-plane (LHP)), and the zero is $1/T_w$ (in the right half-plane (RHP)). Systems with RHP zeros are known as "non-minimum-phase" systems [113]. With one RHP zero, a step increase in the input will result in a temporary decrease of the output, before the output recovers to a higher value. Thus at $t = 0^+$ (initial time), the initial-value theorem yields

$$\Delta \overline{P}_m = \lim_{s \to \infty} s \left(\frac{1 - T_w s}{1 + (T_w/2)s} \right) \frac{\Delta \overline{G}}{s} = -2\Delta \overline{G}$$

As $t \to \infty$ (steady state), the final-value theorem yields

$$\Delta \overline{P}_m = \lim_{s \to 0} s \left(\frac{1 - T_w s}{1 + (T_w/2)s} \right) \frac{\Delta \overline{G}}{s} = \Delta \overline{G}$$

Thus a step change in the command signal for such systems is not desirable. Normally a ramp command is used to reduce the impact of the RHP zero.

For a reader familiar with electric circuits, the simple RL circuit in Figure 12.12 is a useful analog. The power $P_R = i^2 R = E_o i$ consumed by the resistor drops if R is decreased suddenly because the current i in the inductor does not change instantaneously. Eventually P_R will increase because the current i will increase.

12.3.1 Hydraulic Turbine-Governors

As in a steam turbine, a governor for a hydraulic turbine can be used to regulate the machine speed by adjusting the gate to control the amount of water flowing into the turbine and thus the mechanical power. The implementation of a governor in a hydraulic turbine is shown in Figure 12.13, where $\Delta\omega_{\text{ref}}$ (nominally zero) is the change in the reference speed and R_p is the (permanent) droop.

To counter the non-minimum phase effect of the RHP zero due to the penstock, a transient droop is added to modulate the gate ΔG. The implementation of the transient droop is shown in Figure 12.14. Note the transient droop transfer function is like the transient gain reduction block of the voltage regulator (Section 9.3.5). In Figure 12.14, T_R is the reset time ($T_R \simeq 5T_w$) and the transient droop gain is $R_T \simeq 10R_p$. Here the permanent droop R_p can be set to be the droop used in steam turbines.

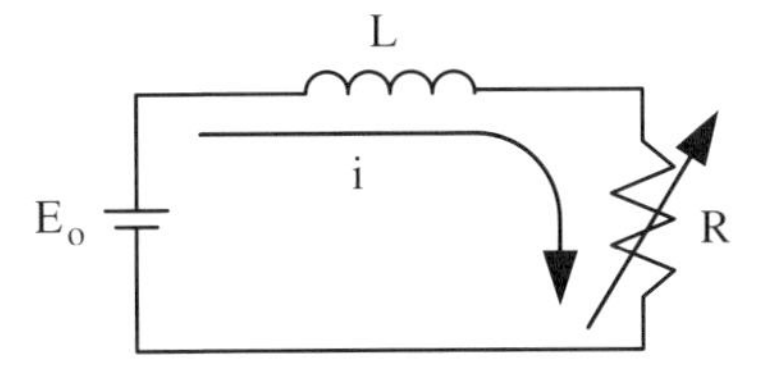

Figure 12.12 Electric circuit analog.

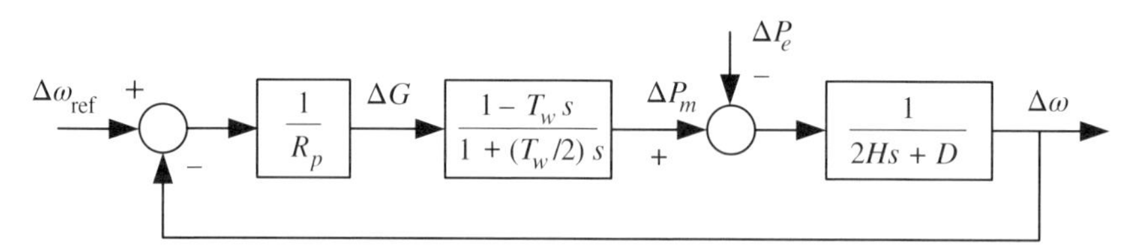

Figure 12.13 Hydraulic-turbine mechanical power control system (deadband not shown).

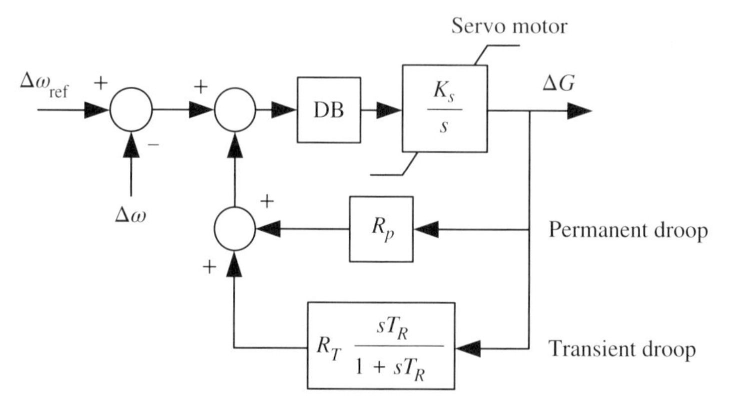

Figure 12.14 Hydraulic turbine-governor with transient droop and deadband (DB) implementation.

Assuming that the servo motor is relatively fast, that is, $K_s > 0$ is large, the transfer function in Figure 12.14 can be simplified to

$$\Delta G = \frac{1}{R_p} \frac{1 + sT_R}{1 + (1 + R_T/R_p)sT_R} (\Delta\omega_{ref} - \Delta\omega) = H_{td}(s)(\Delta\omega_{ref} - \Delta\omega) \tag{12.28}$$

At high frequency ($s \to \infty$), $|H_{td}| = 1/(R_p + R_T)$, and at low frequency ($s \to 0$), $|H_{td}| = 1/R_p$. The governor frequency response of a hydraulic turbine is given in the next example.

Example 12.2

Plot the frequency response of $H_{td}(s)$ given $R_p = 0.05$, $R_T = 10R_p$, and $T_R = 5T_w$, where $T_w = 2$ sec.

Solutions: The frequency response of $H_{td}(s)$ is shown in Figure 12.15, showing a higher gain in steady state than at high frequency. ■

12.3.2 Load Rejection of Hydraulic Turbines

During disturbances such as load rejections, the turbine control system has to close the valves quickly to stop the water from flowing into the turbine or have other means to spill water, to prevent the turbine from overspeeding and suffering damages. Under such a condition in which the water flow is significantly affected, the analysis needs to account for the fact that the concrete wall of the conduit is flexible and the water velocity and the

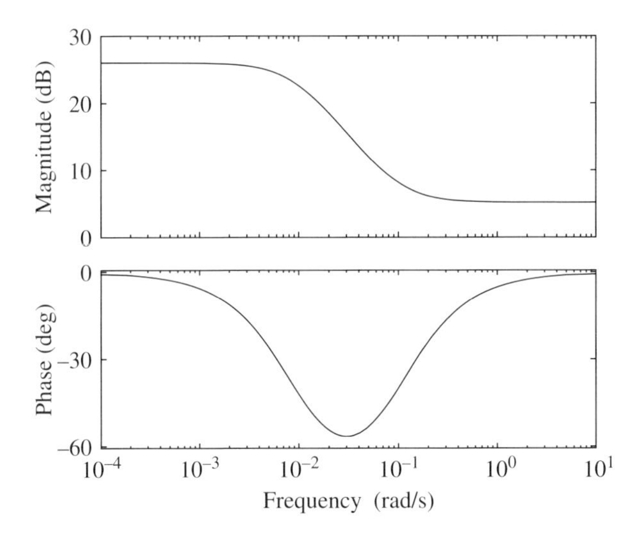

Figure 12.15 Frequency response of $H_{td}(s)$ for Example 12.2.

pressure in the penstock are not uniform throughout the length of the penstock. Thus the water velocity μ should be modeled as a wave equation (which is a second-order linear partial differential equation)

$$\frac{\partial^2 \mu}{\partial t^2} = c^2 \frac{\partial^2 \mu}{\partial x^2} \tag{12.29}$$

where x denotes the distance in the penstock and c is the traveling wave velocity. The solution of the wave equations consists of a wave traveling forward and a wave traveling backward. Such a phenomenon is known as a water hammer and can damage the penstock and the hydraulic turbine.

To protect against traveling waves in the penstock, a surge tank can be incorporated in the design, thus providing an "escape path" for the water to flow into (see Figure 12.16). For projects that cannot provide a surge tank, such as those concerned with earthquakes, a Pelton turbine can be used instead of a Francis turbine. In a Pelton turbine, load rejection can be accomplished by diverting the water jets away from the turbine buckets using deflectors. This is a highly nonlinear process. The paper [144] describes the investigation of the deflector instability of a Pelton turbine and provides a feedforward control design to eliminate the nonlinear stability problem.

12.4 Gas Turbines and Co-Generation Plants

Due to short start-up times, lower emission, and faster construction-to-commission, gas (combustion) turbines and combined-cycle co-generation plants have become an important generation mix for participation as 5- and 15-minute reserves in deregulated electricity markets. Although gas turbines tended to set the market clearing price during the early years of several US deregulated electricity markets, in recent periods, because of lower gas prices, gas turbine and co-generation plants can also compete as base-load units. Models of gas turbines and combined-cycle power plants are quite complex. This

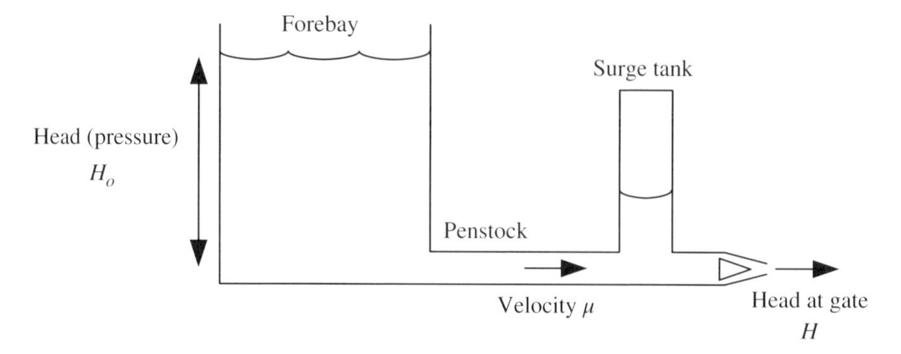

Figure 12.16 A hydraulic turbine configuration showing a surge tank.

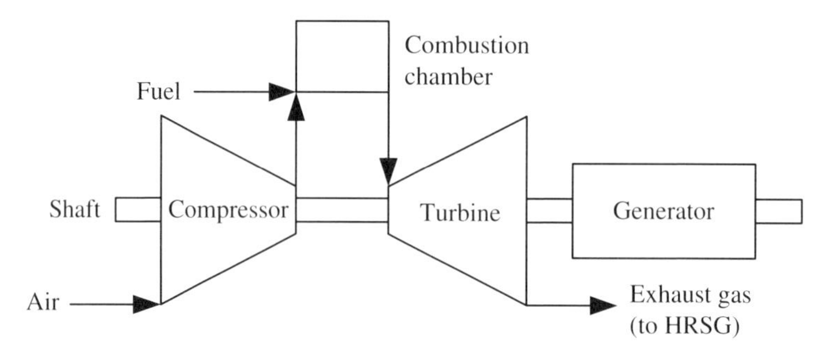

Figure 12.17 Gas turbine model.

section gives a reader a brief overview of such power plants without going into the modeling details, which can be found in [145–147].

A diagram showing the four major components of a gas turbine is given in Figure 12.17. The air for combustion is compressed to a higher pressure by a compressor, which is part of the load of a gas turbine. The compressed air is sent into the combustion chamber, where it is mixed with the fuel and burned to generate high-temperature, high-pressure gas. The energy of the combusted gas is partially captured by the turbine driving a generator. The heat content of the exhaust gas, still at a very high temperature, can be captured separately by a heat-recovery steam-generator (HRSG).

The operation of a gas turbine will depend on the ambient temperature of the inlet air as well as the rotational speed of the shaft. The output power of the turbine is directly proportional to the fuel flow. The maximum power output in steady state is limited by the turbine temperature limit, which is a function of the fuel and air mixture. Generic and vendor-specific gas-turbine models used by industry can be found in [147].

In a combined-cycle power plant (CCPP) with multiple gas turbines, it may be more cost effective to feed all the exhaust gas into a single HRSG. The HRSG and steam turbine may operate at more than one steam pressure level and drive a separate generator.

Figure 12.18 shows the schematic of one of the first HRSGs. This two-pressure-level HRSG produces steam at high pressure (HP) and low pressure (LP), which is then used in a steam turbine [148]. Inside the HRSG is a series of heat exchangers, intended to capture as much useful heat as possible from the hot exhaust gas. The hot exhaust gas

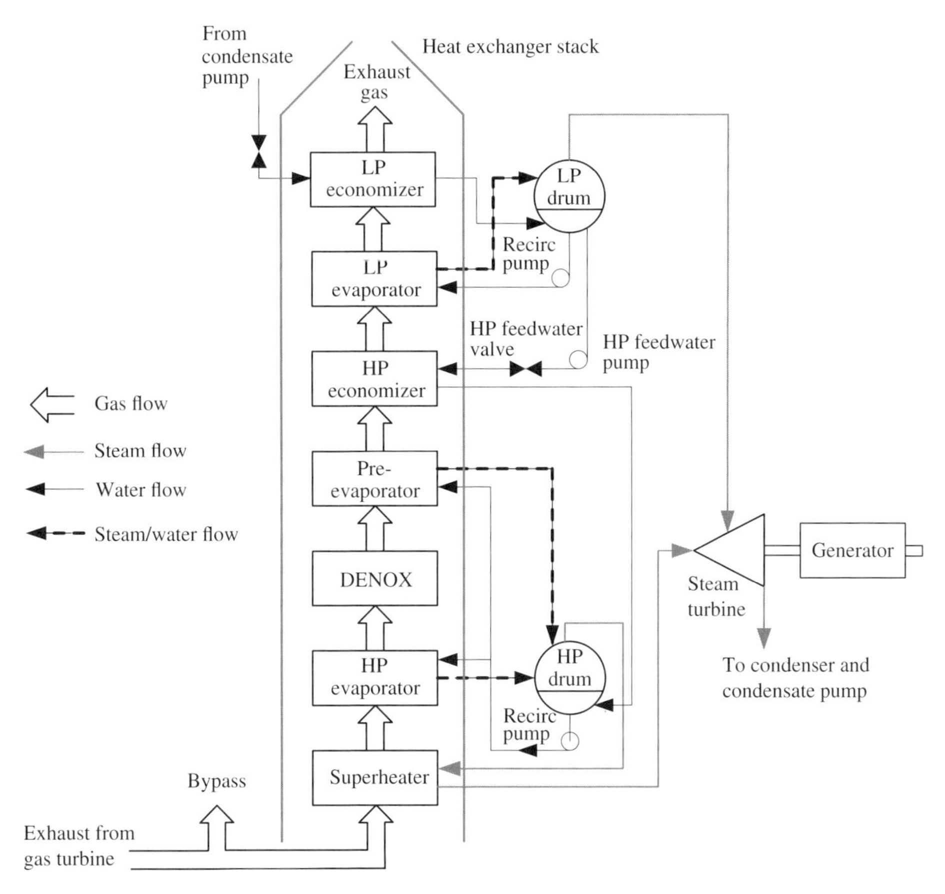

Figure 12.18 Dual-pressure HRSG flow diagram [148] (note the blowdown valves from the drums are omitted) (see color plate section).

goes through the HP heat exchangers before reaching the LP heat exchangers, as the temperature of the exhaust gas is progressively lowered. There are normally three types of heat exchangers: the economizer is for increasing the temperature of the circulating water, the evaporator is to turn water into steam, and the super-heater is to increase the temperature and take moisture out of the HP steam flow.[5] The steam is stored in drums, with one drum corresponding to each pressure level. The flow of water in the fluid stage is controlled by the recirculation pumps, and in the steam stage is a function of the steam pressure.

The dynamics inside a HRSG involve mainly heat transfer between the hot exhaust gas and the steam or water inside the heat exchangers. The rate of heat absorption is approximately modeled by

$$\dot{Q}_e = \eta(T_g - T_i)\dot{m} \tag{12.30}$$

5 Note the component to remove nitrous oxides (DENOX) is included in the model because it uses energy from the exhaust gas, without contributing to steam generation.

where Q_e is the heat absorbed by the heating section, η the heat transfer coefficient, T_g the hot-side gas temperature, T_i the cold-side water/steam temperature, and $\dot{m}$ the flow rate of water/steam. The inputs to the HRSG are the gas-turbine exhaust gas flow rate and temperature, and the feedwater temperature. A detailed HRSG model may have many time constants, one for each heat exchanger and boiler. In [146], a simplified model with two time constants, one of the order of 5 seconds for the heat exchange process and the other of the order 50–200 seconds, is used to represent the boiler storage dynamics. There are continuing investigations of appropriate gas-turbine and co-generation plant models, with efforts to match the simulated power output response against the actual measured power output [147].

An effective power control of the steam turbine in a co-generation plant is the inlet guide vane that can control the amount of gas-turbine exhaust going into the heat-exchanger stack. This control is shown as the bypass in Figure 12.18.

12.5 Primary Frequency Control

A key attribute of a reliable power system is its ability to successfully deal with power imbalances due to sudden changes in generation or load by keeping the system frequency within a tight range.[6] Frequency is a global variable such that generation-load imbalance is visible throughout the entire grid. Power system operators need controllable generation and, as a last resort, load shedding, in order to regulate or restore frequency. The control actions are usually applied in different stages and time frames in a control continuum [149], as described in Table 12.1. The rows in Table 12.1 correspond to the classifications of how the generation and load should respond and the appropriate time frame.

Primary frequency control (or simply frequency response) is provided by generator governor actions responding to frequency deviation from the nominal value and load sensitivities to frequency. Secondary control is normally accomplished with AGC, which deploys frequency regulating generators to reduce the area control error (ACE), consisting of the sum of a frequency bias multiplying the frequency deviation and the difference of the actual interchange from the scheduled interchange between a control area and its neighboring areas. Tertiary control encompasses a new system dispatch to provide new generation and reserves to meet the current loading condition and future contingencies. Time-error correction is applied when needed to correct for the system frequency error over time.

Primary and secondary frequency control mechanisms are addressed here to illustrate the generator dynamic control functions. Tertiary frequency control is more related to power system optimization and is addressed in power system economic operation textbooks such as [11].

In considering primary frequency control, the operation of an isolated system with a single generator is considered first. Then the frequency control of two interconnected units is discussed.

6 Prolonged off-nominal frequency operation may trigger oscillations/resonances in turbine-generators and cause equipment damage.

Table 12.1 Frequency control continuum (taken from Table 1 in [149]).

Control	Ancillary service	Time frame
Primary control	Frequency response	10–60 seconds
Secondary control	Regulation service	1–10 minutes
Tertiary control	Imbalance/reserves	10 minutes to hours
Time control	Time error correction	Hours

Note that in this section on primary frequency control and the next section on secondary frequency control, ω is used to denote the actual machine speed. Machine speed deviation from the nominal frequency will be explicitly denoted by $\Delta\omega$. However, in Section 12.7, ω will again represent the machine speed deviation from the nominal frequency.

12.5.1 Isolated Turbine-Generator Serving Local Load

Figure 12.19 shows an isolated generator with an input mechanical power P_m serving a local load P_e, such that $P_{mo} = P_{eo}$ in steady state. Consider switching on a load of ΔP_e, increasing the load to

$$P_e = P_{eo} + \Delta P_e \tag{12.31}$$

The turbine-governor will respond to this additional load by increasing the mechanical input power to regulate the system frequency.

As an illustration, let the base load be $P_{eo} = 0.7$ pu and the step load increase be $\Delta P_e = 0.2$ pu. The steady-state change in frequency is shown in Figure 12.20, in which two load curves are shown: the vertical curve is due to a load that is independent of frequency, and the curve passing through the origin is due to a load that varies linearly with the frequency. An expanded view of Figure 12.20 is shown in Figure 12.21.

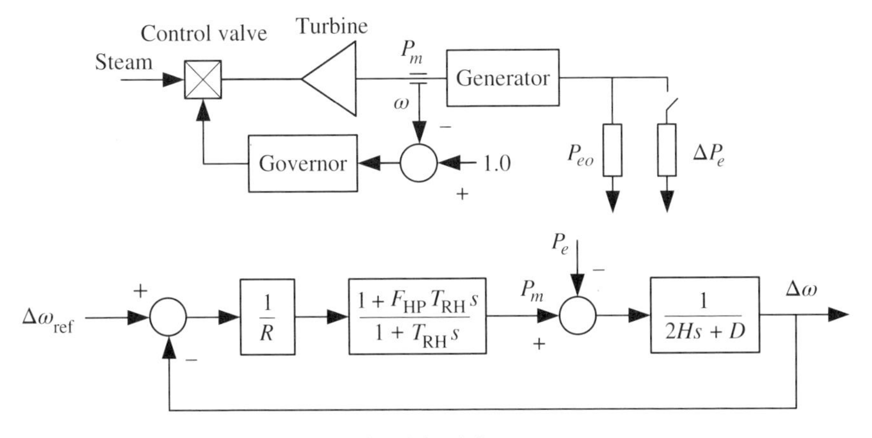

Figure 12.19 Turbine-governor model with load change.

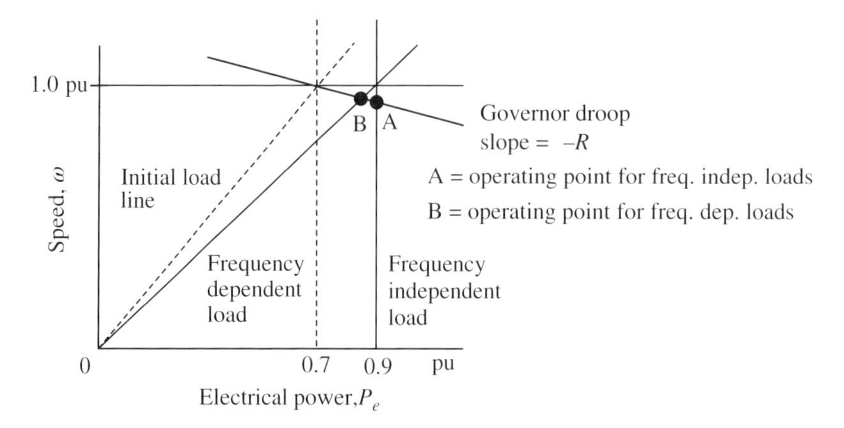

Figure 12.20 Frequency variation for load increase.

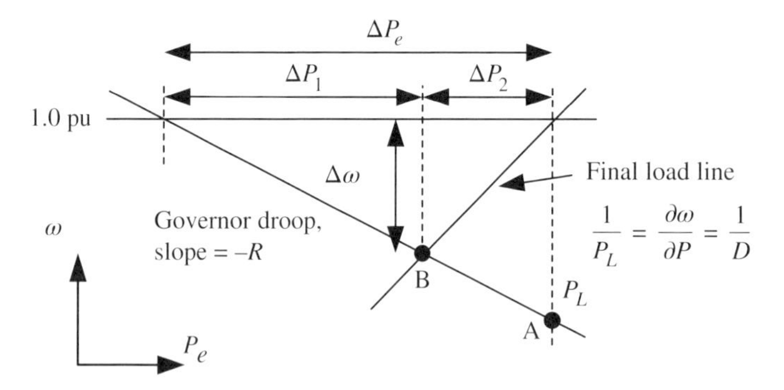

Figure 12.21 Expanded view of the change in operating point.

For frequency-independent load in the isolated system, the change in frequency $\Delta\omega = \omega - \omega_{\text{ref}}$ in steady state is

$$\Delta\omega = -R\Delta P_e \tag{12.32}$$

where ΔP_e is taken to be positive for an increase in load. This situation is denoted by the operating point A. The mechanical input power will be increased to 0.9 pu.

For a load that is linearly dependent on frequency, as denoted by the operating point B in Figure 12.21, the total load is $P_L = P_e$ and the frequency relationship is

$$\frac{1}{P_L} = \frac{-\Delta\omega}{\Delta P_2} \quad \Rightarrow \quad \Delta P_2 = -P_L\Delta\omega \tag{12.33}$$

where ΔP_2 is the load decrement due to frequency dependency. Observe from Figure 12.21 that

$$R = -\frac{\Delta\omega}{\Delta P_1} = -\frac{\Delta\omega}{\Delta P_e - \Delta P_2} \tag{12.34}$$

Combine (12.33) and (12.34) to obtain

$$\Delta\omega = -R(\Delta P_e - \Delta P_2) = -R(\Delta P_e + P_L\Delta\omega) \tag{12.35}$$

that is,

$$\Delta\omega = -\frac{R}{1+RP_L}\Delta P_e = -\frac{1}{(1/R)+P_L}\Delta P_e = -\frac{1}{(1/R)+D}\Delta P_e = -\frac{\Delta P_e}{\beta} \tag{12.36}$$

Note that $\beta = (1/R) + D$ has dual components of governing and load sensitivity, and is commonly known as the system frequency response. It has the unit of pu power/pu frequency.

Example 12.3

An isolated power system with several generators is supplying a load of $L = 1000$ MW. The effective droop of the governors across all the generators is $R = 0.1$, such that $1/R = 10$. Assume that the load is 60% proportional to frequency and 40% constant power. Calculate β and use it to find the steady-state frequency when a load of 50 MW (with the same sensitivity to frequency) is switched on, assuming that the nominal system frequency is 60 Hz. Here it is convenient to use 1000 MVA as the system base such that the load is 1 pu.

Solutions: The load sensitivity to frequency is

$$D = \frac{\partial L}{\partial \omega} = 0.6 \text{ pu power/pu frequency} \tag{12.37}$$

Thus

$$\beta = D + (1/R) = 0.6 + 10 = 10.6 \text{ pu power/pu frequency} \tag{12.38}$$

In a 60 Hz system

$$\beta = \frac{10.6 \times 1{,}000}{60} = 176.67 \text{ MW/Hz} = 17.667 \text{ MW/0.1 Hz} \tag{12.39}$$

Thus an increase in 50 MW of load will cause a frequency decrease of, using (12.38),

$$\Delta\omega = -(50/1000)/10.6 = -0.004717 \text{ pu} \tag{12.40}$$

or, equivalently, using (12.39)

$$\Delta f = -50/176.67 = -0.2830 \text{ Hz} \tag{12.41}$$

■

Example 12.4

In this example, the steady-state and dynamic performance of load frequency control of various types of turbines in an isolated system is shown. In steady state, the load of $P_L = 0.9$ pu is supplied by the turbine providing $P_{mo} = 0.9$ pu. Suppose that all the load is linearly sensitive to the system frequency and the inertia of the generator is $H = 4$ sec. An additional constant power load $\Delta P_L = 0.05$ pu is switched on. Simulate the performance of (a) a turbine with a blocked governor, (b) a steam turbine without reheat and with reheat, using a droop of $R = 0.05$, a reheat time constant of $T_{\text{RH}} = 5$ sec, and a HP power fraction $F_{\text{HP}} = 0.3$, and (c) a hydraulic turbine with $T_w = 2$ sec, using only the permanent droop of $R_p = 0.25$, and then with both the permanent and temporary droop $R_p = 0.05$, $R_T = 10 \times R_p$, and $T_R = 5T_w$.

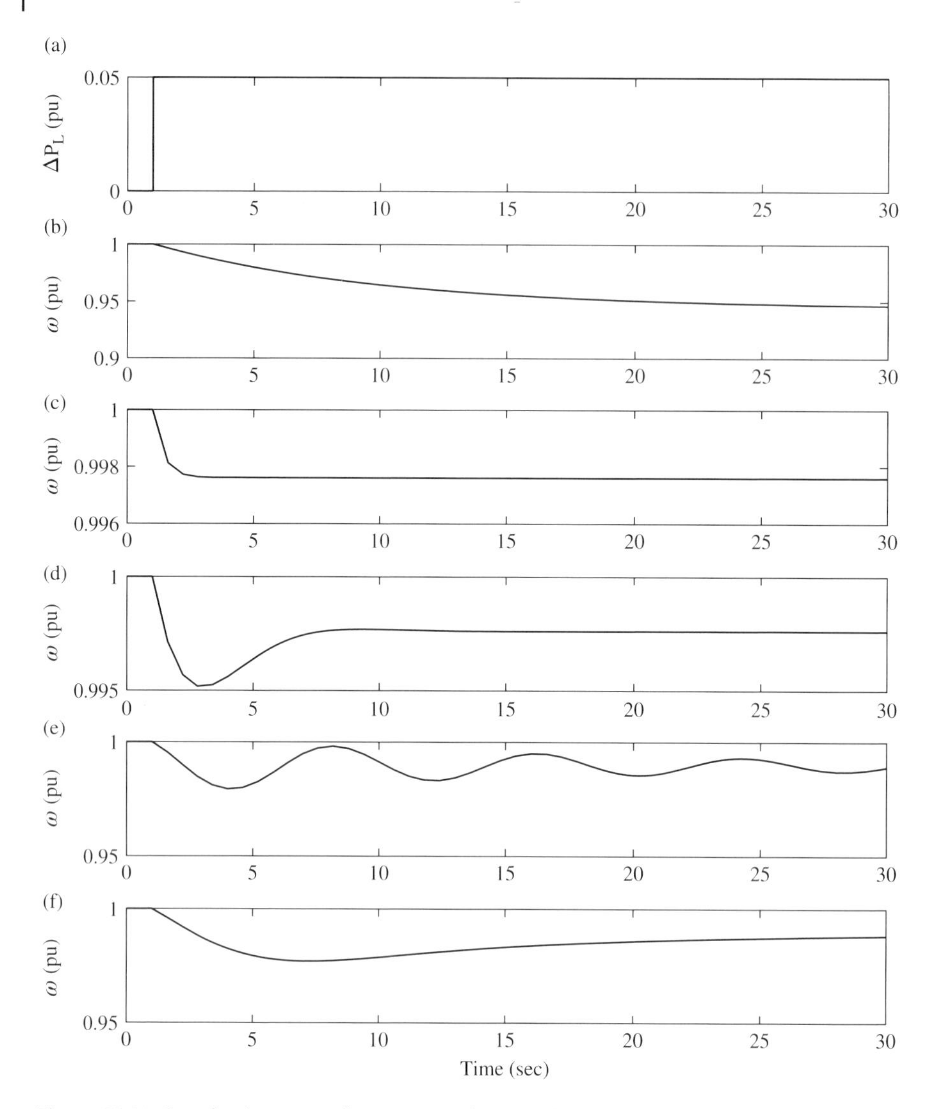

Figure 12.22 Step-load response for various turbines and governors: (a) load step increase, (b) blocked governor response, (c) steam turbine without reheat, (d) steam turbine with reheat, (e) hydraulic turbine with only permanent droop, and (f) hydraulic turbine with temporary droop and permanent droop.

Solutions: The performance of the various load frequency control of the turbines is plotted in Figure 12.22, which is discussed in more detail below.

With the governor blocked, the mechanical input power P_m remains the same. Thus the steady-state change in frequency is

$$\Delta\omega = -0.05/0.9 = -0.0556 \text{ pu} \tag{12.42}$$

The time response (plot (b)) shows a slow exponential decay in frequency. The steam turbine responds quite fast without the reheat function (plot (c)), whereas including the reheat dynamics, the frequency undershoots (plot (d)). In both cases, the frequency change settles to

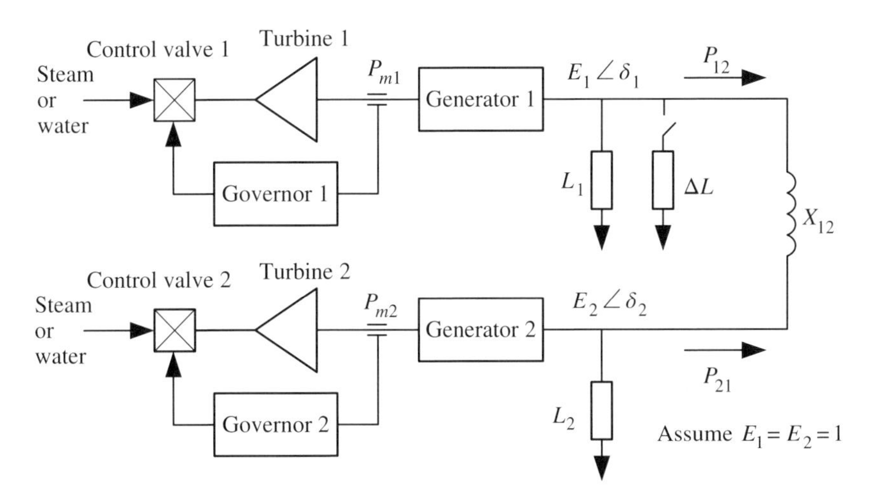

Figure 12.23 Two power systems connected by a reactance X_{12}.

$$\Delta\omega = -0.05/(1/0.05 + 0.9) = -0.0024 \text{ pu} \tag{12.43}$$

Plot (e) shows the response of the hydraulic turbine with a permanent droop of $R_p = 0.25$. Note that a higher value of R_p would yield a larger steady-state $\Delta\omega$. Due to the right-half-plane zero, an oscillatory response ensues, which is problematic as the gate will be opening and closing, producing water wave oscillations in the penstock. A tighter droop will result in a unstable response. By using both the temporary and permanent droop, with the permanent droop the same as the steam turbine droop, a non-oscillatory response is produced, although the settling time is longer. ■

12.5.2 Interconnected Units

Figure 12.23 shows the interconnection of two generators via a tieline with a reactance X_{12}, and the load is assumed to be linearly dependent on the frequency. Also the system voltages are assumed to be 1 pu and line losses are neglected. The power transfer between the two regions is

$$P_{12} = \frac{\sin(\delta_1 - \delta_2)}{X_{12}} = \frac{\sin\delta_{12}}{X_{12}} \tag{12.44}$$

Assuming a linearized power flow model, the power transfer is approximated by

$$P_{12} \simeq A + \frac{\delta_{12}}{X'_{12}} \tag{12.45}$$

where

$$A = \frac{\sin\delta_{12o}}{X_{12}} - \frac{\delta_{12o}}{X'_{12}}, \quad X'_{12} = \frac{X_{12}}{\cos\delta_{12o}} \tag{12.46}$$

and δ_{12o} is the initial condition.

Simplifying the block diagram in Figure 12.24 by combining the integrators and moving A to a different summing junction results in the model shown in Figure 12.25.

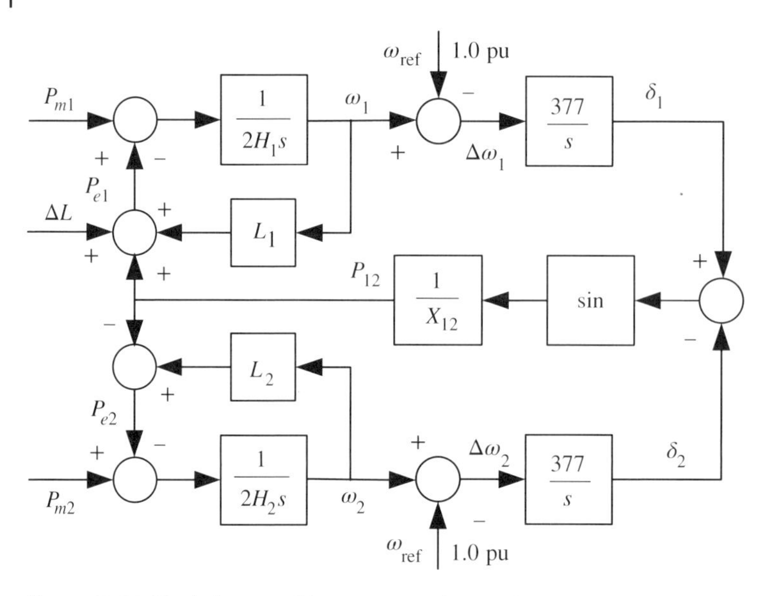

Figure 12.24 Block diagram of interconnected operation.

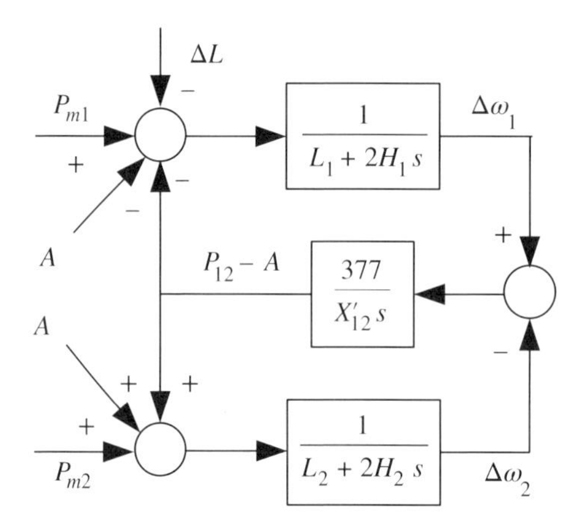

Figure 12.25 Simplified model of the interconnected system.

Neglecting the inputs P_{m1} and P_{m2}, the system in Figure 12.25 can be further simplified to the block diagram shown in Figure 12.26. The transfer function from ΔL to $\Delta\omega_1$ in Figure 12.25 can be readily derived as

$$\Delta\omega_1 = -\frac{T + L_2 s + 2H_2 s^2}{T(L_1 + L_2) + (L_1 L_2 + 2(H_1 + H_2)T)s + 2(H_1 L_2 + H_2 L_1)s^2 + 4H_1 H_2 s^3}\Delta L \tag{12.47}$$

where $T = 377/X'_{12}$.

To develop some additional insight into the transfer function (12.47), three special cases are examined.

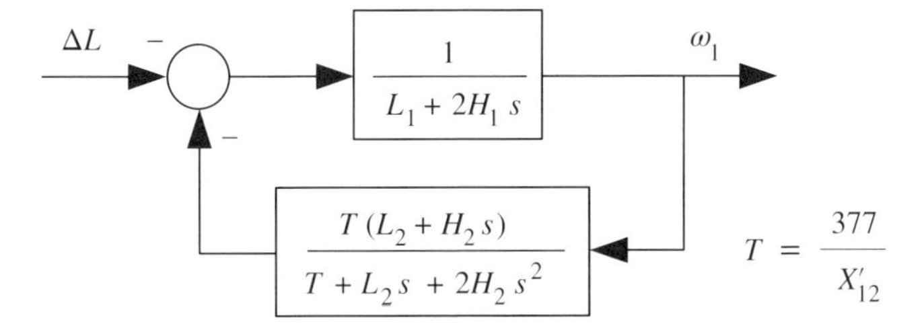

Figure 12.26 Block diagram of transfer function from ΔL to $\Delta\omega_1$.

Case 1: $H_1 = H_2 = H, L_1 = L_2 = L$
In this case, (12.47) becomes

$$\Delta\omega_1 = -\frac{T + Ls + 2Hs^2}{(L + 2Hs)(2T + Ls + 2Hs^2)}\Delta L \tag{12.48}$$

Note that there are two sets of modes in the ω_1 response:

1) The factor $(L + 2Hs)$ gives rise to a real pole $s = -L/2H$, which is responsible for the exponential response of the system frequency.
2) The factor $(2T + Ls + 2Hs^2)$ gives rise to the swing (electromechanical) mode.

Case 2: ω_2 **constant (infinite bus)**
In this case, (12.47) becomes

$$\Delta\omega_1 = -\frac{s}{T + L_1 s + 2H_1 s^2}\Delta L \tag{12.49}$$

Thus only the electromechanical mode due to ω_1 is present. There is no exponential system frequency response, as the system frequency is constant due to the infinite bus.

Case 3: $\omega_1 = \omega_2$ **(lumped inertia model)**
In this case, the transfer function from ΔL to $\Delta\omega$ is

$$\Delta\omega = -\frac{1}{L_1 + L_2 + 2(H_1 + H_2)s}\Delta L \tag{12.50}$$

which has a pole at $s = -(L_1 + L_2)/(2(H_1 + H_2))$, resulting in an exponential decay with no electromechanical mode oscillations.

This analysis illustrates that load and generation imbalances occurring in one control area will be "exported" to other control areas because of the tie line interconnection. This reliance on external systems is beneficial to the overall power system operation as the loss of a large generator in one area is much less significant in an interconnected system with many such large units.

12.5.3 Frequency Response in US Power Grids

The North America Power Grid consists of four asynchronous interconnections: Eastern Interconnection (EI), Western Interconnection (WECC), Texas Interconnection (ERCOT), and Quebec Interconnection. The North America Electric Reliability Corporation (NERC) has been tracking the primary frequency response and β for large generation and load disturbances of each interconnection for many years [149].

As an example, the primary frequency response of the US western power grid (WECC) after a loss of generation event is shown in Figure 12.27. The primary frequency response is measured by three parameters:

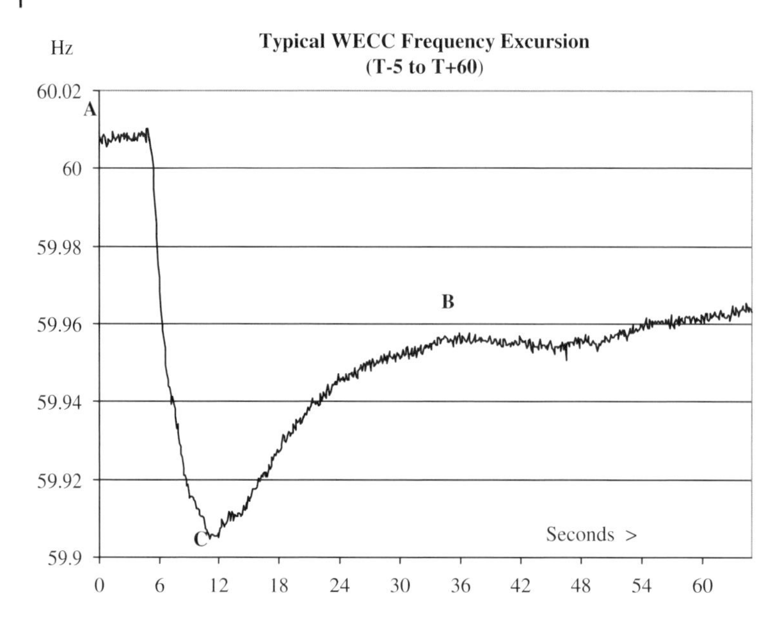

Figure 12.27 WECC frequency response from a generating unit trip at 5 seconds [149].

- Rate of change of frequency (ROCOF) – the rate of change of the frequency at the onset of the disturbance should be lower than certain limits.
- Frequency nadir – the maximum frequency excursion as measured by the frequency difference between Points A and C in Figure 12.27 should also be within certain limits.
- Settling frequency – Point B in Figure 12.27 is the frequency to which the primary frequency control stabilizes within a minute after the onset of the disturbance.

The ROCOF and nadir are strongly affected by the total rotating inertias in the system, whereas the settling frequency is determined by the governor droop functions and the load sensitivities to frequency changes. The combination of governor and load responses to sudden generation and load changes is known as the frequency response, which is described previously as β (12.36), now defined by using measured generation and frequency changes in a large interconnection. This system frequency response is obtained by measuring the frequency decline after the loss of a large generating unit, using the expression

$$\beta = \frac{\text{loss of generation or increase in load (MW)}}{\text{decrease in system frequency (Hz)}} \tag{12.51}$$

For example, if losing a 600 MW generating unit results in a 20 mHz decline in frequency (the difference between Points A and B in Figure 12.27), then

$$\beta = \frac{600}{0.020} = 3000\ \text{MW}/0.1\ \text{Hz} \tag{12.52}$$

where β is normally defined with respect to 0.1 Hz for large interconnections. Table 12.2 shows several values of β reported in [149].

Starting in the 1990s, system operators in the US Eastern Interconnection had observed a steady decline of the so-called system frequency response [150, 151]. With

Table 12.2 Frequency response β in various North America interconnections.

Interconnection	β (MW/0.1Hz)
EI	2760
WECC	1482
ERCOT	650
Quebec	120

many power control regions deregulated, such that only turbine-generators under contract for regulation service will provide frequency control function, a further decline in system frequency response has been observed. Additional concerns have been raised regarding renewable generation as solar panels have no inertia and the inertias of wind turbines are hidden from the grid by their power electronics interface. Thus it is important to continuously monitor the ROCOF, nadir, and β during large system upsets.

12.6 Automatic Generation Control

Most power systems do not operate in isolation. Interconnections with other power systems offer more diversity in energy supply and additional system inertias for reducing system frequency deviation due to large system upsets. Ideally in real-time system operation, a control area[7] should maintain the system frequency and the tie-line active power flow (interchange between neighboring areas) precisely at their optimal dispatch values. Because generators cannot change their power output instantaneously, real-time system dispatch is typically performed every 5–15 minutes, based on real-time actual and predicted loads. As the generator setpoints are changed, regulation units are controlled to minimize frequency and tie-line flow deviations. This automatic generation control (AGC) allows a power system to reach the new operating condition in an orderly manner, despite ramp-rate limits of generators and load uncertainties. Also known as secondary frequency control, AGC is implemented in a distributed manner without communication between the control areas.

To illustrate, consider the simplified two-area system shown in Figure 12.28, in which the two areas are interconnected with a tie-line of reactance X_{12}. In simplified AGC studies, the generators in each area are aggregated into a single unit. There is only one common system frequency f, and the machine speed ω of all the individual generators is the same as the system frequency. Thus interarea oscillations are neglected in this model. Furthermore, the system voltage is assumed to be uniformly at 1 pu, i.e., $E_1 = E_2 = 1$ pu. In Figure 12.28, for Area i, $i = 1, 2$, the power generation is P_{mi} and the load is L_i. The AGC mechanism is to use the system frequency f and the actual interchange power flow IC_i to compute the load control signal $P_{\text{LC}i}$ as the power setpoint of Generator i. The same concept is applicable to systems with more than two areas. The load perturbations ΔL_1 and ΔL_2 are assumed to be independent of the system frequency.

7 Such a control area is also known as a balancing authority.

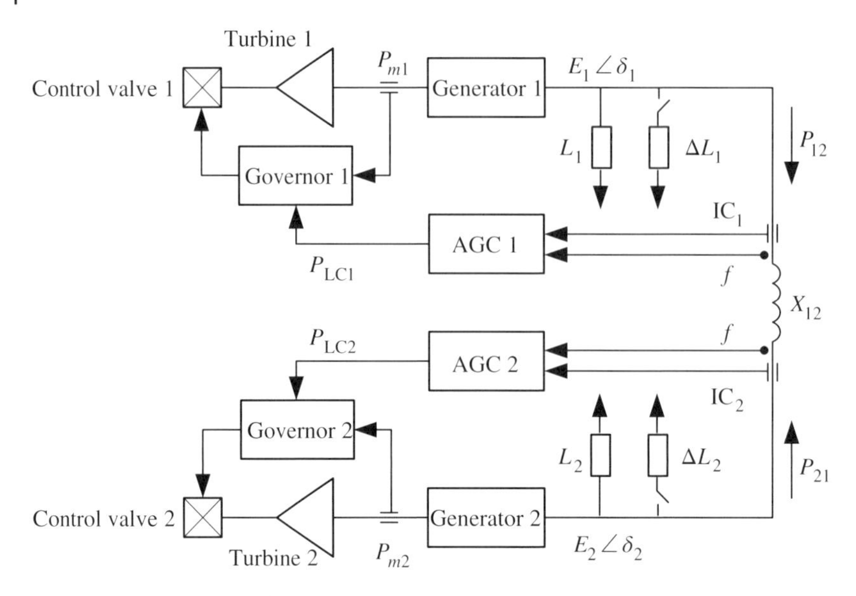

Figure 12.28 Schematic of a two-area automatic generation control system.

For the two-area system, the system frequency is given by, assuming frequency dependent loads,

$$2H\frac{d\omega}{dt} = P_m - \omega L \tag{12.53}$$

where $H = H_1 + H_2$, $P_m = P_{m1} + P_{m2}$, and $L = L_1 + L_2$.

The interchange power flow in this system refers to the power transfer on the tie-line between the two areas. DC power flow will be used such that

$$P_{12} = \frac{\delta_1 - \delta_2}{X_{12}} = -P_{21} \tag{12.54}$$

in which losses are neglected.

The interchange control variable IC_{ref} refers to the scheduled interchange of the tie-line flow, as determined by a security-constrained economic dispatch. A positive IC_{ref} implies that the area is exporting. Without losses, $IC_{ref2} = -IC_{ref1}$. The interchange error in either MW or pu power is defined as

$$\Delta IC_i = IC_i - IC_{refi}, \quad i = 1, 2 \tag{12.55}$$

such that $\Delta IC_i > 0$ implies that Area i is providing tie-line flow above the scheduled value.

The ACE for Area i is defined as [149][8]

$$\begin{aligned} ACE_i &= \text{net interchange deviation} + \text{bias} \times \Delta f \\ &= \Delta IC_i + \beta_i \Delta f \end{aligned} \tag{12.56}$$

The parameter β_i is called the control area (or balancing authority) bias for Area i [149]. The term bias $\times$ Δf is the control area's obligation to support frequency. When assigned

8 In [149], the ACE expression has a negative sign for the second term because β is defined to be negative.

properly, the sum of the control area biases will ideally sum to the system frequency response (12.36) and (12.51), that is,

$$\beta = \beta_1 + \beta_2 \tag{12.57}$$

Note that ACE has the unit of MW or pu power. A control area having a positive ACE would be generating more power than scheduled, and vice versa. A control area having a negative ACE is deficient in scheduled generation.

Due to the stochastic nature of loads, ACE is not always exactly zero for all the control areas. A control area can use the ACE signal to adjust the power settings of the regulating units to reduce ACE. In practice, AGC consists of a sequence of generator setpoint values sent to the regulating units, based on the ramp rate capabilities of the units. These signals become load control commands P_{LCi} to adjust the mechanical input power of the generators in Area i.

Figure 12.29 shows the block-diagram representation of AGC for the two-area system in Figure 12.28, in which the turbine model is represented by a second-order transfer function with a governor time constant T_G and a steam-turbine time constant T_s.

The AGC generator setpoint pulses to form P_{LCi} are implemented as an integration process

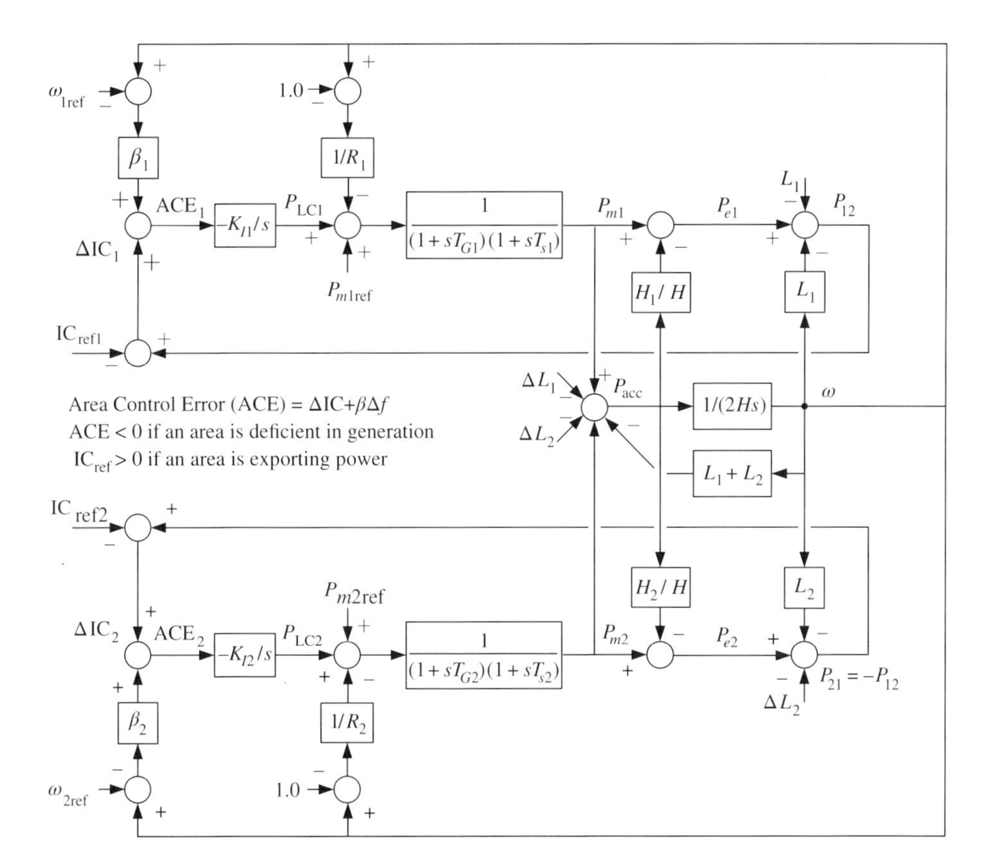

Figure 12.29 Block-diagram representation of a two-area automatic generation control system.

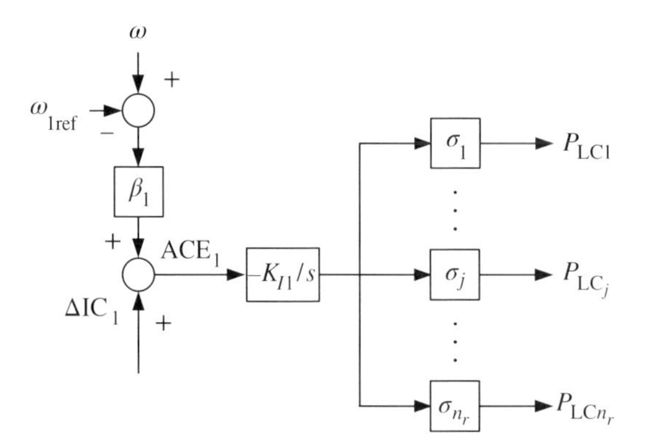

Figure 12.30 Assigning ACE to participating regulation units.

$$P_{\mathrm{LC}i} = K_{Ii} \int -\mathrm{ACE}_i(t)\, dt \tag{12.58}$$

The integral gain K_I of the AGC has a unit of $(d\mathrm{MW}/dt)/\mathrm{MW}$ [139]. Thus it has the meaning of the inverse of a time constant. The AGC loop is intended for secondary frequency control and is supposed to be slower in time than the primary frequency control. In [139], $K_I = 0.05\ \mathrm{sec}^{-1}$ (i.e., $1/K_I = 20$ sec) is used for AGC time response simulation. Also, the time constant should reflect the overall ramp rate of the generating units.

In a practical power system, there is more than one participating regulation unit in a control area. In case of multiple regulating units, the total P_{LC} is broken down into setpoint changes for each generator in the control area, as shown in Figure 12.30. The individual coefficients σ_j, $j = 1, \ldots, n_r$, are known as distribution factors, where n_r is the number of units receiving the new regulation setpoints. The new setpoint calculation can be based on the units' capability to respond, an economic dispatch, or in the case of a deregulated electricity market, the regulating units' bid prices. The AGC signals are normally sent to the generators every 5 seconds.

In a power grid consisting of many interconnected balancing authorities, it is necessary for the power grid to set up some measurable performance indices to ensure that no balancing authority is overly reliant on its neighbors for frequency control, a notion known as "leaning on the tie." In North America, the North America Reliability Council (NERC) has established Control Performance Standard 1 (CSP1) and Control Performance Standard 2 (CPS2) to monitor ACE error over 1-minute intervals and 10-minute intervals, respectively. CPS1 assigns a share of the responsibility for control of the interconnection frequency, and CPS2 intends to limit the unscheduled flows between the control areas. A balancing authority will determine the amount of regulating service to commit for satisfying these two criteria. The details of CPS1 and CPS2 can be found in [149].

Example 12.5

A two-area power system is given in Figure 12.31. The two areas are connected by the tieline between Buses 2 and 3. Each area has multiple generators. The total generation from each area is 0.95 pu, that is, $P_{m10} = P_{m20} = 0.95$ pu. The loads are $P_{L1} = 0.85$ pu and $P_{L2} = 1.05$ pu, such that the total load is $P_L = 1.90$ pu. Assume that for P_{L1} and P_{L2}, 50% is

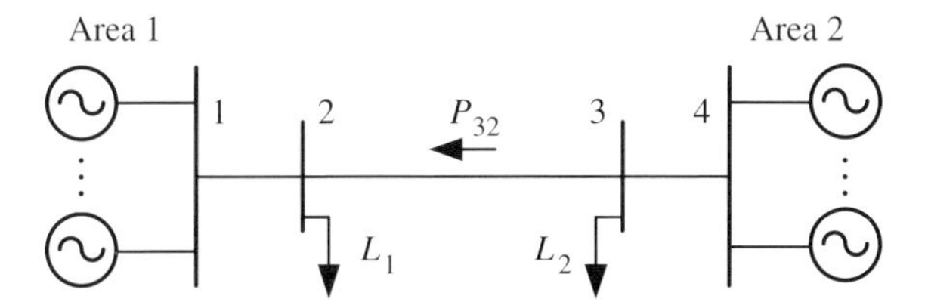

Figure 12.31 An interconnected two-area power system.

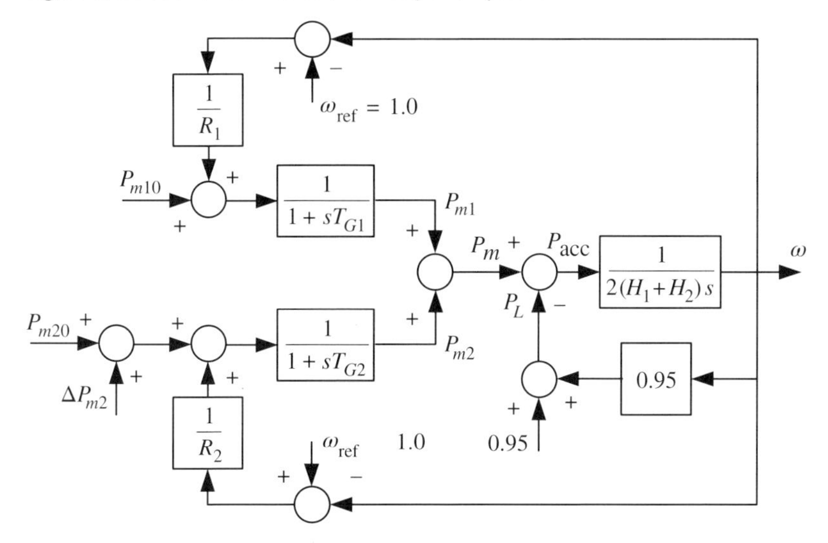

Figure 12.32 Block-diagram representation of the two-area automatic generation control system.

independent of frequency, and the other 50% is frequency-dependent (linearly with ω). To analyze the automatic generation control, the two-area system is represented by the block diagram in Figure 12.32, in which the electromechanical mode oscillation between the machines is neglected.

1) Suppose that a generator with a generation of 0.05 pu in Area 2 is tripped. Find the steady-state values of P_{L1}, P_{L2}, P_{m1}, P_{m2}, the frequency ω, and the power transfer P_{32} 20–30 seconds after the disturbance. Use $R_1 = 0.05$ and $R_2 = 0.10$.
2) Compute the ACE for each area, assuming that the tieline flow before the disturbance is the scheduled flow.

Solutions:

1) The loss of generation in Area 2 is modeled in Figure 12.32 with $\Delta P_{m2} = -0.05$ pu. The accelerating power after the disturbance has decayed is, with the loads separated into frequency dependent and independent loads:

$$\begin{aligned}
P_{acc} &= 0 = P_m - P_L = P_{m1} + P_{m2} - (P_L/2 + (P_L/2)\omega) \\
&= \left(P_{m10} + \frac{1}{R_1}(1-\omega)\right) + \left(P_{m20} + \Delta P_{m2} + \frac{1}{R_2}(1-\omega)\right) - \left(\frac{P_L}{2} + \frac{P_L}{2}\omega\right) \\
&= 1.85 + (20+10)(1-\omega) - 0.95 - 0.95\omega \\
&= (1.85 - 0.95 + 30) - (30 + 0.95)\omega
\end{aligned} \tag{12.59}$$

such that

$$\omega = 30.85/30.95 = 0.9984 \text{ pu} \tag{12.60}$$

The new load, generation, and interchange values are computed as

$$P'_{L1} = P_{L1}/2 + (P_{L1})/2\omega = 0.8493 \text{ pu} \tag{12.61}$$

$$P'_{L2} = P_{L2}/2 + (P_{L2})/2\omega = 1.0492 \text{ pu} \tag{12.62}$$

$$P'_{m1} = P_{m10} + (1 - \omega)/R_1 = 0.9823 \text{ pu} \tag{12.63}$$

$$P'_{m2} = P_{m20} + \Delta P_{m2} + (1 - \omega)/R_2 = 0.9162 \text{ pu} \tag{12.64}$$

The new interchange flow is

$$P'_{32} = P'_{m2} - P'_{L2} = -0.1330 \text{ pu} \tag{12.65}$$

2) The frequency bias factors for the two areas are

$$\beta_1 = 1/R_1 + D_1 = 20 + P_{L1}/2 = 20.425 \text{ pu power/pu frequency} \tag{12.66}$$

$$\beta_2 = 1/R_2 + D_2 = 10 + P_{L2}/2 = 10.525 \text{ pu power/pu frequency} \tag{12.67}$$

The scheduled power flow is $P_{32} = -0.1$ pu.
Thus

$$\begin{aligned} \text{ACE}_1 &= (P_{23\text{act}} - P_{23\text{sched}}) + \beta_1(\omega - \omega_{\text{ref}}) \\ &= (0.133 - 0.1) + 20.425(0.9984 - 1) = 0 \text{ pu} \end{aligned} \tag{12.68}$$

$$\begin{aligned} \text{ACE}_2 &= (P_{32\text{act}} - P_{32\text{sched}}) + \beta_2(\omega - \omega_{\text{ref}}) \\ &= (-0.133 + 0.1) + 10.525(0.9984 - 1) = -0.05 \text{ pu} \end{aligned} \tag{12.69}$$

The ACE calculation shows that Area 1 has no generation deficit, whereas Area 2 has to increase generation by 0.05 pu. ■

12.7 Turbine-Generator Torsional Oscillations and Subsynchronous Resonance

12.7.1 Torsional Modes

In transient stability and frequency regulation studies, the turbine sections and the generators are lumped into a single inertia. However, a more detailed analysis needs to account for the oscillations between the turbine sections and the generator. The analysis of the oscillation among the turbine sections and the generator is known as torsional vibration analysis. The analysis starts with developing a torsional model [152] of the turbine-generator as separate lumped inertias of the turbine sections and the generator connected via rotating shafts. The schematic of a torsional model of a turbine-generator with HP, IP, and LP sections is shown in Figure 12.33 with six separate rotating masses, including a small inertia at the rear end of the shaft to represent a DC1A type excitation system with a rotating amplidyne.[9]

9 No exciter inertia is needed for turbine-generators equipped with solid-state excitation systems.

Figure 12.33 A torsional model for a steam turbine-generator showing lumped masses.

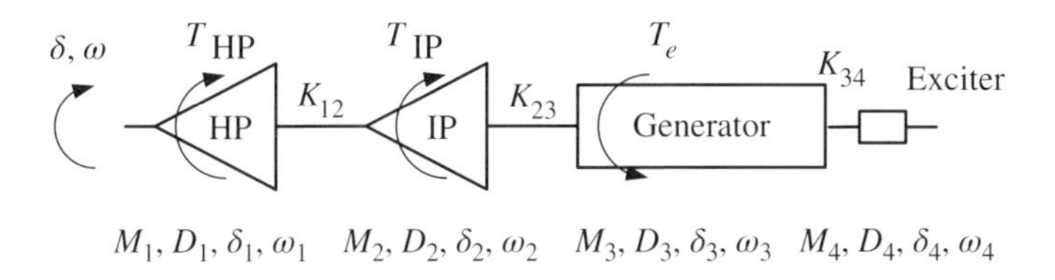

Figure 12.34 A torsional model for a steam turbine-generator showing four lumped masses.

The torsional model of a turbine-generator can be derived by writing the angular accelerator equations for the masses. Consider the four-mass model in [153], shown in Figure 12.34, consisting of the HP, IP, generator, and exciter inertias connected with shafts of various stiffness. The mechanical torques T_{HP} and T_{IP} are applied to the HP and IP sections, respectively, and electrical torque T_e acts on the generator. Note that the positive direction of T_{HP} and T_{IP} is in the same direction as the rotational mass angle θ and speed ω, whereas T_e is defined in the opposite direction. The torque on the exciter is neglected.

Denote the HP, IP, generator, and exciter sections as Masses 1, 2, 3, and 4, respectively. Let M_i, D_i, δ_i, and ω_i be the inertia, damping, angle, and speed, respectively, for Mass i, and denote the shaft stiffness between Masses i and j as K_{ij}. Then the torsional model dynamics for this four-mass system can be described by

$$\begin{aligned}
\dot{\delta}_1 &= \Omega\omega_1, & M_1\dot{\omega}_1 &= T_{\text{HP}} - K_{12}(\delta_1 - \delta_2) - D_1\omega_1 \\
\dot{\delta}_2 &= \Omega\omega_2, & M_2\dot{\omega}_2 &= T_{\text{IP}} - K_{12}(\delta_2 - \delta_1) - K_{23}(\delta_2 - \delta_3) - D_2\omega_2 \\
\dot{\delta}_3 &= \Omega\omega_3, & M_3\dot{\omega}_3 &= -T_e - K_{23}(\delta_3 - \delta_2) - K_{34}(\delta_3 - \delta_4) - D_3\omega_3 \\
\dot{\delta}_4 &= \Omega\omega_4, & M_4\dot{\omega}_4 &= -K_{34}(\delta_4 - \delta_3) - D_4\omega_4
\end{aligned} \tag{12.70}$$

where $\Omega = 2\pi f_o, f_o$ being the nominal system frequency. The turbine section torques can further be modeled as portions of the total mechanical torque given by $T_{\text{HP}} = \rho_1 T_m$ and $T_{\text{IP}} = \rho_2 T_m$, where $\rho_1 + \rho_2 = 1$.

The system (12.70) is linear and thus can be put in the matrix form

$$\dot{x} = \begin{bmatrix} \dot{x}_1 \\ \dot{x}_2 \end{bmatrix} = \begin{bmatrix} 0_{4\times4} & \Omega I_4 \\ M^{-1}K & M^{-1}D \end{bmatrix} \begin{bmatrix} x_1 \\ x_2 \end{bmatrix} + \begin{bmatrix} 0_{4\times2} \\ B_2 \end{bmatrix} u = Ax + Bu \tag{12.71}$$

where the state and control vectors are

$$x_1 = \begin{bmatrix} \delta_1 \\ \delta_2 \\ \delta_3 \\ \delta_4 \end{bmatrix}, \quad x_2 = \begin{bmatrix} \omega_1 \\ \omega_2 \\ \omega_3 \\ \omega_4 \end{bmatrix}, \quad u = \begin{bmatrix} T_m \\ T_e \end{bmatrix} \tag{12.72}$$

$0_{i\times j}$ is a zero matrix of dimension $i \times j$, I_i is an $i \times i$ identity matrix, and the inertia matrix and the damping matrix are, respectively,

$$M = \text{diag}(M_1, M_2, M_3, M_4), \quad D = \text{diag}(D_1, D_2, D_3, D_4) \tag{12.73}$$

the stiffness matrix is

$$K = \begin{bmatrix} -K_{12} & K_{12} & 0 & 0 \\ K_{12} & -(K_{12}+K_{23}) & K_{23} & 0 \\ 0 & K_{23} & -(K_{23}+K_{34}) & K_{34} \\ 0 & 0 & K_{34} & -K_{34} \end{bmatrix} \tag{12.74}$$

and the input matrix is

$$B_2 = \begin{bmatrix} \rho_1 & 0 \\ \rho_2 & 0 \\ 0 & -1 \\ 0 & 0 \end{bmatrix} \tag{12.75}$$

If the front standard speed ω_1 and the rear standard speed ω_4 are measured, the output equation can be set up as

$$y = \begin{bmatrix} \omega_1 \\ \omega_4 \end{bmatrix} = \begin{bmatrix} 0_{2\times 4} & C_2 \end{bmatrix} \begin{bmatrix} x_1 \\ x_2 \end{bmatrix} = Cx \tag{12.76}$$

where

$$C_2 = \begin{bmatrix} 1 & 0 & 0 & 0 \\ 0 & 0 & 0 & 1 \end{bmatrix} \tag{12.77}$$

Note that the dimension of system (12.71) can be reduced by 1 if one of the rotor angles is used as the reference value, as the restoring torque between two adjacent masses $K_{ij}(\delta_i - \delta_j)$ depends on the relative angle difference. Here all the angles are kept so that the modeshapes are simpler to interpret.

The damping of the torsional modes of a steam turbine-generator is very small. For this reason, oscillations associated with these modes take a long time to decay. Figure 12.35 illustrates this phenomenon [154]. The figure shows the measured front standard speed of a large steam turbine-generator following an out-of-step synchronization test. The steam turbine-generator set consists of five masses: one HP section, three LP sections, and one generator. The torsional frequencies are 8.2, 14.7, 19.5, and 21.7 Hz, and the corresponding damping ratios are 0.0030, 0.0014, 0.0009, and 0.0005. The top plot in Figure 12.35 shows the front standard speed over a time frame of 10 seconds; the bottom plot shows how the torsional oscillations still persist, with very little damping, after 8 seconds. Figure 12.36 shows the frequency spectrum in which the lightly damped torsional modes are clearly seen; the sharp peaks at the torsional frequencies indicate low damping. The local mode at a frequency of 1 Hz is also shown.

From the torsional model (12.71) and (12.76), the torsional modes can be computed by finding the eigenvalues of A. The corresponding eigenvectors can be used to study the individual modeshapes, that is, the "twisting" of the shaft for each of the torsional modes. In addition, the torsional model can be used to compute the frequency response and simulate the impulse response. These aspects are illustrated by the next example using the torsional model data from [153].

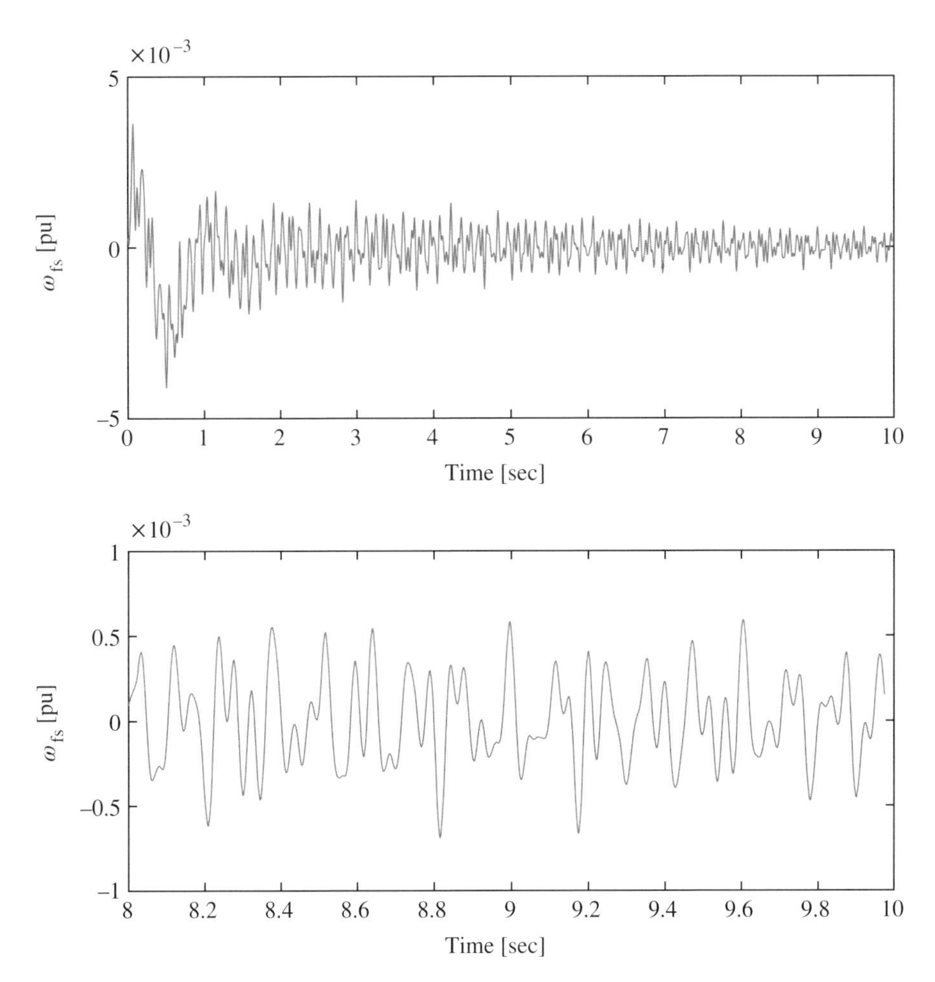

Figure 12.35 Top: Measured front standard speed deviation from nominal speed from a synchronization test [154]; bottom: enlarged plot (see color plate section).

Example 12.6

Consider a four-mass model of the torsional model in Figure 12.33 with the inertias in seconds given by $M_1 = 0.3703$ (HP), $M_2 = 0.6631$ (IP), $M_3 = 1.7313$ (generator), and $M_4 = 0.0629$ (exciter), the stiffness parameters in pu torque/rad given by $K_{12} = 65.51$, $K_{23} = 67.76$, and $K_{34} = 4.94$, and the damping coefficients in pu given by $D_1 = 0.55$, $D_2 = 0.45$, $D_3 = 0.07$, and $D_4 = 0$. The turbine fractions are $\rho_1 = 0.55$ and $\rho_2 = 0.45$. Compute the torsional frequencies and plot the modeshapes.

Solutions: Setting up the A matrix using the given parameters, the eigenvalues of the torsional model are

$$-0.50 \pm j352.44, -0.26 \pm j188.56, -0.15 \pm j167.73, 0, -0.38 \tag{12.78}$$

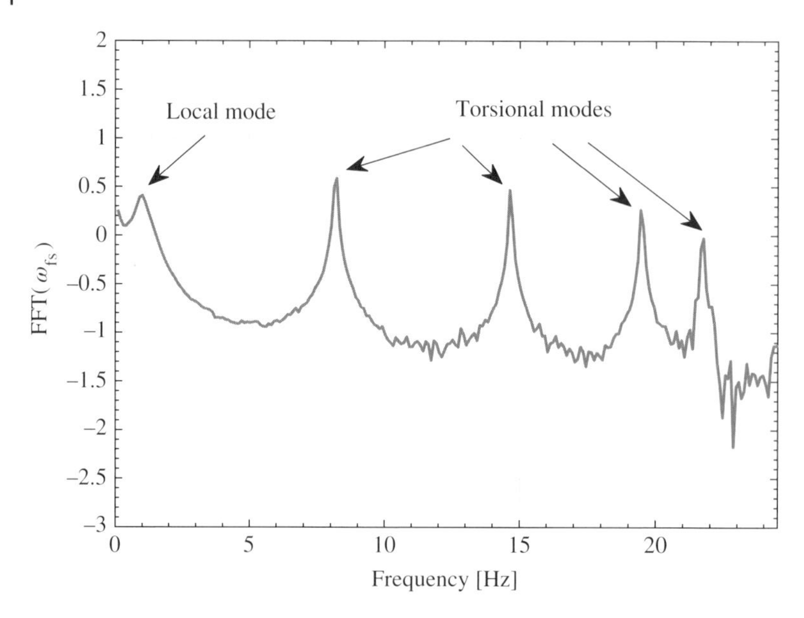

Figure 12.36 Spectrum of the front standard speed in Figure 12.35 [154] (see color plate section).

Note that the 0 eigenvalue exists because of the lack of the reference angle and −0.38 is the average damping on the reference angle motion. The three complex pairs are the torsional modes, with the real part being the damping (in unit of 1/sec) and the imaginary part being the oscillation frequency (in rad/sec). Thus the torsional mode frequencies are 56.10, 30.06, and 26.71 Hz, with very small damping ratios of 0.00142, 0.0000139, and 0.0009, respectively.

The modeshapes can be found from the eigenvectors, which for the first three torsional modes are, respectively,

$$\begin{bmatrix} -0.5505 \pm j0.0000 \\ 0.4751 \mp j0.0014 \\ -0.0657 \pm j0.0004 \\ 0.0206 \mp j0.0002 \\ 0.0007 \mp j0.5147 \\ 0.0007 \pm j0.4442 \\ -0.0003 \mp j0.0614 \\ 0.0002 \pm j0.0192 \end{bmatrix}, \begin{bmatrix} -0.3372 \mp j0.0077 \\ -0.1568 \mp j0.0045 \\ 0.1628 \pm j0.0027 \\ -0.7965 \mp j0.0000 \\ 0.0041 \mp j0.1689 \\ 0.0024 \mp j0.0785 \\ -0.0014 \pm j0.0816 \\ 0.0006 \mp j0.3991 \end{bmatrix}, \begin{bmatrix} 0.1745 \mp j0.0028 \\ 0.1009 \mp j0.0011 \\ -0.0441 \pm j0.0015 \\ -0.8900 \pm j0.0000 \\ 0.0012 \pm j0.0777 \\ 0.0005 \pm j0.0449 \\ -0.0007 \mp j0.0196 \\ 0.0004 \mp j0.3961 \end{bmatrix} \tag{12.79}$$

For completeness, the eigenvectors of the real system modes are

$$\begin{bmatrix} -0.5 \\ -0.5 \\ -0.5 \\ -0.5 \\ 0 \\ 0 \\ 0 \\ 0 \end{bmatrix}, \begin{bmatrix} -0.5000 \\ -0.5000 \\ -0.5000 \\ -0.5000 \\ 0.0005 \\ 0.0005 \\ 0.0005 \\ 0.0005 \end{bmatrix} \tag{12.80}$$

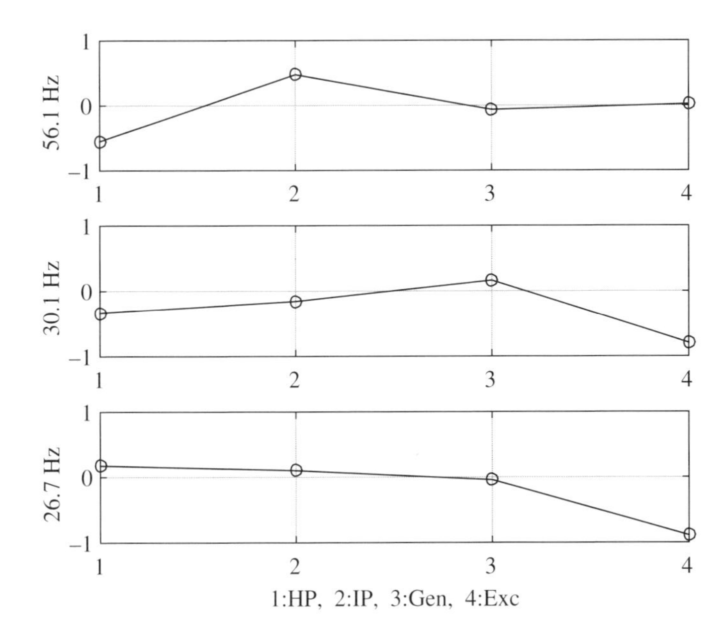

Figure 12.37 The modeshapes of the torsional modes of Example 12.6.

The modeshapes of the three oscillatory modes can be extracted from the real part of the eigenvectors corresponding to the angle states, that is,

$$\begin{bmatrix} -0.5505 \\ 0.4751 \\ -0.0657 \\ 0.0206 \end{bmatrix}, \quad \begin{bmatrix} -0.3372 \\ -0.1568 \\ 0.1628 \\ -0.7965 \end{bmatrix}, \quad \begin{bmatrix} 0.1745 \\ 0.1009 \\ -0.0441 \\ -0.8900 \end{bmatrix} \tag{12.81}$$

These modeshapes are plotted in Figure 12.37.

The modeshape analysis indicates that the 56.10 Hz mode is the oscillation of the HP turbine against the IP turbine, the 30.06 Hz mode is the oscillation of the HP and IP turbines against the generator, and the 26.71 Hz is the oscillation of the exciter against the other masses. ■

Example 12.7

Plot the frequency response of the turbine-generator torsional model in Example 12.6 from the input ΔT_e to the output $\Delta\omega_1$ (front standard speed) from 5 to 60 Hz. Perform an impulse response of the torsional model through the electrical torque input ΔT_e and plot the time response $\Delta\omega_1$ and $\Delta\omega_4$, first for 1 sec and then for 30 sec.

Solutions: The frequency response from T_e to ω_1 is shown in Figure 12.38 with the amplitude plotted in dB. The peaks of the frequency response are due to the lightly damping torsional modes.

The impulse response for the front standard speed is shown in Figure 12.39 and that of the rear standard speed is shown in Figure 12.40. ■

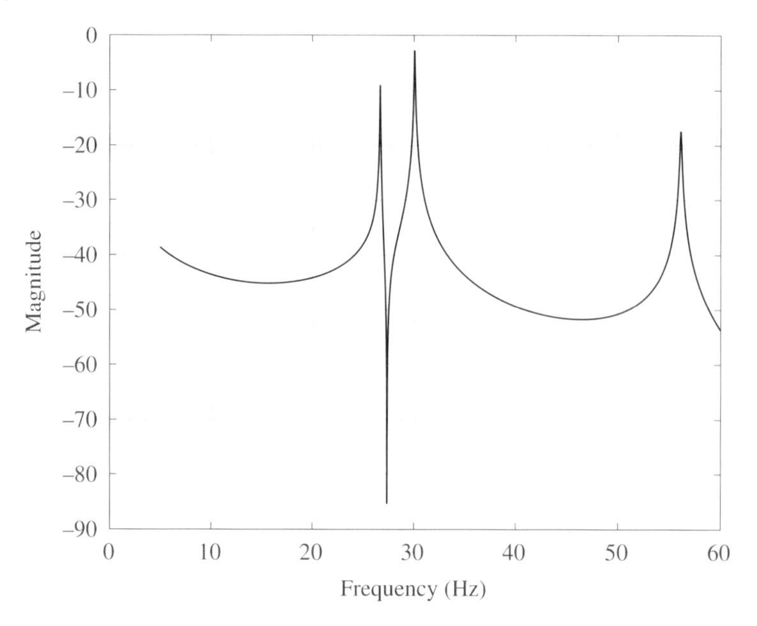

Figure 12.38 Frequency response from T_e to ω_1.

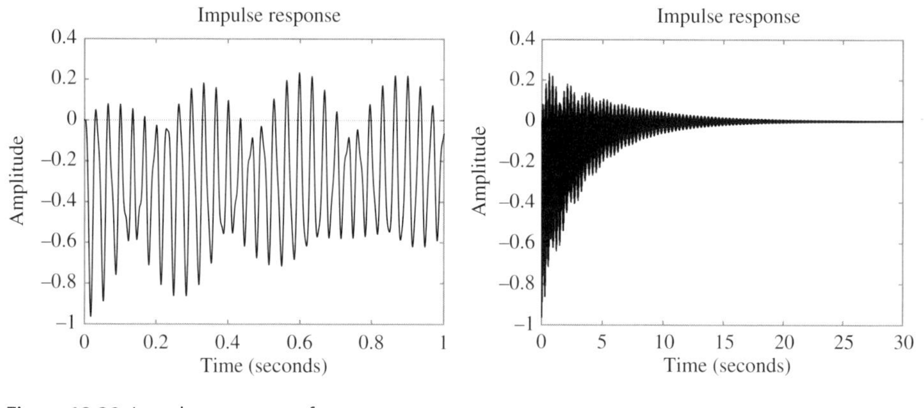

Figure 12.39 Impulse response of ω_1.

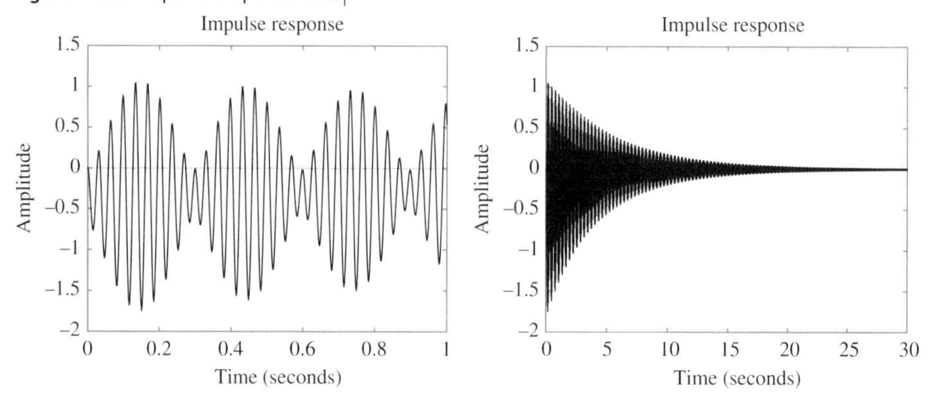

Figure 12.40 Impulse response of ω_4.

For an interpretation of the two real eigenvalues of the torsional model observed in Example 12.6, the equations in (12.70) are added to obtain

$$\begin{aligned}
&\dot{\delta}_1 + \dot{\delta}_2 + \dot{\delta}_3 + \dot{\delta}_4 = \Omega(\omega_1 + \omega_2 + \omega_3 + \omega_4) \\
&M_1\dot{\omega}_1 + M_2\dot{\omega}_2 + M_3\dot{\omega}_3 + M_4\dot{\omega}_4 \\
&\quad = T_{\mathrm{HP}} + T_{\mathrm{IP}} - T_e - (D_1\omega_1 + D_2\omega_2 + D_3\omega_3 + D_4\omega_4)
\end{aligned} \tag{12.82}$$

If the shafts are assumed to be infinitely stiff, then

$$\delta_1 = \delta_2 = \delta_3 = \delta_4 = \delta, \quad \omega_1 = \omega_2 = \omega_3 = \omega_4 = \omega \tag{12.83}$$

As a result, (12.82) reduces to

$$\dot{\delta} = \Omega\omega, \quad M\dot{\omega} = T_m - T_e - D\omega \tag{12.84}$$

where $M = M_1 + M_2 + M_3 + M_4$ and $D = D_1 + D_2 + D_3 + D_4$.

The torsional modes of turbine-generators normally have frequencies from 15 to below 60 Hz. Hence they are called *subsynchronous* frequencies. Because the torsional modes are lightly damped, it is important that they should not be excited by control equipment to avoid shaft fatigue and equipment loss-of-life. A particular concern is that when the generator is connected to transmission lines with series compensation, the electrical network may negatively impact the damping of these modes, resulting in subsynchronous resonance (SSR). A turbine-generator at the Mohave Power Plant was destroyed in 1970, and then again in 1971, until engineers finally attributed the oscillation problem to the SSR phenomenon [152]. The impact of the electrical network on SSR and methods to reduce SSR are discussed next.

12.7.2 Electrical Network Modes

Consider the system shown in Figure 12.41 of a single machine connected to the infinite bus via a series compensated line.[10] A reactance model of the circuit is shown in Figure 12.42, where $X_T = \omega L_T$ is the transformer reactance, $X = \omega L$ is the transmission line reactance, and the series capacitor is inserted at the midpoint of the transmission line. The series compensator level k is defined as

$$X_c = -kX \tag{12.85}$$

At the nominal frequency, $X = \omega_o L$ and $X_c = -1/(\omega_o C)$, where L is the line inductance and C is the series compensation capacitance. Then

$$\frac{1}{\omega_o C} = k\omega_o L \tag{12.86}$$

Because the frequency of concern is in the tens of Hz, the d-axis subtransient reactance is used as the machine reactance, and the machine internal voltage is the subtransient d-axis flux linkage ψ''_d.

10 Additional discussion on series compensator can be found in Chapter 14.

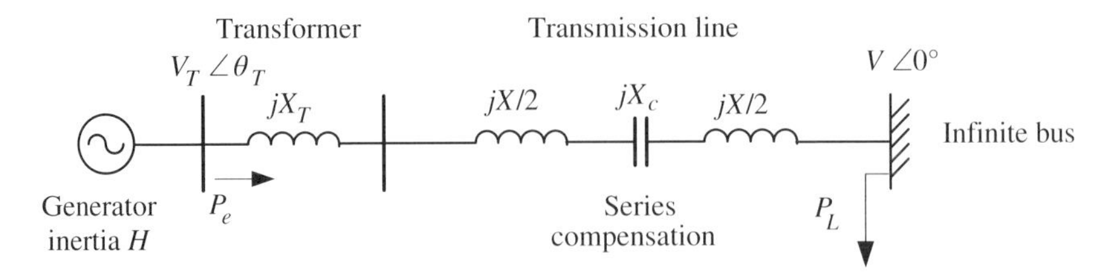

Figure 12.41 Series-compensated single-machine infinite-bus system.

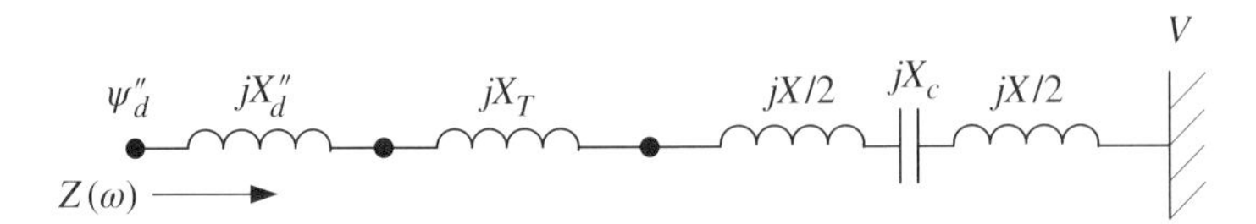

Figure 12.42 Reactance model of the series-compensated SMIB.

The impedance looking into the electrical network from the generator internal voltage as a function of the frequency ω is

$$Z(\omega) = j\omega L''_d + j\omega L_T + j\omega L + \frac{1}{j\omega C}$$

$$= \frac{1}{j\omega C}\left(1 - \omega^2 (L''_d + L_T + L)\frac{1}{k\omega_o^2 L}\right) \tag{12.87}$$

$$= \frac{1}{j\omega C}\left(\frac{k\omega_o^2 L - \omega^2 (L''_d + L_T + L)}{k\omega_o^2 L}\right) \tag{12.88}$$

Resonance occurs when the numerator of (12.88) becomes zero, that is, the electrical network frequency is

$$\omega_e = \omega_o \sqrt{\frac{kL}{L''_d + L_T + L}} = \omega_o \sqrt{\frac{k}{1 + (L''_d + L_T)/L}} \tag{12.89}$$

For $k < 1$, the resonance mode ω is less than ω_o, as $1 + (L''_d + L_T)/L > 1$.

Assuming L''_d and L_T to be smaller than L, the electrical network frequency is approximately

$$\omega_e = \omega_o \sqrt{k} \text{ rad/sec}, \quad \text{or} \quad f_e = f_o \sqrt{k} \text{ Hz} \tag{12.90}$$

For compensation values of $0.05 \leq k \leq 0.75$, the electrical network frequency varies from 13 Hz to 52 Hz for $f_o = 60$ Hz.

Note that the electrical network is connected to the stator of the generator. In the synchronous machine chapters, the fundamental frequency f_o on the stator is generated by a DC current in the rotor field circuit rotating at the fundamental frequency. That is, a frequency f_e on the electrical network is seen in the generator as the f_o complementary frequency

$$f_m = f_o - f_e \tag{12.91}$$

Thus if the frequency f_m is close to a torsional mode frequency, then there would be potential interactions between the electrical and mechanical systems, causing the torsional mode to be unstable, resulting in vibration of the turbine-generator and oscillations in the electrical network.

Example 12.8

For the system in Figure 12.42, the subtransient reactance of the generator is $X_d'' = 0.2$ pu, the transformer reactance is $X_T = 0.15$ pu, and the line reactance is $X = 0.6$ pu, all on the machine base. Suppose a 50% series compensation is used on the transmission line, that is, $X_C = -0.3$ pu. Derive the equivalent impedance $Z(\omega)$ as a function of the frequency ω from the generator looking into the electrical network and compute the electrical network frequency. Assume that the nominal system frequency is 60 Hz. Determine whether there could be possible torsional interactions between the generator torsional modes derived in Example 12.6 and this electrical network.

Solutions: From (12.89), the electrical network frequency is given by

$$f_e = 60\sqrt{\frac{0.5}{1 + (0.2 + 0.15)/0.6}} = 33.717 \text{ Hz} \tag{12.92}$$

The 60 Hz complementary frequency of this electrical network mode is

$$f_m = 60 - 33.717 = 26.283 \text{ Hz} \tag{12.93}$$

The mode is quite close to the 26.71 Hz mode of the four-mass model. Thus there is potential interaction between the torsional mode and the electrical network. ■

12.7.3 SSR Occurrence and Countermeasures

In the power industry, there is consensus on the probability of occurrences of SSR for a generator, summarized as follows [10]:

1) Shunt capacitor banks and static var compensators do not contribute to SSR, as the resulting electrical network frequencies will be super-synchronous, that is, greater than f_o.
2) Generator mechanical resonances are quite sharp (see Figure 12.37) and thus have narrow bandwidths, lessening SSR interaction probability.
3) Hydraulic generators have significantly larger damping than fossil or nuclear turbine-generators, and thus would not have SSR problems.
4) Close-by generators tend to share mechanical damping among the torsional modes. An analysis of this sharing of damping can be found in [155].
5) The series compensation can be adjusted so that the f_o-complement of the electrical network frequency would occur in between the torsional modes.
6) The probability of SSR is highest when a single generating unit is operating in a radial connection with high series compensation. This was the situation leading to the Mohave SSR incidents.

In case series compensation is necessary to provide the proper power transfer, various mechanisms have been developed to mitigate the SSR problem, including the following:

1) SSR filters consisting of capacitors and inductors tuned to the relevant torsional modes can be installed at the generator substation to reduce the interactions between the generator and the electrical network [157].
2) Generator and network variables used in control systems are passed through notched (bandstop) filters to remove the torsional modes. For example, in Chapter 10 on damping control design, the machine speed input signal is passed through a torsional filter before it is used by the power system stabilizer (PSS). The output signal of the PSS can then be injected into the voltage regulator summing junction without feeding the torsional modes back into the electrical network. In PSS design, only the phase impact of the torsional filter is included with a low-order transfer function. The net effect of the torsional filter is that additional phase compensation would be needed.
3) The NGR SSR damper [10] or other power-electronics-based controller can also be used to either add damping to the torsional modes or make the series capacitor look inductive at the torsional frequency. A discussion can be found in Chapter 14 of this text and Chapter 9 of [10].

12.8 Summary and Notes

This chapter discusses models of steam and hydraulic turbines for power system dynamic simulation. Primary and secondary frequency control concepts are also discussed. In addition, torsional models of turbine-generators are provided, which are important for subsynchronous oscillation analysis.

This chapter does not provide a coverage of boiler dynamics for steam turbines. A reader can find more information from [158]. Also various turbine controls have not been discussed. A reader can consult [6] to find out about various control modes, including turbine lead, turbine follow, sliding pressure, and coordinated control. The coverage of gas turbine and co-generation plants is brief. Although gas-turbine models have been developed for quite some time [146, 159], there are still ongoing discussions on appropriate gas turbine models to use. Thus these models are not included here.

The operation of interconnected power systems has a long history. Some historical documentations can be found in [160–162]. Automatic generation control will be a more challenging task with the increasing variable power generation from renewable resources such as wind and solar energy. There is ongoing research on the impact of a lower amount of rotating inertia on the power grid and the need for faster ramping by generating units.

Strict frequency regulation is also important because mechanical oscillatory modes of a turbine-generator not only include the shaft torsional modes, but can also involve steam turbine blades vibrating against the turbine wheels. The blades are lighter; thus the vibrations are at higher frequencies and typically super-synchronous, that is, higher than the system frequency. The geometries of the blades are optimized to extract maximum energy from the steam, with the blades carefully designed to avoid harmonics of the fundamental system frequency. These calculations are typically performed using finite-element analysis programs. If any of the vibrational modes due to the blades is too close to a harmonic, then prolonged periods of off-nominal-frequency operations, such as during light-load periods, may cause resonances with the blades to occur. Such a problem can be avoided with more precise frequency regulation. Hydraulic turbines do not have such an issue and thus can safely operate over a wider frequency range.

The authors would like to acknowledge that the presentations in Sections 12.3, 12.5, and 12.6 follow closely Donald Ewart's lecture notes [138]. The Problems section also includes a few problems from [138].

Problems

12.1 Find the poles and zero of the steam turbine control system in Figure 12.7 using the parameters in (12.2).

12.2 For the hydraulic-turbine system in Example 12.1, simulate its time response due to a ramp increase of the gate G for 0.05 pu in 1 sec.

12.3 For the hydraulic-turbine system in Example 12.1, simulate its time response due to a step decrease of the gate G for -0.05 pu.

12.4 For the hydraulic turbine in Example 12.4 with only the permanent droop, change R_p to 0.1 and show that the frequency governing response is unstable.

12.5 The governor configuration of a hydraulic turbine is shown in Figure 12.43, where $R_p = 0.05$ and $R_T = 0.5$.

1) Find the transfer function from $\Delta\omega$ to the gate position ΔG, with the transient droop function included.
2) The system is operating at steady state with $\omega = \omega_{\text{ref}} = 1.0$ pu. Suppose there is a sudden loss of generation in the system such that the system frequency drops quickly, represented by $\omega(0^+) = 0.995$ pu (300 mHz variation at a nominal frequency of 60 Hz). Calculate the incremental gate opening ΔG and the incremental mechanical output power ΔP_m at the initial time $t = 0^+$ (instantaneous) and at $t = \infty$ (steady state). Assume that ω remains at 0.995 pu in steady state. Repeat the calculation with no transient droop.

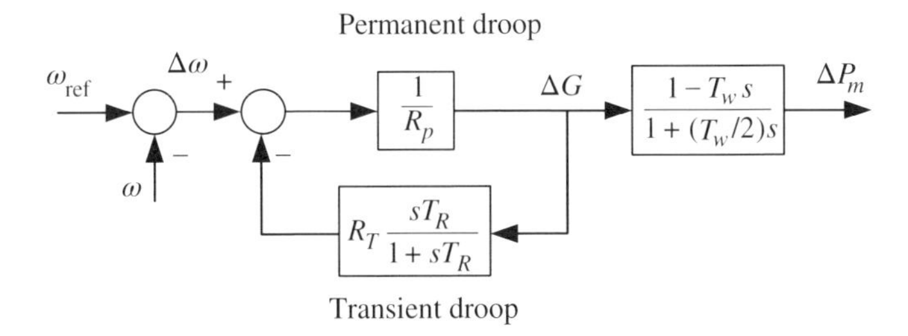

Figure 12.43 Governor block diagram for Problem 12.5.

12.6 Two governor configurations of a hydraulic turbine are shown in Figure 12.44, where $R_p = 0.05$, $T_w = 2$ sec, $T_A = 0.1$ sec, and $T_B = 1.0$ sec. The system is operating in steady state with $\omega = \omega_{\text{ref}} = 1.0$ pu and $P_{mo} = 0.8$ pu. Suppose

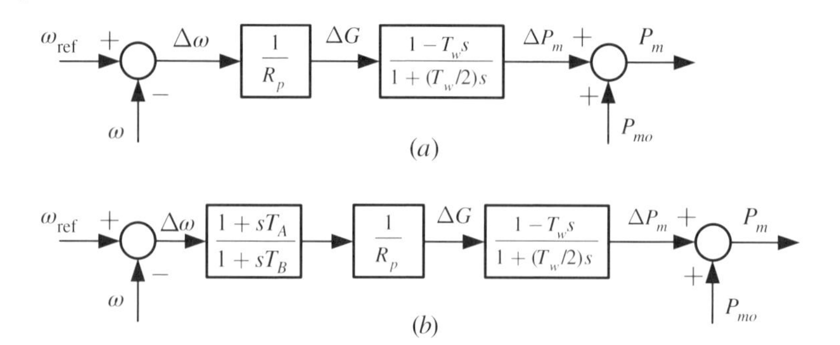

Figure 12.44 Two generator system model block diagrams for Problem 12.6.

there is a sudden loss of generation in the rest of the system such that the system frequency drops quickly, represented by $\omega(0^+) = 0.99$ pu.

1) Calculate the incremental gate opening ΔG and the incremental mechanical output power ΔP_m at the initial time $t = 0^+$ (instantaneous) and at $t = \infty$ (steady-state) for the governor configuration in (a). Assume that ω remains at 0.99 pu in steady state. Also calculate P_m at $t = 0^+$ and $t = \infty$.
2) Repeat the calculation for the configuration in (b).

12.7 A power system with a load of 10,000 MW is supplied by several generators. Assume that the average droop is $R = 0.08$ and 40% of the load varies linearly with the frequency, with the reminder constant power. Determine the frequency response β in pu power/pu frequency and MW/0.1 Hz, using 10,000 MW as the power base. Suppose a constant-power load of 50 MW is added to the system. Calculate the frequency drop.

12.8 In a power grid with a nominal frequency of 60 Hz and a connected generation and load of 700 GW, it is observed that a loss of a 1000 MW generating unit will result in a frequency decline of 33 mHz. (a) Calculate β in MW/0.1 Hz and pu power/pu frequency, using 700 GW as the base power. (b) If the governor droop of all the generators is set to 0.05 and 50% of the load is linearly proportional to the system frequency, calculate the frequency drop for the loss of the same unit.

12.9 Two generators are supplying load as shown in Figure 12.45, where H is the machine inertia, T_G is the turbine time constant, P_m is the generator mechanical power, L is the frequency-dependent load, and R is the frequency regulation constant (droop). Initially, $L = 1.8$ pu, which is supplied by $P_{m10} = P_{m20} = 0.9$ pu. (a) If $\Delta L = 0.1$ pu is inserted, find the individual generator mechanical power and the system frequency in steady state given that $R_1 = R_2 = 0.05$. (b) Repeat Part (a) if $R_1 = 0.05$ and $R_2 = 0.10$.

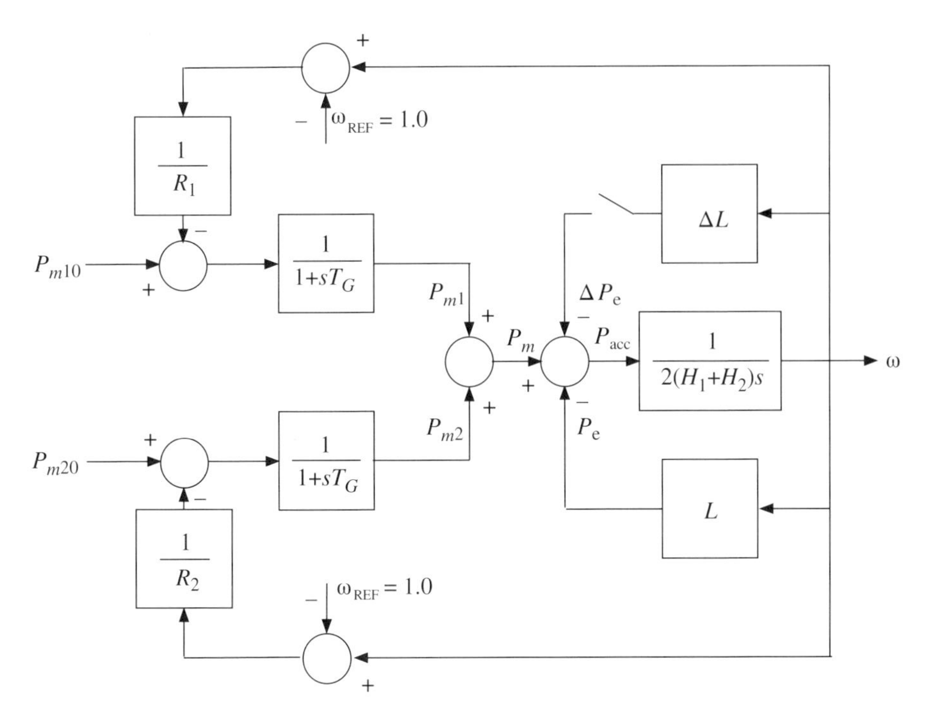

Figure 12.45 Two-generator system model block diagram for Problem 12.9.

12.10 (adapted from Ewart)
The power company which served the ancient Kingdom of Zhou had a total connected capacity of 35 MW. It had no connections to other kingdoms. The power company was divided into three interconnected regions: North Zhou, Middle Zhou, and South Zhou, having a distribution of generation and load shown in Table 12.3.

The power system was running in steady state with a frequency of 1 pu. The regulation $R = 0.1$ was the same for all generators, and the load was proportional to frequency (that is, $D = \partial L/\partial \omega = 1$). Suddenly the Cannon Iron Works in North Zhou switched on an arc furnace, a load of 2 MW. This load was independent of frequency.

Table 12.3 Generation and load in the three regions of Zhou Kingdom.

Region	Capacity of connected generator (MW)	Load (MW)	Actual power being generated (MW)
North Zhou	10	8	9
Middle Zhou	5	5	4
South Zhou	20	17	17
Total	35	30	30

1) Calculate the frequency drop in the Kingdom which would exist in the first 10- or 20-second period after the application of load at the Iron Works.
2) Calculate the net export or import of power for each of the three regions *before* and *after* the application of the arc-furnace load.
3) Calculate the ACE signal for each region.
4) Now assume that the power system broke up into three separate systems at the boundaries of the regions, with the arc-furnace load on. Calculate the resulting frequency in each system. Assume no AGC or supplementary frequency control.

12.11 The rotor mass-spring parameters of the IEEE Benchmark Model for computer simulation of synchronous resonance are given in Table 12.4 [156] (note that $H = M/2$). Build the torsional model of this turbine-generator and compute the torsional modes. Assume the damping parameters to be zero. Calculate and plot the modeshapes of the turbine-generator. Simulate the time response of ω_1 and ω_6 subject to a unit impulse ΔT_e.

Table 12.4 Rotor mass-spring parameters.

Mass	Shaft	Inertia constant H (sec)	Spring constant K (pu torque/rad)
HP		0.092897	
	HP-IP		19.303
IP		0.155589	
	IP-LPA		34.929
LPA		0.858670	
	LPA-LPB		52.038
LPB		0.884215	
	LPB-GEN		70.858
GEN		0.868495	
	GEN-EXC		2.822
EXC		0.0342165	

12.12 Find the electrical mode for the circuit in Example 12.8 for series compensation at $k = 0.1, 0.3, 0.5, 0.7$. For each electrical mode, determine if there is a potential to interact with the torsional modes for the turbine-generator in Problem 12.11.

Part III

Advanced Power System Topics

13

High-Voltage Direct Current Transmission Systems

13.1 Introduction

In the early days of electrification, both direct current (DC) and alternating current (AC) transmission systems were built [164, 165]. However, due to the limited transmission range of DC currents at low voltages, the lack of DC voltage transformers and protection devices, and the invention of the induction motor by Nikola Tesla, DC transmission systems could not compete technically and economically with AC transmission systems.

DC power delivery technology stayed dormant for many years until effective high-voltage, high-power switching devices were invented, first with mercury-arc valves, followed by solid-state thyristors, and more recently voltage-sourced converters (VSCs) such as gate-turnoff (GTO) devices and insulated-gate bipolar transistors (IGBTs) [166]. HVDC systems have found many unique applications, including cost-effective long-distance transmission systems, point-to-point power transmission bypassing congested AC transmission systems, and power transfer between asynchronous neighboring systems. More recently, merchant HVDC systems have also been built in regions with deregulated electricity markets, in which an HVDC system is contracted to deliver a certain amount of active power from low energy price areas to higher energy price areas and be compensated for the energy delivery. In addition, HVDC systems are used for transmission of off-shore wind energy.

This chapter focuses on HVDC transmission systems built using thyristor technology, also known as line-commutated converter (LCC) technology. HVDC systems built on VSC technology are discussed in the next chapter.

The basic function of an HVDC system is to convert AC voltages and currents into DC voltages and currents, known as the rectifying function, and then convert DC voltages and currents back to AC voltages and currents, known as the inverting function.

Several common HVDC configurations are shown in Figure 13.1 [9]. The monopolar link is the basic configuration, in which the rectifier provides a DC voltage and a DC transmission line conducts a DC current to the inverter. The DC current will circulate back to the rectifier via ground. A metallic ground return can also be used. In a bipolar link, there are two sets of converters at each terminal, in which two DC lines are used to complete the circuit. The mid-point of the converters on each terminal is grounded so that the DC voltage on one line is positive and that on the other line is negative. A bipolar HVDC system can operate in a single-pole mode, if one of the poles is out of service. In a homopolar HVDC system, the poles are the mirror image of each other,

Power System Modeling, Computation, and Control, First Edition. Joe H. Chow and Juan J. Sanchez-Gasca.

Companion website: www.wiley.com/go/chow/power-system-modeling

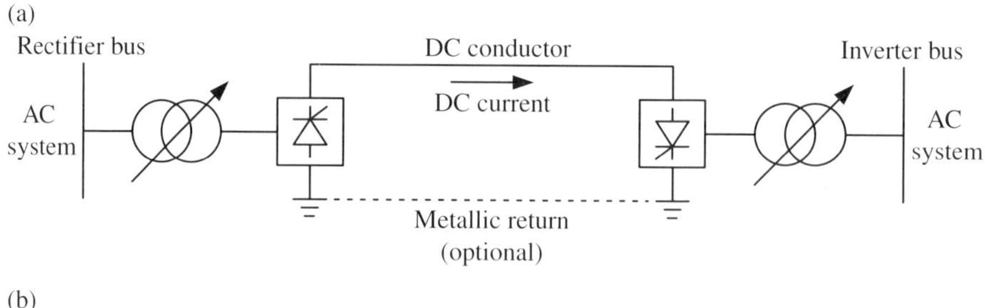

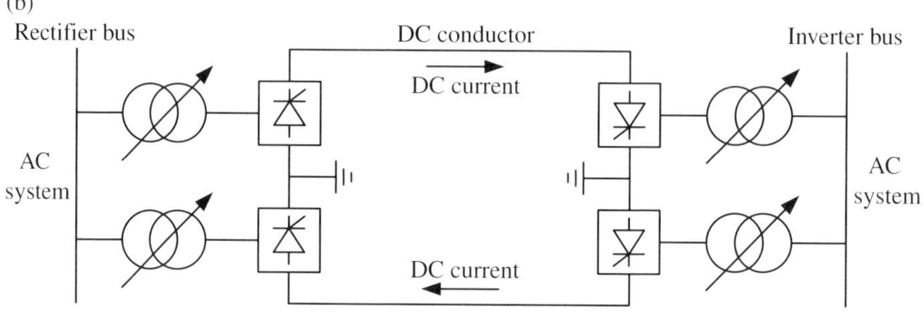

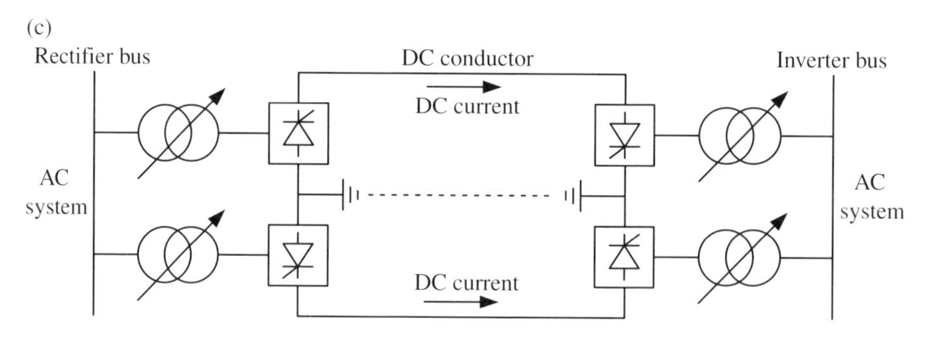

Figure 13.1 Schematics of two-terminal HVDC transmission: (a) monopolar link, (b) bipolar link, and (c) homopolar link [9].

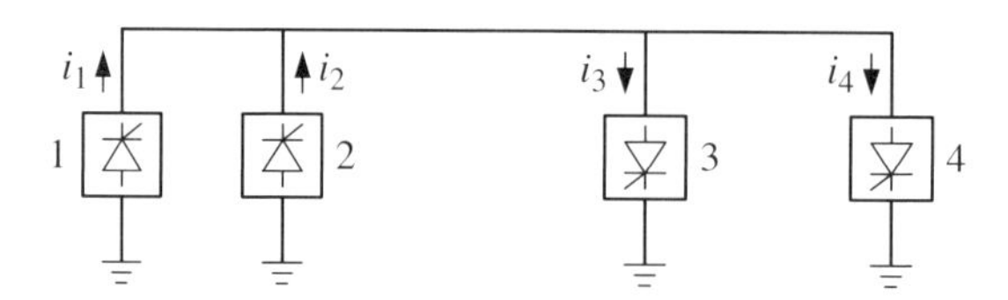

Figure 13.2 Multi-terminal HVDC transmission: a two-rectifier and two-inverter radial system.

such that the DC voltages of both poles are positive. The DC circuit is completed via ground or a metallic return.

The configurations shown in Figure 13.1 are all two-terminal HVDC systems. It is also possible to design multi-terminal HVDC (MTDC) systems with multiple rectifiers and inverters. Figure 13.2 shows the schematic of a four-terminal system in which energy is supplied from Terminals 1 and 2 to Terminals 3 and 4 via a DC link in a radial connection. It is possible to build a loop or ring structure as well, for example constructing another DC line connecting Terminals 1 and 4 directly in the four-terminal system. Although very few land-based MTDC systems have been built, it is anticipated that sea-based wind-turbine farms will most likely be based on MTDC technology.

The organization of this chapter is as follows. Sections 13.2 to 13.4 describe HVDC system operation based on thyristors and line commutation. Section 13.5 discusses operation flexibilities with MTDC. Section 13.6 presents an analysis on HVDC harmonics and filtering requirements. Power flow solution methods for solving AC-DC systems are provided in Section 13.7. Dynamic models and swing mode damping control designs are shown in Sections 13.8 and 13.9.

The materials presented in the first part of this chapter (Sections 13.2–13.4) are primarily based on the information provided in [6, 9]. More updated information can be found in [167].

13.1.1 HVDC System Installations and Applications

There are three main stages in the development of HVDC technologies for DC current control switchings, namely, mercury-arc valves, thyristors, and voltage-sourced converters (VSCs). Tables 13.1 to 13.3 show some of the HVDC system installations, primarily in North America, with one table for each technology area.[1]

1) Mercury-arc valves: These older systems have been mostly phased out or upgraded to newer technologies.
2) Thyristor valves: The use of LCC technology started in the 1970s. This technology will be discussed in this chapter.
3) VSC systems: Many new HVDC installations use the VSC or modular multi-level converter (MMC) technologies. In a VSC system, all converters in series are controlled to switch simultaneously, whereas in a MMC system, each series converter can be switched individually.

Table 13.1 Mercury-arc valve HVDC systems.

Installation	System description
Sweden-Gotland Link (1954)	96 km submarine link from Sweden mainland to Gotland Island; 20 MW, 100 kV monopolar system; alternative to extra thermal generation on island
Pacific DC Intertie (1970, 1985, 1989)	856 mile (1,362 km) bidirection line between Pacific Northwest (Celilo substation) and Southern California (Sylmar substation); 1440 MW, 400 kV mercury-arc valves; upgraded to ±500 kV with thyristors in 1985; additional thyristors added in 1989 for 3100 MW capability; for seasonal diversity of generation and load between Pacific NW and California

1 It should be noted that many ultra-high-voltage HVDC systems have been recently built in the Chinese power system [168].

Table 13.2 LCC HVDC systems.

Installation	System description
Eel River (1972)	First thyristor technology HVDC system, 320 MW at 2 ×80 kV; back-to-back DC interconnection between Hydro Quebec and New Brunswick; connection between two asynchronous systems
Eddie County (1983) Miles City (1985) Oklaunion (1985)	Three similar back-to-back DC connections between US Eastern Interconnection and Western Interconnection; each rated at 200 MW and 82 kV; connections between two asynchronous systems
Itaipu (1986)	Two bipolar DC lines at ±600 kV from Itaipu hydraulic generators to São Paulo; 8 valves at 778 MW/valve
New England HVDC (1986, 1992)	Asynchronous connection between Quebec (HQ) and New England (NE): Phase 1 – Des Cantons (HQ) to Comerford (NE), 690 MW at ±450 kV; Phase 2 – 1500 km three-terminal HVDC system, Radisson (HQ, 2250 MW converter rating) to Nicolet (HQ, 2138 MW) to Sandy Pond (NE, 1800 MW)
Neptune Project (2007)	65 mile undersea/underground DC transmission cable from Sayreville, New Jersey, to Nassau County, Long Island; 660 MW at 500 kV

Table 13.3 VSC and MMC systems.

Installation	System description
Cross Sound Cable (2003)	24 mile (39 km) submarine cable between New Haven, Connecticut, and Shoreham, Long Island; 330 MW, ±150 kV, VSC technology; power transmission contracts from New England to Long Island
Trans Bay Cable (2010)	53 mile undersea DC cable from Pittsburg Substation (East Bay) to Potrero Substation (San Francisco); 400 MW at ±200 kV, MMC technology

There are three main applications for HVDC systems in modern power systems.

1) Long-distance power transfer: For power transfer over long distances, HVDC lines with converters are less expensive to build than AC transmission lines with comparable power ratings. An example is the Pacific DC Intertie in Table 13.1. A discussion on HVDC economics follows in the next section.
2) Connection between asynchronous power systems: HVDC systems can provide power transfer between power systems that operate asynchronously. The Eel River Project and the New England HVDC Project (Table 13.2) allow Hydro Quebec, an asynchronous power system, to send the output power of hydraulic units to the US Eastern Interconnection. The Eddie County, Miles City, and Oklaunion Projects connect the western part of the US Eastern Interconnection to the eastern part of the US Western Power System.
3) Merchant transmission systems: In a deregulated power system, lower price generation may not always be dispatched to load centers if the required transmission paths are constrained by stability or line loading limits. HVDC systems can be built for point-to-point transmission to bypass the congested AC transmission paths. Both

the Cross Sound Cable Project and the Neptune Project deliver power to Long Island using submarine cables. The Trans Bay Cable Project is able to bring additional power into San Francisco. In deregulated electricity markets, such power transmission can command a high premium if the locational marginal price (LMP) of energy at the supply location is much lower than the LMP at the load area. The HVDC system investment and maintenance costs are recovered by multi-year power delivery contracts, most likely negotiated in the project planning phase.

13.1.2 HVDC System Economics

The costs of converters, control systems, transformers, and filters for a DC converter substation can be substantial, in the order of three to four times the cost of an AC substation. However, the cost for the balance of an HVDC system can be much lower than an AC system. They are described as follows:

- For overhead lines, a bipole DC system would require only two lines versus an AC system with three lines. Thus the cost of DC lines is about 60–70% of the cost of AC lines. The cost of tower structures is also lower.
- In underground and submarine applications, the cost of DC cables is about 25–30% of the cost of AC cables. DC cables have lower equivalent resistance from skin effects or pipe losses. Also a higher DC stress on insulation is permissible.
- The active power loss on a 3-phase AC transmission system is $3I^2R$, where R is the line resistance and I is the line current. The active power loss on a bipole DC system is only $2I^2R$. There are also switching losses in the converters of the order of a few percent of the active power transmission on the DC lines.
- The reactive power loss is $3I^2X$ for a 3-phase AC transmission system. There are no reactive power losses on the DC transmission lines. However, reactive power support is needed at both the rectifier and inverter terminals for the LCC technology, and is provided by filters and shunt capacitors.

Thus HVDC systems can be cost competitive for long distance transmission such that the lower cost of the transmission lines more than offsets the higher cost of the HVDC substations. For overhead applications, the capital cost break-even distance is about 500 miles. Above that distance, HVDC systems become more cost effective. For submarine applications, the break-even distance can be as low as 25 miles.

13.2 AC/DC and DC/AC Conversion

In an LCC HVDC system, AC and DC power conversion is enabled by silicon controlled rectifiers (SCRs) called thyristors. The symbol for a controlled thyristor and its ideal characteristics are shown in Figure 13.3. An injection of a small gate current i_G allows a forward-biased thyristor to switch from a very high to a very low impedance device. Thus the ideal VI characteristic of a thyristor resembles a diode, such that when the thyristor is turned on, the voltage drop across the thyristor is zero and the current can go up to a rated limit. This controlled turn-on capability allows for stable operation of HVDC transmission systems.

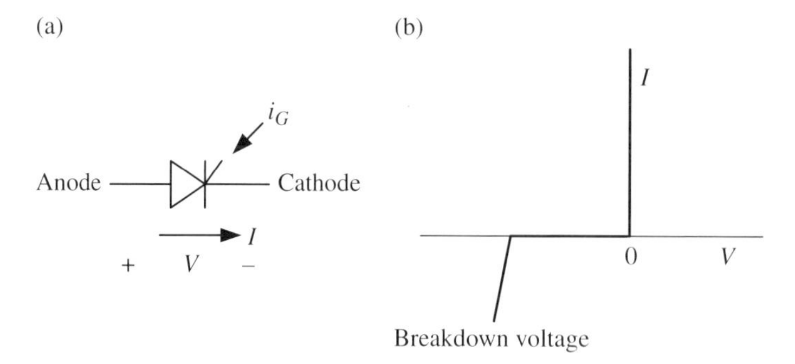

Figure 13.3 Diagrams for (a) a controlled thyristor valve and (b) the ideal thyristor voltage-current characteristics.

A modern thyristor switch can handle several thousand amperes and withstand voltages of several thousand volts. These thyristors can be stacked in series to withstand even higher voltages. They are simultaneously triggered by a light source. In each stack, there are enough thyristors such that if one thyristor in a series malfunctions, it can be bypassed without shutting down the whole system. Thus thyristor-related equipment can be considered quite reliable.

13.2.1 AC-DC Conversion using Ideal Diodes

The illustration of the conversion of AC to DC voltage is first shown using ideal diodes in converter circuits. Consider an AC sinusoidal voltage

$$e = V_m \sin(\omega t) \tag{13.1}$$

supplying a load represented by R, using a half-wave rectifier shown in Figure 13.4.a, where L is a smoothing reactor with a high inductive value. Assume an ideal transformer with turns-ratio (N_t : 1) such that $e = v_{ac}/N_t$. As the diode only conducts if $v_1 > 0$, $v_d = e$ for $e > 0$ and $v_d = 0$ otherwise, as shown in Figure 13.4.b. The average value V_d of v_d can be computed as the average of v_d through one cycle such that

$$V_d = V_m/\pi \tag{13.2}$$

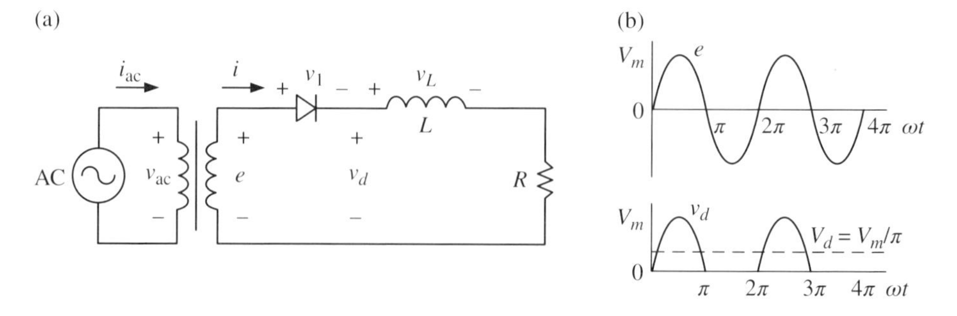

Figure 13.4 (a) Half-wave rectifier circuit and (b) voltage waveforms.

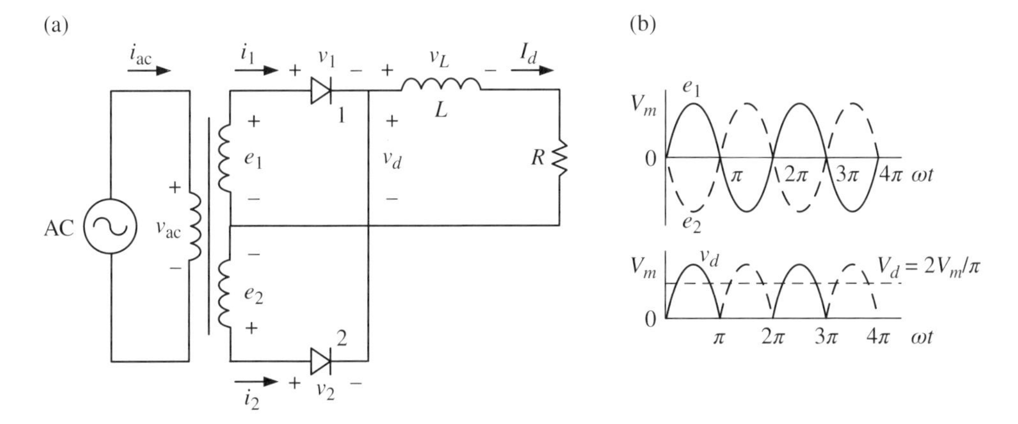

Figure 13.5 Single-phase full-wave rectifier circuit: (a) full-wave rectifier and (b) voltage waveforms.

The variation in the voltage v_d can be reduced by using a full-wave rectifier as shown in Figure 13.5.a. Here

$$e_1 = e_2 = V_m \sin(\omega t) \tag{13.3}$$

but with opposite polarity. Thus the diodes alternate in conducting, resulting in the v_d waveform shown in Figure 13.5.b. The average voltage of v_d is

$$V_d = 2V_m/\pi = 0.637V_m, \quad V_m = 1.571V_d \tag{13.4}$$

For a 3-phase system, a smoother DC voltage can be obtained using three diodes arranged in a 3-phase one-way rectifier circuit as shown in Figure 13.6.a. The AC voltage waveforms e_i, $i = 1, 2, 3$,

$$e_a = V_m \cos(\omega t + 60°), \quad e_b = V_m \cos(\omega t - 60°), \quad e_c = V_m \cos(\omega t - 180°) \tag{13.5}$$

are shown in Figure 13.6.b.

At any one point in time, only one diode will conduct, which corresponds to the phase with the highest voltage, such that the voltage

$$v_d = \max\{e_a, e_b, e_c\} \tag{13.6}$$

is equal to the peak of e_s, $s = a, b, c$, for a 1/3 period of each cycle. The DC voltage average over one period can be computed as

$$V_d = 0.828V_m, \quad V_m = 1.209V_d \tag{13.7}$$

13.2.2 Three-Phase Full-Wave Bridge Converter

A common switching circuit used in practical HVDC systems is the 3-phase full-wave bridge converter, also known as the Graetz circuit [189] shown in Figure 13.7, consisting of three phase-legs driving six ideal diodes to develop a DC voltage. Note that thyristors are used in place of ideal diodes in Figure 13.7 so that the diagram can also be used for illustrating controlled thyristor switching. In addition, the load R is now replaced by a DC voltage source V_{dc}.

(a)

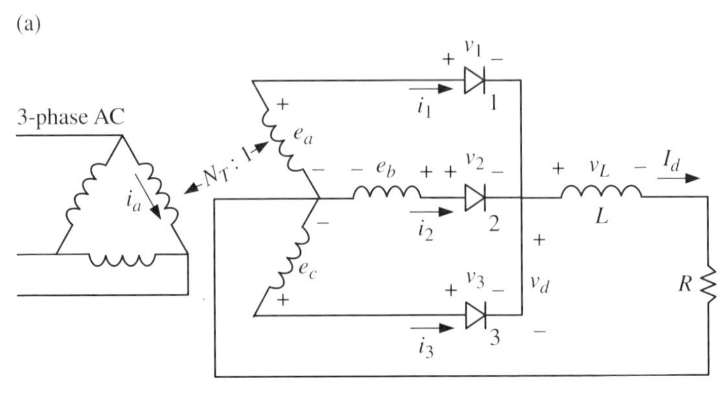

(b)

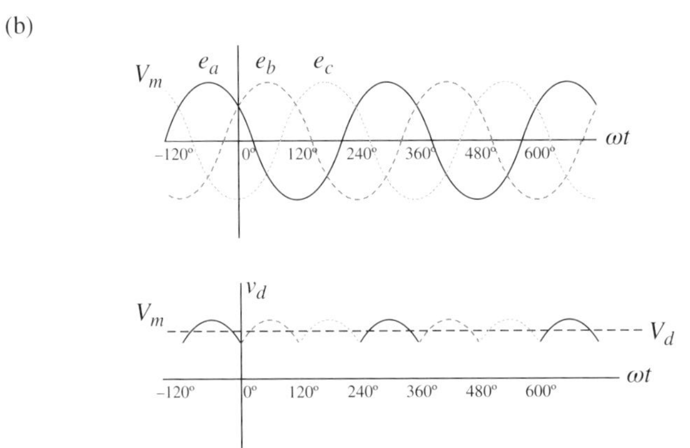

Figure 13.6 Three-phase one-way rectifier: (a) circuit and (b) voltage waveforms (see color plate section).

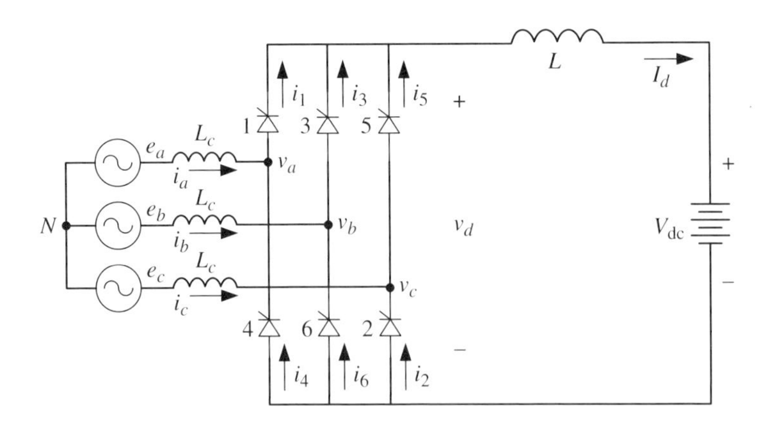

Figure 13.7 Three-phase full-wave bridge rectifier circuit.

Let v_a, v_b, and v_c be balanced 3-phase fundamental frequency sinusoidal voltages spaced 120° apart. In steady-state operation, e_a, e_b, and e_c are also balanced sinusoidal signals:

$$e_a = V_m \cos(\omega t + 60°), \quad e_b = V_m \cos(\omega t - 60°), \quad e_c = V_m \cos(\omega t - 180°) \quad (13.8)$$

The inductance L_c denotes the transformer reactance and is assumed to be the same for all 3 phases. The smoothing reactor L on the DC line is assumed to be large, so that the DC current I_d can be assumed to be constant.

In this circuit, to complete the current flow loop, it is necessary to have two ideal valves conducting, one in the upper leg (Valve 1, 3, or 5) and one in the lower leg (Valve 2, 4, or 6). The conducting pair will depend on the voltage difference between any two phases given by

$$e_{st} = e_s - e_t, \quad s,t = a,b,c, \quad s \neq t \quad (13.9)$$

In the upper row, the conducting diode corresponds to the highest value of e_s, and in the lower row, the conducting diode corresponds to the lowerst value of e_t. The pair of conducting diodes can also be determined by the maximum voltage difference e_{st}. The six possible voltage differences are plotted in Figure 13.8.a.

The voltage e_{ac} is given by

$$e_{ac} = e_a - e_c = V_m(\cos(\omega t + 60°) - \cos(\omega t - 180°)) = \sqrt{3}V_m \sin(\omega t + 30°) \quad (13.10)$$

and is shown in red. It is sinusoidal at the same fundamental frequency with an amplitude of $\sqrt{3}V_m$ and a phase delay of 30° from e_a. The other waveforms e_{st} are also shown in Figure 13.8.a, except that their negative values are not displayed.

Figure 13.8.b shows the voltage

$$v_d = \max\{e_{ab}, e_{ac}, e_{bc}, e_{ba}, e_{ca}, e_{cb}\} \quad (13.11)$$

such that the DC voltage $V_{\rm dr}$ of v_d is the average of the peak waveform over 1/6 of a cycle, which using e_{ac} is

$$\begin{aligned} V_{\rm dr} &= \frac{1}{(\pi/3)}\int_{-\pi/3}^{0} e_{ac} d\theta = \frac{3}{\pi}\int_{-\pi/3}^{0} \sqrt{3}V_m \cos\left(\theta + \frac{\pi}{6}\right) d\theta \\ &= \frac{3\sqrt{3}}{\pi} V_m \sin\left(\theta + \frac{\pi}{6}\right)\Bigg|_{-\pi/3}^{0} = \frac{3\sqrt{3}}{\pi} V_m \cdot 2\sin\left(\frac{\pi}{6}\right) \\ &= \frac{3\sqrt{3}}{\pi} V_m = 1.65 V_m \end{aligned} \quad (13.12)$$

When considering power transfer, it is useful to define the RMS line-to-neutral voltage as $V_{LN} = V_m/\sqrt{2}$ and the RMS line-to-line voltage as $V_{LL} = \sqrt{3}V_m/\sqrt{2}$. The DC voltage can be expressed as

$$V_{\rm dr} = \frac{3\sqrt{6}}{\pi} V_{LN} = \frac{3\sqrt{2}}{\pi} V_{LL} \quad (13.13)$$

Figure 13.8 is also useful for illustrating the conducting valve pairs. For the upper row, as each phase voltage is higher than the other two phase-voltages for 1/3 cycle, the valve for that phase will conduct during that period. For the lower row, as each phase voltage

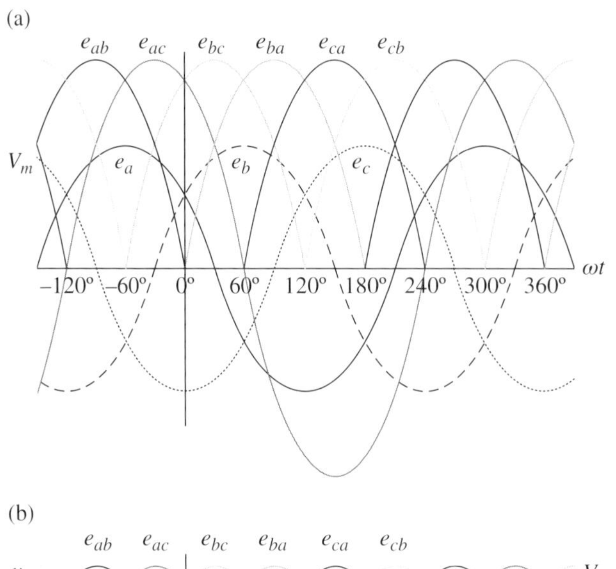

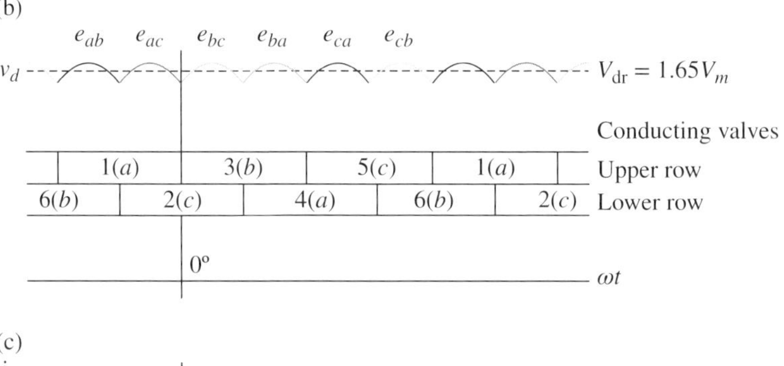

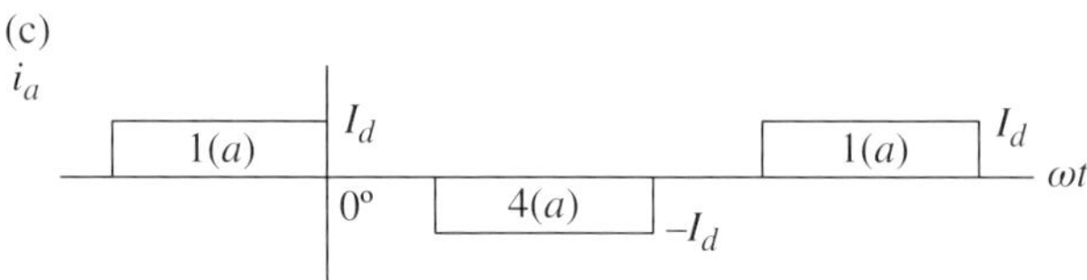

Figure 13.8 Three-phase full-wave bridge rectifier: (a) voltage waveforms, (b) DC voltage waveform and conducting valves, and (c) current in phase a (see color plate section).

is lower than the other two phase-voltages for 1/3 cycle, the valve for that phase will conduct during that period. Furthermore, these conducting periods are staggering by 1/6 cycle, resulting in v_d consisting of the peak waveform of the six phase-voltage differences e_{st}. Thus this is known as a six-pulse bridge circuit. The subscripts in the peak phase-voltage difference inform the conducting valves. For example, when e_{ac} reaches its peak, the phase-a valve in the upper row and the phase-c valve in the lower row will conduct.

With a smoothing reactor, I_d will be kept constant. The current $i_1, \ldots, i_6$ (Figure 13.7) in each valve, when conducting, will also be equal to I_d. For example, when e_{ac} is higher than the other phase voltages, $i_1 = i_2 = I_d$, and all the other currents are zero. Thus $i_a = I_d$, $i_b = 0$, and $i_c = -I_d$. Figure 13.8.c shows the phase-a current i_a.

From the i_a plot, the amplitude of the fundamental frequency component of the current i_a can be extracted using Fourier series as

$$\begin{aligned} I_{LN} &= \frac{1}{\pi}\int_0^{2\pi} I_d \sin\theta d\theta = \frac{2}{\pi}\int_{\pi/6}^{5\pi/6} I_d \sin\theta d\theta \\ &= -\frac{2}{\pi} I_d \cos\theta \Big|_{\pi/6}^{5\pi/6} = -\frac{2}{\pi} I_d \left(-\frac{\sqrt{3}}{2} - \frac{\sqrt{3}}{2}\right) \\ &= \frac{2\sqrt{3}}{\pi} I_d = 1.11 I_d \end{aligned} \tag{13.14}$$

The RMS value of the current magnitude is

$$I_{LI} = \frac{1}{\sqrt{2}}\frac{2\sqrt{3}}{\pi} I_d = \frac{\sqrt{6}}{\pi} I_d = 0.78 I_d \tag{13.15}$$

Thus the DC power delivered by the rectifier is $P_{\mathrm{DC}} = V_{\mathrm{dr}} I_d$. Neglecting losses, the AC active power in the fundamental frequency component provided by all three phases is

$$P_{\mathrm{ac}} = 3\left(\frac{V_m}{\sqrt{2}}\right)\left(\frac{\sqrt{6}}{\pi} I_d\right) = \frac{3\sqrt{3}}{\pi} V_m I_d = V_{\mathrm{dr}} I_d = P_{\mathrm{DC}} \tag{13.16}$$

13.3 Line-Commutation Operation in HVDC Systems

In a real power system, controlled switching using thyristors is necessary for reliable operation of an HVDC system. This section describes the practical operation of HVDC systems and develops an equivalent HVDC circuit for power flow computation, based on the gate control of thyristors.

To ensure successful thyristor switching in light of 3-phase AC system operating with noise and imbalance, the ignition or turn-on of a thyristor is delayed until a phase voltage is either sufficiently larger or smaller than the other two phases. Once a thyristor is turned on, it will not be extinguished unless some other phase voltage has a higher value. The delayed ignition also allows the HVDC power transfer to be regulated and to operate at reduced power if the terminal voltages are depressed. Another aspect that needs to be considered is the commutation effect, as the current flow switching over from one thyristor to another thyristor is not instantaneous as in an ideal diode.

This section first discusses the rectifier operation of converting AC to DC quantities. and then the inverter operation of converting DC back to AC quantities.

13.3.1 Rectifier Operation

13.3.1.1 Thyristor Ignition Delay Angle

Consider the transition of conduction from Valve 1 of phase a to Valve 3 of phase b shown in Figure 13.8. Instead of switching on Valve 3 immediately at $\omega t = 5\pi/6$, the ignition is delayed by an angle α. As an example, for $\alpha = 10^{\circ}$ in a 60 Hz system, the time delay is

$$t = \frac{\alpha}{\omega} = \frac{10 \times \pi/180}{2\pi \cdot 60} = 0.463 \text{ msec} \tag{13.17}$$

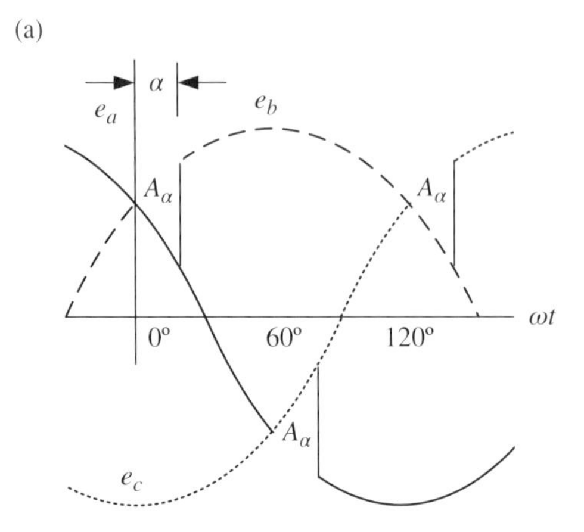

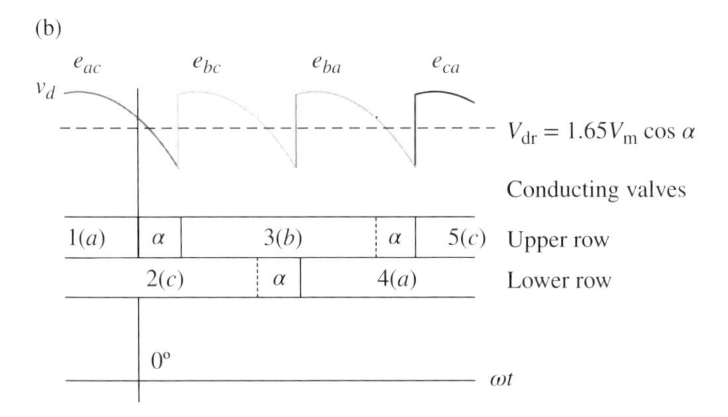

Figure 13.9 Three-phase full-wave bridge converter voltage and current waveforms with ignition delays: (a) voltage waveforms and (b) DC voltage waveform and conducting valves (see color plate section).

The voltage v_d resulting from delaying all valve switchings by α is shown in Figure 13.9. The valve switching sequence is also shown in Figure 13.9.

The average rectifier DC voltage, denoted by V_{dr} over 1/6 cycle, is

$$
\begin{aligned}
V_{dr} &= \frac{1}{(\pi/3)} \int_{-\pi/3+\alpha}^{\alpha} e_{ac} d\theta = \frac{3\sqrt{3}V_m}{\pi} \int_{-\pi/3+\alpha}^{\alpha} \cos\left(\theta + \frac{\pi}{6}\right) d\theta \\
&= -V_{dor} \sin\left(\theta + \frac{\pi}{6}\right)\Bigg|_{-\pi/3+\alpha}^{\alpha} = V_{dor} \cos\alpha
\end{aligned}
\tag{13.18}
$$

where $V_{dor} = 3\sqrt{3}V_m/\pi$ is the DC voltage without ignition delay.

Thus one of the impacts of the ignition delay angle α is that the DC voltage is reduced. Equating the active power transfer between the AC and DC systems yields

$$
3(V_{LN} I_{LI} \cos\phi_r) = V_{dr} I_d = (V_{dor} \cos\alpha) I_d \tag{13.19}
$$

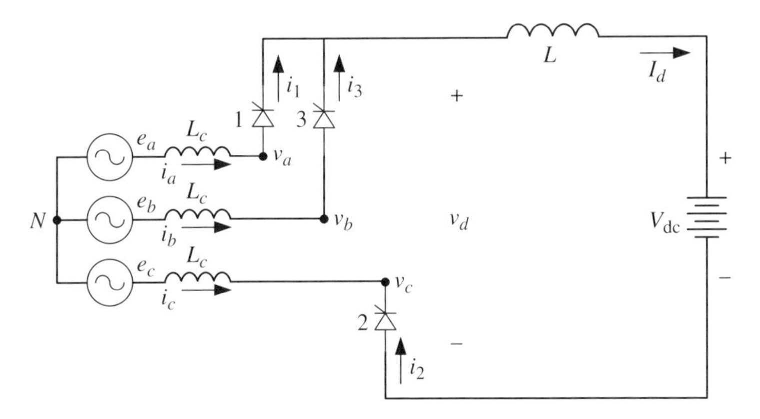

Figure 13.10 Three-phase full-wave bridge converter circuit for commutation from Valves 1 and 2 to Valves 3 and 2 (non-conducting valves not shown).

where ϕ_r is the phase lag of the current waveform relative to the voltage waveform, which is also the power factor angle of the AC power supply. That is, the power factor is given by

$$\cos\phi_r = \cos\alpha \tag{13.20}$$

With ideal switching in which $\alpha = 0$, there is no reactive power requirement from the AC network. However, with ignition delay, the power supply from the AC network to the rectifier is

$$P_{\text{acr}} = V_{\text{dr}} I_d, \quad Q_{\text{acr}} = P_{\text{ac}} \tan\alpha \tag{13.21}$$

in which reactive power is needed to support the ignition delay.

13.3.1.2 Commutation Overlap

In ideal thyristor switching, the current switching between Valves 1 and 3 would occur instantaneously, that is, $i_1 = i_a$ would drop from I_d to 0 and $i_3 = i_b$ would jump from 0 to I_d. In a real power network with inductive elements, i_a and i_b cannot change abruptly.

Consider the circuit representation in Figure 13.10 in which the inductance L_c is an equivalent connection from the rectifier to a voltage source, the voltage drops from the voltage sources to the DC line in the phase-a and phase-b paths can be equated to form

$$e_b - L_c \frac{di_3}{dt} = e_a - L_c \frac{di_1}{dt} \tag{13.22}$$

assuming that the voltage drop across the thyristors is zero. Then

$$e_b - e_a = L_c\left(\frac{di_3}{dt} - \frac{di_1}{dt}\right) = \sqrt{3}V_m \cos(\omega t - 90^\circ) = \sqrt{3}V_m \sin(\omega t) \tag{13.23}$$

Because $I_d = i_1 + i_3$ is constant,

$$\frac{di_3}{dt} + \frac{di_1}{dt} = 0 \tag{13.24}$$

Thus the current in Valve 3 is governed by

$$\frac{di_3}{dt} = \frac{\sqrt{3}V_m}{2L_c}\sin(\omega t) \tag{13.25}$$

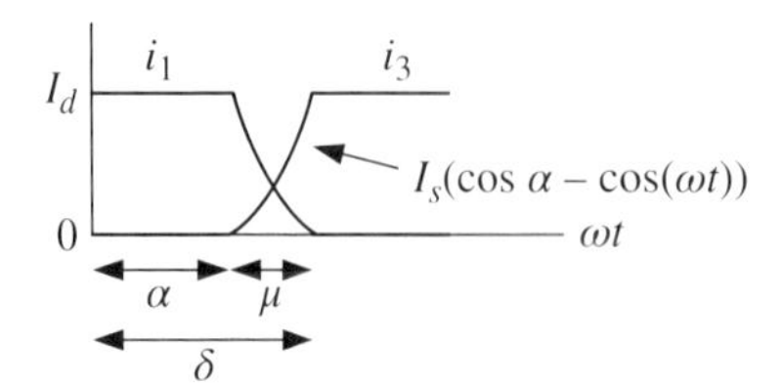

Figure 13.11 Valve currents for commutation from Valves 1 and 2 to Valves 3 and 2.

The build-up of the current i_3 from 0 at $t = \alpha/\omega$ is

$$i_3 = \frac{\sqrt{3}V_m}{2L_c}\int_{\alpha/\omega}^{t}\sin(\omega\tau)d\tau = I_s(\cos\alpha - \cos(\omega t)),\quad I_s = \frac{\sqrt{3}V_m}{2\omega L_c} \tag{13.26}$$

as shown in Figure 13.11.

During this period, i_1 will decrease from I_d to 0 as

$$i_1 = I_d - I_s(\cos\alpha - \cos(\omega t)) \tag{13.27}$$

Let $\delta = \omega t$ when $i_3 = I_d$, that is, when commutation is completed. Then

$$I_d = I_s(\cos\alpha - \cos\delta) \tag{13.28}$$

that is,

$$\delta = \cos^{-1}\left(\cos\alpha - \frac{I_d}{I_s}\right) \tag{13.29}$$

Thus δ is the extinction delay angle of Valve 1. The difference of δ and the ignition delay angle of Valve 3 is the overlap angle

$$\mu = \delta - \alpha \tag{13.30}$$

The commutation also results in a DC voltage reduction. With Valves 1 and 3 conducting at the same time during commutation, the phase-a and phase-b AC voltages are

$$v_a = v_b = e_b - L_c\frac{di_3}{dt} = e_b - \frac{e_b - e_a}{2} = \frac{e_a + e_b}{2} \tag{13.31}$$

which is less than e_b, as shown in Figure 13.12.

The voltage drop is determined by the area A_μ between e_b and $(e_a + e_b)/2$ given by

$$A_\mu = \int_{\alpha}^{\delta}\left(\frac{e_a + e_b}{2}\right)d\theta = \frac{\sqrt{3}V_m}{2}\int_{\alpha}^{\delta}\sin\theta d\theta = \frac{\sqrt{3}V_m}{2}(\cos\alpha - \cos\delta) \tag{13.32}$$

Thus the average voltage drop due to the overlap is

$$\Delta V_{\mathrm{dr}} = \frac{A_\mu}{\pi/3} = \frac{3}{\pi}\frac{\sqrt{3}}{2}V_m(\cos\alpha - \cos\delta) = \frac{V_{\mathrm{dor}}}{2}(\cos\alpha - \cos\delta) \tag{13.33}$$

and thc effective DC voltage on the rectifier side is

$$V_{\mathrm{dr}} = V_{\mathrm{dor}}\cos\alpha - \frac{V_{\mathrm{dor}}}{2}(\cos\alpha - \cos\delta) = \frac{V_{\mathrm{dor}}}{2}(\cos\alpha + \cos\delta) \tag{13.34}$$

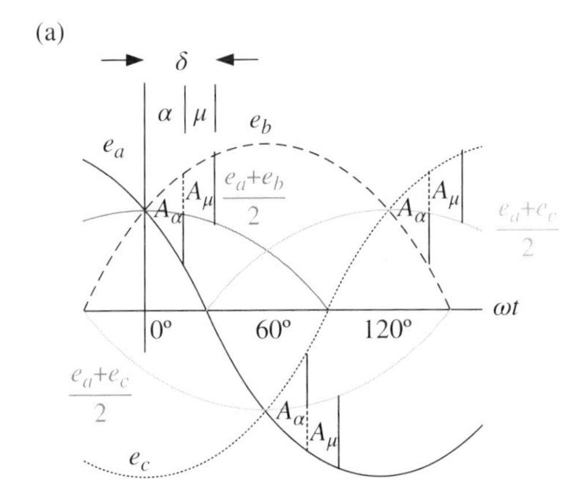

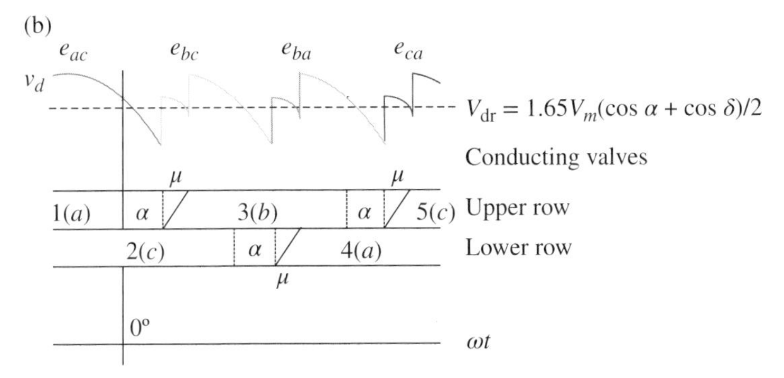

Figure 13.12 Voltage reduction during communtation: (a) voltage waveforms and (b) DC voltage waveform and conducting valves (see color plate section).

Equating the AC active power to the DC power

$$3(V_{LN}I_{LI}\cos\phi) = V_{dr}I_d = \frac{V_{dor}}{2}(\cos\alpha + \cos\delta)I_d \tag{13.35}$$

such that

$$3\left(V_{LN}\frac{\sqrt{6}}{\pi}I_d\right)\cos\phi = \frac{3\sqrt{6}}{\pi}V_{LN}\cdot\frac{1}{2}(\cos\alpha + \cos\delta)\cdot I_d \tag{13.36}$$

that is, the power factor is given by

$$\cos\phi_r = \frac{1}{2}(\cos\alpha + \cos\delta) \tag{13.37}$$

Thus the power supplied from the AC network to the rectifier is

$$P_{acr} = V_{dr}I_d, \quad Q_{acr} = P_{acr}\tan\phi_r \tag{13.38}$$

In power flow formulation, P_{acr} and Q_{acr} are entered as a positive load using the *PQ* bus type.

13.3.2 Inverter Operation

The thyristor circuit for an inverter to convert the DC voltage and current back to 3-phase AC voltages and currents is shown in Figure 13.13. The individual phase quantities at the inverter are denoted by primed (′) variables. For conversion back to AC quantities, the inverter operation requires the phase voltage angle to be 120°. Thus the ignition delay angle would be $\alpha > 120°$.

For inverter operation, the modeling equations can be simplified using the following terminology:

ignition advance angle $\beta = \pi - \alpha$

extinction advance angle $\gamma = \pi - \delta$

overlap angle $\mu = \delta - \alpha = \beta - \gamma$

Following (13.25) and (13.28), the DC current on the inverter side can be represented by

$$I_d = \frac{\sqrt{3}V'_m}{2\omega L'_c}(\cos\alpha - \cos\delta) = I'_s(\cos\gamma - \cos\beta) \tag{13.39}$$

where $I'_s = \sqrt{3}V'_m/(2\omega L'_c)$.

The rectifier voltage equation (13.34) is changed to the inverter voltage equation by changing the sign of the DC voltage

$$V_{\rm di} = -\frac{V_{\rm doi}}{2}(\cos\alpha + \cos\delta) = \frac{V_{\rm doi}}{2}(\cos\gamma + \cos\beta) \tag{13.40}$$

where $V_{\rm doi} = 3\sqrt{3}V'_m/\pi$ and V'_m is the peak amplitude of the AC sinusoidal voltages e'_a, e'_b, and e'_c.

Equating the AC active power transmission from the inverter results in

$$3(V_{LN}I_{LI}\cos\phi_i) = V_{\rm di}I_d = \frac{V_{\rm doi}}{2}(\cos\gamma + \cos\beta)I_d \tag{13.41}$$

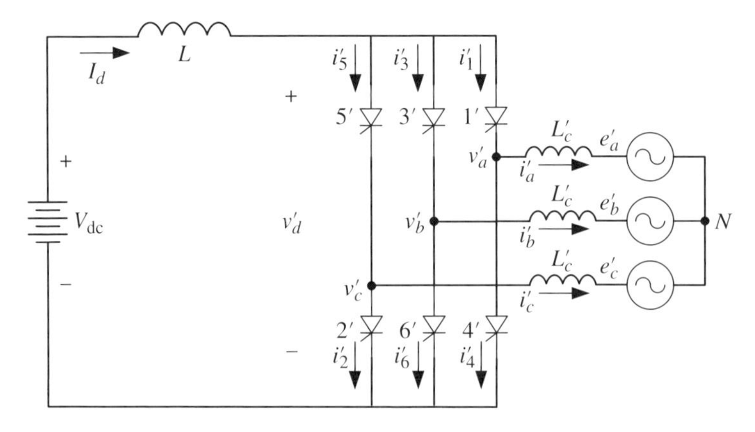

Figure 13.13 Three-phase full-wave bridge inverter circuit.

such that the power factor is given by

$$\cos\phi_i = \frac{1}{2}(\cos\gamma + \cos\beta) \tag{13.42}$$

Thus the power supplied from the AC network to the inverter is

$$P_{\text{aci}} = -V_{\text{di}}I_d, \quad Q_{\text{aci}} = -P_{\text{aci}}\tan\phi_i \tag{13.43}$$

In power flow formulation, P_{aci} is entered as a negative active power load and Q_{aci} a positive reactive power load using the PQ bus type.

13.3.3 Multiple Bridge Converters

The 3-phase bridge converters can be connected in cascade to increase the DC voltage. Also, if the transformers of one bridge are connected Y-Y and those of the second bridge are connected Y-Δ, the amount of harmonics due to the conversion process can be reduced. The computation of the various voltages provided previously is still valid by accounting for the commutation reactance of each bridge. For example, if the rectifier has two bridges, that is, $B_r = 2$, and the commutation inductance for each bridge is L_{cr}, then the total commutation inductance is B_rL_{cr}. However, as the voltage level is double in this case, the per-unit value of L_{cr} will remain the same. With this fact noted, the equivalent circuit of a HVDC link can be developed, which is shown next.

13.3.4 Equivalent Circuit

Including the valve ignition and extinction delays and the commutation overlap, an equivalent circuit based on the DC circuit equations can be established, connecting the rectifier to the inverter and from the converters to the AC buses.

First note that using (13.28) and the expression for I_s, and including multiple rectifier bridges, (13.33) becomes

$$\Delta V_{\text{dr}} = \frac{3}{\pi}\frac{\sqrt{3}}{2}V_mB_r\frac{2\omega L_c}{\sqrt{3}V_m}I_d = B_r\frac{3}{\pi}\omega L_cI_d = B_rR_{cr}I_d \tag{13.44}$$

where $R_{cr} = 3\omega L_{cr}/\pi$ is the equivalent rectifier commutation reactance with the unit of ohms and B_r is the number of bridges. Thus taking both the ignition delay and commutation overlap into account, the DC voltage at the rectifier is given by

$$V_{\text{dr}} = V_{\text{dor}}\cos\alpha - B_rR_{cr}I_d \tag{13.45}$$

The commutation reactance is treated like a resistance in calculating the DC voltage. However, it is not a true resistance and thus does not generate any resistive losses.

Similarly, on the inverter side

$$V_{\text{di}} = \frac{V_{\text{doi}}}{2}(2\cos\gamma + (\cos\beta - \cos\gamma)) = V_{\text{doi}}\cos\gamma - B_i\frac{V_{\text{doi}}}{2}\frac{I_d}{I_s} = V_{\text{doi}}\cos\gamma - B_iR_{ci}I_d \tag{13.46}$$

where $R_{ci} = 3\omega L_{ci}/\pi$ is the inverter commutation reactance in ohms, B_i is the number of inverter bridges, and I_d is taken to be the DC current flowing from the inverter to the rectifier.

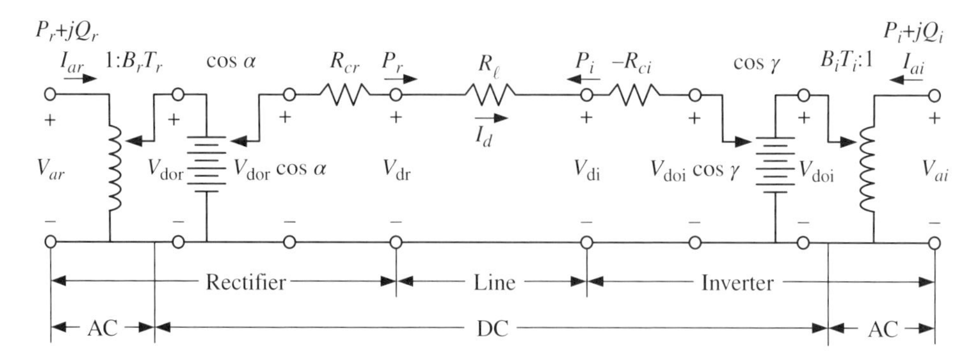

Figure 13.14 Steady-state equivalent circuit for two-terminal DC link, with smoothing reactors, and DC line reactance and capacitance neglected [9].

In addition, the DC voltage drop from the rectifier to the inverter can be modeled as

$$V_{\mathrm{dr}} - V_{\mathrm{di}} = R_{\ell} I_d \tag{13.47}$$

where R_{ℓ} is the DC line resistance.

Combining these equations with the rectifier and inverter equations, the equivalent circuit in Figure 13.14 can be obtained. Note that in Figure 13.14, the rectifier and inverter transformer tap ratios are indicated as T_r and T_i, respectively. These transformers can be used to adjust the DC voltages. Note that the inverter commutation reactance carries a negative sign, as I_d is flowing from the rectifier to the inverter.

Example 13.1: HVDC voltage computation

The parameters of a two-terminal HVDC system are the DC line resistance $R_{\ell} = 5$ ohms, the commutation resistance $R_c = 11.75$ ohms for both converters, the rated DC current is 1000 A, the rated DC voltage is 100 kV at the DC side of the inverter, $\alpha = 18°$, and $\gamma = 18°$. Both converters are 6-pulse converters ($B_r = B_i = 1$).

Calculate for *each* terminal, not considering the AC network and assuming unity transformer taps,

1) the line-to-line voltage V_a on the converter side of the transformer for each converter
2) the active and reactive power flow for each converter, using the power factor angle as $\phi = \cos^{-1}(V_d/V_{do})$
3) the MVA rating of the transformers based on the (P, Q) requirement of the converter at rated condition

Draw the DC system voltage-current diagrams.

Solutions: 1. HVDC system calculation typically starts from the inverter side, as the rated DC voltage defines V_{di}.

Inverter calculations:

$$V_{\mathrm{di}} = 100 \text{ kV}, \quad V_{\mathrm{doi}} = (1/\cos\gamma)(V_{\mathrm{di}} + I_d R_c) = 117.5 \text{ kV}$$

$$V_{\mathrm{aci}} = (\pi/(3\sqrt{2}))V_{\mathrm{doi}} = 87.007 \text{ kV}, \quad V_{\mathrm{doi}} \cos\gamma = 111.75 \text{ kV}$$

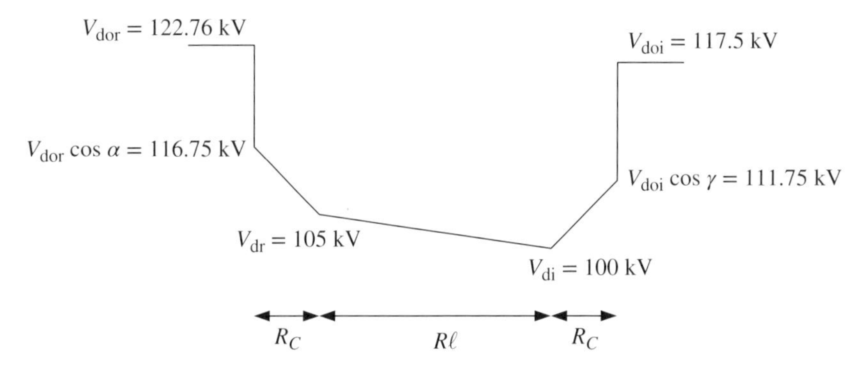

Figure 13.15 Voltage along HVDC link.

Rectifier calculations:

$$V_{dr} = V_{di} + R_\ell I_d = 105 \text{ kV}, \quad V_{dor} = (1/\cos\alpha)(V_{dr} + I_d R_c) = 122.76 \text{ kV}$$
$$V_{acr} = (\pi/(3\sqrt{2})V_{dor} = 90.9 \text{ kV}, \quad V_{dor}\cos\alpha = 116.75 \text{ kV}$$

2) P, Q flow at the rectifier:

$$P_r = V_{dr} I_d = 105 \text{ MW}, \quad Q_r = P_r \sqrt{(V_{dor}/V_{dr})^2 - 1} = 63.597 \text{ MVar}$$

P, Q flow at the inverter:

$$P_i = -V_{di} I_d = -100 \text{ MW}, \quad Q_i = -P_i \sqrt{(V_{doi}/V_{di})^2 - 1} = 61.697 \text{ MVar}$$

3) Transformer ratings

Rectifier: $\sqrt{P_r^2 + Q_r^2} = 122.76$ MVA, Inverter: $\sqrt{P_i^2 + Q_i^2} = 117.5$ MVA

The voltage along the DC line is shown in Figure 13.15. ∎

13.4 Control Modes

An HVDC system is controlled by fast thyristor switching and slower mechanical transformer tap adjustment. These parameters are bounded by their maximum and minimum limits.

The HVDC system control objectives are to:

1) Maintain the DC voltage V_{dc}^{rated} at the inverter terminal, current order I_{ord} at the rectifier terminal, and power P_{dc}^{des} at the rated or desired values.
2) Maintain the power factor at preset values at both the rectifier and inverter terminals.
3) Maintain the ignition and extinction angles and the transformer taps within limits. For proper commutation, the ignition angle α should be higher than 5° to achieve a sufficiently positive voltage before a thyristor is turned on, and the extinction angle γ should be greater than 18° to prevent commutation failure.

The proper operation of an HVDC system requires sufficient voltage support at the converter terminals. As a result, various control modes are developed for HVDC system operation depending on the voltage support capability from the AC system, as described below.

13.4.1 Mode 1: Normal Operation

In normal operation, the *rectifier* will maintain constant DC current at I_{ord} (constant current (CC) operation) and the *inverter* will maintain the DC side of the rectifier voltage at the desired value $V_{\text{dc}}^{\text{rated}}$.

At the rectifier, the current order is determined by the rated DC voltage and power to be

$$I_d = I_{\text{ord}} = P_{\text{dc}}^{\text{des}} / V_{\text{dc}}^{\text{rated}} \tag{13.48}$$

At the inverter, the extinction angle γ is set to maintain V_{di} at the rated DC voltage, with $\gamma \geq \gamma_{\min}$

$$V_{\text{di}} = V_{\text{dc}}^{\text{rated}} = V_{\text{doi}} \cos\gamma - B_i R_{ci} I_d = \frac{3\sqrt{2}}{\pi} B_i T_i V_{ai} \cos\gamma - B_i R_{ci} I_d \tag{13.49}$$

where B_i is the number of inverter bridges, V_{ai} is the phase-*a* voltage RMS value at the inverter terminal, and I_d is set to I_{ord}.

Based on the DC current I_d and the inverter voltage V_{di}, the rectifier voltage becomes

$$V_{\text{dr}} = V_{\text{di}} + R_\ell I_d = V_{\text{dor}} \cos\alpha - B_r R_{rc} I_d, \quad V_{\text{dor}} = \frac{3\sqrt{2}}{\pi} B_r T_r V_{ar} \tag{13.50}$$

where V_{ar} is the phase-*a* voltage RMS value at the rectifier terminal and B_r is the number of rectifier bridges.

The normal operation is illustrated in Figure 13.16.a, which shows the rectifier voltage V_{dr} (13.50) as a function of I_d with constant ignition angle (CIA) α and the inverter voltage V_{di} (13.49) as a function of I_d with constant extinction angle (CEA) γ. The V_{dr} curve is valid up to the current order I_{ord}. The V_{di} curve is extended to $I_d = I_{\text{ord}} - I_m$, where I_m is the current margin, which is typically 10–15% of I_{ord}. The operating point is the intersection of the V_{di} curve and I_{ord}, which is also known as Mode-1 operation.

Figure 13.16.a illustrates the ideal Mode-1 operation in which the DC line current I_d is held fixed at I_{ord}. In practice, I_d is controlled from a current regulator, which with the fast dynamics neglected, determines the rectifier voltage as

$$V_{\text{dor}} \cos\alpha = K(I_{\text{ord}} - I_d) \tag{13.51}$$

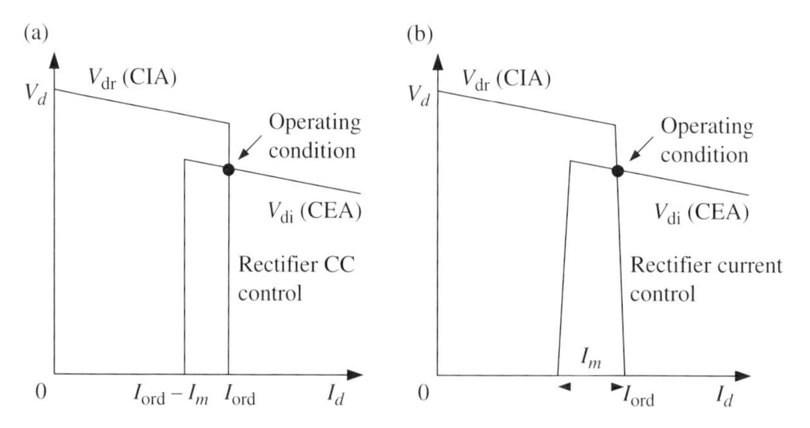

Figure 13.16 HVDC normal operation: (a) Mode 1 and (b) Mode 1 with proportional current control.

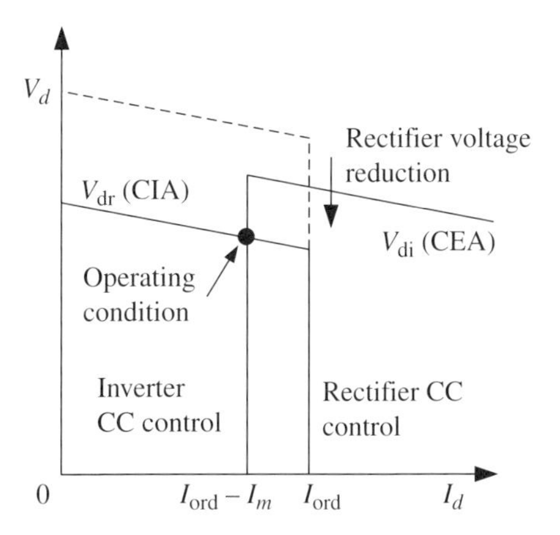

Figure 13.17 HVDC rectifier reduced voltage operation: Mode-2 operation.

where K is the controller proportional gain. Substituting (13.51) into (13.50) yields

$$V_{dr} = K(I_{ord} - I_d) - B_r R_{cr} I_d = K I_{ord} - (K + B_r R_{cr}) I_d \tag{13.52}$$

Thus the ideal constant current characteristic is changed into a straight line regulation with a slope of $-(K + B_r R_{cr})$, as shown in Figure 13.16.b.

13.4.2 Mode 2: Reduced-Voltage Operation

During system operation when the AC bus voltage on the rectifier side is low, V_{dor} will also be low, as the transformer may take some time to adjust its tap position. Even with $\alpha = \alpha_{min}$, the CIA curve will drop, as indicated in Figure 13.17. As a result, the Mode-1 operating condition is infeasible. Under this situation, the inverter will reduce the DC current by the current margin I_m, such that the DC line current becomes $I_d = I_{ord} - I_m$. The inverter DC side voltage V_{di} will then be determined by the rectifier voltage V_{dor}. This is also known as the Mode-2 operation.

Example 13.2: Reduced-voltage operation

Calculate the Mode-2 operating condition of the HVDC system in Example 13.1 when the AC voltage at the rectifier AC bus drops by 20%. Assume the transformer tap ratios remain the same, the current margin is 150 A, and the minimum value of α is 5°. Calculate the voltages on the DC line and the active and reactive power flow for each converter.

Solutions: Reducing the AC voltage at the rectifier by 20% and keeping $V_{dr} = 105$ kV and $I_d = 1$ kA results in

$$\overline{V}_{dor} = V_{dor} \times 0.8 = 98.207 \text{ kV}, \quad \cos\alpha = 116.75/\overline{V}_{dor} = 1.189 > 1 \tag{13.53}$$

implying an infeasible solution. Hence the HVDC system operation is switched to Mode 2.

Reduce the DC current by the current margin $I_m = 150$ A, with $\alpha_{\min} = 5°$

$$\overline{I}_d = I_d - 150 = 850 \text{ A}, \quad \overline{V}_{\text{dor}} \cos(\alpha_{\min}) = 97.83 \text{ kV} \tag{13.54}$$

Using $\alpha_{\min}$ results in the smallest voltage drop for $\overline{V}_{\text{dr}}$, such that

$$\overline{V}_{\text{dr}} = \overline{V}_{\text{dor}} \cos(\alpha_{\min}) - R_c\overline{I}_d = 87.85 \text{ kV} \tag{13.55}$$

$$\overline{V}_{\text{di}} = \overline{V}_{\text{dr}} - R_\ell\overline{I}_d = 83.60 \text{ kV} \tag{13.56}$$

$$V_{\text{doi}} \cos(\overline{\gamma}) = \overline{V}_{\text{di}} + R_c\overline{I}_d = 93.58 \text{ kV} \tag{13.57}$$

$$\cos(\overline{\gamma}) = (\overline{V}_{\text{di}} + R_c\overline{I}_d)/V_{\text{doi}} = 0.7964, \quad \overline{\gamma} = 37.21° \tag{13.58}$$

The P, Q flow at the rectifier is

$$\overline{P}_r = \overline{V}_{\text{dr}}\overline{I}_d = 74.67 \text{ MW}, \quad \overline{Q}_r = \overline{P}_r\sqrt{(\overline{V}_{\text{dor}}/\overline{V}_{\text{dr}})^2 - 1} = 37.32 \text{ MVar} \tag{13.59}$$

and at the inverter is

$$\overline{P}_i = -\overline{V}_{\text{di}}\overline{I}_d = -71.06 \text{ MW}, \quad \overline{Q}_i = -\overline{P}_i\sqrt{(V_{\text{doi}}/\overline{V}_{\text{di}})^2 - 1} = 70.19 \text{ MVar} \tag{13.60}$$

The voltage along along the DC line is shown in Figure 13.15. ■

13.4.3 Mode 3: Transitional Mode

As the HVDC system control transitions from normal-voltage to reduced-voltage operation (Figure 13.18.b), if the slopes of the V_{dr} and V_{di} curves are similar, the intersection of these curves may occur at several points, some of which are undesirable operating conditions. To avoid this situation, a constant ignition advance angle (β) control can be used.

Equations (13.40) and (13.46) can be combined to eliminate $\cos\gamma$ to obtain the inverter DC voltage as

$$V_{\text{di}} = V_{\text{doi}} \cos\beta + B_iR_{ci}I_d \tag{13.61}$$

With β fixed, V_{di} has a positive slope with respect to I_d, as shown in Figure 13.19. Here there is a unique intersection of the DC current I_d with the rectifier voltage V_{dr}. Using the expressions for the rectifier voltage (13.50) in reduced-voltage operation and the inverter voltage (13.61) in constant β operation, the current I_d can be determined as

$$I_d = \frac{V_{\text{dr}} - V_{\text{di}}}{R_\ell} = \frac{V_{\text{dor}} \cos\alpha - V_{\text{doi}} \cos\beta - (B_rR_{cr} + B_iR_{ci})I_d}{R_\ell} \tag{13.62}$$

that is,

$$I_d = \frac{V_{\text{dor}} \cos\alpha - V_{\text{doi}} \cos\beta}{R_\ell + B_rR_{cr} + B_iR_{ci}} \tag{13.63}$$

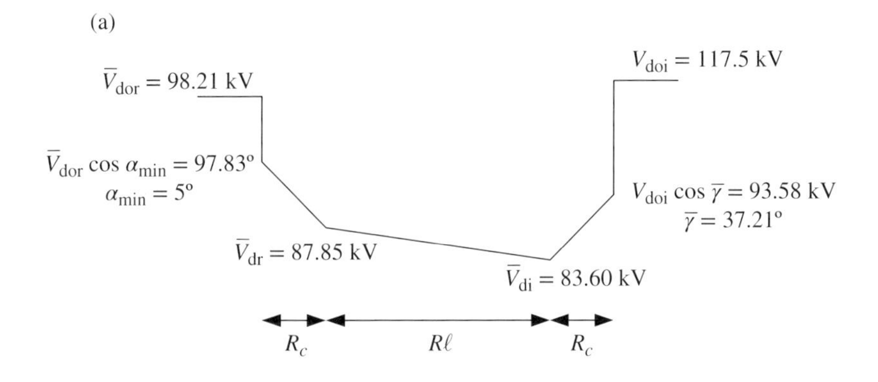

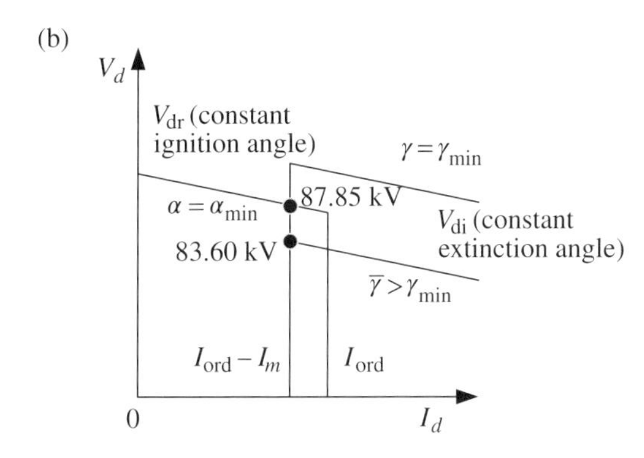

Figure 13.18 Mode-2 operation: (a) voltage profile on DC system and (b) solutions shown on voltage-current diagram.

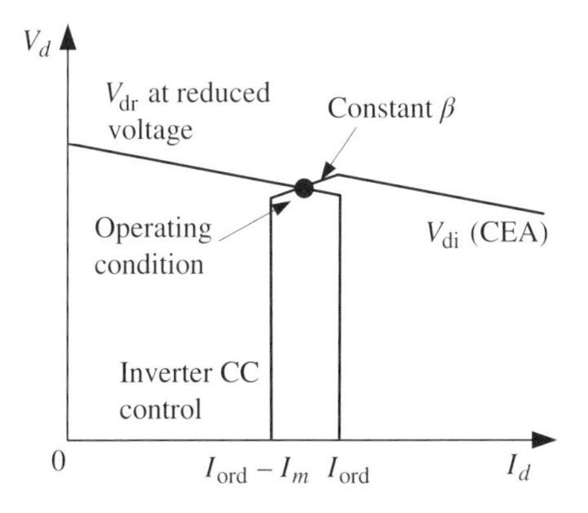

Figure 13.19 Transitional Mode-3 operation.

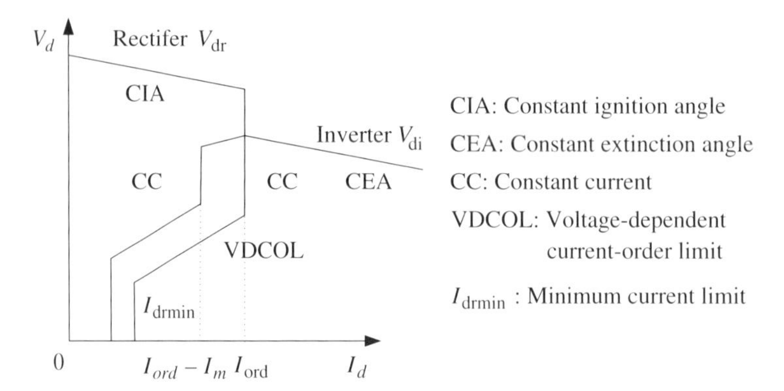

Figure 13.20 HVDC steady-state operating condition with voltage-dependent current-order limiter function.

13.4.4 System Operation Under Fault Conditions

During severe disturbances with short-circuit faults depressing system voltages in the vicinity of an HVDC terminal, it will be beneficial to reduce the DC current and hence active power transfer on the DC system [172]. This approach would reduce the reactive power drawn from the AC system by the DC converters. The reduced current setting is determined by the voltage-dependent current-order limit (VDCOL) shown in Figure 13.20. Such an operation will reduce the possibility of voltage collapse, especially if the terminals are connected to weak AC systems.

13.4.5 Communication Requirements

The converters of an HVDC system are normally coordinated by a centralized controller. For example, in Mode 1, the rectifier receives the current order and the inverter receives the DC voltage reference. The local controllers then implement these setpoints in their control loops. The cental controller also monitors the AC system condition at the converter substations to determine whether for a given abnormal condition, switching to the Mode-2 or Mode-3 operation is required.

The central controller is typically located at one of the converter substations. Communications between the converters were achieved in the past by power line carrier technology, but now are replaced with optical fibers. There are two communication systems in operation: one system is for control and the other for protection.

The types of faults that can be experienced by an HVDC system may include:

1) control failures
2) valve cooling
3) station service failure
4) switching error

The centralized controller is involved in fault recovery. For example, when one pole of the HVDC system is shut down due to AC or DC faults, the system can restart within seconds if the centralized controller has determined that the fault has been cleared. The centralized controller is also involved when reversing power on an LCC HVDC system, as the inverter poles have to be physically reversed.

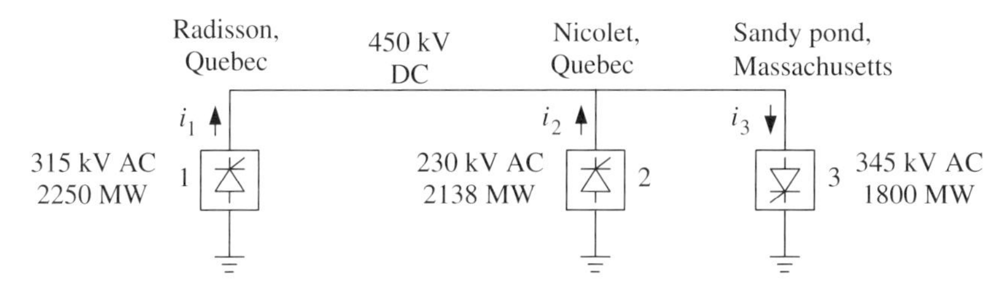

Figure 13.21 Quebec-New England three-terminal HVDC link: the rated values of the DC system and the converters are shown for each terminal.

13.5 Multi-terminal HVDC Systems

An HVDC system is primarily built to transfer power from a generation site (rectifier terminal) to a load center (inverter terminal) as a two-terminal system. However, if the DC line transverses across other generation and load areas, additional terminals can be added to form a multi-terminal HVDC (MTDC) system, such as the radial system shown in Figure 13.2. Instead of providing a general discussion, the operation of one of the world's first high-power MTDC systems is briefly described.[2]

A practical three-terminal HVDC system is the Quebec-New England Phase II system connecting the Province of Quebec (Canada) to New England (USA) and shown in Figure 13.21. This is a biploar HVDC system operating at ±450 kV and capable of transmitting 2000 MW from the hydraulic generators (eight hydraulic power units at 300 MW each) at the Radisson substation to the loads in New England, at a distance of 920 miles, with most of the transmission system located in Canada. HVDC is a superior solution for this power transfer as Quebec and New England are two separate, asynchronous AC systems. The loading on the MTDC is jointly dispatched by the system operators of New England (ISO-NE) and Hydro Quebec (CCR).

The two-terminal system from Radisson to Sandy Pond was commissioned in 1990. The project was built for many New England power companies who signed a long-term power purchase contract with Hydro Quebec. The Nicolet terminal was commissioned in 1993, allowing the power from a nearby nuclear unit to participate in power export to New England. In Figure 13.21, the Nicolet terminal is shown as a rectifier supplying active power to New England. However, the Nicolet terminal may be off-line, so the system is a two-terminal HVDC system, or it can be an inverter terminal if additional power is needed in the Montreal area.

The power rating of this Phase II MTDC system is limited by the trip of the bi-pole system contingency, which will also trip the New England to New Brunswick connection, resulting in a total loss of 2700 MW import to New England. It has been determined from simulation studies that any power loss higher than that amount would cause instability problems in the New York and PJM Interconnection power systems.

Figure 13.22.a shows the Mode-1 operation of Radisson and Nicolet as rectifiers and Sandy Pond as an inverter, whereas Figure 13.22.b shows that of Radisson as a rectifier and Nicolet and Sandy Pond as inverters. In both scenarios, Sandy Pond, being a load bus, will have the lowest voltage, and hence it operates in constant-voltage control to regulate the DC voltage. The other converters are operating in constant-current control.

2 The material in this section was excerpted from presentations by David Bertagnolli, formerly with ISO-NE.

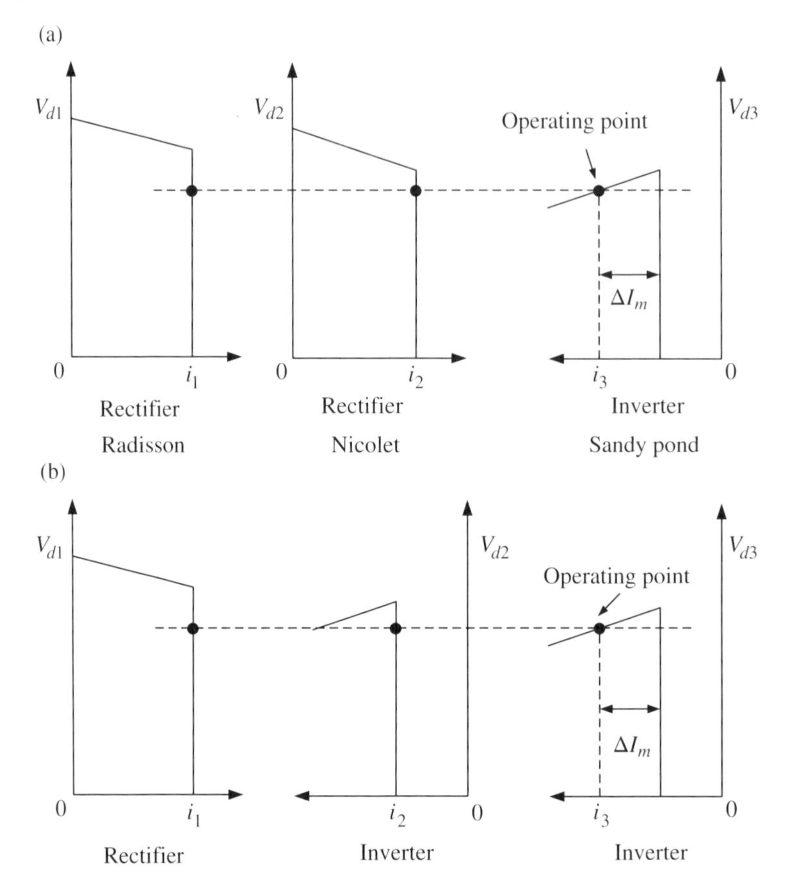

Figure 13.22 Control diagrams for the Quebec-New England three-terminal HVDC link: (a) Radisson and Nicolet in rectifier operation and Sandy Pond in inverter operation, and (b) Radisson in rectifier operation and Nicolet and Sandy Pond in inverter operation.

Thus for the Radisson converter, the DC current i_1 is determined by the desired power generation at its nominal voltage. Similarly, the DC current i_2 for the Nicolet converter is dispatched to achieve the desired supply or load at its nominal voltage.

In recent years with the growth of offshore wind farms using Type-4 wind turbines, MTDC systems connecting the various wind turbines become the defacto methodology, as they achieve the least-cost implementation. Such MTDC systems mostly use the voltage-sourced converter technology, which is discussed in Chapter 14.

13.6 Harmonics and Reactive Power Requirement

13.6.1 Harmonic Filters

From Figure 13.8, the voltage and current waveforms of a full-wave rectifier show that:

- On the AC side, the voltage waveform is sinusoidal (with the assumption of a stiff voltage source), but the current waveform consists of three levels.

- On the DC side, the current is constant (due to a smoothing reactance), but the voltage consists of ripples induced by the AC voltages.

Thus HVDC system operation induces harmonics in a power system, which need to be minimized.

On the AC side, for a 6-pulse Y-Y transformer connection, the phase currents can be expanded in a harmonic series. For phase a, the phase current can be expressed as

$$i_a = \frac{2\sqrt{3}}{\pi} I_d \left(\sin \omega t - \frac{1}{5} \sin 5\omega t - \frac{1}{7} \sin 7\omega t + \frac{1}{11} \sin 11\omega t + \frac{1}{13} \sin 13\omega t + \cdots \right) \tag{13.64}$$

Expressions with similar harmonic contents can be written for i_b and i_c. Note that i_a does not contain any even or $3n$ harmonics, $n = 1, 2, \ldots$. Thus the harmonics are of $6n \pm 1$, $n = 1, 2, \ldots$.

On the other hand, if a 6-pulse Δ-Y transformer connection is used, the harmonic representation of the phase-a current is

$$i_a = \frac{2\sqrt{3}}{\pi} I_d \left(\sin \omega t + \frac{1}{5} \sin 5\omega t + \frac{1}{7} \sin 7\omega t + \frac{1}{11} \sin 11\omega t + \frac{1}{13} \sin 13\omega t + \cdots \right) \tag{13.65}$$

in which some of the harmonics are time-shifted from (13.64).

Thus for a 12-pulse bridge with one Y–Y connection and one Δ–Y connection in series, in the so-called multiple-bridge converter configuration [6], the phase-a current is now given by

$$i_a = \frac{2\sqrt{3}}{\pi} 2I_d \left(\sin \omega t + \frac{1}{11} \sin 11\omega t + \frac{1}{13} \sin 13\omega t + \frac{1}{23} \sin 23\omega t + \frac{1}{25} \sin 25\omega t + \cdots \right) \tag{13.66}$$

such that only $12n \pm 1$, $n = 1, 2, \ldots$, harmonic components remain.

The derivation of the harmonics in (13.66) assumes no commutation delay. From Figure 13.11, it can be seen that commutation delay would reduce somewhat the amount of harmonic components in the phase currents.

For a 12-pulse HVDC system, an appropriate filter bank is shown in Figure 13.23, consisting of the parallel connections of tuned 11th and 13th harmonic filters and a highpass filter for all the other harmonics. The frequency response of a particular filter bank is shown in Figure 13.24. The tuned filters provide low-impedance paths for the harmonic currents of those frequencies to pass through, without going into the AC system.

As evident from Figure 13.8, the harmonics on the DC voltage have components of $6n$, $n = 1, 2, \ldots$, for a 6-pulse converter. For a 12-pulse configuration, the DC voltage will have harmonic components at $12n$, $n = 1, 2, \ldots$. Thus on the DC side, it is adequate to have filters for the 12th and 24th harmonics to eliminate the most significant harmonics in the DC voltage.

13.6.2 Reactive Power Support

Because of the firing and commutation delays in the thyristors, both the rectifier and inverter consume reactive power from the AC system. Example 13.1 shows that the

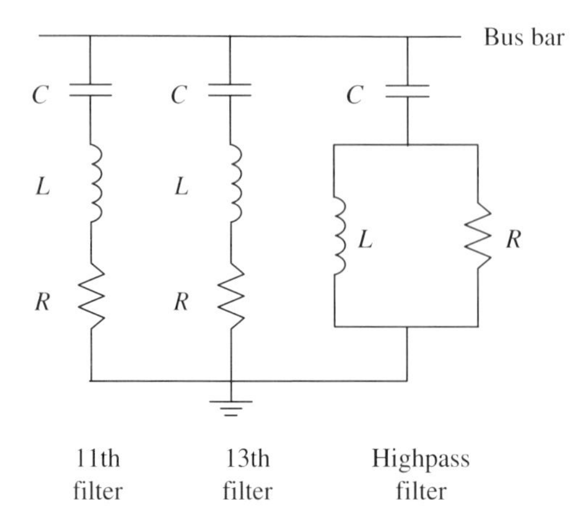

Figure 13.23 Components of single-phase HVDC AC filters.

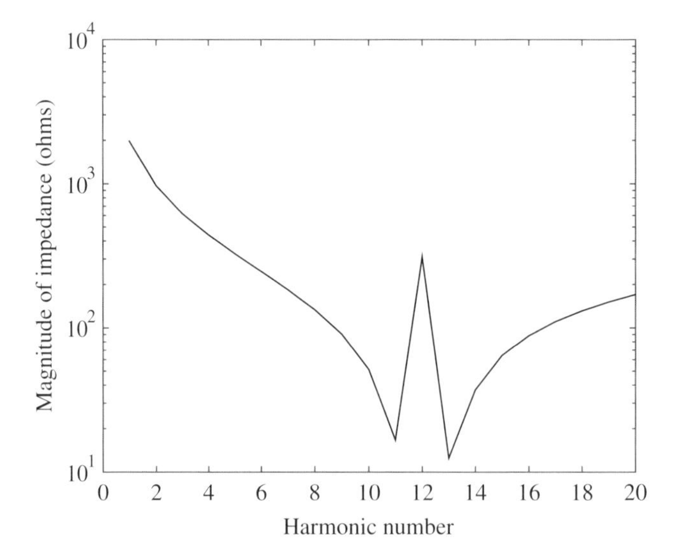

Figure 13.24 Frequency response of the impedance an HVDC AC filter bank vs harmonic number.

reactive power support is about 60% of the active power transfer. The thyristors are not capable of producing reactive power themselves. For HVDC systems with rectifiers close to generators, the reactive power can be supplied directly from the generators, and no additional reactive power compensation may be needed.

If a converter is located in an area with limited reactive power support, shunt capacitor banks will be required. Note that part of the reactive power supply can come from the AC filter banks. At the fundamental frequency, the filter banks are capacitive. For example, a tuned filter has an impedance of

$$H(j\omega) = R + j\omega L - j/(\omega C) \tag{13.67}$$

For low harmonic values, the imaginary part of $H(j\omega)$ is negative.

In some cases, dynamic voltage support such as static var compensation may also be needed for maintaining stability after severe disturbances. This is particularly true if either the rectifier or the inverter is connected to a weak power system.

Reactive power management is also important during the start-up period of an HVDC system. With the filter banks connected, the AC side voltage may be high before the power transfer level is sufficiently increased. As a result, some reactor banks may also be needed.

13.7 AC-DC Power Flow Computation

If the dynamics of the DC system are not important, then the rectifier can be modelled as a generator bus with negative active power and the inverter as a generator bus with positive active power. As generator buses, their bus voltages can be specified which will give an indication of the amount of reactive power support needed. The inverter bus generation is typically at rated power, whereas the active power at the rectifier terminal is higher by an amount equal to the losses on the DC line.

If the HVDC system dynamics are important, then the DC line current and converter voltage variables need to be consistent with the AC terminal bus voltages and currents, so that the proper reactive power support is provided. For a two-terminal HVDC system, the DC bus voltage V_d, the DC line current I_d, the firing angle α, the extinction angle γ, and the transformer taps at the rectifier and inverter need to be solved from a power flow solution involving both the AC and DC systems.

There are two general approaches to solve an AC-DC power flow solution. One approach is called the unified method [169] in which the combined AC-DC power flow is solved simultaneously. In this approach, the DC system variables are added to the Newton-Raphson AC power flow solution by including the required equations to represent the HVDC system power flow and control variables. This method expands the dimension of the Jacobian matrix.

The approach discussed in this chapter is the sequential method in which the AC power flow is solved first. Then the AC terminal bus voltages are used to compute the DC variables and determine the active and reactive power (P, Q) settings at the rectifier and inverter terminals. The updated (P, Q) values of the DC terminals are then used to determine the current injections at the HVDC terminal buses and obtain a new AC power flow solution. This process is repeated until both the AC and DC power flow solutions converge to within a pre-specified tolerance. In this approach, the AC power flow solution uses a Newton-Raphson method and the DC power flow solution is based on the equations developed in the earlier sections of this chapter. This is the method used in the Power System Toolbox as developed by Graham Rogers [60].

Consider the single-machine infinite-bus (SMIB) system with parallel AC and DC lines shown in Figure 13.25. The rectifier is at Bus 4 and is connected to Bus 3 via a branch with a small reactance and a variable tap transformer. The inverter is at Bus 5, which is connected to the infinite bus also via a small reactance and a variable tap transformer.

An expanded view of the rectifier components is shown in Figure 13.26. Following Figure 13.7, the transformer model is represented by the commutation reactance X_c and the variable tap T_r. The AC voltage on the thyristor side is commonly called $\tilde{V}_{\mathrm{HT}}$ (for haute (high) tension (potential)) and the AC voltage on the bus terminal is called $\tilde{V}_{\mathrm{LT}}$

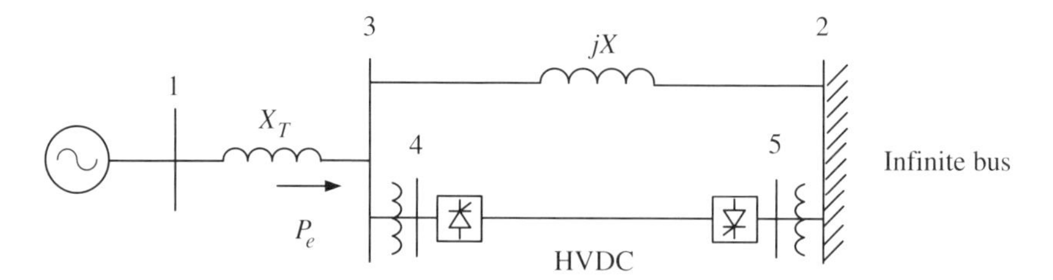

Figure 13.25 SMIB with parallel AC and DC lines.

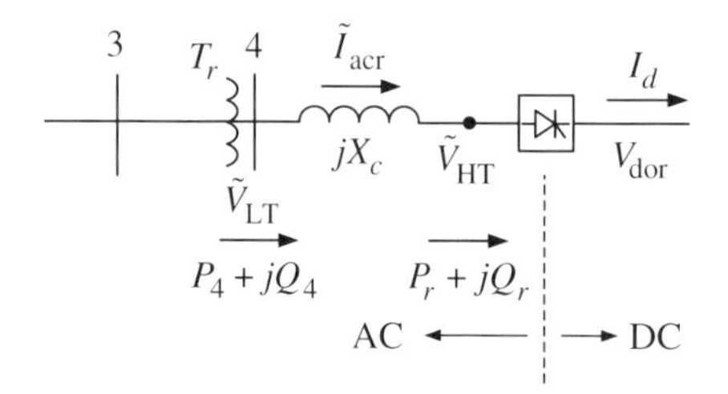

Figure 13.26 Rectifier terminal showing AC-DC interface.

(for low tension) which is equal to $\tilde{V}_4$. Bus 4 is designated as a nonforming load bus, allowing its load $P_4 + jQ_4$ to be adjusted as the DC solution is updated. As there are no losses on the commutation reactance, $P_r = P_4$. Furthermore, Q_r is determined by the rectifier power factor (13.37), and can be used to compute Q_4 by accounting for the reactive power flow on X_c.

Note that there are "redundant" control variables in an HVDC system. The DC voltage at the inverter can be set by either the transformer tap ratio or the extinction angle γ, which are independent variables. The rectifier side also has two independent control variables: the ignition angle α and the transformer tap ratio. The two controls at each terminal are needed because the firing and extinction angles are range limited. The transformer taps can be used if the firing or extinction angle alone cannot achieve the desired DC voltage.

The following solution strategy is suggested for the iterative AC-DC solution approach [60]:

1) The inverter DC voltage is set by the specification, and the inverter extinction angle is set to its minimum value.
2) The inverter transformer tap ratio is required to maintain the DC voltage to within 1% of the desired DC voltage value.
3) The DC current is determined from the rated DC power and the inverter voltage.
4) The rectifier DC voltage is calculated from the DC line current and the inverter DC voltage, and is used to compute the rectifier firing angle.
5) If necessary, the rectifier transformer tap ratio is adjusted to maintain the firing angle to within limits.

At the completion of each DC power flow solution iteration, the new transformer tap ratios and the active and reactive power flow modeled as loads on the rectifier and inverter buses are returned to the AC power flow solution step.

For the remainder of this section, the DC power flow computation is illustrated by an example using the system in Figure 13.25.

Example 13.3: HVDC system power flow computation

Consider the SMIB system with the parallel AC and DC lines in Figure 13.25. The rated voltage levels of the rectifier Bus 4 is 115 kV, and the inverter Bus 5 is 100 kV.[3] The generator is supplying 900 MW to the infinite bus acting as the load. The DC line is rated at 500 kV and 500 MW. The DC line resistance is 20 ohms. There are four rectifier and inverter bridges each, such that $B_r = B_i = B = 4$. The commutation reactances are $X_{cr} = X_{ci} = X_c = 3$ ohms. Thus the commutation resistances are

$$R_{cr} = R_{ci} = (3/\pi)BX_c = 11.459 \text{ ohms} \tag{13.68}$$

and the equivalent commutation reactances are

$$X_{equr} = X_{equi} = \sqrt{3}X_c/B = 1.299\text{ohms} \tag{13.69}$$

In addition, the firing angle α is in the range $[5°, 30°]$ and the extinction angle γ is in the range $[18°, 25°]$. The DC voltage should be in the range of $\pm1\%$ of the nominal voltage of 500 kV. The DC current is

$$I_d = (\text{rated DC power})/(\text{rated DC voltage}) = 500 \text{ MW}/500 \text{ kV} = 1 \text{ kA} \tag{13.70}$$

The converged power flow solution using the data file *datasmib_dc.m* with a system base of 100 MVA is given by

			GENERATION		LOAD	
BUS	VOLTS	ANGLE	REAL	REACTIVE	REAL	REACTIVE
1.0000	1.0500	16.3752	9.0000	4.5750	0	0
2.0000	1.0000	0	-8.8000	2.2719	0	0
3.0000	0.9926	8.8645	0	0	0.0000	-0.0000
4.0000	0.9700	8.5482	0	0	5.2106	3.0261
5.0000	0.9834	0.2963	0	0	-5.0106	1.7647

Note that the loads on Buses 4 and 5 are initially given as $5.0 + j2.5$ pu and $-5.0 + j2.5$ pu, respectively.

The AC power flow computation is quite routine and it will be skipped as it cannot be computed manually anyway. Only the DC power flow is illustrated here, showing how the AC solution variables are used to compute the DC variables, which then are provided as inputs to the AC power flow solution. For further simplification, the converged AC power flow solution variables are used.

Solutions: The computation will be performed first for the inverter and then for the rectifier. Also, this illustration will be based on the DC system operating in the normal mode.

---

3 In an AC power flow, the voltages would have all been set to pu values, and hence the voltage levels are not needed for the solution. In DC systems, the AC bus voltage values are used to set the DC voltage levels.

Inverter Terminal Computation

The current flowing into the inverter is

$$\tilde{I}_{\text{aci}} = \frac{(P_5 + jQ_5)^*}{\tilde{V}_5^*} = \frac{-5.0106 - j1.7647}{0.9834\angle -0.2963^\circ} = 5.4019\angle -160.3014^\circ \text{ pu} \tag{13.71}$$

which is equal to

$$\tilde{I}_{\text{aci}} = \frac{5.4019\angle -160.3014^\circ}{\sqrt{3}} \times \frac{\text{base MVA}}{\text{base kV}}$$

$$= \frac{5.4019\angle -160.3014^\circ}{\sqrt{3}} \frac{100 \times 10^6}{100 \times 10^3} = 3.1188\angle -160.3014^\circ \text{ kA} \tag{13.72}$$

Then the DC current is

$$I_d = |\tilde{I}_{\text{aci}}|/(B_i\sqrt{6}/\pi) = 1 \text{ kA} \tag{13.73}$$

which is the desired value.

The voltage at the HT side of the inverter is

$$\tilde{V}_{\text{HT}i} = \tilde{V}_i + jX_{\text{equ}i}\tilde{I}_{\text{aci}}$$
$$= 0.9834\angle 0.2963^\circ \times 100 \times 10^3 + j1.299 \times 3.1188\angle -160.3014^\circ \times 10^3$$
$$= 99.760\angle -1.899^\circ \text{ kV} \tag{13.74}$$

The DC voltage is

$$V_{\text{doi}} = 3\sqrt{(2)}|\tilde{V}_{\text{HT}i}|B_i/\pi = 538.90 \text{ kV} \tag{13.75}$$

Thus the DC line voltage V_{di} on the inverter side is given by

$$V_{\text{doi}} \cos\gamma = V_{\text{di}} + I_d R_{ci} \tag{13.76}$$

Setting $\gamma = 18^\circ$, the minimum value, the DC line voltage is

$$V_{\text{di}} = V_{\text{doi}} \cos\gamma - I_d R_{ci} = 538.90 \cos 18^\circ - 1 \times 11.459 = 501.061 \text{ kV} \tag{13.77}$$

This value is within $\pm 1\%$ of the desired voltage of 500 kV, and thus no adjustment of the transformer tap is necessary. Note if it is desired for V_{di} to be exactly at the rated voltage of 500 kV, one can increase γ to 18.3630°.

The power carried by the DC transmission is

$$P_i = V_{\text{di}} \times I_d = 501.064 \times 1 = 501.064 \text{ MW} = P_5 \tag{13.78}$$

By obtaining

$$\phi_i = \cos^{-1}(V_{\text{di}}/V_{\text{doi}}) = \cos^{-1}(501.064/538.90) = 21.5975^\circ \tag{13.79}$$

the reactive power requirement is, using pu values in the computation,

$$Q_5 = Q_i - X_{\text{equ}i}|\tilde{I}_{\text{aci}}|^2 = P_i \tan\phi_i - X_{\text{equ}i}|\tilde{I}_{\text{aci}}|^2$$
$$= 5.01064 \tan 21.5975^\circ - 0.0075 \times 5.4019^2$$
$$= 1.9836 - 0.2189 = 1.7647 \text{ pu} \tag{13.80}$$

Note that the reactive power demand Q_5 is less than Q_i (1.9836 pu) because the voltage $V_{\text{HT}i} = 99.760/100 = 0.9976$ pu is higher than $V_{\text{LT}i} = V_4 = 0.9834$ pu. Note that the P_5

and Q_5 values agree with the solved power flow values, except that P_5 carries a minus sign in the AC power flow data as the inverter terminal provides positive active power injection.

Rectifier Terminal Computation

The current flowing into the rectifier is

$$\tilde{I}_{\text{acr}} = \frac{(P_4 + jQ_4)^*}{\tilde{V}_4^*} = \frac{5.2106 - j3.0261}{0.9700\angle 8.5482^\circ} = 6.2122\angle -21.5984^\circ \text{ pu} \tag{13.81}$$

whose magnitude is 3.1188 kA, the same as the current magnitude computed on the inverter terminal.

The voltage of the HT side of the rectifier is

$$\begin{aligned}\tilde{V}_{\text{HT}r} &= \tilde{V}_r + jX_{\text{equr}}\tilde{I}_{\text{acr}} \\ &= 0.9700\angle 8.5482^\circ \times 115 \times 10^3 + j1.299 \times 6.2122\angle -21.5984^\circ \times 10^3 \\ &= 113.64\angle 10.3149^\circ \text{ kV}\end{aligned} \tag{13.82}$$

The DC voltage is

$$V_{\text{dor}} = 3\sqrt{(2)}|\tilde{V}_{\text{HT}r}|B_r/\pi = 613.84 \text{ kV} \tag{13.83}$$

Thus the DC line voltage V_{dr} on the rectifier side is given by

$$V_{\text{dr}} = V_{\text{di}} + I_d R_\ell = 501.061 + 1 \times 20 = 521.061 \text{ kV} \tag{13.84}$$

The ignition angle α is determined by the relationship

$$V_{\text{dor}} \cos\alpha = V_{\text{dr}} + I_d R_{cr} \tag{13.85}$$

Thus α can be solved as

$$\alpha = \cos^{-1}\left(\frac{V_{\text{dr}} + I_d R_{cr}}{V_{\text{dor}}}\right) = \cos^{-1}\left(\frac{521.061 - 1 \times 11.459}{613.84}\right) = 29.8287^\circ \tag{13.86}$$

This value is within the range of $[5^\circ, 30^\circ]$, and thus no adjustment of the transformer tap is necessary. Therefore the power carried by the DC transmission is

$$P_r = V_{\text{dr}} \times I_d = 521.061 \times 1 = 521.061 \text{ MW} = P_4 \tag{13.87}$$

The 20 MW difference between P_r and P_i is the losses incurred on the DC transmission line resistance.

By obtaining

$$\phi_r = \cos^{-1}(V_{\text{dr}}/V_{\text{dor}}) = \cos^{-1}(521.061/613.84) = 31.9133^\circ \tag{13.88}$$

the reactive power requirement is, using per-unit values in the computation,

$$\begin{aligned}Q_4 &= Q_r - X_{\text{equr}}|\tilde{I}_{\text{acr}}|^2 = P_r \tan\phi_r - X_{\text{equr}}|\tilde{I}_{\text{acr}}|^2 \\ &= 5.21061 \tan(31.9133^\circ) - 0.0057 \times 5.4019^2 \\ &= 3.2450 - 0.2189 = 3.0262 \text{ pu}\end{aligned} \tag{13.89}$$

Note that the P_4 and Q_4 values agree with the power flow solution values. Here P_r is a positive active power load.

Transformer Tap Setting

The transformer tap setting is used to ensure that on the inverter terminal, $V_{\rm di}$ is within some tolerance of the desired DC voltage, and on the rectifier terminal, α is within limits. The tap adjustment to obtain a desirable $V_{\rm di}$ is illustrated here.

At the conclusion of the initial AC power flow solution, the inverter bus AC voltage when converted into the DC voltage $V_{\rm doi}$ is 548.96 kV. With $\gamma = 18°$,

$$V_{\rm di} = V_{\rm doi} \cos\gamma - I_d R_{ci} = 548.96 \cos 18° - 1 \times 11.459 = 510.63 \text{ kV} \tag{13.90}$$

which is beyond the 1% tolerance range of 500 kV. If $V_{\rm di}$ is set to the maximum value of 505 kV, then the desired $V_{\rm doi}$ would be

$$V_{\rm doi} = \frac{V_{d\rm imax} + I_d R_{ci}}{\cos\gamma} = \frac{505 + 1 \times 11.459}{\cos 18°} = 543.04 \text{ kV} \tag{13.91}$$

Thus the transformer tap is set to

$$T_i = 548.96/543.04 = 1.0109 \tag{13.92}$$

For a transformer with a discrete step of 0.005, the next higher setting of $T_i = 1.015$ is used. The transformer tap will remain at this value until an AC power flow iteration yields a $V_{\rm di}$ outside its desired range.

The computed transformer tap ratio T_i will be used to solve the next AC power flow. ■

As illustrated by this example, the converters require certain amounts of reactive power support. Here the converter terminals are close to a generator and the infinite bus, which can readily supply the reactive power. For an initial analysis of an HVDC system whose terminals are not close to generators, synchronous condensers are temporarily put close to the converter terminal buses. Once a power flow solution has converged, a designer then can examine the reactance power outputs of the synchronous condensers to see how many switched shunt capacitor banks would be required.

13.8 Dynamic Models

In dynamic simulation of HVDC systems using positive-sequence power networks, the converter switching dynamics are not included because such dynamics are fast. However, states are used to model the converter control for achieving the desired voltage and current flow of the HVDC system. In addition, voltage and current states are used to model the DC line dynamics, such that the DC line current can be used as the injection to interface the HVDC system with the AC network.

13.8.1 Converter Control

The steady-state operation of HVDC systems has been discussed in Section 13.4 with optimal α and γ control to maintain the desired DC voltage and current. In disturbance conditions, these desired operating conditions are enabled by feedback control systems, with the rectifier regulating the firing angle α and hence the DC line current, and the inverter regulating the extinction angle γ and hence the DC voltage, as illustrated by

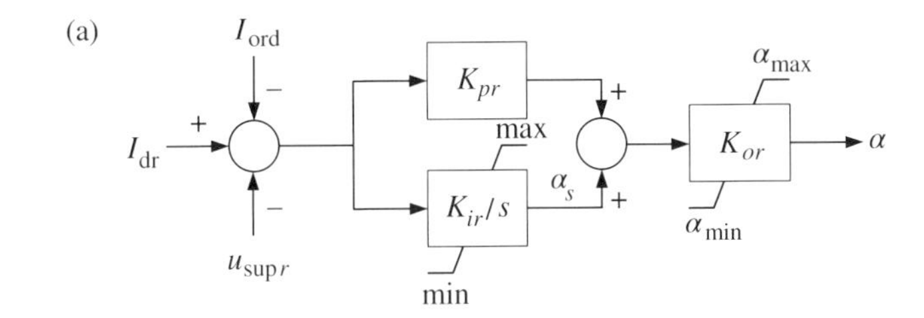

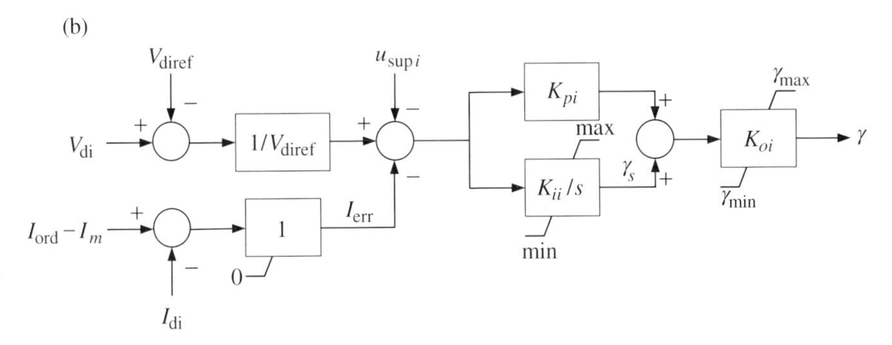

Figure 13.27 HVDC control systems: (a) rectifier control and (b) inverter control.

the control diagram in Figure 13.27. It is worth mentioning again that during dynamic conditions, the transformer tap ratios will remain fixed.

In normal rectifier operation, α responds to the error between the actual DC current I_{dr} and the desired current order I_{ord}. This regulation is achieved through a proportional gain K_{pr} and an integral gain K_{ir}, with an overall output gain K_{or}, as determined by

$$\alpha = K_{or}(\alpha_s + K_{pr}(I_{dr} - I_{ord} - u_{sup\,r})) \tag{13.93}$$

$$\dot{\alpha}_s = K_{ir}(I_{dr} - I_{ord} - u_{sup\,r}) \tag{13.94}$$

where α_s is the state which keeps the steady-state value of α for $K_{or} = 1$. The input u_{supr} is a supplementary control signal, which can be used to provide electromechanical (or interarea) mode damping [170, 171]. The limits $[\alpha_{min}, \alpha_{max}]$ are imposed on the control system to keep the values of α within a reasonable range. During initialization, I_{ord} is set to the I_{dr} value and α_s is set so that the α value matches the solved power flow. Using this rectifier controller, if I_{dr} is below I_{ord}, a negative error signal will lower α, increasing V_{dr} and allowing a larger voltage drop across the DC line to increase I_{dr}. The inverse situation holds if I_{dr} is higher than I_{ord}.

In normal inverter operation, γ responds to the error between the actual DC voltage V_{di} and the desired current order V_{diref}. This regulation is achieved through a proportional gain K_{pi} and an integral gain K_{ii}, with an overall output gain of K_{oi}, as given by

$$\gamma = K_{oi}(\gamma_s + K_{pi}((V_{di} - V_{diref})/V_{diref} - I_{err} - u_{supi})) \tag{13.95}$$

$$\dot{\gamma}_s = K_{ii}((V_{di} - V_{diref})/V_{diref} - I_{err} - u_{supi}) \tag{13.96}$$

where γ_s is the state (and also the output) of the integrator, which keeps the steady-state value of γ for $K_{oi} = 1$, u_{supi} is a supplementary control signal, and I_{err} is a current error signal. The current error signal is given by

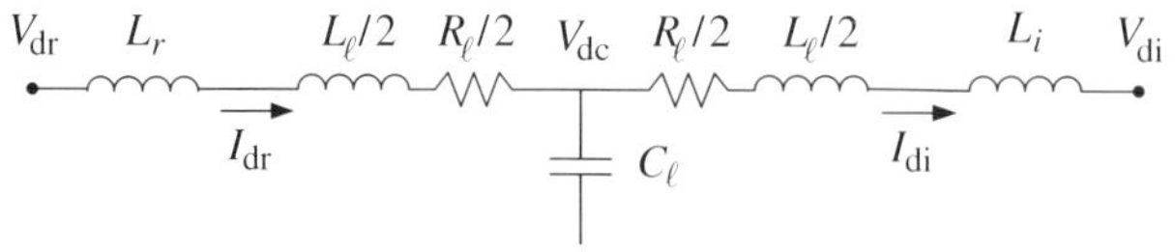

Figure 13.28 HVDC line model with inductance and capacitance components.

$$I_{\text{err}} = \begin{cases} I_{\text{ord}} - I_m - I_{\text{di}} & \text{if } I_{\text{ord}} - I_m - I_{\text{di}} > 0 \\ 0 & \text{otherwise} \end{cases} \tag{13.97}$$

In normal operation, I_{err} is zero. The control logic in (13.97) is activated in low-voltage operation during disturbances and will reduce the DC current to achieve the Mode-2 operation.

The limits $[\gamma_{\min}, \gamma_{\max}]$ are imposed on the control system to keep the values of γ within a reasonable range. During initialization, V_{diref} is set to the V_{di} value and γ_s is set to the γ value from the solved power flow. Using this inverter controller, if V_{di} is below V_{diref}, a negative error signal will lower γ, and thus increase V_{di}. The inverse situation holds if V_{di} is higher than V_{diref}. In addition, γ may be further reduced during low voltage conditions.

13.8.2 DC Line Dynamics

Note that in (13.93) and (13.95), different DC currents I_{dr} and I_{di} are used for the rectifier and the inverter, respectively. In steady-state power flow calculations, these values are assumed to be the same as I_d. In an HVDC dynamic simulation, the line current will be determined by the DC line resistance, inductance, and shunt capacitance, and the smoothing reactors.

A T-equivalent representation of the DC line is shown in Figure 13.28, in which the smoothing reactors, line inductance, and line capacitance are added to the DC line shown in Figure 13.14. In this representation, the line capacitance C_ℓ is located in the middle of the DC line. The line inductance L_ℓ and resistance R_ℓ are divided into halves and located on each arm of the T-equivalent. The smoothing reactors on the rectifier and inverter sides are represented by L_r and L_i, respectively.

The circuit equations of the currents I_{dr} and I_{di} of the DC line are

$$\left(L_r + \frac{L_\ell}{2}\right)\frac{dI_{\text{dr}}}{dt} = V_{\text{dr}} - V_{\text{dc}} - \frac{R_\ell}{2}I_{\text{dr}} \tag{13.98}$$

$$\left(L_i + \frac{L_\ell}{2}\right)\frac{dI_{\text{di}}}{dt} = V_{\text{dc}} - V_{\text{di}} - \frac{R_\ell}{2}I_{\text{di}} \tag{13.99}$$

The capacitor voltage V_{dc} is governed by

$$C_\ell \frac{dV_{\text{dc}}}{dt} = I_{\text{dr}} - I_{\text{di}} \tag{13.100}$$

Thus the HVDC dynamic model with the T-equivalent DC-line representation is represented by five states: α, γ, I_{dr}, I_{di}, and V_{dc}.

If the dynamics due to the DC-line capacitance can be neglected, the T-equivalent DC-line representation simplifies to the radial representation shown in Figure 13.29.

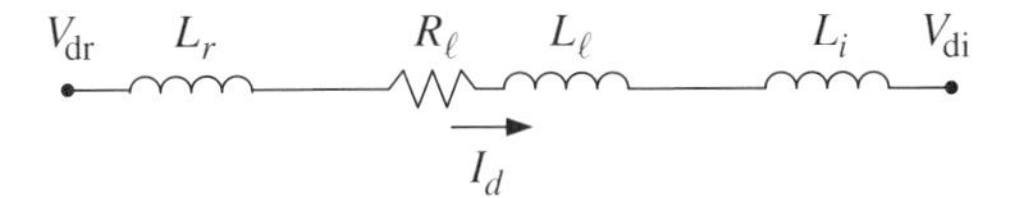

Figure 13.29 HVDC line model with inductance.

Here there is only a single DC current state variable I_d, which is determined by the circuit equation

$$(L_r + L_i + L_\ell)\frac{dI_d}{dt} = V_{dr} - V_{di} - R_\ell I_d \tag{13.101}$$

Thus the HVDC dynamic model with the radial DC-line representation is represented by three states: α, γ, and I_d.

The time constants associated with the DC lines are quite small, and thus would require a smaller integration step size compared to the positive-sequence AC system dynamics. Typically, the DC line dynamics are computed using an integration stepsize 1/10 of the AC system dynamics integration stepsize. As the number of states for each DC line is not large, the HVDC dynamics computation would not impose a significant computation burden.

13.8.3 AC-DC Network Solution

At each step of the dynamic simulation, for each new condition of the HVDC system, the DC quantities are used by the AC network to determine the power flow in the AC system. Given V_{dr}, I_{dr}, and α, the power demand from the AC network by the rectifier bus is given by (13.38)

$$P_{acr} = V_{dr} I_{dr}, \quad Q_{acr} = P_{acr} \tan\phi_r \tag{13.102}$$

In the network equation $Y\tilde{V} = \tilde{I}$, the current injection due to the rectifier operation is

$$\tilde{I}_{ar} = -\frac{(P_{acr} + jQ_{acr})^*}{\tilde{V}_{ar}^*} \tag{13.103}$$

where $\tilde{V}_{ar}$ is the rectifier bus AC voltage phasor and the negative sign is used to point the injection into the AC network.

Similarly, given V_{di}, I_{di}, and γ, the power supply to the AC network from the inverter bus is

$$P_{aci} = V_{di} I_{di}, \quad Q_{aci} = P_{aci} \tan\phi_i \tag{13.104}$$

where P_{aci} is the negative active power load and Q_{aci} a positive reactive power load. Thus the current injection due to the inverter operation is

$$\tilde{I}_{ai} = \frac{(P_{aci} - jQ_{aci})^*}{\tilde{V}_{ai}^*} \tag{13.105}$$

where $\tilde{V}_{ai}$ is the inverter bus AC voltage phasor.

Note that both $\tilde{V}_{ar}$ and $\tilde{V}_{ai}$ are unknown variables in the voltage phasor $\tilde{V}$ of the network solution. Thus the network solution has to be solved iteratively, in which $\tilde{I}_{ar}$ and $\tilde{I}_{ai}$ are updated when new values of $\tilde{V}_{ar}$ and $\tilde{V}_{ai}$ are available.

As this is a nonlinear network solution, the entries of the Jacobian used for the solution need to be updated with the partial derivatives of $(P_{\text{acr}}, Q_{\text{acr}})$ with respect to $\tilde{V}_{ar}$, and of $(P_{\text{aci}}, Q_{\text{aci}})$ with respect to $\tilde{V}_{ai}$.

The following example shows the HVDC control performance during a disturbance.

Example 13.4: HVDC system disturbance response

Consider the SMIB system with the parallel AC and DC lines, operating in the steady-state condition given in Example 13.3. The generator is modeled with subtransient effects and an excitation system, but without a PSS. An impedance-to-ground fault, with a fault reactance of 0.2 pu, is applied at Bus 3 and self-cleared after 50 msec. Simulate the AC-DC system, and plot the system trajectories of the rectifier and inverter voltages, and the control variables α and γ. The data for the simulation is given in the file *datasmib_dc_fault.m*. Some of the parameters important for this simulation are:

$$R_\ell = 20 \text{ ohms}, \quad L_\ell = 100 \text{ mH}, \quad C_\ell = 0\ \mu\text{F}, \quad L_r = L_i = 1000 \text{ mH}$$
$$K_{pr} = 1, \quad K_{ir} = 1, \quad K_{or} = 1, \quad K_{pi} = 1, \quad K_{ii} = 1, \quad K_{oi} = 0.1$$
$$\alpha_{\min} = 5^\circ, \quad \alpha_{\max} = 90^\circ, \quad \gamma_{\min} = 15^\circ, \quad \gamma_{\max} = 90^\circ$$
$$I_{\text{ord}} = 1\text{pu}, \quad V_{\text{diref}} = 1\text{pu}, \quad I_m = 0.15\text{pu}$$

Solutions: In this system, as the inverter bus is connected to the infinite bus, its voltage would not be affected very much. However, the rectifier voltage may drop significeantly during the fault. Thus the simulation is to see how α reacts to the disturbance. The simulation results are shown in Figures 13.30 and 13.31.

For the inverter, the DC and AC voltages stay roughly the same. The inverter γ decreases by a small amount, and I_{err} is active during the fault-on period. For the rectifier, the DC and AC voltage both drop by 15%. As a result, γ is decreased, thereby increasing the DC voltage. However, The fault is still on, and the AC voltage is still depressed. During the disturbance, the DC current has dropped below 85%. The DC current does not recover until the fault is cleared. Note that in this example, no DC

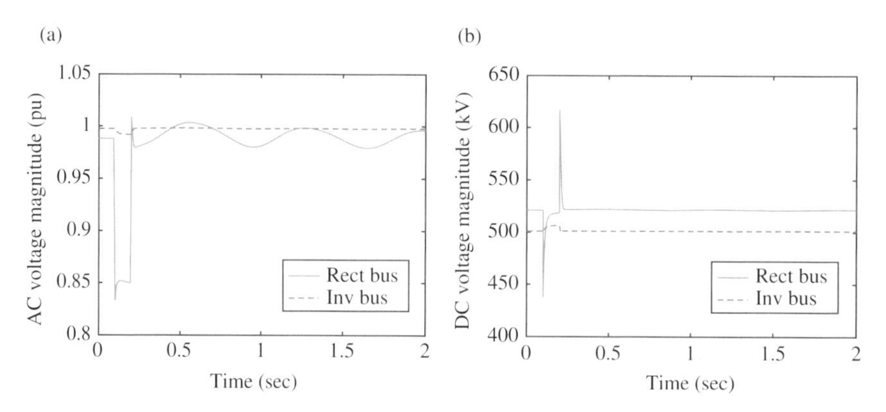

Figure 13.30 Disturbance response for Example 13.4: (a) AC voltages and (b) DC voltages (see color plate section).

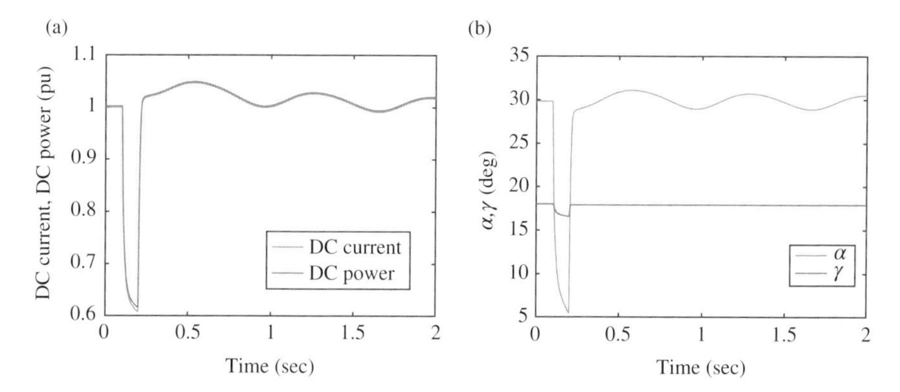

Figure 13.31 Disturbance response for Example 13.4: (a) DC current and power, and (b) firing and extinction angles (see color plate section).

line shunt susceptance is modeled, and thus there is only a single DC current. The DC power drops during the fault period, due to the drop in current. There is no damping control in the system. As a result, there is an undamped oscillation after the fault is cleared. In the next section, the design of an HVDC controller to improve the damping is discussed. ■

13.9 Damping Control Design

HVDC systems can be used to provide additional damping to electromechanical modes. This section illustrates the damping control design using the methodology previously presented in Chapter 10 for power system stabilizer design. Linearized models are needed to perform the control design. However, developing linearized models involving HVDC systems would be cumbersome to do manually. Thus it is assumed that a computer linearization function is available to generate the required linearized model with appropriate inputs and outputs.

An effective way for an HVDC system to add damping to power swing modes is to modulate the DC power flow. This action can be achieved by controlling the DC line current, via the signal u_{supr} into the rectifier current summing junction shown in Figure 13.27.a. The choice of the input signal used in the damping controller depends on a number of factors. For the SMIB system shown in Figure 13.25, good candidate input signals for damping the oscillation of the generator against the infinite bus include the voltage-angle difference between the rectifier and inverter terminal buses, the frequency at the rectifier bus, and the power flow on the parallel AC line. For example, if the bus voltage angles θ_4 and θ_5 at Buses 4 and 5, respectively, are used, the controller transfer function (Figure 13.32) can take the form

$$u_{\mathrm{supr}} = G_{\mathrm{ldlg}}(s)G_d(s)G_{\mathrm{wo}}(s)G_{\mathrm{lp}}(s)\theta_d \tag{13.106}$$

where $\theta_d = \theta_4 - \theta_5$, the lowpass filter

$$G_{\mathrm{lp}}(s) = \frac{1}{1 + T_{\mathrm{lp}}s} \tag{13.107}$$

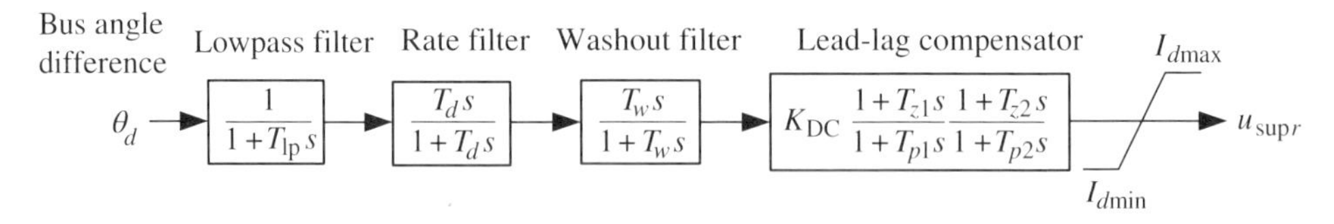

Figure 13.32 HVDC damping controller using rectifier current order.

with a time constant of T_{lp} is used for reducing noise at high frequency, the washout filter

$$G_{\text{wo}}(s) = \frac{T_w s}{1 + T_w s} \tag{13.108}$$

with a time constant of T_w would render the controller inactive in steady state, the rate filter

$$G_d(s) = \frac{T_d s}{1 + T_d s} \tag{13.109}$$

with a time constant of T_d would convert the input angle signal into a frequency signal, and the lead-lag compensator

$$G_{\text{ldlg}}(s) = K_{\text{DC}} \frac{1 + T_{z1}s}{1 + T_{p1}s} \frac{1 + T_{z2}s}{1 + T_{p2}s} \tag{13.110}$$

where T_{zi} and T_{pi}, $i = 1, 2$, are the lead and lag time constants, respectively, and K_{DC} is the control gain. The controller $G_{\text{ldlg}}(s)$ provides the required gain and phase compensation to make the damping control effective. Note that the signal u_{supr} is limited to prevent the damping controller from taking over the HVDC system regulation control.

Example 13.5 shows the process of designing a damping controller using the rectifier supplemental current input signal.

Example 13.5: HVDC system damping control design

Consider again the SMIB system in Figure 13.25 with the parallel AC and DC lines, operating in the steady-state condition given in Example 13.3. Obtain the linearized model of the system and compute the swing mode of the system. Then design a damping controller in the form of (13.106) to increase the damping ratio of the closed-loop system to 0.1, using the angle difference $\theta_d = \theta_4 - \theta_5$ as the input signal to the controller. Apply the designed controller to the linearized model and simulate the closed-loop linearized model to show the response of $\Delta\theta_d$ subject to a 0.05 pu step in the excitation system reference voltage value.

Solutions: After removing the rotor angle and speed of the large machine representing the infinite bus, there are 11 states in the following order: the two generator mechanical states for the rotor angle δ and speed ω; the four generator electrical states E'_q, ψ_{kd}, E'_d, and ψ_{kq}; the two exciter states V_R and E_{fd}; and three DC states α, γ, and I_d. The entries of the 11×11 system A matrix are given by

```
Columns 1 through 6
          0   3.7699e+02            0            0            0            0
-2.2138e-01  -5.8734e-04  -1.1884e-01  -1.0702e-01  -1.6010e-02  -6.9250e-02
 1.1124e-01  -4.4138e-04  -2.1738e+00   2.3382e+00  -1.1894e-02  -5.1077e-02
-3.5075e+00   2.3003e-02   3.0169e+01  -3.4136e+01   3.1049e-02   7.4233e-02
-2.4341e-01  -3.4554e-03  -3.0690e-02  -2.8001e-02  -8.4423e+00   6.3190e+00
-2.5771e+00  -3.6583e-02  -3.2492e-01  -2.9646e-01   1.5185e+01  -2.1623e+01
 2.8696e+00  -6.6513e-02  -9.0935e+00  -8.1915e+00   2.5309e+00   1.1307e+01
 2.2109e+01  -1.0098e+00  -1.5145e+02  -2.0588e+02   4.5756e+01   2.0389e+02
          0            0            0            0            0            0
-4.8992e-05   9.6388e-07   4.4008e-05   3.9713e-05  -1.8464e-05  -8.3461e-05
-2.8199e+01   8.1414e-01   5.2452e+01   4.7296e+01  -1.6444e+01  -7.4507e+01

Columns 7 through 11
          0            0            0            0            0
-2.2234e-05  -5.8734e-05   4.2564e-02  -1.8689e-04   2.2012e-03
-1.6709e-05   1.4555e+00  -2.5957e-03  -1.4015e-04   2.5927e-02
 8.7077e-04   2.3003e-03  -1.4998e-01   7.3064e-03  -1.1509e+00
-1.3080e-04  -3.4554e-04   1.1849e-01  -1.0984e-03   1.0554e-01
-1.3849e-03  -3.6583e-03   1.2545e+00  -1.1629e-02   1.1174e+00
-1.5244e-01  -6.6513e-03   1.3396e+00  -2.1134e-02   2.6921e+00
 1.4923e+01  -1.3069e+02   2.3010e+01  -3.2087e-01   3.8995e+01
          0            0            0            0   1.0000e+00
          0            0  -2.3268e-05  -3.0392e-02  -3.6324e-05
 3.0819e-02   8.1414e-02  -1.6107e+02   7.5103e+00  -1.8906e+02
```

The eigenvalues of the system are given by

$$\begin{bmatrix} -187.19 \\ -129.90 \\ -33.826 \\ -0.19695 \pm j8.9470 \\ -4.2655 \\ -1.3128 \pm j0.61897 \\ -0.88614 \\ -0.030393 \end{bmatrix} \tag{13.111}$$

The complex pair $-0.19695 \pm j8.9470$ is the swing mode with a frequency of 1.424 Hz and a damping ratio of 0.022.

The controller (13.106) will be used to add damping to this lightly damped mode. First, the sensitivity due to the supplementary signal u_{supr} is used to form the input matrix B (a column vector), whose transpose is given by

```
Columns 1 through 6
          0  -5.8734e-04  -4.4138e-04  2.3003e-02  -3.4554e-03   -3.6583e-02

Columns 7 through 11
-6.6513e-02  -1.0098e+00            0  1.0000e+00   8.1414e-01
```

Second, the difference between the sensitivities due to the voltage angles of Buses 4 and 5 is used as the output matrix C (a row vector), whose entries are

```
Columns 1 through 6
   5.6402e-01  -2.7146e-05   1.9530e-01   1.7577e-01   6.3880e-02   2.8073e-01

Columns 7 through 11
            0            0   4.0337e-02  -1.2954e-03  -2.6829e-02
```

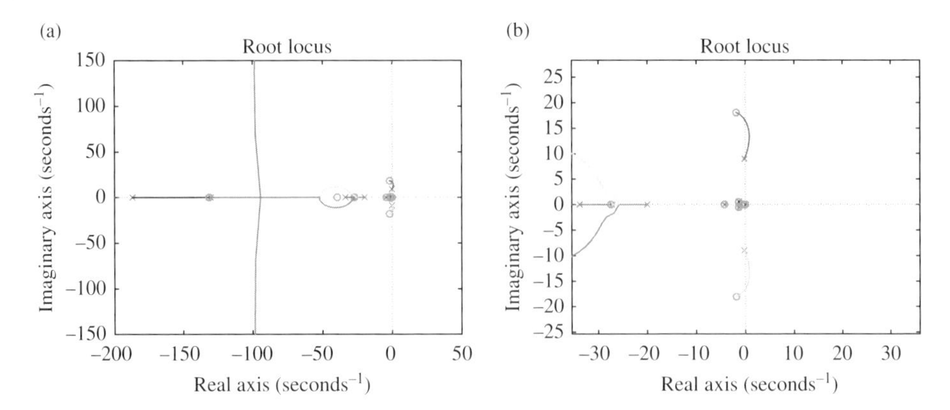

Figure 13.33 (a) Root-locus plot of uncompensated system and (b) enlarged plot around the swing mode (see color plate section).

The matrices $(A, B, C, 0)$ form the system $G(s)$.

The filter parameters are chosen as

$$T_{\text{lp}} = 20 \text{ sec}, \quad T_w = T_d = 10 \text{ sec} \tag{13.112}$$

Then the root-locus design method used for PSS design in Chapter 10 is applied to obtain the gain and phase of the lead-lag compensator. The root-locus plot of the uncompensated system

$$G(s)G_d(s)G_{\text{wo}}(s)G_{\text{lp}}(s) \tag{13.113}$$

is shown in Figure 13.33. Note that the root-locus branch leaving the swing mode at $-0.19695 + j8.9470$ almost goes vertically up, implying that the compensator phase requirement is about 90° at the swing mode frequency. This phase-lead requirement is divided equally among the two lead-lag sections.

For each lead-lag section to achieve a phase lead of 45°, the pole-zero ratio is selected as 5.8284. With a center frequency of 8.9470 rad/sec, the time constants are computed to be $T_{z1} = T_{z2} = 0.046024$ sec and $T_{p1} = T_{p2} = 0.26825$ sec.

Next a root-locus analysis is performed for the compensated system

$$G(s)G_d(s)G_{\text{wo}}(s)G_{\text{lp}}(s)G_{\text{ldlg}}(s) \tag{13.114}$$

which is shown in Figure 13.34. Note that the direction of the root-locus branch leaving the swing mode has been rotated about 90° counterclockwise as compared to the uncompensated system. In Figure 13.34.b, the MATLAB® data cursor function shows that if $K_{\text{DC}} = 7.85$ is used, the swing mode of the close-loop system will have a damping ratio of 0.112.

A gain of $K_{\text{DC}} = 8$ is used to obtain the closed-loop system, with the swing modes at $-1.0626 \pm j9.2940$. To simulate the step response, the input B matrix is changed to the one for the excitation system reference voltage signal. The resulting time response of the incremental voltage angle difference between Buses 4 and 5 is shown in Figure 13.35. Note that the steady-state value of $\Delta\delta_d$ is about −0.4°. This observation is consistent with the reasoning that as the bus voltages become higher, a smaller angle difference between the buses would be needed to achieve the same AC power transfer. ■

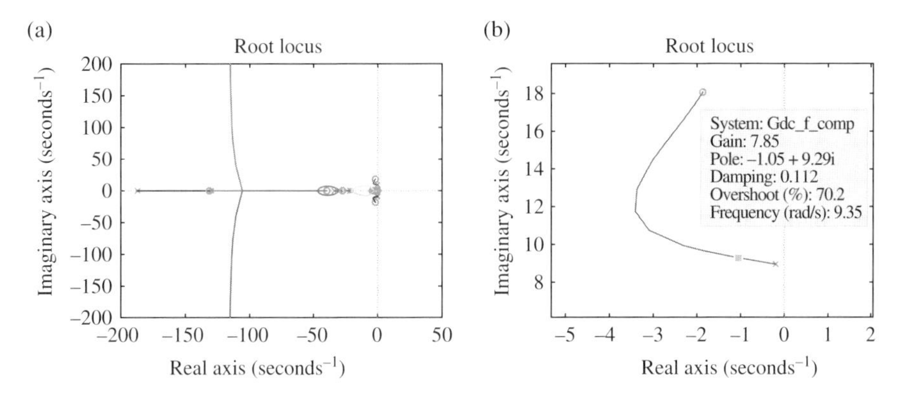

Figure 13.34 (a) Root-locus plot of compensated system and (b) enlarged plot around the swing mode (see color plate section).

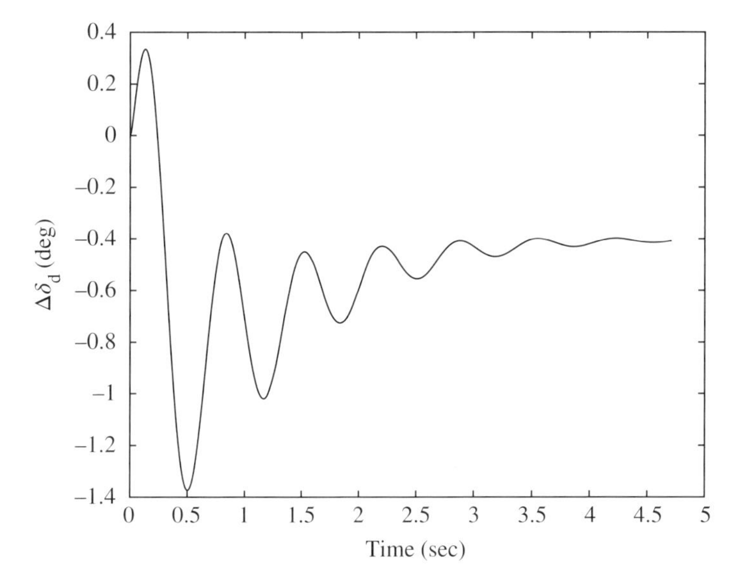

Figure 13.35 Incremental voltage angle difference between Buses 4 and 5 in response to a 0.05 pu step in excitation system reference voltage value.

To implement the angle-signal controller (13.106), equipment such as synchrophasor measurement units (PMUs) [8, 3] would be needed to compute the bus voltage angles. In addition, the controller is located at the rectifier terminal so that the inverter AC terminal bus angle is a remote signal. The communication time required to acquire the remote signal needs to be accounted for in the design of damping controllers. Essentially a time delay translates to a phase lag in the frequency domain. Thus the phase of the damping controller needs to be increased to compensate for this additional phase lag [173]. When using remote signals for control, the reliability of the communication system should also be considered. The practical implementation of the interarea mode damping controller for the US Pacific DC Intertie considers an ensemble of remote signals [171] in case of an individual PMU signal dropout.

13.10 Summary and Notes

This chapter describes HVDC system operation based on the thyristor or line-commutated converter technology. From the AC-DC conversion process, an equivalent steady-state DC line model is developed. The HVDC control modes are presented and illustrated with numerical examples. The DC-AC interface calculation in the power flow solution process is discussed and also illustrated with numerical examples. Harmonics and filter requirement are discussed. HVDC system dynamic control models are developed and used for the design of swing mode damping controllers with the converter AC bus voltage angle difference as the input signal. Problem 13.12, which is on the HVDC damping controller design using line active power flow, can be assigned as a design project.

Additional topics on HVDC systems, such as the use of multiple bridge converter configurations to further reduce harmonic components, system operation under faults on the HVDC system, and commutation failure, can be found in [9], [6], and [163].

The coverage of MTDC systems in this chapter is quite brief. A reader can consult papers such as [175] and [176] for more operational details of thyristor-based MTDC systems.

The HVDC code in PST was developed by Graham Rogers, to whom the authors are very grateful. Some of the discussions on DC power flow computation and operating condition are based on his code.

Problems

13.1 (AC versus DC system cost comparison) Because DC transmission systems with two lines are cheaper than AC transmission systems with three lines, HVDC systems are more economical than AC transmission systems over long distances. For each comparison below, find the break-even line length for the AC and DC systems, based only the capital cost consideration.

1) Overhead line

	AC	DC
Line cost	$0.20/kW-Mile	$0.12/kW-Mile
Substation equipment	$20/kW	$70/kW

2) Cable 1

	AC	DC
Line cost	$1.20/kW-Mile	$0.40/kW-Mile
Substation equipment	$20/kW	$70/kW

3) Cable 2

	AC	DC
Line cost	$3.00/kW-Mile	$1.00/kW-Mile
Substation equipment	$20/kW	$60/kW

13.2 Derive equations (13.2) and (13.4).

13.3 Calculate the HVDC system operating condition in Example 13.2 if the AC rectifier bus voltage drops by 15%.

13.4 An HVDC system has the following parameters: rated voltage of 500 kV at the inverter, rated current of 2 kA, $R_\ell = 20$ ohms, $R_c = 3$ ohms/bridge, $B = 4$ bridges, ignition angle $5^\circ \leq \alpha \leq 30^\circ$, extinction angle $18^\circ \leq \gamma \leq 25^\circ$, and current margin of 15%. The transformer ratio limits are sufficiently large so they do not come into consideration. The base voltage for the converter AC buses is 115 kV. Determine the operating condition (including the transformer tap ratios) and the active and reactive power transfer at the rectifier and the inverter, assuming that the converter AC buses are at rated voltage. For the rectifier, set the transformer ratio at unity.

13.5 Calculate the HVDC system operating condition in Problem 13.4 if the AC rectifier bus voltage drops by 20%.

13.6 For the SMIB system in Example 13.3, use the PST function *lfdc* or any other AC-DC power flow program to compute the solution.

13.7 For the SMIB system in Example 13.4, first use the PST function *s_simu* or any other power system simulation program to compute the system response to the disturbance. Then reduce the fault reactance from 0.2 pu to 0.17 pu and repeat the disturbance simulation. Comment on the response of α and γ.

13.8 The Klein-Rogers-Kundur 2-area, 4-machine system with a DC line model is shown in Figure 13.36. The DC line is in parallel with the AC lines, connecting the two areas. The system data can be found in the data file *d_testdc_115kv.m*. Use the PST function *lfdc* to run the loadflow for this data set. Observe the iteration process and the solution. The transformer tap ratios can be found in *line_sol* or *tapi* and *tapr* after the AC-DC power flow has converged.

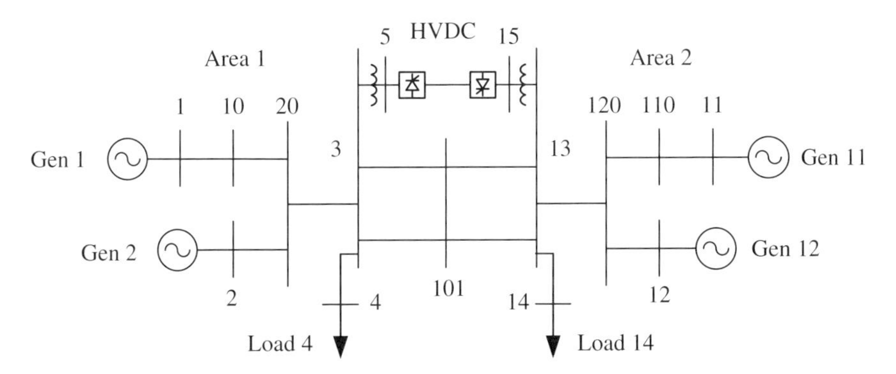

Figure 13.36 Two-area, four-machine power system with HVDC line added.

13.9 Repeat Problem 13.8 by specifying the voltage on Bus 13 to be 1.01 pu.

13.10 Use *s_simu* to simulate the disturbance listed in the file *d_testdc_115kv.m*. Plot and comment on the converter voltages (both AC and DC) and DC current time response. Note that the time array for the AC variables is *t*, and the time array for DC variables is *t_dc*. The PST variables needed are *bus_v, Vdc,* and *i_dcr*.

13.11 (HVDC filters) A set of HVDC filter parameters for the Three Gorges–Changzhou HVDC project is given in [174]. The filter parameters for the 11th and 13th harmonic AC filters at the Longquan substation and the 12th and 24th harmonic DC filters at the Zhengping substation are given in Table 13.4 corresponding to the filter configuration in Figure 13.37. The nominal system frequency is 50 Hz. Plot the magnitude of the frequency response of these filters. Use logarithmic scale for magnitude and linear scale for frequency. For the AC filters at Longquan (rated at 525 kV), find the reactive power supplied by the filters at the nominal system voltage, assuming that the AC filter parameters in Table 13.4 are given for each phase.

Table 13.4 AC and DC filter parameters used in the China Three Gorges-Changzhou HVDC project.

	11/13 AC filters	12/24 DC filters
C_1, μF	1.6	2.8
L_1, mH	43.8	13.0
C_2, μF	57.8	5.1
L_2, mH	1.2	7.1
R_1, ohms	2000	300

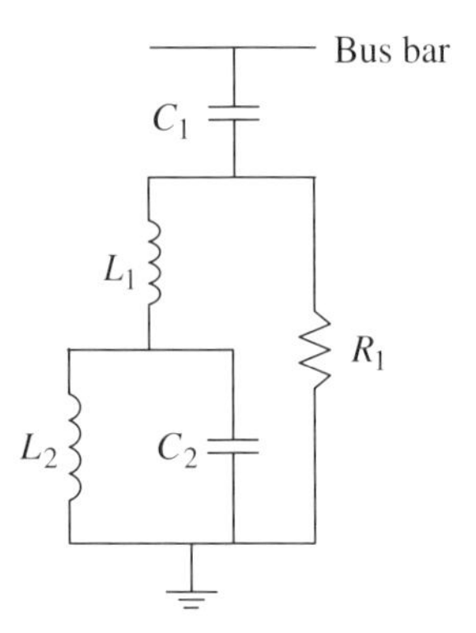

Figure 13.37 Simplified AC and DC filter configuration used in the China Three Gorges-Changzhou HVDC project.

13.12 (Design project: HVDC damping control design) Redesign the damping controller in Example 13.5 using the active power flow P_{32} from Bus 3 to Bus 2 as the controller input signal. The same control performance specification in Example 13.5 is used, that is, to increase the damping ratio of the swing mode to 0.1. This problem requires the C matrix corresponding to ΔP_{32}, which is not provided by the PST linearization function *svm_mgen*. However, it can be obtained by taking the partial derivatives of

$$P_{32} = \frac{V_3 V_2 \sin(\theta_3 - \theta_2)}{X} \tag{13.115}$$

to obtain the perturbation equation

$$\Delta P_{32} = \left.\frac{\partial P_{32}}{\partial V_3}\right|_o \Delta V_3 + \left.\frac{\partial P_{32}}{\partial V_2}\right|_o \Delta V_2 + \left.\frac{\partial P_{32}}{\partial \theta_3}\right|_o \Delta \theta_3 + \left.\frac{\partial P_{32}}{\partial \theta_2}\right|_o \Delta \theta_2 \tag{13.116}$$

where the subscript "o" denotes evaluation at the equilibrium condition from the power flow solution. PST provides C matrices for the bus voltage magnitude and angle in the variables *c_v* and *c_ang*, respectively, from which the C matrix for ΔP_{32} can be obtained.

14

Flexible AC Transmission Systems

14.1 Introduction

An AC transmission system that is robust at low power transfer may become stressed at higher power transfer level, such that it may not be able meet the required stability margins. HVDC systems can be used to relieve the stress on parallel AC transmission paths. However, this will require the installation of transmission lines, which are costly and may incur a lengthy permitting process, unless the routing uses mostly submarine cables.

Another approach to improving the strength or stability margins of an AC transmission system is to apply flexible AC transmission systems (FACTS), which are power-electronic-based controllers such as static var compensators (SVCs) and thyrsitor-controlled series compensators (TCSCs) [10]. These substation-based controllers can be a quicker and less expensive way for a utility company to upgrade its existing AC transmission system.

FACTS controllers are supported by two power electronics technologies.[1] The thyristor technology and its operation have been covered in Chapter 13 on HVDC systems. The voltage-sourced converter (VSC) technology has recently become more popular. Gate-turnoff thyristors (GTOs) were used in earlier applications. Newer applications use insulated-gate bipolar transistors (IGBTs), which are also used in the power-electronic interface for energy storage units and renewable resources such as wind turbine-generators and photo-voltaic panels. Similar to thyristors, the VSCs can be turned on when a gate signal is applied. However, the VSCs can also be turned off without waiting for their current flow be zero. This capability allows VSC-based FACTS controllers to also provide reactive power support. In addition, VSC-based FACTS controllers can be coupled to allow for active power circulation.

FACTS controllers can be installed in either the shunt or series configuration. When there are multiple VSCs, they can be coupled to provide more versatile configurations. Table 14.1 lists some of the FACTS controllers and their common acronyms [10].

The main objective of FACTS controllers is to improve system stability: transient, voltage, and small-signal, such that the AC transmission system becomes more reliable or

1 Mechanical switching technology has also been suggested, but the time response of such switching systems would be slower and the number of switchings limited.

Power System Modeling, Computation, and Control, First Edition. Joe H. Chow and Juan J. Sanchez-Gasca.

Companion website: www.wiley.com/go/chow/power-system-modeling

Table 14.1 FACTS controllers.

Configurations	Thyristors	GTOs/IGBTs
Shunt controllers	Static var compensator (SVC)	Static compensator (STATCOM)
Series controllers	Thyristor-controlled series compensator (TCSC)	Static synchronous series compensator (SSSC)
Coupled DC buses		Back-to-back STATCOM (B2B or VSC HVDC), Unified power flow controller (UPFC), Interline power flow controller (IPFC)

additional power flow can be transferred on critical paths. This chapter discusses the use of FACTS controllers to accomplish these goals. The discussion starts with SVCs in Section 14.2 and TCSCs in Section 14.3. The VSC-based FACTS controllers are covered in subsequent chapters.

14.2 Static Var Compensator

14.2.1 Circuit Configuration and Thyristor Switching

An SVC nominally consists of a fixed shunt capacitor in parallel with a reactor controlled by two thyristors connected to form a bidirectional switch, as shown in Figure 14.1, such that the current flowing into or out of the SVC would be continuous because the current flow in a reactor is continuous. If the thyristor pair is connected in series with the capacitor, the current flowing through the SVC would be discontinuous during switching, as the current in a capacitor can change abruptly.

With one of the thyristors conducting, the current i_L will be 90° lagging the sinusoidal voltage waveform $v(t) = V\cos(\omega t)$ as

$$i_L(t) = \frac{1}{L}\frac{dv(t)}{dt} = \frac{V}{\omega L}\sin(\omega t) \tag{14.1}$$

By delaying the ignition angle α of the forward-biased (FB) thyristor, the current flow in the reactor when $v(t) > 0$ is controlled as

$$i_L(\alpha, t) = \frac{1}{L}\int_{\alpha}^{\omega t} v(t)\,dt = \frac{V}{\omega L}(\sin(\omega t) - \sin\alpha), \qquad \frac{\alpha}{\omega} \le t \le \frac{\pi - \alpha}{\omega} \tag{14.2}$$

Note that at $t = (\pi - \alpha)/\omega$, $i_L(\alpha, t)$ is zero and the FB thyristor will cease to conduct. When $v(t) < 0$, the reverse-biased (RB) thyristor will conduct with a firing angle delay of α, producing a reactor current $i_L(\alpha, t)$ that is opposite in sign to the current given by (14.2).

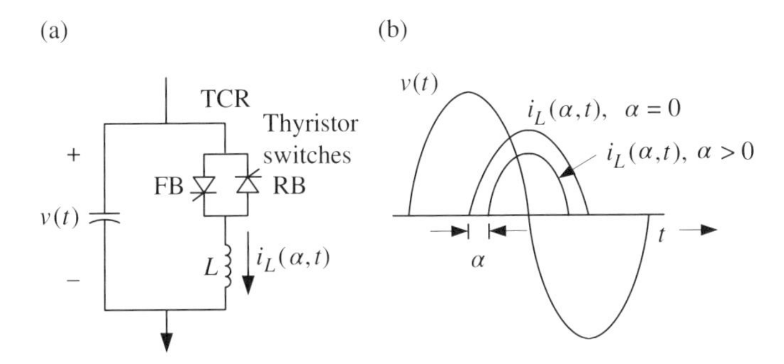

Figure 14.1 (a) An SVC showing a thyristor-controlled reactor in parallel with a fixed capacitor. (b) The inductor current i_L as controlled by the firing delay angle.

For a fixed α, $i_L(\alpha, t)$ is periodic. Applying Fourier series expansion, it can be shown that the fundamental component of $i_L(\alpha, t)$ in the thyristor-controlled reactor (TCR) at the frequency ω is

$$i_{LF}(\alpha, t) = I_{LF}(\alpha)\sin(\omega t) = \frac{V}{\omega L}\left(1 - \frac{2}{\pi}\alpha - \frac{1}{\pi}\sin(2\alpha)\right)\sin(\omega t) \tag{14.3}$$

Thus the effective susceptance of the TCR, using the fundamental component of $i_L(\alpha, t)$ is

$$B_L(\alpha) = -\frac{I_{LF}(\alpha)}{V} = -\frac{1}{\omega L}\left(1 - \frac{2}{\pi}\alpha - \frac{1}{\pi}\sin(2\alpha)\right) \tag{14.4}$$

Thus the total susceptance of the SVC is

$$B_{\mathrm{SVC}} = \omega C + B_L(\alpha) \tag{14.5}$$

Note that if $\alpha = 90°$, the thyristors will not conduct. Hence $I_{LF} = 0$ and $B_L = 0$. In this situation, the SVC is providing maximum reactive power via the capacitor.

In an SVC, the reactive power is provided by capacitors and inductors, that is, if an SVC is rated from 0 to 100 MVar capacitive, both the capacitor and the inductor are rated at 100 MVar. This would not be the case for VSC-based FACTS controllers as the rating of the DC bus capacitor is much smaller.

For a single-phase reactor current i_L, the harmonic contents include all the odd harmonics, that is, $3, 5, 7, \ldots$. For balanced 3-phase operation in delta connection, the $3n$, $n = 1, 2, \ldots$, harmonics will circulate only within the TCRs. For appropriate 12-pulse arrangements [10], cancellation of the $6n \pm 1$, $n = 1, 2, \ldots$, harmonics will be possible. Thus the only harmonics left are the $12n \pm 1$, $n = 1, 2, \ldots$. These harmonic frequencies are the same as those in HVDC systems. Figure 14.2 shows the 5th and 7th harmonic filters installed in parallel to the TCR.

14.2.2 Steady-State Voltage Regulation and Stability Enhancement

This section discusses the SVC steady-state operation, in which the SVC would regulate its terminal voltage precisely to a desired value in negligible time. Two applications will be analyzed: one for voltage stability enhancement and one for transient stability enhancement.

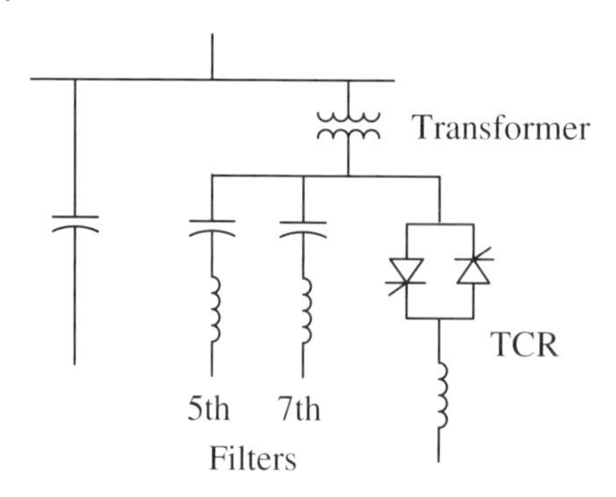

Figure 14.2 SVC configuration showing 5th and 7th harmonic filters.

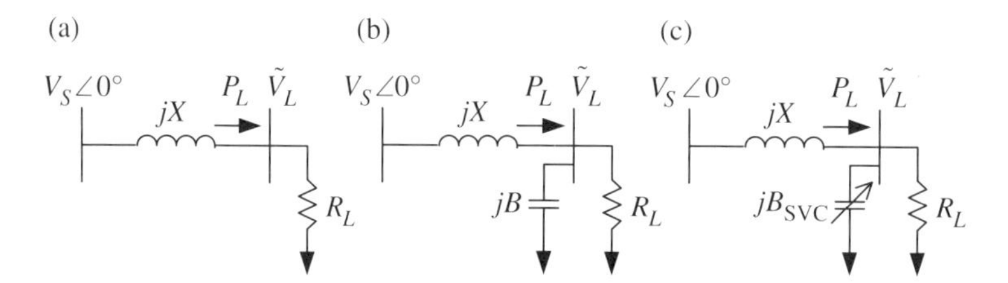

Figure 14.3 A stiff-source-to-load system: (a) no compensation, (b) with fixed capacitor, and (c) with SVC.

14.2.2.1 Voltage Stability Enhancement

Consider a power system with a stiff voltage source of magnitude $\tilde{V}_S$ supplying power to a load at unity power factor via a transmission line of impedance jX (Figure 14.3.a). The unity power factor load is modeled as a variable resistance R_L. Note that the load is zero as $R_L \to \infty$ and the power consumption increases as R_L is lowered.

Assuming the voltage angle of the stiff source is zero, the load bus voltage $\tilde{V}_L$ across R_L and the load current are given by

$$\tilde{V}_L = \frac{R_L}{R_L + jX} V_S, \quad \tilde{I}_L = \frac{\tilde{V}_L}{R_L} = \frac{V_S}{R_L + jX} \tag{14.6}$$

Thus the load bus voltage magnitude is

$$V_L = \frac{R_L}{\sqrt{R_L^2 + X^2}} V_S \tag{14.7}$$

The active power P_L delivered to the load is

$$P_L = \tilde{V}_L \tilde{I}_L^* = \frac{R_L V_S^2}{R_L^2 + X^2} \tag{14.8}$$

As mentioned in Chapter 3, for each value of P_L, (14.8) is quadratic in R_L and thus yields two solutions. The maximum power is delivered when the two roots of R_L are equal, which is also the voltage collapse point. From Chapter 3, these equations can be used to generate the PV curve for this system in per unit quantities, which is shown as the solid curve in Figure 14.4, with $X = 0.1$ pu and $V_S = 1.0$ pu.

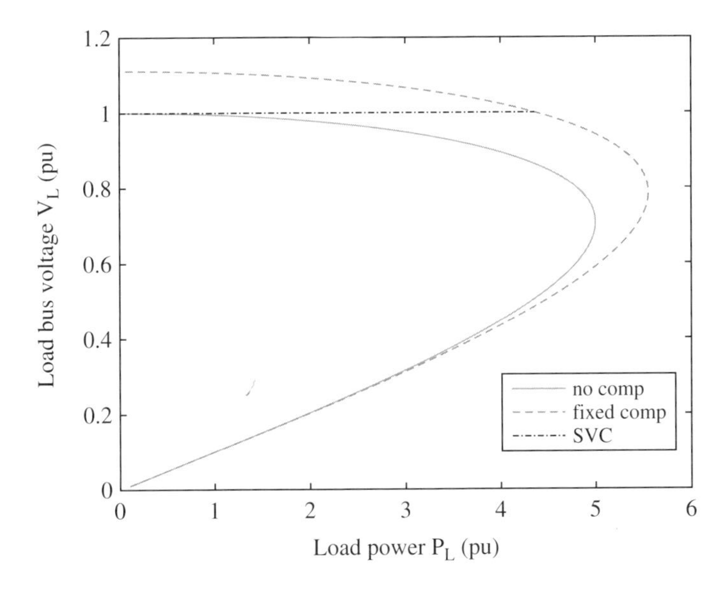

Figure 14.4 *PV* curves for the stiff-source-to-load system with no compensation, fixed capacitor, and SVC at the load bus (see color plate section).

Now consider adding a fixed shunt capacitor with susceptance B at the load bus (Figure 14.3.b). The combined impedance Z_L at the load bus is the parallel combination of R_L and the capacitor, that is,

$$Z_L = R_L \parallel \frac{1}{jB} = \frac{R_L}{1 + jR_L B} \tag{14.9}$$

Thus the load bus voltage and current in R_L are

$$\tilde{V}_L' = \frac{Z_L}{Z_L + jX} V_S = \frac{R_L}{R_L(1 - BX) + jX} V_S, \quad \tilde{I}_L' = \frac{\tilde{V}_L}{R_L} = \frac{V_S}{R_L(1 - BX) + jX} \tag{14.10}$$

The load bus voltage magnitude is

$$V_L' = \frac{R_L}{\sqrt{(R_L(1 - BX))^2 + X^2}} V_S \tag{14.11}$$

and the active power P_L delivered to the load is

$$P_L' = \tilde{V}_L' \tilde{I}_L'^* = \frac{R_L V_S^2}{(R_L(1 - BX))^2 + X^2} \tag{14.12}$$

Comparing (14.7) and (14.11), it can be observed that $V_L' > V_L$ for the same value of R_L because the denominator in (14.11) is smaller. For a similar reason, $P_L' > P_L$ for the same value of R_L. Thus adding fixed capacitive power compensation to the load improves voltage stability and power transfer to the load. The *PV* curve with the fixed capacitor compensator is shown as the dashed curve in Figure 14.4, where $B = 1.0$ pu, showing a higher voltage and extending the power transfer of the *PV* curve without compensation. In practice, the capacitor will be switched on only when the load bus voltage falls below a threshold. Leaving the capacitor on at low power transfer level may lead to over voltage.

Further consider replacing the fixed capacitor by an SVC with the same capacitor rating. Suppose the SVC would regulate the load bus voltage to be the same as the stiff source voltage V_S. Thus the upper part of the PV curve is fixed at V_S, until the maximum capacitive power compensation is reached, that is, the thyristors in the TCR are no longer firing. From that point on, the SVC and fixed capacitor compensation PV curves would be the same. This is illustrated also in Figure 14.4 with the flat voltage profile shown as the dot-dashed curve.

Although the fixed shunt capacitor and the SVC seem to have the same increase in the voltage stability margin, for fast acting disturbances, the mechanically switched shunt capacitor may not be switched on immediately. For example, if the pre-contingency power flow is at $P_L = 3.0$ pu, the shunt capacitor may not be on, as otherwise the voltage would exceed 1 pu. If suddenly an active power load of 1.0 pu is switched on at the load center, the voltage may drop to 0.9 pu before the shunt capacitor can be switched on. With the SVC connected, the voltage will still be maintained at about 1.0 pu.

Many urban load centers are decommissioning fossil units in or near their commercial and residential areas. As a result, dynamic reactive power support provided by synchronous generators would no longer be available. In addition, the import of additional active power requires more reactive power support. A good solution is to install an SVC or a STATCOM (see Section 14.4) in the footprint vacated by the decommissioned power plant. An example is the STATCOM installation replacing four old fossil units at the Holly substation in the city of Austin, Texas [177].

Example 14.1: Voltage stability computation

For the stiff-source-to-load system in Figure 14.3, let $V_S = 1$ pu, $X = 0.1$ pu, and $B = 1$ pu on the system base of 100 MVA. Assuming a resistive load, compute the maximum power transfer values for the system without compensation, with fixed capacitor compensation and for an SVC with a maximum capacitance of $B_{max} = B = 1$ pu. Also compute the voltage values at the point of collapse in each of these cases. Use the VIP method discussed in Section 3.4, which points to the fact that voltage collapse occurs when the combined load impedance is the same as the line reactance.

Solutions: For the no-compensation case, it follows that maximum power transfer occurs at $R_L = X = 0.1$ pu. Substituting $R_L = 0.1$ pu into (14.7) and (14.8) yields the familiar solution at the voltage collapse point

$$V_L = 0.7071 \text{ pu}, \quad P_L = 5 \text{ pu} \tag{14.13}$$

With fixed capacitor compensator of $B = 1$ pu, the maximum power transfer occurs at

$$|Z_L| = X, \quad \text{or} \quad R_L/\sqrt{1 + R_L^2 B^2)} = X \tag{14.14}$$

such that

$$R_L = X/\sqrt{1 - B^2 X^2} = 0.1005 \text{ pu} \tag{14.15}$$

At the voltage collapse point

$$V_L' = 0.7454 \text{ pu}, \quad P_L' = 5.5277 \text{ pu} \tag{14.16}$$

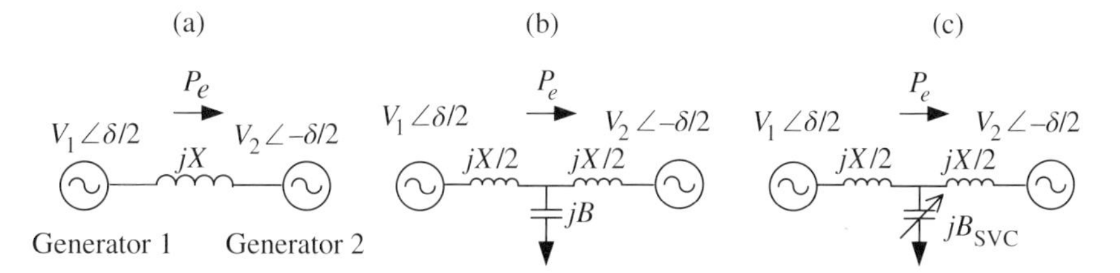

Figure 14.5 Two machines connected via a reactance: (a) no compensation, (b) with fixed capacitor, and (c) with SVC.

Thus the addition of a 100 MVA shunt capacitor improves the maximum power transfer by about 53 MW.

Because the maximum capacitance of the SVC is the same as the fixed shunt compensation, the SVC has the same voltage collapse point.

The PV curves in Figure 14.4 are generated using the parameter values from this example. For the SVC, the power transfer level in which the SVC saturates is given by $R_L = 0.2294$ pu and $P_L'' = 4.3589$ pu.

Some utilities put a lower operating voltage limit at 0.9 pu. According to Figure 14.4, the additional stability margin provided by the shunt compensator would be more than 100 MW. ■

14.2.2.2 Transient Stability Enhancement

The ability to control reactive power to regulate bus voltages can also be applied to enhance transient stability. Consider the two-machine system connected via an equivalent reactance jX as shown in Figure 14.5.a. For a simplified treatment, classical models are used to represent the machines which are assumed to be identical, and the reactance X consists of the transmission line reactances, transformer reactances, and the machine direct-axis transient reactances. The machine internal voltages are $\tilde{V}_1 = V_1\angle\delta/2$ and $\tilde{V}_2 = V_2\angle-(\delta/2)$ where $V_1 = V_2 = V_m$ and the active power flows from Generator 1 to Generator 2.

As shown in Chapter 4, the power transfer $P\delta$-curve without shunt compensation has the form

$$P_e = (V_m^2/X)\sin\delta = P_o\sin\delta \tag{14.17}$$

where $P_o = V_m^2/X$ is the maximum power transfer occurring when $\delta = 90°$. Note that the system is steady-state stable if $\delta < 90°$. This $P\delta$-curve is shown in Figure 14.6 as the solid curve.

Consider the situation when a fixed capacitor with susceptance B is installed at the midpoint of the interconnection. In such a case, the equivalent reactance between the two generators becomes (see Problem 14.2)

$$X_{eq} = X(1 - BX/4) \tag{14.18}$$

Thus the $P\delta$ curve for this system with a capacitor at the midpoint is

$$P = \frac{V_m^2}{X_{eq}}\sin\delta = \frac{P_o}{1 - BX/4}\sin\delta > P_o\sin\delta \tag{14.19}$$

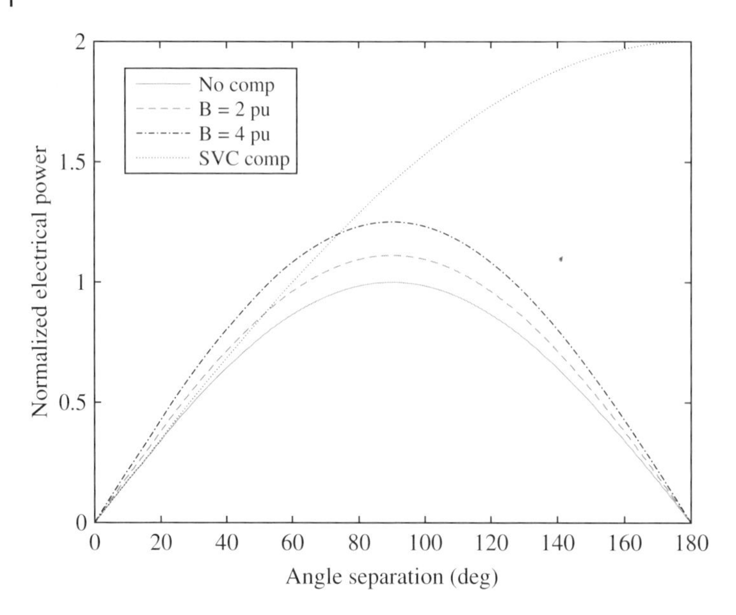

Figure 14.6 $P\delta$-curves for systems in Figure 14.5 with $X = 0.2$ pu and $V_m = 1$ pu on 100 MVA base - electrical power is normalized with respect to P_o (see color plate section).

The $P\delta$-curves for two different capacitors with $B = 2$ pu and $B = 4$ pu are plotted in Figure 14.6 as the dashed and dash-dot curves. Note the increase of the maximum power transfer.

Suppose an SVC is installed at the midpoint of the transmission line and set to control the voltage at the midpoint to V_m. As the two sides of the system are symmetrical, the voltage angle at the midpoint is zero. The power transfer on the transmission line is now

$$P = \frac{V_m^2}{X/2} \sin(\delta/2) = 2P_o \sin(\delta/2) \tag{14.20}$$

Thus the maximum power transfer is reached when $\delta = 180^\circ$ at a power level of $2P_o$. This system is steady-state stable if $\delta < 180^\circ$. The $P\delta$-curves for (14.17) and (14.20) are shown as the dotted curve in Figure 14.6.

The amount of reactive power to maintain the voltage at the midpoint may be substantial at high power transfer levels. For the parameters used in Figure 14.6, the SVC is providing 585.8 MVA of reactive power when the angle separation is 90°. Thus the SVC may be loaded up to its limit before getting to $\delta = 180^\circ$. At the point of saturation, the SVC operates as a fixed capacitor, and thus the $P\delta$-curve will switch to a fixed-capacitor characteristic. For example, if the maximum output of the SVC is 200 MVA, the $P\delta$-curve will follow the (black) dotted curve until δ reaches about 50°, at which point it will switch to the (red) dashed curve.

By exercising voltage control, the SVC will improve transient stability because the restoring torque after a fault is cleared will be higher. Let the power transfer be at $0.8P_o$ for the system configurations in Figure 14.5. Consider a short-circuit fault at one of the generator buses, cleared shortly after without tripping any lines. Following the discussion in Chapter 5, in the no-compensation case, the accelerating energy (kinetic energy) is given by the area A_1 between $P_e = 0$ and $P_e = 0.8$ from the initial rotor angle to the

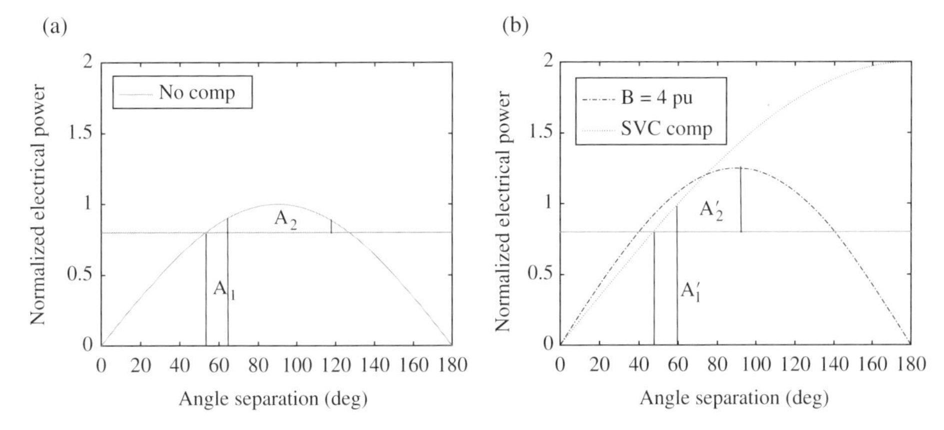

Figure 14.7 Equal area criterion applied to the $P\delta$ curves: (a) no compensation and (b) SVC compensation (see color plate section).

rotor angle when the fault is cleared, as shown in Figure 14.7.a. The restoring energy (potential energy) is the area A_2 between $P = 0.8$ pu and the $P\delta$ curve. Stability is maintained if Area A_2 is greater than Area A_1. Figure 14.7.a shows that there is not much stability margin left.

Consider an SVC compensation scheme with $B_{max} = 4.0$ pu. With the same disturbance, the area A'_1 denotes the kinetic energy accumulated when the fault is on. After the fault is cleared, the restoring energy is represented by the area between the SVC $P\delta$ curve and $P_e = 0.8$ pu, starting at the angle separation at the instant of fault clearing as shown in Figure 14.7.b. When the SVC reaches saturation at 400 MVar, the area will then be bounded by the area between the fixed-compensation $P\delta$ curve for $B = 4$ pu and $P_e = 0.8$, as indicated in Figure 14.7.b by the area A'_2. Note that there is still a substantial stability margin left. Hence the application of an SVC will enhance transient stability of the system.

14.2.3 Dynamic Voltage Control and Droop Regulation

So far it is assumed that the SVC will control its terminal voltage exactly to a desired value, which is a good assumption in steady state. For dynamic control, a droop function is implemented in the voltage-responsive control block diagram as shown in Figure 14.8.

The control system acts on the voltage error $V_{err} = V_{ref} - V_{SVC}$, where V_{ref} is the reference voltage and V_{SVC} is the SVC bus voltage. The transfer function $G_f(s) = (1 + sT_1)/(1 + sT_2)$ provides some filtering of the error signal and transient gain reduction if necessary. For this discussion $G_f(s)$ is set to unity with $T_2 = T_1$. Without the feedback loop, the integral control in the K_{SVC}/s block, where K_{SVC} is the SVC control gain, would regulate V_{SVC} to exactly V_{ref}. This control action may not be practical as the SVC may make up for the reactive power deficit all by itself, as nearby voltage control systems may back off and not contribute any reactive power support.

To allow other voltage control systems to provide reactive power support, the control system in Figure 14.8 incorporates a droop function α. In steady state,

$$V_{SVC} = V_{ref} - \alpha I_C \tag{14.21}$$

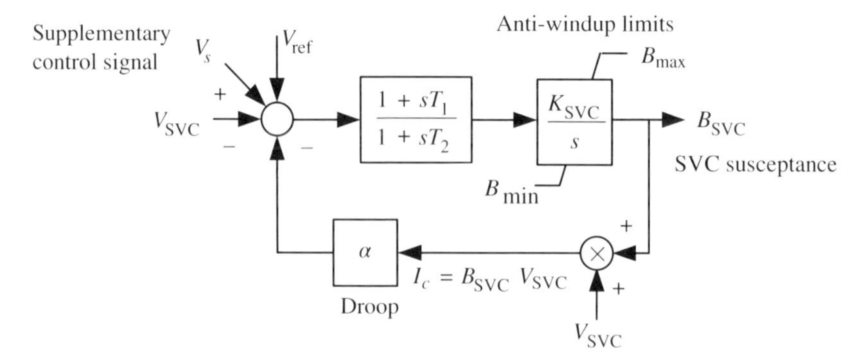

Figure 14.8 A SVC dynamic control diagram with droop regulation.

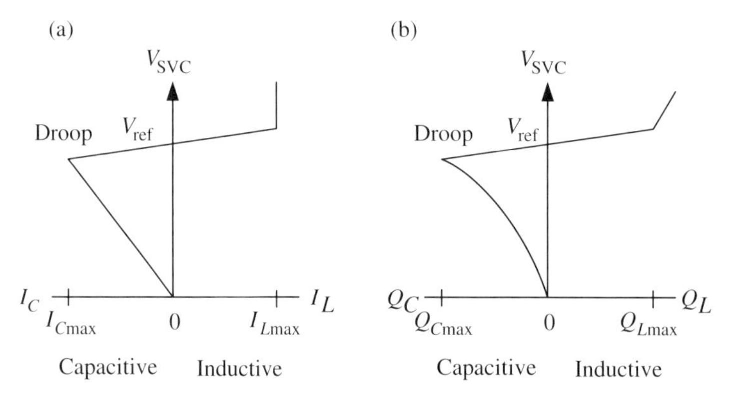

Figure 14.9 SVC droop control curves: (a) V_{SVC} versus I_C and (b) V_{SVC} versus Q_C.

where V_{SVC} is the SVC bus voltage magnitude. This droop function is shown in Figure 14.9.a, in which the slope of the droop line is α. It is customary to represent a positive capacitive current on the negative real axis and positive inductance current on the positive real axis. Figure 14.9.b shows the variation of V_{SVC} versus Q_C, in which $Q_C = B_{SVC} I_C^2$. The SVC droop regulation is typically set to $\alpha = 3$–5%.

The transfer function of the SVC controller for controlling the shunt susceptance is

$$B_{SVC} = \frac{K_{SVC}/s}{1 + \alpha(K_{SVC}/s)V_{SVC}}(V_{ref} - V_{SVC}) = \frac{K_{SVC}}{s + \alpha K_{SVC} V_{SVC}}(V_{ref} - V_{SVC}) \quad (14.22)$$

which is nonlinear due to the V_{SVC} term in the denominator. However, as V_{SVC} is close to 1 pu, the shunt susceptance can be simplified to

$$B_{SVC} = \frac{K_{SVC}}{s + \alpha K_{SVC}}(V_{ref} - V_{SVC}) \quad (14.23)$$

Note that in the control system, anti-windup is required for the integral control block K_{SVC}/s as B cannot physically exceed either B_{max} or B_{max}.

To see how the droop regulation would determine the voltage operating point, consider the stiff-source-to-load system with an SVC installed at the load bus in Figure 14.10.a. Let the load be fixed at $P + jQ$ and the voltage at the load bus be V_1

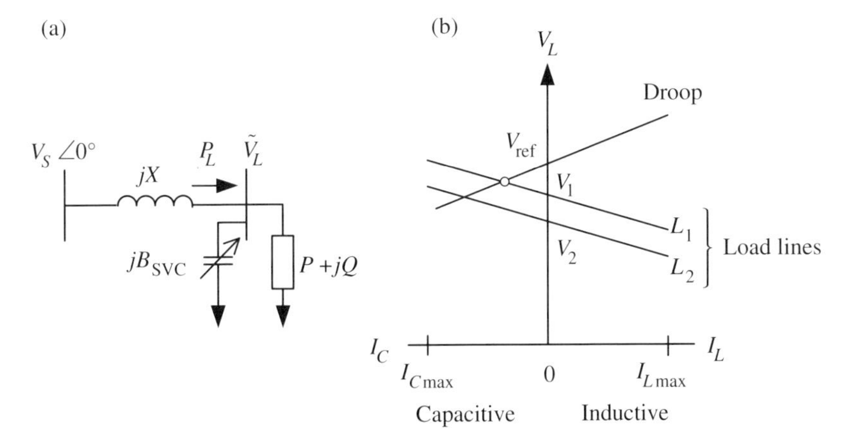

Figure 14.10 Droop regulation and load lines for an SVC: (a) SVC at load bus and (b) voltage-current diagram.

when $I_C = 0$. If the SVC injects positive capacitive current I_C into the system, the bus voltage will increase. Conversely, if the SVC injects positive inductive current I_L into the system, then the bus voltage V_L will decrease. For small changes in I_C, this variation is linear, which is commonly known as a load line (voltage versus capacitive current characteristic), denoted by L_1 in Figure 14.10.b. If there is a disruption of capacitive power supply to the load bus, the load bus voltage V_L may drop to V_2. The load line L_2 in Figure 14.3.b shows the variation of the bus voltage V_L as a function of the SVC current injection.

Let the linear load line A be represented by

$$V_L = V_1 + \beta I_C \tag{14.24}$$

where $\beta > 0$ is the sensitivity of the load bus voltage to capacitive current injection. The current I_C is negative if the SVC is operating in the inductive region. Let the droop regulation characteristics be given by (14.21) with a reference voltage of V_{ref}. Then the operating point will be determined by the intersection of these curves. Equating I_C from these two equations results in

$$\frac{V_{ref} - V_L}{\alpha} = \frac{V_L - V_1}{\beta} \quad \Rightarrow \quad V_L = \frac{\alpha V_1 + \beta V_{ref}}{\alpha + \beta} \tag{14.25}$$

which is denoted by the marker "∘" in Figure 14.10.b. Note that V_L is the weighted average of the SVC reference voltage and the load line zero-injection voltage.

An important consideration in SVC steady-state operation is to preserve the dynamic range of the SVC for disturbance response. It is unwise to use up all the capacitive current of an SVC during normal operation. In case of a disturbance (generator or line trip), the SVC will no longer have any capacitive response capability left. Thus SVCs are normally operated in the var reserve mode, that is, the SVC would operate within 10–20% of its dynamic range by adjusting the value of V_{ref}.

Figure 14.11.a illustrates the var reserve concept. Suppose the system is operating at Point a on the droop line with V_{ref1} (Figure 14.11.b), which is outside the var reserve region. One of the causes of this condition could be that V_{ref1} is set too high. Thus to

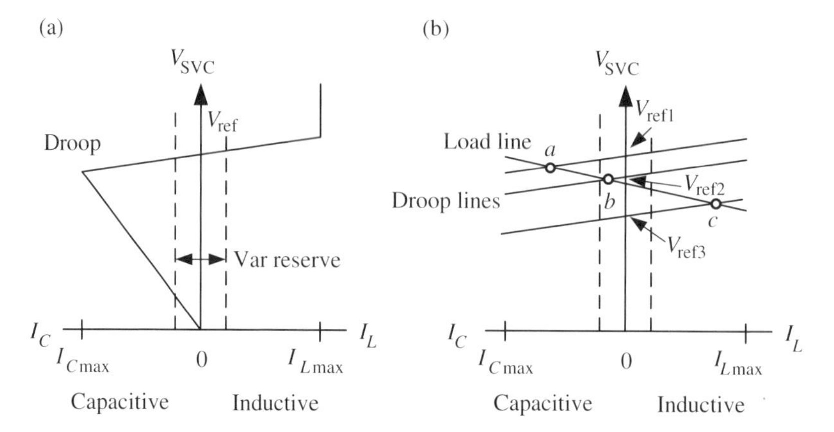

Figure 14.11 Var reserve mode operation of an SVC: (a) var reserve region and (b) impact of adjusting V_{ref}.

reduce I_C, the reference voltage can be reduced to an appropriate value so that the intersection of the load line through Point a intersects the new droop line with V_{ref2} lying inside the var reserve region, as illustrated by Point b in Figure 14.11.b. Note that if the system is operating at Point c with inductive current injection, then the SVC reference voltage V_{ref3} is set too low. The remedy is to increase the SVC reference voltage.

Var reserve is not enforced when the system is subjected to a large disturbance. Thus locally at the SVC substation, there is instrumentation to detect disturbances based on the deviation from a desired voltage range or the rate of change in bus voltage frequency (df/dt), which will trigger the disabling of the var reserve logic and allow a full response from the SVC. After the system has settled to a steady state, the system operator can reset the reference voltage such that the new operating point will fall inside the var reserve region again. In a later section on STATCOM (which has a similar behavior to an SVC), recorded PMU data will be used to illustrate droop regulation and var reserve operation of a STATCOM.

Example 14.2: Var reserve operation

Without any reactive power compensation, the voltage at a substation under some loading condition is 0.99 pu. If a capacitor rated at 1.0 pu reactive power is switched on at the bus, the voltage would go back up to 1.0 pu. Calculate the expression for the load line. Suppose an SVC is installed at the substation. Its voltage reference value (V_{ref}) is set at 1.02 pu, and the slope of the voltage regulation droop line is 0.03 pu voltage/pu reactive power injection. Find the resulting voltage on the bus, if the SVC voltage regulation loop is turned on (without the switched capacitor). Note that as the voltage is about 1.0 pu, the reactive power injection can be set equal to the reactive current injection. In addition, if the var reserve limit is enforced at 0.2 pu capacitive current, calculate the new operating point.

Solutions: The load line is

$$V_{SVC} = V_L + \beta I_C, \quad V_L = 0.99 \text{ pu}, \quad \beta = \frac{1.00 - 0.99}{1.0} = 0.01$$

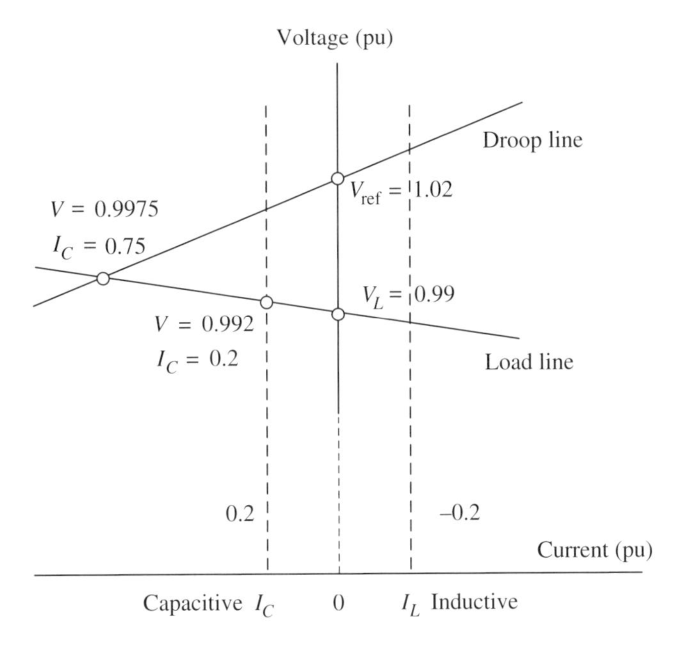

Figure 14.12 Figure for Example 14.2.

where V_L is the intercept of the load line with the $I_C = 0$ axis. The droop regulation line is

$$V_{\text{ref}} - V_{\text{SVC}} = \alpha I_C, \quad V_{\text{ref}} = 1.02, \quad \alpha = 0.03$$

To solve the operating condition, substitute the load line equation into the droop regulation equation to obtain

$$V_{\text{ref}} - V_L - \beta I_C = \alpha I_C$$

that is,

$$I_C = \frac{V_{\text{ref}} - V_L}{\alpha + \beta} = \frac{1.02 - 0.99}{0.03 + 0.01} = 0.75 \text{ pu}$$

which is capacitive. Thus the SVC bus voltage is

$$V_{\text{SVC}} = 1.02 - 0.03 \times 0.75 = 0.9975 \text{ pu}$$

As $I_C = 0.75$ pu is higher than the var reserve limit of 0.2 pu, the operating point for var reserve at $I_C = 0.2$ pu is

$$V_{\text{SVC}} = 0.99 + 0.01 \times 0.2 = 0.992 \text{ pu}$$

■

14.2.4 Dynamic Simulation

In case of a power system disturbance, an SVC is able to inject either capacitive or reactive power to maintain its terminal voltage, and thus able to assist in maintaining the system voltage and facilitate electrical power transfer. A faster recovery of electrical power implies that the accelerating torques on the affected generators will be reduced and is beneficial to system stability.

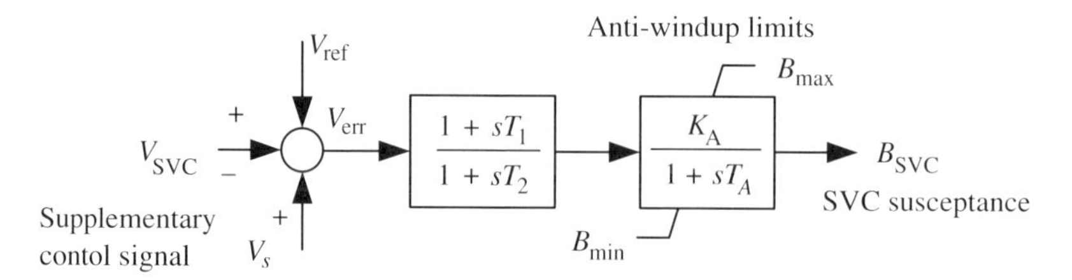

Figure 14.13 A SVC dynamic control diagram with $K_A = 1/\alpha$.

To perform dynamic simulation involving SVC, the simulation code needs to implement a control model such as the one in Figure 14.13. The terminal voltage V_{SVC} is obtained from the network solution. The output of the controller is the SVC susceptance B_{SVC} which is used for solving the system voltages and currents. To accommodate the time varying B_{SVC}, the SVC bus is treated as a conforming load bus. At each integration step, the network Y matrix is updated with the most recent value of B_{SVC}. Note that B_{SVC} is a linear element, such that if the system does not have any nonconforming loads, the network solution is direct and does not require iterations. However, adding an entry into the Y matrix may require a refactorization of the new Y matrix. Alternatively, the impact of the SVC can be modeled as a reactive current injection of $B_{SVC}V_{SVC}$ into the power system. As V_{SVC} is not known ahead of time, this method requires iterations, similar to the HVDC converters discussed in Chapter 13. In this approach, the factorization of the Y matrix can be reused without change.

The following example illustrates how an SVC can enhance system stability.

Example 14.3: Transient stability

An SVC is used to control the voltage on Bus 4 in the Single-Machine Infinite-Bus (SMIB) system shown in Figure 14.14. The generator, with a rating of 1000 MVA, is supplying 950 MW to another system modeled by the infinite bus with a terminal voltage of 1.05 pu. The line reactances are $X_T = 0.15$ pu, $X_{L1} = 0.6$ pu, $X_{L2} = 0.2$ pu, and $X_{L3} = 0.4$ pu on the system base of 1000 MVA. In steady state, the voltage of Bus 4 is controlled to 1.008 pu.

1) Use the power flow solution to determine the reference voltage used in the SVC controller.
2) A 3-phase short-circuit fault is applied on Line 3-2 close to Bus 3, and is cleared by removing Line 3-2. Use the Power System Toolbox to find out whether the system is transiently stable for fault clearing times of 4, 5, and 6 cycles (60 Hz system) for
   a) no SVC compensation
   b) an SVC with a control range of ±200 MVar.

In the dynamic SVC model, the gain K_A (which is the inverse of droop) is $1/0.03 = 33.3$ and the control time constant is 0.05 sec. The time constants T_1 and T_2 are both set to 0.2 sec and thus the filter block is bypassed.

The data file for this simulation is *datasmib_svc_example_14_3.m*, where the subtransient model is used for the generator, which is controlled by an excitation system and a governor. In the power flow formulation, the SVC is modeled as a generator bus. In

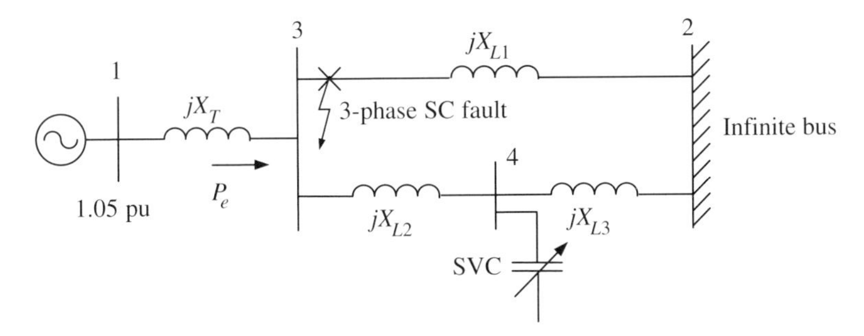

Figure 14.14 Figure for Example 14.3.

Table 14.2 Simulation results for various clearing time.

Reactive power control	4-cycle	5-cycle	6-cycle
No SVC	Stable	Unstable	Unstable
+/- 200 MVar SVC	Stable	Stable	Stable

dynamic simulation, the SVC bus needs to be declared as a non-conforming load bus because its admittance will change during a simulation. Initiate the fault at $t = 0.1$ sec so that a short time span of steady state will be evident in the time response.

Solutions: (1) From the power flow solution, the shunt susceptance provided by the SVC to hold the SVC bus voltage at 1.009 pu is 0.0030134 pu (3.01 MVar). Then following (14.23), the SVC reference voltage is

$$V_{\text{ref}} = V_{\text{SVC}} + (1/K_A)B_{\text{SVC}} = 1.003 + (1/33.3) \times 0.0030134/1.009^2 = 1.0299 \text{ pu} \tag{14.26}$$

(2) The simulation results are summarized in Table 14.2. Without the SVC, the system is transiently stable for a 4-cycle fault and unstable for a 5-cycle fault. The time responses for these two faults are shown in Figure 14.15. For the 4-cycle fault, the voltage at Bus 4 recovers into a negatively damped oscillation as shown in Figure 14.15.a. For the 5-cycle fault, the voltage at Bus 4 swings widely, due to the generator rotor angle going unstable (also known as pole slipping).

With the addition of a ±200 MVar SVC at Bus 4, the system is stable for a 6-cycle fault. The local-mode oscillation is still lightly damped, as shown in Figure 14.16.a. It is also of interest to see in Figure 14.16.b that the SVC undergoes a bang-bang-like control as the VSC voltage exceeds or falls below the SVC reference voltage value. Note that the SVC susceptance eventually settles to 0.168 pu. ■

14.2.5 Damping Control Design using SVC

Example 14.3 shows that an SVC is able to enhance the transient stability in a stressed system with high levels of power transfer. Such high power transfer level often causes

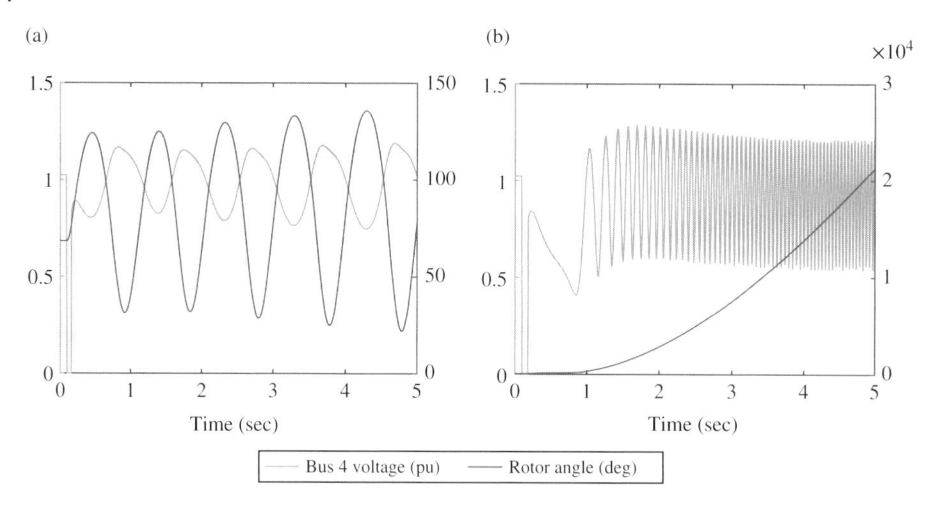

Figure 14.15 Bus 4 voltage and rotor angle response for system without SVC: (a) 4-cycle fault clearing and (b) 5-cycle clearing (see color plate section).

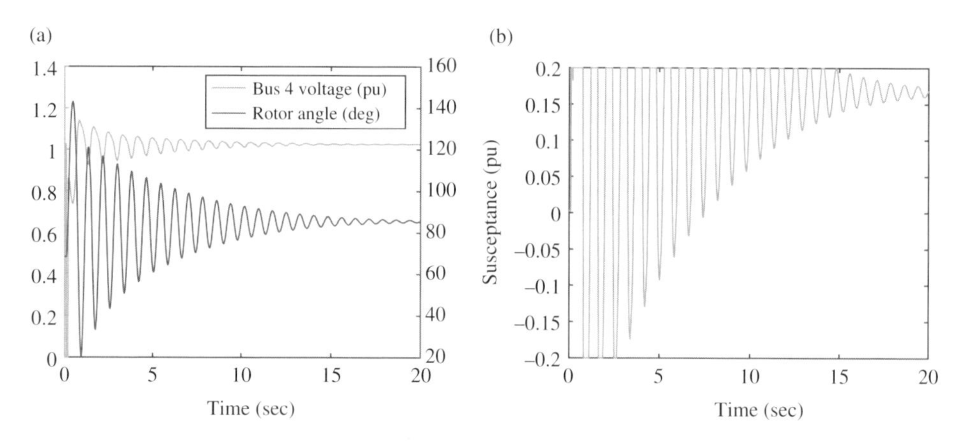

Figure 14.16 Time response of system with SVC for 6-cycle fault clearing: (a) Bus 4 voltage and (b) SVC susceptance response to Bus 4 voltage (see color plate section).

the electromechanical modes to be lightly damped. Because of its fast response, an SVC can also accommodate a damping control signal (V_s in Figure 14.13) to modulate the SVC susceptance for improving the damping of the critical electromechanical modes [178].

SVC damping controller design consists of two stages: first, the analysis of the linearized model and the selection of measured local signals as inputs to the damping controller; and second, the selection of the controller structure and the design steps. These steps are illustrated in the following example. Note that remote signals can also be used for damping control design, as shown in Example 13.5 for the HVDC system damping controller. Using local signals allows the illustration of additional control design aspects.

Example 14.4: SVC damping controller design

This example is in two parts:

1) Generate the linearized state-space model of the system in Example 14.3, using the same model parameters, except that Line 3-2 is disconnected.[2] Compute the eigenvalues of the linearized model. Select the voltage magnitude V_4 for Bus 4, the active-power P_e on Line 4-2, and the current magnitude I_m on Line 4-2 as possible output signals. Analyze the output matrix corresponding to these signals and determine the suitability of the signals for use by the damping controller.
2) Design a damping controller to improve the damping ratio of the electromechanical mode to about 0.15. Use the linearized closed-loop system to show the response of ΔI_m subject to a 0.05 pu step in the excitation system reference voltage value.

Solutions: (1) After removing the rotor angle and speed of the large machine representing the infinite bus, the linear model has 10 states in the following order: the two generator mechanical states for the rotor angle δ and speed ω, the four generator electrical states E'_q, ψ_{kd}, E'_d, and ψ_{kq}, the two exciter states V_R and E_{fd}, and two SVC states B_{SVC} (the SVC susceptance) and x_f (the filter state). The entries of the 10×10 system A matrix are given by:

```
Columns 1 through 6
          0   3.7699e+02            0            0            0            0
-1.5451e-01  -3.4775e-01  -1.0670e-01  -9.6032e-02  -2.6415e-03  -1.1622e-02
 9.4892e-02            0  -2.1839e+00   2.3138e+00  -1.5910e-02  -7.0002e-02
-3.5731e+00            0   3.0518e+01  -3.3824e+01            0            0
-3.4293e-02            0            0            0  -8.4117e+00   6.4454e+00
-3.6307e-01            0            0            0   1.5509e+01  -2.0285e+01
 4.2937e+00            0  -8.2488e+00  -7.4239e+00   3.2463e+00   1.4284e+01
 4.8264e+01            0  -1.3384e+02  -1.8857e+02   5.8535e+01   2.5755e+02
 1.4597e+02            0  -7.1663e+01  -6.4497e+01   5.1295e+01   2.2570e+02
          0            0            0            0            0            0

Columns 7 through 10
          0            0            0            0
          0            0  -3.8764e-02            0
          0   1.3843e+00  -1.0634e-02            0
          0            0   6.3727e-01            0
          0            0  -1.7686e-01            0
          0            0  -1.8725e+00            0
-1.4993e-01            0  -2.7225e+00            0
 1.4961e+01  -1.2431e+02  -4.3600e+01            0
          0            0  -1.7613e+02   6.6600e+02
          0            0            0  -5.0000e+00
```

2 The SVC would be more effective in control because without Line 3-2, all the power has to flow through Bus 4.

The eigenvalues of the system are given by

$$\begin{bmatrix} -172.74 \\ -123.97 \\ -33.610 \\ -28.128 \\ -0.29605 \pm j8.2490 \\ -4.4126 \\ -1.0934 \pm j0.70265 \\ -5.0000 \end{bmatrix} \tag{14.27}$$

The complex pair $-0.29605 \pm j8.2490$ is the swing mode with a frequency of 1.313 Hz and a damping ratio of 0.359.

The input matrix B for the signal V_s is a 10×1 column vector, whose only none-zero entry is for the B_{SVC} state, having a value of $K_A/T_A = 666$. Normally the SVC control would come through the filter state, except that $T_1 = T_2$, which makes the filter state unnecessary (i.e., the eigenvalue $-1/T_2 = -5$ is uncontrollable).

A good damping control design requires controllability of the actuation and observability of the input signal. For an SVC, the actuation is fast and thus there is sufficient controllability. The observability part requires finding a signal in which the lightly damped electromechanical mode is readily observable. For an SVC, there are plenty of local voltage and current signals in which the electromechanical oscillations are dominant. A third element of good control design is to use an input signal which can accommodate a sufficiently high gain design without forcing some other modes to be unstable. For an SVC, the local signals are network variables that are functions of the states. Any changes in network susceptance values would "immediately" cause the network variables to change. This is known as the "inner-loop effect" in [178]. Instead of using a modal decomposition as in [178], the entries of the C matrix from various signals are analyzed below to show the constraint on controller gain.

This example studies three local signals: the SVC bus voltage V_4, the active-power flow on Line 4-2 P_e, and the magnitude of the current flow on Line 4-2 I_m.[3] The C matrix entries of these local signals are given by:

V_4

```
Columns 1 through 6
-2.1918e-01        0    1.0760e-01    9.6842e-02   -7.7020e-02   -3.3889e-01

Columns 7 through 10
          0        0    2.3442e-01            0
```

P_e

```
Columns 1 through 6
 8.8060e-01        0    6.0814e-01    5.4733e-01    1.5055e-02    6.6242e-02

Columns 7 through 10
          0        0    2.2093e-01            0
```

3 The active power part of current flow I_a and bus frequency signals are also considered in [178]. The active power current flow signal I_a is similar to P_e. To analyze the bus frequency, a frequency state needs to be created from the bus angle.

I_m

```
Columns 1 through 6
 1.0310e+00              0    5.9169e-01    5.3252e-01    5.1665e-02    2.2733e-01

Columns 7 through 10
          0              0    1.2422e-01             0
```

Observability can be analyzed by examining the entry corresponding to generator angle $\Delta\delta$, which is the first entry. A higher value means that the mode is more observable. However, it should be noted that the scaling/variation of the signal also needs to be accounted for. For example, a voltage signal is typically about 1 pu, but a current signal can be 5–6 pu, and thus the first entries of the current signals are higher than the voltage signal.

Another important entry in the C matrix is the value corresponding to the state ΔB_{SVC}, which is the 9th entry. This entry signifies the impact B_{SVC} has on the signal. This entry for V_4 at 0.2344 says that for every 1 pu change in B_{SVC}, the SVC bus voltage increases by 0.2344 pu. This is clearly an undesirable effect as the control object is to damp the oscillation of rotor angles. Closing this loop with a high gain may result in destabilizing the voltage regulation function. On the other hand, the entry in I_m is smaller at 0.1242. This is due to the current flow being mostly determined by bus angle separations, although it is also a function of bus voltages.

A simple indicator of the suitability of a signal for damping control is the ratio of the entries of $\Delta\delta$ and ΔB_{SVC}. The ratios for the three signals are given by:

Signal	**Observability and inner-loop gain ratio**
V_4	$0.2192/0.2344 = 0.9350$
P_e	$0.8806/0.2209 = 3.986$
I_m	$1.0310/0.1242 = 8.300$

The higher values of the ratios for P_e and I_m make these signals more effective than V_4 for damping control. For this example, the control design will make use of the I_m signal. The current magnitude signal control design is independent of the power flow direction. For the P_e signal, if the power flow changes direction, the sign of the control gain also needs to be changed to ensure positive damping effect. This design is posed as Problem 14.12.

(2) The root-locus method will be used to design the damping controller using I_m as the input signal. Denote $G(s)$ as the linearized model of the power system with V_s as the input and ΔI_m as the output. The controller structure is shown in Figure 14.17. The rate filter $G_{\mathrm{wo}}(s) = sT_w/(1 + sT_w)$ will render the controller inactive when the input signal is constant. The lead-lag compensator $G_{\mathrm{ldlg}}(s) = (1 + sT_a)/(1 + sT_b)$ provides phase compensation to achieve enhanced damping. The controller gain is K_d, which is implemented with a time constant T_d.

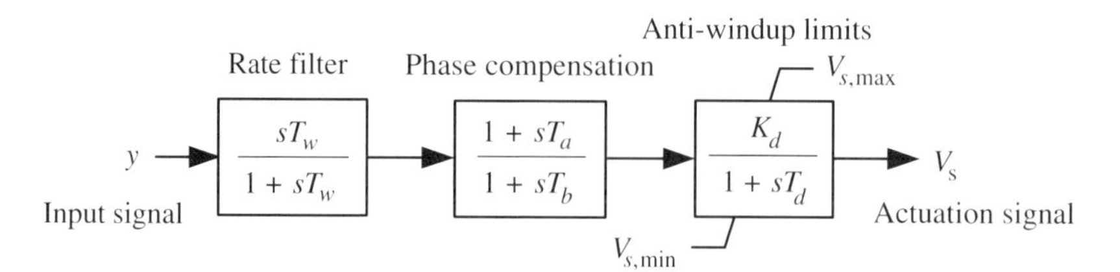

Figure 14.17 SVC damping controller structure.

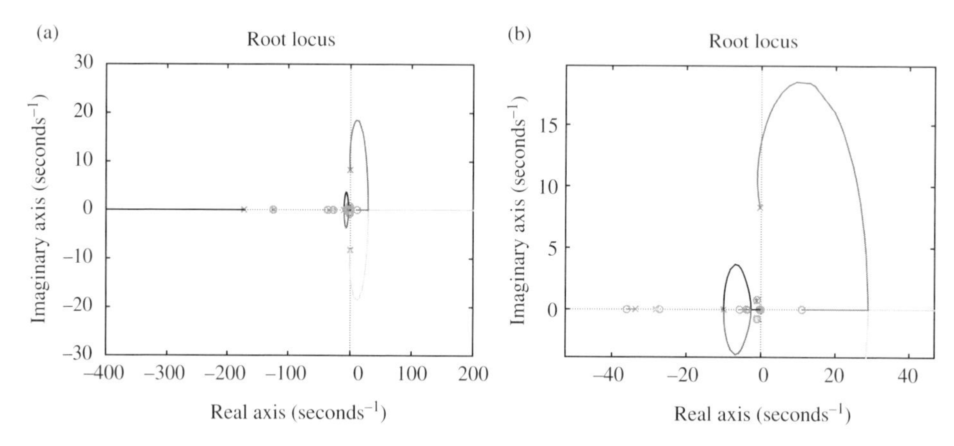

Figure 14.18 (a) Root-locus plot of uncompensated system and (b) enlarged plot around the swing mode (see color plate section).

Given that the frequency of the electromechanical mode is about 10 rad/sec, the time constants T_w and T_d are both set to $1/10 = 0.1$ sec. The root-locus plot of the transfer function of the uncompensated system

$$G(s)\frac{1}{1+sT_d}G_{\text{wo}}(s) \tag{14.28}$$

is shown in Figure 14.18. Note that V_s comes into the controller summing junction as a positive sign.

The angle of departure of the swing mode is about 117°. To change the direction of the angle of departure pointing in the direction of damping improvement, a lead-lag compensator with 45° phase-lead compensation at 8.249 rad/sec is designed, resulting in $T_a = 0.2929$ sec and $T_b = 0.05021$ sec.

Next a root-locus analysis is performed for the compensated system

$$G(s)\frac{1}{1+sT_d}G_{\text{ldlg}}G_{\text{wo}}(s) \tag{14.29}$$

which is shown in Figure 14.19.a. In the enlarged RL plot of Figure 14.19.b, the MATLAB® data cursor function shows that if $K_d = 0.2$ is used, the swing mode of the close-loop system will have a damping ratio of 0.184.

Figure 14.19.c shows that if K_d is increased to about 0.9, there will be a pair of complex eigenvalues crossing into the right-half plane. Thus using a gain of $K_d = 0.2$ would provide a gain margin of $20\log_{10}(0.9/0.2) = 13.06$ dB, which is quite adequate. The resulting

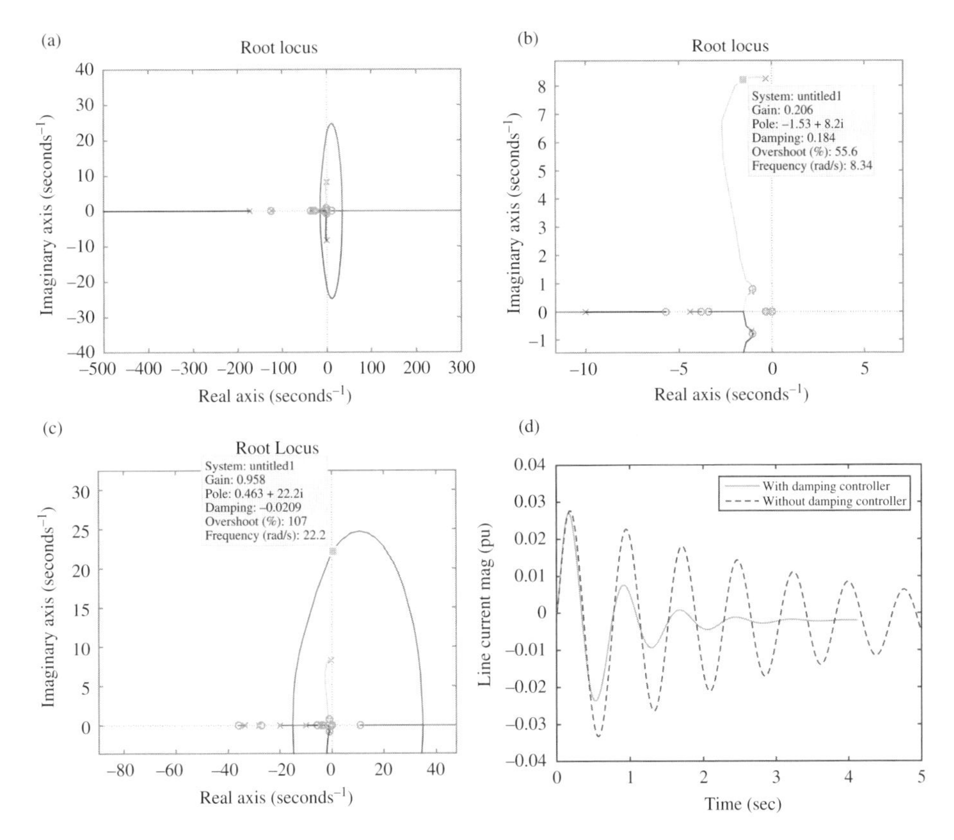

Figure 14.19 (a) Root-locus plot of compensated system, (b) enlarged plot around the swing mode, (c) unstable gain, and (d) comparison of compensated and uncompensated systems subject to a 0.05 pu step increase in the excitation system voltage reference (see color plate section).

line-current magnitude response due to a 0.05 pu step increase in V_{ref} of the excitation system is shown in Figure 14.19.d. Note the line current magnitude is slightly lower as the voltage is increased on the generator bus. As compared to the same voltage step in the uncompensated system, the oscillation in the compensated system is damped much more quickly. ■

14.3 Thyristor-Controlled Series Compensator

The second part of this chapter is on series compensation with fixed series capacitors and thyristor-controlled series compensators for improving power system transfer capability and stability. The fast-acting TCSCs can dynamically change the effective impedance of a transmission line and thus perform other tasks such as damping of swing modes.

In contrast to SVCs, the number of TCSC installations is still limited. The first TCSC was installed at Kayenta [179], followed by the installations at Slatt [180], in Sweden [181], in Brazil [182], and more recently in China [183].

A typical TCSC installation will also consist of fixed series capacitor compensation so as to reduce cost. Thus the discussion here first starts with fixed series compensation.

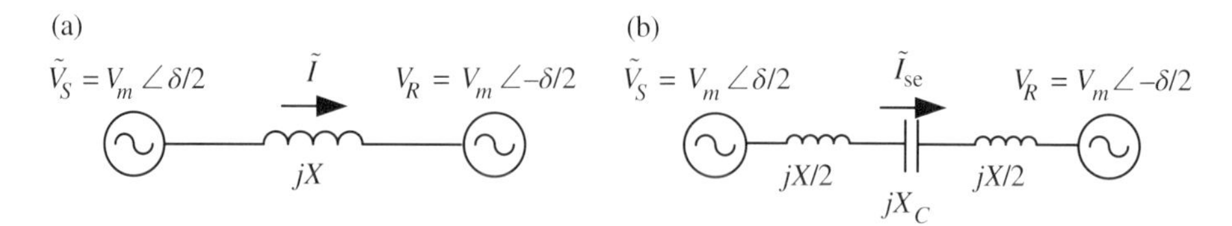

Figure 14.20 Two-machine system with (a) no compensation and (b) fixed series compensation.

14.3.1 Fixed Series Compensation

Consider the two-machine system in Figure 14.20.a, where the sending-end voltage and the receiving-end voltage are, respectively,

$$\tilde{V}_s = V_m \angle(\delta/2), \quad \tilde{V}_r = V_m \angle(-\delta/2) \tag{14.30}$$

The power transfer between the two bus terminals is

$$P = \frac{V_m^2}{X} \sin \delta \tag{14.31}$$

Suppose a fixed capacitor with reactance $X_C = -kX$ is inserted in the middle of the transmission line, as shown in Figure 14.20.b. The parameter k, $0 \leq k < 1$, is called the compensation ratio. Then the effective reactance between the two machines is

$$X_{\text{eff}} = X + X_C = (1 - k)X \tag{14.32}$$

Thus the power transfer for a series compensation ratio k is

$$P_k = \frac{V_m^2}{X_{\text{eff}}} \sin \delta = \frac{V_m^2}{(1 - k)X} \sin \delta \tag{14.33}$$

Thus series compensation can significantly improve power transfer of long transmission lines. For high k, series compensation is broken into smaller series capacitors installed at various locations along a long transmission line to prevent a large difference in voltage across any particular capacitor and help level out the voltage profile on the line.

Series compensation also improves voltage-stability limited power transfer, due to the fact that the Thèvenin equivalent reactance will be reduced by the series compensation. The impact of the series compensation on voltage stability can be studied using the techniques introduced earlier in Chapter 3.

As mentioned in the discussion of torsional oscillations in Chapter 12 on power generation, a potential problem with series compensation is that it may destabilize the torsional modes of turbine-generators, requiring the use of torsional oscillation protection equipment such as torsional filters [114].

14.3.2 TCSC Circuit Configuration and Switching

Instead of a fixed series compensation, there are additional benefits in using a fast variable series compensation by combining a thyristor-controlled reactance (TCR) in parallel with a fixed series capacitor, as shown in Figure 14.21. The TCSC was proposed by Dr. John Vithayathil [184].

As in an SVC, the TCSC changes its equivalent reactance by adjusting the ignition delay angle α of the thyristor pair. Let the reactance value be $X_L = \omega L$. Then the effective

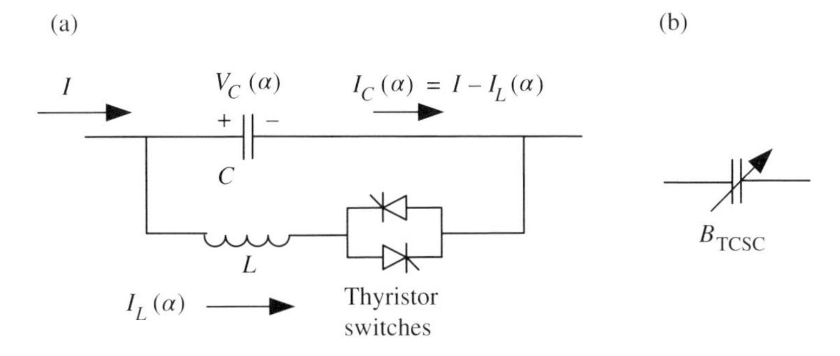

Figure 14.21 TCSC: (a) circuit representation and (b) symbolic representation.

reactance as a function of the ignition delay angle α is, considering only the fundamental frequency component,

$$X_L(\alpha) = X_L \frac{\pi}{\pi - 2\alpha - \sin(2\alpha)}, \quad 0 \leq \alpha \leq \frac{\pi}{2} \tag{14.34}$$

Note that TCSC operation does not require harmonics filters as the harmonic currents will mostly circulate within the TCSC.

Thus the combined reactance $X_C = -1/(\omega C)$ of the fixed capacitor and the TCR is

$$X_{\text{TCSC}} = \frac{X_C X_L(\alpha)}{X_C + X_L(\alpha)} \tag{14.35}$$

It should be noted that for values of α such that $X_C + X_L(\alpha)$ is small, X_{TCSC} can be very large (resonance) and can cause operating difficulties. Thus α normally operates away from the resonance condition. For control purposes, it can be assumed that the TCSC can provide continuous variation from $X_L > 0$ to $X_C < 0$, which is translated to $B_{\min} = -1/X_L$ and $B_{\max} = -1/X_C$ as the limits of B_{TCSC}.

For closed-loop control, a TCSC is normally put in the reactance modulation mode. That is, the reactance X_{TCSC} is maintained to a desired value. There are other control functions that can be added to the TCSC overall control scheme.

One such function is for a TCSC to provide transient stability augmentation, in which the TCSC rapidly switches to $B_{\max}$ by a signal, most likely a breaker status indicating line or generator trips. Consider the system shown in Figure 14.22, where the generating area is supplying power to the load area with four parallel lines, two at 765 kV and two at 500 kV. The system may readily handle the simultaneous loss of the two 500 kV lines. However, the simultaneous loss of the two 765 kV lines would require tripping some generators and shedding some loads to maintain the stability of the rest of the system. However, if TCSCs are installed on the 500 kV lines, they can be immediately switched to maximum capacitance upon the trip of the 765 kV lines, improving the power transfer on the 500 kV lines, reducing the number of generators tripped, and eliminating the need to disconnect some loads [185]. Note that the TCSC on the 500 kV lines should not be switched to $B_{\max}$ in normal operating conditions because more current will flow on the 500 kV lines, resulting in higher resistive losses.

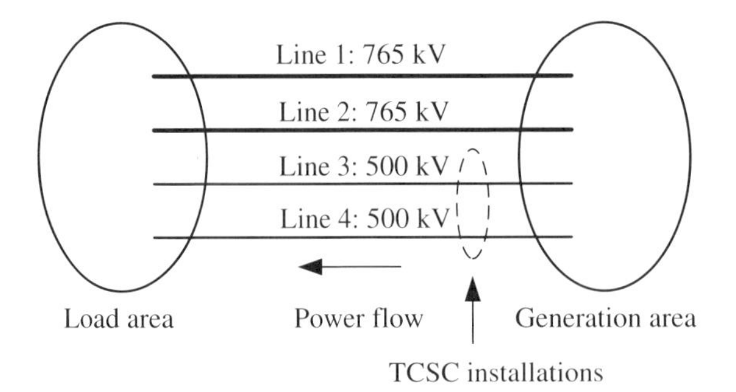

Figure 14.22 Two-area system with two 765 kV lines and two 500 kV lines.

14.3.3 Voltage Reversal Control

An added advantage of a TCSC is its short-term overload capability. This is known as the voltage reversal or vernier control [180], and is illustrated for a single phase in Figure 14.23. In steady-state operation, the current I is assumed to be periodic at the nominal frequency. For full-insertion operation, the reactor is switched off by blocking the thyristors (Figure 14.23.a). The capacitor voltage $V_C = jX_C I$ also is sinusoidal at the nominal frequency, but 90° out of phase with the current I.

Now consider activating the thyristor pairs when V_C is sightly positive before becoming negative. Then the capacitor voltage will produce a current I_L in the inductor L, allowing it to discharge faster, as illustrated with the voltage polarity and current direction in Figure 14.23.a. When V_C turns negative, the current I_L will be continuous and flow back into the capacitor. In this period, the capacitor will be charged by two currents I and I_L, and as a result, will have a more negative voltage value. Eventually I_L will decay to zero and the thyristor switches are turned off. The process repeats when the capacitor voltage reaches a small negative value and the thyristor switches are turned on, with a phase β inducing a current I_L to flow through the inductor and discharging the voltage faster. The resulting capacitor voltage waveform $\hat{V}_C$ due to the voltage-reversal control will have a higher amplitude than V_C. Thus the effective capacitive reactance will have a value of

$$\hat{X}_C = \max\{\hat{V}_C\}/\max\{I\} \tag{14.36}$$

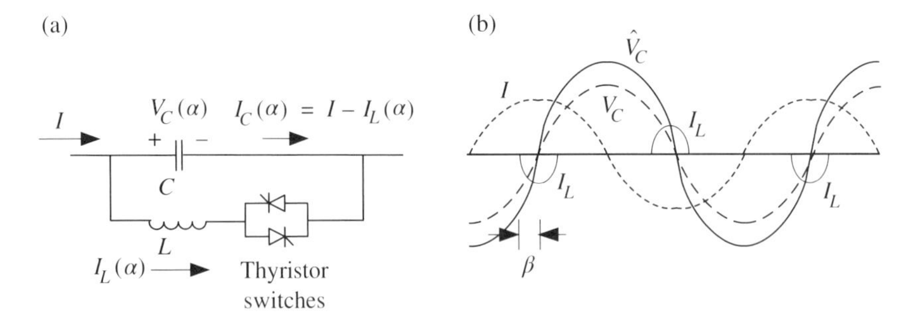

Figure 14.23 Steady-state wave shapes of TCSC voltages and currents during constant-reactance vernier control: (a) TCSC circuit configuration and (b) increase of V_C to $\hat{V}_C$ through voltage reversal [180].

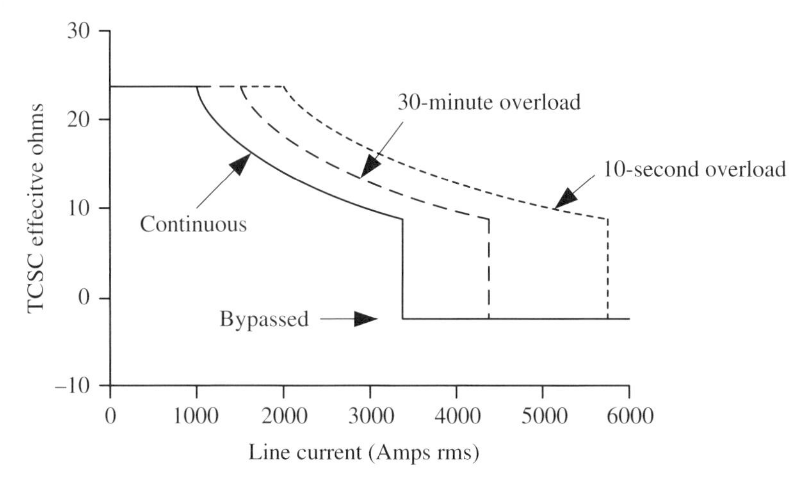

Figure 14.24 Reactance capability of TCSC [180].

It was reported in [180] that the normal rating of the Slatt TSCS of 8 ohms can be increased to 9.2 ohms in steady state for nominal current, and 16 ohms for 10 seconds and 12 ohms for 30 minutes utilizing the overload capability (Figure 14.24). If the TCSC is primarily used for transient stability enhancement, then it should be sized according to its short-term limit, resulting in a lower nominal rating and cost.

Note that the temporary overload limit of a TCSC is determined by the dielectric stress and total rms current through the capacitor, with the rms current value being more restrictive. As seen in Figure 14.23, the voltage-reversal control will generate harmonics. However, the harmonic currents mainly circulate in the reactor-switched-capacitor loop, with very little entering the transmission system. Thus harmonic filters are not needed.

14.3.4 Mitigation of Subsynchronous Oscillations

When the TCSC is operating in the maximum capacitive mode, there is a potential for interactions with the torsional modes of nearby generators. There are several approaches to alleviating such SSR phenomena. A modulation function can be added to the thyristor control to add damping to the relevant torsional modes [186]. Voltage reversal control can also be used to make the frequency response of the TCSC to be inductive at the generator torsional mode frequency. In doing so, torsional interactions can be avoided.

A separate torsional oscillation damping scheme for fixed series compensation is the NGH SSR Damper proposed by Dr. Narain G. Hingorani [10] in 1981. This is a relatively low-rating device, functioning as a thyristor-controlled impedance, as shown in Figure 14.25 with the thyristors in series with a resistance and a reactance. The thyristors will force the voltage of the series capacitor to zero at the end of each half-cycle if it exceeds the value associated with the fundamental voltage component of the synchronous power frequency. This action will remove undesirable frequencies, including torsional frequencies, from the capacitor voltage, thus preventing torsional interactions with turbine-generator rotor shafts.

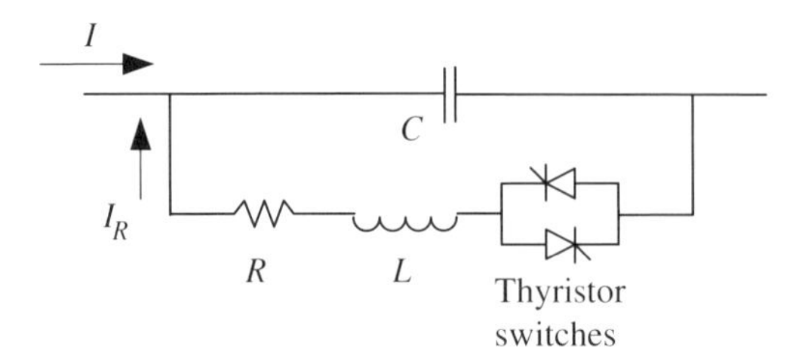

Figure 14.25 NGH SSR damper [10].

14.3.5 Dynamic Model and Damping Control Design

The actuation of a TCSC is represented by the dynamic model shown in Figure 14.26. The regulator contains a gain K_A (normally set to unity if the input signal u_{TCSC} is a desired TCSC susceptance B_{TCSC}) and a small time constant T_A to represent thyristor switching. The filter time constants T_1 and T_2 provide capability to shape the input control signals. This controller has two states: the TCSC susceptance B_{TCSC}) and the filter state.

There are two input signals into the TCSC control model. The signal u_{TCSC} is used for large disturbances. A TCSC is normally not used to regulate current flows or bus voltages. Instead, in a severe contingency such as line or generator tripping, a discrete control signal such as a step increase can be issued allowing the TCSC to reach its emergency rating and facilitate increased flow on the transmission line. By this action, the synchronizing torque across the transmission line will be increased. This would be a suitable control action for the system in Figure 14.22.

In a dynamic simulation program, u_{TCSC} is used to hold the initial value of the TCSC insertion. The output B_{TCSC} is used by the network solution to solve for the bus voltages and line currents. Because changing B_{TCSC} requires changing the network admittance matrix Y, both the connecting buses of the TCSC have to be designated as non-conforming buses. The adjustment in Y will occur in the entries Y_{ii}, Y_{ij}, Y_{ji}, and Y_{jj}, where i and j are TCSC buses. Note that the output B_{TCSC} is bounded by its ratings B_{max} and B_{min}. Anti-windup limit is applied to the B_{TCSC} state (see Appendix 9.A).

The second signal u_{sup} is for implementing a damping controller to improve the damping on the relevant swing modes. Similar to an SVC, a TCSC is useful to provide interarea damping enhancement, in situations when such interarea damping cannot be properly provided by power system stabilizers installed on synchronous generators. An example is the Brazil North-South Interconnection, connecting the previously asynchronous North and South regions of the Brazilian power system [182], as illustrated

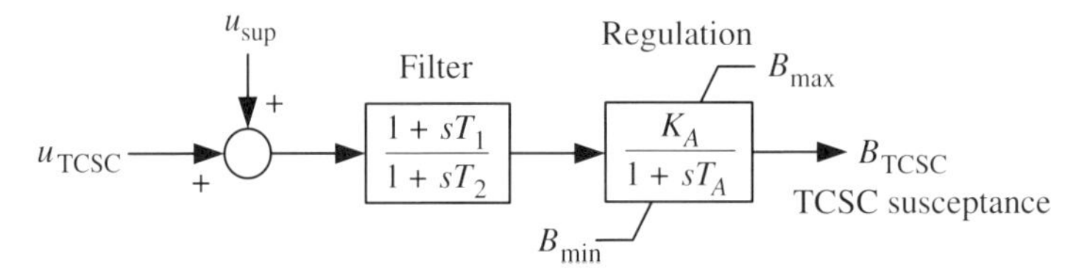

Figure 14.26 TCSC dynamic model representation.

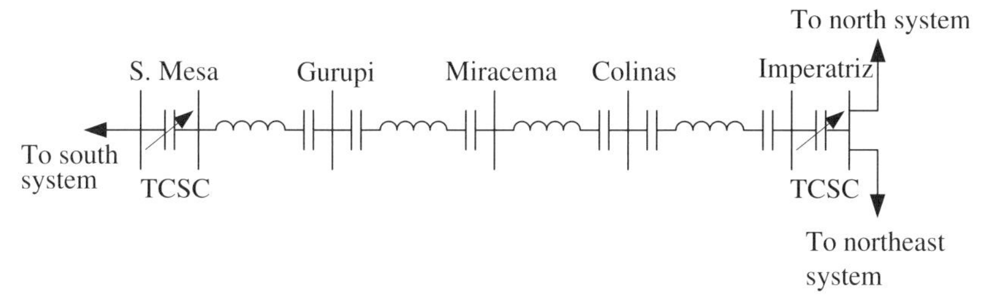

Figure 14.27 Brazil power system North-South Interconnection schematic [182].

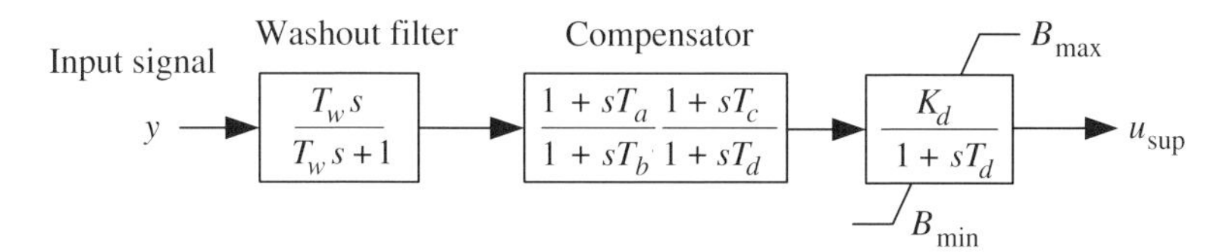

Figure 14.28 TCSC damping controller.

in Figure 14.27. The interconnection is at 500 kV and 1020 km long, starting from Imperatriz in the north and Serra da Mesa in the south. There are six separate series capacitors providing 54% compensation. The TCSCs are installed at Imperatriz and S. Mesa, each providing 6% of compensation. The main function of the two TCSCs is to provide damping enhancement to the 0.2 Hz N-S interarea mode, as otherwise, the system is small-signal unstable for many contingencies.

The block diagram of a TCSC damping controller is shown in Figure 14.28. As in the SVC damping control design, the input signal y can be a local signal such as bus voltage and line current flow, a measured remote signal such as the bus angles obtained from synchrophasor measurement units, or a remote signal estimated using local measurements [187]. A washout filter with time constant T_w is used to stop the damping controller action if the input signal y has settled to a steady state. The damping control gain is K_d and is rolled off by the time constant T_d. The second-order compensator is designed to provide the proper phase lead or lag for swing-mode damping enhancement.

The damping control design for a TCSC is illustrated in the following example. The bus voltage magnitude signal will be used as input signal [188].[4]

Example 14.5: TCSC damping controller design

Consider the SMIB system in Figure 14.29, which is modified from Figure 14.14 by replacing the SVC with a TCSC connecting Bus 4 to a new Bus 5. In addition, Line 3-2 is disconnected by opening the circuit breaker (CB). This example is in two parts:

1) Generate the linearized state-space model of the system, using the same model parameters as in Example 14.3. Note that the reactance of the TCSC is set at

4 The bus voltage magnitude signal has not been used previously in this text for damping control. Recall that the machine speed signal is used in Example 10.3 for PSS design, the remote bus angle signals in Example 13.5 for HVDC system damping design, and the line-current magnitude in Example 15.4 for SVC damping design.

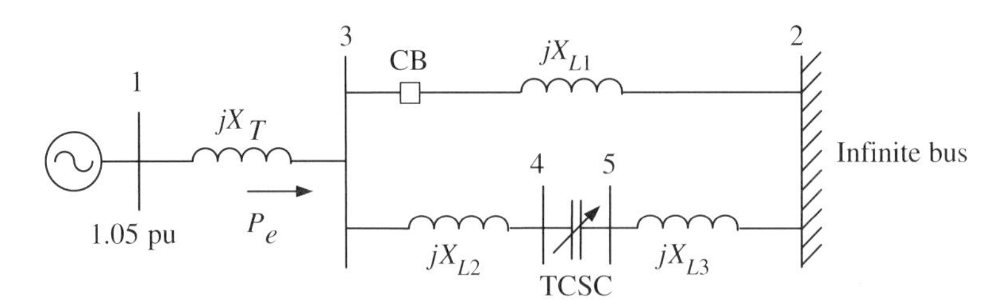

Figure 14.29 Figure for Example 14.5.

$X_C = -0.01$ pu initially. The TCSC parameters are $B_{\max} = 0.1$ pu, $B_{\mathrm{mim}} = -0.1$ pu, $K_A = 1$, and $T_A = 0.05$ sec. The filter states are not used in the model. Compute the eigenvalues of the linearized model.

2) Design a damping controller using the Bus 4 voltage magnitude V_4 to improve the damping ratio of the electromechanical mode to about 0.15. Use the linearized closed-loop system to show the response of ΔV_4 subject to a 0.05 pu step in the excitation system reference voltage value.

Solutions: (1) After removing the rotor angle and speed of the large machine representing the infinite bus, the nine states for the linearized model are in the following order: the two generator mechanical states for the rotor angle δ and speed ω, the four generator electrical states E'_q, ψ_{kd}, E'_d, and ψ_{kq}, the two exciter states V_R and E_{fd}, and the TCSC state B_{TCSC}. The entries of the 9×9 system A matrix are given by:

```
Columns 1 through 6
          0   376.9911          0          0          0          0
    -0.1476    -0.3478    -0.1017    -0.0915    -0.0042    -0.0187
     0.0977          0    -2.1839     2.3298    -0.0140    -0.0614
    -3.4048          0    30.4535   -33.8822          0          0
    -0.0551          0          0          0    -8.4148     6.4319
    -0.5837          0          0          0    15.4764   -20.4285
     4.2500          0    -8.8227    -7.9404     2.9903    13.1573
    48.1389          0  -144.2781  -199.3731    53.8576   236.9727
          0          0          0          0          0          0

Columns 7 through 9
          0          0          0
          0          0          0
          0     1.4572          0
          0          0     0.0003
          0          0          0
          0          0     0.0002
    -0.1499          0    -0.0003
    14.9613  -130.7363    -0.0022
          0          0   -20.0000
```

The eigenvalues of the system are given by

$$\begin{bmatrix} -129.81 \\ -33.117 \\ -26.078 \\ 0.090693 \pm j7.3574 \\ -3.8930 \\ -1.7904 \\ -1.6323 \\ -20.000 \end{bmatrix} \tag{14.37}$$

The complex pair $0.090693 \pm j7.3574$ is the unstable swing mode with a frequency of 1.171 Hz and a damping ratio of -0.0137.

The input matrix B for the signal u_{sup} is a 9×1 column vector, whose only none-zero entry is for the B_{TCSC} state, having a value of $K_A/T_A = 20$.

The row of C matrix corresponding to the voltage magnitude V_4 at Bus 4 is

```
Columns 1 through 6
-2.1699e-01            0   1.1123e-01   1.0011e-01  -6.9884e-02  -3.0749e-01

Columns 7 through 9
          0            0   6.5684e-06
```

The swing mode is clearly observable in the ΔV_4 signal.

(2) The root-locus method will be used to design the damping controller using V_4 as the input signal. Denote $G(s)$ as the linearized model of the power system with u_{sup} as the input and ΔV_4 as the output.

The time constant for the washout filter is set at $T_w = 0.1$ sec, and the time constant for the control gain is set at $T_d = 0.1$ sec. The root-locus plot of the transfer function of the uncompensated system

$$G(s)\frac{1}{1+sT_d}G_{wo}(s) \tag{14.38}$$

is shown in Figure 14.30. Note that u_{sup} comes into the controller summing junction with a positive sign.

The angle of departure of the swing mode is about 90°. To change the direction of the angle of departure pointing in the direction of damping improvement, a combined lead compensation G_{ldlg} of 70° is desirable. This phase lead is split into two lead-lag compensators, each with a 35° phase-lead compensation at 7.357 rad/sec, resulting in $T_a = T_c = 0.2611$ sec and $T_b = T_d = 0.07076$ sec.

Next a root-locus analysis is performed for the compensated system

$$G(s)\frac{1}{1+sT_d}G_{ldlg}G_{wo}(s) \tag{14.39}$$

which is shown in Figure 14.31.a. In the enlarged RL plot of Figure 14.31.b, the MATLAB® data cursor function shows that if $K_d = 5530$ is used, the swing mode of the close-loop system will have a damping ratio of 0.134. Thus the gain K_d is set to 6000, which achieves a damping ratio of 0.1499.

Figure 14.31.c shows that if K_d is increased to about 23500, there will be a pair of complex eigenvalues crossing into the right-half plane. Thus using a gain of $K_d = 6000$

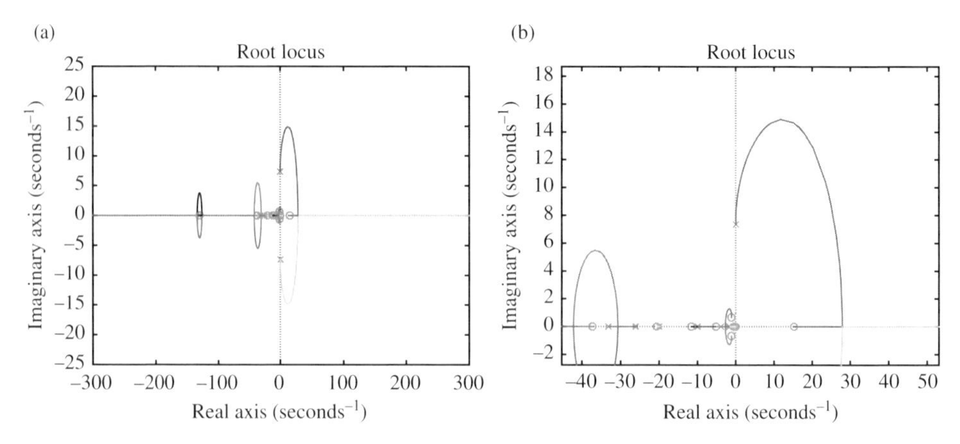

Figure 14.30 (a) Root-locus plot of uncompensated system and (b) zoomed-in plot around the swing mode (see color plate section).

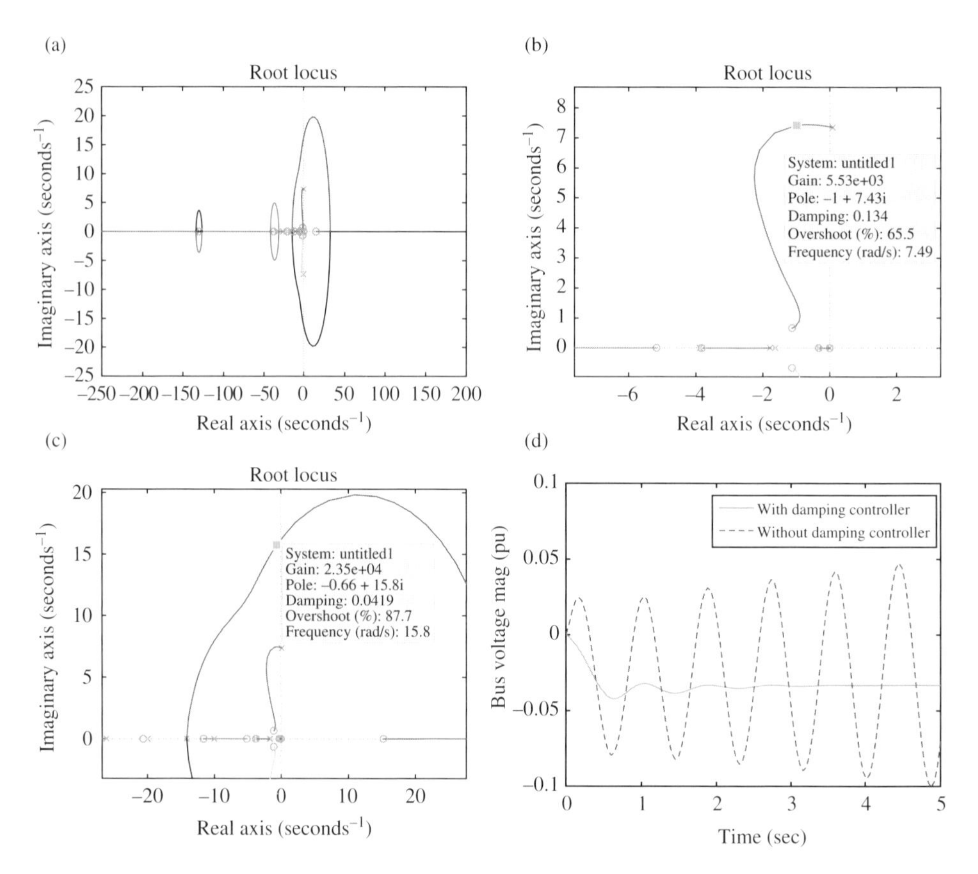

Figure 14.31 (a) Root-locus plot of compensated system, (b) zoomed-in plot around the swing mode, (c) unstable gain, and (d) comparison of compensated and uncompensated systems subject to a 0.05 pu step in excitation system voltage reference (see color plate section).

would provide a gain margin of $20\log_{10}(23500/6000) = 11.86$ dB, which is quite adequate. Using a gain of $K_d = 6000$, the response of the Bus 4 voltage magnitude V_4 in the compensated linearized model due to a 0.05 pu step in the excitation system is shown in Figure 14.31.d. Note that the time response of the uncompensated system is unstable, but the oscillations in the compensated system damped out quite rapidly. ■

14.4 Shunt VSC Controllers

This section and the next section discuss the use of controllers based on a more recent voltage-sourced converter (VSC) technology for AC-DC conversion, allowing them to perform both reactive and active power control. There are several unique properties of the VSC technology, including the following advantages:

1) Unlike a thyristor, a VSC can turn off the conducting current without waiting for the current to cross zero, thus allowing it to perform power factor control.
2) Without requiring full-size capacitors and reactors to provide reactive power support, VSC-based controllers can fit into a smaller footprint such as a substation in an urban area.
3) The DC bus of a VSC-based controller can be connected to the DC bus of another VSC-based controller, allowing active power to circulate between the two controllers. Several shunt-to-shunt and shunt-to-series configurations are possible, allowing for more controllability of the AC power system.

The power-electronics aspect of the components of VSCs can be found in detail in many power electronics books [10, 189] and will not be discussed here. Only operational level descriptions essential for power system applications are provided. This section considers shunt controllers, whereas the next section considers series and coupled controllers. The presentation of these two sections relies on the materials in [10], as it provides clear explanations to this highly complex subject matter.

14.4.1 Voltage-Sourced Converters

The basic component of a VSC consists of a valve with both turn-on and turn-off capabilities. The valve is connected in parallel with a reverse diode, as shown in Figure 14.32. Early implementation of VSC uses a gate turned-off thyristor (GTO), which is a silicon-based thyristor, and more recently, GTO devices have been replaced by insulated gate bipolar transistors (IGBTs) having also turn-on and turn-off capabilities [190]. IGBTs have faster turn-on and turn-off capabilities, thus resulting in smaller losses. A VSC allows bidirectional flow of currents: from A to B if the valve is activated by a gate current, and from B to A through the reverse diode.

The steady-state operation of a VSC is illustrated by the single leg of a full-wave converter circuit connected to the AC grid via a transformer, as shown in Figure 14.33. The DC voltage V_{dc} across the capacitor (also known as the DC bus) is assumed to be constant. Suppose Valve 1 is on, that is, either GTO 1 or Reverse Diode 1′ is on.[5] Under ideal operation, as the voltage drop across the valve is zero, $v_a = V_{dc}/2$ as referenced to

5 Here GTO is used to denote a generic turn-off device controlled by a gate current.

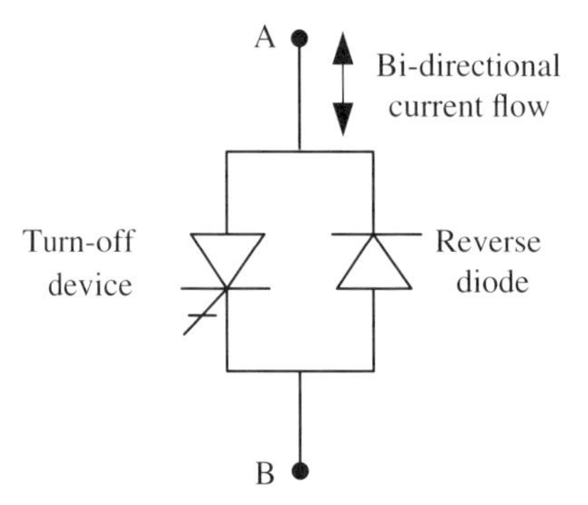

Figure 14.32 VSC valve.

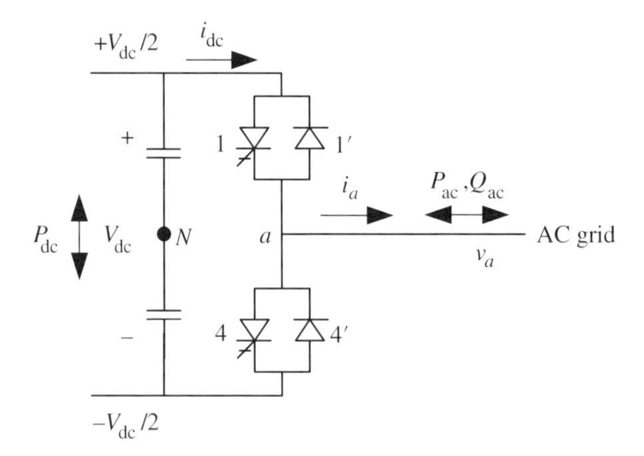

Figure 14.33 A single-phase-leg VSC connected to AC power grid.

the neutral N. Conversely, suppose Valve 4 is on. Then $v_a = -V_{dc}/2$, also referenced to N. Thus as Valves 1 and 4 turn on every half cycle, v_a is a square wave (Figure 14.34.a).

Let i_a be the sinusoidal current flowing to the AC grid through the conducting valves (Figure 14.34.b) and i_{dc} be the DC current leaving the capacitor, with their directions as indicated in Figure 14.33. When i_a is positive and GTO 1 is turned on, i_{dc} is positive. When i_a crosses zero, GTO 1 will no longer conduct and i_a will start to flow in Diode 1′. In this operation, both the currents $i_a = i_{dc}$ are negative.

After half of a cycle, Valve 4 is turned on. This action results in polarity reversal on the voltage v_a to $-V_{dc}/2$. The current i_a will flow in GTO 4 until i_a becomes zero and turns positive. Then the positive current will flow through Diode 4′, until Valve 1 is turned in the next cycle. In this operation, $i_{dc} = -i_a$ as the direction of i_a is opposite to that of i_{dc} (Figure 14.34.c). The voltage across Valve 1 is shown in Figure 14.34.d.

The time period that the diodes are conducting is known as the rectifier mode, as DC current i_{dc} is flowing into the DC voltage source, thus extracting energy from the AC grid. When the GTOs are conducting, the VSCs are in the inverter mode, as i_{dc} is leaving the DC voltage source, thus providing energy to the AC grid. If the DC source voltage is not supported by some kind of energy supply (like energy storage), the rectifying and inverting modes should result in zero energy exchange with the AC grid every cycle.

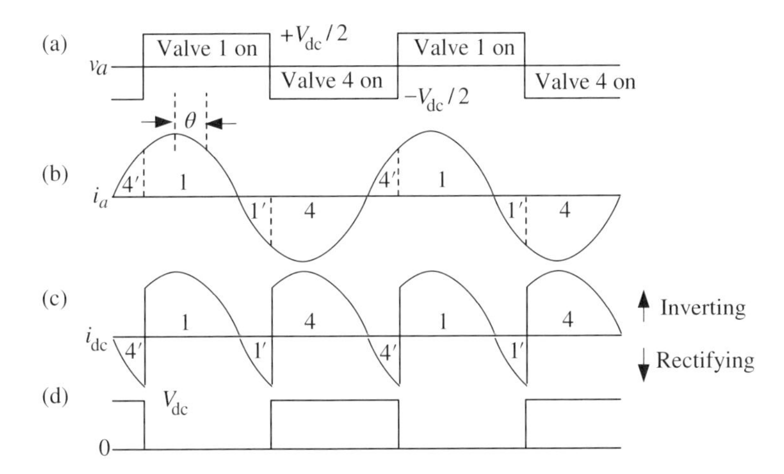

Figure 14.34 Single-phase-leg full-wave VSC operational waveforms: (a) AC voltage with respect to *N*, (b) AC current, (c) DC current, and (d) voltage across Valve 1 [10].

The control of Valves 1 and 4 will determine the power factor of the VSC operation. Consider the phase relationship of v_a and i_a in Figure 14.34. From the square-wave, a sinusoidal component of the fundamental frequency can be extracted for v_a, which peaks at t_{pv}. The peak of i_a at t_{pi} leads v_a by a phase of

$$\theta = (t_{pi} - t_{pv})/T_f \tag{14.40}$$

where T_f is the fundamental period of the AC grid. As $\theta > 0$, the VSC is supplying capacitive power to the AC grid.

Additional reactive power control of VSCs is illustrated via Figure 14.35 using the single phase-leg of the circuit in Figure 14.33. Let the current i_a be sinusoidal and unaffected by the VSC switching. Suppose v_a is controlled such that its fundamental frequency component is in phase with i_a, as in Figure 14.35.b. In other words, the phasors $\tilde{I}_a$ and $\tilde{V}_a$ point in the same direction. The VSC is acting as an inverter, converting DC power to AC active power with unity power factor. Suppose that the VSC is controlled such that v_a is advanced by 1/4 cycle (90°) (Figure 14.35.c). In this case, the phasor $\tilde{V}_a$ leads $\tilde{I}_a$ by 1/4 cycle, such that only inductive power is supplied, that is, the power factor is zero.

Further advancing the voltage v_a by 1/4 cycle (Figure 14.35.d), the voltage phasors $\tilde{V}_a$ is in the direction opposite to $\tilde{I}_a$. In this case, the VSC is acting as a rectifier by converting AC grid active power into DC power. Finally, when v_a is advanced by another 1/4 cycle (Figure 14.35.e), the phasor $\tilde{V}_a$ now lags $\tilde{I}_a$ by 1/4 cycle. In this case, the VSC provides only capacitive power to the AC grid, that is, the power factor is zero.

It is possible to control the valves to provide a full four-quadrant (P, Q) operation provided that there is an energy supply at the DC bus.

14.4.1.1 Three-Phase Full-Wave VSCs

The configuration of VSCs to enable 3-phase operation is shown in Figure 14.36, utilizing six VSCs. The GTOs at each valve pair (1,4), (3,6), and (5,2) are turned on 120° apart, with two GTOs conducting at any one point in time, forming a two-level phase voltage

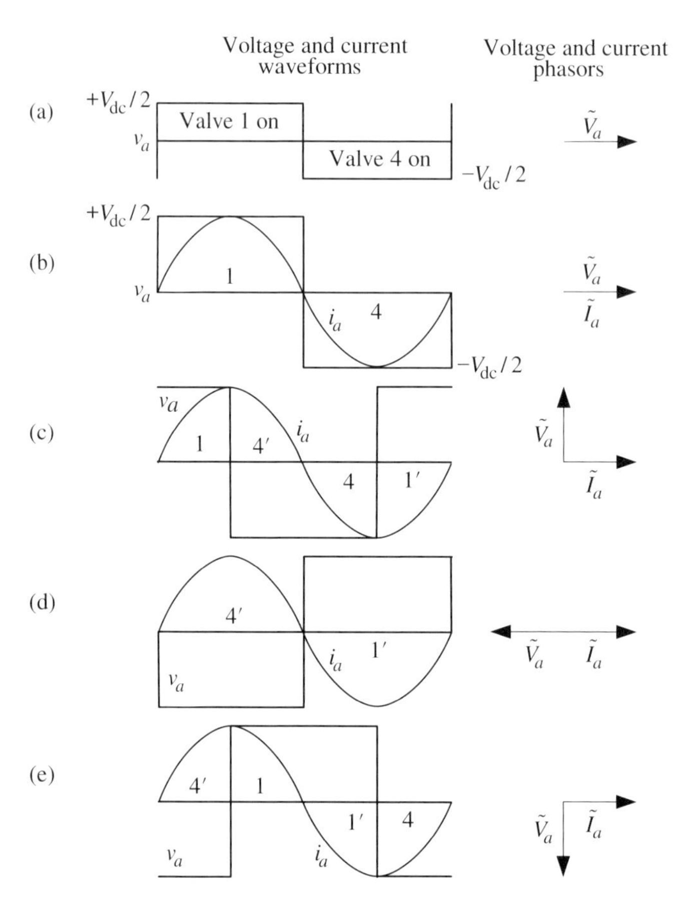

Figure 14.35 Operation of VSC to achieve various power factors: (a) AC voltage with respect to N as controlled by the valves, (b) voltage and current waveforms for inverter unity PF, (c) inductive power injection, (d) rectifier unity PF, and (e) capacitive power injection [10].

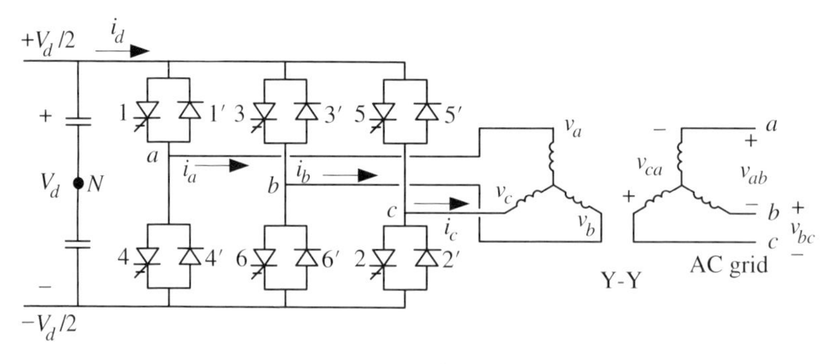

Figure 14.36 Three-phase full-wave VSC circuit diagram.

as shown Figure 14.37.a, in which the voltage drop across an enabled GTO is taken to be zero. Then the phase-to-neutral voltages

$$v_{ab} = v_a - v_b, \quad v_{bc} = v_b - v_c, \quad v_{ca} = v_c - v_a \tag{14.41}$$

the ungrounded neutral voltage

$$v_n = (v_a + v_b + v_c)/3 \tag{14.42}$$

and the phase-to-neutral voltages

$$v_{an} = v_a - v_n, \quad v_{bn} = v_b - v_n, \quad v_{cn} = v_c - v_n \tag{14.43}$$

can be computed, as shown in Figure 14.37.b–d.

Example 14.6

Suppose v_a is controlled such that it lags i_a by 45°. Compute the phase and DC currents for the 3-phase full-wave converter.

Solutions: Assuming that the AC phase current is sinusoidal. The current i_a is shown in the top plot of Figure 14.38. The DC current i_{dc}^a due to i_a is shown in the second plot, which is similar to that in Figure 14.34. Similarly, the DC currents i_{dc}^b and i_{dc}^c due to i_b and i_c, respectively, are shown in the third and fourth plots of Figure 14.38. The net DC current given by

$$i_{\text{dc}} = i_{\text{dc}}^a + i_{\text{dc}}^b + i_{\text{dc}}^c \tag{14.44}$$

is shown in the last plot of Figure 14.38. As the average of i_{dc} is positive, the VSC is providing active power to the AC grid. As the current leads the voltage, the VSC is also providing capacitive power to the AC grid. ■

14.4.1.2 Three-Level Converters

As seen from the AC voltage and DC current waveforms, the harmonics on the waveforms can be quite significant. One way of reducing harmonics is to use 3-level converter circuits, as shown in Figure 14.39.a. The circuit consists of two converter levels per phase-leg, with a clamping diode connecting the midpoint of these two converters to the midpoint of the DC bus denoted by N.

During the time when the GTOs 1A and 4A are turned on while the GTOs 1 and 4 are blocked, v_a will be equal to the neutral voltage. When the GTOs 1 and 4 are turned on during the periods denoted by σ, v_a will have three voltage levels, as shown in Figure 14.39.b. Thus the phase-to-phase voltage v_{ab} will have 5 levels, more closely resembling a sinusoidal waveform.

14.4.1.3 Harmonics

Using Fourier analysis, it can be shown that for a 3-phase, full-wave, 2-level converter, with the DC bus voltages at $\pm V_{\text{dc}}/2$, the phase a to neutral voltage is

$$v_a = v_{aN} = \frac{4}{\pi}\frac{V_{\text{dc}}}{2}\left[\cos(\omega t) - \frac{1}{3}\cos(3\omega t) + \frac{1}{5}\cos(5\omega t) - \frac{1}{7}\cos(7\omega t) + \cdots\right] \tag{14.45}$$

which has all the odd harmonics. The harmonics in phase-b and -c voltages are displaced by $\mp 2\pi/3$ rad, respectively.

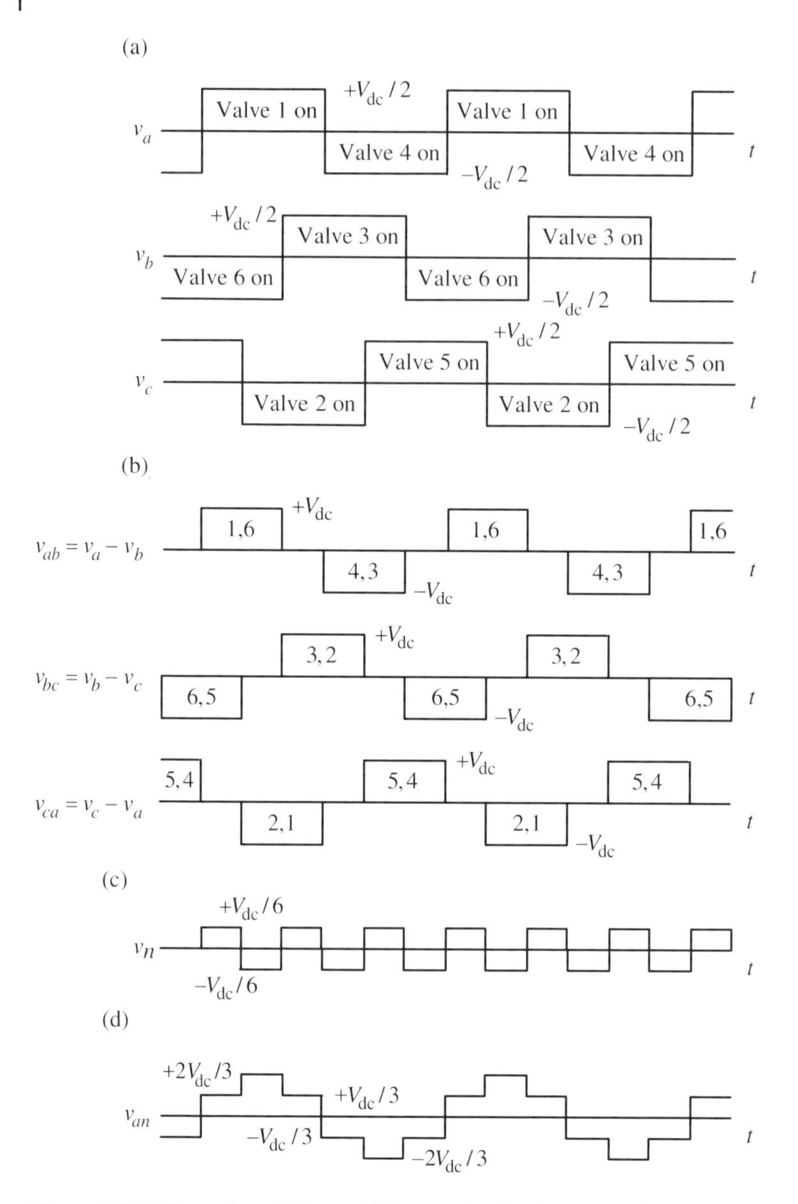

Figure 14.37 Three-phase full-wave VSC operational voltage waveforms: (a) phase voltages, (b) phase-to-phase voltages, (c) neutral voltage, and (d) phase-to-neutral voltage [10].

The phase-to-phase voltage is given by

$$v_{ab} = \frac{2\sqrt{3}}{\pi} V_d \left[\cos(\omega t) - \frac{1}{5}\cos(5\omega t) + \frac{1}{7}\cos(7\omega t) + \cdots - \frac{1}{11}\cos(11\omega t) + \cdots \right] \tag{14.46}$$

in which only the $6n \pm 1$ (triplen) harmonics are left.

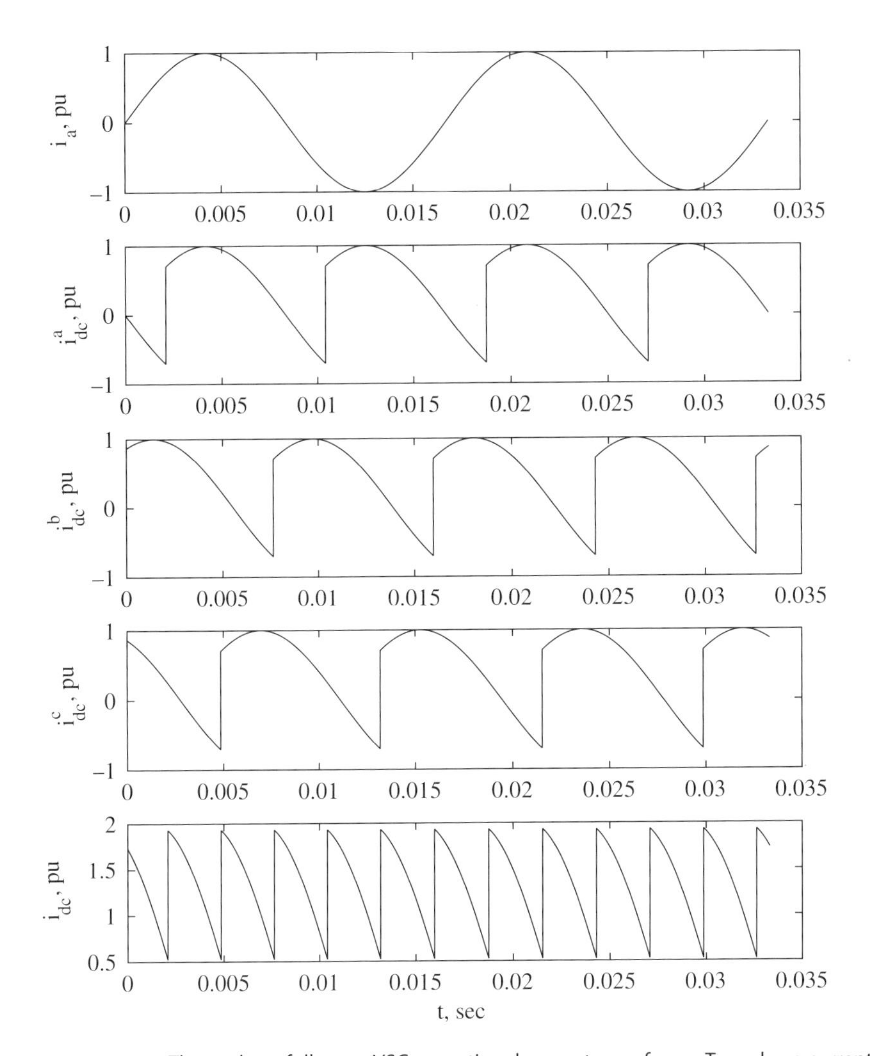

Figure 14.38 Three-phase full-wave VSC operational current waveforms. Top: phase current, (2-4) DC currents due to individual phases. Bottom: total DC bus currents [10].

On the DC side, the DC current i_{dc}, which is the sum of i_a, i_b, and i_c, can be expressed as

$$i_{\mathrm{dc}} = I_{\mathrm{dc}} + \text{harmonics of } 6k, \quad k = 1, 2, \dots \tag{14.47}$$

where

$$I_{\mathrm{dc}} = \frac{2\sqrt{3}}{\pi} I_{\mathrm{ac}} \cos\phi \tag{14.48}$$

where I_{ac} is the rms value of the AC current and ϕ is the power factor. The kth harmonic current is a function of the power factor. It is minimum for unity power factor, given by

$$I_k/I_{dm} = \sqrt{2}/(k^2 - 1) \tag{14.49}$$

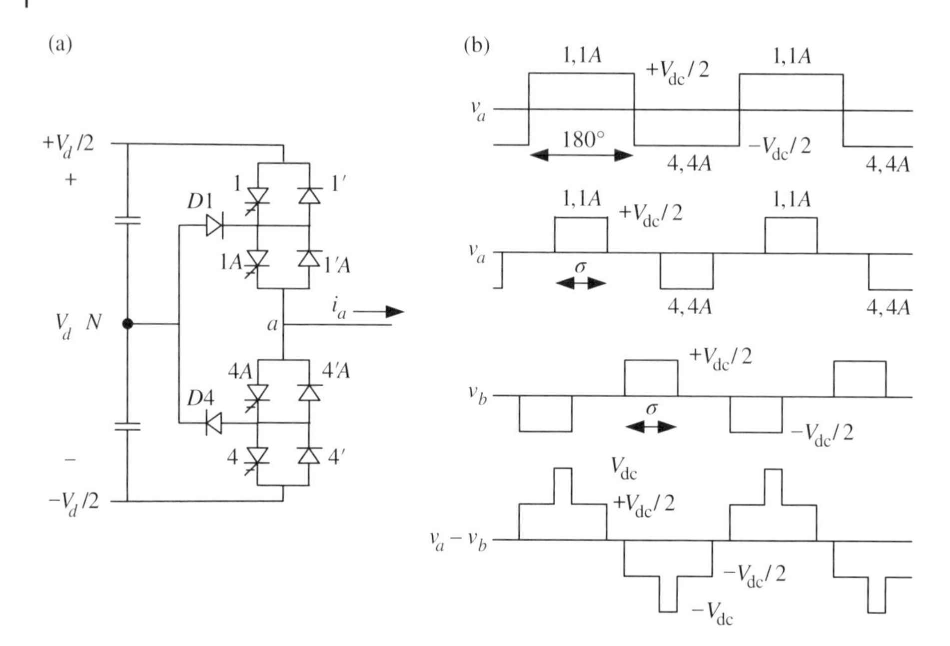

Figure 14.39 Three-level VSC converters: (a) circuit diagram and (b) AC voltage waveforms.

where I_{dm} is the peak value of DC bus current. It increases to the maximum

$$I_k/I_{dm} = \sqrt{2}k/(k^2 - 1) \tag{14.50}$$

when the power factor is zero.

Harmonics reduction can be achieved using a three-level converter circuit and 24- and 48-pulse operations. It is also possible to use pulse-width modulation (PWM), instead of line commutation, to control the VSC and reduce harmonic components.

14.4.2 Static Compensator

14.4.2.1 Steady-State Analysis

The shunt connection of a VSC to a high-voltage bus via a step-up transformer is known as a static compensator (STATCOM) and is shown in Figure 14.40.a. The STATCOM acts as a current source $\tilde{I}_{\text{sh}}$ to support the bus voltage $\tilde{V}_{\text{sh}}$, as shown in Figure 14.40.b, where X_T is the transformer reactance. It can inject either capacitive or inductive currents to control the bus voltage V_{sh} on the low-side of the transformer or the voltage V at the high-side of the transformer. Thus STATCOM functions like an SVC, except that the size of the capacitor to maintain the DC bus voltage of the STATCOM is much smaller than the capacitor used in an SVC. Thus a STATCOM has a smaller footprint than an SVC and can fit into urban substations with limited space.

Assuming that the shunt transformer is lossless and there is no active power injection by the STATCOM, the angle θ_{sh} of the injected voltage phasor $\tilde{V}_{\text{sh}} = V_{\text{sh}}\angle\theta_{\text{sh}}$ is equal to the angle θ of the bus voltage phasor $\tilde{V} = V\angle\theta$, such that the injected current is

$$\tilde{I}_{\text{sh}} = (\tilde{V}_{\text{sh}} - \tilde{V})/(jX_T) = ((V_{\text{sh}} - V)/X_T)\angle(\theta - \pi/2) = I_{\text{sh}}\angle(\theta - \pi/2) \tag{14.51}$$

that is, $\tilde{I}_{\text{sh}}$ is in quadrature to the bus voltage phasor $\tilde{V}$.

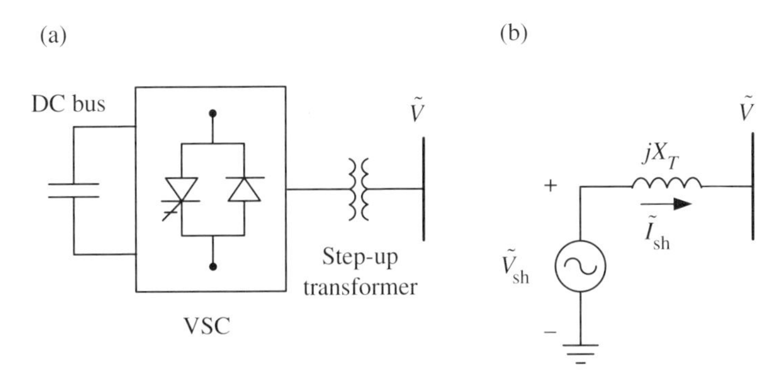

Figure 14.40 STATCOM: (a) schematic and (b) equivalent circuit diagram.

For the STATCOM to inject capacitive power into the system, the magnitude of the voltage source V_{sh} is increased to be greater than the bus voltage V. This is accomplished by increasing the voltage on the DC bus in the VSC. On the other hand, the voltage V_{sh} of the STATCOM can be lowered to below V to inject inductive power into the system.

As in the SVC, the regulation of the terminal bus voltage V is accomplished by

$$V_{sh} = V_{ref} - \alpha I_{sh} \tag{14.52}$$

where V_{ref} is the reference voltage value for the VSC and α is the droop, nominally between 3% and 5%.

With droop control, the steady-state operational diagrams of the STATCOM with respect to the reactive current and power are shown in Figure 14.41. As in the SVC, capacitive current and power would point to the left of the origin, and inductive current and power would point to the right. The current I_{sh} is taken to be positive if capacitive current is injected into the system. In this notation, the droop curve is a straight line with a positive slope. At maximum capacitive power injection, the STATCOM becomes a constant current device until the system voltage drops so low that constant current can no longer be maintained (Figure 14.41.a). With saturation at constant current, the reactive power output of the STATCOM decreases linearly as shown in Figure 14.41.b. Comparing the STATCOM capability in Figure 14.41 and the SVC capability in Figure 14.9, one can observe that a STATCOM is more effective than an SVC in low-voltage situations.

In power flow calculation, STATCOM is modeled as a generator bus with active power equal to zero. Its voltage V_{sh} is specified and its reactive power output limited by its rating. The droop is not taken into account in the power flow solution, as it is assumed that V_{ref} can be set properly to satisfy (14.51). If the maximum reactive power is reached, then the STATCOM bus is set to a *PQ* bus with zero active power and the reactive power set equal to its limit.

14.4.2.2 Dynamic Model

To assess the impact of a STATCOM on the stability of a power system, a dynamic model is needed to capture the response of its control system. In positive-sequence dynamic

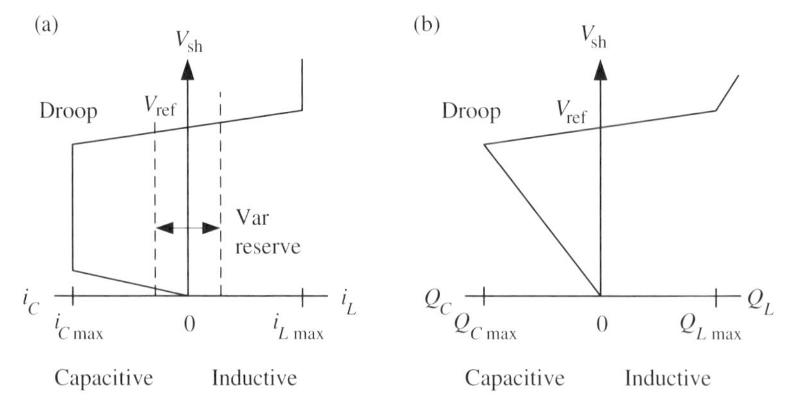

Figure 14.41 STATCOM operational diagram as a function of (a) injected current and (b) injected reactive power.

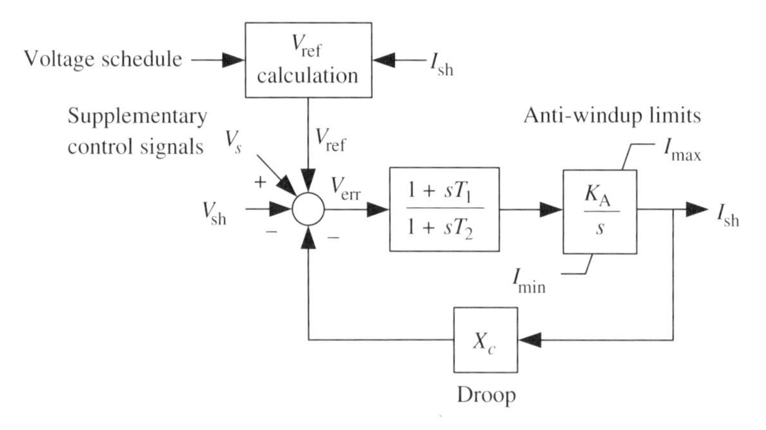

Figure 14.42 STATCOM dynamic model.

simulation, the action of the converters can be assumed to be instantaneous, as otherwise the small time constants associated with the converter switchings may require a very small integration stepsize.

A simplified control model is shown in Figure 14.42, in which the error signal

$$V_{err} = V_{ref} - V_{sh} - X_c I_{sh} \tag{14.53}$$

is filtered and used to set the STATCOM current I_{sh} through an integrator with a gain K_A, where V_{ref} is the reference voltage value set to provide the desired voltage V_{sh} at the STATCOM bus. Note that $X_c = \alpha$ plays the role of droop regulation, as the droop is interpreted as a reactance. The signal V_s allows the input of a damping control signal, similar to that in the SVC. In addition, this control diagram also shows a mechanism, using proportional and integral gains and var reserve limits, to adjust V_{ref} according to the STATCOM current output I_{sh} and the voltage schedule provided by the system dispatch.

Note that I_{sh} is a real quantity and is positive when capacitive current is provided by the STATCOM and negative for inductive current. In the dynamic simulation process, the STATCOM current, denoted as a phasor $\tilde{I}_{sh}$, is denoted as a constant current source, so that the network solution part will treat it as a non-conforming load. Iterations are needed to obtain the network solution.

Neglecting the time constants T_1 and T_2 of the filter, a small-signal transfer function from the bus voltage to the STATCOM current can be obtained as

$$\Delta I_{\text{sh}} = -\frac{K_A/s}{1 + X_c K_A/s}\Delta V_{\text{sh}} = -\frac{1/X_c}{1 + s/(X_c K_A)}\Delta V_{\text{sh}} = -\frac{K_{\text{sc}}}{1 + T_{\text{sc}} s}\Delta V_{\text{sh}} \tag{14.54}$$

which is a first-order system with a gain of $K_{\text{sc}} = 1/X_c$, the inverse of droop, and a time constant of $T_{\text{sc}} = 1/(X_c K_A)$, which is usually small. Thus when installed at a critical location in a power grid, a STATCOM is expected to provide a fast response to maintain the voltage at that location. As pointed out in the SVC discussion, STATCOM can also improve transient stability, voltage stability, and damping enhancement of swing modes. In particular, the supplementary input signal V_s in Figure 14.42 can be used to accommodate a damping signal.

To illustrate the performance of a STATCOM rated at 200 MVA, the phasor measurement recording of the high-side transformer voltage and the current injection during a disturbance will be shown and used to develop the transfer function (14.54) [191]. The phasor data rate is 30 samples per second.

The disturbance was a generating unit trip, which resulted in a rise in voltages at substations around the STATCOM installation. Figure 14.43 shows the response of the voltage magnitude and current injection after the disturbance, which occurred at about 69 sec.[6] Note that the voltage response showed an increase in magnitude in the post-fault condition, in addition to a few oscillations due to electromechanical swings. Before

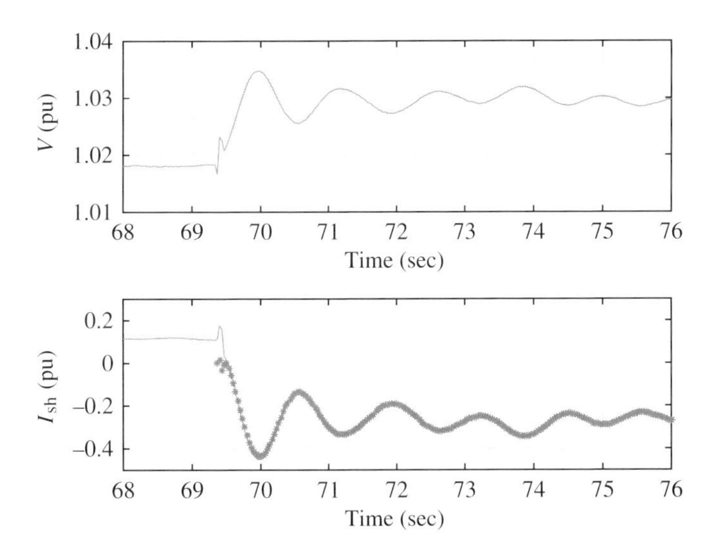

Figure 14.43 STATCOM disturbance response: (upper plot) measured high-side bus voltage magnitude, and (lower plot) measured STATCOM reactive current injection (solid curve): capacitive current is positive and inductive current is negative. The lower plot also shows the approximation achieved by the estimated transfer function (dotted curve) (see color plate section).

6 The time of the disturbance has no particular significance, as the first time tag of data set is arbitrarily set to zero.

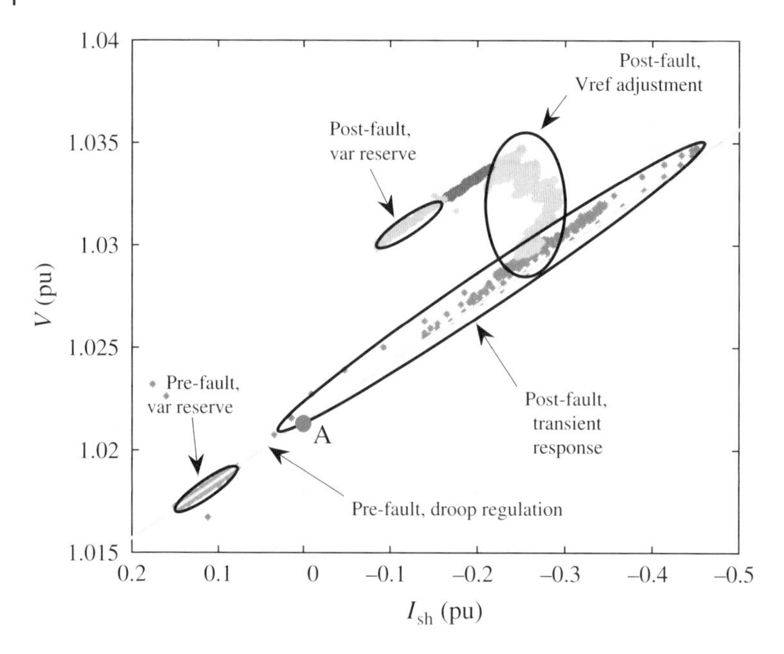

Figure 14.44 Pre-fault and post-fault STATCOM droop characteristics (see color plate section).

the disturbance, the STATCOM was providing capacitive power support with $I_{sh} > 0$. However, in the post-fault period, because of the increase in voltage magnitude, the STATCOM started to absorb capacitive power, such that $I_{sh} < 0$, that is, the current is inductive.

To show the droop regulation, the disturbance-response bus voltage magnitude is plotted against the STATCOM current injection, as shown in Figure 14.44, for the pre-fault and post-fault operating conditions. Initially in the pre-fault region, the STATCOM was responding to random system loading variations by supplying a small amount of capacitive current to the power network. A droop regulation line computed from the pre-fault PMU data is shown in Figure 14.44, with a slope of 0.0311 pu power/pu current. Point A is the value of the pre-fault $V_{ref} = 1.0214$ pu. After the generating unit was stripped, the STATCOM reacted to the higher voltage by injecting inductive current I_{sh}, which followed closely the droop regulation curve. One can also observe from the power swings that when the bus voltage increased, the current I_{sh} became more negative.

As the post-fault I_{sh} was outside of the Var reserve region (about ±15% of the STATCOM rating), the value of V_{ref} was readjusted until a new post-fault droop regulation line was formed, which was raised from the pre-fault droop regulation line because the post-fault V_{ref} is higher than the pre-fault value. The region of V_{ref} adjustment is also indicated in Figure 14.44.

The measured voltage and current response can be used to estimate the transfer function (14.54). Using the transfer function estimation function *tfest* in MATLAB®, a simplified STATCOM control system transfer function was found to be

$$-\frac{K_{sc}}{1+T_{sc}s} = -\frac{32.2}{1+0.0329s} \tag{14.55}$$

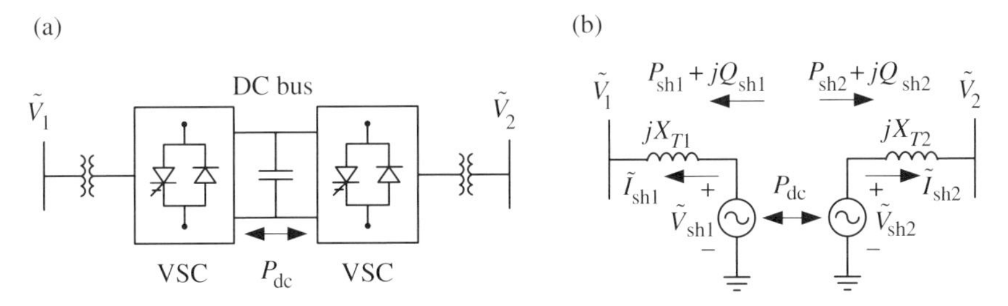

Figure 14.45 B2B HVDC system: (a) schematic and (b) equivalent circuit.

Thus for this particular STATCOM, the droop is $1/32.2 = 3.11\%$, which agrees with the pre-fault droop curve. The time constant $T_{sc} = 32.9$ msec signifies a fast response time (about 2 cycles), as compared to generator excitation system control.

The voltage magnitude perturbation from the measured disturbance data is passed through the estimated transfer function. The resulting current is shown in Figure 14.43.

14.4.3 VSC HVDC Systems

Chapter 13 describes the operation of LCC HVDC systems based on thyristors. An alternative HVDC system can be developed by coupling the DC buses of shunt VSCs. These systems are called VSC HVDC systems, back-to-back (B2B) HVDC systems, or B2B STATCOMs. They are commercially known as HVDC Light[7] or HVDC Plus[8].

The schematic and equivalent circuit of a VSC HVDC system are shown in Figure 14.45, illustrating the coupling of the DC buses of the two shunt VSCs. With the ability to exchange active power P_{dc} from one VSC to the other VSC, the injected currents I_{sh1} and I_{sh2} no longer have to be reactive only. The two VSCs can be located in nearby substations, or the DC buses can be connected via a DC transmission line.[9] In the latter case, the DC line resistance needs to be considered in loss calculation.

14.4.3.1 Steady-State Operation

With an active power source, the injected current $\tilde{I}_{shi}$, $i = 1, 2$, will no longer have to be in quadrature to its terminal bus voltage $\tilde{V}_i$. Figure 14.46 illustrates the relationship between $\tilde{I}_{shi}$ and $\tilde{V}_i$, with $\tilde{V}_i$ drawn to point upward [10]. The voltage drop across the transformer

$$\tilde{V}_{\ell i} = jX_{Ti}I_{shi} = \tilde{V}_{shi} - \tilde{V}_i \tag{14.56}$$

is still in quadrature to $\tilde{I}_{shi} = I_{shi}\angle\phi_{shi}$, where $\tilde{V}_{shi}$ is the output voltage phasor of VSC i. The apparent power injected into the terminal bus of VSC i is given by

$$\begin{aligned} P_i + jQ_i &= \tilde{V}_i\tilde{I}^*_{shi} = V_iI_{shi}\angle(\pi/2 - \phi_{shi}) \\ &= V_iI_{shi}\cos(\pi/2 - \phi_{shi}) + jV_iI_{shi}\sin(\pi/2 - \phi_{shi}) \end{aligned} \tag{14.57}$$

---

7 Trademark of ABB.

8 Trademark of Siemens.

9 The 36 MVA VSC Eagle Pass HVDC system is an asynchronous link connecting the Eagle Pass substation in Texas with the Peidras Negras substation in Mexico. The two substations are literally right next to each other. The 330 MW VSC Cross Sound Cable HVDC system connects the Halvarsson substation in Connecticut to the Shoreham substation on Long Island. The DC link is a 40 km submarine cable.

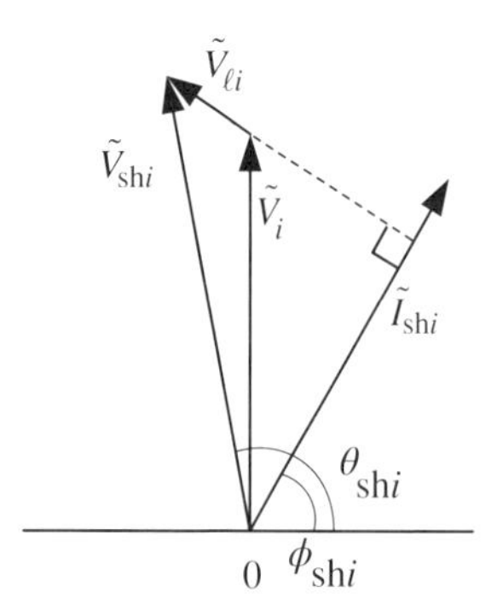

Figure 14.46 Voltage and current relationship of a shunt VSC with active power injection.

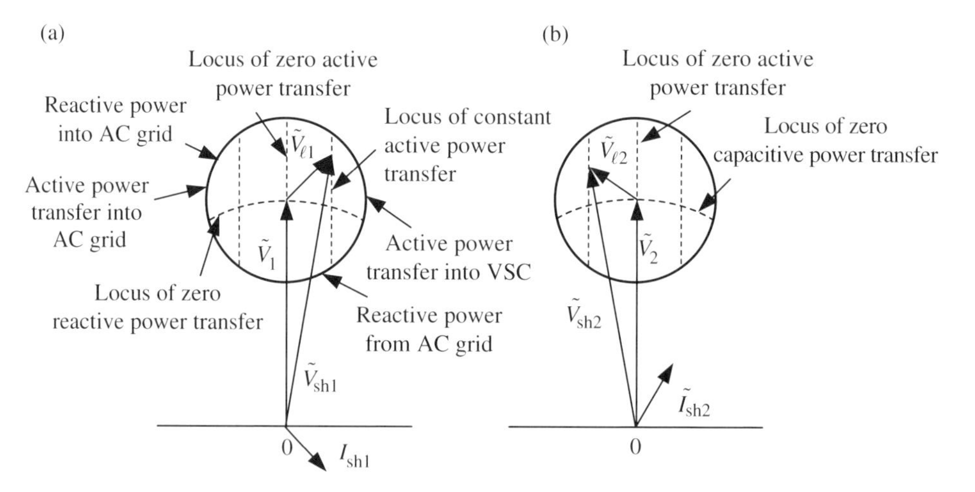

Figure 14.47 Operational regions of a VSC HVDC system: (a) VSC in rectifier mode and (b) VSC in inverter mode [10].

such that the VSC is injecting active power into the AC network (inverter mode) if $|\pi/2 - \phi_{shi}| < \pi/2$ (i.e., $0 < \phi_{shi} < \pi$) and injecting capacitive power if $0 < \pi/2 - \phi_{shi} < \pi$ (i.e., $|\phi_{shi}| < \pi/2$). Figure 14.47 shows the relationship of $\tilde{I}_{shi}$ with respect to $\tilde{V}_{shi}$ in the rectifier and inverter mode for the HVDC system in Figure 14.45. For convenience the rectifier and inverter voltage phasors $\tilde{V}_{shi}$ are both plotted pointing upward, which may not be the case if the rectifier and inverter nodes are synchronously connected.

An important consideration in shunt VSC operation is its operating constraints, one of which is the current limit $|I_{max}|$ of the VSC. The current limit constraint places a bound on the voltage drop across the transformer

$$|\tilde{V}_{\ell i}| = |jX_{Ti}I_{shi}| \leq X_{Ti}I_{max} \tag{14.58}$$

The feasible regions of operation of a VSC HVDC system are illustrated in Figure 14.47. Due to the maximum current constraints (14.58), the voltage difference between $\tilde{V}_{shi}$ and $\tilde{V}_i$ lies in a circle centered at $\tilde{V}_i$, $i = 1, 2$. The circle limiting the voltage difference is bisected into two halves by a line collinear with $\tilde{V}_i$. On this line there is no active power transfer because the current in the VSC is in quadrature to the bus voltage. In the right half-circle, power is supplied to the VSC from the AC grid, that is, a rectifying operation. In the left half-circle, power is supplied from the VSC to the AC grid, an inverting operation. In Figure 14.47, Bus 1 is in the rectifier mode and Bus 2 is in the inverter mode. These modes can also be determined from the direction of $\tilde{I}_{shi}$. One can also readily determine the direction of power transfer by comparing $\tilde{V}_i$ and $\tilde{V}_{shi}$: if $\tilde{V}_i$ leads $\tilde{V}_{shi}$, then VSC i is in the rectifier mode; otherwise, it is in the inverter mode.

Figure 14.47 shows that $V_{shi} > V_i$, $i = 1, 2$. Thus both VSCs are injecting capacitive power into the AC network.[10]

The above discussion can be further generalized by observing in Figure 14.47 that the actual power transfer from the VSC bus to the terminal bus is

$$P_{dci} = V_{shi} V_i \sin(\theta_{shi} - \pi/2)/X_T \tag{14.59}$$

Thus, assuming that V_i is fixed (a stiff-voltage-source assumption), P_{dc} will remain the same as long as $V_{shi} \sin(\theta_{shi} - \pi/2)$ remains the same. This expression becomes a constant active power transfer line shown inside the operating region in Figure 14.47, which is parallel to $\tilde{V}_i$.

For steady-state operation of a VSC HVDC system, although the two VSCs can control the magnitude and phase of the individual injected currents $\tilde{I}_{sh1}$ and $\tilde{I}_{sh2}$, the degree of freedom is reduced from four to three because of the active power circulation constraint. If the two VSCs are geographically apart and connected by a DC line, the losses on the DC line should be accounted for in the circulating power calculation. Once the DC power transfer is scheduled, the magnitude and phase of the injected currents are determined to provide the reactive power support and voltage control at the rectifier and inverter terminal buses.

Compared to thyristor-based HVDC systems, VSC HVDC systems not only control the DC power, but also provide voltage regulation of the terminal buses. In addition, power reversal on a VSC HVDC system can be accomplished very quickly, simply by controlling the phase and sequence of the firing of the individual VSCs.

Power flow computation

The impact of a VSC HVDC system on the steady-state power flow can be computed by specifying both the rectifier and inverter buses as generator (*PV*) buses. The active power injection of the rectifier bus is equal to the desired power transfer and carries a negative sign, and the bus magnitude is a desired voltage value controlled by the VSC to a reference value. The active power injection at the inverter bus is positive, signifying a power import, and the bus magnitude is also a desired voltage value controlled by the VSC to a reference value. The active power injected into the AC network by the inverter may be set at a lower value than the rectifier power injection to account for DC line resistive losses, which can be expressed as a function of the DC power transfer. In addition, the upper and lower limits of the reactive power generation for the rectifier and inverter buses can be entered into the bus data. These limits will be dependent on the active power loading on the converters. If any of these limits is violated, the PV bus type will be converted to a *PQ* bus type.

---

10 An arc with V_i as the radius in the operating circle serves as the zero capacitive power injection line (see Figure 14.47). If V_{shi} is above the arc, then the VSC is providing capacitive power. Otherwise, the VSC is providing inductive power.

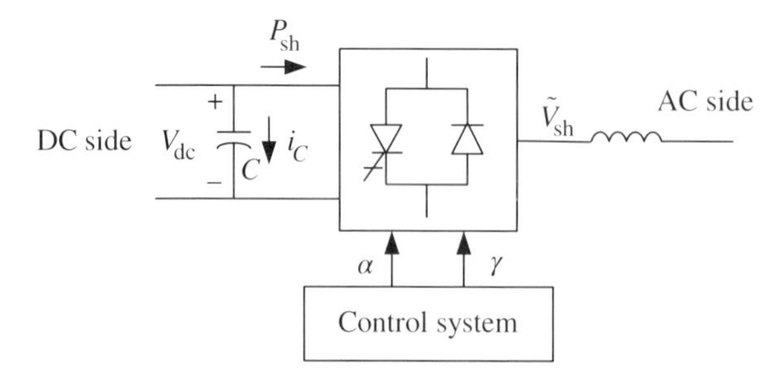

Figure 14.48 VSC control schematic.

Note that in contrast to an LCC HVDC system power flow, the VSC HVDC system power flow does not require an iterative AC-DC solution method because such VSC HVDC systems can control the terminal bus voltage magnitude.

The solution of the power flow can be used to initialize the dynamic VSC HVDC system models, including computing the reference values using droop regulation. These aspects are discussed in the next section.

14.4.3.2 Dynamic Model

The dynamic model of the VSC HVDC system is used to capture the transient behavior of its control system during a disturbance. In a positive-sequence network simulation environment, VSC switching effects will be neglected. A high-level block diagram of the VSC (Figure 14.48) shows the DC-AC interface and the control system. The key VSC control variables include the power angle α and the voltage amplitude gain γ, such that the terminal voltage phasor becomes

$$\tilde{V}_{\text{sh}} = \gamma V_{\text{dc}} e^{j\alpha} \tag{14.60}$$

The operation of the two VSCs is coordinated with regard to active power transmission but they work independently on reactive power control. For active power control, the objective is to maintain the desired power transfer. For reactive power control, each converter can either supply a constant level of reactive power or regulate the converter bus voltage. The operation modes of a VSC HVDC system are described next.

DC Capacitor Model

During transients, the DC capacitor voltage would not be constant. From Figures 14.45 and 14.48, its current I_C is given by

$$I_C = \frac{-(P_{\text{sh1}} + P_{\text{sh2}})}{V_{\text{dc}}} \tag{14.61}$$

where $P_{\text{sh1}} + P_{\text{sh2}}$ is the net AC power injected into the HVDC system. In steady state, $P_{\text{sh2}} = -P_{\text{sh1}}$ if losses are neglected. Then $I_C = 0$. Following (14.61), the change in the DC capacitor can be modeled by the nonlinear equation

$$\frac{dV_{\text{dc}}}{dt} = \frac{1}{C}\frac{-(P_{\text{sh1}} + P_{\text{sh2}})}{V_{\text{dc}}} \tag{14.62}$$

The block diagram of (14.62) is shown in Figure 14.49.

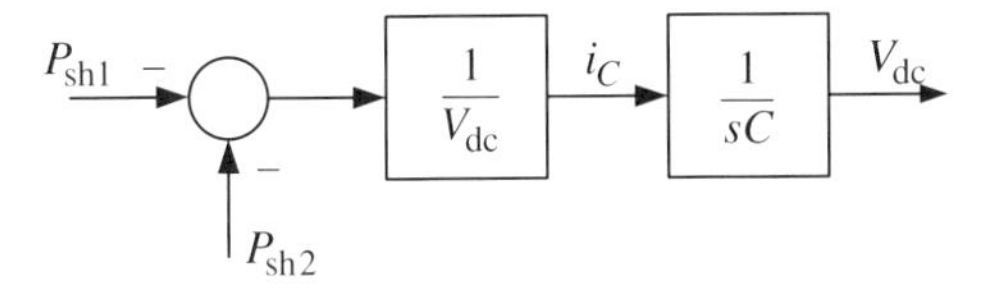

Figure 14.49 DC link dynamics.

Regulation Modes

The two VSCs are coordinated with regard to active power transmission but work independently on reactive power control. For active power control, one of the VSCs controls the active power injection and the other VSC regulates the DC capacitor voltage. The control modes are listed as follows.

Active Power Flow Regulation: One of the DC converter terminals will regulate the scheduled active power transfer, which is achieved by a proportional-integral control. Figure 14.50 shows the block diagram of the rectifier (VSC1) in the active power regulation mode. In Figure 14.50, P_{sh1ref} and P_{sh1} are the desired and actual (measured) DC power flow, respectively, K_{p1} and K_{i1} are the proportional and integral control gains, respectively, T_1 is a time constant representing filtering and commutating effects, and α_{sh1} is the control angle, which when combined with the high-side of the step-up transformer bus voltage angle θ_1 will be the phase angle α_1 of the VSC voltage $\tilde{V}_{\text{sh1}}$. This feedback loop will control the active power P_{sh1} on the transformer to the desired value P_{sh1ref}. Note that both P_{sh1ref} and P_{sh1} are taken to be positive values. If the DC power transfer P_{sh1} is less than the desired power transfer level P_{sh1ref}, α_{sh1} will increase such that $\theta_1 - \alpha_1$ will have a higher value, allowing more AC active power to flow into the DC system.

DC Voltage Regulation: The other DC converter terminal will regulate the scheduled voltage on the DC link, which can also be achieved by a proportional-integral control. Figure 14.51 shows the block diagram of the inverter (VSC2) in the DC voltage regulation mode. In Figure 14.51, V_{dcref} and V_{dc} are the desired and actual (measured) DC voltages, respectively, K_{p2} and K_{i2} are the proportional and integral control gains, respectively, T_2 is a time constant representing filtering and commutating effects, and α_{sh2} is the control angle, which when combined with the high-side of the step-up transformer bus voltage angle θ_2 will be the phase angle α_2 of the VSC voltage $\tilde{V}_{\text{sh2}}$. This feedback loop will control the DC link voltage V_{dc} to the desired value V_{dcref}. If $V_{\text{dc}} < V_{\text{dcref}}$, α_{sh2} will increase. Thus $\alpha_2 - \theta_2$ will be smaller, such that less DC power is transmitted, allowing the DC capacitor to accumulate additional charges.

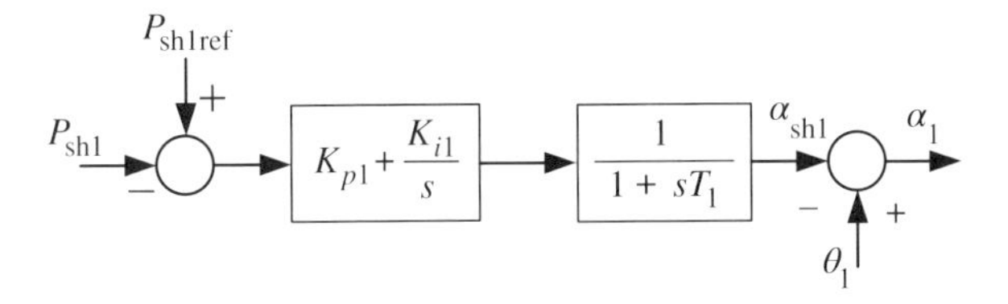

Figure 14.50 Rectifier VSC in active power control mode.

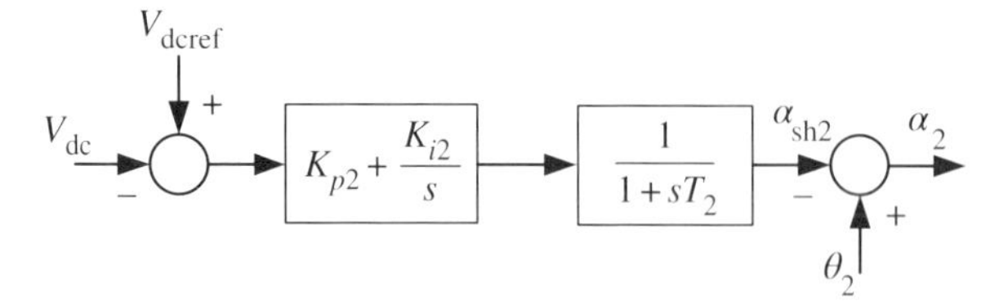

Figure 14.51 Inverter VSC in DC voltage control mode.

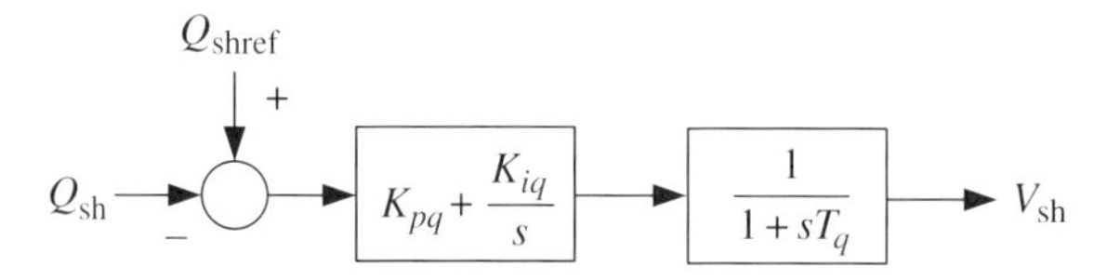

Figure 14.52 VSC in reactive power control mode.

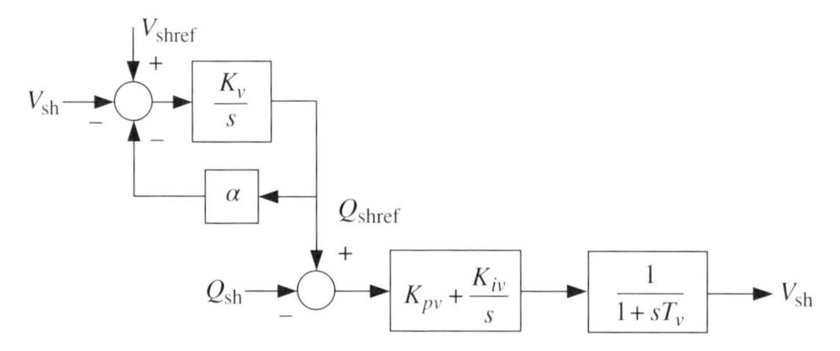

Figure 14.53 VSC in bus voltage control mode.

Reactive Power Regulation: The two VSCs are also able to control the reactive power output to the AC system. There are typically two modes of operation. The first mode is to regulate the reactive power output using a proportional-integral control, as shown in Figure 14.52. In Figure 14.52, Q_{shref} and Q_{sh} are the desired and actual (measured) reactive power output, respectively, K_{pq} and K_{iq} are the proportional and integral control gains, respectively, T_q is a time constant representing filtering and commutating effects, and V_{sh} is the VSC voltage on the AC side. This feedback loop will control the reactive power output Q_{sh} to the desired value Q_{shref}. If $Q_{sh} < Q_{shref}$, V_{sh} will increase, implying that the VSC will provide additional reactive power to the AC system.

Converter Bus Voltage Regulation: Instead of just keeping reactive power output to a fixed value, the VSC can also be used to control the converter bus on the AC side. The control block diagram is shown in Figure 14.53, in which the deviation of V_{sh} from V_{shref} is used to generate a new reference Q_{shref} through an integral gain K_v and a droop of α. As in the case of other droop functions, α allows the sharing of reactive power control within a certain voltage control region. If $V_{sh} < V_{shref}$, Q_{shref} will increase, which in turn will increase V_{sh}.

14.5 Series and Coupled VSC Controllers

14.5.1 Static Synchronous Series Compensation

14.5.1.1 Steady-State Analysis

When a VSC is inserted in series in a transmission line, it can control the power flow by providing a voltage source into the transmission line. This is in contrast to series capacitor compensation in which the power flow is controlled by adjusting the equivalent reactance of the line. This series VSC controller is known as a static synchronous series compensator (SSSC) consisting of a standalone VSC connecting into a transmission line via a series transformer, as shown in Figure 14.54.a, in which the transmission line resistance is neglected. The equivalent circuit of an SSSC is shown in Figure 14.54.b in which the VSC is represented by a voltage source $\tilde{V}_q$ in series with the series transformer winding reactance X_s, which is much smaller than the leakage reactance of a shunt transformer. For a simplified analysis, X_s can be neglected.

Because the DC bus of an SSSC is not connected to an energy supply, the inserted voltage $\tilde{V}_q$ is in quadrature to the line current $\tilde{I}_{se}$. Assume that Buses 1 and 2 are stiff voltage buses with equal voltage magnitude, shown in Figure 14.55, with $\tilde{V}_1 = V\angle\theta/2$ and $\tilde{V}_2 = V\angle-\theta/2$. The angles of these phasors are measured relative to the vertical axis. Then the line current $\tilde{I}_{se}$ will be pointing upward with a phase of zero and is a function of the inserted voltage $\tilde{V}_q$. The inserted voltage $\tilde{V}_q$ will be collinear with the voltage $\tilde{V}_L$ across the transmission line. It has a phase of $\pm\pi/2$. As drawn in Figure 14.55, $\tilde{V}_q$ leads $\tilde{I}_{se}$ by a phase of $\pi/2$ and effectively increases the magnitude of $\tilde{V}_1$, hence also increasing the magnitude of the voltage $\tilde{V}_L$.

With the SSSC and the inserted voltage $\tilde{V}_q$ in phase with $\tilde{V}_L$, the current $\tilde{I}_{se}$ becomes

$$\tilde{I}_{se} = \frac{(\tilde{V}_1 + \tilde{V}_q) - \tilde{V}_2}{jX_L} = \frac{2V\sin(\theta/2) + V_q}{X_L} \tag{14.63}$$

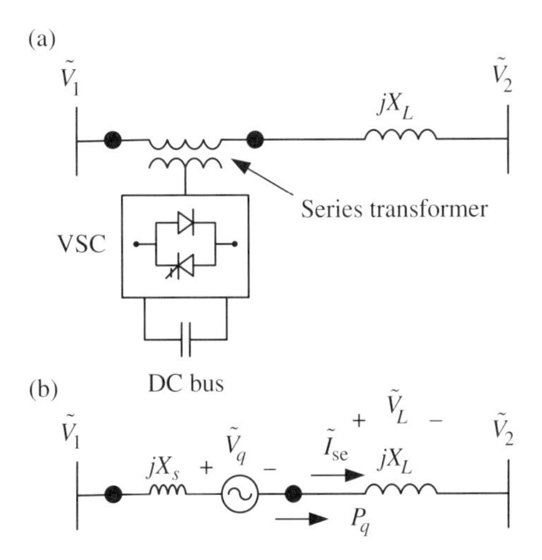

Figure 14.54 SSSC model: (a) schematic and (b) equivalent circuit.

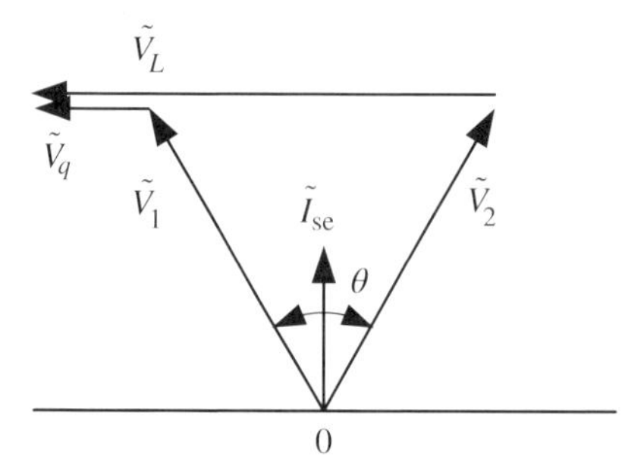

Figure 14.55 Voltage and current diagram of an SSSC.

Thus the active power transfer on the transmission line becomes

$$P = \text{Re}(\tilde{V}_1 \tilde{I}_{se}^*) = \frac{V^2 \sin\theta + VV_q \cos(\theta/2)}{X_L} = \frac{V^2 \sin\theta}{X_L} + P_q \tag{14.64}$$

an increase of $P_q = VV_q \cos(\theta/2)/X_L$ due to the presence of V_q. Thus during a power system disturbance, if an SSSC is turned on in the post-fault period to facilitate additional power transfer, the system stability margin can be improved.

The reactive power injected into the transmission line by the SSSC is

$$Q_q = \text{Im}(\tilde{V}_q \tilde{I}_{se}^*) = \frac{V_q^2 + 2VV_q \sin(\theta/2)}{X_L} \tag{14.65}$$

When applied to Figure 14.55, (14.65) indicates that the SSSC is providing capacitive power into the transmission line.

The operation of the SSSC is limited by the magnitude of the voltage insertion $V_q < V_{max}$. The voltage limit V_{max} is of the order of 10% of the nominal transmission line voltage level.

Example 14.7: SSSC insertion

Consider the SSSC circuit model shown in Figure 14.54. The transmission line reactance is $X_L = 0.035$ pu. The SSSC transformer reactance at $X_s = 0.0034$ pu will be neglected. Without the SSSC insertion, the from- and to-bus voltages are both at $V = 1.02$ pu, and the active power flow on the transmission line is $P = 700$ MW flowing from Bus 1 to Bus 2. Calculate the change in active power flow and the capacitive power injected by the SSSC for a voltage insertion of $V_q = \pm 0.1$ pu. Assume that the system base is 100 MVA and Buses 1 and 2 maintain their voltage magnitudes and angles with or without the SSSC.

Solutions: Without the SSSC, the angular separation between the two buses is

$$\theta = \sin^{-1}\left(\frac{PX_L}{V^2}\right) = 13.62° \tag{14.66}$$

With $V_q = 0.1$ pu, (14.64) yields

$$P = \frac{0.1^2 \sin 13.62° + 1.02 \times 0.1 \cos 6.81°}{0.035} = 7.0 + 2.8937 = 9.8937 \text{ pu} \tag{14.67}$$

Thus the SSSC enables an additional power transfer of 289.27 MW between the two buses. The reactive power provided by the SSSC is

$$Q_q = \frac{0.1^2 + 1.02 \times 0.1 \sin 6.81^\circ}{0.035} = 0.9769 \text{ pu} \tag{14.68}$$

which is in fact capacitive.

With $V_q = -0.1$ pu, (14.64) yields

$$P = \frac{0.1^2 \sin 13.62^\circ - 1.02 \times 0.1 \cos 6.81^\circ}{0.035} = 7.0 - 2.8937 = 4.1063 \text{ pu} \tag{14.69}$$

Thus the SSSC reduces the power transfer by 289.37 MW between the two buses. The reactive power provided by the SSSC is

$$Q_q = \frac{(-0.1)^2 - 1.02 \times 0.1 \sin 6.81^\circ}{0.035} = -0.4054 \text{ pu} \tag{14.70}$$

which is inductive.

These calculations reveal that the SSSC has to be rated at least 98 MVA. This rating also has to account for the variations of the voltage magnitude of Buses 1 and 2 and the power transfer on the transmission line. ■

14.5.2 Unified Power Flow Controller

14.5.2.1 Steady-State Analysis

As in the case of the back-to-back HVDC system, the DC bus of a series VSC can also be connected to the DC bus of either a shunt or series VSC to enhance the flexibility of the FACTS controller. Two such power exchange configurations are discussed here: a unified power flow controller (UPFC) and an interline power flow controller (IPFC).

If the DC bus of an SSSC is coupled to that of a STATCOM, the two VSCs will become a UPFC (Figure 14.56) [192]. An equivalent circuit diagram is shown in Figure 14.57, in which instead of injecting only reactive power into the transmission line, active power P_{dc} can be circulated from the shunt VSC into the series VSC, and vice versa.

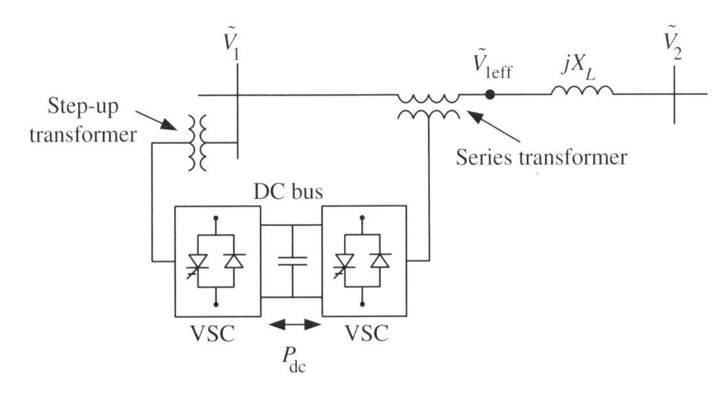

Figure 14.56 Schematic of a UPFC.

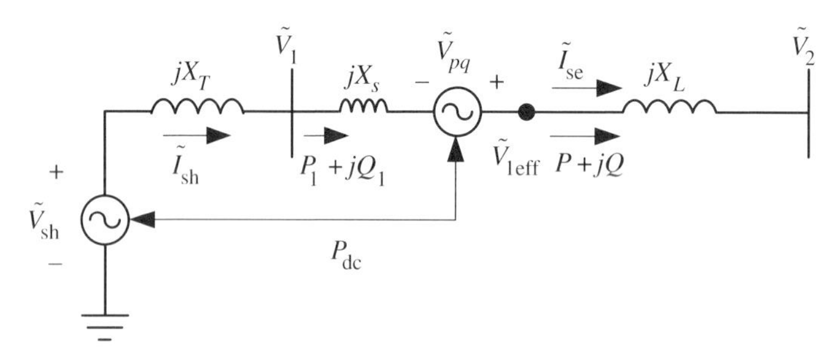

Figure 14.57 An equivalent circuit for a UPFC.

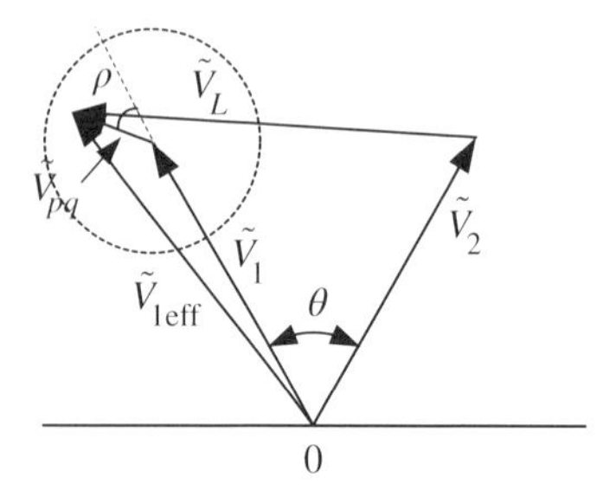

Figure 14.58 Controllability of $\tilde{V}_{pq}$ of a UPFC.

The controllability of a UPFC can be further illustrated by letting $\tilde{V}_{pq}$[11] pointing from $\tilde{V}_1$, forming a circle of radius V_{max} shown in 14.58 [10]. There are several modes of UPFC control that are of interest.

1) In the voltage-regulation mode, $\tilde{V}_{pq}$ is inserted in the same direction as $\tilde{V}_1$, as illustrated in Figure 14.59.a. In this case the effective sending-end voltage becomes

$$\tilde{V}_{1eff} = \tilde{V}_1 + \tilde{V}_{pq} = \left(1 + \frac{V_{pq}}{V}\right)\tilde{V}_1 \tag{14.71}$$

   This is the voltage support mode using series voltage insertion.
2) In the SSSC mode (Figure 14.59.b), $\tilde{V}_{pq}$ is inserted in quadrature to the line current $\tilde{I}_{se}$. There is no power circulation in this mode. Note that the shunt VSC can still regulate its terminal bus voltage V_1.
3) In the phase-shifting mode (Figure 14.59.c), $\tilde{V}_{pq}$ is inserted such that the magnitude of the effective voltage

$$\tilde{V}_{1eff} = \tilde{V}_1 + \tilde{V}_{pq} = \tilde{V}_1 e^{j\sigma} \tag{14.72}$$

   remains the same but the phase of the voltage is shifted.
4) Figure 14.59.d shows the general case in which the UPFC provides simultaneous active and reactive power (P, Q) flow control.

11 The inserted voltage is now denoted by V_{pq} as it affects both active and reactive power injection from the VSC, as opposed to V_q for an SSSC, which only affects reactive power injection.

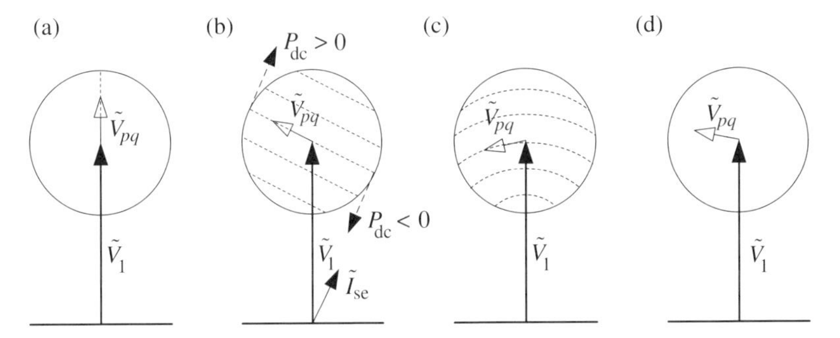

Figure 14.59 Various modes of UPFC operation: (a) voltage regulation, (b) SSSC, (c) phase shifting, and (d) simultaneous voltage, reactance, and angle control [10].

With the insertion of the voltage $\tilde{V}_{pq}$, the current flowing from Bus 1 to Bus 2 is given by

$$I_{se} = \frac{\tilde{V}_1 + \tilde{V}_{pq} - \tilde{V}_2}{jX_L} \tag{14.73}$$

Then the apparent power flowing from Bus 1 to Bus 2 is

$$P_1 + jQ_1 = \tilde{V}_2 I_{se}^* = P_o + \frac{VV_{pq}\sin\rho}{X_L} + jQ_o + \frac{VV_{pq}\cos\rho}{X_L} \tag{14.74}$$

where $\tilde{V}_1 = Ve^{j\theta/2}$, $\tilde{V}_2 = Ve^{-j\theta/2}$, $\tilde{V}_{pq} = V_{pq}e^{j(\theta/2+\rho)}$, $P_o = V^2\sin\theta/X_L$, and $Q_o = V^2(1-\cos\theta)/X_L$.

The power injected into the transmission line by the UPFC is

$$S_{se} = \tilde{V}_{pq}\tilde{I}_{se}^* = P_{pq} + jQ_{pq} \tag{14.75}$$

where

$$P_{pq} = P_{dc} = \frac{2VV_{pq}\sin(\theta/2)\cos(\theta/2+\rho)}{X_L} \tag{14.76}$$

is the circulating DC power and

$$Q_{pq} = \frac{V_{pq}^2}{X_L} + \frac{2VV_{pq}\sin(\theta/2)\sin(\theta/2+\rho)}{X_L} \tag{14.77}$$

is the capacitive power provided by the series VSC of the UPFC.

From the P_{dc} expression (14.76), it follows that $P_{dc} > 0$ (meaning that DC power is circulating from the shunt VSC to the series VSC) if $\cos(\theta/2+\rho) > 0$, that is, $-\pi/2 < (\theta/2+\rho) < \pi/2$. The regions of positive and negative P_{dc} are shown in Figure 14.59.b.

Thus the power flow from $\tilde{V}_{1eff}$ to $\tilde{V}_2$ is given by

$$P + jQ = (P_1 + jQ_1) + (P_{pq} + jQ_{pq}) \tag{14.78}$$

Thus for a UPFC to be in the (P, Q) control mode (Mode 4), the voltage magnitude V_{pq} and phase ρ need to satisfy the following two expressions for the desired power flow:

$$P_{des} = \frac{V^2\sin\theta}{X_L} + \frac{VV_{pq}\sin\rho}{X_L} + \frac{2VV_{pq}\sin(\theta/2)\cos(\theta/2+\rho)}{X_L} \tag{14.79}$$

$$Q_{\text{des}} = \frac{V^2(1-\cos\theta)}{X_L} + \frac{VV_{pq}\cos\rho}{X_L} + \frac{V_{pq}^2}{X_L} + \frac{2VV_{pq}\sin(\theta/2)\sin(\theta/2+\rho)}{X_L} \tag{14.80}$$

Note that these two expressions assume that $X_s = 0$.

However, the additional power delivered must also observe

$$\sqrt{(P_{pq}(\theta,\rho))^2 + (Q_{pq}(\theta,\rho))^2} = \frac{VV_{\max}}{X} \tag{14.81}$$

Thus when the operational limit is reached, P_{pq} and Q_{pq} must be jointly dispatched.

The active power $P_{\text{dc}} = P_{\text{se}}$ supplied by the series VSC must come from the shunt VSC. The apparent power $P_{\text{sh}} + jQ_{\text{sh}}$ supplied by the shunt VSC to the AC power system is

$$P_{\text{sh}} + jQ_{\text{sh}} = \tilde{V}_{\text{sh}}I_{\text{sh}}^*, \quad I_{\text{sh}} = (\tilde{V}_{\text{sh}} - \tilde{V}_1)/(jX_T) \tag{14.82}$$

Here $P_{\text{sh}} = -P_{\text{se}}$ is determined by the series VSC. The remaining computation is to determine Q_{sh} to maintain V_{sh} at V_{ref}.

In normal operation, the voltage insertion by the series VSC is close to zero, akin to the Var reserve mode of a STATCOM. When a disturbance occurs, the series VSC voltage V_{pq} can be deployed to control the flow on the compensated line to maintain a higher level of stability. The shunt VSC can provide voltage support. The power circulation between the shunt and series VSCs provides some additional flexibility. The UPFC installed at the Inez 230 kV substation with the series VSC inserted in the Inez-Big Sandy line is part of an overall solution to prevent voltage instability in the region by limiting the flow on the Inez-Big Sandy line during disturbances [10].

Example 14.8: UPFC insertion

Consider the UPFC circuit model shown in Figure 14.57 in which the transmission line reactance is $X_L = 0.035$ pu. The reactance of the shunt transformer is $X_T = 0.15$ pu and the series transformer reactance $X_s = 0.0014$ pu is neglected. Without the UPFC insertion, the from- and to-bus voltages are both at $V = 1.02$ pu, and the active power flow on the transmission line is $P = 700$ MW flowing from Bus 1 to Bus 2. Calculate the (P, Q) flow after the insertion of the series voltage of $\tilde{V}_{pq} = 0.05$ pu (Mode 1 in Figure 14.58 with $\rho = 0°$) and the shunt VSC current $\tilde{I}_{\text{sh}}$ if $V_{sh} = 1.03$ pu. Assume that the system base is 100 MVA and Buses 1 and 2 maintain the same voltage magnitudes and angles with or without the UPFC.

Solutions: From Example 14.7, the voltages in pu are given by

$$\tilde{V}_1 = 1.02e^{j6.810°}, \quad \tilde{V}_2 = 1.02e^{-j6.810°}, \quad \tilde{V}_{pq} = 0.05e^{j6.810°} \tag{14.83}$$

such that the series current in pu is given by

$$\tilde{I}_{\text{se}} = \frac{\tilde{V}_1 + \tilde{V}_{pq} - \tilde{V}_2}{jX_L} = 7.0809 - j1.4185 \tag{14.84}$$

Thus the line power flow values in pu are given by

$$P_1 + jQ_1 = \tilde{V}_1\tilde{I}_{\text{se}}^* = 7.0000 + j2.2931 \tag{14.85}$$

and

$$P + jQ = (\tilde{V}_1 + \tilde{V}_{pq})\tilde{I}_{se}^{*} = 7.3431 + j2.4055 \tag{14.86}$$

The power in pu injected by the series VSC is

$$P_{pq} + jQ_{pq} = \tilde{V}_{pq}\tilde{I}_{se}^{*} = 0.3431 + j0.1124 \tag{14.87}$$

Although the series VSC does not inject much capacitive power Q_{pq} into the transmission line, the inserted $\tilde{V}_{pq}$ facilitates additional reactive power transfer on the transmission line. In practical systems, this application of $\tilde{V}_{pq}$ could increase the voltage magnitude of Bus 2 by 0.5–1.0%.

The active power circulation of $P_{dc} = P_{pq} = 34.31$ MW has to be supported by the shunt VSC. From the power transfer function over a lossless branch, the phase of $\tilde{V}_{sh}$ is computed from

$$\theta_{sh} = \frac{\theta}{2} - \sin^{-1}\left(\frac{P_{pq}X_T}{V_{sh}V}\right) = 4.0020° \tag{14.88}$$

Thus

$$\tilde{I}_{sh} = \frac{\tilde{V}_{sh} - V_1}{jX_T} = -0.3271 - j0.0979 \text{ pu} \tag{14.89}$$

such that the power provided by the shunt VSC is

$$P_{sh} + jQ_{sh} = \tilde{V}_{sh}I_{sh}^{*} = -0.3431 + j0.0771 \text{ pu} \tag{14.90}$$

■

14.5.3 Interline Power Flow Controller

14.5.3.1 Steady-State Analysis

Another configuration of coupling the DC buses of two VSCs is the interline power flow controller (IPFC) in which the DC buses of two series VSCs on two different lines are connected. The schematic of an IPFC is shown in Figure 14.60 [193]. The equivalent

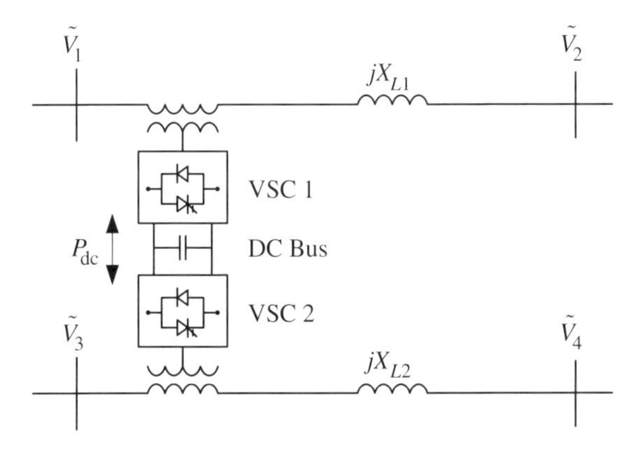

Figure 14.60 IPFC schematic.

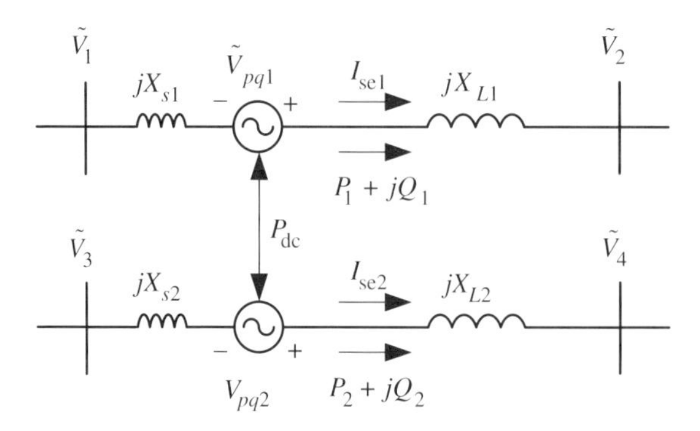

Figure 14.61 IPFC equivalent circuit.

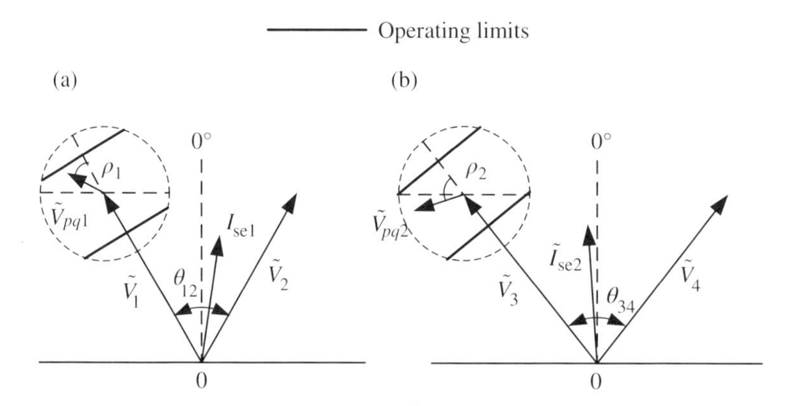

Figure 14.62 IPFC operating condition: (a) primary series VSC and (b) secondary series VSC [10].

circuit in Figure 14.61 shows the possibility of the two series VSCs to exchange active power P_{dc}.

As in a UPFC, although there are four controlled variables $V_{pq1}\angle\rho_1$ and $V_{pq2}\angle\rho_2$, there are only three quantities that can be independently controlled. A typical choice is to control the two active power flows P_1 and P_2, and one reactive power flow, say, Q_1, as shown in Figure 14.61. The fourth degree of control freedom is used to satisfy the DC power circulation P_{dc}. The transmission line whose P, Q flow is controlled is called the primary line, and that whose P flow is controlled is called the secondary line.

The operation of the two connected VSCs is shown in Figure 14.62, where the circles indicate maximum VSC voltage insertion. Following the notation used for the IPFC in Figure 14.59, $\theta_{12} = \theta_1 - \theta_2$ and $\theta_{34} = \theta_3 - \theta_4$ are the angular differences between Buses 1 and 2, and between Buses 3 and 4, respectively. Here the bus voltages $\tilde{V}_i$, $i = 1, \ldots, 4$, are assumed to be stiff. For consistency with the derivation in the UPFC section, the phases of $\tilde{V}_1$ and $\tilde{V}_2$ are taken as $\theta_{12}/2$ and $-\theta_{12}/2$, and the phases of $\tilde{V}_3$ and $\tilde{V}_4$ are taken as $\theta_{34}/2$ and $-\theta_{34}/2$. In an interconnected power system, the 0° direction in Figure 14.62.b is displaced from the 0° direction in Figure 14.62.a, say by θ_{rot}. In the expressions below for the secondary VSC, all the angles can be rotated by the amount θ_{rot}.

The operation of the primary series VSC is similar to that of the series VSC of a UPFC, shown in Figures 14.59 and 14.58. As shown in Figure 14.62.a, the primary VSC is simultaneously increasing the active and reactive power flow between Buses 1 and 2. In addition, DC power is also flowing from the secondary VSC to the primary VSC. In Figure 14.62.b, the secondary VSC is operating by absorbing active power and sending it to the primary VSC. It is also simultaneously increasing the active and reactive power flow between Buses 3 and 4.

For simplicity, assume that all the bus voltage magnitudes are equal to V. Then the desired (P, Q) expressions for the primary line are given by

$$P_{1\text{des}} = \frac{V^2 \sin\theta_{12}}{X_{L1}} + \frac{VV_{pq1}\sin\rho_1}{X_{L1}} + \frac{2VV_{pq1}\sin(\theta_{12}/2)\cos(\theta_{12}/2+\rho_1)}{X_{L1}} \tag{14.91}$$

$$Q_{1\text{des}} = \frac{V^2(1-\cos\theta_{12})}{X_{L1}} + \frac{VV_{pq1}\cos\rho_1}{X_{L1}} + \frac{V_{pq1}^2}{X_{L1}} + \frac{2VV_{pq1}\sin(\theta_{12}/2)\sin(\theta_{12}/2+\rho_1)}{X_{L1}} \tag{14.92}$$

The DC power circulating from the secondary VSC to the primary VSC is

$$P_{pq1} = P_{\text{dc1}} = \frac{2VV_{pq1}\sin(\theta_{12}/2)\cos(\theta_{12}/2+\rho_1)}{X_{L1}} \tag{14.93}$$

and reactive power provided by the primary VSC is

$$Q_{pq1} = \frac{V_{pq1}^2}{X_{L1}} + \frac{2VV_{pq1}\sin(\theta_{12}/2)\sin(\theta_{12}/2+\rho_1)}{X_{L1}} \tag{14.94}$$

Similar expressions can be obtained for the secondary series VSC. The desired P expression for the secondary line is given by

$$P_{2\text{des}} = \frac{V^2 \sin\theta_{45}}{X_{L2}} + \frac{VV_{pq2}\sin\rho_2}{X_{L2}} + \frac{2VV_{pq2}\sin(\theta_{34}/2)\cos(\theta_{34}/2+\rho_2)}{X_{L2}} \tag{14.95}$$

The DC power circulating from the primary VSC to the secondary VSC is

$$P_{pq2} = P_{\text{dc2}} = \frac{2VV_{pq2}\sin(\theta_{34}/2)\cos(\theta_{34}/2+\rho_2)}{X_{L2}} \tag{14.96}$$

and reactive power provided by the secondary VSC is

$$Q_{pq2} = \frac{V_{pq2}^2}{X_{L2}} + \frac{2VV_{pq2}\sin(\theta_{34}/2)\sin(\theta_{34}/2+\rho_2)}{X_{L2}} \tag{14.97}$$

The IPFC operation is then determined by (14.91), (14.92), (14.95), and

$$P_{pq1} + P_{pq2} = 0 \tag{14.98}$$

One of the main functions of an IPFC is to control the flow on parallel paths. At present, the convertible static compensator (CSC) at the Marcy substation in central New York State is the only FACTS controller that can operate as an IPFC [194]. Although the CSC is mostly operated in the shunt mode for maximum power transfer benefit, it has been used occasionally in the IPFC or 2-SSSC mode to optimize the flow in the area north of New York City during line and breaker maintenance.

The DC power circulation between the two VSCs places a constraint on the operation of the IPFC. That is, the largest amount of P_{dc} supply demanded by the primary is limited by the amount of P_{dc} that can be provided by the secondary VSC. Suppose the two VSCs are of the same rating and the reactance X_{L2} of the second is significantly higher than that of X_{L1}. Then $\tilde{I}_{se2}$ will be smaller than that of $\tilde{I}_{se1}$. As P_{dc} is proportional to the magnitude of the line currents, a P_{dc} disparity between the two VSCs exists. However, this imbalance can be overcome by installing an energy storage system on the DC bus.

The next example continues from Example 14.8 to compute the operating condition of the secondary VSC.

Example 14.9: IPFC insertion

Consider the IPFC circuit model shown in Figure 14.61 in which Bus 1 and Bus 3 are the same bus and the transmission line reactances are $X_{L1} = 0.035$ pu and $X_{L2} = 0.068$ pu. The series transformer reactance X_s is neglected. Without the IPFC insertion, $P_1 = 700$ MW on Line 1 and $P_2 = 570$ MW. Assume that Buses 1, 2, and 4 are stiff buses with the same voltage magnitude $V = 1.02$ pu, and the bus voltage angles remain the same with or without the IPFC. Calculate (P_1, Q_1) after the insertion of the series voltage of $\tilde{V}_{pq1} = 0.05$ pu with $\rho_1 = 0°$. Assuming $V_{pq2} = 0.08$ pu, calculate ρ_2 and the resulting P_2 on Line 2.

Solutions: The solutions for the Line 1 quantities are already given in Example 14.8 as

$$P_1 + jQ_1 = 7.3431 + j2.4055 \text{ pu} \tag{14.99}$$

The DC power in pu injected into the line by VSC 1 is

$$P_{pq1} = P_{dc} = 0.3431 \text{ pu} \tag{14.100}$$

For the secondary line, because Buses 1 and 3 are the same bus, the angle difference for the secondary line is changed to θ_{14}. Without the IPFC,

$$\theta_{14} = \sin^{-1}\left(\frac{P_2 X_{L2}}{V^2}\right) = 21.87° \tag{14.101}$$

Thus with an inserted voltage magnitude of $V_{pq2} = 0.08$ pu, the required phase to facilitate the transfer of 34.31 MW of power from the secondary VSC to the primary VSC is

$$\rho_2 = -\frac{\theta_{14}}{2} + \cos^{-1}\left(\frac{-P_{pq1} X_{L2}}{2VV_{pq2}\sin(\theta_{14}/2)}\right) = 127.97° \tag{14.102}$$

■

In Example 14.9, if the primary VSC voltage is set at $V_{pq1} = 0.1$ pu, there is no solution to this problem if V_{pq2} is limited to within 0.1 pu.

14.5.4 Dynamic Model

In this section, several series-VSC dynamic models for different control modes are shown. The control models shown here are by no means exhaustive. The purpose is illustrate series voltage insertion, DC power circulation, and line (P, Q) flow regulation. These dynamic models are much simpler than the corresponding models in commercial software. However, they do contain the main feedback components.

14.5.4.1 Series Voltage Insertion

Series voltage insertion is a common SSSC mode of operation. Following the earlier notation, the inserted series voltage is defined as

$$\tilde{V}_q = V_q e^{j\rho} \tag{14.103}$$

where V_q is allowed to take a positive or negative value and ρ is measured relative to the Bus 1 angle. When $V_q > 0$, the VSC is injecting capacitive power into the transmission line. When $V_q < 0$,

$$\tilde{V}_q = V_q e^{j\rho} = |V_q| e^{j(\rho-\pi)} \tag{14.104}$$

that is, the VSC is injecting inductive power into the transmission line.

Keeping the notation in (14.103), the voltage insertion control is shown in Figure 14.63. Figure 14.63.a shows the voltage control on the DC bus, where K_v is the conversion factor from the DC bus voltage to the voltage across the series transformer. The V_{dc} dynamic is given by

$$C\frac{d|V_q|}{dt} = I_{dc} = \frac{\Delta P_{dc}}{V_{dc}} \tag{14.105}$$

where the difference between $|V_{qref}|$ and $|V_q|$ is taken to be proportional to the circulating power ΔP_{dc} required to adjust the VSC voltage $|\tilde{V}_q|$.

The control of the phase ρ to maintain a desired power circulation is shown in Figure 14.63.b. For an SSSC, the desired power circulation would always be $P_{dcref} = 0$. The control system consists of a filter and a proportional-integral (PI) controller. Because the series VSC can operate in either the inductive or the capacitive mode, the signs S_1 and S_2 need to be adjusted accordingly, as indicated in the caption of Figure 14.63.

Note that the control diagram Figure 14.63.b is also suitable for modeling the power circulation control of the secondary VSC in an IPFC and the shunt VSC in a UPFC.

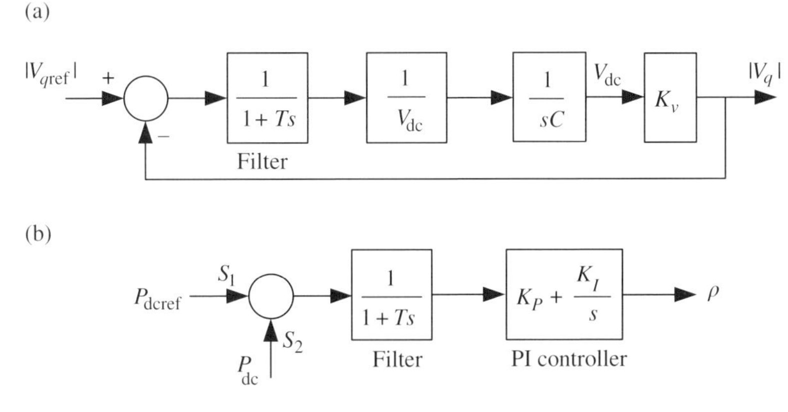

Figure 14.63 Voltage insertion control for SSSC application: (a) voltage magnitude control and (b) power circulation control. In (b), if $V_q > 0$, then S_1 = "−" and S_2 = "+", and if $V_q < 0$, then S_1 = "+" and S_2 = "−".

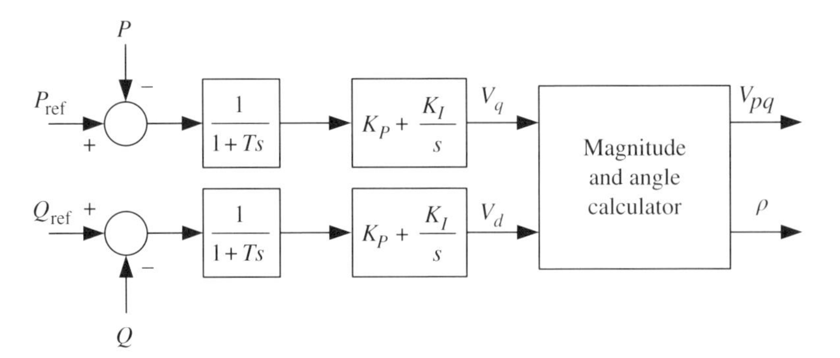

Figure 14.64 Line active and reactive power control of a UPFC and the primary VSC of an IPFC.

14.5.4.2 Line Active and Reactive Power Flow Control

For the series VSC in a UPFC and the primary VSC in an IPFC, there are two degrees of controllability to regulate the line (P,Q) flow. A control diagram is shown in Figure 14.64 with separate P and Q regulation against the desired values P_{ref} and Q_{ref}, respectively, using PI control, with the time constant T, the proportional gain K_P, and integral gain K_I denoting generic parameters. Their values may be different in the two control paths.

As shown in Figure 14.64, each PI controller is used to build up a voltage. The P loop contributes a voltage V_q in phase with the line current $\tilde{I}_{se}$ and the Q loop contributes a voltage V_d in quadrature to $\tilde{I}_{se}$. Hence this control scheme needs to measure the phase angle of $\tilde{I}_{se}$, using, for example, a phase-lock-loop (PLL) circuit. The expressions used in the "magnitude and angle calculator" box are

$$V_{pq} = \sqrt{V_d^2 + V_q^2}, \quad \rho = \tan^{-1}(V_q/V_d) \tag{14.106}$$

Note that ρ ranges from 0° to 360°, depending on the (P,Q) values with respect to (P_{ref},Q_{ref}).

14.6 Summary and Notes

This chapter describes FACTS based on both the thyristor and voltage-sourced converter technologies. These controllers can be applied in shunt configuration to support voltage and in series configuration to control flows. With the voltage-sourced converter technology, the DC buses of the converters can be connected such that active power can circulate from one converter to another converter, thus increasing the versatility of FACTS controllers. Dynamic control models are discussed and damping controllers for an SVC and a TCSC are illustrated with examples. Note that both the controller designs in Problems 14.12 and 14.13 can be assigned as design projects.

Converter technology has been evolving. The latest technology is based on the modular multi-level converter (MMC) topology, in which each converter switch has its own DC bus supported by a capacitor. This technology would reduce the amount of harmonics.

Some of the materials in this chapter were developed by several PhD students in their research work on FACTS. The authors would like to acknowledge Xuan Wei [195], Xia Jiang (UPFC) [196], Xinghao Fang (back-to-back STATCOM) [197], and Wei Li (KTH) for the STATCOM voltage-current plot in Figure 14.44.

Problems

14.1 Solve for P'_L in Example 14.1 analytically using (14.11) and (14.12).

14.2 For the network connection shown in Figure 14.65, if Bus C is eliminated, the equivalent reactance between Buses A and B is given by, from the star-delta transformation [198],

$$X_{eq} = X_1 X_2 \left(\frac{1}{X_1} + \frac{1}{X_2} + \frac{1}{X_3} \right) \tag{14.107}$$

Use this expression to verify X_{eq} given by (14.18).

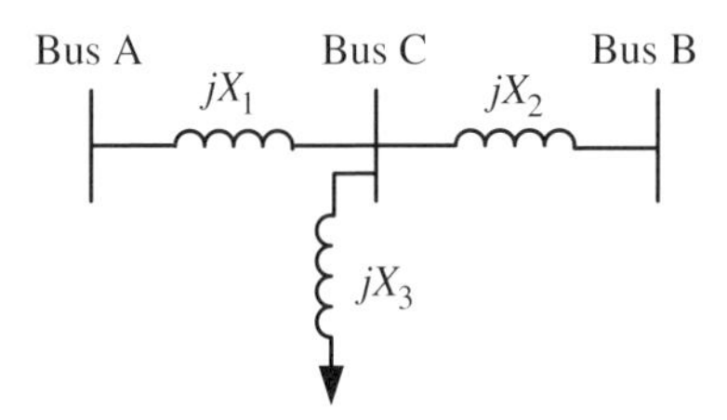

Figure 14.65 Figure for Problem 14.2.

14.3 Consider the radial two-machine system shown in Figure 14.66. The voltages at the machine terminal buses are $\tilde{V}_1 = V_1 e^{j\delta/2}$ and $\tilde{V}_2 = V_2 e^{-j\delta/2}$. The transmission line with reactance X is compensated at regular intervals by $(n-1)$ static var compensator. Suppose $V_1 = V_2 = V$ and all the SVCs are regulating their voltages to be also equal to V. Find the active power flow P from Bus 1 to Bus 2. Verify that in the limit as $n \to \infty$, $P = (V^2/X)\delta$.

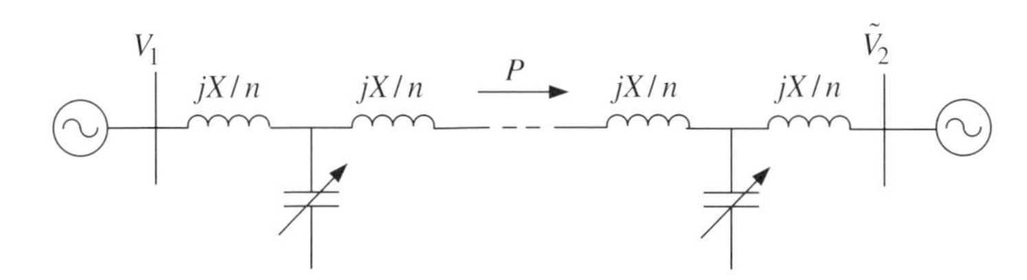

Figure 14.66 Transmission path with SVC compensation at regular intervals.

14.4 A load $P + jQ$ is served by a stiff voltage source $E = 1.0$ pu via a transmission line with impedance $Z = R + jX = 0.01 + j0.1$ pu, all on 100 MVA base as shown in Figure 14.67.

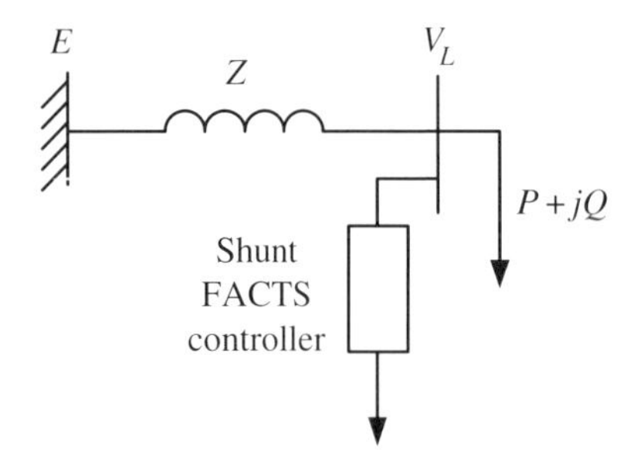

Figure 14.67 A stiff-source-to-load system with shunt compensation at the load bus.

Use the PST power flow solver to generate the upper part of the *PV*-curves for the load bus (1) with no compensation, (2) with an SVC with a rating of B_c = 1.0 pu, and (3) a STATCOM with a rating of 100 MVar. For the SVC, model it as a susceptance and set any voltage V_L higher than 1.0 pu as 1.0 pu. For the STATCOM, model it as a generator with zero active power, $Q_{\max} = 100$ MVar, and $Q_{\min} = -100$ MVar (also known as a synchronous condenser) and set the desired voltage on the load bus at 1.0 pu. Perform this analysis for the load with unity power factor, and leading and lagging 0.9 power factor. Comment on the similarities and differences of the SVC and STATCOM.

14.5 An SVC is used to regulate the voltage V on a load bus. The sensitivity of the load bus voltage with respect to the capacitive current injection I_C is

$$V = V_o + \beta I_C, \quad \beta = 0.02 \tag{14.108}$$

where V_o is the bus voltage without compensation. The SVC voltage regulation is represented by

$$V_{\rm SVC} = V = V_{\rm ref} - \alpha I_C, \quad \alpha = 0.04 \tag{14.109}$$

Find V and I_C for the following operating conditions and provide explanations for the results:
1) $V_o = 1.0$ pu and $V_{\rm ref} = 1.03$ pu
2) $V_o = 1.04$ pu and $V_{\rm ref} = 1.03$ pu
If the SVC Var reserve of ± 0.2 pu is exceeded, redispatch $V_{\rm ref}$ so that the operating point is within Var reserve.

14.6 An SVC is used to control the voltage on Bus 4 for the SMIB system shown in Figure 14.14. The generator is supplying 900 MW to the load modeled by the infinite bus with a terminal voltage of 1.05 pu. The line reactances are $X_T = 0.1$ pu, $X_{L1} = 0.4$ pu, and $X_{L2} = X_{L3} = 0.2$ pu on the system base of 1000 MVA. In steady state, the voltage of Bus 4 is controlled to 1.01 pu. A 3-phase short-circuit fault is applied on Line 3-2 close to Bus 3, and is cleared by removing Line 3-2. Use PST to determine whether the system is transiently stable for a fault clearing times of 0.13, 0.14, and 0.15 sec for (1) no SVC compensation, (2) an SVC with a control range of ± 200 MVar, and (3) an SVC with a control range of ± 800 MVar. The breaker opening at the remote end occurs at 0.01 sec after the fault is cleared on the faulted side. The gain of the dynamic SVC model is $1/0.03 = 33.3$ and the time constant is 0.05 sec.

The data file for this simulation is *datasmib_svc.m*. The synchronous machine is represented by the subtransient model and is controlled by an excitation system and a governor. In the power flow solution, the SVC is modeled as a generator bus. In the simulation, the SVC bus needs to be declared as a non-conforming load bus because the admittance will change during the simulation. To emulate no SVC, set the control gain to a very small number. Apply the fault at $t = 0.1$ sec so a short time span of steady state will be evident in the time response. Plot the machine angles for each clearing time.

14.7 The benefit of series compensation for a transmission line is investigated in this problem.

The transmission line with no series compensation is shown in Figure 14.68.a. Given the voltage phasors $\tilde{V}_S = 1.0\angle 110°$ pu and $\tilde{V}_R = 1.0\angle 70°$ pu, and the transmission line reactance $X_L = 0.1$ pu, calculate the current phasor $\tilde{I}$ in the transmission line and the power transfer. Plot all the voltage and current phasors, including $\tilde{V}_L$. In Figure 14.68.b, the line is 50% compensated with $X_C = -0.05$ pu. Assume that the network is stiff so that $\tilde{V}_S$ and $\tilde{V}_R$ remain unchanged. Calculate and plot all voltage and current phasors, and compute the power transfer. In Figure 14.68.c, an SSSC is applied to the transmission line and is represented by a voltage injection $\tilde{V}_q$ of 0.1 pu in magnitude. Note that this voltage injection is in quadrature with the line current. Calculate and plot all voltage and current phasors, and compute the power transfer.

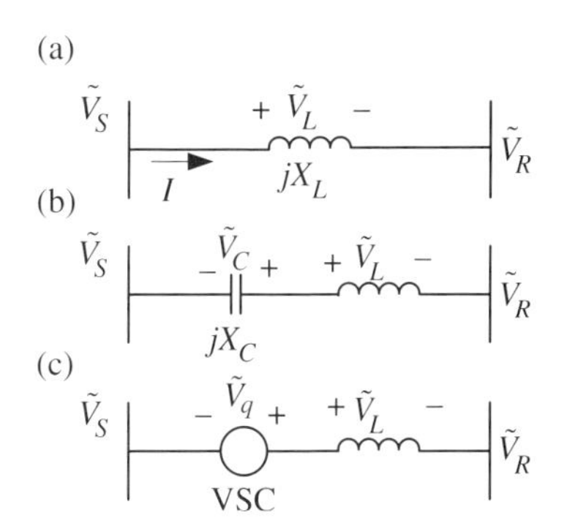

Figure 14.68 Series compensation schemes: (a) no compensation, (b) fixed series compensation, and (c) VSC series compensation.

14.8 Normally series compensation is inserted into the middle of a transmission line, as shown in Figure 14.69.b, and not at one end of the transmission line, as shown in 14.69.a.

Given the voltage phasors $\tilde{V}_S = 1.0\angle 110°$ and $\tilde{V}_R = 1.0\angle 70°$, $X_L = 0.1$ pu and $X_C = -0.05$ pu, calculate the voltage magnitudes at A and the mid-point of X_L in Figure 14.69.a, and the voltage magnitudes at B and C in Figure 14.69.b. Plot the voltage profiles on the transmission line for both cases from $\tilde{V}_S$ to $\tilde{V}_R$, assuming that the capacitor takes up negligible distance.

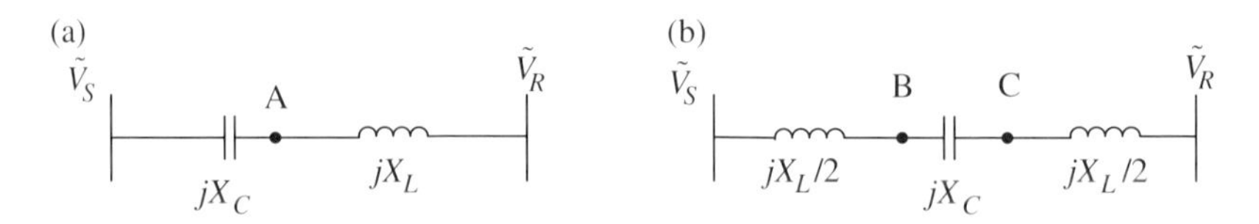

Figure 14.69 Series compensation at two different locations: (a) at the end of a transmission line and (b) in the middle of a transmission line.

14.9 Show that the power delivered to the receiving bus in a UPFC configuration (Figure 14.58) is given by

$$P = \frac{V^2}{X}\sin\theta + \frac{VV_{pq}}{X}\sin(\theta+\rho), \quad Q = -\frac{V^2}{X}(1-\cos\theta) + \frac{VV_{pq}}{X}\cos(\theta+\rho) \tag{14.110}$$

14.10 The injected voltage source model of a UPFC is shown in Figure 14.70.a. Note that the series transformer reactance is neglected, and the shunt transformer does not impact the calculation. Assume that the voltage phasors are fixed at $\tilde{V}_S = 1.0\angle 110°$ and $\tilde{V}_R = 1.0\angle 70°$, and the line reactance is $X_L = 0.1$ pu. The maximum series voltage insertion $\tilde{V}_{pq}$ magnitude is 0.1 pu. Thus the voltage phasor diagram can be represented in Figure 14.70.b. Consider the operation of the UPFC at maximum $\tilde{V}_{pq}$ for

a) voltage regulation (in the direction of capacitive power injection)
b) SSSC mode (in the direction of increasing power transfer)
c) phase shifting (in the direction of increasing power transfer).

For each case, calculate the required active power circulation P_{pq} between the series and shunt VSCs and the reactive power injection by the series VSC.

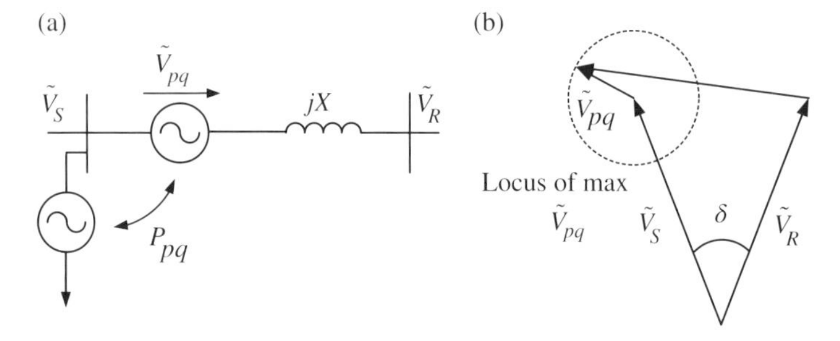

Figure 14.70 Figure for Problem 14.10: (a) UPFC circuit diagram and (b) voltage diagram.

14.11 Consider the injected voltage source model of a STATCOM shown in Figure 14.71. The STATCOM is rated at 100 MVar, and the system base is 100 MVA. The transformer reactance is $X_{t1} = 0.15$ pu. The voltage at Bus 1 is fixed at $\tilde{V}_1 = 1.0e^{j\pi/2}$ pu.

1) Suppose the STATCOM is operating as a standalone unit. Find $\tilde{V}_{sh}$ and $\tilde{I}_{sh}$ if $S = j1.0$ pu, that is, the STATCOM is injecting 1.0 pu capacitive power into Bus 1. Draw the voltage and current vector diagram.

2) Suppose the DC bus of the STATCOM is connected to the DC bus of another VSC FACTS controller. Find $\tilde{V}_{sh}$ and $\tilde{I}_{sh}$ if $S = -0.6 + j0.8$ pu, that is, the STATCOM is injecting 0.8 pu capacitive power into Bus 1 but drawing 0.6 pu active power into its DC bus. Draw the voltage and current vector diagram.

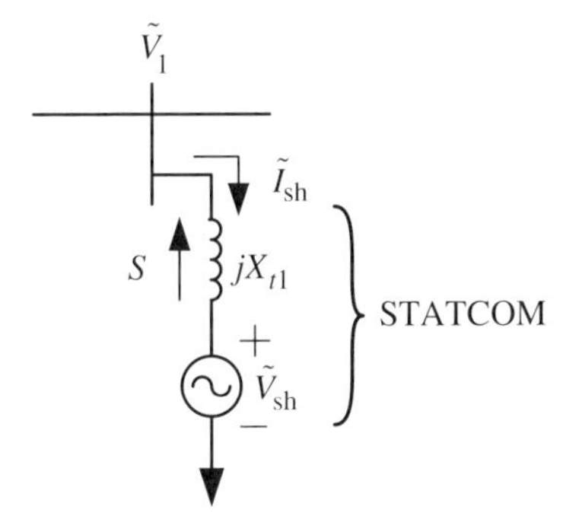

Figure 14.71 Figure for Problem 14.11.

14.12 (Design project: SVC damping controller design) Redesign the SVC damping controller in Example 14.4, using the line active power flow from Bus 4 to Bus 2 as the input signal to the damping controller. Also assume that Line 3-2 has been disconnected. Comment on the performance of the design for a 0.05 pu step increase in the excitation system voltage reference value.

14.13 (Design project: TCSC damping controller design) Redesign the TCSC damping controller in Example 14.5, using the bus voltage angle difference between Buses 2 and 3 as the input signal (assuming synchronized phasor measurements are available).[12] Comment on the performance of the design for a 0.05 pu step increase in the excitation system voltage reference value.

12 In the TCSC damping controller design paper [187], synthesized remote angles are used as the input signal. A remote angle estimation scheme is also given in [187].

15

Wind Power Generation and Modeling

The subject of this chapter is a topic of great interest to the power industry. This attention arises from the need to address the ever-growing demand for electric power while meeting very different types of requirements aimed at addressing environmental, economic, and operational concerns. It is also a subject related to different areas of science and engineering, e.g., materials sciences, electrical and mechanical engineering, etc. Consequently, it is a subject that cannot be completely described, even lightly, in a single book, let alone in a single chapter. For these reasons, the objective of this chapter is restricted to providing a reader with an understanding of fundamental concepts related to the modeling, simulation, and control of wind power plants (WPPs) in bulk (large) power systems.

Sections 15.1–15.3 provide a brief historical background, describe wind turbine components, and introduce several fundamental analytical relations used to quantify wind power, followed by Section 15.4 with a brief description of the main types of wind turbines used by the power industry. Recognizing that most wind turbines use an induction machine to generate power, the steady-state characteristics of such machines are discussed in Section 15.5. Section 15.6 describes the overall representation of a WPP for planning studies, and Section 15.7 outlines the control criteria for variable-speed wind turbines. Section 15.8 presents a wind turbine dynamic model similar to that used in several commercial transient stability programs. Section 15.9 describes a plant controller for coordinating the individual wind turbines in a wind farm.

Important engineering aspects associated with wind power are still in a state of flux and are bound to change and evolve as the penetration of wind power in existing systems continues to increase. This chapter provides models and control systems that represent the state-of-the-art wind turbine-generator (WTG) systems, allowing a reader to readily follow new wind technology developments.

15.1 Background

Wind power has become an important part of the generation resources in several countries, and its relevance is likely to increase as environmental concerns become more prominent. Wind is not only an abundant resource of power but the power extracted from the wind does not involve the emission of pollutants into the atmosphere, i.e., wind

Power System Modeling, Computation, and Control, First Edition. Joe H. Chow and Juan J. Sanchez-Gasca.

Companion website: www.wiley.com/go/chow/power-system-modeling

power is considered "green power." Its abundance and minimum environmental impact make wind power a very attractive form of power generation.

The use of wind power in the United States began toward the end of the nineteenth century. By 1888, the Brush wind turbine in Cleveland, Ohio had produced 12 kW of DC power for charging batteries at variable speed [199]. Related developments using DC turbines continued through the early years of the twentieth century. However, as AC became the dominant form of generation and transmission of electric power, the use and development of variable-speed DC wind turbines came to an end [199]. The "modern" use of wind energy began in the late 1970s as one consequence of the oil crises of 1973 and 1979 [200]. Following these events, the wind industry has developed very rapidly since the 1980s.

A confluence of mature technologies, environmental awareness, and economic conditions makes it now possible to manufacture and install large, utility-scale wind turbines to the extent that in some countries the electric power derived from the wind accounts for a large percentage of the total generated power. Although the wind speed is never constant and the areas with good wind resources are often located far from the centers of power consumption, the power industry has demonstrated that these are very surmountable obstacles for the development and use of wind power.

A wind turbine is a device that converts the kinetic energy in the wind into electrical energy. Succinctly, the energy in the wind is used to turn a set of blades, the blades turn a shaft which, depending on the turbine design, is either coupled directly or through a gearbox to a generator, and the generator then transforms the mechanical power into electrical power.

Depending on the position of the axis of rotation, wind turbines are classified as either horizontal axis wind turbines (HAWTs) or vertical axis wind turbines (VAWTs), as shown in Figure 15.1. In the horizontal axis configuration, the wind turbine is aligned with the wind and the generator resides at the top of a tower. Conversely, in a vertical axis wind turbine, the generator resides near the ground and the turbine does not have to be aligned with the wind.

Other commonly used classifications of wind turbines include variable-versus fixed-pitch rotor blades, and fixed- versus variable-speed machines. Variable-pitch wind turbines have the capability of turning their blades along their longitudinal axes

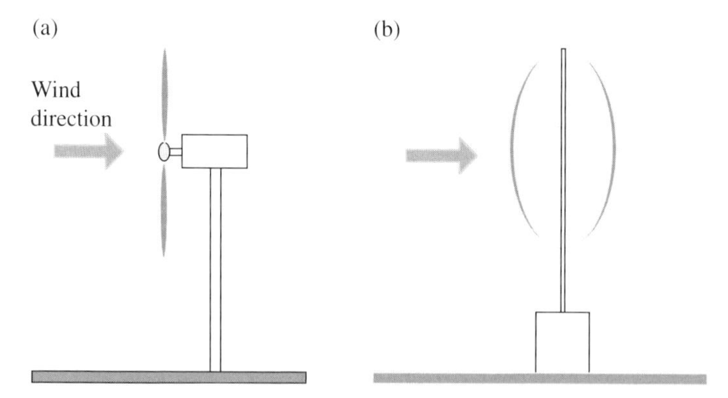

Figure 15.1 Wind energy conversion: (a) horizontal axis wind turbine (HAWT) and (b) vertical axis wind turbine (VAWT) (see color plate section).

Figure 15.2 Horizontal axis wind turbine (courtesy of GE Renewable Energy) (see color plate section).

whereas fixed-pitch machines do not. Fixed-pitch wind turbines are typically smaller machines and operate essentially at a constant speed whereas variable-speed wind turbines can operate efficiently over a much wider range of speeds. Modern large utility-scale wind turbines are of the variable-speed type. A critical technology that allows variable-speed wind turbines to be integrated into an AC power system at synchronous frequency is the application of voltage-sourced converters, allowing the circulation of active power between the power grid and the WTG, such that a wind turbine can remain online during wind gusts and system fault conditions. Note that voltage-sourced converters have already been discussed in Chapter 14.

Commercial WTG designs have converged to a three-blade, horizontal axis configuration where the blades are located in front of the unit (Figure 15.2). This particular configuration is assumed in this chapter.

15.2 Wind Turbine Components

The largest and most noticeable components of a wind turbine are the nacelle, rotor, and tower. The nacelle is located at the top of the tower and houses the generator as well as other equipment including the yaw drives and the gearbox (Figures 15.3 and 15.4). The yaw drives are used to align the nacelle with the wind direction; the gearbox

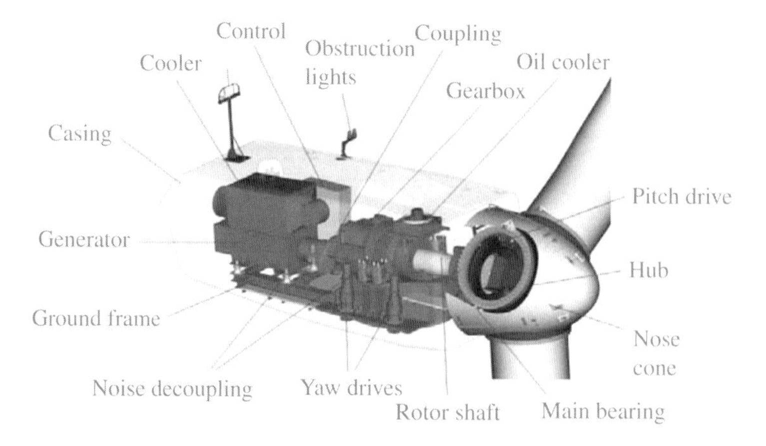

Figure 15.3 Nacelle (courtesy of GE Renewable Energy) (see color plate section).

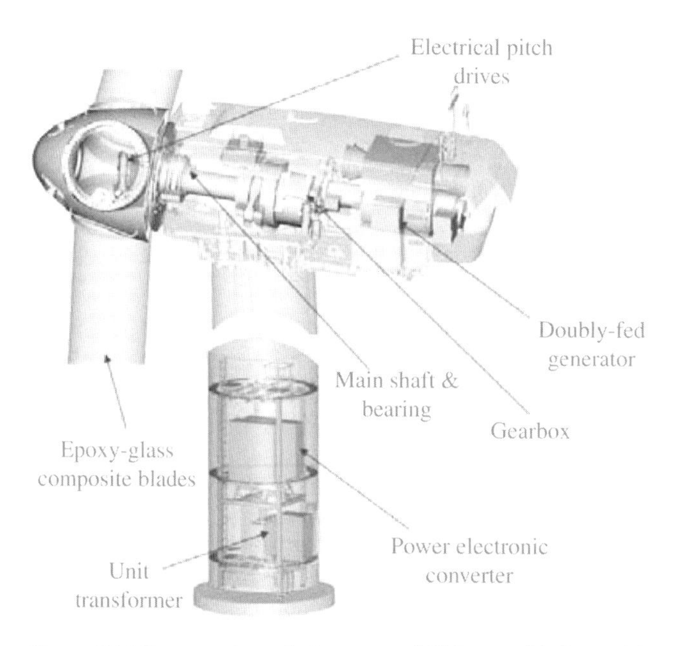

Figure 15.4 Tower and nacelle (courtesy of GE Renewable Energy) (see color plate section).

couples the rotor shaft (low-speed shaft) with the generator shaft. The rotor includes the turbine blades and the hub. Figures 15.3 and 15.4 show the electrical pitch drives located in the hub. The function of the pitch drives is to turn (pitch) the blades along their longitudinal axis. The tower is the structure that supports the nacelle and rotor, and houses other electrical equipment, e.g., power-electronic converters and transformers. Figure 15.4 shows a power electronic converter and a transformer located at the bottom of the tower. The transformer (unit transformer) is used to connect the wind turbine to the power plant electric network; power electronic converters are key components of modern wind turbines and are used for multiple control functions.

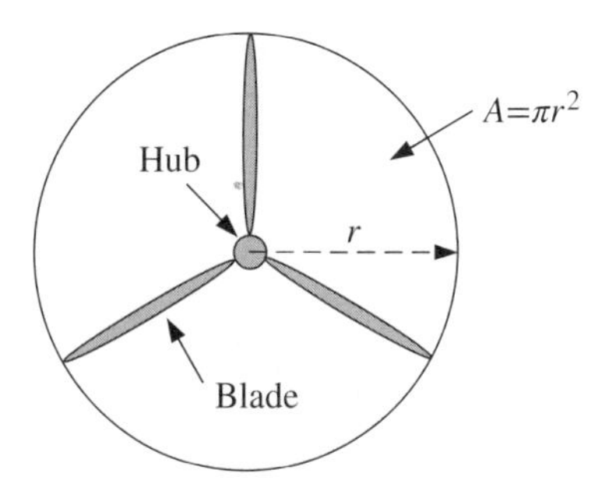

Figure 15.5 Area swept by wind turbine blades (see color plate section).

15.3 Wind Power

The total power available in the wind, P_{wind}, is a function of three factors: the air density, the wind speed raised to the third power, and the area swept by the blades, and is given by [201]

$$P_{wind} = \frac{1}{2}\rho A v^3 \tag{15.1}$$

where ρ is the air density (kg/m^3), A is the area swept by the blades (m^2), and v is the wind speed (m/sec). The wind power P_{wind} is not constant since the air speed varies continuously and the air density is a function of temperature and humidity. The air density also decreases with increasing altitude (at sea level, at a temperature of 15°C, the air density is approximately 1.225 kg/m^3 [201]). The area swept by the blades is the circle generated by the blade tips and is given by $A = \pi r^2$ where r is the rotor radius (Figure 15.5). Hence, (15.1) can also be written as

$$P_{wind} = \frac{1}{2}\rho\pi r^2 v^3 \tag{15.2}$$

Equation (15.2) shows that, with everything else being equal, a wind turbine with longer blades will produce a significantly larger amount of power than a turbine with shorter blades. Figure 15.6 shows the wind power as a function of the wind speed for two turbines of different radius (30 m and 50 m). The capability to generate larger amounts of power with longer blades is one of the reasons behind the trend toward building larger utility-size turbines.

The maximum amount of power that can in theory be extracted from the wind is 59.26% of P_{wind} [201]

$$P_{max} = 0.593 P_{wind} = 0.593\frac{1}{2}\pi\rho r^2 v^3 \tag{15.3}$$

This limit is known as the Betz limit after Albert Betz (1885–1968), the German engineer who discovered it. In practice, the amount of power extracted by utility-size wind turbines is in the order of 40% of the available power [202].

Example 15.1

Compute the theoretical maximum amount of power that can be extracted from the wind by a wind turbine whose rotor diameter is 70 m when the wind speed is (a) 11 m/sec and (b) 12 m/sec. Assume that the air density is 1.225 kg/m^3.

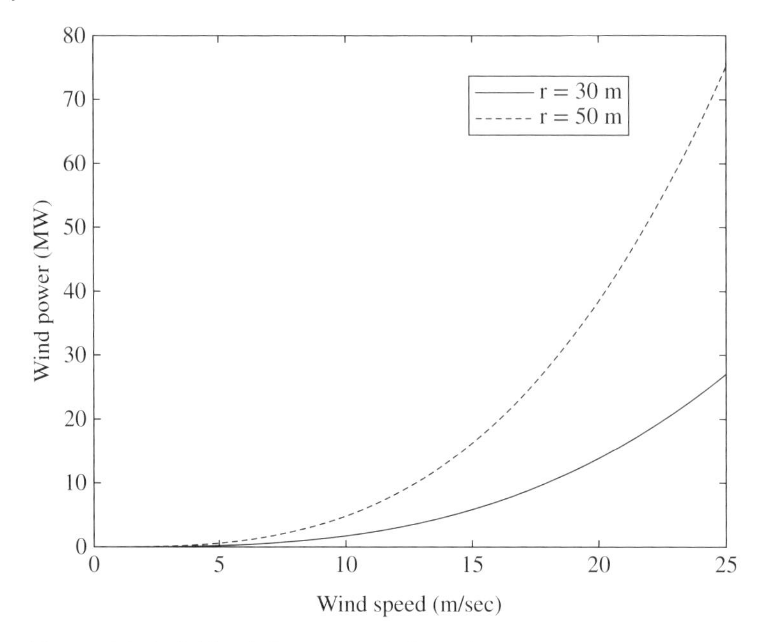

Figure 15.6 Available wind power vs wind speed.

Solutions: Substituting the given data in (15.3)

$$11\ \text{m/sec: } P_{max} = 0.593\frac{1}{2}\pi 1.225(70/2)^2 11^3 = 1.86\ \text{MW} \tag{15.4}$$

$$12\ \text{m/sec: } P_{max} = 0.593\frac{1}{2}\pi 1.225(70/2)^2 12^3 = 2.42\ \text{MW} \tag{15.5}$$

The results obtained clearly show the significant effect on P_{max} caused by relatively small changes in the wind speed. ■

15.3.1 Blade Angle Orientation

Early wind turbine blade shapes were derived from the experience accumulated in the aircraft industry, but it became evident that airfoil designs that were suitable for aircraft wings were not the best for wind turbines [203]. This led to the development of wind turbine blades specially designed for horizontal axis wind turbines [203, 205]. Several terms are used to describe an airfoil, including those shown in Figure 15.7:

- Chord line: the straight line that connects the leading and trailing edges of the airfoil.
- Angle of attack: the angle α between the relative wind and the chord line. Relative wind is the movement of the air relative to the airfoil.
- Pitch angle: the angle between the plane of rotation of the airfoil and the chord line. In other words, the pitch angle refers to the angle a blade is rotated around its longitudinal axis. Pitch angle control is an effective way of regulating power in variable-speed wind turbines. Relatively small changes in the pitch angle can have a significant impact on the generated power [203].

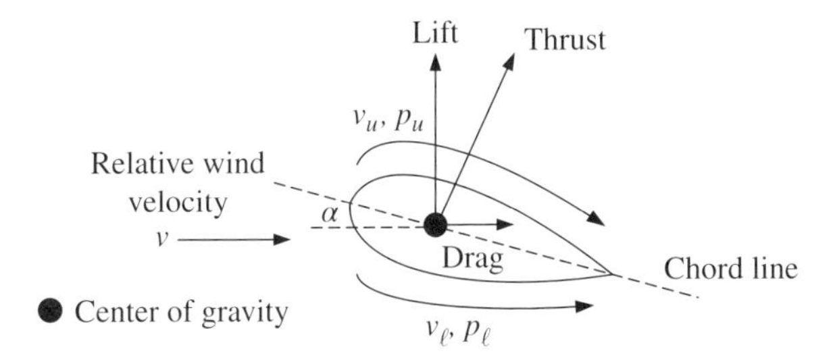

Figure 15.7 Cross-section of an airfoil: (v_u,p_u) and (v_ℓ,p_ℓ) are the air velocities and pressures on the upper and lower sides of an airfoil, with α the angle of attack.

Also shown in Figure 15.7 are two forces induced on the airfoil: lift and drag. Drag force on the airfoil is the resistive force in the direction of the relative wind; the lift force is perpendicular to the relative wind. These forces arise due to the pressure difference between the upper and lower surfaces of an airfoil. In Figure 15.7, the velocity v_u is higher than v_ℓ such that the air pressure p_u is lower than p_ℓ, thus creating a lift force. The blades of a HAWT are propelled by the lift force. However, the rotation of a VAWT is enabled by the drag force acting on the blades, and as a result a VAWT is less efficient than a HAWT.

For maximal wind energy capture, the optimal pitch of the blade varies along the blade, as the rotational velocity $v_t = \omega r$, where the angular velocity ω is in rad/sec, increases when measured from the hub (smallest) to the tip (highest). As Figure 15.8 indicates, close to the hub, v_t is small such that the angle θ_r between the relative velocity v_r and the wind velocity v_1 is small. Thus the optimal pitch angle is

$$\theta = \theta_r + \alpha \tag{15.6}$$

where the optimal angle of attack α is approximately 5°.

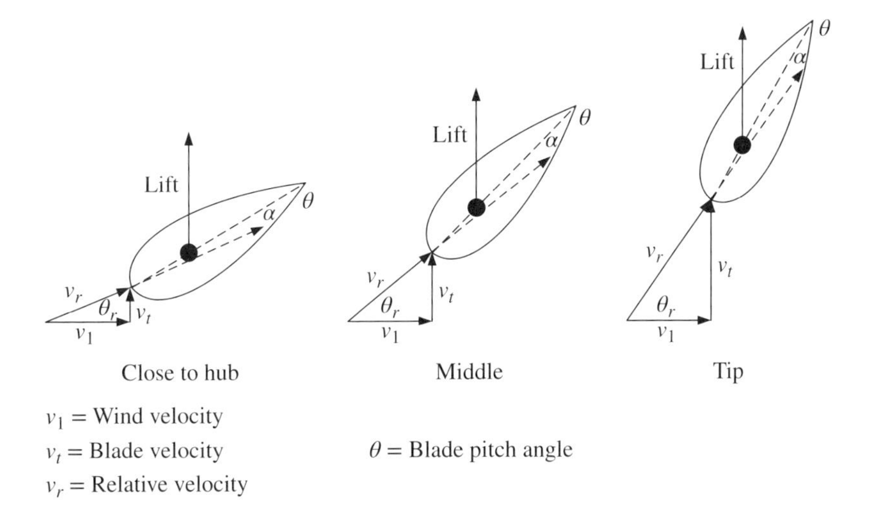

Figure 15.8 Variation of blade pitch along a wind turbine blade [204].

Notice that the pitch angle also increases, reaching its largest value at the blade tip, where $v_t/v_1 = \lambda$. The parameter λ is known as the tip-speed ratio, which is defined as the ratio of the rotor blade tip speed and the wind speed

$$\lambda = \omega r/v \tag{15.7}$$

where ω is the rotor angular speed (rad/sec), r is the rotor radius (m), and v is the wind speed (m/sec). The optimal value of λ is approximately 8, from which the pitch can be calculated as

$$\theta_{\text{tip}} = \tan^{-1}(8/1) + 5^\circ = 82.88^\circ + 5^\circ = 87.88^\circ \tag{15.8}$$

that is, the blade orientation is almost perpendicular to the wind direction. The pitch angle will be higher for smaller values of r. Thus the wind turbine blade has a twist by design. To avoid confusion with the variable pitch requirement, the optimal pitch configuration is defined by $\theta = 0^\circ$. For low wind speed condition, the pitch will be set to $\theta = 0^\circ$ to achieve maximum wind energy capture. When the wind speed exceeds its rated value, θ is increased to reduce the amount of wind energy capture and maintain the turbine output power at its rated value (1 pu).

15.3.2 Power Coefficient

The amount of mechanical power $P_{\text{mech}} = P_{\text{WT}}$ that can actually be transferred to a wind turbine is P_{wind} (15.2) times a dimensionless quantity known as the power coefficient $C_p(\lambda, \theta)$

$$P_{\text{WT}} = C_p(\lambda, \theta) P_{\text{wind}}, \quad \text{or} \quad C_p(\lambda, \theta) = \frac{P_{\text{WT}}}{P_{\text{wind}}} \tag{15.9}$$

where $C_p(\lambda, \theta)$ represents the fraction of the power available in the wind that can be extracted by the wind turbine. Note that λ and C_p are dimensionless quantities and are not a function of the turbine size. Figure 15.9 is a plot of C_p versus λ for several pitch angles obtained from test data. The figure shows that C_p reaches a single maximum point at a specific tip-speed ratio. This is an important characteristic that illustrates the motivation for using variable-speed wind turbines in order to maximize the power extracted from the wind. For instance, if a wind turbine can only operate at a fixed rotational speed, the maximum C_p can only be achieved at one specific wind speed (the wind speed for which $\lambda = \lambda_{\text{max}}$). However, if the turbine can adjust its rotational speed, it can then operate at maximum C_p over a wider range of wind speeds.

For simulation purposes, storing the C_p curves in Figure 15.9 and then interpolating between wind velocity and pitch angle is cumbersome. For this reason a polynomial expression to approximately represent the C_p curves for λ between 3 and 15 is often used instead.

Example 15.2

Assuming a wind speed of 12 m/sec and a pitch of 1°, compute the rotor angular speed that maximizes P_{WT} for a wind turbine with a rotor diameter of 90 m using the power coefficient versus wind speed curve shown in Figure 15.9.

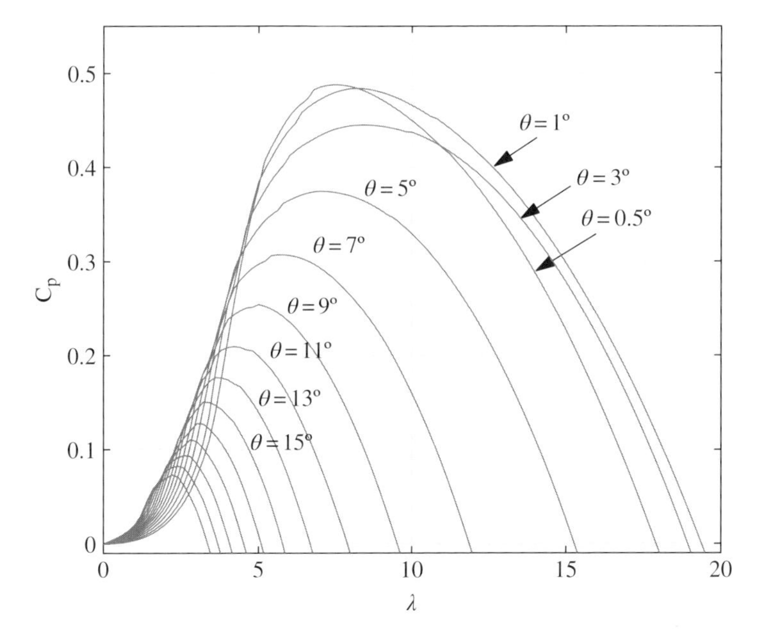

Figure 15.9 Power coefficient versus wind speed at different pitch angles (courtesy of GE Renewable Energy) (see color plate section).

Solutions: From Figure 15.9, for a pitch of 1°, C_p reaches its maximum value when the tip speed ratio λ is approximately 8. Hence, from (15.7)

$$\omega = \frac{\lambda v}{r} = \frac{8 \times 12}{90/2} = 2.13 \text{ rad/sec} = 20.4 \text{ rpm} \tag{15.10}$$

that is, the rotor completes one revolution in about 3 sec. ■

An important plot used to describe the performance of a wind turbine at different wind speeds is the power curve. This curve is the plot of the wind power output as a function of wind speed at the hub height (Figure 15.10). The curve consists of four regions, each defined by the following three wind speeds:

1) Cut-in wind speed: The minimum wind speed at which the wind turbine generates power (typical cut-in wind speeds are in the range of 3 to 5 m/sec).
2) Rated wind speed: The minimum wind speed at which the turbine generates its rated power.
3) Cut-out wind speed: The maximum wind speed at which the wind turbine delivers power. By design, to prevent equipment damage, wind turbines do not operate when the wind speed exceeds the cut-out wind speed.

Wind turbines are designed to generate power while operating in Regions 2 and 3. In Region 2, the control objective of a variable-speed wind turbine is to maximize the power extraction of the wind by maximizing the power coefficient C_p. In Region 3, when the wind speed is greater than its rated value, the output power of the turbine is limited to its rated power. This is done in order to prevent equipment damage and is accomplished by pitching the turbine blades. In general, the objective of the turbine controls is to maximize the power generation within the design constraints of the equipment. It is not

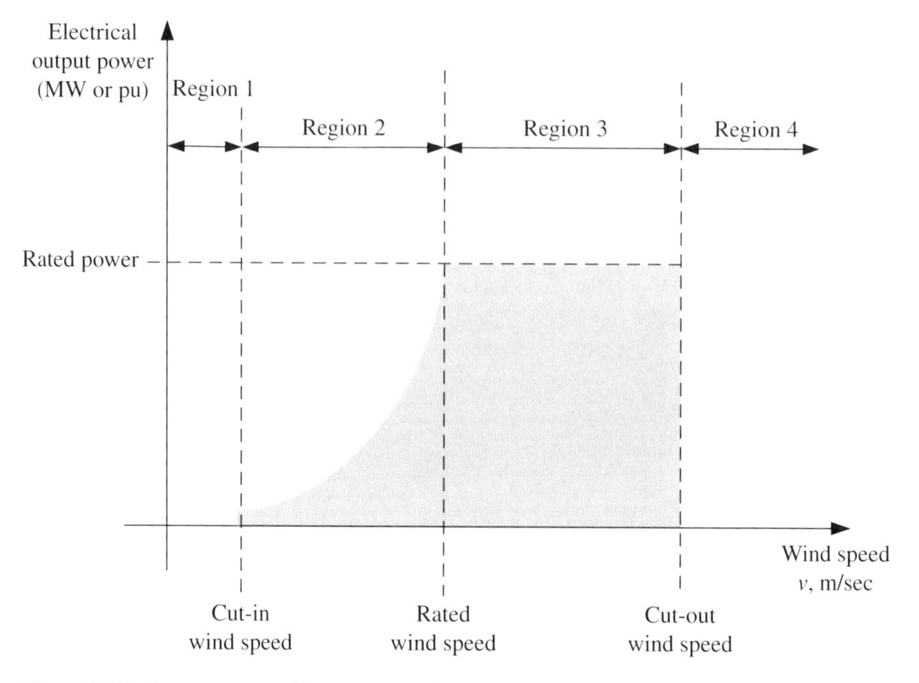

Figure 15.10 Power curve and four regions of operation (see color plate section).

economical to generate power for wind speeds below the cut-in wind speed and wind turbines are not designed to operate at wind speeds in excess of the cut-out wind speed.

15.4 Wind Turbine Types

Depending on the method used to control the speed of the wind turbine, wind turbines are classified as follows [206, 214]:

1) Type 1 - Fixed speed
2) Type 2 - Limited variable speed
3) Type 3 - Variable speed with rotor-side converter
4) Type 4 - Variable speed with full-converter interface

15.4.1 Type 1

The Type-1 wind turbine is the simplest and also the oldest design. Its basic configuration is shown in Figure 15.11. It consists of a squirrel-cage induction generator connected to the network with the generator coupled to the turbine through a gearbox. As discussed in Chapter 11, induction generators, not unlike induction motors, consume reactive power. Typically, mechanically switched capacitors are used to provide the reactive power consumed by the induction generator. Induction machines also draw large amounts of reactive power and current when they operate at low speeds. For this reason, an electronic "soft starter" is often used to mitigate these effects. The Type-1

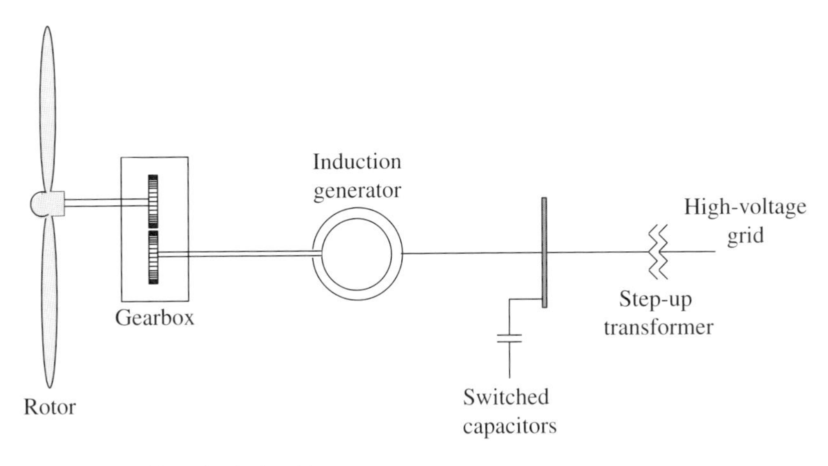

Figure 15.11 Type-1 wind turbine.

wind turbine is simple and robust but operates at essentially a constant speed due to the steep torque-speed characteristics of the induction generator. The nearly constant speed operation means that the energy capture from the wind cannot be optimized.

15.4.2 Type 2

The Type-2 wind turbine is very similar to the Type-1 machine. The difference is that in the Type-2 turbine, the resistance of the rotor can be rapidly adjusted externally, allowing the turbine to operate over a larger speed range than the Type-1 wind turbine (Figure 15.12). An electronic "soft starter" may be included to mitigate large currents and reactive power consumption when the machine is started.

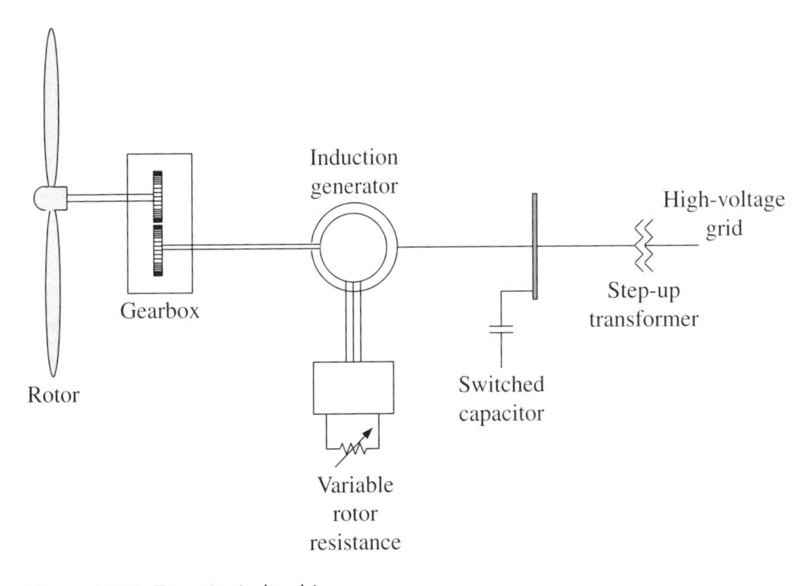

Figure 15.12 Type-2 wind turbine.

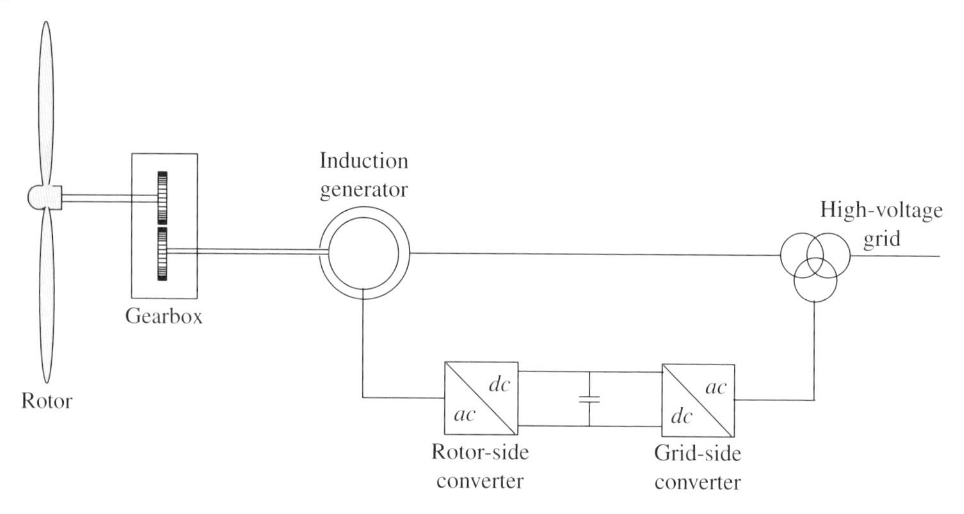

Figure 15.13 Type-3 wind turbine.

15.4.3 Type 3

The Type-3 wind turbine consists of a wound-rotor induction generator with slip rings. The slip rings are used to interface the rotor windings to a voltage-sourced converter (VSC) as shown in Figure 15.13.[1] With this arrangement, excitation frequency can be adjusted very rapidly, allowing for the control of the rotor speed over a wide range of speeds; the practical speed range may vary from 0.7 pu to 1.1 pu of system synchronous frequency [210]. Wind turbines with this structure are termed doubly-fed induction generators (DFIGs) and doubly-fed asynchronous generators (DFAGs). An intrinsic characteristic of the DFIG configuration is that only a fraction of the total turbine power is processed by the converters; typically, this amount is approximately 30%. Consequently, the converters are rated accordingly.

15.4.4 Type 4

In Type-4 wind turbines the generator is connected to the power grid through a full converter, i.e., the converter carries all the generated power (Figure 15.14). This arrangement completely decouples the generator speed from the power system frequency and allows operation over a wider range of speeds. Different types of generators have been used in this type of wind turbine, including permanent magnet rotor synchronous machines, wound rotor synchronous machines, and squirrel cage induction machines [214].

WTGs of Types 1 and 2 reflect earlier technologies of wind power generation. Today, the great majority of large utility-scale wind turbines are of Types 3 and 4, i.e., variable-speed type machines. Type-4 wind turbines are particularly attractive for offshore wind farms.

1 The operation of VSCs has been discussed in Chapter 14.

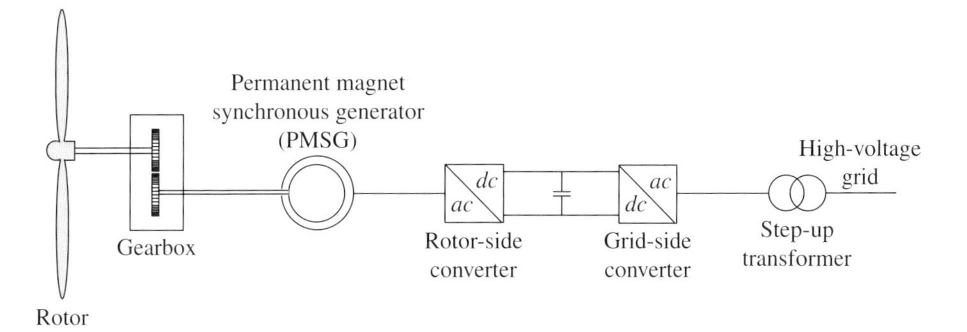

Figure 15.14 Type-4 wind turbine.

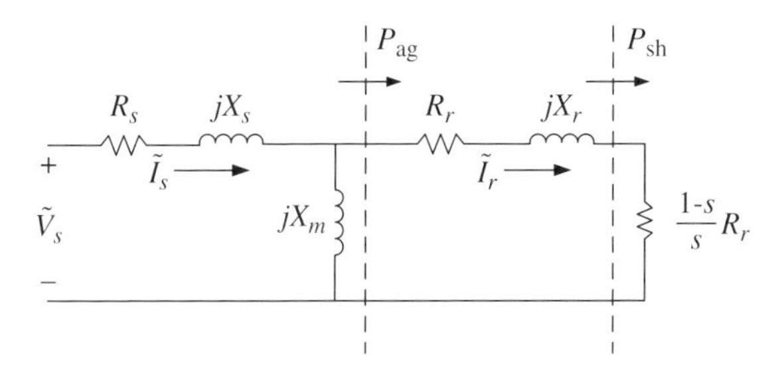

Figure 15.15 Per-phase equivalent circuit of an induction machine.

15.5 Steady-State Characteristics

In this section, a brief description of the steady-state characteristics of different types of wind turbines is given. This is done by analyzing the torque versus speed (or slip) curve of an induction machine. Some basics of induction machines have already been covered in Chapter 11. Here the discussion describes those machines in more detail in the context of wind power generation.

15.5.1 Type-1 Wind Turbine

Under steady-state conditions, the torque-slip characteristics of the wind turbines of Types 1 and 2 can be derived from the equivalent circuit of an induction machine. Figure 15.15 shows the steady-state single-phase equivalent circuit of an induction machine with all quantities referred to the stator [6], where R_s and R_r are the rotor and stator resistances, respectively, X_s and X_r are the rotor and stator leakage reactances, respectively, X_m is the magnetizing reactance, and s is the slip. Due to its familiarity, a motor convention for current flow and slip is adopted here.

From the equivalent circuit in Figure 15.15, the air-gap power, P_{ag}, and the power losses in the rotor resistance, $P_{r\text{loss}}$, are given by

$$P_{ag} = \frac{R_r}{s} I_r^2, \quad P_{r\text{loss}} = R_r I_r^2 \tag{15.11}$$

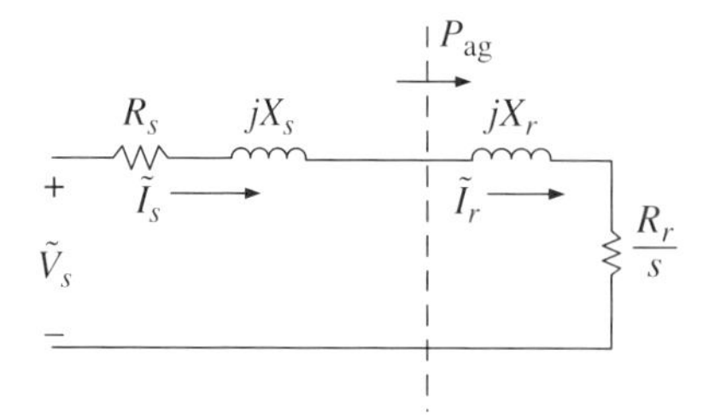

Figure 15.16 Per-phase equivalent circuit of an induction machine used to compute its torque-slip equation.

where I_r is the rotor current magnitude. The air-gap power minus the rotor power losses gives the power P_{sh} transferred to the shaft

$$P_{\text{sh}} = R_r \frac{1-s}{s} I_r^2 \tag{15.12}$$

In generator operation, $s < 0$ and thus $P_{\text{sh}} < 0$. The negative P_{sh} is consistent with the fact that the rotor is providing power to the stator. To compute the torque developed by a three-phase machine, P_{sh} is multiplied by 3 and divided by the rotational speed ω_m (in mechanical rad/sec)

$$T_e = 3\frac{P_{\text{sh}}}{\omega_m} = 3\frac{1}{\omega_m} R_r \frac{1-s}{s} I_r^2 \tag{15.13}$$

The magnitude of the magnetizing reactance X_m is much larger than the magnitude of the stator impedance. This allows for the use of the equivalent circuit shown in Figure 15.16 in which X_m is neglected to derive the torque-slip relationship.[2]

From Figure 15.16, the rotor current phasor $\tilde{I}_r$ as a function of the stator voltage phasor $\tilde{V}_s$ is given by

$$\tilde{I}_r = \frac{\tilde{V}_s}{(R_s + R_r/s) + j(X_s + X_r)} \tag{15.14}$$

Hence, the magnitude of $\tilde{I}_r$ and $\tilde{V}_s$ are related by

$$I_r^2 = \frac{V_s^2}{(R_s + R_r/s)^2 + (X_s + X_r)^2} \tag{15.15}$$

The rotational speed ω_m in mechanical rad/sec is related to the rotational speed ω_r in electrical rad/sec by $2/N_p$, where N_p is the number of poles, such that

$$\omega_m = \frac{2}{N_p}\omega_r = \frac{2}{N_p}(1-s)\omega_o \tag{15.16}$$

In (15.16), $\omega_o = 2\pi f_o$, where f_o is the system nominal frequency, typically 50 or 60 Hz.

Substituting (15.15) and (15.16) in the torque equation (15.13) gives the relationship between the torque and the slip as

$$T_e = 3\frac{N_p}{2}\frac{R_r}{s\omega_o}\frac{V_s^2}{(R_s + R_r/s)^2 + (X_s + X_r)^2} \tag{15.17}$$

2 Alternatively, a Thèvenin equivalent incorporating X_m can be used (see Chapter 11).

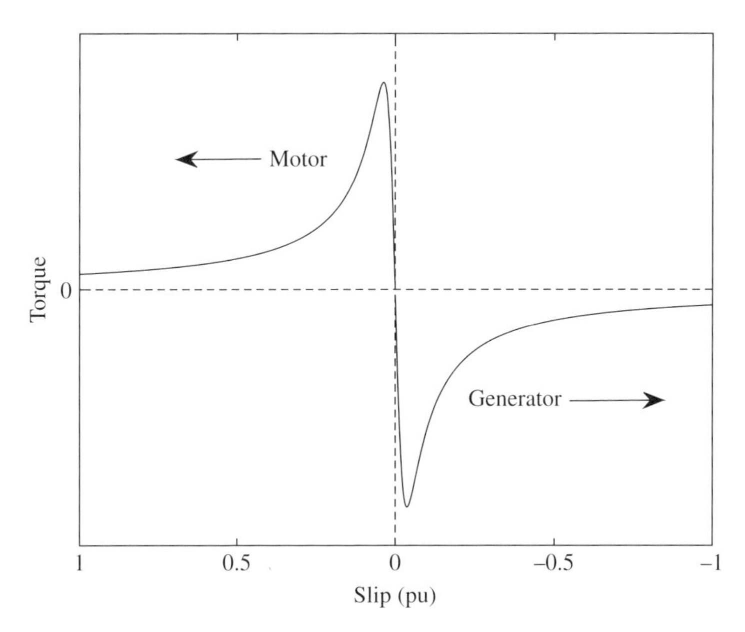

Figure 15.17 Torque vs slip characteristic of an induction machine.

A typical torque versus slip curve for an induction machine is shown in Figure 15.17. When the slip is greater than minus one and less than zero, the machine functions as a generator; for slip values greater than zero and less than one, the machine behaves as a motor. The figure shows that when the slip is zero, i.e., when the machine is running at synchronous speed, the torque is also zero.

A Type-1 wind turbine does not include the means to control its speed, i.e., the turbine operates at essentially a constant speed regardless of the wind speed. This is an inherent characteristic of this type of machine and is due to the steep slope of the torque-slip curve of the induction generator in the neighborhood of the rated operating point (approximately 1% slip at rated torque) [208]. The small range of operation is shown in Figure 15.18.

15.5.2 Type-2 Wind Turbine

Unlike Type-1 WTGs, a Type-2 machine allows for a limited adjustment of the rotor resistance R_r via electronic controls. This permits the turbine to operate over a wider speed range than the simpler Type-1 design, which translates into an increased capability for extracting more energy from the wind. However, the rotor resistance also increases the losses. The torque generated by a Type-2 WTG with an external resistance of $(\alpha - 1)R_r$ is given by

$$T_e = 3\frac{N_p}{2}\frac{\alpha R_r}{s\omega_o}\frac{V_s^2}{(R_s + \alpha R_r/s)^2 + (X_s + X_r)^2} \tag{15.18}$$

In this arrangement, the maximum torque remains constant as the rotor resistance varies. Note that for $\alpha = 1$, (15.18) corresponds to a Type-1 wind turbine.

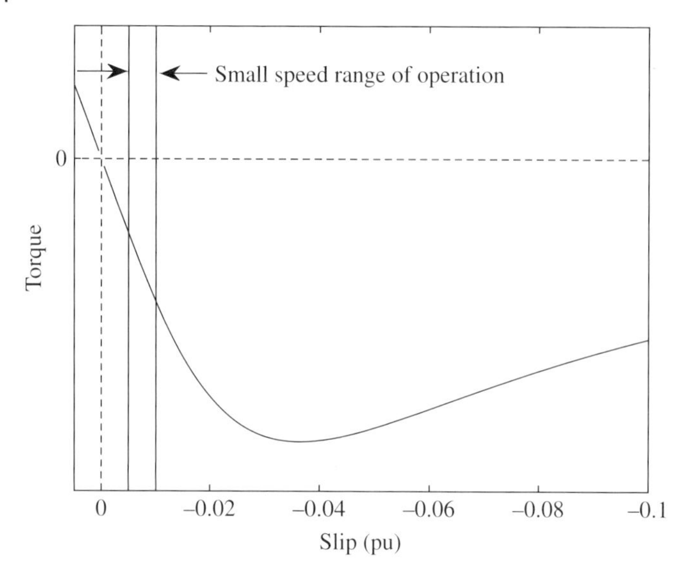

Figure 15.18 Torque vs slip characteristic of an induction generator in the neighborhood of its rated operating point.

Example 15.3

Plot the torque-slip curve of a Type-2 WTG of a single phase for an induction generator with the parameters $R_s = 0.00488$ pu, $X_s = 0.09241$ pu, $R_r = 0.007$ pu, and $X_r = 0.09955$ pu [207], for $\alpha = 1$, 4, and 8, and $s = -0.5$ to 0.5 pu. Assume that $V_s = 1.0$ pu, $N_p = 2$, and $\omega_o = 1$ pu.

Solutions: The torque-slip equation for a single phase is

$$T_e = \frac{\alpha R_r}{s\omega_o} \frac{V_s^2}{(R_s + \alpha R_r/s)^2 + (X_s + X_r)^2} \tag{15.19}$$

The torque-slip curve is shown in Figure 15.19. ■

15.5.3 Type-3 Wind Turbine

A DFIG consists of a wound-rotor induction machine equipped with a back-to-back voltage-sourced converter connected to the rotor circuit through slip rings, as shown in Figures 15.20 and 15.21. The converter allows power exchange between the rotor circuit and the power network. A consequence of this arrangement is that the active power is divided between the stator and rotor circuits approximately in proportion to the slip frequency, i.e., $P_{\text{rotor}} = -sP_{\text{stator}}$ (neglecting losses).

For negative slip values (rotor speeds greater than synchronous speed), the rotor active power is injected into the network through the converter. Conversely, for rotor speeds less than synchronous speed, active power flows from the network to the rotor, as indicated in Figure 15.21. In both cases the stator is feeding active power into the grid. There is no converter power flow between the grid and the rotor when the machine operates at synchronous speed ($s = 0$) and losses are neglected.

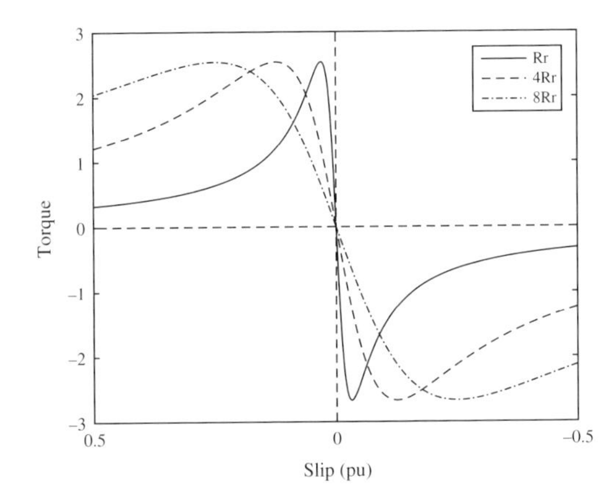

Figure 15.19 Torque-slip curves for a Type-2 WTG with several external resistance values.

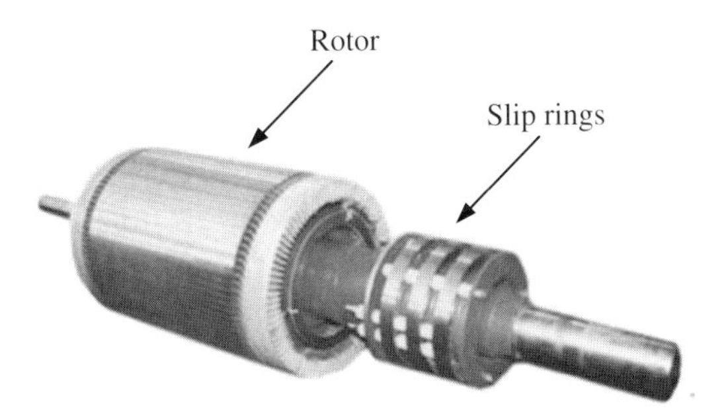

Figure 15.20 Rotor with slip rings (courtesy of GE Renewable Energy) (see color plate section).

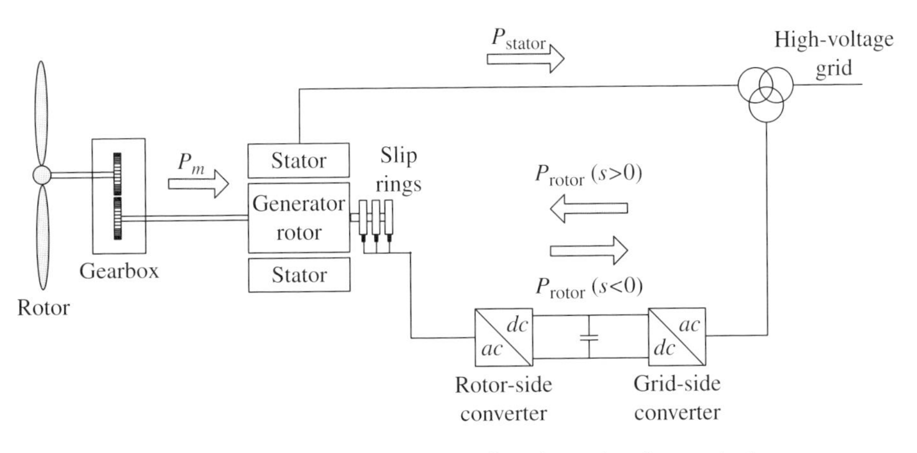

Figure 15.21 DFIG configuration and active power flows (see color plate section).

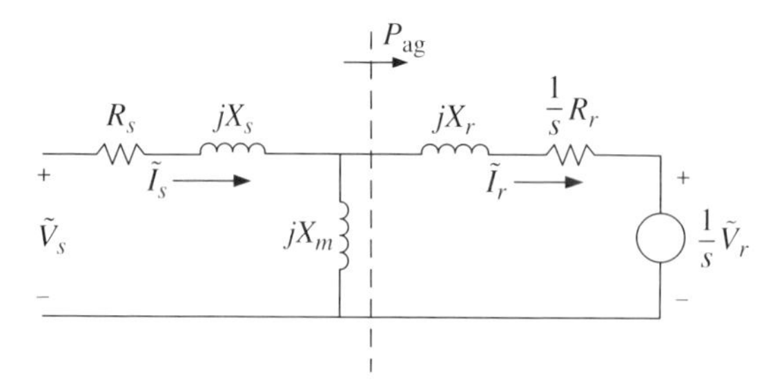

Figure 15.22 DFIG per-phase equivalent circuit.

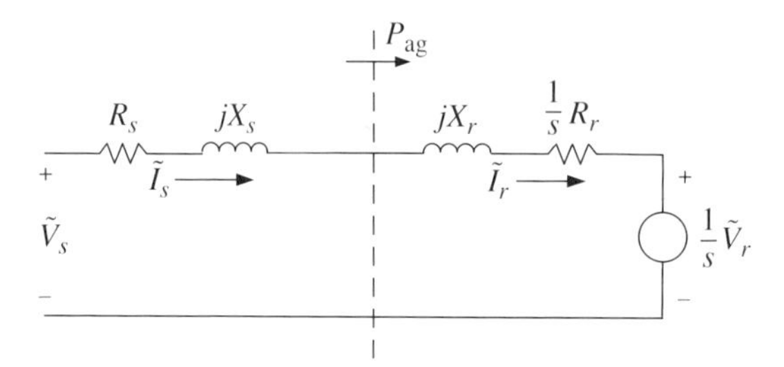

Figure 15.23 Per-phase equivalent circuit of a DFIG used to compute its torque-slip equation.

The torque-slip characteristics of a Type-3 wind turbine can also be derived from the modified equivalent circuit of an induction machine shown in Figure 15.22. This circuit includes an injected voltage that varies with the slip to represent the effect of the rotor-side converter control $\tilde{V}_r/s$ [207]. In Figure 15.22, the rotor voltage and parameters are referred to the stator.

Because the magnitude of the magnetizing reactance X_m is much larger than the magnitude of the stator impedance, the simpler equivalent circuit shown in Figure 15.23, with the magnetizing reactance neglected, can be used to derive the desired torque-slip relationship.

From the circuit in Figure 15.23, the rotor current is given by [208]

$$\tilde{I}_r = \frac{\tilde{V}_s - \tilde{V}_r/s}{(R_s + R_r/s) + j(X_s + X_r)} \tag{15.20}$$

A comparison with (15.14) shows that the injected voltage $\tilde{V}_r/s$ in the numerator of (15.20) is the only difference.

To understand the impact of the insertion of $\tilde{V}_r$, consider the voltage insertion of $\tilde{V}_r$ to be in phase with $\tilde{V}_s$. This can be readily accomplished by controlling the voltage insertion and measuring the phase angle of the terminal bus voltage, using for example, a phase-locked-loop circuit. Then the phasor notation can be dropped by assuming the phases of $\tilde{V}_s$ and $\tilde{V}_r$ to both be zero. The resulting electrical torque expression can be simplified to

$$T_e = \frac{3N_p}{2} \frac{V_s(V_s - V_r/s)(R_s + R_r/s)}{(R_s + R_r/s)^2 + (X_s + X_r)^2} \tag{15.21}$$

As an illustration, the torque-slip curve (15.21) using the induction generator parameters from Example 15.3 for values of $V_r = -0.02$, 0, and 0.02 pu is shown in Figure 15.19.

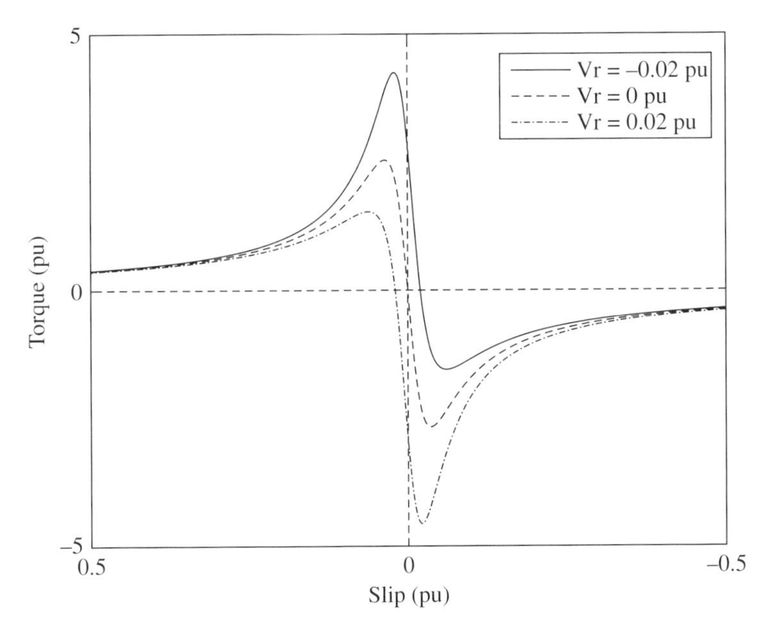

Figure 15.24 Torque-slip curve of Type-3 WTG for Example 15.3.

Note that the electrical torque can be made nonzero even when $s = 0$ pu, due to the fact that the rotor current $\tilde{I}_r$ becomes nonzero when $V_r \neq 0$.

15.6 Wind Power Plant Representation

A large WPP may consist of hundreds of relatively small WTGs deployed over a large geographical area. The individual WTGs are equipped with a step-up transformer and are interconnected via an electrical network termed "the collector system." The WPP is connected to the power grid at a bus called the point of interconnection (POI) or point of common coupling (PCC), as shown in Figure 15.25.

Within the WPP, each individual wind turbine operates at slightly different conditions due to the wind variability and the different impedance between each turbine and the POI. This is illustrated in Figure 15.26, which shows simulation results for the terminal voltage of individual wind turbines in a WPP responding to a sudden voltage drop in the bulk power network. The figure shows the similarity of the responses of the units as well as their slightly different operating conditions.

Although each individual WTG in a WPP operates at slightly different conditions, from the standpoint of the bulk power system, it is the dynamic performance of the WPP at the POI that really matters. Most WPPs include a plant-level controller that processes measurements taken at the POI and coordinates the operation of the different plant components, such as switched capacitors, SVCs, and the individual WTGs, in order to meet the operational requirements imposed at the POI. The plant-level controller also processes commands issued by the transmission system operator [206]. The concept of a plant-level controller is illustrated in Figure 15.27.

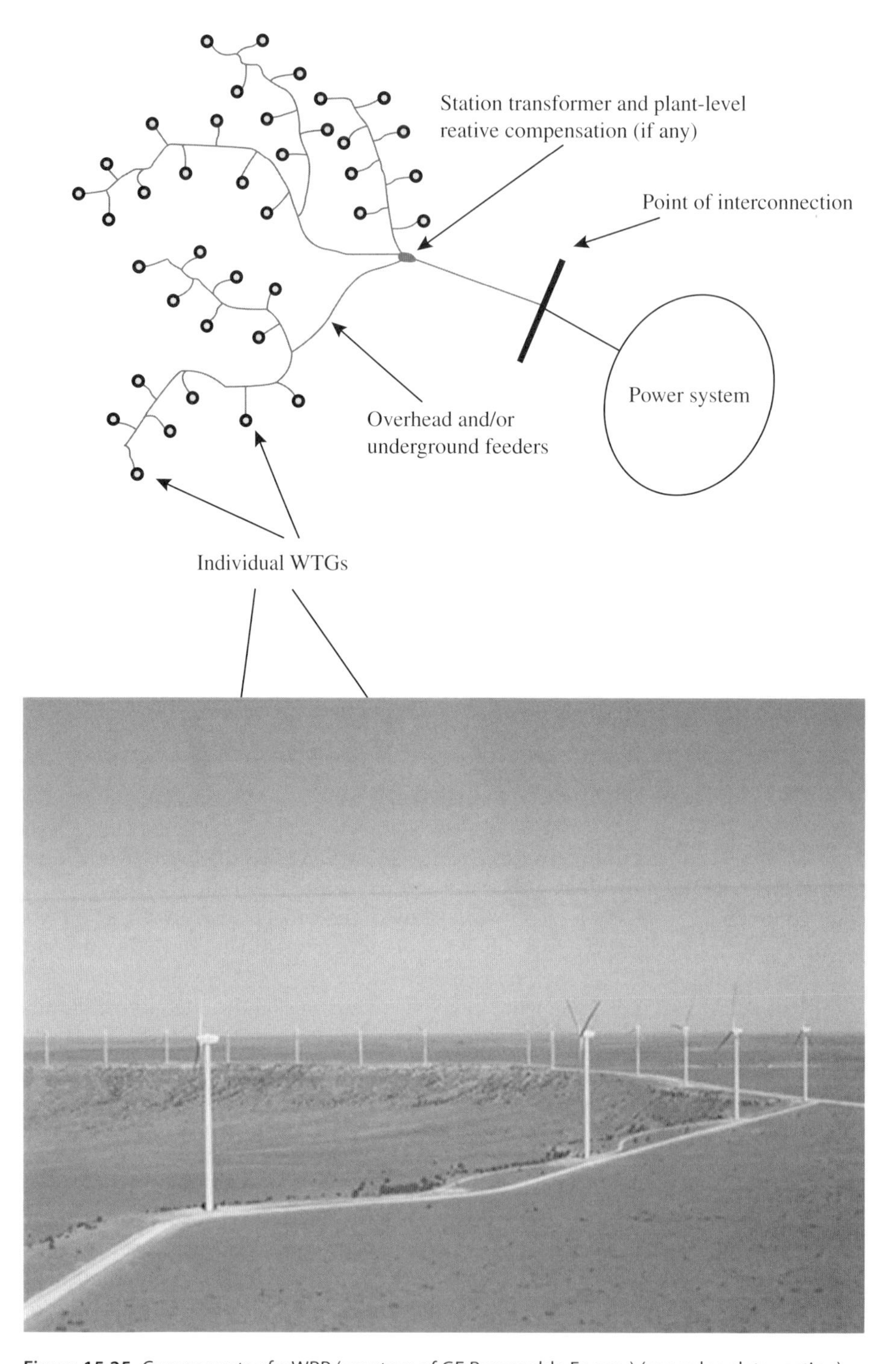

Figure 15.25 Components of a WPP (courtesy of GE Renewable Energy) (see color plate section).

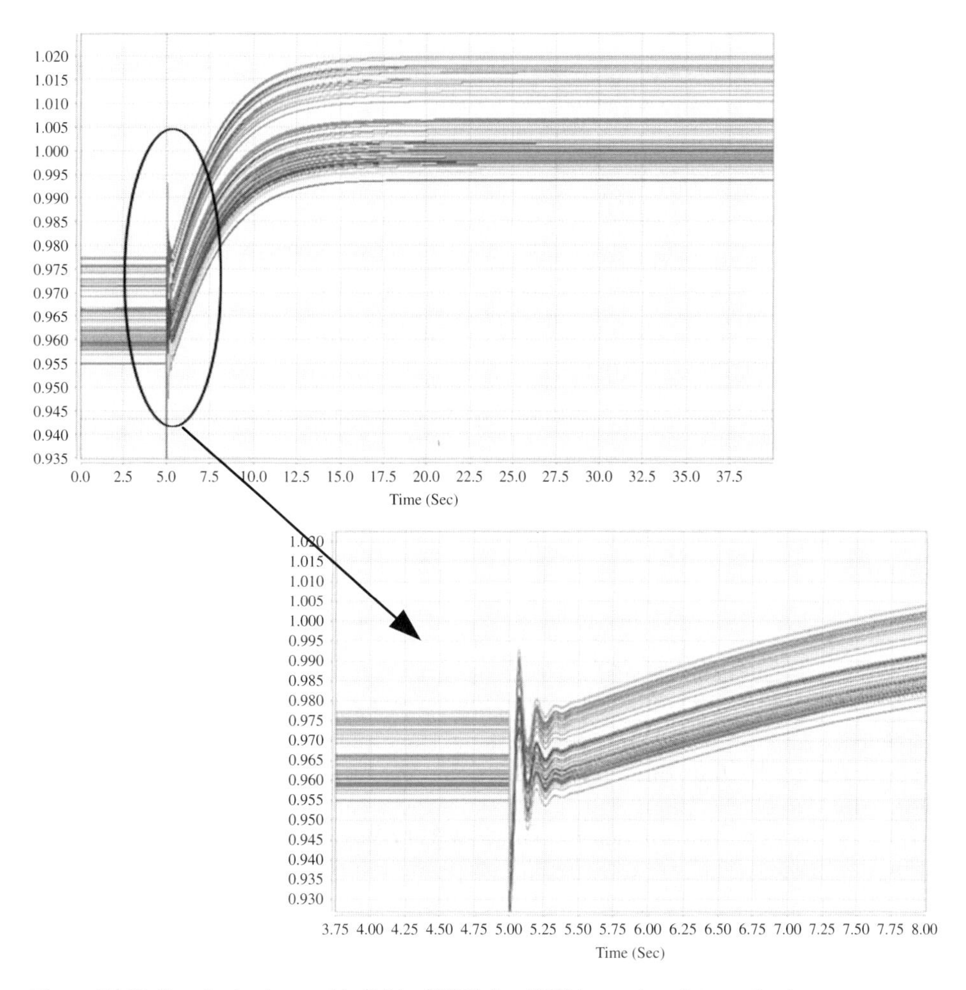

Figure 15.26 Terminal voltage of individual WTGs in a WPP (see color plate section).

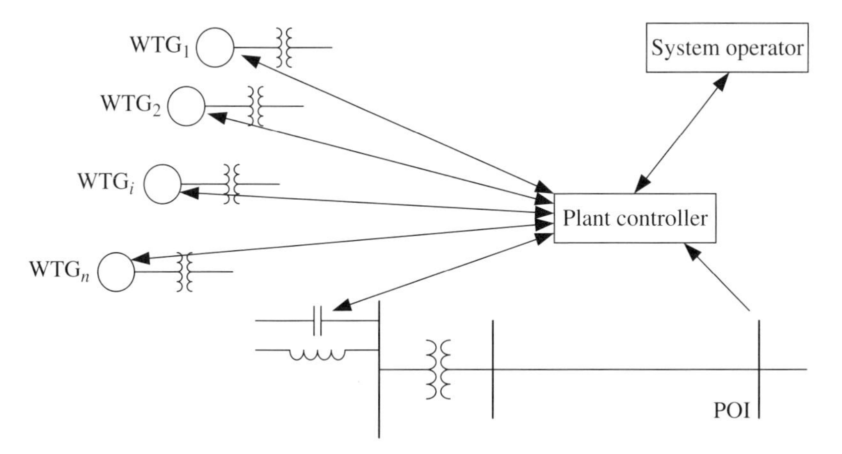

Figure 15.27 Plant controller interface with reactive support devices.

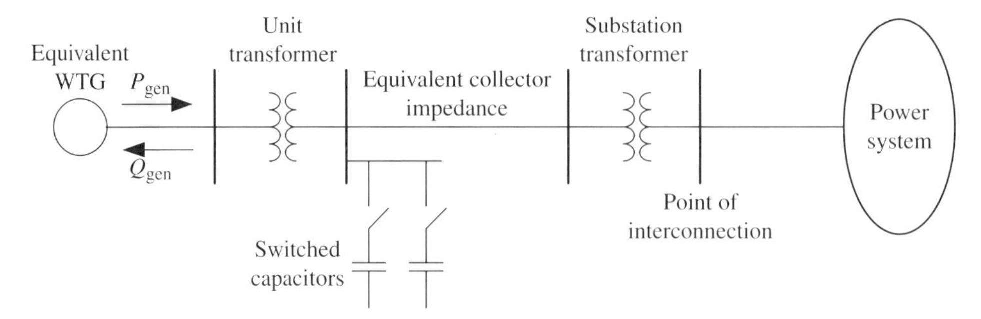

Figure 15.28 Aggregate model for WTG of Types 1 and 2.

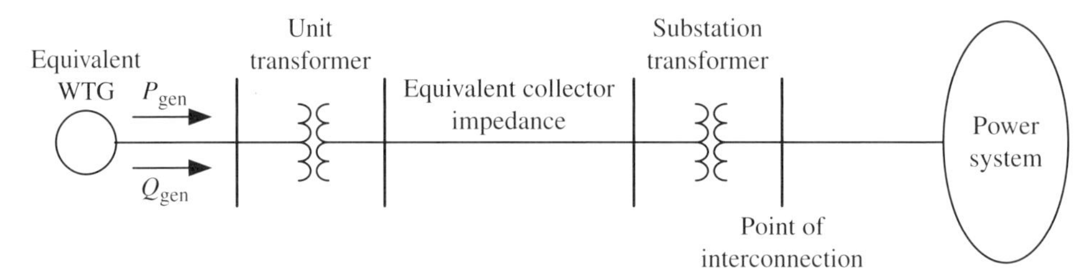

Figure 15.29 Aggregate model for WTG of Types 3 and 4. For Type-4 WTG, the interconnection of the WTG to the grid uses a DC transmission system (see Figure 15.14).

Typical planning studies of large power systems involve the execution of a large number of simulations to study different contingencies and system operating conditions. To perform this type of study, it is impractical and unnecessary to include a model of each individual generating unit. Rather, the single WTG equivalent representations shown in Figures 15.28 and 15.29 are used [208]. This configuration has been found to provide the level of detail required in power system planning studies and allow for the efficient execution of transient stability simulations. In power plants that consist of more than a single type of WTG, or when the plant contains feeders of very different impedances, more than a single equivalent can be used.

The equivalent WPPs shown in Figures 15.28 and 15.29 consist of an equivalent WTG, an equivalent unit transformer, an equivalent collector system, and the substation transformer. The rating of the equivalent wind turbine is equal to the rating of an individual machine times the number of machines in operation. The unit transformer is typically a delta-wye connected transformer, and its rating is also equal to the rating of an individual unit times the number of units in service. The equivalent collector system is a simple impedance. Notice that the equivalent WPP shown in Figure 15.28 is for wind turbines of Types 1 and 2. Because the generator in these types of turbines is essentially a conventional induction generator, reactive power must be provided by a local source such as switched capacitors. Figure 15.29 refers to wind turbines of Types 3 and 4, although additional reactive compensation might be required in a particular installation due to system conditions, Type-3 and Type-4 wind turbines do not inherently require reactive support and can generate or absorb reactive power. The stator of a Type-4 wind turbine is connected directly to DC rectifiers. The wind power is then transmitted over a DC line and converted back to AC via inverters connected to an AC grid. The inverters can also perform reactive power control.

The equivalent collector impedance is a very simple representation of a rather intricate electrical network designed to minimize voltage drops and power losses. A straightforward method was introduced in [211] and further illustrated in [206] to compute the equivalent impedance of a collector system. The resultant impedance captures the active and reactive power losses as well as voltage drops. The method is based on several simplifying assumptions and leads to two simple equations to compute the collector equivalent impedance

$$Z_{\mathrm{eq}} = R_{\mathrm{eq}} + jX_{\mathrm{eq}} = \left(\sum_{k=1}^{K} Z_k n_k^2 \right) / N^2, \quad B_{\mathrm{eq}} = \sum_{k=1}^{K} B_k \tag{15.22}$$

where k denotes the individual lines, Z_k and B_k are the impedance and charging of line k, respectively, Z_{eq}, R_{eq}, X_{eq}, and B_{eq} are the equivalent impedance, resistance, reactance, and charging of the equivalent line, respectively, K is the total number of branches in the collectors system, and N is the total number of wind turbines in operation. The parameter n_k is the number of wind turbines connected between a given bus and the end of the collector branch(es). The use of (15.22) is illustrated in the following example.

Example 15.4

A small WPP consisting of 14 wind turbines is shown in Figure 15.30. Each wind turbine is rated at 1.7 MVA and equipped with a 0.6 kV/34.4 kV unit transformer rated at 1.75 MVA. The collector impedance data is given in the R, X, and B columns in Table 15.1. Use (15.22) to derive a single WTG equivalent of the WPP.

Solutions: Consider the collector section between Buses 4 and 5 (row 4 in Table 15.1). The number of wind turbines between Bus 5 and the end of the collector branch is 4 (the

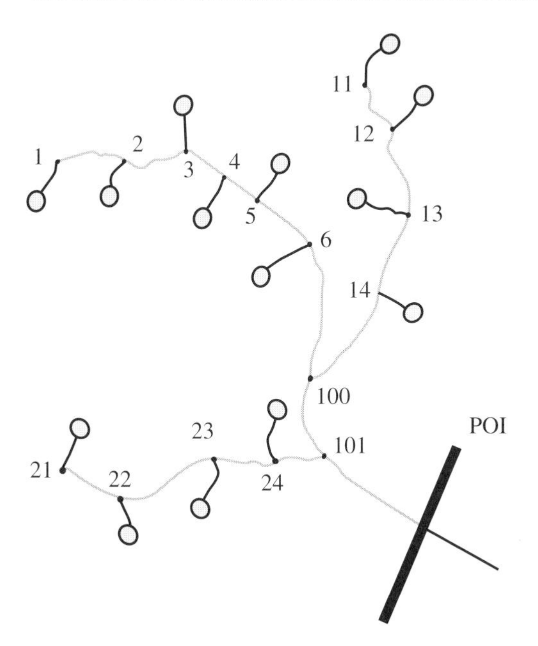

Figure 15.30 Wind power plant (see color plate section).

Table 15.1 System data and collector impedance computation.

	From bus	To Bus	*R*	*X*	*B*	*n*	n^2	n^2R	n^2X
1	1	2	0.0023	0.0322	0.0005	1	1	0.0023	0.0322
2	2	3	0.0047	0.0448	0.0009	2	4	0.0188	0.1792
3	3	4	0.0073	0.0350	0.0002	3	9	0.0657	0.3150
4	4	5	0.0056	0.0461	0.0004	4	16	0.0896	0.7376
5	5	6	0.0019	0.0275	0.0008	5	25	0.0475	0.6875
6	6	100	0.0029	0.0092	0.0012	6	36	0.1044	0.3312
7	11	12	0.0033	0.0321	0.0000	1	1	0.0033	0.0321
8	12	13	0.0096	0.0400	0.0000	2	4	0.0384	0.1600
9	13	14	0.0033	0.0373	0.0003	3	9	0.0297	0.3357
10	14	100	0.0039	0.0364	0.0006	4	16	0.0624	0.5824
11	100	101	0.0080	0.0265	0.0009	10	100	0.8000	2.6500
12	21	22	0.0060	0.0237	0.0006	1	1	0.0060	0.0237
13	22	23	0.0022	0.0135	0.0008	2	4	0.0088	0.0540
14	23	24	0.0093	0.0336	0.0003	3	9	0.0837	0.3024
15	24	101	0.0077	0.0274	0.0004	4	16	0.1232	0.4384
16	101	POC	0.0059	0.0412	0.0007	14	196	1.1564	8.0752
					Col. sum $B_{eq} = 0.0086$			Col sum 2.6402	Col. sum 14.9366
								$R_{eq} = 0.01347$	$X_{eq} = 0.07621$

number of wind turbines connected at Buses 1, 2, 3, and 4). Hence $n_k = 4$. Similarly, consider the collector section between Buses 100 and 101 (row 11 in Table 15.1). In this case $n_k = 10$ (the number of wind turbines connected at Buses 1, 2, 3, 4, 5, 6, 11, 12, 13, and 14).

Accordingly, the n_k^2R and n_k^2X values are tabulated in the last two columns of Table 15.1. The equivalent resistance and reactance are obtained by dividing the column sums of n_k^2R and n_k^2X by $N^2 = 14^2 = 196$. ■

15.7 Overall Control Criteria for Variable-Speed Wind Turbines

The overall control criteria for a variable-speed WTG can be described in terms of the wind power versus speed curve first introduced in Section 15.3. Two main regions of operation are defined by this curve: Regions 2 and 3. In Region 2 where the wind speed is below its rated value, the control objective is to maximize power extraction from the wind by adjusting the turbine speed in response to changes in the wind. In this region, the pitch angle is fixed. Since the power extracted from the wind is proportional to the power coefficient C_p, which is a function of both the pitch angle θ and the tip speed

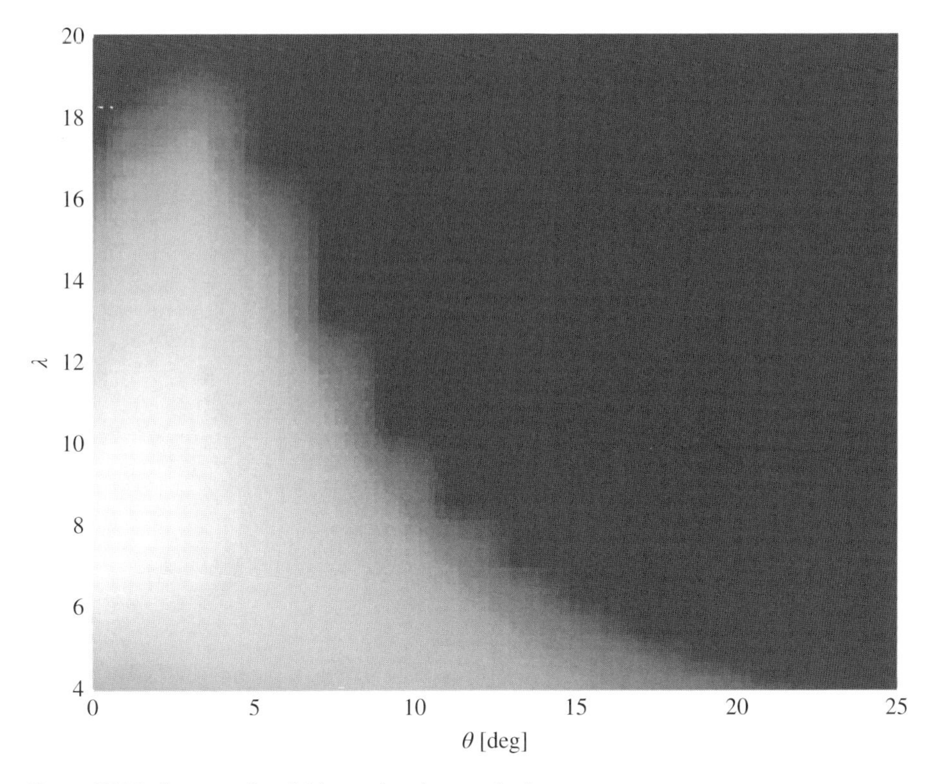

Figure 15.31 C_p versus θ and λ (see color plate section).

ratio λ, the control objective in Region 2 is to hold the tip speed ratio at a value that corresponds to maximum C_p. This concept is illustrated in Figure 15.31, which shows a heat map plot of C_p versus θ and λ. With the turbine operating in Region 2 and θ held at its minimum value of 0°, the turbine speed would be adjusted to operate at a value of λ of approximately 7.7 where C_p reaches its maximum value.

In Region 3, where the wind speed is above its rated value, the main control objective is to limit the turbine output power to its nominal value. In this region, the pitch angle is adjusted to shed the extra power in the wind and meet the control objective to prevent electrical and/or mechanical power overloads. Figure 15.32 is a conceptual view of the main control functions associated with WTGs of Types 3 and 4; Figure 15.33 summarizes the control actions in Regions 2 and 3.

Although the final objective of all wind turbines is to produce electrical power efficiently and economically within their design constraints, the manner in which this goal is achieved is different for each wind turbine manufacturer. Utility-size wind turbines have different controllers designed to function in different time scales. The rotor-side converter is typically used for torque and terminal voltage or power factor control; the grid side converter is used to hold the DC-link capacitor voltage at a set value and to maintain converter operation at a given power [207]. The reactive power output of the generator can be controlled by varying the magnitude of the rotor currents. This gives the Type-3 and Type-4 WTGs voltage regulation capabilities not unlike a synchronous generator but with greater speed of response [217]. The dynamic characteristics of the

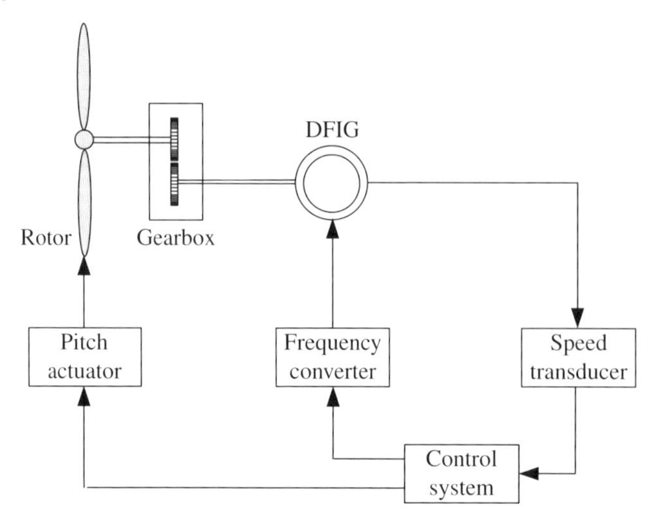

Figure 15.32 Main control functions.

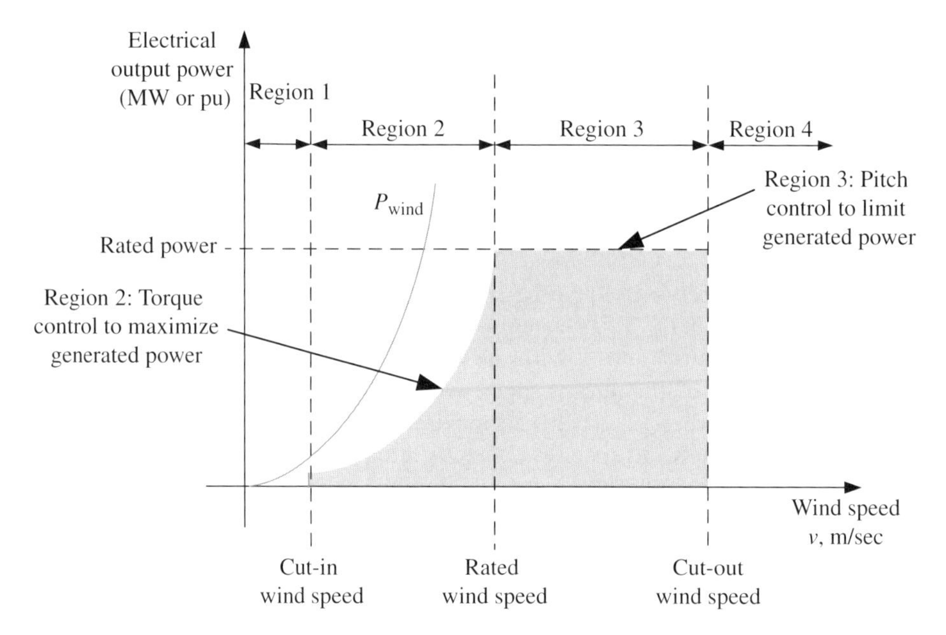

Figure 15.33 Power vs speed curve (see color plate section).

Type-3 and Type-4 WTGs are completely dominated by the very fast acting converters and are very different than the dynamic characteristics of the Type-1 and Type-2 WTGs.

The control actions outlined in the previous paragraphs have been represented in many simulation models of varying levels of detail. Very detailed models are used by wind turbine manufacturers to study and develop their products. These models essentially mimic the performance of the actual equipment and include the representation of very fast electronic components. At the other end of the spectrum, in terms of complexity and detail, are the models used to perform transient stability planning studies. Planning studies are performed to demonstrate compliance with power system performance criteria and for planning system upgrades. They are typically conducted using very large data sets and involve the execution of a large number of transient stability simulations. For these reasons, dynamic models used in planning studies need to be simple enough for computational efficiency and, at the same time, detailed enough to represent the dynamic phenomena of interest. The next section describes one such model.

15.8 Wind Turbine Model for Transient Stability Planning Studies

The dynamic wind turbine models for planning studies reflect the practical constraints associated with the representation of advanced equipment whose dynamic characteristics span a wide time frame. In this type of model, the turbine mechanical controls are simplified and the very fast dynamics associated with the converters are modeled by algebraic equations. The wind turbine model described in this section is used to represent Type-3 WTGs. The model is suitable for representing individual WTGs or an equivalent WPP, and has been implemented in commercial transient stability programs used to conduct planning studies in North America [215].

The model conforms to various modeling and performance assumptions, including [216]:

- The model is intended for the simulation of events in a time associated with typical transient stability simulations, i.e., 10–20 seconds.
- It is assumed that for fault condition studies in the simulation time frame (10–20 seconds), the wind speed remains constant.
- The model is not intended for use in simulations that involve severe frequency excursions or weak systems.

Wind turbine control systems are more complex than those for synchronous machines. In a synchronous machine, the mechanical power setpoint for the governor and the voltage setpoint for the excitation system are determined by optimal dispatch and can be set externally by a control center. For a wind turbine, the active and reactive power setpoints are determined by the wind speed and interconnection conditions. Their reference values are computed using additional control functions.

15.8.1 Overall Model Structure

Figure 15.34 shows the overall model structure and connectivity between the model components used to represent a Type-3 wind turbine. The model required to represent a Type-4 wind turbine is a subset of those used to represent a Type-3 wind turbine [215]. The models are generator/converter, electrical control, aerodynamic conversion (aero), pitch controller, drive-train, and torque controller. An additional model, a plant-level controller, may also be included to represent WPPs where remote measurements are used to regulate variables outside the power plant. This case is discussed later.

The wind turbine model shown in Figure 15.34 is a simplified representation of a complex electro-mechanical system. The model represents system dynamics and

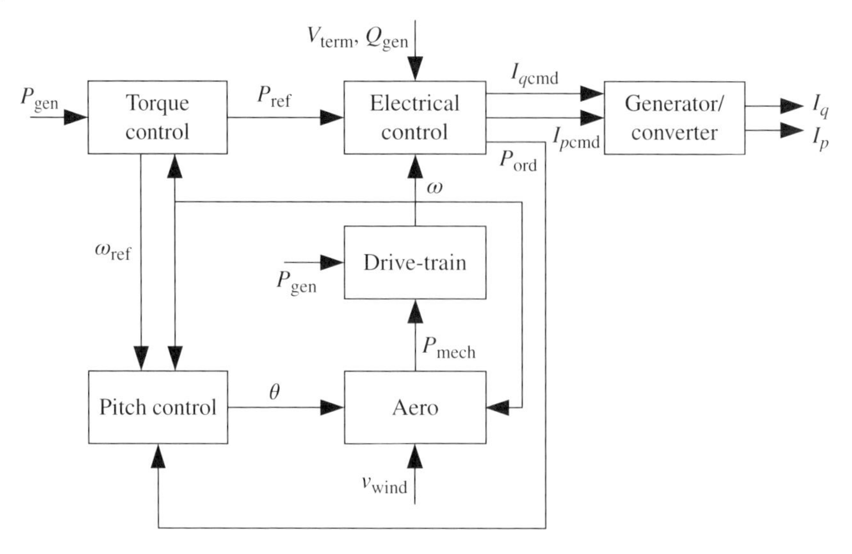

Figure 15.34 Structure of a Type-3 WTG.

control actions that are pertinent for transient stability planning studies. The individual models shown in Figure 15.34 include several features whose description is beyond the scope of this chapter. These features include means for reconfiguring the model in different topologies, power factor control, and the specification of a prescribed reactive control response when the generator terminal voltage drops below a given threshold [215].

15.8.2 Generator/Converter Model

The generator/converter model described in this section is suitable for power system planning studies of the type performed by power system planners. The model is essentially an algebraic model where the fast flux dynamics have been eliminated to reflect the rapid response to the higher level commands from the electrical controls through the converter. The only dynamic components in the model represent small time delays in signal processing. Figure 15.35 shows the basic structure of the model. The model injects active and reactive components (I_p and I_q) of the inverter current into the external network in response to the current commands (I_{pcmd} and I_{qcmd}) from the Electrical Controller.[3] Not unlike other voltage source converters, e.g. STATCOM, the WTG converter synthesizes a voltage behind a reactance which results in the desired current injection [217].

The generator/converter model includes functions to limit the reactive current injected into the network such that the terminal voltage of the machine does not

3 The active component I_p of the current is in phase with the terminal voltage phasor and the reactive component I_q is in quadrature to the terminal voltage phasor.

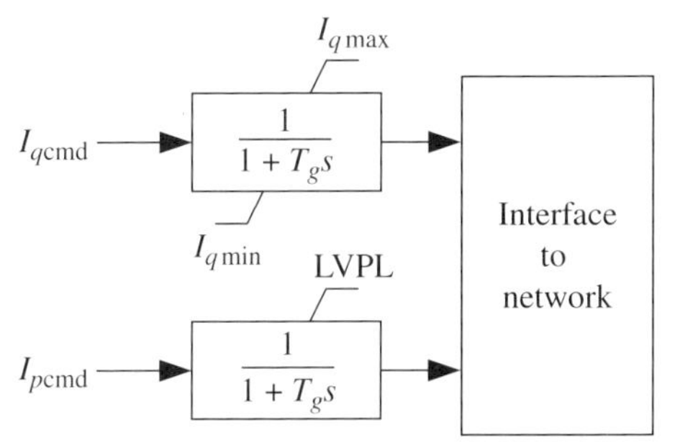

Figure 15.35 Generator/converter model (see color plate section).

exceed a given threshold, as long as the converter is within the current limits (see [215] for a more detailed description of this and other related functions). Furthermore, unlike a conventional generator or motor model, the generator/converter model does not include the state variables associated with the rotor dynamics; these states are included in the drive-train model.

The actual implementation of a model of this nature in a transient stability program may require program-specific logic to mitigate numerical issues that arise due to the approximation of high-bandwidth hardware components by a simple model. Figure 15.35 simply shows a box to denote the presence of these functions in a commercial implementation.

From a practical standpoint, a simplified model of this nature exhibits limitations in its range of validity and should not be used to represent a WPP connected to a very weak grid or to study system performance under severe system frequency excursions.

15.8.3 Electrical Control Model

The electrical control model emulates active and reactive power control actions. Its function is to compute the real and reactive current commands $I_{p\mathrm{cmd}}$ and $I_{q\mathrm{cmd}}$, respectively, that go into the generator/converter model (Figure 15.36). The electrical control model has three inputs: the generator terminal voltage V_{term}, the generator reactive power Q_{gen}, and the real power reference P_{ref} set by the torque controller. If a plant level controller model is included in the representation of the WPP, then the reactive power reference, Q_{ref}, is set by this model.

The electrical control model consists of two essentially independent controllers. One controller computes the reactive current command and the other the active current command. The former controller includes two proportional-integral (PI) blocks for reactive power and local voltage control. The latter computes the active current command based on the active power reference from the torque controller and the terminal voltage of the generator (the time constant T_{rv} represents measurement delays and is typically small).

The model also includes current limits to prevent the current commands from exceeding the equipment rating. These limits are voltage dependent and can be set to enforce reactive or active power priority as follows:

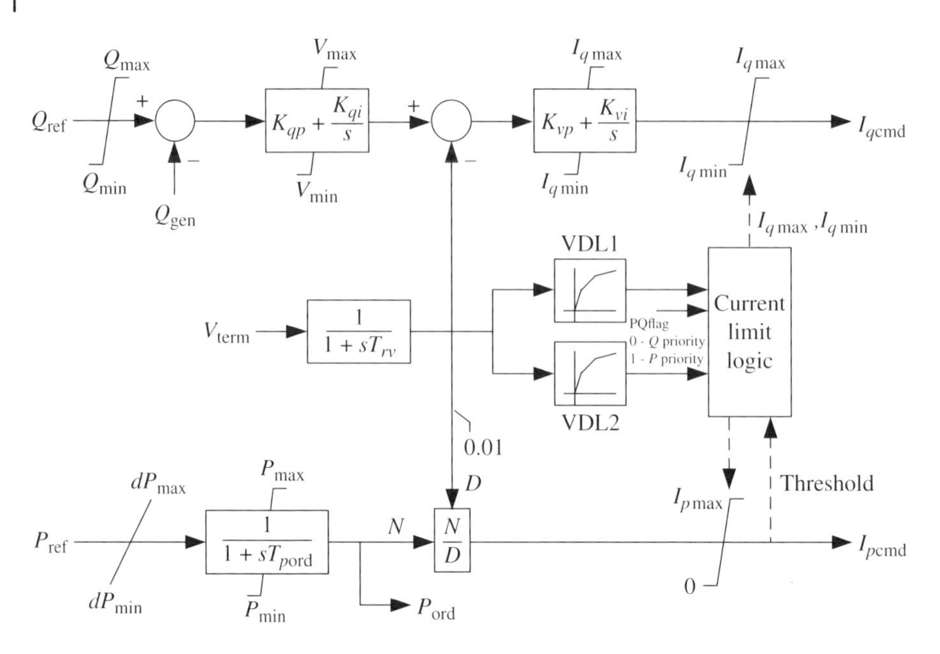

Figure 15.36 Electrical control model (see color plate section).

For Q priority:

$$I_{qmax} = \min\{\text{output of VDL1}, I_{max}\} \tag{15.23}$$

$$I_{qmin} = -I_{qmax} \tag{15.24}$$

$$I_{pmax} = \min\{\text{output of VDL2}, \sqrt{I_{max}^2 - I_{qcmd}^2}\} \tag{15.25}$$

$$I_{pmin} = 0 \tag{15.26}$$

For P priority:

$$I_{qmax} = \min\left\{\text{output of VDL1}, \sqrt{I_{max}^2 - I_{pcmd}^2}\right\} \tag{15.27}$$

$$I_{qmin} = -I_{qmax} \tag{15.28}$$

$$I_{pmax} = \min\{\text{output of VDL2}, I_{max}\} \tag{15.29}$$

$$I_{pmin} = 0 \tag{15.30}$$

VDL1 and VDL2 represent limits as a function of the terminal voltage. In addition to the reactive and active current commands, the electrical controller also computes the power order processed by the pitch controller.

Low-voltage ride-through

WPPs include protective functions that define the voltage levels at which the plant is to remain on-line to meet low- and high-voltage ride-through (LVRT and HVRT) requirements. By meeting these requirements, a WPP remains connected to the power grid during conditions of low voltage. Such a capability reduces the risk of a major incident detrimental to the system operation such as a voltage collapse. Figure 15.37

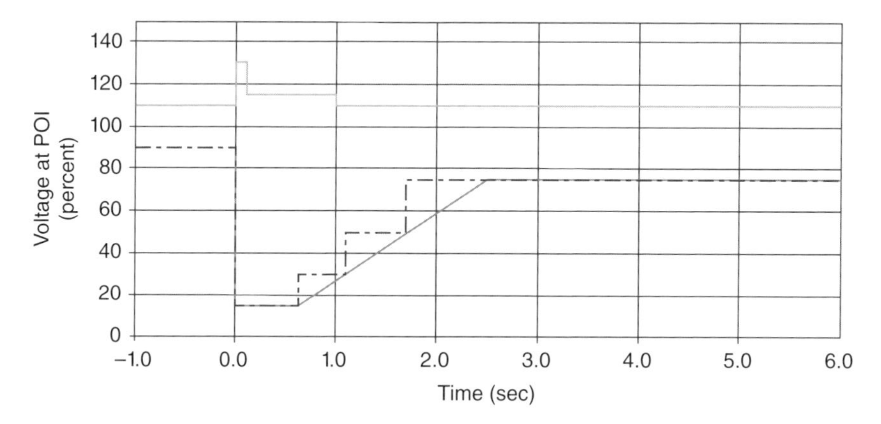

Figure 15.37 LVRT specifications [217] (see color plate section).

is an example of LVRT specifications [217], which shows both high and low thresholds (denoted by the upper and lower red curves) within which a power plant must operate for at least the specified amount of time. For example, the plant is to remain on-line for at least 2.0 seconds when the voltage at the POI is 0.6 pu. In Figure 15.37 the dashed line is a stepwise fit to the curve that defines the equipment minimum performance. The thresholds and associated times are set to meet specific grid codes and/or interconnection requirements [217].

15.8.4 Drive-Train Model

The drive-train model represents the dynamics of the inertial components of the WTG. Although the drive train of the WTG set consists of several masses, including those of the hub, turbine blades, gearbox, and generator, as well as associated damping coefficients and spring constants, a one- or two-mass model is adequate for planning studies.

In the two-mass model, the turbine and generator inertia constants, H_t and H_g, respectively, are represented as two separate entities as shown in Figures 15.38 and 15.39. In the single-mass model (Figure 15.40) the turbine and generator inertia constants are combined into an equivalent inertia constant $H = H_t + H_g$. The torsional damping coefficient D_{tg} approximates the damping provided by a damping function in the actual turbine controller. The spring constant K_{shaft} is the stiffness constant. The initial speed is represented by ω_o.

The gearbox in a wind turbine is a complex device designed to withstand the sudden changes in load and transform the speed of rotation of the very slowly rotating turbine to the shaft speed needed to operate the generator [212].

The state equations for the two-mass model are given by

$$\begin{aligned}
\frac{d\Delta\omega_t}{dt} &= \frac{1}{2H_t}(T_{\text{mech}} - D_{\text{shaft}}(\Delta\omega_t - \Delta\omega_g) - K_{\text{shaft}}\delta_{tg}) \\
\frac{d\Delta\omega_g}{dt} &= \frac{1}{2H_g}(-T_{\text{elec}} + D_{\text{shaft}}(\Delta\omega_t - \Delta\omega_g) + K_{\text{shaft}}\delta_{tg}) \\
\frac{d\Delta\delta_{tg}}{dt} &= \Delta\omega_t - \Delta\omega_g
\end{aligned} \tag{15.31}$$

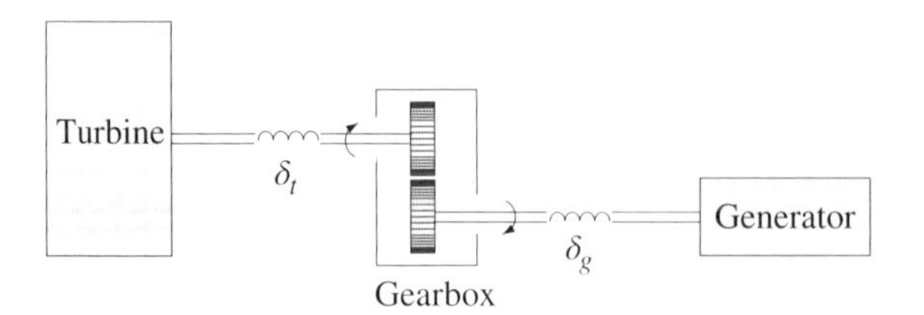

Figure 15.38 Two-mass turbine-generator model (see color plate section).

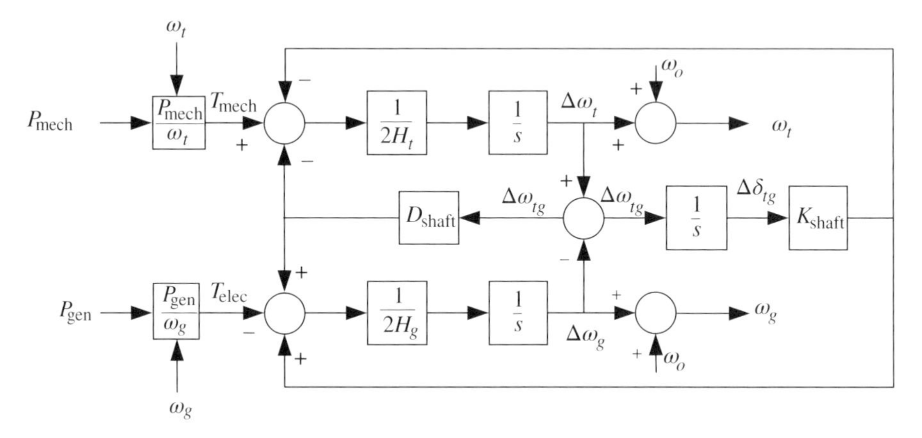

Figure 15.39 Block diagram representation of the two-mass turbine-generator model.

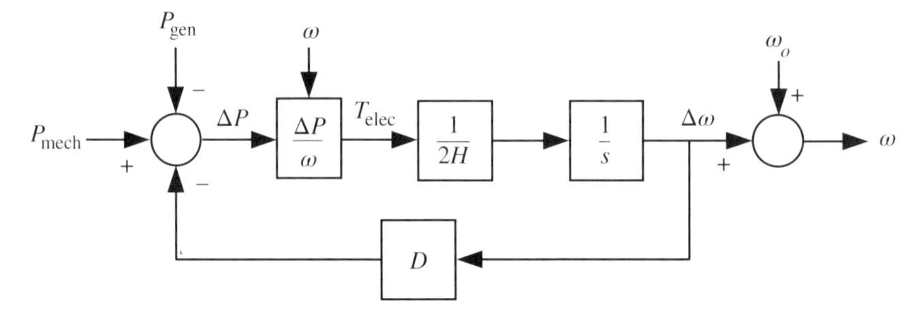

Figure 15.40 Block-diagram representation of the one-mass turbine-generator model.

where

$$T_{\text{mech}} = P_{\text{mech}}/\omega_t, \quad T_{\text{elec}} = P_{\text{gen}}/\omega_g \tag{15.32}$$

Example 15.5

Consider the following parameters for the two-mass WTG model shown in Figure 15.39, given in per unit on the generator base:

$$H_t = 4.33 \text{ sec}, \quad H_g = 0.62 \text{ sec}, \quad D_{\text{shaft}} = 1.5 \text{ pu}, \quad K_{\text{shaft}} = 139.48 \text{ pu} \tag{15.33}$$

Compute the initial values of the state variables in (15.31) assuming that P_{mech} and P_{gen} are equal to 1 pu, and $\omega_t = \omega_g = \omega_o = 1.2$ pu.

Solutions: Three state variables describe the model (the outputs of the integrators): $\Delta\omega_t$, $\Delta\omega_g$, and δ_{tg}. A dynamic model for transient stability analysis is typically initialized assuming steady-state conditions, i.e., the derivatives of the state variables are zero. Furthermore, in steady state, the speed deviations, $\Delta\omega_t$ and $\Delta\omega_g$, are zero. The initial value of the third state variable, δ_{tg}, can be obtained from either the first or second equation in (15.31) (with $P_{\text{mech}} = P_{\text{gen}}$) as

$$\delta_{tg} = \frac{1}{K_{\text{shaft}}}\frac{P_{\text{gen}}}{\omega_o} = 0.006 \text{ rad} \tag{15.34}$$

■

Example 15.6

For the wind turbine considered in Example 15.5, verify that the torsional frequency of oscillation is approximately 1.8 Hz.

Solutions: From (15.31), the state matrix of the two-mass turbine model is given by

$$A = \begin{bmatrix} -D_{\text{shaft}}/2H_t - P_{\text{mech}}/(2H_t\omega_o^2) & D_{\text{shaft}}/2H_t & -K_{\text{shaft}}/2H_t \\ D_{\text{shaft}}/2H_g & -D_{\text{shaft}}/2H_g + P_{\text{gen}}/(2H_g\omega_o^2) & K_{\text{shaft}}/(2H_g) \\ 1 & -1 & 0 \end{bmatrix} \tag{15.35}$$

Substituting the given numerical values gives

$$A = \begin{bmatrix} -0.2534 & 0.1732 & -16.106 \\ 1.2097 & -0.6496 & 112.48 \\ 1 & -1 & 0 \end{bmatrix} \tag{15.36}$$

The oscillatory mode of A is $-0.4515 \pm j11.3288$ whose frequency is $11.3288/(2\pi) = 1.803$ Hz.

■

15.8.5 Torque Control Model

The objective of the torque control model is to compute the required power reference for optimal power generation. The torque controller is shown in Figure 15.41. It is essentially a PI controller whose function is to set the power reference P_{ref} for the electrical controller, based on a pre-determined speed versus power relation $\omega_t = f(P_e)$. A typical power versus speed curve is shown in Figure 15.42. The curve follows the maximum value of the power versus the speed loci of the wind turbine for different values of the wind speed when the pitch angle is fixed at a value of zero (Region 2 operation). The difference between ω_{ref} and the actual generator speed ω is the input signal into a PI controller to produce a torque reference. The product of the torque reference and the generator speed forms the power reference fed into the electrical controller. The time constant $T_{\omega\text{ref}}$ is relatively large, e.g., 5 seconds, which allows for the speed reference to smoothly track changes in power.

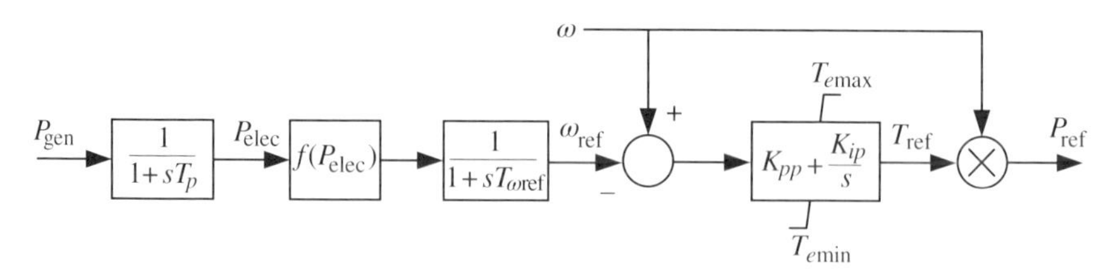

Figure 15.41 Torque controller.

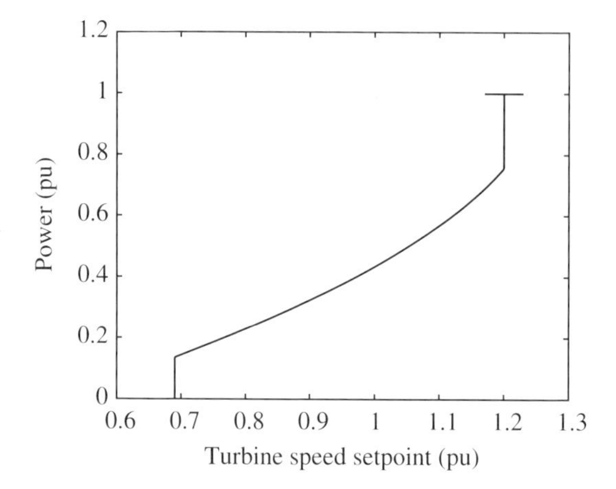

Figure 15.42 Power versus speed curve.

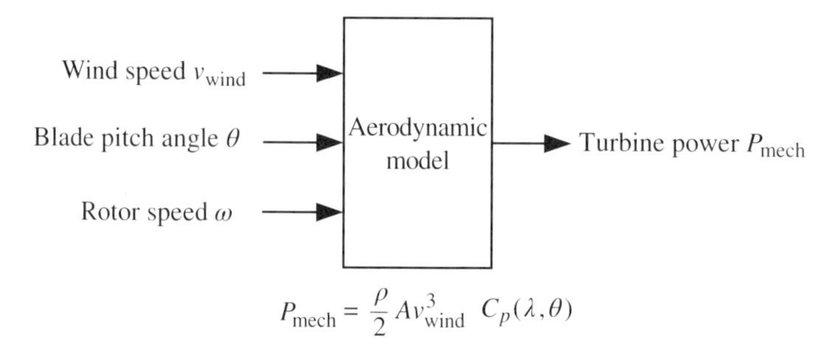

Figure 15.43 Aerodynamic model.

15.8.6 Aerodynamic Model

The aerodynamic model (aero model) computes the mechanical power generated by the wind turbine, P_{mech}, from the energy contained in the wind (Figure 15.43). Since this power is a function of the wind speed and the power coefficient C_p, the complexity of the aerodynamic model depends on the level of detail used to represent C_p.

The coefficient C_p is a nonlinear function of the tip speed ratio λ and the pitch angle θ. A representative three-dimensional plot for C_p derived from measured data is shown in Figure 15.44. The use of a three-dimensional C_p curve in a transient stability program requires the interpolation of λ and θ to compute C_p – a time consuming process. This level of detail, however, is not required for transient stability analyses.

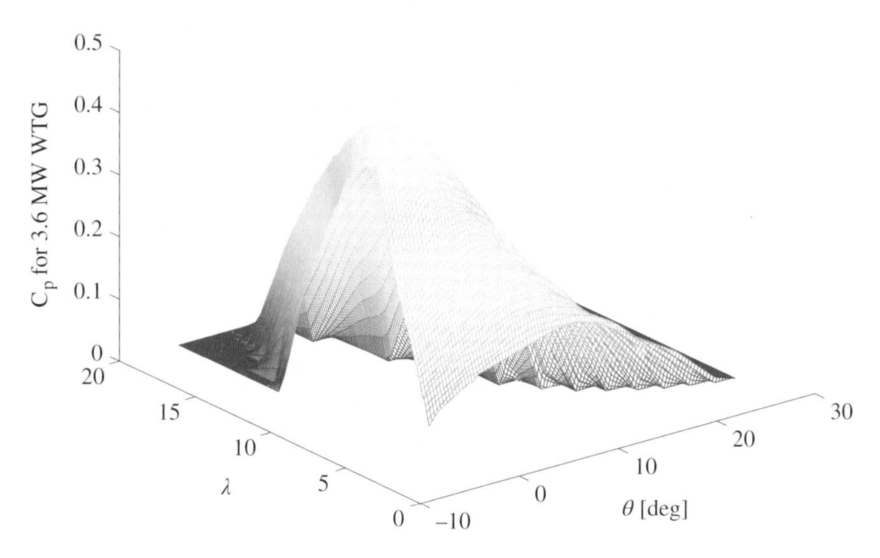

Figure 15.44 Three-dimensional C_p surface (see color plate section).

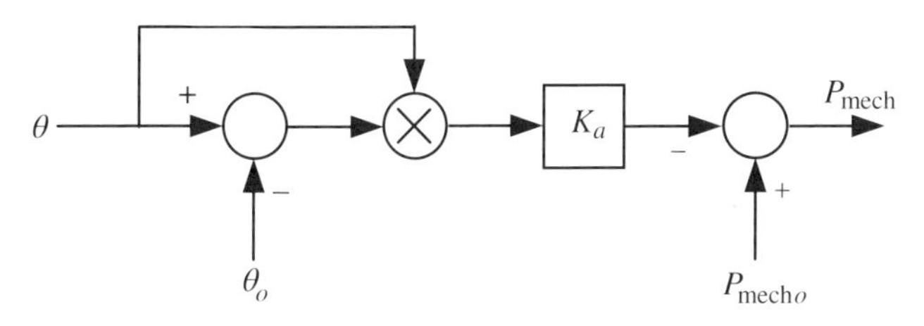

Figure 15.45 Simplified aerodynamic model.

Simpler C_p representations have been proposed which consist of an algebraic relationship between C_p, λ, and θ [214]. The aero model used in the WPP model shown in Figure 15.45 is even simpler and is based on the assumption that wind speed is constant [213]. This assumption is reasonable for the level of detail required in power system planning simulations involving grid disturbances and spanning a time frame of simulation of 10–20 seconds. The simplified aerodynamic model, shown in Figure 15.45 has a single input θ, the pitch angle, and its output is the mechanical power P_{mech}. Their initial values are P_{mecho} and θ_o.

From Figure 15.45, the equation that describes the aero model is given by

$$P_{\mathrm{mech}} = P_{\mathrm{mecho}} - K_a(\theta - \theta_o)\theta \tag{15.37}$$

The model is based on empirical observations of the performance of a more detailed model which showed linear relations between different variables including the rate of change of power $dP/d\theta$ with respect to the pitch angle θ (Figure 15.46) as well as the wind speed and generated power (Figure 15.47). The red dots in the figures are values obtained from the more detailed model, and the lines are straight-line-fit to those points.

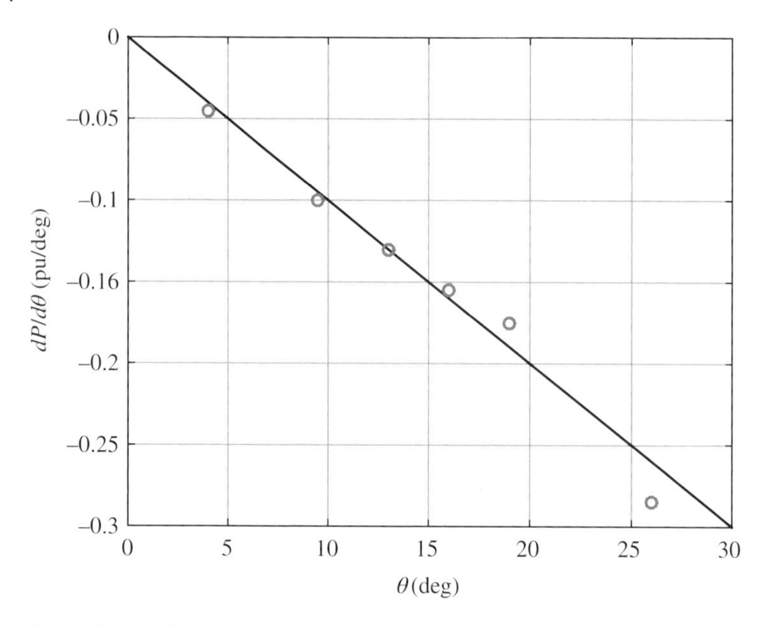

Figure 15.46 $dP/d\theta$ versus θ (see color plate section).

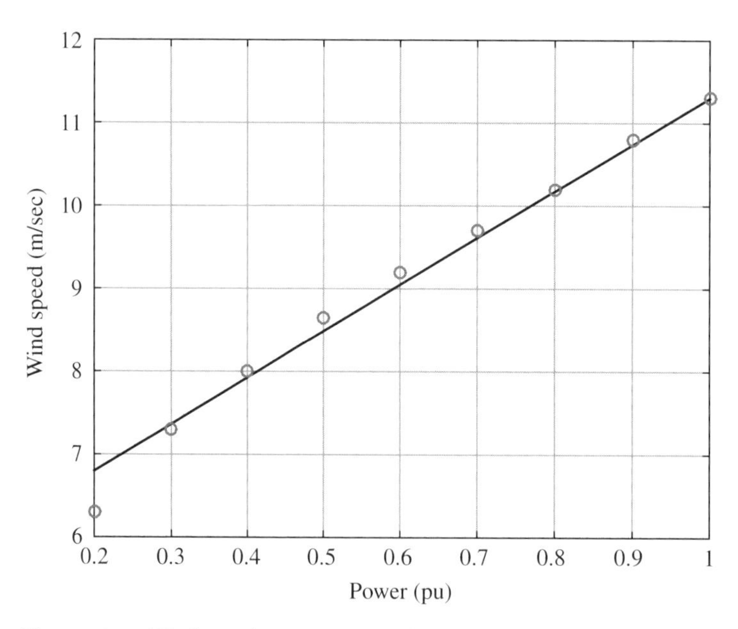

Figure 15.47 Wind speed versus generated power (see color plate section).

15.8.7 Pitch Controller

The function of the pitch controller model is to limit the power generated by the turbine to its rated value at 1.0 pu when operating in Region 3. In this region the available wind power is above the equipment rating and the blades have to be pitched to limit

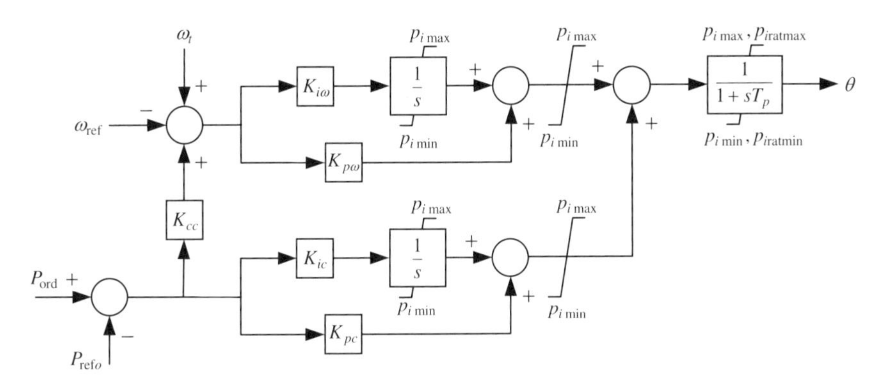

Figure 15.48 Pitch controller.

the mechanical power delivered to the shaft to 1.0 pu. The dynamics of the pitch controller can have a significant impact on dynamic simulation results. When the available wind power is less than rated, the blades are held at minimum pitch, which in the model implementation of Figure 15.48 is at zero degrees.

The pitch controller depicted in Figure 15.48 consists of two PI controllers with limits at (p_{imin}, p_{imax}), a time constant, and rate limits $(p_{iratmin}, p_{iratmax})$ associated with the blade angle (pitch). The input to one of the PI controllers is power deviation. The input to the other PI controller is speed deviation. This signal may be augmented with a signal proportional to the power deviation.

The pitch controller shown in Figure 15.48 is a functional representation of the actual control system. In other words, this controller is not intended to be a replica of the actual controller; rather it emulates the dynamic characteristics of an actual controller.

Example 15.7

This example illustrates the dynamic response of a WPP connected to a large system modeled as an infinite bus (Figure 15.49). The rated wind speed of the equivalent WTG is 12 m/sec, the rated turbine power is 162 MW, and the rated generator power is 180 MVA. The steady-state wind speed is 14 m/sec, such that the blades are pitched to 9.58° so as not to exceed the wind-turbine power rating. Perform the simulation for the following disturbances:

1) Wind gust: a sinusoidal component starting at 5 sec and ending at 10 sec is added to the wind speed, as shown in the top plot Figure 15.50. The wind speed drops to 10 m/sec at its lowest value.
2) Fault condition: a 0.1 sec, 3-phase short-circuit fault is applied at Bus 5 and cleared without tripping any lines.

The Simulink® diagram to simulate these disturbances is contained in the folder *WT_model*. This model is based on an earlier version of the GE WTG model described in [217]. A two-mass model is used for the drive train. The masses are $H_t = 4.33$ pu and $H_g = 0.62$ pu. The other parameters can be found in the example folder.

Solutions:

1) Wind gust: The simulation results are shown in Figure 15.50. As the wind speed starts to drop at 5 sec, the turbine mechanical power P_{mech} and the rotor speed also drop,

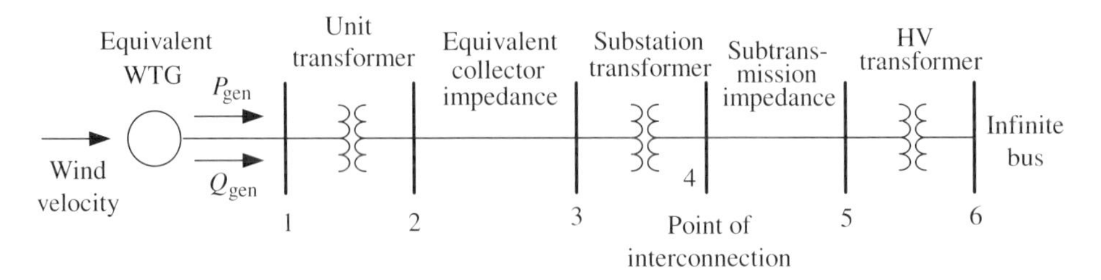

Figure 15.49 Simple test system.

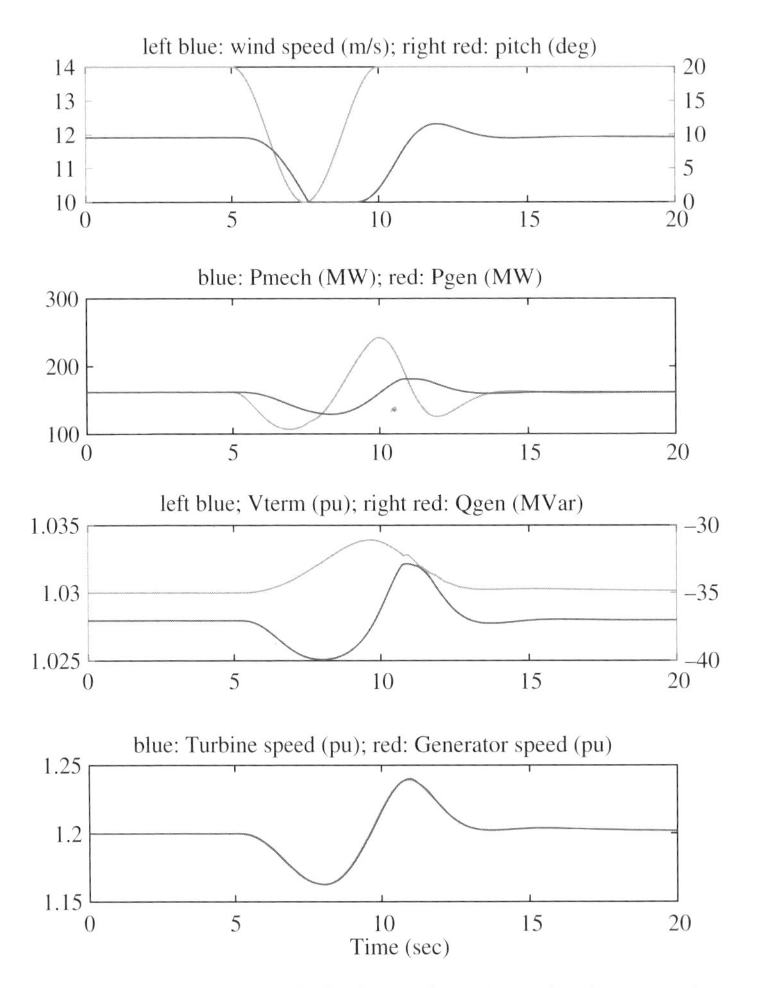

Figure 15.50 Simulation results for the wind gust (see color plate section).

followed by the generator electric power P_{gen}. With reduced active power output, the terminal voltage starts to rise and the amount of reactive power required by the generator drops. The pitch controller starts to reduce the pitch angle θ to extract more power from the wind. As the wind speed drops below the rated speed of 12 m/sec, the pitch angle saturates at the minimum value of 0°. As the wind speed begins to increase,

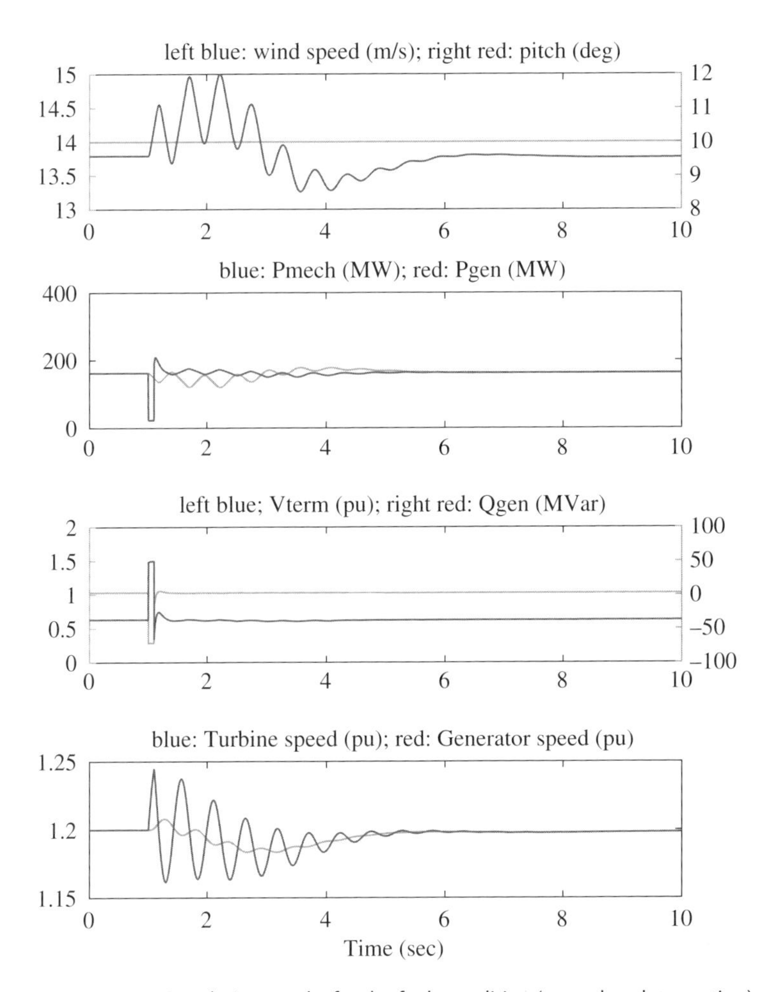

Figure 15.51 Simulation results for the fault condition (see color plate section).

with θ still at 0°, the mechanical power temporarily exceeds its rating of 162 MW. As a result, the pitch angle starts to increase, causing the turbine to return to its original operating condition. Note that in this simulation, the torsional oscillation between the two wind-turbine masses is not observed because the wind variation is relatively slow and first impacts the turbine mass, which is heavy enough not to transmit the transient impact to the generator mass.

2) Fault condition: The short-circuit fault simulation results are given in Figure 15.51. The fault disrupts the network power flow and causes the electric power output of the generator to drop significantly, until the fault is cleared. During the fault-on period, the rotor speed starts to increase, causing the pitch angle to increase, reducing the power extracted from the wind. As this disturbance affects the generator mass directly, the torsional oscillation between the turbine mass and the generator mass at 1.8 Hz is clearly visible. The pitch controller also reacts to this oscillation. The system, however, is stable and returns to the original operating condition. ■

15.9 Plant-Level Control Model

The plant-level controller model processes remote measurements, e.g., bus voltage, branch power flow, and frequency, and issues commands to the electrical controller. Referring to Figure 15.36, the commands into the electrical controller from the plant-level controller would be Q_{ref} and P_{ref}. The measurements are often taken at the POI. A simplified rendition of the model is shown in Figure 15.52. The complete block diagram for this model can be found in [215]. The control system depicted in Figure 15.52 is essentially a simple PI controller whose function is to regulate a measured voltage, V_{meas}. The time constant T_{fltr} may be used to represent a transducer. The deadband prevents unwarranted control actions due to small deviations in the neighborhood of the reference voltage V_{ref}.

15.9.1 Simulation Example

As part of the interconnection studies for the WPP shown in Figure 15.53, it is required to assess its voltage regulation capability at the POI, following a sudden voltage drop in the "Main system" ("Large system"). The "Existing system" represents a small network. The WPP consists of 203 units rated at 1.64 MVA each.

To perform the assessment, the response of the voltage at the POI to a sudden voltage drop from 1.05 pu to 1.00 pu at the bus labeled "Main system" is simulated with and without the plant controller.

The results of the simulations are shown in Figure 15.54. The solid and dashed lines show the POI voltage, with and without the plant controller, respectively. With the plant controller enabled, the POI voltage is brought back to its pre-disturbance value. This example highlights the effect of the plant controller in a simple system. Other system components and/or control criteria are often included in more realistic scenarios.

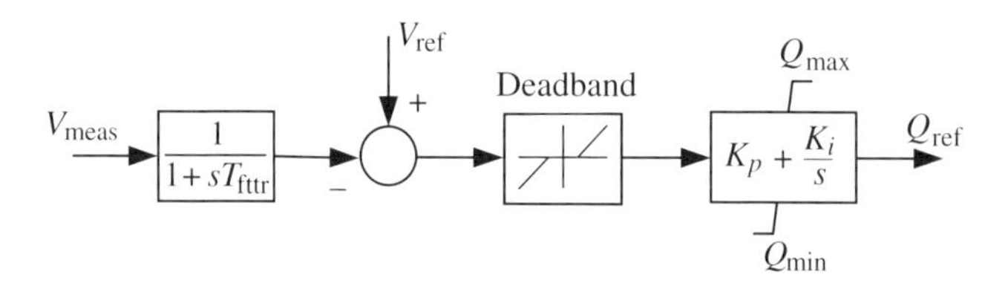

Figure 15.52 Plant-level controller.

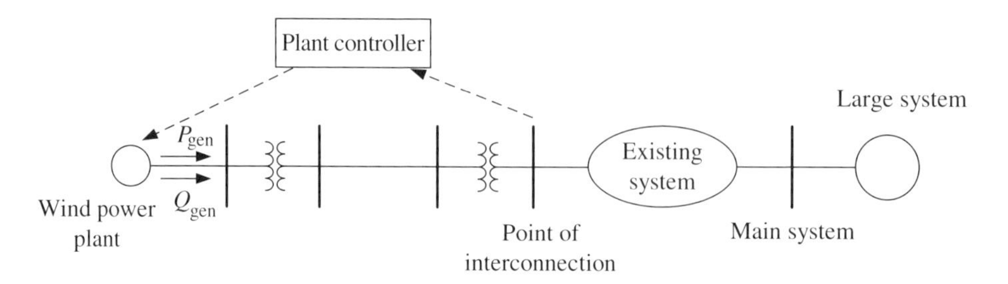

Figure 15.53 Test system.

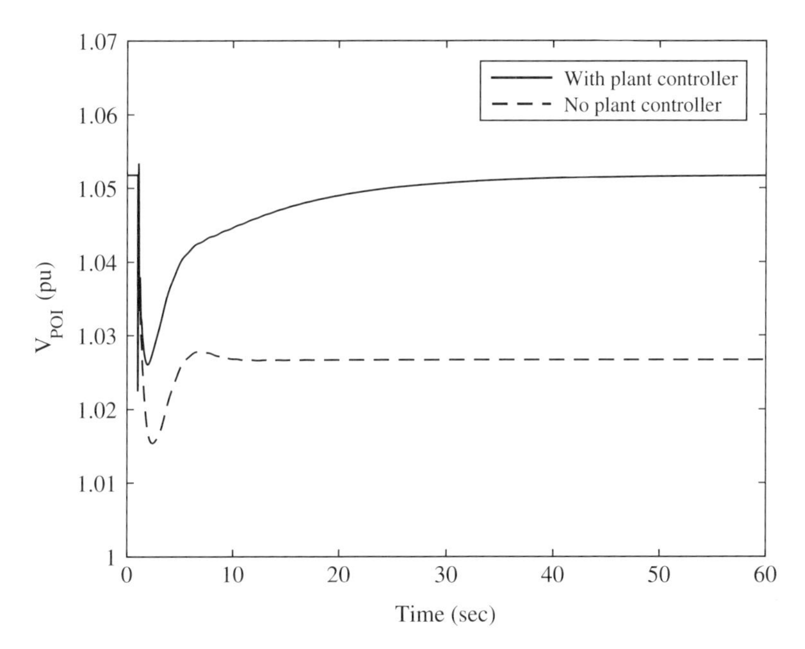

Figure 15.54 POI voltage with and without the plant controller (see color plate section).

15.10 Summary and Notes

This chapter discusses wind energy conversion to electrical energy. The wind turbine types are presented, with this chapter focusing on Type-3 DFIG wind turbines. The equivalent circuit of the induction generator with converter voltage insertion in the rotor winding is used for a simplified calculation of the inserted voltage. Equivalent wind plant models are developed, and active and reactive power control models for Type-3 wind turbines are presented. A Simulink model is used to demonstrate the typical control performance for Type-3 wind turbines.

The Type-3 control models can also be applied to Type-4 wind turbines. Here the design calls for the active power control to deal with only the wind energy conversion, and the reactive power control to deal only with the converters, as the DC link between the rectifier and the inverter separates the active and reactive power control dynamics.

Subjects related to transient stability analyses of wind turbines not covered here include protection schemes, and the tuning and coordination of controllers. Also, since the wind turbine model covered in Section 15.8 does not explicitly include the flux state equations, a description of the WTG in terms of differential equations is not presented, but can be found in publications such as [202, 207, 209, 214].

The frequency in a power system deviates from its nominal value, e.g., 60 Hz, when there is a power imbalance between the generated power and the power consumed by the loads. For instance, following a sudden loss of generation, the frequency will drop as shown in Figure 15.55 [219], which is similar to the Western Electricity Coordinating Council (WECC) frequency response displayed in Figure 12.27. An inherent dynamic characteristic of synchronous generators is that they mitigate the rate of decay of the frequency following the contingency by transferring some of the kinetic energy in the

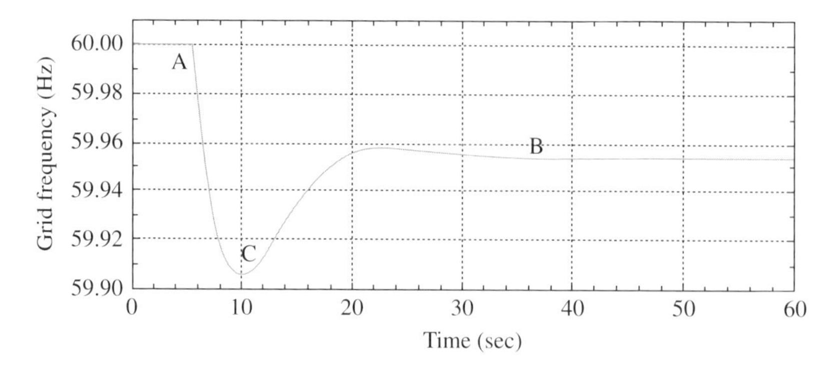

Figure 15.55 Typical frequency response of a power system for a loss of generation event (see color plate section).

rotors to the grid. This is known as the inertial response. Following the inertial response, additional frequency control is then provided by the governors.

Unlike the dynamic response of a conventional power plant, modern WPPs do not inherently provide inertial response or governor-type frequency control. Without inertial response, unacceptable frequency deviations become a real possibility in systems where the amount of wind power is significant. To address this problem, using the control capabilities of the electronic converters, several methods to emulate inertial response have been proposed in the literature and implemented by wind-turbine manufacturers [220–223]. These methods emulate the frequency response by control actions and can be roughly divided into two categories: methods designed to extract extra power from the rotor inertia when a frequency drop occurs and methods that de-load the wind turbine to provide headroom for frequency regulation [219].

Besides wind energy, another important renewable resource is solar energy. Installations of small roof-top solar photo-voltaic (PV) panels and large-scale solar PV farms with several MW at peak power output have been growing at an exponential pace. Time constants of solar PV systems affecting energy output are very small as solar energy generation does not required a rotating inertia. In addition, such systems operate with maximum power output strategies and thus are not controlled based on power grid conditions. If necessary, the wind energy frequency response and reactive power control strategies can also be applied to solar energy systems.

The Wind Turbine Simulink® model used in several examples has additional features that are not used in this chapter. A reader is encouraged to download the computer code and try out the various topologies and disturbances.

Problems

15.1 (Betz law) Given an upstream wind speed of v_1, show that the maximum wind power that can be extracted is

$$P_{\max} = \frac{16}{27} A \frac{\rho}{2} v_1^3$$

where ρ is the air density and A is the area swept by the wind turbine blades. Note that you have to use the expression

$$v_2 = \frac{1}{2}(v_1 + v_3)$$

where v_2 is the wind velocity at the blades and v_3 is the downstream wind velocity.

15.2 Consider a wind turbine with a blade radius of 50 m. The turbine is designed to be optimal at a wind speed of 15 m/sec and a tip speed ratio of 8. Assuming that the optimal angle of attack for this particular design of the blade is 6° throughout the length of the blade, calculate the blade pitch angle at the tip (50 m), mid point (25 m), and close to the hub (3 m).

15.3 Calculate the power coefficient C_p curve for a GE DFIG WTG using the data in Table 4-11 of the GE Wind Turbine Modeling Report [217] under the condition of constant pitch angle of 1° and λ ranging from 0 to 18. The C_p coefficients can be found in the m file *cp.m*. Note that the pitch angle is in degrees for the C_p calculation.

15.4 Plot the per-phase torque-slip curve of a Type-2 wind turbines for the parameters $R_s = 0.004888$ pu, $X_s = 0.09241$ pu, $R_r = 0.00549$ pu, $X_r = 0.09955$ pu [218] ranging from $s = -0.5$ to 0.5 pu using (15.19). Assume $V_s = 1.0$ pu. Show the results for $\alpha = 1$, 5, and 10 on the same plot.

15.5 Derive (15.21) for a Type-3 WTG for a V_r voltage insertion in-phase with V_s.

15.6 Consider a radial connection (like a daisy chain) of four wind turbines, as shown in Figure 15.56.a. The wind turbines are identical with the same rating S and inertia H. Under identical wind and operating conditions for each turbine, the currents satisfy $I_1 = I_2 = I_3 = I_4 = I$. By matching the output current I_S and the active and reactive power losses within the daisy chain, derive an aggregate model as shown in Figure 15.56.b. Note that the total active and reactive losses in an impedance Z subject to a current I can be expressed as $S_{\text{loss}} = I^2 Z$.

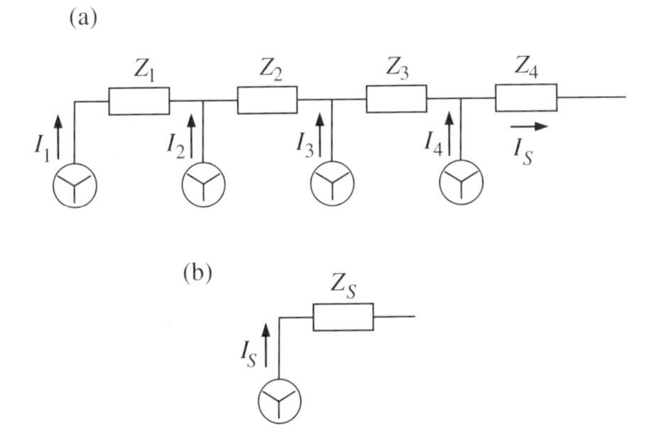

Figure 15.56 (a) Radial connection of four wind turbines and (b) equivalent model.

15.7 The turbine inertia and the generator inertia for a 1.6 MW Type-3 wind turbine are given by $H_t = 4.06$ sec and $H_g = 0.58$ sec, respectively. The damping constant is $D_{\text{shaft}} = 1.5$ pu for each section, and the shaft stiffness constant between the turbine and the generator is $K_{\text{shaft}} = 130.69$ pu/rad. Assume that the wind turbine is operating at $P_{\text{mech}} = P_{\text{gen}} = 1.0$ pu and $\omega_t = \omega_g = \omega_o = 1.2$ pu. Find the torsional frequency of oscillation between the turbine and the generator. Also compute the steady-state shaft twist δ_{tg} between the turbine and generator.

15.8 The MATLAB® files for a wind farm in a 6-bus radial system (Figure 15.49) used in Example 15.7 can be found in the *WT_Model* folder. The WTG is located at Bus 1 and the infinite bus is taken as Bus 6. The Simulink® file *WTG* can be used to simulate various disturbance responses. It is necessary to first edit the *asyst5_init.m* file to set up the disturbance condition. Then run the *asyst5_master.m* file to set up the parameters and initial condition in the MATLAB workspace before starting the simulation.

a) Simulate a wind gust response using the wind gust profile in the file *asyst5_init.m*. The wind gust drops from 14 m/sec to 8 m/sec and then goes back up to 14 m/sec as a sinusoidal function. (Adjust *wmag* in *asyst5_init.m*.)
b) Simulate a wind gust response using a wind gust profile similar to Part (a) except that the wind gust goes from 14 m/sec to 18 m/sec and then goes back to 14 m/sec.
c) Simulate the disturbance response for a 200 msec fault on Bus 5 starting at 1 sec, keeping the wind speed constant at 14 m/sec. Clear the fault without tripping a line (there are no parallel lines to trip) (Adjust *Tfault* and *Tclear* in *asyst5_init.m*.)
d) Continuing from Part (c), simulate the disturbance response for a sufficiently long fault so that the WTG becomes unstable.
e) At 1 sec, reduce the voltage at the infinite bus by 0.04 pu in a step function. This can be accomplished using the *Vref* input of the excitation system of the synchronous generator used to model the infinite bus. (The *Vref* input can be found in the Simulink® diagram. Also adjust the parameter *Step_time* in *asyst5_init.m*.)

For each disturbance, plot the wind speed, pitch angle, mechanical power, generator active and reactive power, bus voltage magnitude, turbine speed, and generator speed. When appropriate, plot some of the variables in the same plot. Comment on the wind turbine dynamic response in each case. Make sure the MATLAB® workspace is cleared before starting a new simulation.

16

Power System Coherency and Model Reduction

16.1 Introduction

Dynamic simulation on a digital computer is the main tool for a power system engineer to study system planning as well as contingency analysis for real-time operation. There are two main objectives in such nonlinear dynamic simulations: (1) the power system dynamics should be accurately captured in the simulation and (2) the model size should not be too large so that all the necessary simulation cases can be completed expeditiously, allowing the results to be used in a timely manner.

To achieve the second objective, a common practice of a power system control region is to retain its own system in full detail, also known as the study system, and reduce the model of the neighboring systems, also known as the external system (Figure 16.1). In general, the contingencies to be studied occur in the study region. The external system is needed because of parallel flow effects. For example, a short-circuit fault cleared by a line trip between Buses A and B in the study system shown in Figure 16.2 will redistribute the pre-fault flow on the tripped line to other paths, some of which may circulate through the external system. The incremental flow ΔP_1 from the study region to the external system will be accompanied by an incremental flow ΔP_2 from the external system back to the study region at a different location.[1] Note that because of losses, ΔP_1 and ΔP_2 are not necessarily equal. These parallel flows between the boundary buses need to be modeled accurately in order for the stability simulation to be valid.

Because disturbances will not be applied in the external system, a less detailed model of the external system can be used as long as it can model the parallel flow sufficiently accurately. As the reduced external system needs to be connected to the study system and be used in a dynamic simulation program, it would be desirable for the reduced external system to have the structure of a conventional power system. This nonlinear model reduction process is the main focus of this chapter. The treatment is based on the coherency concept, in which synchronous machines that are coherent can be aggregated into larger equivalent machines, thus reducing the number of differential equations to be simulated. Algorithms have also been developed to optimize the dynamic parameters of the aggregated machines.

1 Such parallel flow is sometimes also called loop flow.

Power System Modeling, Computation, and Control, First Edition. Joe H. Chow and Juan J. Sanchez-Gasca.

Companion website: www.wiley.com/go/chow/power-system-modeling

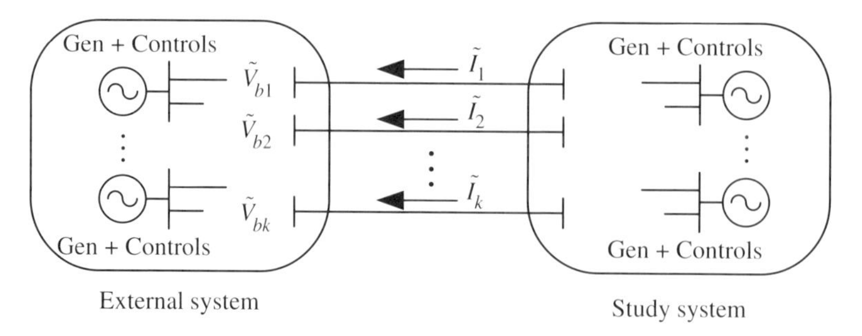

Figure 16.1 Separation of a power system into a study system and an external system.

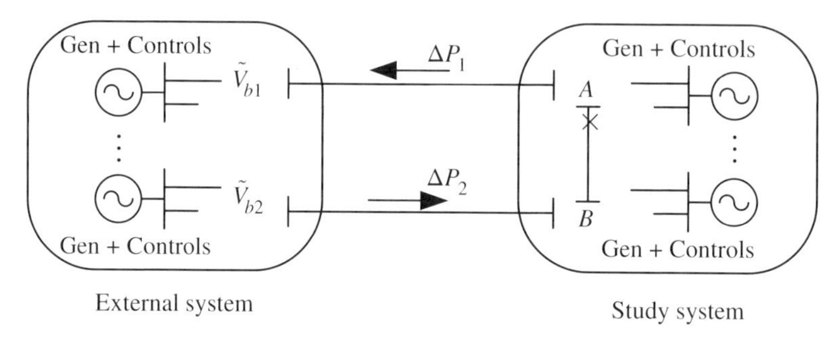

Figure 16.2 Parallel flow through the external system.

Other approaches have also been developed to obtain reduced models of the external system. Most notable is the use of linear reduced models, starting from the modal equivalent work in [224–227], the selective modal analysis method in [86, 87], and more recently the input-output model reduction techniques such as the Krylov method and the balanced truncation technique [228]. Another input-output model approach is to use a nonlinear neural network to represent the external system [229]. These methods have not yet been adopted for practical use by control centers and utility companies.

The reminder of the chapter presents the power system dynamic model reduction as used in the industry, with Section 16.2 on coherency and Section 16.3 on generator aggregation and model reduction. These two sections form the basis for the power system model reduction tools used in the power industry. Section 16.4 shows a comparison of the accuracy of the simulated time responses from the methods discussed in Section 16.3. Section 16.5 is on linear model reduction methods and Section 16.6 is a brief discussion on the coherency and aggregation tools available in the Power System Toolbox.

16.2 Interarea Oscillations and Slow Coherency

In disturbance simulations, it is common to observe many generators swinging together as a group, which are called coherent machines. Figure 16.3 shows the rotor angle swings of 21 generators in Arizona subject to a disturbance in the US Western Electricity Coordinating Council (WECC) system [231]. Note that the steady-state values of the individual angles have been subtracted from all the rotor angles so that the initial rotor angles all

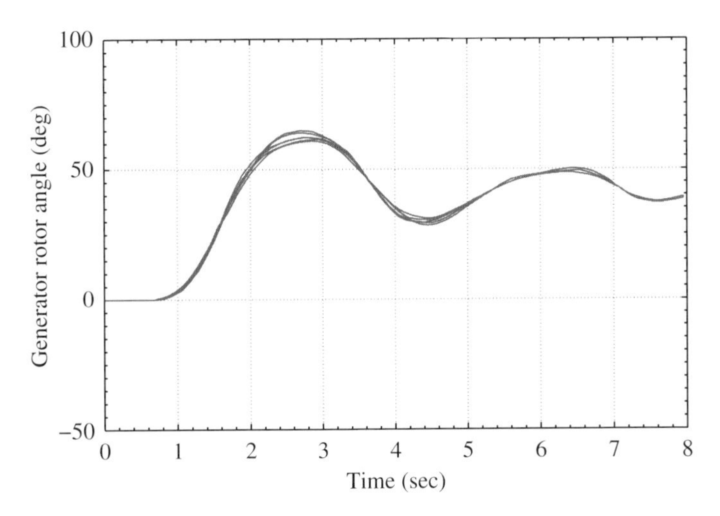

Figure 16.3 Swing curves of a group of 21 generators in Arizona [231] (see color plate section).

start at 0°. For all practical purposes, the response of each of the machines in this coherent group is indistinguishable from the other machines in the same coherent group. If their dynamics are similar, it would be logical to combine them into a single aggregate machine. Thus the first step in power system model reduction is to identify the coherent machines.

The authors of [230, 231] recognized that it is not necessary to simulate the nonlinear dynamics of a power system to find the coherent machines. Instead, it is adequate to simulate the linearized model, which would be much faster. Furthermore, disturbances can also be approximated to further reduce computation time. After a disturbance has been simulated, a generator clustering algorithm based on the generator rotor angle response is used to identify the coherent machine groups. A machine is coherent with a group of other machines if

$$|\Delta\delta_i(t) - \Delta\delta_r(t)| \leq \varepsilon, \quad \text{for } t_o \leq t \leq t_f \tag{16.1}$$

where t_o and t_f are the initial and final times of the simulation, respectively, δ denotes the machine rotor angle, i is the index for the machine being clustered, r is the index for the reference machine of the group under consideration, and ε is the specified tolerance in degrees.

Application experience has shown that the coherent groups obtained from criterion (16.1) are dependent on the disturbance applied. As a result, multiple disturbances have been suggested for finding coherent groups that are valid for all the selected disturbances.

This empirical method using (16.1) is simple but does not offer much insight into this coherency phenomenon in many large power systems. In the remainder of this section, a slow-coherency approach is developed to link the concepts of weak and strong connections and interarea mode oscillation between power system control regions. More advanced coherency identification methods have also been developed in [232].

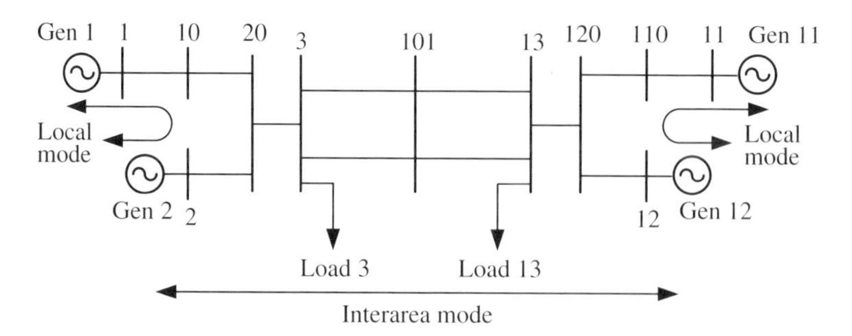

Figure 16.4 Two-area, 4-machine power system [21].

16.2.1 Slow Coherency

A large power system usually consists of tightly connected synchronous machines in each control region with tie-lines located strategically for power exchange and reserve sharing between the control regions. As discussed in Chapter 6, the oscillations between these groups of strongly connected machines are the interarea modes. These interarea modes are lower in frequency than the local modes and intra-plant modes. To develop the slow coherency concept between machines in different coherent groups, the singular perturbations technique can be used to show this time-scale separation between the interarea and local modes. The following example shows the key ideas of slow coherency, that is, the coherency with respect to the interarea modes.

Example 16.1

For the 2-area, 4-machine system shown in Figure 16.4, a 3-phase-to-ground fault is applied at Bus 3 and cleared by removing one of the lines between Buses 3 and 101. Plot the time response of the machine rotor speeds and identify the coherent machines.

Solutions: Figure 16.5 shows the machine rotor speed response to the disturbance. Note that the machine speeds exceed 1 pu and thus the absolute machine rotor angles will be increasingly unbounded. As a result, the swing modes are harder to discern from the absolute machine rotor angles. Because $\omega = (1/\Omega)\, d\delta/dt$, where Ω is the conversion factor from pu speed to radians, coherent machines also have similar speed response. Figure 16.5 shows that the disturbance excites the 1.3 Hz local mode between Generators 1 and 2, and the 0.6 Hz interarea mode between Generators 1 and 2, and Generators 11 and 12. From the time response, one can see that Generators 11 and 12 are clearly coherent, and disregarding the presence of the local mode, Generators 1 and 2 are also coherent with respect to the interarea mode. ■

To analyze the interarea mode coherency shown in Example 16.1, a linear electromechanical power system model first shown in Chapter 6 will be used. An N-bus, n-machine power system with classical electromechanical model representation and constant impedance loads can be expressed as

$$m_i\ddot{\delta}_i = P_{mi} - P_{ei} = P_{mi} - \frac{E_i' V_j \sin\left(\delta_i - \theta_j\right)}{X_{di}'} = f_i(\delta, V) \tag{16.2}$$

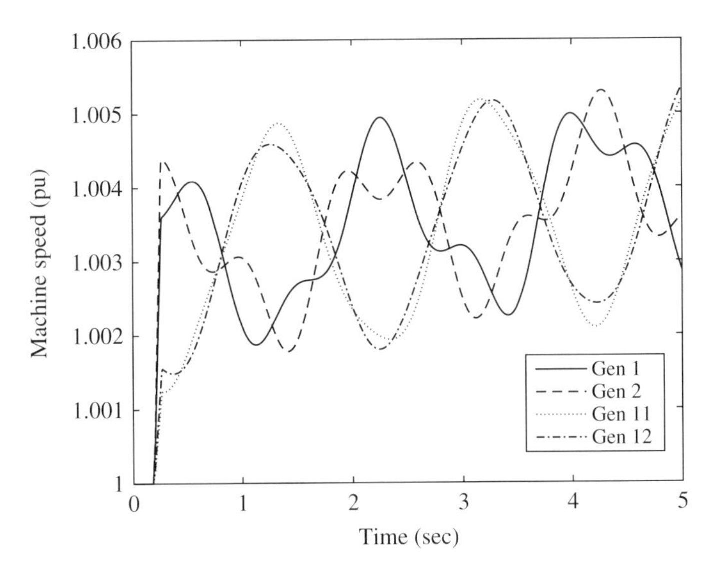

Figure 16.5 Two-area, 4-machine power system disturbance response.

where the synchronous machine i is modeled as a constant voltage E'_i behind a transient reactance X'_{di}, $m_i = 2H_i/\Omega$ where H_i is the inertia of Machine i in seconds and $\Omega = 2\pi f_o$ where f_o is the nominal system frequency in Hz, δ_i is the machine rotor angle in radians, P_{mi} is the input mechanical power, and P_{ei} is the output electrical power. Damping in this electromechanical model is neglected.

The electrical output power P_{ei} is determined from the power flow in the network. Let V_{jre} and V_{jim} be the real and imaginary parts of the bus voltage phasor at Bus j, the terminal bus of Machine i. The voltage magnitude and angle of Bus j is given by

$$V_j = \sqrt{V_{jre}^2 + V_{jim}^2}, \quad \theta_j = \tan^{-1}\left(\frac{V_{jim}}{V_{jre}}\right) \tag{16.3}$$

Then for Bus j, the active power flow balance is[2]

$$P_{ej} - \text{Real}\left\{ \sum_{k=1,k\neq j}^{N} \left(V_{jre} + jV_{jim}\right)\left(\frac{V_{jre} + jV_{jim} - V_{kre} - jV_{kim}}{R_{Ljk} + jX_{Ljk}}\right)^* \right\} \\ - V_j^2 G_j = g_{2j-1} = 0 \tag{16.4}$$

and the reactive power flow balance is

$$Q_{ej} - \text{Imag}\left\{ \sum_{k=1,k\neq j}^{N} \left(V_{jre} + jV_{jim}\right)\left(\frac{V_{jre} + jV_{jim} - V_{kre} - jV_{kim}}{R_{Ljk} + jX_{Ljk}}\right)^* \right\} \\ - V_j^2 B_j + V_j^2 \frac{B_{Ljk}}{2} = g_{2j} = 0 \tag{16.5}$$

where R_{Ljk}, X_{Ljk}, and B_{Ljk} are the resistance, reactance, and line charging, respectively, of the Line j-k, G_j and B_j are the load conductance and susceptance at Bus j. If Bus j is

2 Note that j denotes the imaginary number if it is not used as a bus index.

a generator terminal bus, P_{ej} and Q_{ej} are generator active and reactive electrical output power, respectively. Note that if Machine i is located at Bus j and the generator circuit resistances are neglected, then $P_{ej} = P_{ei}$.

Thus the electromechanical model with the power network explicitly represented is given by

$$M\ddot{\delta} = f(\delta, V), \quad 0 = g(\delta, V) \tag{16.6}$$

where M is the diagonal matrix of machine inertias m_i, f is a vector of acceleration torques, g is the vector containing network power flow, δ is the n-vector of machine rotor angles, and V is the $2N$-vector of the real and imaginary parts of the bus voltages.

The model (16.6) is linearized about a nominal power flow equilibrium (δ_o, V_o) to obtain

$$M\Delta\ddot{\delta} = \left.\frac{\partial f(\delta, V)}{\partial \delta}\right|_{\delta_o, V_o} \Delta\delta + \left.\frac{\partial f(\delta, V)}{\partial V}\right|_{\delta_o, V_o} \Delta V = K_1\Delta\delta + K_2\Delta V \tag{16.7}$$

$$0 = \left.\frac{\partial g(\delta, V)}{\partial \delta}\right|_{\delta_o, V_o} \Delta\delta + \left.\frac{\partial g(\delta, V)}{\partial V}\right|_{\delta_o, V_o} \Delta V = K_3\Delta\delta + K_4\Delta V \tag{16.8}$$

where $\Delta\delta$ is an n-vector of machine angle deviations from δ_o and ΔV is a $2N$-vector of the real and imaginary parts $(\Delta V_{\mathrm{re}}, \Delta V_{\mathrm{im}})$ of the load bus voltage deviations from V_o, with the subscript o denoting the equilibrium values. The matrices K_1, K_2, and K_3 are partial derivatives of the power transfer between the machines and their terminal buses, in which K_1 is diagonal, and K_4 is the nonsingular network admittance matrix. The sensitivity matrices K_i can be derived analytically or from numerical perturbations as discussed in Chapter 6.

The algebraic equation (16.8) is solved to obtain

$$\Delta V = -K_4^{-1}K_3\Delta\delta \tag{16.9}$$

such that its substitution in (16.7) results in

$$M\Delta\ddot{\delta} = \left(K_1 - K_2K_4^{-1}K_3\right)\Delta\delta = K\Delta\delta \tag{16.10}$$

where the (i, j) entry of K is

$$K_{ij} = E_i'E_j'\left(B_{ij}\cos\left(\delta_i - \delta_j\right) - G_{ij}\sin\left(\delta_i - \delta_j\right)\right)\Big|_{\delta_o, V_o}, \quad i \neq j \tag{16.11}$$

and $G_{ij} + jB_{ij}$ is the equivalent admittance between Machines i and j. Furthermore,

$$K_{ii} = -\sum_{j=1, j\neq i}^{n} K_{ij} \tag{16.12}$$

Thus the row sum of K equals to zero. The entries K_{ij} are the synchronizing torque coefficients discussed in Chapter 6.

16.2.2 Slow Coherent Areas

Assume a power system has r areas of slow coherent machines and the load buses that interconnect these machines. Define

$\Delta\delta_i^\alpha$ = deviation of the rotor angle of Machine i in Area α from its equilibrium value
m_i^α = the inertia of Machine i in Area α

The machines are ordered such that each $\Delta\delta_i^\alpha$ from the same coherent areas appears consecutively in $\Delta\delta$.

The slow coherency phenomenon is attributed primarily to the connections between the machines in the same coherent areas being stiffer than those between different areas, which can be due to two reasons:

1) The admittances of the external connections B_{ij}^E are much smaller than the admittances of the internal connections B_{pq}^I, such that their ratio becomes a small parameter

$$\varepsilon_1 = B_{ij}^E / B_{pq}^I \tag{16.13}$$

where the superscript E denotes external and I denotes internal, and i, j, p, q are bus indices. This situation also includes heavily loaded high-voltage, long-distance transmission lines between two coherent areas.
2) The number of external connections is much less than the number of internal connections, such that their ratio becomes a second small parameter

$$\varepsilon_2 = \overline{\gamma}^E / \underline{\gamma}^I \tag{16.14}$$

where

$$\overline{\gamma}^E = \max_\alpha \{\gamma_\alpha^E\}, \quad \underline{\gamma}^I = \min_\alpha \{\gamma_\alpha^I\}, \quad \alpha = 1, \ldots, r \tag{16.15}$$

γ_α^E = (the number of external connections of Area α)/N^α
γ_α^I = (the number of internal connections of Area α)/N^α
and N^α is the number of buses in Area α.

Thus for a large power system, the weak connections between coherent areas can be scaled by the product of two small parameters, represented by

$$\varepsilon = \varepsilon_1 \varepsilon_2 \tag{16.16}$$

Thus the network admittance matrix is separated into

$$K_4 = K_4^I + \varepsilon K_4^E \tag{16.17}$$

where K_4^I is due to the internal connections and K_4^E the external connections.

Based on (16.17), the synchronizing torque or connection matrix K is

$$\begin{aligned} K &= K_1 - K_2\left(K_4^I + \varepsilon\left(K_4^E\right)\right)^{-1} K_3 \\ &= K_1 - K_2\left(K_4^I\left(I + \varepsilon\left(K_4^I\right)^{-1} K_4^E\right)\right)^{-1} K_3 = K^I + \varepsilon K^E \end{aligned} \tag{16.18}$$

where

$$K^I = K_1 - K_2\left(K_4^I\right)^{-1} K_3, \quad K^E = K_2\left(K_4^I\right)^{-1} K_4^E \left(K_4^I\right)^{-1} K_3 + O(\varepsilon) \tag{16.19}$$

In the decomposition (16.18), the property that each row of K^I sums to zero is preserved.

Define in each area an inertia-weighted *aggregate variable*

$$y^\alpha = \sum_{i=1}^{n_\alpha} m_i^\alpha \Delta\delta_i^\alpha / m^\alpha, \quad \alpha = 1, 2, \ldots, r \tag{16.20}$$

where m_i^α is the inertia of Machine i in Area α, n_α is the number of machines in Area α, and

$$m^\alpha = \sum_{i=1}^{n_\alpha} m_i^\alpha, \quad \alpha = 1, 2, \ldots, r \tag{16.21}$$

is the aggregate inertia of area α.

Denote by y the r-vector whose αth entry is y^α. The matrix form of (16.20) is

$$y = C\Delta\delta = M_a^{-1} U^T M \Delta\delta \tag{16.22}$$

where

$$U = \text{blockdiag}\left(u_1, u_2, \ldots, u_r\right) \tag{16.23}$$

is the grouping matrix with $n_\alpha \times 1$ column vectors of 1s

$$u_\alpha = \begin{bmatrix} 1 & 1 & \ldots & 1 \end{bmatrix}^T, \quad \alpha = 1, 2, \ldots, r \tag{16.24}$$

and

$$M_a = \text{diag}\left(m^1, m^2, \ldots, m^r\right) = U^T M U \tag{16.25}$$

is the $r \times r$ diagonal aggregate inertia matrix.

Select in each area a reference machine, say the first machine, and define the motions of the other machines in the same area relative to this reference machine by the *local variables*

$$z_{i-1}^\alpha = \Delta\delta_i^\alpha - \Delta\delta_1^\alpha, \quad i = 2, 3, \ldots, n_\alpha, \quad \alpha = 1, 2, \ldots, r \tag{16.26}$$

Denote by z^α the $\left(n_\alpha - 1\right)$-vector of z_i^α and consider z^α as the αth subvector of the $(n - r)$-vector z. Equation (16.26) in matrix form is

$$z = G\Delta\delta = \text{blockdiag}\left(G_1, G_2, \ldots, G_r\right) \Delta\delta \tag{16.27}$$

where G_α is the $\left(n_\alpha - 1\right) \times n_\alpha$ matrix

$$G_\alpha = \begin{bmatrix} -1 & 1 & 0 & . & 0 \\ -1 & 0 & 1 & . & 0 \\ . & . & . & . & . \\ -1 & 0 & 0 & . & 1 \end{bmatrix} \tag{16.28}$$

The transformation of the original state $\Delta\delta$ into the aggregate variable y and the local variable z is given by

$$\begin{bmatrix} y \\ z \end{bmatrix} = \begin{bmatrix} C \\ G \end{bmatrix} \Delta\delta \tag{16.29}$$

The inverse of this transformation is

$$\Delta\delta = \begin{pmatrix} U & G^+ \end{pmatrix} \begin{bmatrix} y \\ z \end{bmatrix} \tag{16.30}$$

where

$$G^+ = G^T \left(GG^T\right)^{-1} \tag{16.31}$$

is block-diagonal.

Apply the transformation (16.29) to the model (16.10) to obtain

$$M_a\ddot{y} = \varepsilon K_a y + \varepsilon K_{ad} z$$
$$M_d\ddot{z} = \varepsilon K_{da} y + (K_d + \varepsilon K_{dd}) z \tag{16.32}$$

where

$$M_d = (GM^{-1}G^T)^{-1}, \quad K_a = U^T K^E U$$
$$K_{da} = U^T K^E M^{-1} G^T M_d, \quad K_{da} = M_d G M^{-1} K^E U$$
$$K_d = M_d G M^{-1} K^I M^{-1} G^T M_d, \quad K_{dd} = M_d G M^{-1} K^E M^{-1} G^T M_d \tag{16.33}$$

Note that in (16.32), the connection matrices K_a, K_{ad}, and K_{da} are independent of the internal connections in K^I. The system (16.32) is in the *standard singularly perturbed form* [100], showing y as the slow variable and z as the fast variable. What is special about (16.32) is that ε is both the weak connection parameter and the singular perturbation parameter, giving rise to slow coherency.

Neglecting the fast dynamics in (16.32), the slow subsystem is approximated by

$$M_a\ddot{y} = \varepsilon K_a y \tag{16.34}$$

The fast subsystem of (16.32) is approximated by

$$M_d\ddot{z} = K_d z \tag{16.35}$$

in which ε is set to 0.

It is also possible to formulate the slow-fast dynamics separation including the power network. Apply (16.29) directly to the model (16.7) and (16.8) to obtain

$$M_a\ddot{y} = K_{11} y + K_{12} z + K_{13}\Delta V$$
$$M_d\ddot{z} = K_{21} y + K_{22} z + K_{23}\Delta V$$
$$0 = K_{31} y + K_{32} z + (K_4^I + \varepsilon K_4^E)\Delta V \tag{16.36}$$

where

$$K_{11} = U^T K_1 U, \quad K_{12} = U^T K_1 G^+, \quad K_{13} = U^T K_2, \quad K_{21} = (G^+)^T K_1 U$$
$$K_{22} = (G^+)^T K_1 G^+, \quad K_{23} = (G^+)^T K_2, \quad K_{31} = K_3 U, \quad K_{32} = K_3 G^+ \tag{16.37}$$

Eliminating the fast variables, the slow subsystem is approximated by

$$M_a\ddot{y} = K_{11} y + K_{13}\Delta V$$
$$0 = K_{31} y + K_4 \Delta V \tag{16.38}$$

which consists of both the machine angles and the network bus voltages as variables. This is the *inertial aggregation model* [83], which is equivalent to linking the internal nodes of the coherent machines by infinite admittances.

For a more accurate model, consider z to vary with y. As a first-order approximation, from (16.36) the *quasi-steady state* of z is obtained as

$$z = -K_{22}^{-1}(K_{21} y + K_{23}\Delta V) \tag{16.39}$$

Eliminating z from (16.36) results in the *slow-coherency (aggregate) model*

$$M_a\ddot{y} = K_{11s} y + K_{13s}\Delta V$$
$$0 = K_{31s} y + K_{4s}\Delta V \tag{16.40}$$

where

$$K_{11s} = K_{11} - K_{12}K_{22}^{-1}K_{21}, \quad K_{13s} = K_{13} - K_{12}K_{22}^{-1}K_{23}$$
$$K_{31s} = K_{31} - K_{32}K_{22}^{-1}K_{21}, \quad K_{4s} = K_4 - K_{32}K_{22}^{-1}K_{23} \tag{16.41}$$

The difference between this model and the inertial aggregation model is that in (16.40), the internal nodes of the coherent machines are no longer connected by infinite admittances. Instead, the singular perturbation method introduces impedance correction terms to the connection matrices K_{11}, K_{13}, K_{31}, and K_4.

These models are illustrated with Example 16.2, which can also be found in [83].

Example 16.2

For the 2-area, 4-machine system shown in Figure 16.4, find the swing modes and their eigenvectors. Then develop the slow and fast subsystems, and compare the swing modes computed from these subsystems to the ones obtained from the full system.

Solutions: Using a linearization method, the connection matrix is obtained as

$$K = \begin{bmatrix} -9.4574 & 8.0159 & 0.5063 & 0.9351 \\ 8.7238 & -11.3978 & 0.9268 & 1.7472 \\ 0.6739 & 0.9520 & -9.6175 & 7.9917 \\ 1.3644 & 1.9325 & 8.1747 & -11.4716 \end{bmatrix} \tag{16.42}$$

The eigenvalues of the $M^{-1}K$ are

$$\lambda\left(M^{-1}K\right) = 0, -14.2787, -60.7554, -62.2531 \tag{16.43}$$

with the corresponding eigenvector vectors

$$v_1 = \begin{bmatrix} 0.5 \\ 0.5 \\ 0.5 \\ 0.5 \end{bmatrix}, \quad v_2 = \begin{bmatrix} 0.4878 \\ 0.4031 \\ -0.5672 \\ -0.5271 \end{bmatrix}, \quad v_3 = \begin{bmatrix} 0.6333 \\ -0.7446 \\ 0.1924 \\ -0.0863 \end{bmatrix}, \quad v_4 = \begin{bmatrix} 0.1102 \\ -0.1494 \\ -0.8098 \\ 0.5566 \end{bmatrix} \tag{16.44}$$

Because $M^{-1}K$ represents a system of second-order differential equations, the swing modes can be computed as the square roots of the eigenvalues in (16.43), that is,

- Interarea mode: $\sqrt{-14.279} = \pm j3.779$ rad/s
- Local modes: $\sqrt{-60.755} = \pm j7.795$ rad/s and $\sqrt{-62.253} = \pm j7.890$ rad/s

Decompose K into the internal and external connections as

$$K^I = \begin{bmatrix} -8.0159 & 8.0159 & 0 & 0 \\ 8.7238 & -8.7238 & 0 & 0 \\ 0 & 0 & -7.9917 & 7.9917 \\ 0 & 0 & 8.1747 & -8.1747 \end{bmatrix} \tag{16.45}$$

$$\varepsilon K^E = \begin{bmatrix} -1.4414 & 0 & 0.5063 & 0.9351 \\ 0 & -2.6739 & 0.9268 & 1.7472 \\ 0.6739 & 0.9520 & -1.6258 & 0 \\ 1.3644 & 1.9325 & 0 & -3.2969 \end{bmatrix} \tag{16.46}$$

Table 16.1 Electromechanical modes computed from various models.

	Exact model	Inertial aggreg.	Error	Slow coherency	Error
Interarea mode	3.779 rad/s	3.816 rad/s	0.98%	3.799 rad/s	0.53%
Area 1 local mode	7.795 rad/s	7.344 rad/s	−5.79%		
Area 2 local mode	7.890 rad/s	7.393 rad/s	−6.30%		

The slow subsystem is represented by (16.34)

$$M_a = \frac{1}{2\pi \times 60}\begin{bmatrix} 234 & 0 \\ 0 & 234 \end{bmatrix}, \quad \varepsilon K_a = \begin{bmatrix} -4.1154 & 4.1154 \\ 4.9227 & -4.9227 \end{bmatrix} \tag{16.47}$$

The eigenvalues of $M_a^{-1}K_a$ are 0 and −14.561, implying that the interarea mode frequency is $\sqrt{-14.561} = \pm j3.816$ rad/s.

The fast local dynamics (16.35) are represented by

$$M_d = \frac{1}{2\pi \times 60}\begin{bmatrix} 58.500 & 0 \\ 0 & 55.611 \end{bmatrix}, \quad K_d = \begin{bmatrix} -8.3699 & 0 \\ 0 & -8.0628 \end{bmatrix} \tag{16.48}$$

The eigenvalues of $M_d^{-1}K_d$ are −53.939 and −54.660 such that the frequencies of the local modes are $\pm j7.3443$ and $\pm j7.3932$ rad/s.

Applying a correction term to the slow subsystem results in

$$\varepsilon K_{as} = K_{11s} - K_{13s}K_{4s}^{-1}K_{31s} = \begin{bmatrix} -4.0698 & 4.0699 \\ 4.8861 & -4.8860 \end{bmatrix} \tag{16.49}$$

so that the approximated interarea mode becomes $\pm j3.799$ rad/s, which is very close to the interarea mode of the full system.

A summary of the interarea mode and local mode frequencies from the various models is given in Table 16.1. It is of interest to note that the approximated interarea mode from the inertial aggregation is higher in frequency than that of the exact model because the coherent machines are connected with infinite admittances and the network becomes stiffer (see Section 16.3 for additional discussions). On the other hand, the local modes from the initial aggregation are lower in frequencies than those of the exact model because impedances connecting two nodes in a local area through external connections are neglected. That is, the effective stiffness of the connections within an area is reduced, resulting in lower frequencies of oscillations.

Note that the local modes of the full system are approximated within 6.5% by those from the subsystems. ■

16.2.3 Finding Coherent Groups of Machines

The eigenvectors of the matrix MK^{-1} are known as the modeshapes of the machine rotor angles. Consider the eigenvectors (16.44) in Example 16.2. The eigenvector v_1 shows that all the rotor angles are moving together in the system mode, which is due to generation-load imbalance. The eigenvector v_2 shows that for the interarea mode, Generators 1 and 2 swing together against Generators 11 and 12, representing the power

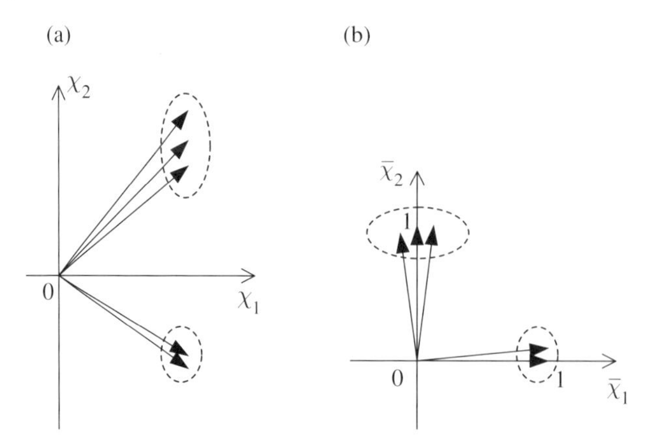

Figure 16.6 Row of slow eigenvectors V_s of the interarea modes: (a) original coordinates and (b) new coordinates.

oscillations on the tielines. There are two local modes: the dominant parts in v_3 show the oscillation between Generators 1 and 2, and those in v_4 the oscillation between Generators 11 and 12. These eigenvectors provide a clear picture that there are two slow coherent areas in this system: Area 1 containing Generators 1 and 2, and Areas 2 containing Generators 11 and 12.

For systems with r coherent areas, the slow-coherent machines will have similar components in all the $(r - 1)$ interarea mode eigenvectors. Let V_s be the matrix containing the system mode and the interarea mode eigenvectors. Then coherent machines must have similar rows in V_s. Figure 16.6.a shows that these rows would form clusters. As these clusters may be close to each other, a Gaussian-elimination procedure can be used to find the most linearly independent row vectors in V_s and use them as reference vectors to find coherent machines in the same group.

The main idea is to find the r most linearly independent rows of V_s and reorder V_s so that these rows are the first part of V_s, forming the matrix V_{s1}. Then the following transformation can be performed:

$$V_s = \begin{bmatrix} V_{s1} \\ V_{s2} \end{bmatrix} \Rightarrow \begin{bmatrix} V_{s1} \\ V_{s2} \end{bmatrix} V_{s1}^{-1} = \begin{bmatrix} I \\ L \end{bmatrix} = \begin{bmatrix} I \\ L_g \end{bmatrix} + \begin{bmatrix} 0 \\ O(\varepsilon) \end{bmatrix} \tag{16.50}$$

In ideal coherency, the matrix L would be a grouping matrix L_g consisting of only zeros and ones. In practice, L will be close to L_g. The clusters of coherent machines in the V_{s1} coordinates are shown in Figure 16.6.b.

In addition to the machines, it is also useful to find the network buses in the coherent areas. Extending the machine rotor angle idea, the bus voltage angle θ can be computed as a sensitivity matrix C_θ of the rotor angles. If a linearization program is available, the bus voltage angles can be specified as the output variables. Then the part of the output matrix C of the linearized dynamic system for these rows of θ corresponding to the rotor angles δ would form C_θ. It follows that the modeshapes of the bus angles with respect to the interarea modes are given by $V_\theta = C_\theta V_s$. The rows of V_θ can also be clustered with the reference machines using the transformation $V_\theta V_{s1}^{-1}$.

To summarize, a grouping algorithm for finding the coherent machines and buses is given as follows:

Coherency Grouping Algorithm [50]

1) Compute the electromechanical modes of an n-machine power system.
2) Select the (interarea) modes, which are the modes with frequencies less than 1 Hz.
3) Compute the eigenvectors (mode shapes) of these slower modes.
4) Use the V_s matrix to identify the machines with similar mode shapes to form slow coherent groups.
5) Compute $V_\theta V_{s1}^{-1} = C_\theta V_s V_{s1}^{-1}$ to identify the buses for the coherent groups.

The use of this algorithm is illustrated in the following example.

Example 16.3

Find the coherent machines and buses in the power system in Example 16.2 using the Coherency Grouping Algorithm.

Solutions: With two coherent areas, there is only one interarea mode, which has the frequency 3.779 rad/sec = 0.601 Hz. The eigenvectors v_1 and v_2 in (16.44) form the V_s matrix

$$V_s = \begin{bmatrix} 0.5 & 0.4878 \\ 0.5 & 0.4031 \\ 0.5 & -0.5672 \\ 0.5 & -0.5271 \end{bmatrix} \begin{matrix} \text{Gen 1} \\ \text{Gen 2} \\ \text{Gen 11} \\ \text{Gen 12} \end{matrix} \qquad V_{s1} = \begin{bmatrix} 0.5 & 0.4878 \\ 0.5 & -0.5672 \end{bmatrix} \begin{matrix} \text{Gen 1} \\ \text{Gen 11} \end{matrix} \tag{16.51}$$

Using Gaussian elimination techniques, the rows in V_s due to Generators 1 and 11 are the most linearly independent pair, resulting in V_{s1} as shown in (16.51). In the transformed coordinates, the largest entry in each row of L, underlined in (16.52), is used for assigning the other machines to Generators 1 and 11:

$$V_s' V_{s1}^{-1} = \begin{bmatrix} 1 & 0 \\ 0 & 1 \\ \underline{0.9198} & 0.0802 \\ 0.0380 & \underline{0.9620} \end{bmatrix} \begin{matrix} \text{Gen 1} \\ \text{Gen 11} \\ \text{Gen 2} \\ \text{Gen 12} \end{matrix} \tag{16.52}$$

Continuing with the network buses, $C_\theta V_s$ and $V_\theta V_{s1}^{-1}$ are computed as

$$V_\theta = C_\theta V_s = \begin{bmatrix} 0.5 & 0.4283 \\ 0.5 & 0.3535 \\ 0.5 & 0.2556 \\ 0.5 & 0.3844 \\ 0.5 & -0.5018 \\ 0.5 & -0.4667 \\ 0.5 & -0.3556 \\ 0.5 & 0.3128 \\ 0.5 & -0.0523 \\ 0.5 & -0.4671 \\ 0.5 & -0.4125 \end{bmatrix} \begin{matrix} \text{Bus 1} \\ \text{Bus 2} \\ \text{Bus 3} \\ \text{Bus 10} \\ \text{Bus 11} \\ \text{Bus 12} \\ \text{Bus 13} \\ \text{Bus 20} \\ \text{Bus 101} \\ \text{Bus 110} \\ \text{Bus 120} \end{matrix} \qquad V_\theta V_{s1}^{-1} = \begin{bmatrix} \underline{0.9436} & 0.0564 \\ \underline{0.8727} & 0.1273 \\ \underline{0.7800} & 0.2200 \\ \underline{0.9020} & 0.0980 \\ 0.0620 & \underline{0.9380} \\ 0.0953 & \underline{0.9047} \\ 0.2006 & \underline{0.7994} \\ \underline{0.8342} & 0.1658 \\ 0.4880 & \underline{0.5120} \\ 0.0949 & \underline{0.9051} \\ 0.1466 & \underline{0.8534} \end{bmatrix} \tag{16.53}$$

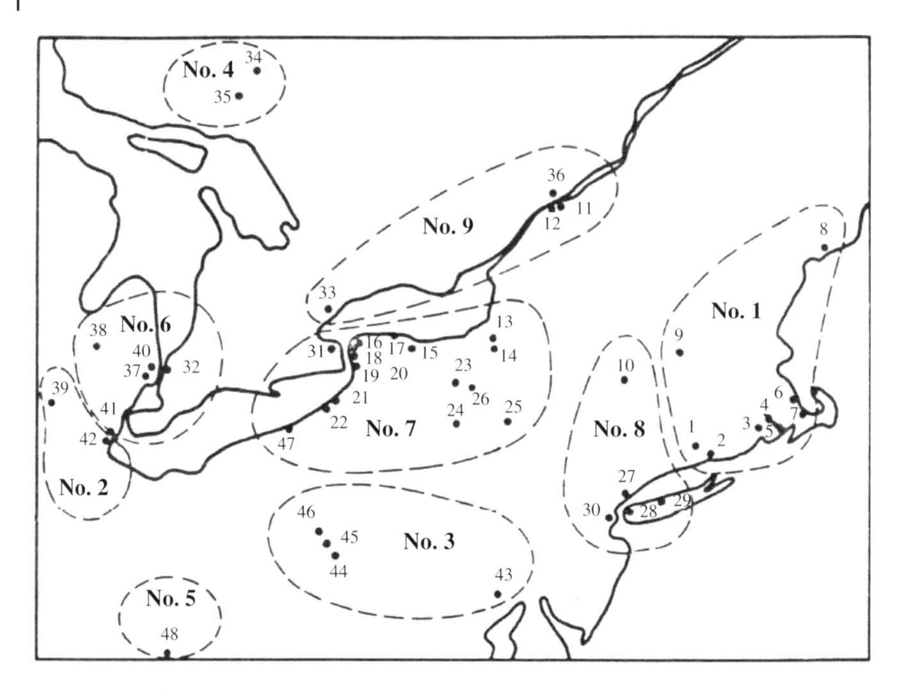

Figure 16.7 Nine-area partition of the NPCC system.

Then the largest entry in each row of $V_\theta V_{s1}^{-1}$, which is underlined, assigns the bus with either Generators 1 and 11. Thus the two coherent areas are:

Area 1 – Generators: 1, 2 Buses: 1, 2, 3, 10, 20
Area 2 – Generators: 11, 12 Buses: 11, 12, 13, 110, 120

Note that Bus 101 is not included in either area, as the two entries 0.4880 and 0.5120 are about equal. Although it can be assigned to Area 2, in practice, it is a boundary bus between the two areas, as can be seen from Figure 16.4. Boundary buses are important to retain if in the model reduction stage, it is desired to keep the areas separate. ■

For a larger system, a 9-area partition of a 48-machine, 140-bus NPCC system using the Grouping Algorithm is shown in Figure 16.7 [50].

A more advanced tolerance-based grouping algorithm that uses the angles between the row vectors in V_s can be found in [232]. There are also techniques that use the weak and strong admittances in a power system to identify the slow coherent areas [64, 233].

16.3 Generator Aggregation and Network Reduction

The slow subsystem in (16.34) is a linear reduced model useful for interarea mode analysis. For application to nonlinear power system dynamic simulation, a reduced model should have the following desirable characteristics:

1) The reduced model needs to have a power system structure, with equivalent generators and their control dynamics connected via a power network. As a result, the reduced model can be analyzed using existing power system simulation software.

2) The study system, in which disturbances are applied, should be kept in full detail.
3) The reduced model maintains the same power flow solution as the base-case full model. In particular, a bus in the study system of the reduced model will have the same voltage and current flows to the connecting buses as in the full model. However, the reduced model may not keep precisely the same power flow if a line is removed from the study system.

The model reduction technique proposed by Dr. Robin Podmore and others [230, 231, 235] addresses the above constraints and is carried out in three stages, which are described in the following subsections.

16.3.1 Generator Aggregation

The premise of this technique is that generators in the same coherent group can be represented by an aggregate generator because the dynamics due to the differences between the individual generators in the same coherent group are no longer important. Two methods are discussed here.

The first method is the Podmore aggregation [230].

Podmore Aggregation Algorithm:

1) Create an equivalent generator bus for a group of coherent generators.
2) Connect the equivalent generator bus to the individual generator terminal buses.
3) Remove the individual generator terminal buses.

This algorithm will be illustrated using the two coherent generators on Buses 1 and 2 in Figure 16.8, which are connected to the reminder of the power network via the transformers represented by the leakage reactances X_1 and X_2.

Step 1: In the power flow solution, let the voltages at Buses 1 and 2 and power output of the generators be $\tilde{V}_i = V_i\angle\theta_i$ and $P_i + jQ_i$, $i = 1, 2$, respectively. These two generators are moved to a new equivalent bus, called Bus 5, as shown in Figure 16.9. The voltage at

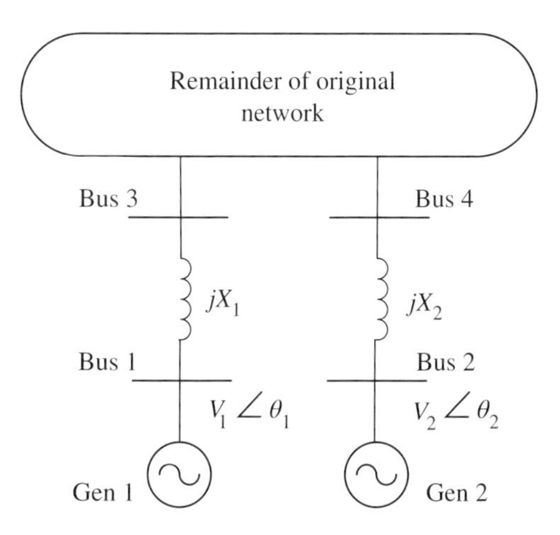

Figure 16.8 Coherent generators in original network.

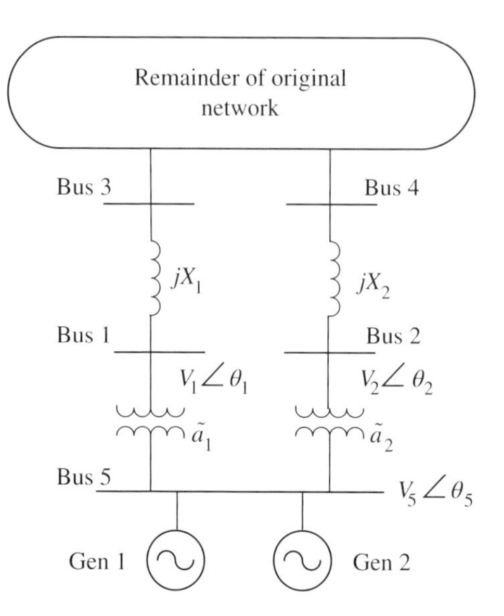

Figure 16.9 Coherent generator buses are connected to an equivalent bus, and generators are moved to the equivalent bus.

Bus 5 should resemble the voltages at Buses 1 and 2. One way to define $\tilde{V}_5$ is to set it to be the average voltage of Buses 1 and 2

$$\tilde{V}_5 = V_5\angle\theta_5 = \left(V_1\angle\theta_1 + V_2\angle\theta_2\right)/2 \tag{16.54}$$

Alternatively, it can be set as the weighted average of the generator MVA ratings, or simply the bus voltage of the generator with the largest active power output.

Step 2: To preserve the power flow from Buses 1 to 3 and Buses 2 to 4, Bus 5 is connected to Bus 1 via an ideal transformer[3] with a tap ratio and phase shift of

$$\tilde{a}_1 = a_1\angle\phi_1 = \tilde{V}_5/\tilde{V}_1 = \left(V_5/V_1\right)\angle\left(\theta_5 - \theta_1\right) \tag{16.55}$$

Similarly, the complex transformer tap ratio from Bus 5 to Bus 2 is

$$\tilde{a}_2 = a_2\angle\phi_2 = \tilde{V}_5/\tilde{V}_2 = \left(V_5/V_2\right)\angle\left(\theta_5 - \theta_2\right) \tag{16.56}$$

Step 3: Buses 1 and 2 are removed so that Bus 5 is connected to Bus 3 with a reactance of X_1 and a complex transformer ratio of $\tilde{a}_1$, and connected to Bus 4 with a reactance of X_2 and a complex transformer ratio of $\tilde{a}_2$, as shown in Figure 16.10. This new network configuration aggregates the coherent generators and preserves the power flow into the remainder of the original network.

This generation aggregation can be extended to more than two coherent generators. The method can also handle additional complexity in the network. For example, if Buses 1 and 2 in Figure 16.8 are connected with a branch, then this branch is replaced by shunt admittances on Buses 1 and 2. Also all the shunt admittances and loads on Buses 1 and 2 are moved to the equivalent generator bus. Details of such operations can be found in [231].

3 An ideal transformer has no leakage fluxes and hence its leakage reactance is zero.

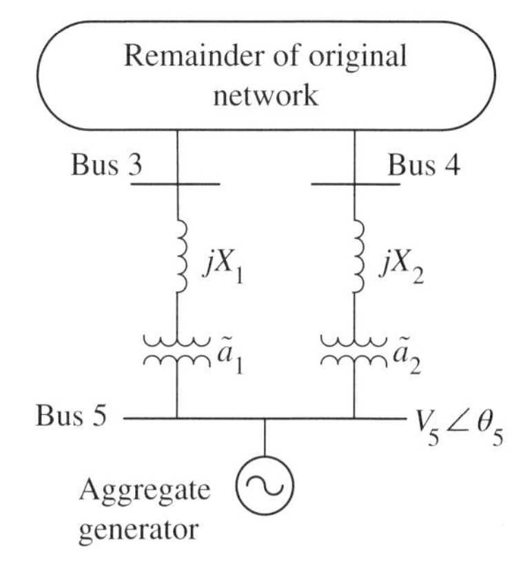

Figure 16.10 Original generator terminal buses eliminated and aggregate generator formed.

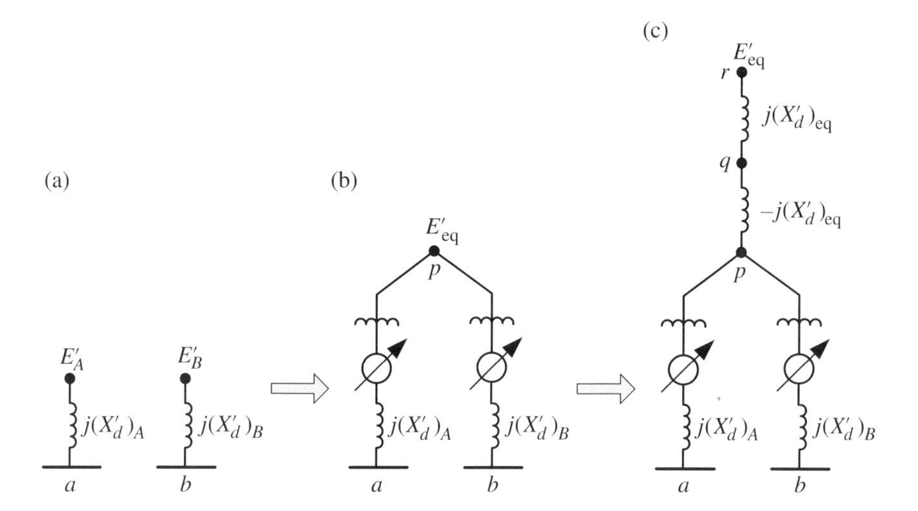

Figure 16.11 Inertial aggregation: (a) original network, (b) internal nodes of generators connected, and (c) internal node of the aggregate generator created (see color plate section).

The second method is the inertial aggregation method utilizing the model (16.38). Here the dynamics inside a coherent area are assumed to be infinitely fast and thus neglected. This is akin to connecting the states of (16.38), which are the internal nodes of the machines, with infinite impedances. To illustrate, the inertial aggregation for two machines, A and B, is shown in Figure 16.11, where a and b are the generator terminal buses. In the inertial aggregation technique, the machine internal node voltages E'_A and E'_B are computed (Figure 16.11.a). These two generator internal nodes are tied together to a common Bus p with infinite admittances (Figure 16.11.b). The voltage at Bus p, E'_{eq}, can be set either to an average of E'_A and E'_B or a weighted average with respect to the active and reactive power generation. To preserve the power flow, ideal transformers with complex turns ratios $\alpha_a \angle\phi_a$ and $\alpha_b \angle\phi_b$ and zero impedances link Buses a and b to Bus p.

The linking of the internal nodes creates an equivalent generator with multiple terminal buses, which is not a conventional power system network representation. Taking $\left(X'_d\right)_{\text{eq}}$ from (16.57) as an equivalent transient reactance, the network is extended beyond node p by two buses with reactances of $-\left(X'_d\right)_{\text{eq}}$ and $\left(X'_d\right)_{\text{eq}}$, as shown in Figure 16.11.c. The node r serves as the internal node of the equivalent machine, and the node q the generator terminal bus. The node p can be eliminated if desired.

There is also a third aggregation method for the construction of the slow-coherency model (16.40), which is based on two important observations. First recall that (16.40) requires a singular perturbation correction involving only the network within the coherent area, which is referred to as the "individual coherent area" aggregation. This property does not hold if additional corrections are made to (16.40). The second observation is that a nonlinear power system model has to be derived from the linearized model (16.40). This method is quite involved [234] and will not be elaborated here.

16.3.2 Dynamic Aggregation

If the generators are represented by the classical model, the parameters of the aggregate generator can be obtained by summing the MVA bases and inertias, and paralleling the transient reactances. Given n coherent machines, each machine having an MVA base of S_i, inertia of H_i, and transient reactance of X'_{di}, $i = 1, 2, .., n$, the parameters of the aggregate generator will be represented by

$$\begin{aligned} \text{MVA rating} &: S = \sum_{i=1}^{n} S_i \\ \text{Inertia} &: H = (1/S) \sum_{i=1}^{n} H_i S_i \\ \text{Transient reactance} &: X'_d = 1/\left[S \sum_{i=1}^{n} \left(S_i X'_{di}\right)^{-1} \right] \end{aligned} \tag{16.57}$$

in which the inertia and transient reactance are both weighted by the individual machine ratings.

If the individual generators are modeled with detailed machine models, excitation systems, and governors, a dynamic aggregation process has to be used. The aggregate models of the detailed machine model and the control equipment can be obtained with various methods. For example, let the turbine-governors for coherent machines $1, 2, \ldots, n$, be represented by the individual transfer functions $G_i(s)$ such as

$$P_{mi} = G_i(s)\,\omega_i(s) \tag{16.58}$$

where P_{mi} is the mechanical power driving Machine i and ω_i is the speed of Machine i. Combine the coherent turbine-governors into an aggregate turbine generator

$$P_m = \sum_{i=1}^{n} P_{mi} = \left(\sum_{i=1}^{n} G_i(s) \right) \omega(s) = G(s)\,\omega(s) \tag{16.59}$$

in which individual machine speeds ω_i are assumed to be the same as the speed of the aggregate machine ω. Then an optimization is used to approximate the frequency response of $G(s)$ by $G_{agg}(s)$ having the structure of physical turbine governors as discussed in Chapter 12. The use of such optimization methods to tune the parameters of the aggregate model for frequency response matching with structurally constraints is quite involved and will be not discussed here. Further discussions can be found in [235].

If the control models of the machines in the same coherent group are of different types, it may be necessary to form separate aggregate machines. For example, if some machines are hydraulic units and the other machines are steam units, it would be necessary to aggregate the hydraulic units into one aggregate machine, and the steam units into another aggregate machine. In other words, there will be an aggregate steam unit and an aggregate hydraulic unit on the equivalent bus.

Example 16.4

In Example 16.1, Machines 11 and 12 (shown in Figure 16.12.a) of the 2-area, 4-machine system are found to be coherent. This example shows how these two machines can be combined into an aggregate machine. The power flow solution for Buses 11 and 12 in Example 16.1 is given in Table 16.2. The machine parameters are:

Gen 11: base MVA = 1100, $H = 6.5$ sec, $X'_d = 0.25$ pu
Gen 12: base MVA = 700, $H = 6.5$ sec, $X'_d = 0.25$ pu

Use the power flow solution of the 2-area system to combine Generator 11 and Generator 12 into a single equivalent machine on a common bus.

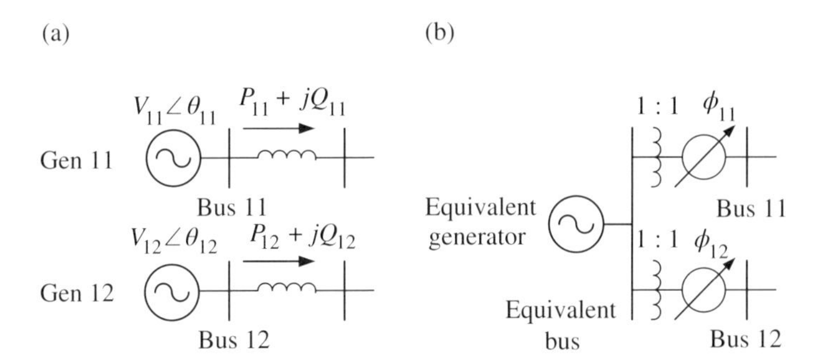

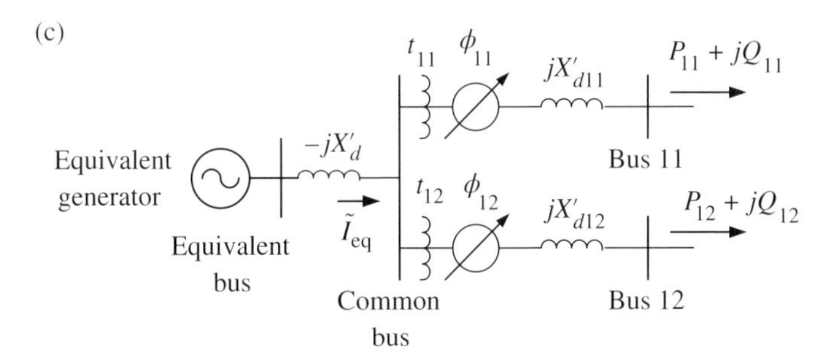

Figure 16.12 One-line diagrams showing the steps of generator aggregation for Example 16.4: (a) original system, (b) Podmore aggregation, and (c) inertial aggregation.

Table 16.2 Bus data for Example 16.4.

Bus	Voltage magnitude pu	Voltage angle degrees	*P* gen pu	*Q* gen pu
11	1.05	0	9.2471	2.3435
12	1.05	−16.6	5.0000	3.2249

1) Use the Podmore method: The common bus is taken to be a new bus whose voltage magnitude is the average of the voltage magnitudes of Buses 11 and 12, and the voltage angle is the average of the voltage angles at these two buses.
2) Use the inertial aggregation method: The common internal node is taken as the average of the internal node voltages of these two machines, whose angle is the average of the internal node voltage angles.

Solutions:

a) Podmore aggregation method: The common bus voltage magnitude and angle are set as

$$V_{eq} = \left(V_{11} + V_{12}\right)/2 = 1.05 \text{ pu} \tag{16.60}$$

$$\angle\tilde{V}_{eq} = \left(0^\circ - 16.60^\circ\right)/2 = -8.30^\circ \tag{16.61}$$

Because $V_{eq} = V_{11} = V_{12}$, the tap ratios are set at $t_{11} = 1$ and $t_{12} = 1$. The phase-shifting transformers are set up as

$$\phi_{11} = 0^\circ - \left(-8.30^\circ\right) = 8.30^\circ, \quad \phi_{12} = -16.60^\circ - \left(-8.30^\circ\right) = -8.30^\circ \tag{16.62}$$

The aggregate generator base is

$$1100 + 700 = 1800 \text{ MVA} \tag{16.63}$$

The aggregate inertia on 1800 MVA base is

$$H = (6.5 \times 1100 + 6.5 \times 700) / (1100 + 700) = 6.5 \text{ sec} \tag{16.64}$$

The equivalent X'_d on 1800 MVA base is

$$\left(\frac{X'_{d11}}{1100} \parallel \frac{X'_{d12}}{700}\right) \times 1800 = \frac{\left(X'_{d11}/1100\right)\left(X'_{d12}/700\right)}{\left(X'_{d11}/1100\right) + \left(X'_{d12}/700\right)} = 0.25 \text{ pu} \tag{16.65}$$

The aggregated generator network is shown in Figure 16.12.b.

b) Inertial aggregation method: The generator output currents are given by

$$\tilde{I}_{11} = \left(\frac{P_{11} + jQ_{11}}{\tilde{V}_{11}}\right)^*, \quad \tilde{I}_{12} = \left(\frac{P_{12} + jQ_{12}}{\tilde{V}_{12}}\right)^* \tag{16.66}$$

Assuming classical machine models, the Generator 11 and 12 internal voltages are, respectively,

$$\tilde{E}'_{11} = \tilde{V}_{11} + \tilde{I}_{11}\frac{jX'_{d11}}{11} = 1.1188\angle 10.306^\circ \tag{16.67}$$

$$\tilde{E}'_{12} = \tilde{V}_{12} + \tilde{I}_{12}\frac{jX'_{d12}}{7} = 1.1721\angle - 8.2571^\circ \tag{16.68}$$

The common bus voltage magnitude and phase are

$$V_{\text{cb}} = (1.1188 + 1.1721)/2 = 1.1454 \text{ pu} \tag{16.69}$$

$$\angle\tilde{V}_{\text{cb}} = \left(10.306^\circ - 8.2571^\circ\right)/2 = 1.0244^\circ \tag{16.70}$$

The transformer ratio and phase shift angles are

$$\phi_{11} = 10.306^\circ - 1.0244^\circ = 9.2816^\circ, \quad t_{11} = 1.1188 : 1.1454 = 1 : 1.0238 \tag{16.71}$$

$$\phi_{12} = -8.2571^\circ - 1.0244^\circ = -9.2816^\circ, \quad t_{12} = 1.1721 : 1.1454 = 1 : 0.9772 \tag{16.72}$$

Thus the generator output current is

$$\tilde{I}_{\text{eq}} = \tilde{I}_{11}\angle - 9.2816^\circ/1.0238 + \tilde{I}_{12}\angle 9.2816^\circ/0.9772 = 12.570 - j7.2767 \tag{16.73}$$

The aggregate generator bus voltage is

$$\tilde{V}_{\text{eq}} = \tilde{V}_{\text{cb}} - j\tilde{I}_{\text{eq}} \times 0.25/18 = 1.0442 - j0.15411 = 1.0555\angle - 8.3955^\circ \tag{16.74}$$

Note that the MVA base and the inertia of the equivalent generator are the same as that of the Podmore method (16.63) and (16.64). The aggregated generator network is shown in Figure 16.12.c. ■

16.3.3 Load Bus Elimination

In addition to the reduction of the generators and the generation buses, the load buses can also be reduced using a variety of traditional methods such as the Ward-Hale reduction and the REI method [17]. Here the Ward-Hale method is described.

Consider the 4-node star connection subnetwork in Figure 16.13.a. Disregarding the flows into the Nodes 1, 2, and 3, the network equation is represented by

$$\begin{bmatrix} Y_{14} & 0 & 0 & -Y_{14} \\ 0 & Y_{24} & 0 & -Y_{24} \\ 0 & 0 & Y_{34} & -Y_{34} \\ -Y_{14} & -Y_{24} & -Y_{34} & Y_{44} \end{bmatrix} \begin{bmatrix} \tilde{V}_1 \\ \tilde{V}_2 \\ \tilde{V}_3 \\ \tilde{V}_4 \end{bmatrix} = \begin{bmatrix} 0 \\ 0 \\ 0 \\ \tilde{I}_4 \end{bmatrix} \tag{16.75}$$

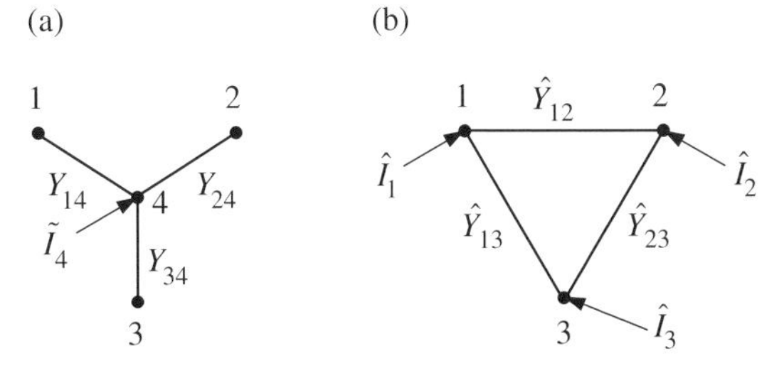

Figure 16.13 Load bus reduction: (a) 4-node star connection and (b) 3-node delta connection.

where Y_{ij} is the admittance of the line connecting Nodes i and j,

$$Y_{44} = Y_{14} + Y_{24} + Y_{34} \tag{16.76}$$

$\tilde{V}_i$ is the voltage phasor on Node i, and $\tilde{I}_4$ is the current injection into Node 4. The voltage at Node 4 can be computed in terms of the other variables as

$$\tilde{V}_4 = \left(\tilde{I}_4 + \tilde{V}_1 Y_{14} + Y_{24}\tilde{V}_2 + Y_{34}\tilde{V}_3\right) / Y_{44} \tag{16.77}$$

Eliminating $\tilde{V}_4$ from (16.75) results in the subnetwork described by

$$\begin{bmatrix} \left(1-\alpha_{14}\right) Y_{14} & -\alpha_{14}Y_{24} & -\alpha_{14}Y_{34} \\ -\alpha_{24}Y_{14} & \left(1-\alpha_{24}\right) Y_{24} & -\alpha_{24}Y_{34} \\ -\alpha_{34}Y_{14} & -\alpha_{34}Y_{24} & \left(1-\alpha_{34}\right) Y_{34} \end{bmatrix} \begin{bmatrix} \tilde{V}_1 \\ \tilde{V}_2 \\ \tilde{V}_3 \end{bmatrix}$$

$$= \begin{bmatrix} \hat{Y}_{11} & \hat{Y}_{12} & \hat{Y}_{13} \\ \hat{Y}_{12} & \hat{Y}_{22} & \hat{Y}_{23} \\ \hat{Y}_{13} & \hat{Y}_{23} & \hat{Y}_{33} \end{bmatrix} \begin{bmatrix} \tilde{V}_1 \\ \tilde{V}_2 \\ \tilde{V}_3 \end{bmatrix} = Y_{\mathrm{red}} \begin{bmatrix} \tilde{V}_1 \\ \tilde{V}_2 \\ \tilde{V}_3 \end{bmatrix} = \begin{bmatrix} \hat{I}_1 \\ \hat{I}_2 \\ \hat{I}_3 \end{bmatrix} \tag{16.78}$$

where $\alpha_{i4} = Y_{i4}/Y_{44}$, $i = 1, 2, 3$. The network (16.78) is shown in Figure 16.13.b. Note that this 3-node subnetwork has the same number of branches as the original 4-node subnetwork.

If the transmission lines are modeled with line charging and if constant impedance loads are added to the diagonal entries of the Y matrix, then the row sums of Y_{red} will no longer be equal to zero. Instead, the row-sum difference will become the G and B shunts at the buses. For example, the shunts at Bus 1 in (16.78) are given by

$$\hat{G}_1 + j\hat{B}_1 = \hat{Y}_{11} + \hat{Y}_{12} + \hat{Y}_{13} \tag{16.79}$$

The line reactance can be reconstructed from

$$\hat{R}_{ij} + j\hat{X}_{ij} = 1/\hat{Y}_{ij} \tag{16.80}$$

The equivalenced lines will not have line charging. In an equivalenced system with generator aggregation, transmission lines can have negative values of $\hat{R}_{ij}$ and buses can have negative values of $\hat{G}_i$. These values are typically small and can be a result of a generator output represented as negative load because no dynamic generator data is available. The line parameters $\hat{R}_{ij}$ and $\hat{X}_{ij}$ and the shunts $\hat{G}_i$ and $\hat{B}_i$ are used as the line and bus data in the power flow solution for the reduced model.

Example 16.5

Consider Buses 12, 13, 110, and 120 in the 2-area, 4-machine system, in which Bus 120 is connected to the other three buses (Figure 16.4). Develop the network equation (16.75) using the line parameters given in Table 16.3. Note that there are no injections into Bus 120. Eliminate Bus 120 and obtain the line and bus parameters of the remaining three buses.

Solutions: With the bus voltage vector arranged in the order $\tilde{V}_{12}$, $\tilde{V}_{13}$, $\tilde{V}_{110}$, and $\tilde{V}_{120}$, the admittance matrix is formed as

$$Y = \begin{bmatrix} 0-j59.88 & 0 & 0 & 0+j59.88 \\ 0 & 9.90-j99.00 & 0 & -9.90+j99.01 \\ 0 & 0 & 3.96-j39.58 & -3.96+j39.60 \\ 0-j59.88 & -9.90+j99.01 & -3.96+j39.60 & 13.86-j198.46 \end{bmatrix} \tag{16.81}$$

Table 16.3 Line parameters for Example 16.5.

From bus	To bus	*R*, pu	*X*, pu	*B*, pu
12	120	0	0.0167	0
13	120	0.0010	0.0100	0.0175
110	120	0.0025	0.0250	0.0437

Eliminating the last row and column of Y results in

$$Y_{\text{red}} = \begin{bmatrix} 1.2557 - j41.9009 & -0.8965 + j29.9358 & -0.3586 + j11.9743 \\ -0.8965 + j29.9358 & 3.4689 - j49.6516 & -2.5728 + j19.7398 \\ -0.3586 + j11.9743 & -2.5728 + j19.7398 & 2.9313 - j31.6862 \end{bmatrix} \tag{16.82}$$

The new line impedances are (all in pu):

Line 12-13: $-1/\left(-0.8965 + j29.9358\right) = 0.0010 + j0.0334 = R_{12,13} + jX_{12,13}$
Line 12-110: $-1/\left(-0.3586 + j11.9743\right) = 0.0025 + j0.0834 = R_{12,110} + jX_{12,110}$
Line 13-110: $-1/\left(-2.5728 + j19.7398\right) = 0.0065 + j0.0498 = R_{13,110} + jX_{13,110}$

The shunt loads on these buses are obtained by adding each row of Y_{red} such that the G_{shunt} and B_{shunt} are (all in pu):

Bus 12: $G_{12} = 0.0006$, $B_{12} = 0.0092$
Bus 13: $G_{13} = -0.0005$, $B_{13} = 0.0240$
Bus 110: $G_{110} = -0.0002$, $B_{110} = 0.0280$

Note that positive G consumes real power and positive B generates capacitive power. Here two of the loads actually generate real power (albeit a very small amount) to preserve the power balance. ■

An important issue in load bus reduction is the number of new branches being generated during the reduction process, as it is desirable to minimize this number. Several fundamental network configurations are shown in the left column of Figure 16.14. The resulting networks and the connections after the reduction of one node are shown in the right column of Figure 16.14.

In the reduction of the radial 3 nodes to 2 nodes, the number of branches is decreased by 1. For the reduction of the 4-node star connection to the 3-node delta connection, the number of lines remains the same. However, in the reduction of Bus 5 from the 5-node star connection, the number of branches has grown from 4 to 6. In fact, if a node connected to N other nodes is eliminated, then $N\,(N-1)\,/2$ new lines will be created.

In the reduction of a large power system, it is necessary to order the load-bus elimination in such a way such that the least number of branches will be created in each step. This is akin to a look-forward step. As a result, the 3-node radial configuration in Figure 16.14.a will be reduced first. In this stage, nodes with only two connections each will be identified and eliminated, resulting in reducing both the number of buses and branches. Next nodes with three connections each will be identified and reduced, corresponding to Figure 16.14.b. Once the reduction process starts eliminating nodes with four or more connections, the number of branches will increase, although the number of buses will continue to decrease.

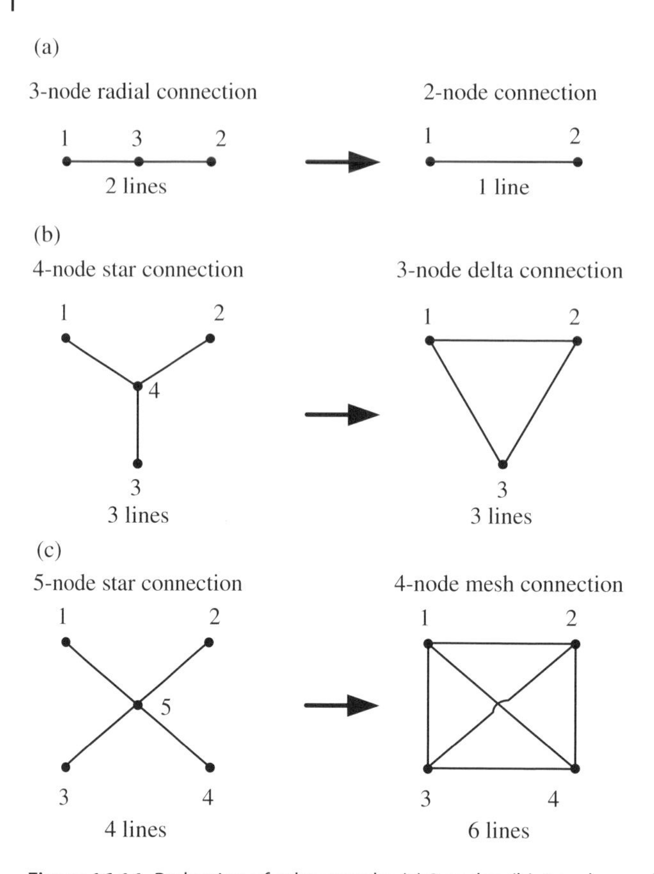

Figure 16.14 Reduction of subnetworks: (a) 3 nodes, (b) 4 nodes, and (c) 5 nodes.

Thus a practical issue in load bus reduction is to prevent an excessive number of branches. In general, power system model reduction consists of keeping the study region intact without reduction, and reduces the external system. The external system would consist of many control regions. The connections within a control region are typically more than the connections between the control regions. Thus a good strategy is to retain the boundary buses between the control regions. With this option, the internal buses of a control area will not have connections to any buses outside of the control area during the reduction process, thus helping to limit the growth of branches. In addition, the flows and oscillations on the tielines can be monitored in a simulation study using the reduced model. After the load bus reduction process has been completed, high-impedance branches can be eliminated. For example, a line with an impedance higher than 10 pu (on 100 MVA base) can be removed as the flow error would not be more than 10 MW, which is also approximately the maximum amount of power that can flow on the line.

A factor to be considered in load-bus reduction is the presence of phase shifters. When a line with a phase shifter is eliminated in conjunction to a load bus elimination, one may not be able to recover a "physical" line with the usual set of line parameters from the reduced admittance matrix. As a result, the power flow solution will not be "exact" anymore. Thus it may not be desirable to eliminate phase-shifter lines with low impedance values.

16.4 Simulation Studies

The NPCC 48-machine, 140-bus system [50] will be used to provide a comparison of the nonlinear aggregate models obtained from the two reduction techniques, and to illustrate the model reduction process. Note that the New England system model of the 48-machine system is the same as that of the NPCC 16-machine system shown in Figure 4.25, except that the buses and generators have been renumbered. A disturbance will be applied in the New England part of the system. Thus in the reduced model, all nine generators and all the buses in New England will be kept in detail, and only the power system external to New England will be reduced.

The coherent machine groups external to the New England system are found by applying a tight-coherency criterion to the machine groups obtained from the slow-coherent algorithm. The details of finding the 15 machine groups (some of which are single-machine areas) outside of New England are described in Chapter 3 of [83]. The Podmore and inertial aggregation algorithms described earlier will be used to obtain 15 aggregate generators outside of New England. Thus in addition to the 9 original generators in the study region, the reduced model contains a total of 24 generators. The accuracy of these reduced models are illustrated in the following example.

Example 16.6

For the NPCC 48-machine system, a 6-cycle short-circuit fault is applied at the Medway bus (Bus 7 in the NP48 data set *datanp48.m*, which is Bus 16 in Figure 4.25), and is cleared by removing the line from Medway to the Sherman Road bus (Bus 6 in *datanp48.m*, which is Bus 15 in Figure 4.25). The example folder contains all the MATLAB files to build the two aggregate models from the coherent groups.[4] Simulate the disturbance on the original model and the two reduced models and evaluate the accuracy of the reduced models from the time responses of Generator 8 (Maine Yankee, connected to Bus 29 in Figure 4.25) and Generator 4 (Brayton Point unit 3, connected to Bus 20 in Figure 4.25).

Solutions: Using the files provided in the example folder, this Medway disturbance is simulated with the resulting time responses of Generators 8 and 4 shown in Figures 16.15 and 16.16. These machine rotor angle time responses are plotted using Generator 48, which is a large machine with its own area, as the reference.

The full-order model time responses show a dominant interarea-type oscillation at slightly lower than 0.4 Hz. There is also a higher frequency oscillation in the Generator 4 response, which is its oscillation against Brayton Point unit 4 on the same bus. This is known as an intra-plant mode.

The Podmore aggregation model provides good approximations up to 4 seconds, after which the error in the slow interarea mode frequency approximation becomes

4 Note that the aggregation has to start from a solved power flow solution.

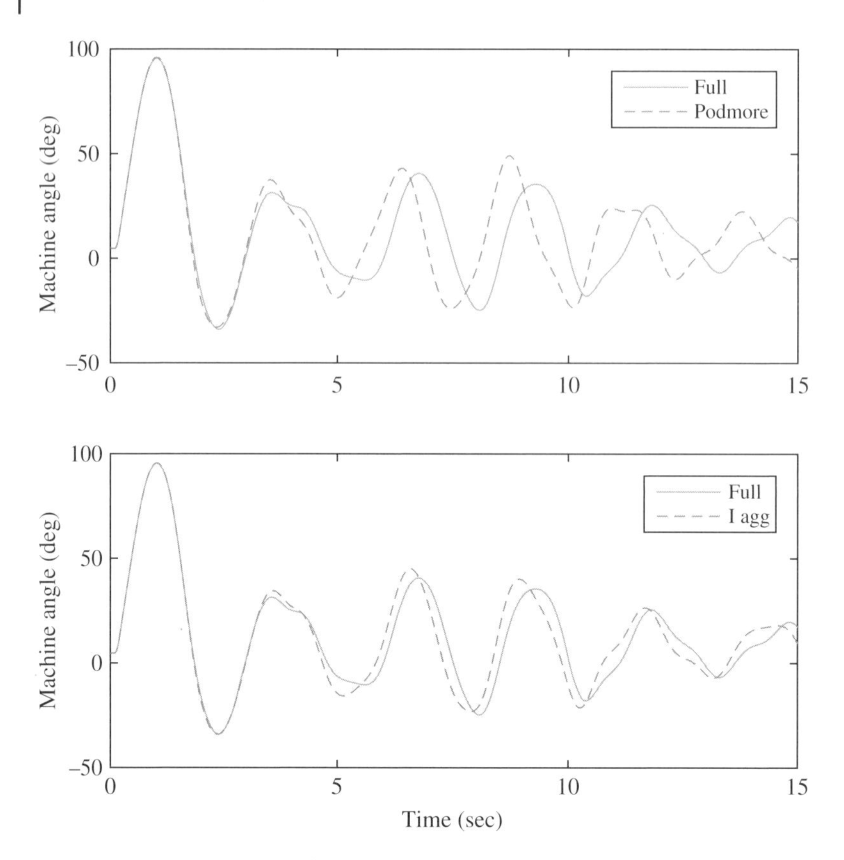

Figure 16.15 Time response of Machine 8 for the Medway disturbance (see color plate section).

evident. One can interpret that the Podmore model shows an oscillation frequency close to 0.4 Hz. As explained earlier, the Podmore aggregation "stiffens" the power network, resulting in a higher synchronizing torque and thus a higher frequency of oscillations.[5]

The inertial aggregation model captures the interarea oscillatory modes more accurately. In so doing, the differences in the time responses between the full mode and the inertial aggregation model are smaller.

Note that none of the load buses in the reduced models have been reduced, so that issues with load bus reduction connected with lines with phase shifters do not affect the model accuracy. ■

16.4.1 Singular Perturbations Method

Once slow coherency has been identified for a power system, one can obtain the quasi-steady-state approximation from the singular perturbations method as presented in [236]. In this approach, the intra-area or local modes within each coherent area are assumed to be fast and have settled to their quasi-steady-state values. As a result, the differential equations describing the intra-area modes are solved as algebraic equations.

5 An eigenvalue analysis of the reduced models can be found in Chapter 3 of [83].

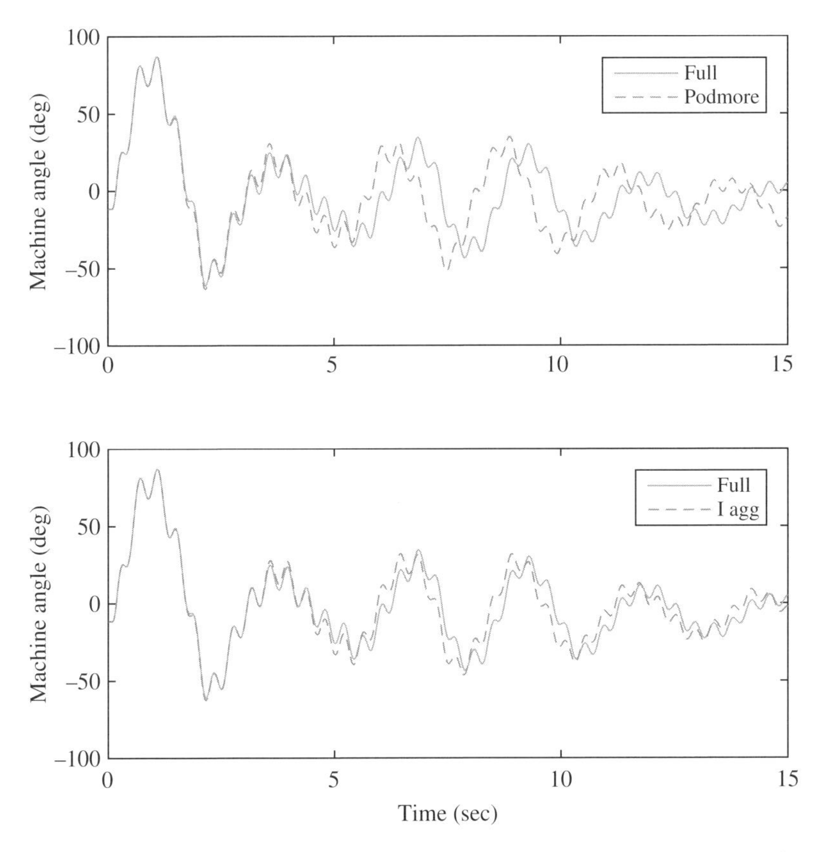

Figure 16.16 Time response of Machine 4 for the Medway disturbance (see color plate section).

This approach would require the development of special computer code to calculate the quasi-steady solution in a conventional power system simulation program [236].

16.5 Linear Reduced Model Methods

Instead of the dynamic aggregation method discussed in the earlier sections, which produces a reduced model with a power system structure, methods producing a linear reduced model for the external systems have also been proposed. When represented by a linear model, the external system can be detached from the study system, in which the tielines to the study system are represented by current injections, as shown in Figure 16.17.

The input-output model of the external system can be represented by

$$\frac{d\overline{x}_e}{dt} = \overline{f}_e\left(\overline{x}_e, \overline{V}_e, \overline{I}\right), \quad \overline{V}_e = \overline{g}_e\left(\overline{x}_e, \overline{I}\right), \quad \overline{V}_b = \overline{g}_b\left(\overline{x}_e, \overline{I}\right) \tag{16.83}$$

in which the tieline current injections $\overline{I}$ are the inputs to the model and the boundary bus voltages $\overline{V}_b$ are the outputs. The dynamic equation $\overline{f}_e$ is driven by the bus voltages $\overline{V}_e$ and the input $\overline{I}$. Note that the network equation $\overline{g}_e$ for the internal buses and $\overline{g}_b$ for the

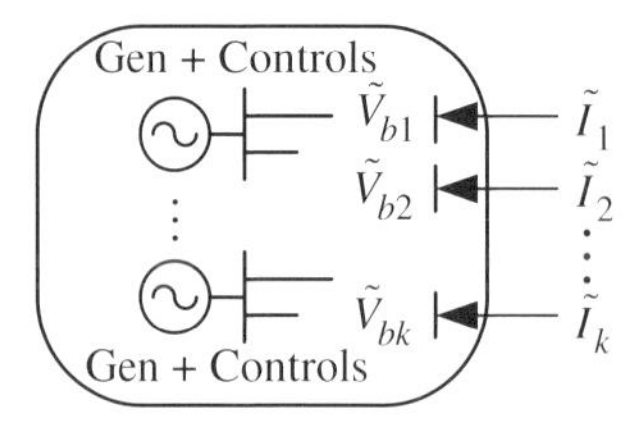

Figure 16.17 Input-output model of the external system.

boundary buses are no longer dependent on the study network variables. In particular, the current injection $\bar{I}$ is an independent variable and no longer a function of both the study and external systems.

This input–output model formulation conforms to the procedure in power system dynamic simulation, as described in Chapter 4. Note that it is also possible to use a formulation with the boundary bus voltages as input variables and the currents injected into the study system as the output variables [224]. This formulation is equivalent to (16.83) if the boundary buses are extended into the study system.

One of the first model reduction ideas is to linearize the external model (16.83) as

$$\frac{d\Delta\bar{x}_e}{dt} = A\Delta\bar{x}_e + B\Delta\bar{I}, \quad \Delta\bar{V}_b = C\Delta\bar{x}_e + D\Delta\bar{I} \tag{16.84}$$

at the pre-fault power flow condition. The network equation and the feedback control loops have been included in (16.84) without showing them explicitly. Nonlinear model linearization has been discussed in Chapter 6.

Representing the external system with a linear model would decrease the amount of computation needed for simulating a disturbance in the study system. The linear model will be able to account for the loop flow shown in Figure 16.1. To achieve further computation reduction, various approaches have been proposed for reducing the size of the linear model of the external system. A recent survey of linear model reduction approaches can be found in [237].

16.5.1 Modal Truncation

In the early 1970s, John Undrill, William Price, and others developed a modal truncation approach [224–227] following such work as [238] by selecting the modes to be retained in the reduced external systems. In this approach, (16.84) is represented by a reduced model in the modal form

$$\Delta\dot{\bar{x}}_m = A_m\Delta\bar{x}_m + B_m\Delta\bar{I}, \quad \Delta\bar{V}_b = C_m\Delta\bar{x}_m + D\Delta\bar{I} \tag{16.85}$$

in which only the dominant modes, including the electromechanical modes, are retained. The dimension of $\Delta\bar{x}_m$ is less than that of $\Delta\bar{x}_e$. Furthermore, if the state matrix A_m is expressed as a diagonal matrix, in which 2×2 real diagonal blocks are used for complex eigenvalues, the computation needs during simulation can be further reduced. Special computer code needs to be developed to interface the reduced linear model of the external system to the study system in a nonlinear power system simulation program.

16.5.2 Balanced Model Reduction Method

Another modal elimination approach is based on balanced truncation methods, which eliminate modes that are less controllable or observable [239, 240]. With balanced truncation, the reduced linear model of the external system will have the form

$$\Delta\dot{\bar{x}}_{\text{br}} = A_{\text{br}}\Delta\bar{x}_{\text{br}} + B_{\text{br}}\Delta\bar{I}, \quad \Delta\overline{V}_b = C_{\text{br}}\Delta\bar{x}_{\text{br}} + D\Delta\bar{I} \tag{16.86}$$

where the reduced matrices $(A_{\text{br}}, B_{\text{br}}, C_{\text{br}})$ are obtained using controllability and observability Gramians. Such methods do not keep the external system modes exactly but provide a better frequency response approximation of the input-output model compared to the modal truncation method. Balanced truncation methods are supported by efficient computation algorithms [241].

Note that the inputs and outputs of the linear models (16.85) and (16.86) are incremental variables. To couple them to the nonlinear study model, the pre-fault steady-state solution for the tieline currents $\bar{I}_o$ will provide the nominal values such that incremental input is

$$\Delta\bar{I} = \bar{I} - \bar{I}_o \tag{16.87}$$

Similarly, the pre-fault steady-state solution of the boundary bus voltages $\overline{V}_o$ will provide the actual input values to the study system as

$$\overline{V} = \overline{V}_o + \Delta\overline{V} \tag{16.88}$$

Equations (16.87) and (16.88) allow the initial condition of the linear models to be set to zero.

If the post-fault steady-state tieline flows are different from the initial conditions $\bar{I}_o$, then $\bar{x}$ will converge to a nonzero steady-state value. As a result, the steady-state boundary bus voltages will also be different from $\overline{V}_o$. Thus disturbances involving line trips and loss of generation can be accommodated by linear external models.

16.6 Dynamic Model Reduction Software

The DYNRED program is a commercial software capable of obtaining reduced models from large power systems with upwards of 30,000 buses. In this software, the coherency algorithm described in Section 16.3 is one of the methods used to find the coherent generators. For large systems, it would be practical to compute only the interarea modes. The Arnoldi method can be used efficiently to compute these eigenvalues [89]. The generator aggregation and load bus reduction are based on the algorithms described in Section 16.4.

The Power System Toolbox contains several functions for identifying the coherent generators and for aggregating coherent generators. The main PST EQUIV functions are listed as follows:

- *L_group*: grouping algorithm based on eigenvectors
- *ex_group*: tolerance-based grouping algorithm used in obtaining the 17-coherent areas of the NPCC system.
- *podmore*: R. Pormore's algorithm of aggregating generators at terminal buses
- *i_agg*: inertial aggregation at generator internal buses

- *s_coh3*: slow-coherency aggregation, with additional impedance corrections
- *reduce*: load bus elimination

These functions use the same power system power flow input data file as the *loadflow* function and the same machine data file.

16.7 Summary and Notes

This chapter discusses the concepts supporting the various aspects of power system model reduction. It first provides the analytical justification for slow coherency due to the sparse and weak connection between coherent areas. Then methods of aggregating coherent generators are provided, including the singular perturbation methods to improve on the reduced models. Power engineers tasked with obtaining 5-year or 10-year system planning models may find in this chapter the rationales for many of the steps that they need to take in coming up with their desired reduced models.

Research and development of methods for constructing reduced-order models of large power systems are still continuing. Here are some comments on topics not covered in this chapter and on some recent developments.

The modal truncation method [238] and the singular perturbations method [100] eliminate the modes with fast decaying transients and as such are less important. The selective modal analysis technique (Section 6.5.2) designates the modes with high modal participating factors as relevant modes that will be retained, and those modes with lower modal participating factors as less relevant modes that will be eliminated [86, 87]. From a state-space model form, the relevant reduced model can be computed iteratively. The reduced models are also suitable for designing damping controllers, such as power system stabilizers.

Besides modal truncation and balanced model reduction, the Krylov method [228, 242] has also been studied. Here the Markov parameters of the linear input–output model (16.84) are preserved up to a certain index. This matching of part of the controllability-observability subspace is also known as moment matching. The Krylov method is less computationally expensive than the balanced truncation method. It does not provide any error bounds but, in general, seems to work well.

In addition to being the basis of analyzing interarea oscillations, slow coherency has also been proposed for intentional islanding of large power systems in case of severe contingencies [243]. The idea here is to separate a connected system into islands along weak connections, so that interarea oscillations after islanding can be eliminated.

With the availability of synchronized phasor measurements, [70] has proposed a measurement-based method to construct a nonlinear 2-machine equivalent model for radial transmission systems, using a dominant interarea mode. The nonlinear model readily allows for transient energy analysis (see Chapter 5).

Problems

16.1 The transformation from the original variable x to the aggregate variable y and the local variable z is defined as

$$\begin{bmatrix} y \\ z \end{bmatrix} = \begin{bmatrix} C \\ G \end{bmatrix} x$$

Show that the inverse of this transformation is

$$x = \begin{bmatrix} U & G^{+} \end{bmatrix} \begin{bmatrix} y \\ z \end{bmatrix}$$

where

$$G^{+} = M^{-1}G^{T}\left(GM^{-1}G^{T}\right)^{-1}$$

16.2 Applying the transformation in Problem 16.1 to the second-order electromechanical model $M\ddot{x} = Kx$, show that the resulting model is

$$M_a\ddot{y} = \varepsilon K_a y + \varepsilon K_{ad} z \tag{16.89}$$

$$M_d\ddot{z} = \varepsilon K_{da} y + \left(K_d + \varepsilon K_{dd}\right) z \tag{16.90}$$

16.3 Consider the 2-area, 4-machine system (Figure 16.4). The data of the system is given in *d2a_em_coh_HW.m*. Note that in the original system data, all 4 machines have the same MVA rating (900 MVA) and inertia (6.5 sec on machine base). In this data file, the ratings of Generators 11 and 12 have been changed to 1,100 and 700 MVA, respectively. In this system, Generators 1 and 2 are coherent and Generators 11 and 12 are coherent.

1) Use the linearization function *svm_mgen* to generate the 8×8 A matrix of the system. (Do not set any reference generators.) Find the eigenvalues and identify the interarea and local modes.
2) Find the synchronizing coefficient (or stiffness) matrix K. Decompose K into the internal connection matrix K_I and the external connection matrix K_I. Note that there is no need to identify a singular perturbation or weak coupling parameter ε.
3) Find M_a, K_a, and $A_{i,j}$, $i, j = 1, 2$.
4) Find the eigenvalues of the slow aggregate model and the fast local models, without any corrections.
5) Find the eigenvalues of the slow coherency (corrected) aggregate model.

16.4 Use the eigenvectors of the interarea mode and the grouping algorithm to find the coherent areas of the 2-area system in Problem 16.3.

16.5 (Generator bus aggregation) Consider Example 16.4 with a new power flow solution given in Table 16.4. Aggregate Generator 11 and Generator 12 into a single equivalent machine on a common bus.

1) Use the Podmore method. The common bus is taken to be a new bus whose voltage phasor is given by the weighted power average according to

$$V_{\text{eq}} = \frac{P_{11}V_{11} + P_{12}V_{12}}{P_{11} + P_{12}} \tag{16.91}$$

$$\theta_{\text{eq}} = \frac{P_{11}\theta_{11} + P_{12}\theta_{12}}{P_{11} + P_{12}} \tag{16.92}$$

2) Use the inertial aggregation method based on (16.91) and (16.92).

Do not simplify the resulting power network.

Table 16.4 Bus data for Problem 16.5.

Bus	V pu	θ deg.	P gen, pu	Q gen, pu
11	1.03	0	9.2471	1.2000
12	1.05	−16.6	5.0000	3.2249

16.6 (Load bus reduction) Example 16.5 illustrates the elimination of Bus 120. Suppose a shunt capacitor of 100 MVar (1 pu susceptance) is added to Bus 120. Repeat Example 16.5 to compute the resulting 3-bus network.

16.7 (Load bus reduction) In the NPCC 16-machine model, Bus 1 is connected to five other buses with line resistance R, reactance X, and charging B shown in Table 16.5. The load on Bus 1 is $2.527 + j1.1856$ pu, and the voltage magnitude of Bus 1 is 1.0590 pu. Eliminate Bus 1 by representing the load at Bus 1 as a constant impedance load. Use the reduced Y matrix to reconstruct the lines connecting Buses 27, 30, 31, and 47, and the loads moved from Bus 1 to these buses.

Table 16.5 Line parameters for Problem 16.7.

From Bus	To Bus	R, pu	X, pu	B, pu
1	2	0.0035	0.0411	0.6987
1	27	0.0320	0.3200	0.41
1	30	0.0008	0.0074	0.48
1	31	0.0016	0.0163	0.25
1	47	0.0013	0.0188	1.31

16.8 (Load bus reduction) Repeat Problem 16.7 by using the *reduce* function on the NPCC 16-machine system in the file *data16em.m*. Solve the system power flow before applying *reduce*. Compare the results to those from Problem 16.7. Note that *reduce* is a function in the Aggregation Toolbox folder *aggreg*.

16.9 (NPCC 48-machine System) The data for the NPCC 48-machine power system is given in *datanp48.m*. Use the *L_group* function in the Equivalencing Toolbox folder *equiv* to find the coherent groups of machines using the first nine slowest eigenvalues (including the system mode). For this problem you need to solve the power flow, obtain a linearized model, and extract the eigenvector matrix for the nine slowest modes.

References

1 C. C. Liu, S. McArthur, and S.-J. Lee, *Smart Grid Handbook*, Wiley, 2016.
2 C. Concordia and R. P. Schulz, "Appropriate Component Representation for the Simulation of Power System Dynamics," IEEE Publication 75CH0970-4-PWR, 1975.
3 P. W. Sauer, M. A. Pai, and J. H. Chow, *Power System Dynamics and Stability with Synchrophasor Measurement and Power System Toolbox,* second edition, Wiley-IEEE Press, 2017.
4 A. Greenwood, *Electrical Transients in Power Systems,* second edition, Wiley-Interscience, 1991.
5 P. M. Anderson and A. A. Fouad, *Power System Control and Stability,* second edition, IEEE-Wiley Press, 2002.
6 P. Kundur, *Power System Stability and Control*, McGraw-Hill, 1994.
7 J. Machowski, J. Bialek, and J. Bumby, *Power System Dynamics: Stability and Control,* second edition, Wiley, 2008.
8 A. G. Phadke and J. S. Thorp, *Synchronized Phasor Measurements and Their Applications*, Springer, 2008.
9 E. W. Kimbark, *Direct Current Transmission*, vol. 1, Wiley, 1971.
10 N. Hingonari and L. Gyugyi, *Understanding FACTS: Concepts and Technology of Flexible AC Transmission Systems*, Wiley-IEEE Press, 1999.
11 A. J. Wood, B. F. Wollenberg, and G. B. Sheble, *Power Generation, Operation, and Control,* third edition, Wiley-Interscience, 2013.
12 J. J. Grainger and W. D. Stevenson, *Power System Analysis*, McGraw-Hill, New York, 1994.
13 J. H. Chow and K. W. Cheung, "A Toolbox for Power System Dynamics and Control Engineering Education and Research," *IEEE Trans. on Power Systems*, vol. 7, no. 4, pp. 1559–1564, 1992.
14 IEEE Standard C57.135, *Guide for the Application, Specification, and Testing of Phase-Shifting Transformers*, 2012.
15 M. A. Pai, *Power Circuits and Electromagnetics*, Stipes Publishing, 2012.
16 B. Stott, "Review of Load-Flow Calculation Methods," *Proc. of IEEE*, vol. 62, no. 7, pp. 916–929, 1974.
17 G. W. Stagg and A. H. El-Abiad, *Computer Methods in Power System Analysis*, McGraw-Hill, 1968.
18 Y. Saad, *Iterative Methods for Sparse Linear Systems*, SIAM, 2003.
19 G. H. Golub and C. F. Van Loan, *Matrix Computation*, John Hopkins University Press, 1983, Chapter 4.

Power System Modeling, Computation, and Control, First Edition. Joe H. Chow and Juan J. Sanchez-Gasca.

Companion website: www.wiley.com/go/chow/power-system-modeling

20 W. W. Price, *Power System Computation Course Notes*, Power Systems Engineering Course, General Electric Company, 1980.
21 G. Rogers, *Power System Oscillations*, Springer, 2012.
22 W. F. Tinney and J. W. Walker, "Direct Solutions of Sparse Network Equations by Optimally Ordered Triangular Factorization," *Proc. of IEEE*, vol. 55, pp. 1801–1809, 1967.
23 M. L. Crow, *Computational Methods for Electric Power Systems,* third edition, CRC Press, 2015.
24 L. Powell, *Power System Load Flow Analysis*, McGraw-Hill, 2004.
25 C. W. Taylor, *Power System Voltage Stability*, McGraw-Hill, 1994.
26 T. Van Cutsem and C. Vournas, *Voltage Stability of Electric Power Systems*, Springer, 1998.
27 V. Ajjarapu, *Computational Techniques for Voltage Stability Assessment and Control*, Springer, 2006.
28 C. Barbier and J. P. Barret, "Analysis of Phenomena of Voltage Collapse on a Transmission System," *Revue Generale de l'electricite*, vol. 89, pp. 672–690, Oct. 1980.
29 A. Kurita and T. Sakurai, "The Power System Failure on July 23, 1987 in Tokyo," *Proc. IEEE Conf. on Decision and Control*, pp. 2093–2097, 1988.
30 T. Ohno and S. Imai, "The 1987 Tokyo Blackout," *Proc. of Power System Computation Conf.*, 2006.
31 C. W. Taylor and D. C. Erickson, "Recording and Analyzing the July 2 Cascading Outage," *IEEE Computer Applications in Power*, pp. 26–30, Jan. 1997.
32 L. Warland, *A Voltage Instability Predictor Using Local Area Measurements, VIP++*, PhD Dissertation, Department of Electric Power Systems, Norwegian University of Science and Technology, 2002. Available: http://www.diva-portal.org/smash/get/diva2:121795/FULLTEXT01.pdf.
33 D. E. Julian, R. P. Schulz, K. T. Vu, W. H. Quaintance, N. B. Bhatt, and D. Novosel, "Quantifying Proximity to Voltage Collapse using the Voltage Instability Predictor (VIP)," *Proc. IEEE PES Summer Power Meeting*, vol. 2, pp. 931–936, July 2000.
34 B. Gao, G. K. Morison, and P. Kundur, "Voltage Stability Evaluation using Modal Analysis," *IEEE Trans. on Power Systems*, vol. 7, no. 4, pp. 1529–1542, 1992.
35 V. Ajjarapu and C. Christy, "The Continuation Power Flow: A Tool for Steady State Voltage Stability Analysis," *IEEE Trans. on Power Systems*, vol. 7, no. 1, pp. 416–423, 1992.
36 H.-D. Chiang, A. J. Flueck, K. S. Shah, and N. Balu, "CPFLOW: A Practical Tool for Tracing Power System Steady-State Stationary Behavior due to Load and Generation Variations," *IEEE Trans. on Power Systems*, vol. 10, no. 2, pp. 623–634, 1995.
37 S. G. Ghiocel and J. H. Chow, "A Power Flow Method using a New Bus Type for Computing Steady-State Voltage Stability Margins," *IEEE Trans. on Power Systems*, vol. 29, no. 2, pp. 958–965, 2014.
38 S. G. Ghiocel, *Applications of Synchronized Phasor Measurements for State Estimation, Voltage Stability, and Damping Control*, PhD Dissertation, Rensselaer Polytechnic Institute, 2013.
39 J. H. Chow, R. Galarza, P. Accari, and W. Price, "Inertial and Slow Coherency Aggregation Algorithms for Power System Dynamic Model Reduction," *IEEE Trans. on Power Systems*, vol. 10, no. 2, pp. 680–685, 1995.

40 B. Otomega, V. Sermanson, and T. Van Cutsem, “Reverse-Logic Control of Load Tap Changers in Emergency Voltage Conditions,” *Proc. IEEE Power Tech Conf*., Bologna, Italy, 2003.
41 S. Corsi, *Voltage Control and Protection in Electric Power Systems*, Springer, 2015.
42 M. Ilić and J. Zaborszky, *Dynamics and Control of Large Electric Power Systems*, Wiley-IEEE Press, 2000.
43 D. H. Boteler, Q. Bui-Van, and J. Lemay, “Directional Sensitivity to Geomagnetically Induced Currents of the Hydro-Quebec 735 kV Power System,” *IEEE Trans. on Power Delivery*, vol. 9, no. 4, pp. 1963–1971, 1994.
44 M. Parniani, J. H. Chow, L. Vanfretti, B. Bhargava, and A. Salazar, “Voltage Stability Analysis of a Multiple-Infeed Load Center Using Phasor Measurement Data,” *Proc. of IEEE Power System Conf. and Expo.*, October 2006.
45 S. M. Abdelkader and D. J. Morrow, “Online Tracking of Thévenin Equivalent Parameters using PMU Measurements,” *IEEE Trans. on Power Systems*, vol. 27, no. 2, pp. 975–983, 2012.
46 S. M. Burchett, D. Douglas, S. G. Ghiocel, M. Liehr, J. H. Chow, D. Kosterev, A. Faris, E. Heredia, and G. H. Matthews, “An Optimal Thévenin Equivalent Estimation Method and its Application to the Voltage Stability Analysis of a Wind Hub,” *IEEE Trans. on Power Systems*, vol. 33, no. 4, pp. 3644–3652, 2018.
47 M. S. Calović, “Modeling and Analysis of Under-Load Tap Changing Transformer Control Systems,” *IEEE Trans. on Power Apparatus and Systems*, vol. PAS-103, no. 7, pp. 1909–1915, 1984.
48 J. Medanić, M. Ilić, and J. Christensen, “Discrete Models of Slow Voltage Dynamics for Under Load Tap-Changing Transformer Coordination,” *IEEE Trans. on Power Systems*, vol. 2, no. 4, pp. 873–880, 1987.
49 IEEE PES Voltage Stability Working Group, *Test Systems for Voltage Stability Analysis and Security Assessment*, Technical Report PES-TR19, 2015.
50 B. Avramović, J. H. Chow, P. V. Kokotović, G. M. Peponides, and J. R. Winkelman, *Time-Scale Modeling of Dynamic Networks with Applications to Power Systems*, Springer, 1982.
51 B. L. Agrawal, P. M. Anderson, C. Concordia, R. G. Farmer, A. A. Foaud, P. Kundur, W. W. Price, and C. W. Taylor, “Damping Representation for Power System Stability Studies,” *IEEE Trans. on Power Systems*, vol. 14, no. 1, pp. 151–157, 1999.
52 A. R. Bergen and V. Vittal, *Power System Analysis*, second edition, Pearson, 1999.
53 A. Friedman, *Advanced Calculus*, Dover, 2007.
54 C. W. Gear, *Numerical Initial Value Problems in Ordinary Differential Equations*, Prentice Hall, 1971.
55 P. R. Benyon, “A Review of Numerical Methods for Digital Simulation,” *Simulation*, pp. 219–238, 1968.
56 J. J. Sanchez-Gasca, R. D’Aquila, W.W. Price, and J. J. Paserba, “Variable Time Step, Implicit Integration for Extended-Term Power System Dynamic Simulation,” *Proc. 1995 PICA Conf*., pp. 948–956, 1995.
57 A. Kurita, H. Kubo, K. Oki, S. Agematsu, D. B. Klapper, N. W. Miller, W.W. Price, J. J. Sanchez-Gasca, K. A. Wirgau, and T. D. Younkins, “Multiple Time-Scale Power System Dynamic Simulation,” *IEEE Trans. on Power Systems*, vol. 8, no. 1, pp. 216–223, 1993

58 J. F. Luini, R. P. Schulz, and A. E. Turner, "A Digital Computer Program for Analyzing Long Term Dynamic Response of Power Systems," *Proc. 1975 IEEE PICA Conf.*, 75CH0962-I PWR, pp. 136–143, 1975.
59 F. Milano, *Power System Modelling and Scripting*, Springer, 2010.
60 J. H. Chow and R. J. Rogers, *Power System Toolbox*, 2000. Available http://www.ecse.rpi.edu/ ˜chowj.
61 Z. Huang, Y. Chen, and J. Nieplocha, "Massive Contingency Analysis with High Performance Computing," *Proc. of IEEE PES General Meeting*, Calgary, Canada, 2009.
62 M. A. Pai, *Energy Function Analysis for Power System Stability*, Kluwer, 1989.
63 A. A. Fouad and V. Vittal, *Power System Transient Stability Analysis using Transient Energy Method*, Prentice Hall, 1992.
64 J. Zaborszky, G. Huang, B. Zheng, and T.-C. Leung, "On the Phase Portrait of a Class of Large Nonlinear Dynamic Systems such as the Power System," *IEEE Trans. on Automatic Control*, vol. 33, no. 1, pp. 4–15, 1988.
65 H.-D. Chiang, M. W. Hirsch, and F. F. Wu, "Stability Regions of Nonlinear Autonomous Dynamical Systems," *IEEE Trans. on Automatic Control*, vol. 33, no. 1, pp. 16–27, 1988.
66 H. K. Khalil, *Nonlinear Systems,* third edition, Pearson, 2014.
67 J. P. LaSalle, "Some Extensions of Lyapunov's Second Method," *IRE Trans. on Circuit Theory*, vol. CT-7, no. 4, pp. 520–527, 1960.
68 J. H. Chow, A. Chakrabortty, M. Arcak, B, Bhargava, and A. Salazar, "Synchronized Phasor Data Based Energy Function Analysis of Dominant Power Transfer Paths in Large Power Systems," *IEEE Trans. on Power Systems*, vol. 22, no. 2, pp. 727–734, May 2008.
69 A. Chakrabortty, *Estimation, Analysis and Control Methods for Large-Scale Electric Power Systems using Synchronized Phasor Measurements*, PhD Dissertation, Rensselear Polytechnic Institute, August 2008.
70 J. H. Chow, A. Chakrabortty, L. Vanfretti, and M. Arcak, "Estimation of Radial Power System Transfer Path Dynamic Parameters using Synchronized Phasor Data," *IEEE Trans. on Power Systems*, vol. 23, no. 2, pp. 564–571, May 2008.
71 V. I. Arnold and R. A. Silverman, *Ordinary Differential Equations*, MIT Press, 1978.
72 M. W. Hirsch, S. Smale, and R. L. Devaney, *Differential Equations, Dynamical Systems, and an Introduction to Chaos,* third edition, Academic Press, 2012.
73 H.-D. Chiang, *Direct Methods for Stability Analysis of Electric Power Systems: Theoretical Foundation, BCU Methodologies, and Applications*, Wiley, 2010.
74 H.-D. Chiang and L. F. C. Alberto, *Stability Regions of Nonlinear Dynamical Systems: Theory, Estimation, and Applications*, Cambridge University Press, 2015.
75 A. H. El-Abiad and K. Nagappan, "Transient Stability Regions for Multi-Machine Power Systems," *IEEE Trans. on Power Apparatus and Systems*, vol. PAS-85, pp. 169–179, 1966.
76 N. Kakimoto, N. Y. Ohsawa, and M. Hayashi, "Transient Stability Analysis of Large-Scale Power Systems by Lyapunov's Direct Method," *IEEE Trans. on Power Apparatus and Systems*, vol. PAS-103, pp. 160–167, 1984.
77 T. Athay, R. Podmore, and S. Virmani, "A Practical Method for the Direct Analysis of Transient Stability," *IEEE Trans. on Power Apparatus and Systems*, vol. PAS-98, pp. 573–584, 1979.

78 P. S. Prabhakara and A. H. El-Abiad, "A Simplified Determination of Stability Regions for Lyapunov Method," *IEEE Trans. on Power Apparatus and Systems*, vol. PAS-94, pp. 672–689, 1975.
79 M. Ribbens-Pavella and B. Lemal, "Fast Determination of Stability Regions for On-Line Power System Studies," *Proc. of IEE*, vol.123, pp. 689–969, 1976.
80 NERC Technical Report, *Reliability Concepts*, 2007.
81 D. Halliday, R. Resnick, and J. Walker, *Fundamentals of Phyics*, vol. 1, tenth edition, Wiley, 2013.
82 M. Klein, G. J. Rogers, and P. Kundur, "A Fundamental Study of Inter-Area Oscillations in Power Systems," *IEEE Trans. on Power Systems*, vol. 6, no. 3, pp. 914–921, August 1991.
83 J. H. Chow, ed., *Power System Coherency and Model Reduction*, Springer, 2013.
84 P. M. DeRusso, R. J. Roy, C. M. Close, and A. A. Desrochers, *State Variables for Engineers*, Wiley-Interscience, 1997.
85 C. Moler and C. Van Loan, "Nineteen Dubious Ways to Compute the Exponential of a Matrix, Twenty-Five Years Later," *SIAM Review*, vol. 45, no. 1, pp. 1–46, 2003.
86 I. J. Pérez-Arriaga, G. C. Verghese, and F. C. Scheweppe, "Selective Modal Analysis with Applications to Electric Power Systems, Part I: Heuristic Introduction," *IEEE Trans. on Power Apparatus and Systems*, vol. PAS-101, pp. 3117–3125, 1982.
87 G. C. Verghese, I. J. Pérez-Arriaga, and F. C. Scheweppe, "Selective Modal Analysis with Applications to Electric Power Systems, Part II: The Dynamic Stability Problem," *IEEE Trans. on Power Apparatus and Systems*, vol. PAS-101, pp. 3126–3134, 1982.
88 N. Martins and H. J. C. P. Pinto, "Modern Tools for Small-Signal Stability Analysis and Design of FACTS Assisted Power Systems," available online: http://www.nelsonmartins.com/pdf/publications/powertech.pdf.
89 L. Wang and A. Semlyen, "Application of Sparse Eigenvalue Techniques to the Small-Signal Stability Analysis of Large Power Systems," *IEEE Trans. on Power Systems*, vol. PWRS-5, no. 6, pp. 635–642, 1990.
90 J. H. Chow, J. Cullum, and R. Willoughby, "A Sparsity-Based Technique for Identifying Slow-Coherent Areas in Large Power Systems," *IEEE Trans. on Power Apparatus and Systems*, vol. PAS-103, no. 3, pp. 463–473, 1984.
91 M. A. Pai, D. P. Sen Gupta, and K. R. Padiyar, *Small Signal Analysis of Power Systems*, Alpha Science International, UK, 2004.
92 R. J. Piwko, H. A. Othman, O. A. Alvarez, and C. Y. Wu, "Eigenvalue and Frequency-Domain Analysis of the Intermountain Power Project and the WSCC Network," *IEEE Trans. on Power Systems*, vol. 6, no. 1, pp. 238–244, 1991.
93 C. Concordia, *Synchronous Machine*, John Wiley and Sons, New York, 1951.
94 D. A. Swann, *Synchronous Machines Course Notes*, Power System Engineering Course, General Electric Company, 1979.
95 IEEE Standard 1110-2002, *IEEE Guide for Synchronous Generator Modeling Practices and Applications in Power System Stability Analysis*, IEEE Power Engineering Society, 2002.
96 R. H. Park, "Two-Reaction Theory of Synchronous Machines – Generalized Method of Analysis – Part I," *AIEE Transactions*, vol. 48, pp. 716–727, 1929.
97 C. C. Young, *Synchronous Machine Theory*, Power System Engineering Course, General Electric Company, 1963.

98 IEEE Standard C37.102-2006, *IEEE Guide for AC Generator Protection*, IEEE Power Engineering Society, 2006.
99 R. P. Schulz, "Synchronous Machine Modeling," IEEE Publication 75 CH0970-4-PWR, 1975.
100 P. Kokotovic, H. K. Khalil, and J. O'Reilly, *Singular Perturbation Methods In Control: Analysis and Design*, Academic Press, 1986.
101 IEEE Standard 115-2009, *IEEE Guide for Test Procedures for Synchronous Machines: Part I – Acceptance and Performance Testing, and Part II – Test Procedures and Parameter Determination for Dynamic Analysis*, IEEE Power Engineering Society, 2009.
102 S. J. Salon, *Finite Element Analysis of Electrical Machines*, Springer, 1995.
103 Z. Huang, P. Du, D. Kosterev, and S. Yang, "Generator Dynamic Model Validation and Parameter Calibration using Phasor Measurements at the Point of Connection," *IEEE Trans. on Power Systems*, vol. 28, no. 2, pp. 1939–1949, 2013.
104 J. H. Chow, M. T. Glinkowski, R. J. Murphy, T. W. Cease, and N. Kosaka, "Generator and Exciter Parameter Estimation of Fort Patrick Henry Hydro Unit 1," *IEEE Trans. on Energy Conversion*, vol. 14, pp. 923–929, 1999.
105 IEEE Standard 421.5-1992, *IEEE Recommended Practice for Excitation System Models for Power System Stability Studies*, IEEE Power Engineering Society, 1992.
106 IEEE Standard 421.1-2007, *IEEE Standard Definitions for Excitation Systems for Synchronous Machines*, Power Engineering Society, 2007.
107 IEEE Committee Report, "Computer Representation of Excitation Systems," *IEEE Trans. on Power Apparatus and Systems*, vol. PAS-87, no. 6, pp. 1460–1464, 1968.
108 F. P. DeMello and C. Concordia, "Concepts of Synchronous Machine Stability as Affected by Excitation Control," *IEEE Trans. on Power Apparatus and Systems*, vol. PAS-88, no. 4, pp. 16–329, 1969.
109 P. M. Anderson and R. G. Farmer, *Series Compensation of Power Systems*, IPBLSH! Inc., 1996.
110 W. G. Heffron and R. A. Phillips, "Effect of a Modern Amplidyne Voltage Regulator in Underexcited Operation of Large Turbine Generators," *AIEE Trans., Part III: Power Apparatus and Systems*, vol. 71, pp. 692–697, Aug. 1952.
111 A. Murdoch and G. Boukarim, "Performance Criteria and Tuning Techniques," Chapter 3, IEEE PES *Tutorial on Power System Stabilization via Excitation Control*, June 2007.
112 J. H. Chow, G. E. Boukarim, and A. Murdoch, "Power System Stabilizers as Undergraduate Control Design Projects," *IEEE Trans. on Power Systems*, vol. 19, no. 1, pp. 144–151, 2004.
113 G. F. Franklin, J. D. Powell, and A. Emami-Naeini, *Feedback Control of Dynamic Systems*, seventh edition, Prentice Hall, 2014.
114 R. A. Lawson, D. A. Swann, and G. F. Wright, "Minimization of Power System Stabilizer Torsional Interaction on Large Steam Turbine Generators," *IEEE Trans. on Power Apparatus and Systems*, vol. PAS-97, no. 1, pp. 183–190, 1978.
115 I. Kamwa, R. Grondin, and G. Trudel, "IEEE PSS2B versus PSS4B: The Limits of Performance of Modern Power System Stabilizers," *IEEE Trans. on Power Systems*, vol. 20, no. 2, pp. 903–915, 2005.
116 E. V. Larsen and D. A. Swann, "Applying Power System Stabilizers, Part I: General Concepts; Part II: Performance Objectives and Tuning Concepts; Part III: Practical

Considerations," *IEEE Trans. of Power Apparatus and Systems*, vol. PAS-100, pp. 3017–3046, 1981.

117 D. C. Lee, R. E. Beaulieu, and J. R. R. Service, "A Power System Stabilizer using Speed and Electrical Power Inputs – Design and Field Experience," *IEEE Trans. of Power Apparatus and Systems*, vol. PAS-100, pp. 4151–4157, 1981.

118 A. Murdoch, S. Venkataraman, R. A. Lawson, and W. R. Pearson, "Integral of Accelerating Power Type PSS, Part 1 – Theory, Design, and Tuning Methodology," *IEEE Trans. on Energy Conversion*, vol. 14, no. 4, pp. 1658–1663, 1999.

119 A. Murdoch, S. Venkataraman, and R. A. Lawson, "Integral of Accelerating Power Type PSS, Part 2 – Field Testing and Performance Verification," *IEEE Trans. on Energy Conversion*, vol. 14, no. 4, pp. 1664–1672, 1999.

120 S. J. Chapman, *Electric Machinery and Power System Fundamentals*, McGraw-Hill, 2002.

121 NERC Report, *Reliability Guideline: Developing Load Model Composition Data*, March 2017. Available: www.nerc.com.

122 A. Keyhani, *Design of Smart Power Grid Renewable Energy Systems*, Wiley-IEEE Press, 2011.

123 C. Concordia and S. Ihara, "Load Representation in Power System Stability Studies," *IEEE Trans. on Power Apparatus and Systems*, vol. PAS-101, pp. 969–977, 1982.

124 IEEE Task Force Report, "Load Representation for Dynamic Performance Analysis," *IEEE Trans. on Power Systems*, vol. 8, no. 2, pp. 472–482, 1993.

125 IEEE Task Force Report, "Standard Load Models for Power Flow and Dynamic Performance Simulation," *IEEE Trans. on Power Systems*, vol. 10, no. 3, pp. 1302–1313, 1995.

126 D. Kosterev, A. Meklin, J. Undrill, B. Lesieutre, W. Price, D. Chassin, R. Bravo, and S. Yang, "Load Modeling in Power System Studies: WECC Progress Update," *Proc. IEEE PES General Meeting*, 2008.

127 B. R. Williams, W. R. Schmus, and D. C. Dawson, "Transmission Voltage Recovery Delayed by Stalled Air Conditioner Compressors," *IEEE Trans. on Power Systems*, vol. 7, no. 3, pp. 1173–1181, August 1992.

128 L. Y. Taylor, R. A. Jones, and S. M. Halpin, "Development of Load Models for Fault Induced Delayed Voltage Recovery Dynamic Studies," *Proc. of IEEE PES General Meeting*, pp. 1–7, 2008.

129 R. J. Bravo and D. P. Chassin, "Fault Induced Delayed Voltage Recovery (FIDVR) Model Validation," *Proc. IEEE/PES Transmission and Distribution Conference and Exposition*, 2016.

130 Federal Power Commission Report, *Prevention of Power Failures, Vol. III, Studies of the Task Groups on the Northeast Power Interruption*, June 1967.

131 J. V. Milanović, K. Yamashita, S. Martinez Villanueva, S. Ž. Djokić, and L. M. Korunović, "International Industry Practice on Power System Load Modeling," *IEEE Trans. on Power Systems*, vol. 28, no. 3, pp. 3038–3046, 2013.

132 J. R. Smith, G. J. Rogers, and G. W. Buckley, "Application of Induction Motor Simulation Models to Power System Auxiliary Pump Drives," *IEEE Trans. on Power Apparatus and Systems*, vol. PAS-98, no. 5, pp. 1824–1831, 1979.

133 G. J. Rogers, J. Di Manno, and R. T. H. Alden, "An Aggregate Induction Motor Model for Industrial Plants," *IEEE Trans. on Power Apparatus and Systems*, vol. PAS-103, no. 4, pp. 683–690, 1984.

134 G. J. Rogers, "Demystifying Induction Motor Behavior," *IEEE Computer Applications in Power*, pp. 20–33, Jan. 1994.
135 A. E. Fitzgerald, C. Kingsley, and S. D. Umans, *Electric Machinery,* seventh edition, McGraw-Hill, 2014.
136 D. Ruiz-Vega, T. I. Asiaín Olivares, and D. Olguín Salinas, "An Approach to the Inititalization of Dynamic Induction Motor Models," *IEEE Trans. on Power Systems*, vol. 17, no. 3, pp. 747–751, 2002.
137 P. Aree and E. Acha, "Power Flow Initialisation of Dynamic Studies with Induction Motor Loads," *IET Generation, Transmission & Distribution*, vol. 5, iss. 4, pp. 417–424, 2011.
138 D. N. Ewart, *System Operation and Control Course Notes*, Power System Engineering Course, General Electric Company, 1980.
139 D. N. Ewart, "Automatic Generation Control: Performance under Normal Conditions," *Proc. Engineering Foundation Conference: Systems Engineering for Power*, Henniker, NH, 1975.
140 H. G. Stoll, *Least-Cost Electric Utility Planning*, John Wiley, Hoboken, 1989.
141 T. D. Younkins, J. H. Chow, A. S. Brower, J. Kure-Jensen, and J. B. Wagner, "Fast Valving with Reheat and Straight Condensing Steam Turbines," *IEEE Trans. on Power Systems*, vol. PWRS-2, pp. 397–405, 1987.
142 I. Bremmer, *The J Curve: a New Way to Understand Why Nations Rise and Fall*, Simon and Schuster, 2007.
143 R. Dornbusch, S. Fischer, and R. Startz, *Macroeconomics*, McGraw-Hill Economics, 12th Edition, 2013.
144 R. M. Johnson, J. H. Chow, and M. V. Dillon, "Pelton Turbine Deflector Overspeed Control in Small Power Systems," *IEEE Trans. on Power Systems*, vol. 19, pp. 1032–1037, May 2004.
145 W. L. Rowen, "Simplified Mathematical Representations of Heavy-Duty Gas Turbines," *ASME Journal of Engineering for Power,* Series A, pp. 865–869, Oct. 1983.
146 IEEE Committee Report, "Dynamic Models for Combined Cycle Plants in Power System Studies," *IEEE Trans. on Power Systems*, vol. 9, no. 3, pp. 1698–1708, 1994.
147 IEEE PES Task Force report, *Dynamic Models for Turbine-Governors in Power System Studies*, Special Publication TP 538, 2013.
148 T. D. Younkins and J. H. Chow, "A Multivariable Feedwater Control Design For a Dual Drum HRSG," *IEEE Control Systems Magazine*, pp. 77–80, April 1988.
149 NERC Technical Report, *Balancing and Frequency Response*, 2011. Available: www .nerc.com.
150 J. W. Ingleson and M. Nagle, "Decline of Eastern Interconnection Frequency Response," *Proc. of Georgia Tech Fault and Disturbance Conf.*, 1999.
151 J. W. Ingleson and E. Allen, "Tracking the Eastern Interconnection Frequency Governing Characteristic," *Proc. of IEEE PES General Meeting*, 2010.
152 P. M. Anderson, B. L. Agrawal, and J. E. Van Ness, *Subsynchronous Resonance in Power Sytems*, Wiley, 1999.
153 J. H. Chow, S. H. Javid, J. J. Sanchez-Gasca, C. E. J. Bowler, and J. S. Edmonds, "Torsional Model Identification for Turbine-Generators," *IEEE Trans. on Energy Conversion*, vol. 4, no. 4, pp. 83–91, 1986.

154 J. J. Sanchez-Gasca, "Computation of Turbine-Generator Subsynchronous Torsional Modes from Measured Data using the Eigensystem Realization Algorithm," *Proc. IEEE PES Winter Power Meeting*, Columbus, OH, 2001.
155 B. Agrawal and R. G. Farmer, "Effective Damping for SSR Analysis of Parallel Turbine-Generators," *IEEE Trans. on Power Systems*, vol. 3, pp. 1441–1448, 1988.
156 IEEE Subsynchronous Resonance Task Force, "First Benchmark Model for Computer Simulation of Subsynchronous Resonance," *IEEE Trans. on Power Apparatus and Systems*, vol. PAS-96, pp. 1565–1572, 1977.
157 R. G. Farmer, A. L. Schwalb, and E. Katz, "Navajo Project Report on Subsynchronous Resonance Analysis and Solutions," *IEEE Trans. on Power Apparatus and Systems*, vol. PAS-96, pp. 1226–1232, 1977.
158 F. P. de Mello, *Boiler Dynamics and Control*, XLIBRIS, 2013.
159 P. Pourbeik, *Dynamic Models for Turbine-Governors in Power System Studies*, IEEE Task Force on Turbine-Governor Modeling, 2013.
160 N. Cohn, *Control of Generation and Power Flow on Interconnected Power Systems*, Wiley, 1966.
161 L. K. Kirchmeyer, *Economic Operation of Power Systems*, General Electric Series, John Wiley, 1958.
162 L. K. Kirchmeyer, *Economic Control of Interconnected Systems*, General Electric Series, John Wiley, 1959.
163 J. Arrillaga, *High Voltage Direct Current Transmission*, second edition, IEE Power and Energy Series, 1998.
164 M. Guarnieri, "The Begining of Electric Energy Transmission: Part One," *IEEE Industrial Electronics Magazine*, vol. 7, no. 1, pp. 50–52, Mar. 2013.
165 M. Guarnieri, "The Begining of Electric Energy Transmission: Part Two," *IEEE Industrial Electronics Magazine*, vol. 7, no. 2, pp. 52–59, June 2013.
166 D. Tiku, "DC Power Transmission: Mercury-Arc to Thyrsitor HVDC Valves," *IEEE Power and Energy Magazine*, vol. 12, no. 2, pp. 76–96, 2014.
167 EPRI, *High Voltage Direct Current (HVDC) Transmission Reference Book*, 2012.
168 D. Huang, Y. Shu, J. Ruan, and Y. Hu, "Ultra High Voltage Transmission in China: Developments, Current Status and Future Prospects," *Proc. of IEEE*, vol. 97, no. 3, pp. 555–583, 2009.
169 J. Arrillaga and N. R. Watson, *Computer Modelling of Electrical Power Systems*, second edition, Wiley, 2001.
170 R. Cresap, W. Mittelstadt, D. Scott, and C. Taylor, "Operating Experience with Modulation of the Pacific HVDC Intertie," *IEEE Trans. on Power Apparatus and Systems*, vol. PAS-97, no. 4, pp. 1053–1059, July 1978.
171 D. Trudnowski, B. Pierre, F. Wilches-Bernal, D. Schoenwald, R. Elliott, J. Neely, R. Byrne, and D. Kosterev, "Initial Closed-Loop Testing Results for the Pacific DC Intertie Wide Area Damping Controller," Proc. IEEE PES General Meeting, 2017.
172 R. Bunch and D. Kosterev, "Design and Implementation of AC Voltage Dependent Current Order Limiter at Pacific HVDC Interite," *IEEE Trans. on Power Systems*, vol. 15, no. 1, pp. 293–299, 2000.
173 J. H. Chow and S. Ghiocel, "An Adaptive Wide-Area Power System Damping Controller using Synchrophasor Data," in *Control and Optimization Theory for Electric Smart Grids*, A. Chakrabortty and M. Ilic, eds, Springer, 2012.

174 B. Bergdahl and R. Dass, "AC-DC Harmonic Filters for Three Gorges-Changzhou ±500 kV HVDC Project," available online: library.e.abb.com.

175 J. Reeve, "Multiterminal HVDC Power Systems," *IEEE Trans. on Power Apparatus and Systems*, vol. PAS-99, no. 2, pp. 729–737, 1980.

176 F. Nozari, C. E. Grund, and R. L. Hauth, "Current Order Coordination in Multiterminal DC Systems," *IEEE Trans. on Power Apparatus and Systems*, vol. PAS-100, no. 11, pp. 4628–4635, 1981.

177 A. Oskoui, B. Matthews, J.-P. Hasler, M. Oliveira, T. Larsson, Å. Petersson, and E. John, "Holly STATCOM – FACTS to Replace Critical Generation, Operational Experience," *Proc. IEEE PES Transmission and Distribution Conference and Exhibition*, 2006.

178 E. V. Larsen and J. H. Chow, "SVC Control Design Concepts for System Dynamic Performance," IEEE Power Engineering Society Publication 87TH0187-5-PWR *Application of Static Var Systems for System Dynamic Performance*, 1987.

179 B. L. Agarwal, R. A. Hedin, R. K. Johnson, A. H. Montoya, and V. A. Vossler, "Advanced Series Compensator (ASC): Steady-State, Transient Stability and Subsynchronous Resonance Studies," *Proc. of Flexible AC Transmission Systems Conf.*, Boston, MA, May 1992.

180 J. Urbanek, R. J. Piwko, E. V. Larsen, B. L. Damsky, B. C. Furumasu, W. Mittlestadt, and J. D. Eden, "Thyristor Controlled Series Compensation Prototype Installation at the Slatt 500 kV Substation," *IEEE Trans. on Power Delivery*, vol. 8, pp. 1460–1469, 1993.

181 D. Holmberg, M. Danielson, P. Halvarsson, and L. Ängquist, "The Stöde Thyrsitor Controlled Series Capacitor," CIGRE Session 1998, paper 14-105, Paris, 1998.

182 C. Gama. L. Anquist, G. Ingestrom, amd M. Noroozian, "Commissioning and Operative Experience of TCSC for Damping Power Oscillation in the Brazilian North-South Interconnection," CIGRE Session 2000, Paris, France, 2000.

183 Y. Fan and B. Quan, "Electrical Design Aspects of Pingguo TCSC Project," *Proc. IEEE PES Transmission and Distribution Conference and Exhibition: Asia and Pacific*, 2005.

184 J. J. Vithayathil, *Scheme for Rapid Adjustment of Network Impedance*, US Patent 5032738A, 1991.

185 H.-J. Lee, S.-H. Kim, K. Hur, J.-S. Choi, H.-J. Oh, B.-J. Lee, G. Jang, and J. H. Chow, "Integrating TCSC to Enhance Transmission Capability and Security: Feasibility Studies for Korean Electric Power System," *Proc. IEEE PES General Meeting*, pp. 1–6, 2016.

186 E. V. Larsen and K. Clark, *Thyristor Controlled Series Capacitor Vernier Control System*, US patent US5202583 A, 1993.

187 E. V. Larsen, J. J. Sanchez-Gasca, and J. H. Chow, "Concepts for Design of FACTS Controllers to Damp Power Swings," *IEEE Trans. on Power Systems*, vol. 10, pp. 948–956, 1995.

188 K. Clark, B. Fardanesh, and R, Adapa, "Thyristor Controlled Series Compensation Application Study – Control Interaction Consideration," *IEEE Trans. on Power Delivery*, vol. 10, pp. 1031–1037, 1995.

189 N. Mohan, T. M. Undeland, and W. P. Robbins, *Power Electronics: Converters, Applications, and Design*, Wiley, 2003.

190 B. J. Baliga, *Fundamentals of Power Semiconductor Devices*, Springer, 2008.

191 W. Li, *PMU-based State Estimation for Hybrid AC and DC Grids*, PhD Dissertation, KTH Royal Institute of Technology, 2018.

192 L. Gyugyi, C. D. Schauder, S. L. Williams, T. R. Rietman, D. R. Torgerson, and A. Edris, “The Unified Power Flow Controller: A New Approach to Power Transmission Control,” *IEEE Trans. on Power Delivery*, vol. 10, no. 2, pp. 1085–1097, 1995.

193 L. Gyugyi, K. K. Sen, and C. D. Schauder, “The Interline Power Flow Controller Concept: A New Approach to Power Flow Management in Transmission Systems,” *IEEE Trans. on Power Delivery*, vol. 14, no. 3, pp. 1115–1123, 1999.

194 X. Jiang, X. Fang, J. H. Chow, A.-A. Edris, E. Uzunovic, M. Parisi, and L. Hopkins, “A Novel Approach for Modeling Voltage-Sourced Converter-Based FACTS Controllers,” *IEEE Trans. on Power Delivery*, vol. 23, no. 4, pp. 2591–2598, 2008.

195 X. Wei, *Voltage-Sourced Converter based FACTS Controllers, Modelling, Dispatch, Computation, and Control*, PhD Dissertation, Rensselaer Polytechnic Institute, 2004.

196 X. Jiang, *Operating Modes and their Regulations of Voltage-Sourced Converter based FACTS Controllers*, PhD Dissertation, Rensselaer Polytechnic Institute, 2007.

197 X. Fang, *Rated-Capacity Dispatch, Sensitivity Analysis, amd Controller Design of VSC-based FACTS Controllers*, PhD Dissertation, Rensselaer Polytechnic Institute, 2008.

198 J. D. Glover, T. Overbye, and M. S. Sarma, *Power System Analysis and Design*, sixth edition, CL Engineering, 2016.

199 P. W. Carlin, A. S. Laxson, and E. B. Muljadi, *The History and State of the Art of Variable-Speed Wind Turbine Technology*, NREL Technical Report, NREL/TP-500-28607, Feb. 2001.

200 J. Vestergaard, L. Brandstrup, and R. D. Goddard, “A Brief History of the Wind Turbine Industries in Denmark and the United States,” *Academy of International Business Conf. Proc.*, pp. 322–327, Nov. 2004.

201 J. F. Manwell, J. G. McGowan, and A. L. Rogers, *Wind Energy Explained: Theory, Design and Applications*, Wiley, 2002.

202 CIGRE Brochure 328, *Modeling and Dynamic Behavior of Wind Generation as It Relates to Power System Control and Dynamic Performance*, WG C4.601, August 2007.

203 T. Burton, N. Jenkins, D. Sharpe, and E. Bossanyi, *Wind Energy Handbook*, second edition, Wiley, 2011.

204 P. Jain, *Wind Energy Engineering*, McGraw-Hill, 2011.

205 J. L. Tangier and D. M. Somers, “NREL Airfoil Families for HAWTs,” NREL/TP-442-7109, 1995.

206 A. Ellis and E. Muljadi, “Wind Power Plant Representations in Large-Scale Power Flow Simulations in WECC,” *Proc. IEEE PES General Meeting*, pp. 20–24, July 2008.

207 O. Anaya-Lara, N. Jenkins, J. Ekanayake, P. Cartwright, and M. Hughes, *Wind Energy Generation, Modeling and Control*, Wiley, 2009.

208 A. Ellis, E. Muljadi, J. J. Sanchez-Gasca, and Y. Kazachkov, “Generic Models for Simulation of Wind Power Plants in Bulk System Planning Studies,” *Proc. IEEE PES General Meeting*, July 2011.

209 V. Akhmatov, *Induction Generators for Wind Power*, Multi-Science Publishing Company, Ltd, 2005.

210 B. Fox, D. Flynn, L. Bryans, N. Jenkins, D. Milborrow, M. O’Malley, R. Watson, and O. Anaya-Lara, *Wind Power Integration*, IET Power and Energy Series 50, 2007.

211 E. Muljadi, A. Ellis, J. Mechenbier, J. Hochheimer, R. Young, N. Miller, R. Delmerico, R. Zavadil, and J. C. Smith, “Equivalencing the Collector System of a Large Wind Power Plant,” *Proc. IEEE PES General Meeting*, June 2006.

212 F. Oyague, *Gearbox Modeling and Load Simulation of a Baseline 750-kW Wind Turbine using State-of-the-Art Simulation Codes*, Technical Report NREL/TP-500-41160, Feb. 2009.

213 W. W. Price and J. J. Sanchez-Gasca, "Simplified Wind Turbine Generator Aerodynamic Models for Transient Stability Studies," *Proc. IEEE PES Power Systems Conf. and Expo.* (PSCE), pp. 986–992, Oct. 2006.

214 L. Fan and Z. Miao, *Modeling and Analysis of Doubly Fed Induction Generator Wind Energy Systems*, Elsevier, 2015.

215 A. Ellis, P. Pourbeik, J. J. Sanchez-Gasca, J. Senthil, and J. Weber, "Generic Wind Turbine Generator Models for WECC – A Second Status Report," *Proc. IEEE PES General Meeting*, July 2015.

216 A. Ellis, Y. Kazachkov, E. Muljadi, P. Pourbeik, and J. J. Sanchez-Gasca, "Description and Technical Specifications for Generic WTG Models – A Status Report," *Proc. IEEE PES Power Systems Conf. and Expo. (PSCE)*, Mar. 2011.

217 M. Shao, N. Miller, J. J. Sanchez-Gasca, and J. MacDowell, *Modeling of GE Wind Turbine-Generator for Grid Studies*, version 4.6, Technical Report, General Electric International, 2013.

218 Z. Lubosny, *Wind Turbine Operation in Electric Power Systems*, Springer-Verlag, Berlin, 2003.

219 F. Wilches-Bernal, J. H. Chow, and J. J. Sanchez-Gasca, "A Fundamental Study of Applying Wind Turbines for Power System Frequency Control," *IEEE Trans. on Power Systems*, vol. 31, no. 2, pp. 1496–1505, 2016.

220 N. W. Miller, K. Clark, M. E. Cardinal, and R. W. Delmerico, "GE Wind Plant Dynamic Performance for Grid and Wind Events," *AEE TECHWINDGRID 09*, Madrid, Spain, April 2009.

221 M. Fischer, S. Engelken, N. Mihov, and A. Mendonca, "Operational Experiences with Inertial Response Provided by Type 4 Wind Turbines," *IET Renewable Power Generation*, vol. 10, no. 1, pp. 17–24, 2016.

222 M. Asmine and C. E. Langlois, "Field Measurements for the Assessment of Inertial Response for Wind Power Plants based on Hydro-Québec TransÉnergie Requirements," *IET Renewable Power Generation*, vol. 10, no. 1, pp. 25–32, 2016.

223 A. Mullane and M. O'Malley, "The Inertial Response of Induction-Machine-Based Wind Turbines," *IEEE Trans. on Power Systems*, vol. 20, no. 3, pp. 1496–1503, 2005.

224 J. M. Undrill and A. E. Turner, "Construction of Power System Electromechanical Equivalents by Modal Analysis," *IEEE Trans. on Power Apparatus and Systems*, vol. PAS-90, pp. 2049–2059, 1971.

225 J. M. Undrill, J. A. Casazza, E. M. Gulachenski, and L. K. Kirchmayer, "Electromechanical Equivalents for Use in Power System Stability Studies," *IEEE Trans. on Power Apparatus and Systems*, vol. PAS-90, pp. 2060–2071, 1971.

226 W. W. Price, E. M. Gulachenski, P. Kundur, F. J. Lange, G. C. Loehr, B. A. Roth, and R. F. Silva, "Testing of the Modal Dynamic Equivalents Technique," *IEEE Trans. on Power Apparatus and Systems*, vol. PAS-97, pp. 1366–1372, 1978.

227 W. W. Price and B. A. Roth, "Large-Scale Implementation of Modal Dynamic Equivalents," *IEEE Trans. on Power Apparatus and Systems*, vol. PAS-100, pp. 3811–3817, 1981.

228 S. Liu, P. W. Sauer, D. Chaniotis, and M. A. Pai, "Krylov Subspace and Balanced Truncation Methods for Power Sytem Model Reduction," Chapter 6, *Power System Coherency and Model Reduction*, J. H. Chow, ed., Springer, New York, 2013.
229 V. Vittal and F. Ma, "A Hybrid Dynamic Equivalent using ANN-Based Boundary Matching Technique," Chapter 5, *Power System Coherency and Model Reduction*, J. H. Chow, ed., Springer, New York, 2013.
230 R. W. deMello, R. Podmore, and K. N. Stanton, "Coherency-Based Dynamic Equivalents: Applications in Transient Stability Studies," *PICA Conf. Proc.*, pp. 23–31, 1975.
231 R. Podmore, "Coherency in Power Systems," Chapter 2, *Power System Coherency and Model Reduction*, J. H. Chow, ed., Springer, 2013.
232 J. H. Chow, "New Algorithms for Slow Coherency Aggregation of Large Power Systems," in *Systems and Control Theory for Power Systems*, IMA Volumes in Mathematics and its Applications, Volume 64, J. H. Chow, R. J. Thomas, and P. V. Kokotović, eds, Springer-Verlag, 1994.
233 R. Nath, S. S. Lamba, and K. S. P. Rao, "Coherency Based System Decomposition into Study and External Areas using Weak Coupling," *IEEE Trans. on Power Apparatus and Systems*, vol. PAS-104, pp. 1443–1449, 1985.
234 R. A. Date, *Dynamic Model Reduction of Large Scale Systems: Application to Power Systems*, MS Thesis, Rensselaer Polytechnic Institute, 1989.
235 A. J. Germond and R. Podmore, "Dynamic Aggregation of Generating Unit Models," *IEEE Trans. on Power Apparatus and Systems*, vol. PAS-97, no. 4, pp. 1060–1069, 1978.
236 J. R. Winkelman, J. H. Chow, B. C. Bowler, B. Avramovic, and P. V. Kokotović, "An Analysis of Interarea Dynamics of Multi-Machine Systems," *IEEE Trans. on Power Apparatus and Systems*, vol. PAS-100, pp. 754–763, 1981.
237 S. D. Dukić and A.T. Sarić, "Dynamic Model Reduction: An Overview of Available Techniques with Application to Power Systems," *Serbian Journal of Electrical Engineering*, vol. 9, no. 2, pp. 131–169, 2012.
238 E. J. Davison, "A Method for Simplifying Dynamic Systems," *IEEE Trans. on Automatic Control*, vol. AC-11, pp. 93–101, 1966.
239 B. C. Moore, "Principal Component Analysis in Linear Systems: Controllability, Observability, and Model Reduction," *IEEE Trans. on Automatic Control*, vol. AC-26, pp. 17–32, 1981.
240 K. Glover, "All Optimal Hankel-Norm Approximations of Linear Multivariable Systems and their L^{∞} Norms," *International Journal of Control*, vol. 39, pp. 1115–1193, 1984.
241 P. Benner, V. Mehrmann, and D. C. Sorensen, *Dimension Reduction of Large-Scale Systems*, Lecture Notes in Computational Sciences and Engineering, vol. 45, Springer, Berlin, 2005.
242 A. C. Antoulas, *Approximation of Large-Scale Dynamical Systems*, SIAM, Philadelphia, 2005.
243 H. You, V. Vittal, and X. Wang, "Slow Coherency-based Islanding," *IEEE Trans. on Power Systems*, vol. 19, no. 1, pp. 483–491, 2004.

Index

Power System Modeling, Computation, and Control, First Edition. Joe H. Chow and Juan J. Sanchez-Gasca.

Companion website: www.wiley.com/go/chow/power-system-modeling

t

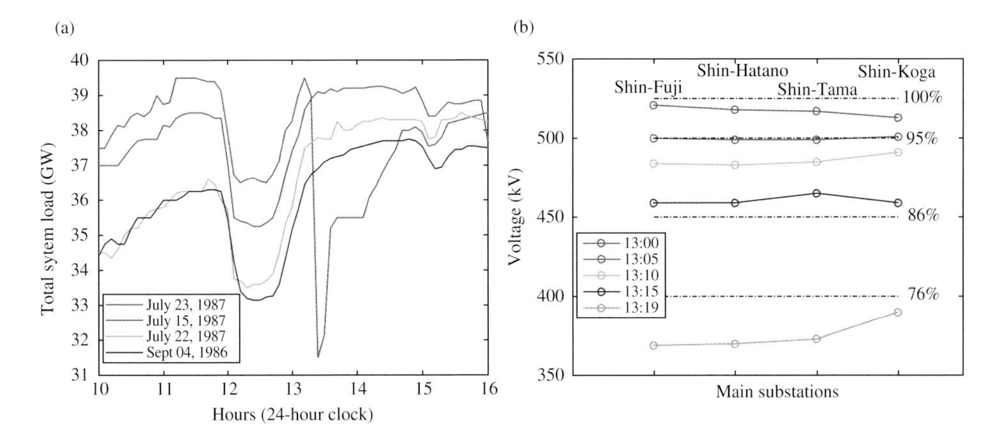

Figure 3.1 TEPCO system: (a) system load profile around noon time for several high-load days and (b) voltage drops at main substations with time progression indicated by different colors [29] (data courtesy of Teruo Ohno, TEPCO).

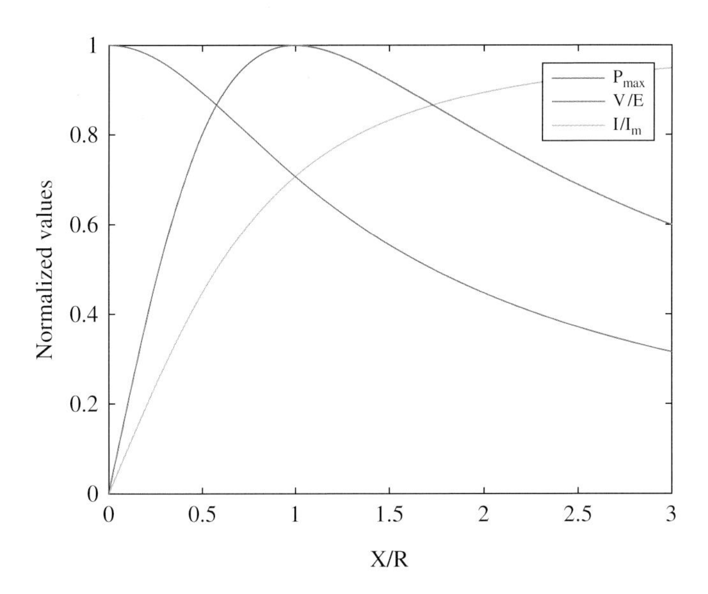

Figure 3.6 Voltage, current, and power profile for various resistive loads.

Power System Modeling, Computation, and Control, First Edition. Joe H. Chow and Juan J. Sanchez-Gasca.

Companion website: www.wiley.com/go/chow/power-system-modeling

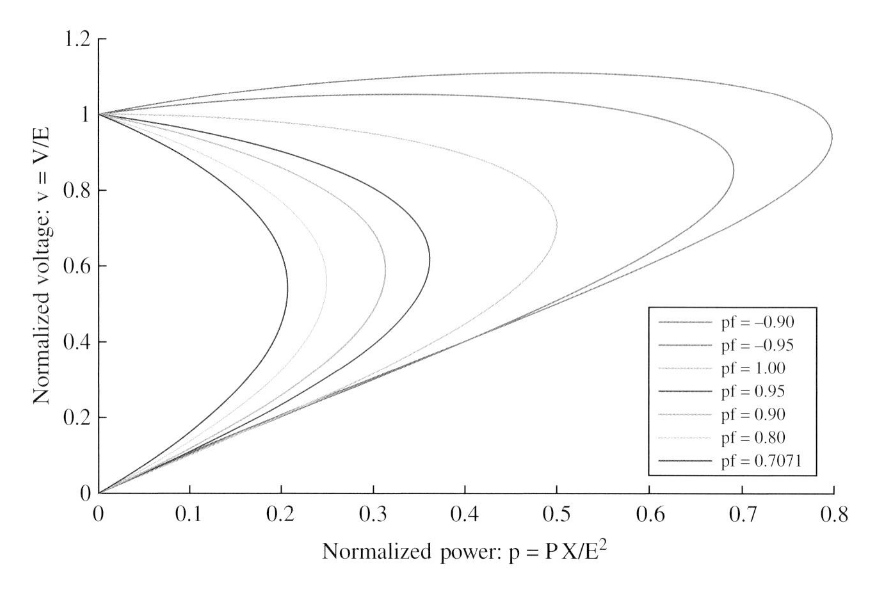

Figure 3.7 *PV* curves for various power factors.

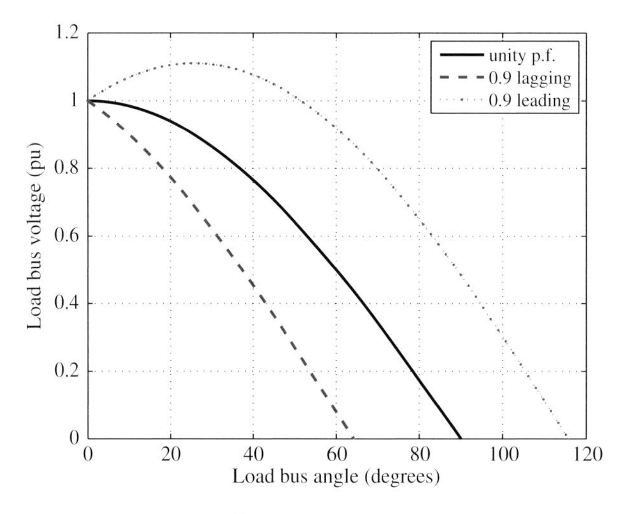

Figure 3.10 Variation of V versus θ.

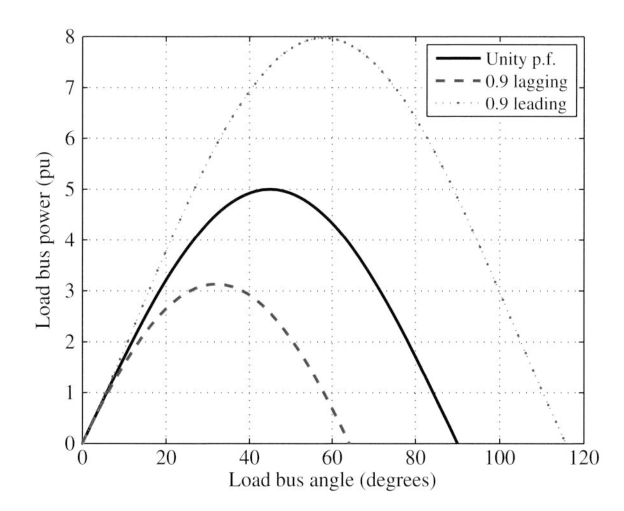

Figure 3.11 Variation of P_L versus θ.

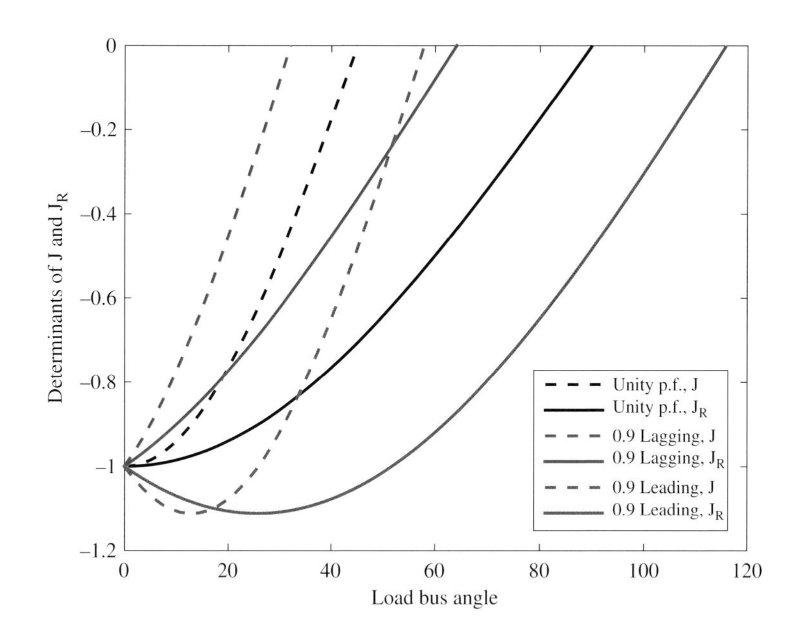

Figure 3.12 Determinants of J and J_R as a function of θ in degrees.

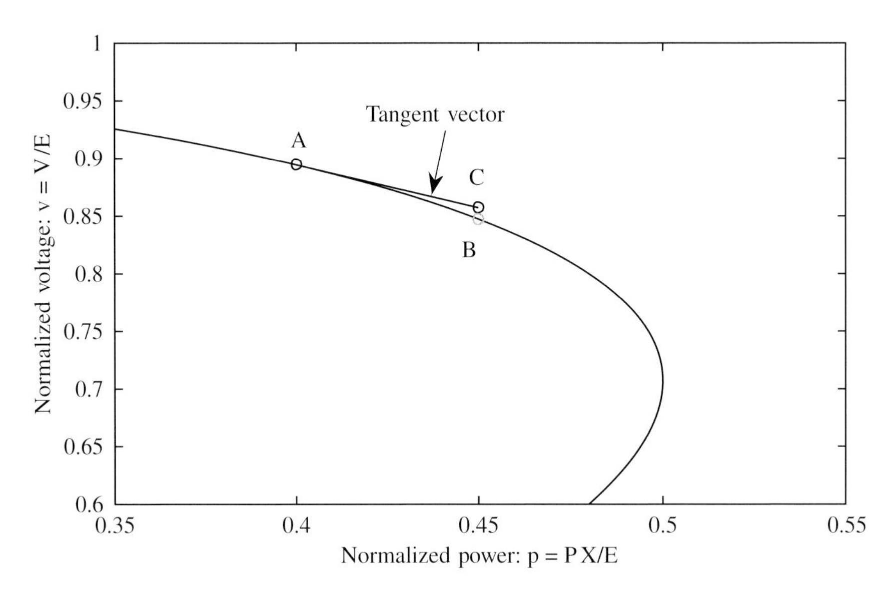

Figure 3.14 Tangent vector at $p = 0.4$ pu to find the initial guess of the power flow solution for $p = 0.45$ pu.

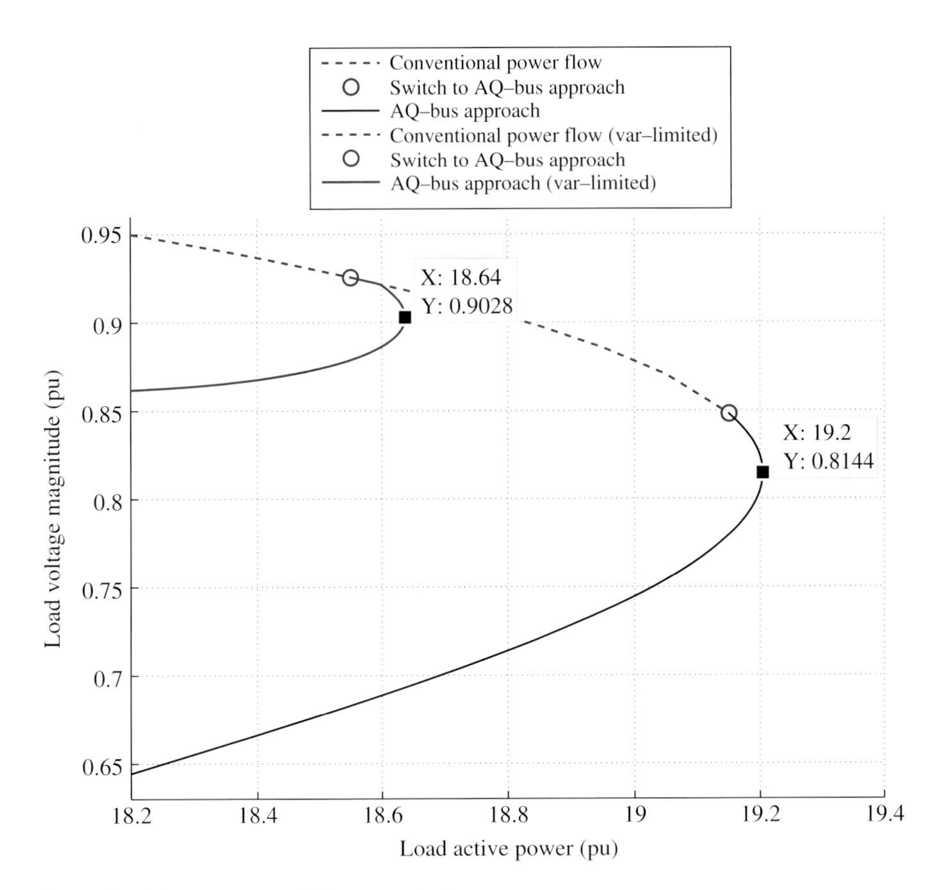

Figure 3.15 Power-voltage (*PV*) curves of a 2-area system.

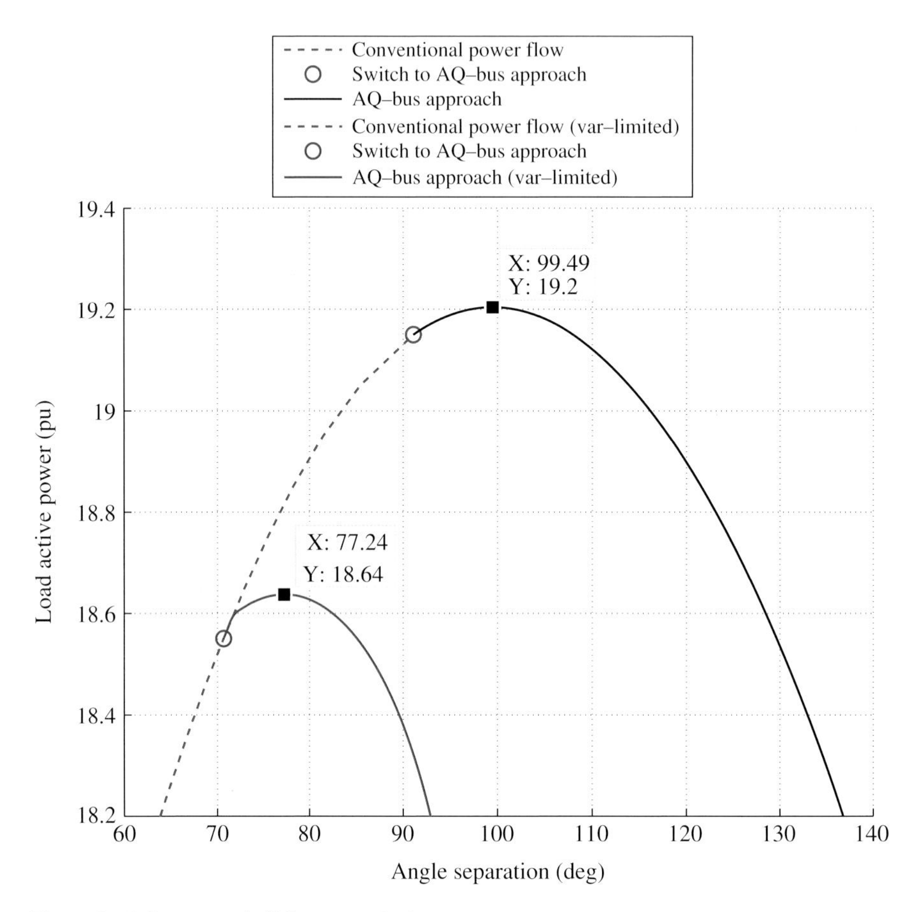

Figure 3.16 Power-angle ($P\theta$) curves of a 2-area system.

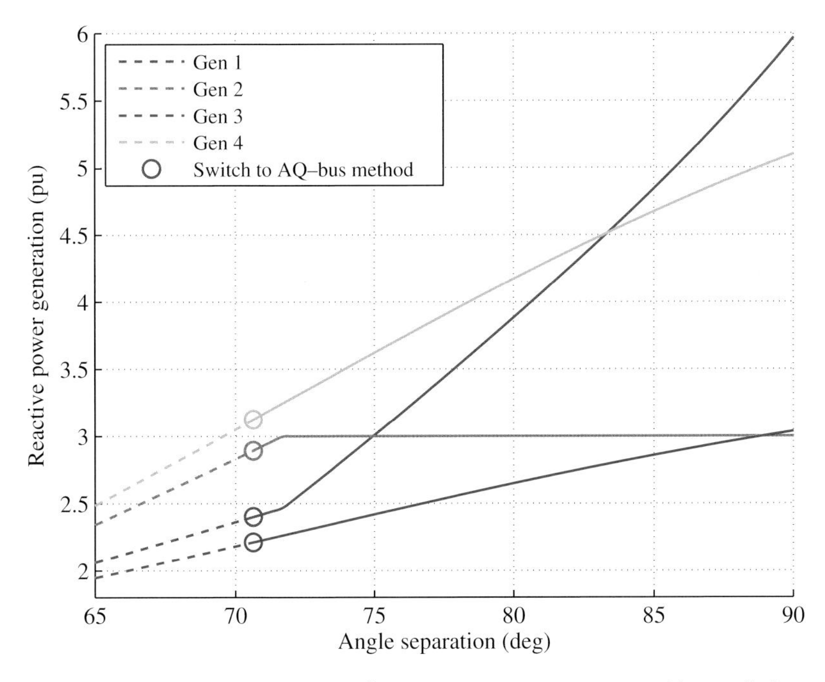

Figure 3.17 Reactive power output of generators in a 2-area system with a var limit.

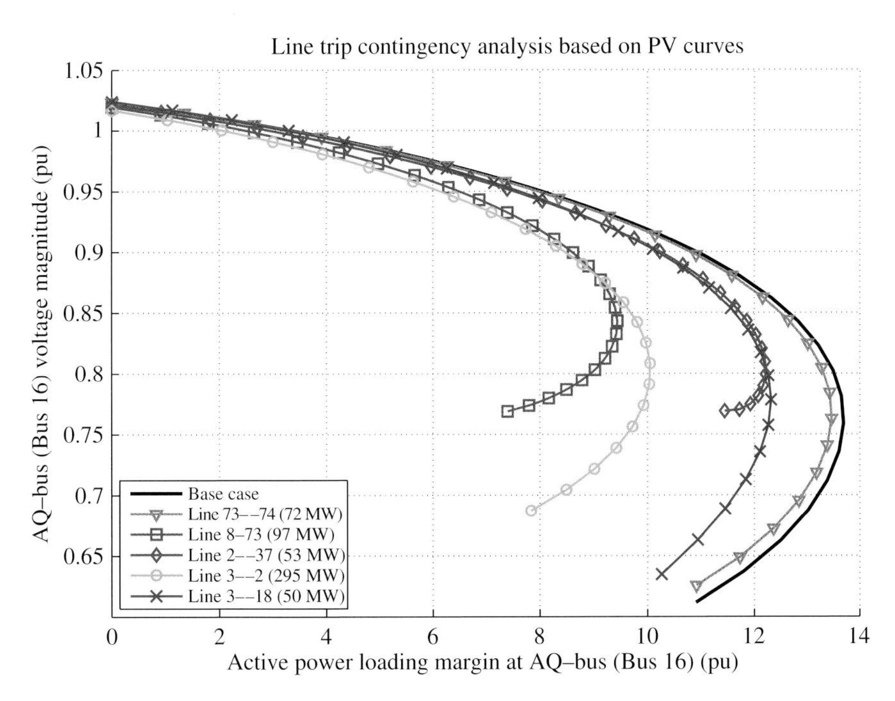

Figure 3.18 Contingency-based voltage stability analysis.

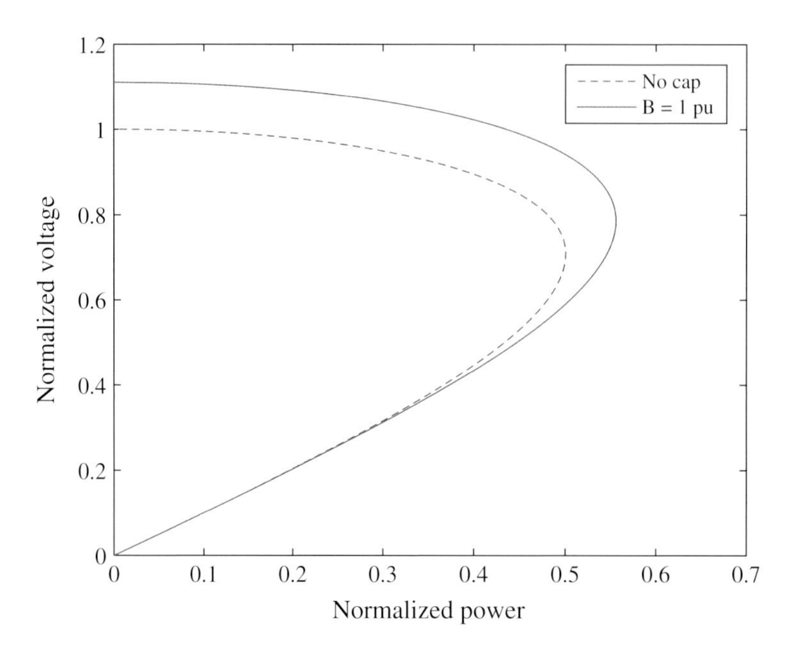

Figure 3.20 *PV* curves with and without shunt capacitor at the load bus.

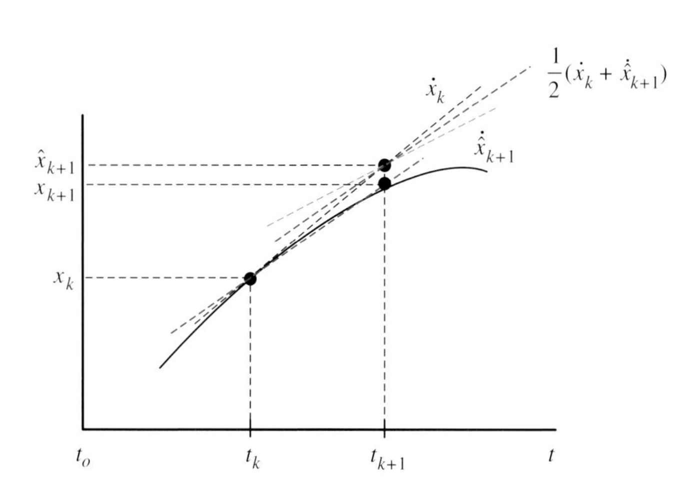

Figure 4.14 Euler method with full-step modification.

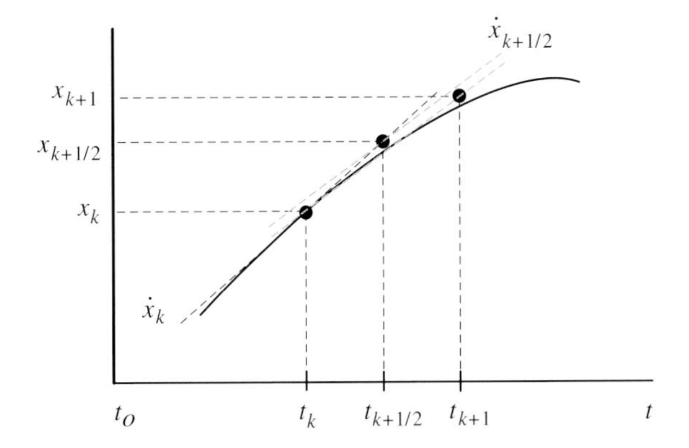

Figure 4.15 Euler method with half-step modification.

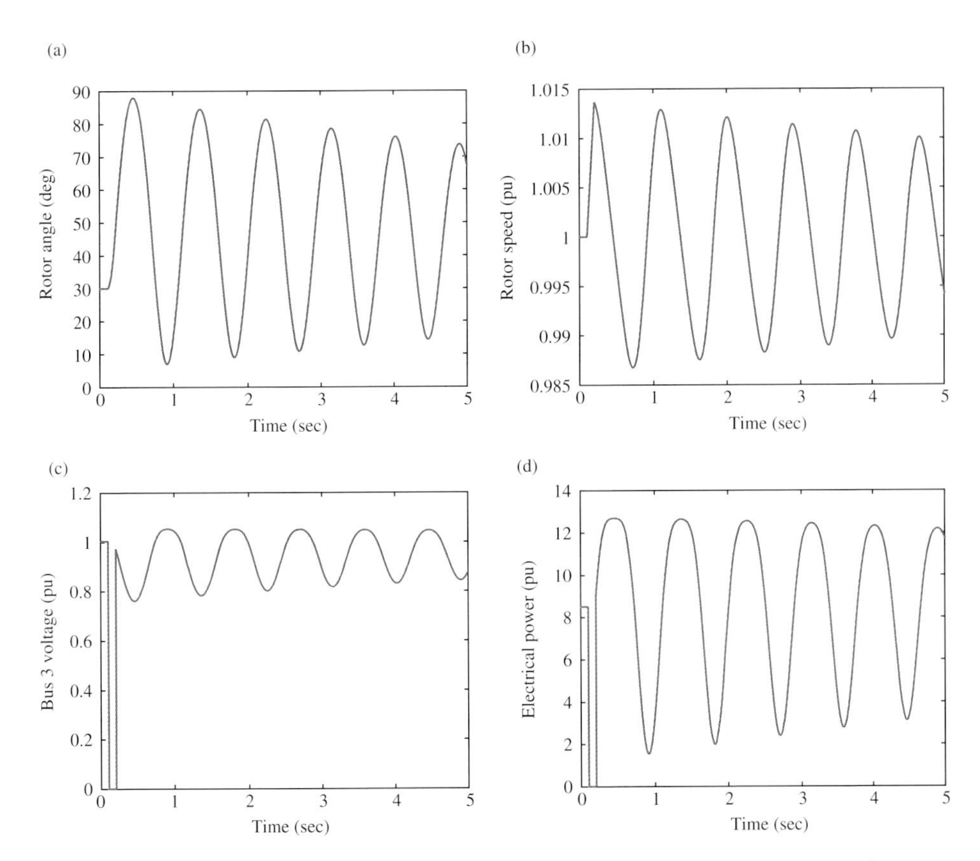

Figure 4.18 Generator 1 disturbance responses: (a) rotor angle, (b) rotor speed, (c) Bus 3 voltage magnitude, and (d) electrical power.

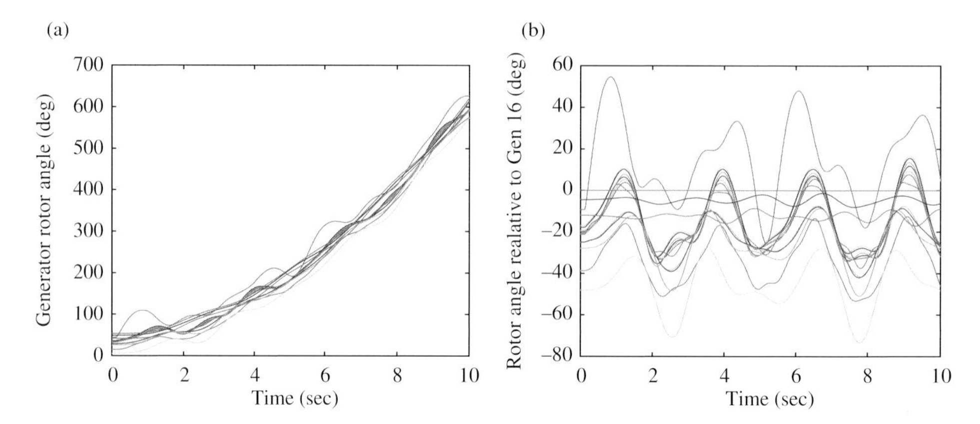

Figure 4.26 Generator disturbance responses for a 3-cycle fault: (a) absolute rotor angles and (b) rotor angles relative to Generator 16. Generator 9 has the highest angle variations.

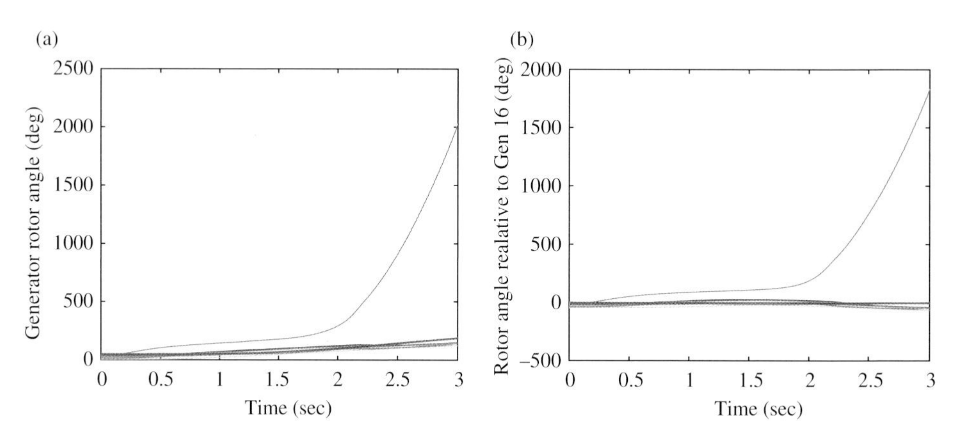

Figure 4.27 Generator disturbance responses for a 4-cycle fault: (a) absolute rotor angles and (b) rotor angles relative to Generator 16. Note that Generator 9 is unstable.

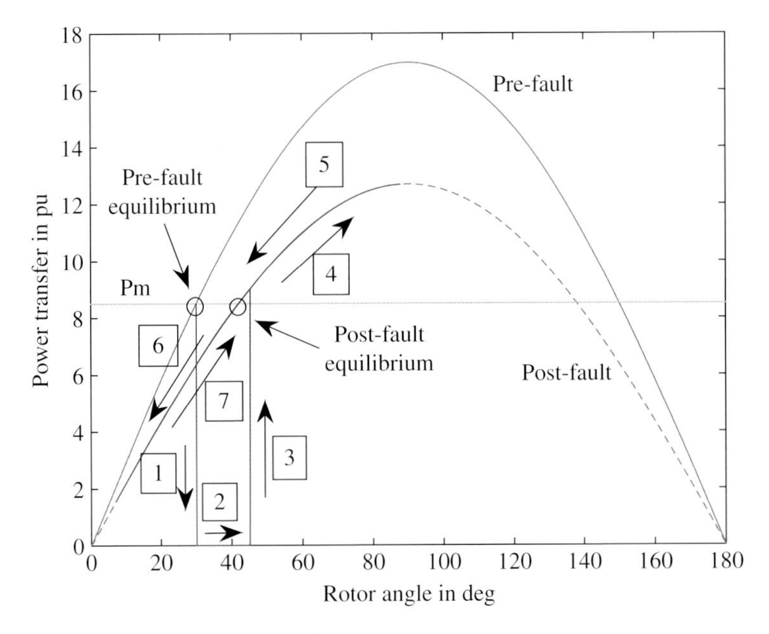

Figure 5.3 *P*-δ curve for Example 5.1.

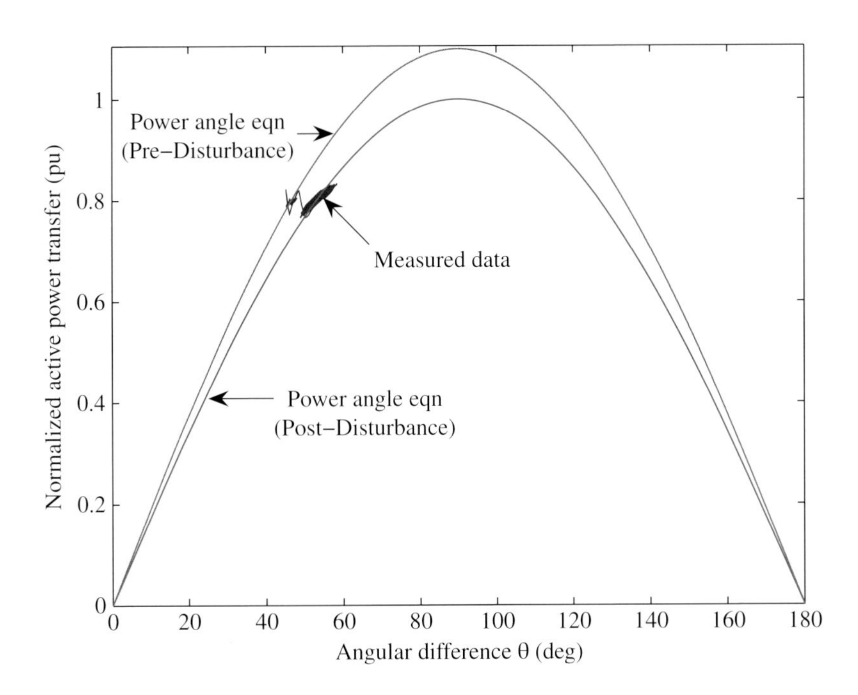

Figure 5.10 Power-angle curves.

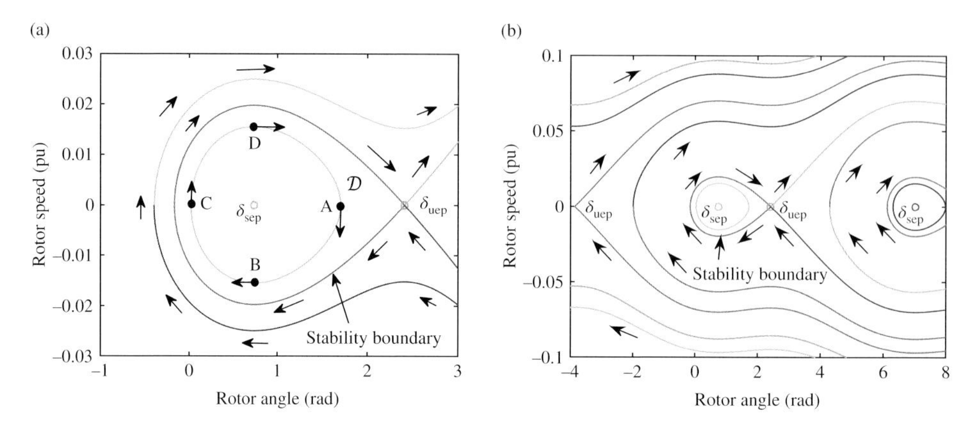

Figure 5.13 (a) Phase portraits and region of stability and (b) multiple equilibrium points.

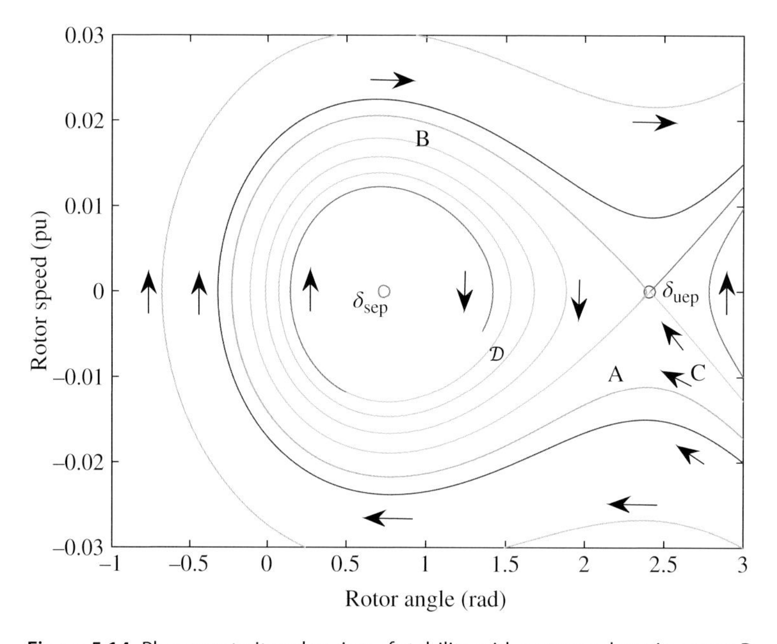

Figure 5.14 Phase portrait and region of stability with nonzero damping term *D*.

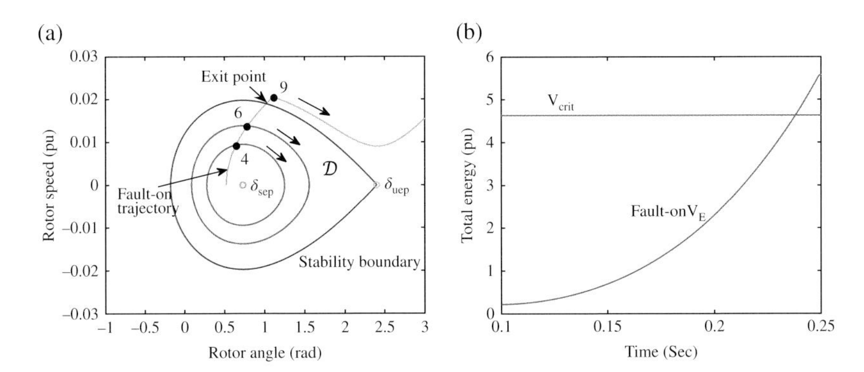

Figure 5.15 (a) Phase portraits for various fault clearing times (the numbers indicate the clearing times in cycles) and (b) potential energy at the unstable equilibrium point δ_{uep} and fault-on total energy.

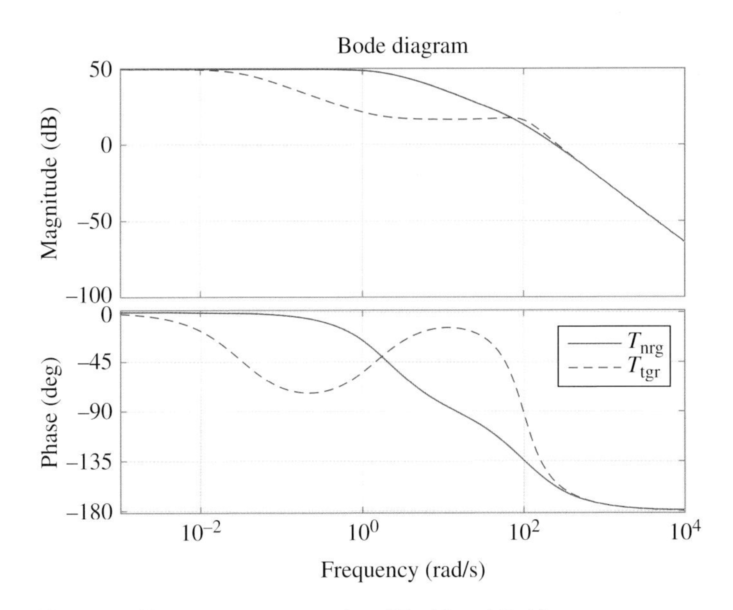

Figure 9.8 Frequency response plot of $T_{ngr}(s)$ and $T_{tgr}(s)$.

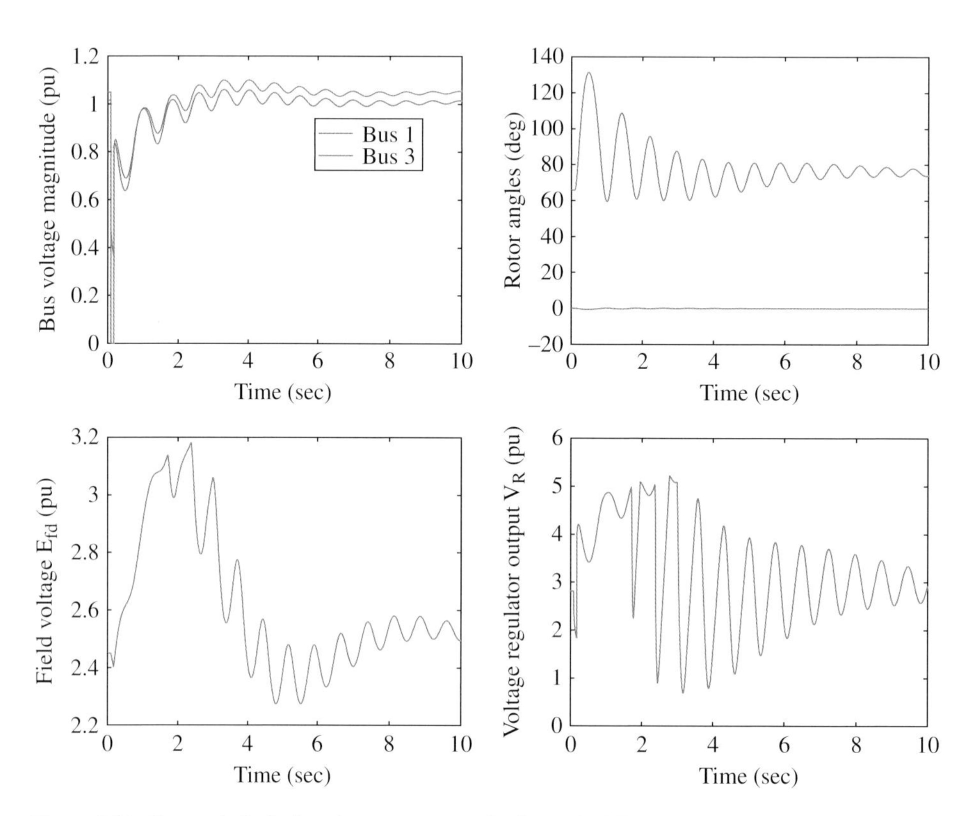

Figure 9.11 Five-cycle fault disturbance response for Example 9.5.

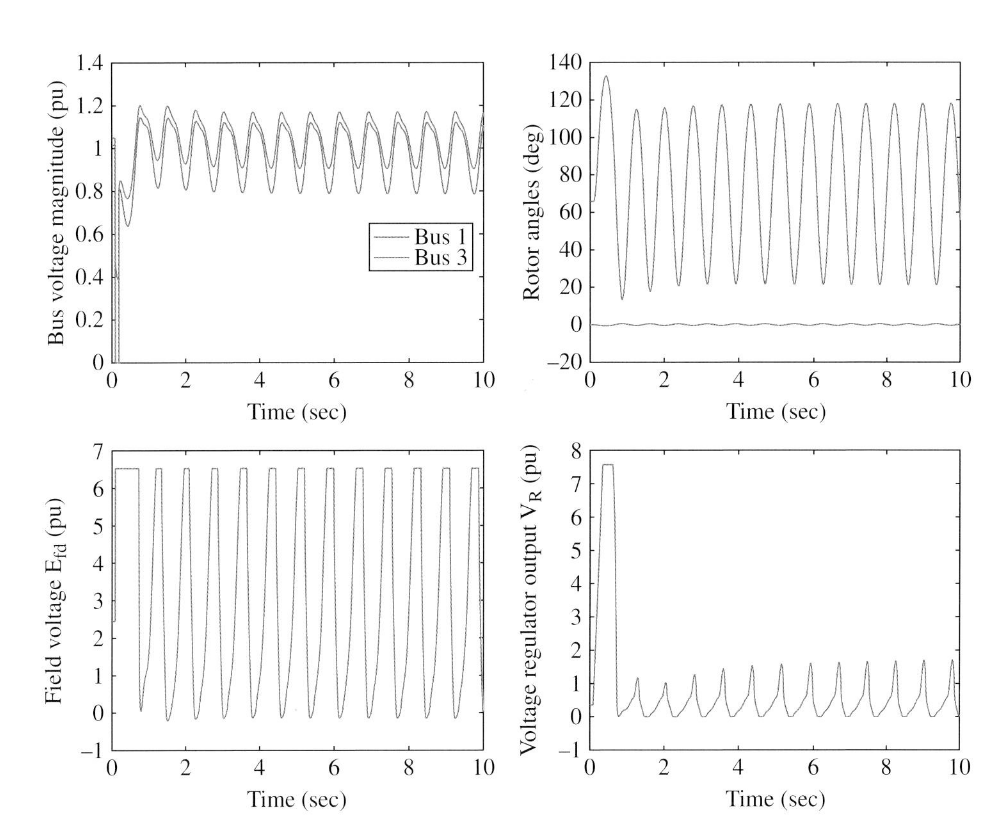

Figure 9.16 Disturbance response for Example 9.8.

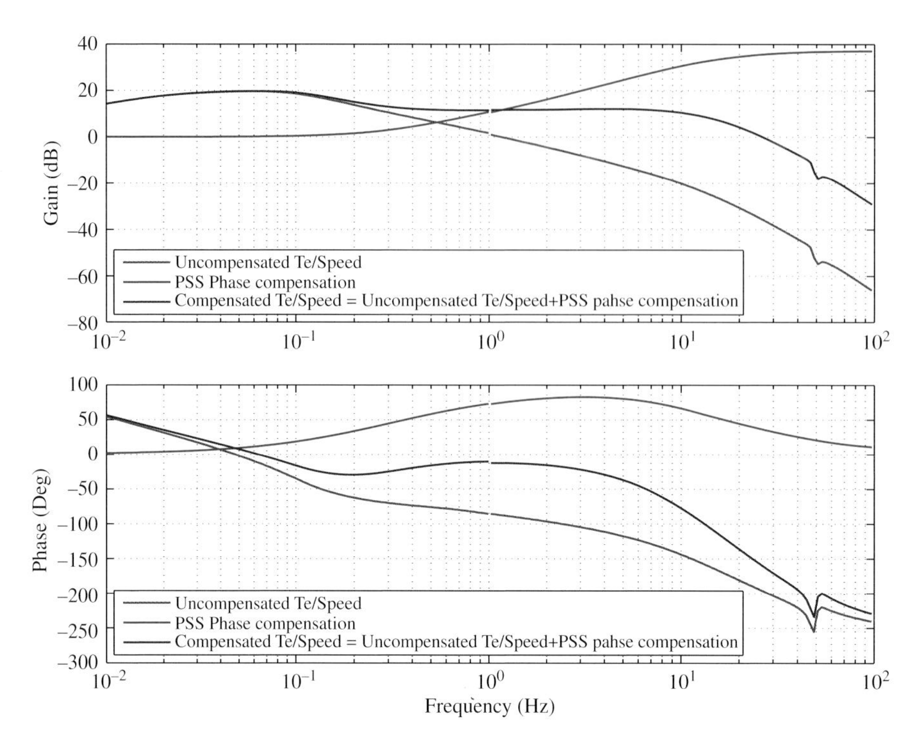

Figure 10.9 Uncompensated, compensated, and PSS phase plot (courtesy of GE Energy Consulting).

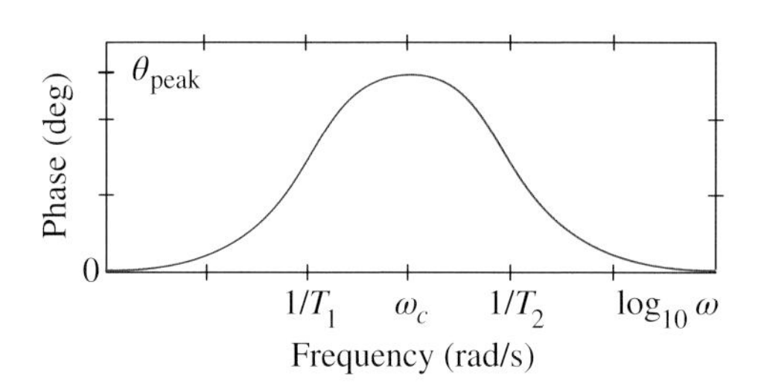

Figure 10.10 Phase characteristic of a phase-lead compensator versus frequency.

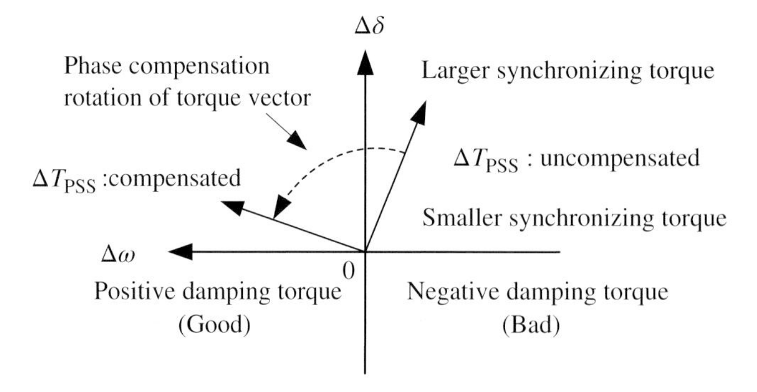

Figure 10.11 Rotation of ΔT_{PSS} through phase compensation.

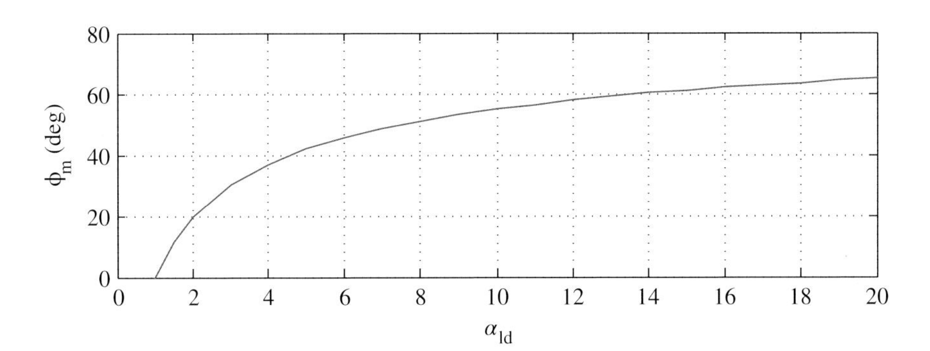

Figure 10.14 The maximum phase angle θ_{peak} versus α_{ld} for a lead compensator.

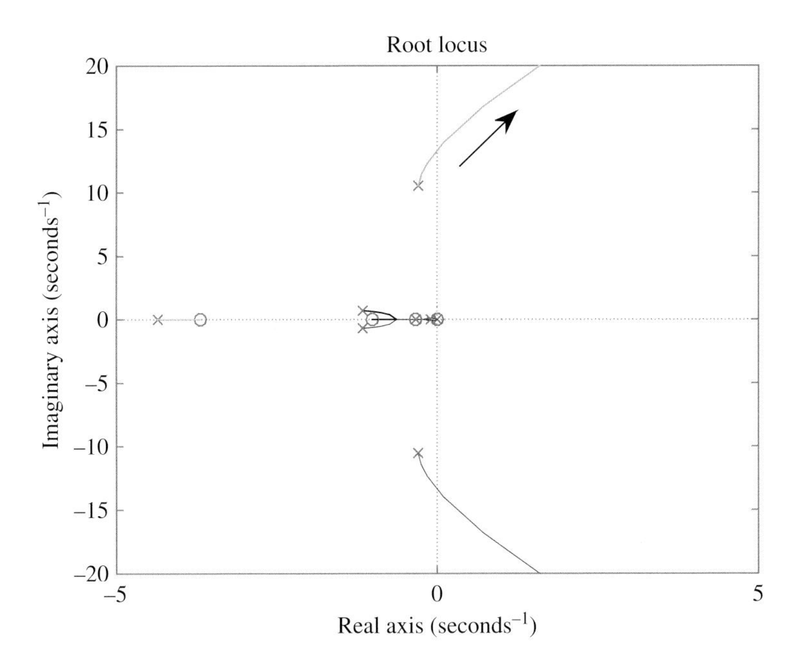

Figure 10.15 Root-locus plot of the uncompensated system (note that the scales of the horizontal and vertical axes are not the same).

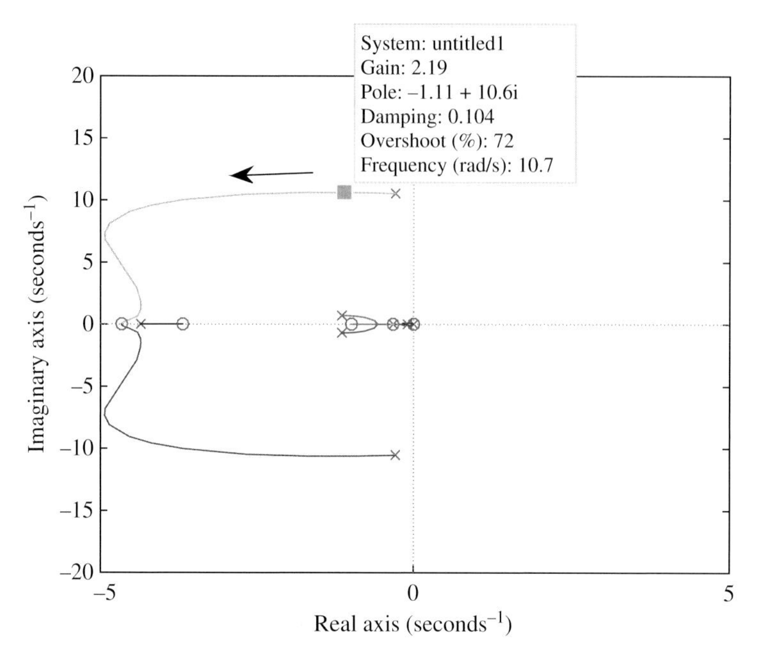

Figure 10.16 Root-locus plot of the compensated system.

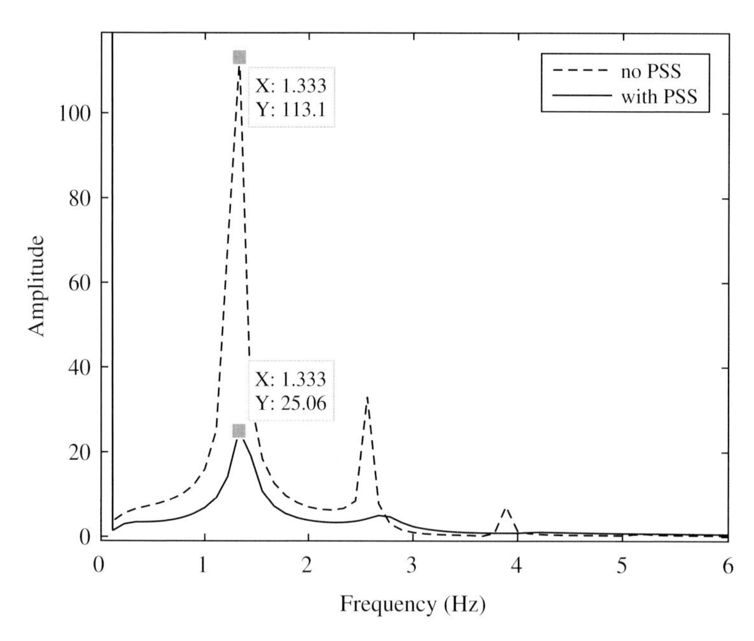

Figure 10.18 Spectral analysis of the post-fault generator terminal voltage response with and without power system stabilizer.

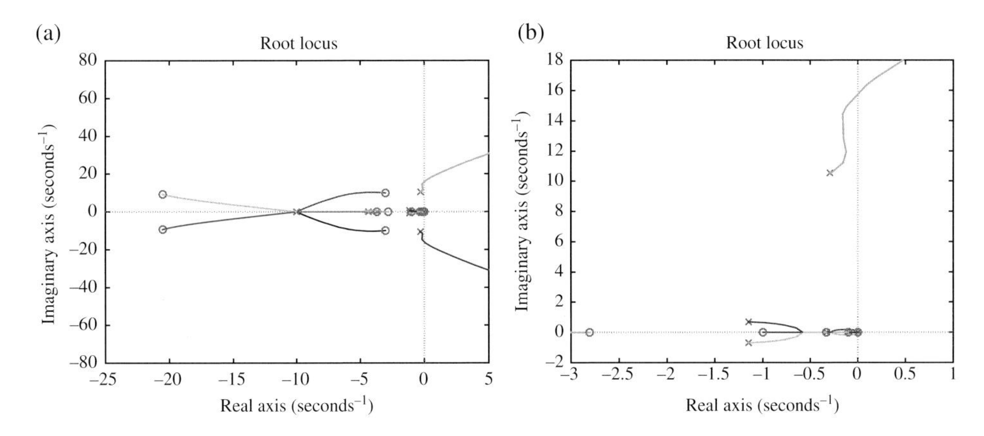

Figure 10.25 (a) Uncompensated root-locus plot of Example 10.4 and (b) zoomed-in plot.

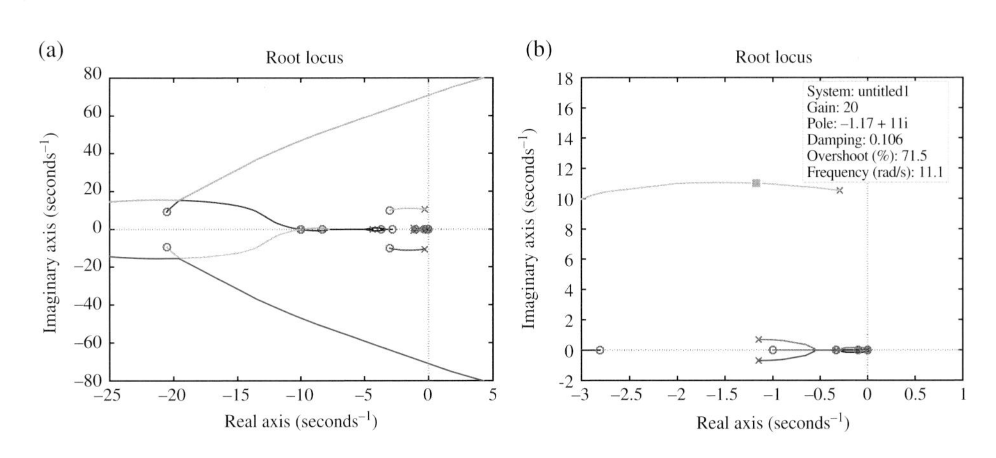

Figure 10.26 (a) Compensated root-locus plot of Example 10.4 and (b) zoomed-in plot.

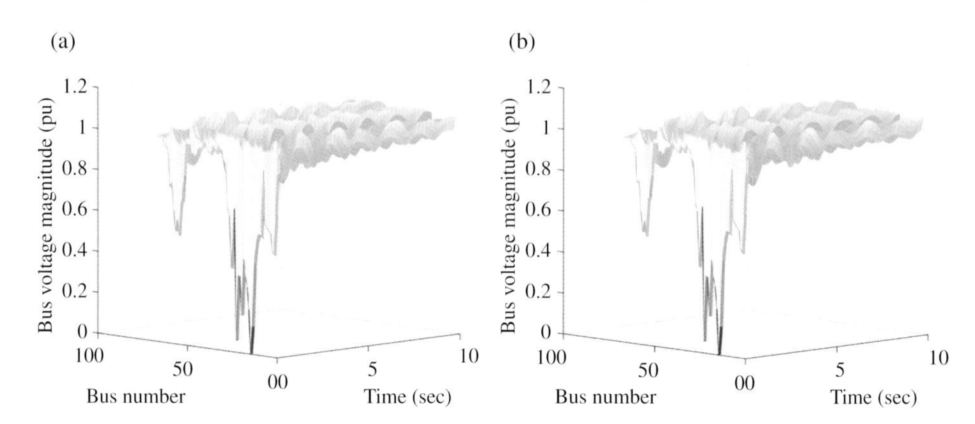

Figure 11.2 Eight-cycle fault voltage magnitude response for (a) constant-Z active power load model and (b) constant-I active power load model.

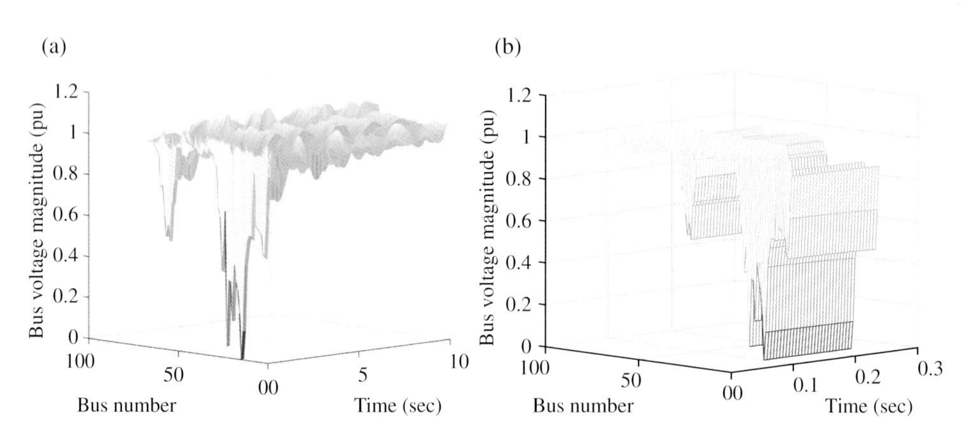

Figure 11.3 Nine-cycle fault voltage magnitude response for (a) constant-Z active power load model, and (b) constant-I active power load model.

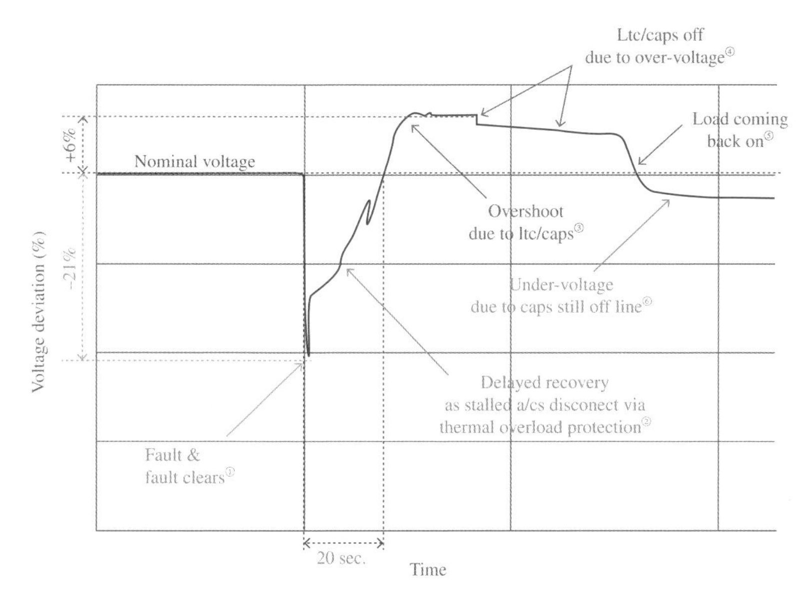

Figure 11.14 Fault-induced delayed voltage recovery [129].

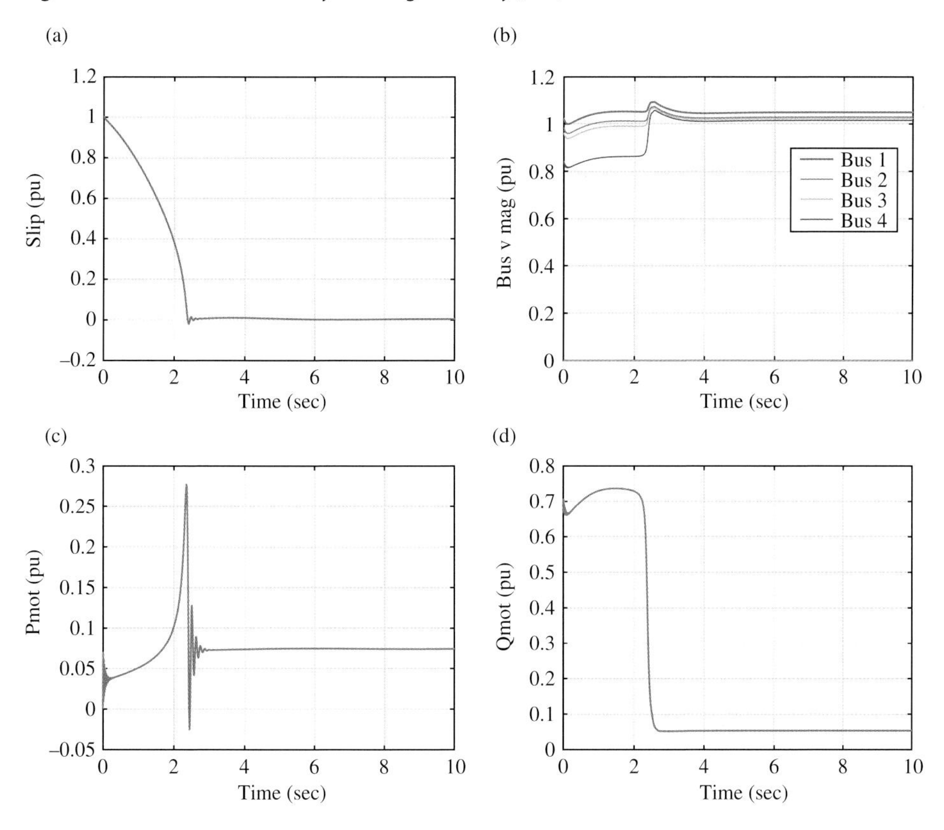

Figure 11.16 Induction motor startup response (a) slip, (b) bus voltage magnitude, (c) active power consumption, and (d) reactive power consumption.

Figure 12.2 Low-pressure sections of a steam turbine (Courtesy of General Electric Company).

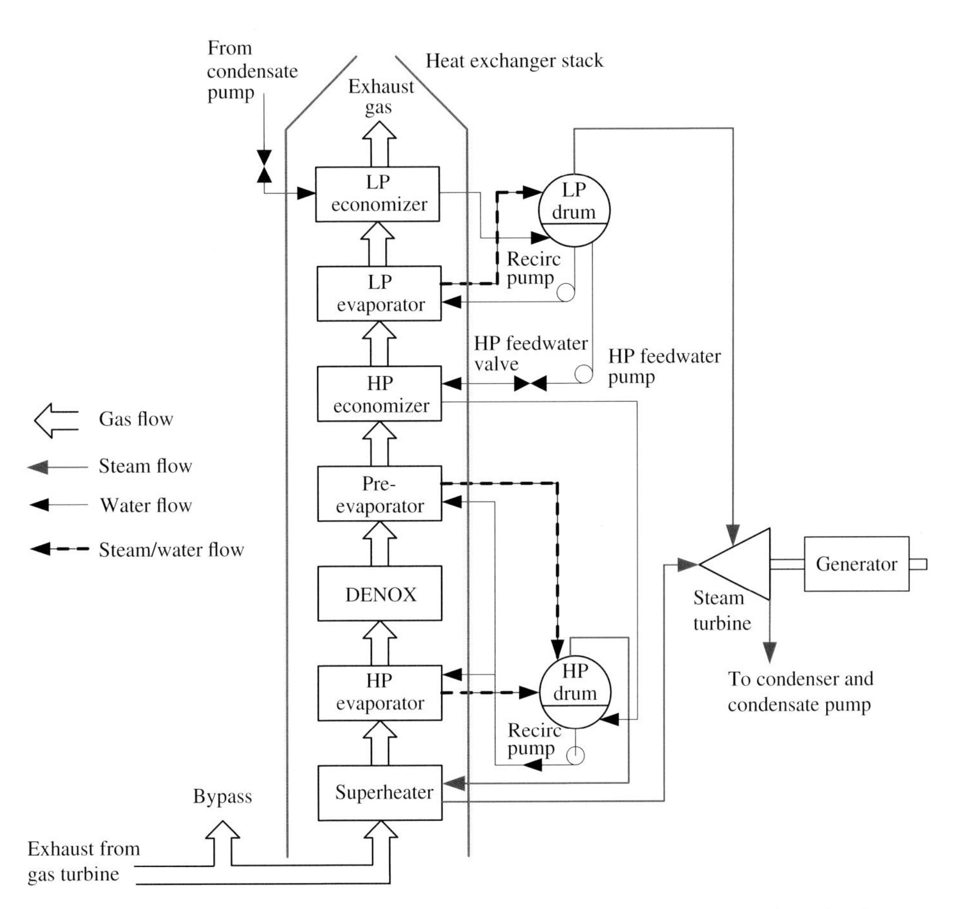

Figure 12.18 Dual-pressure HRSG flow diagram [148] (note the blowdown valves from the drums are omitted).

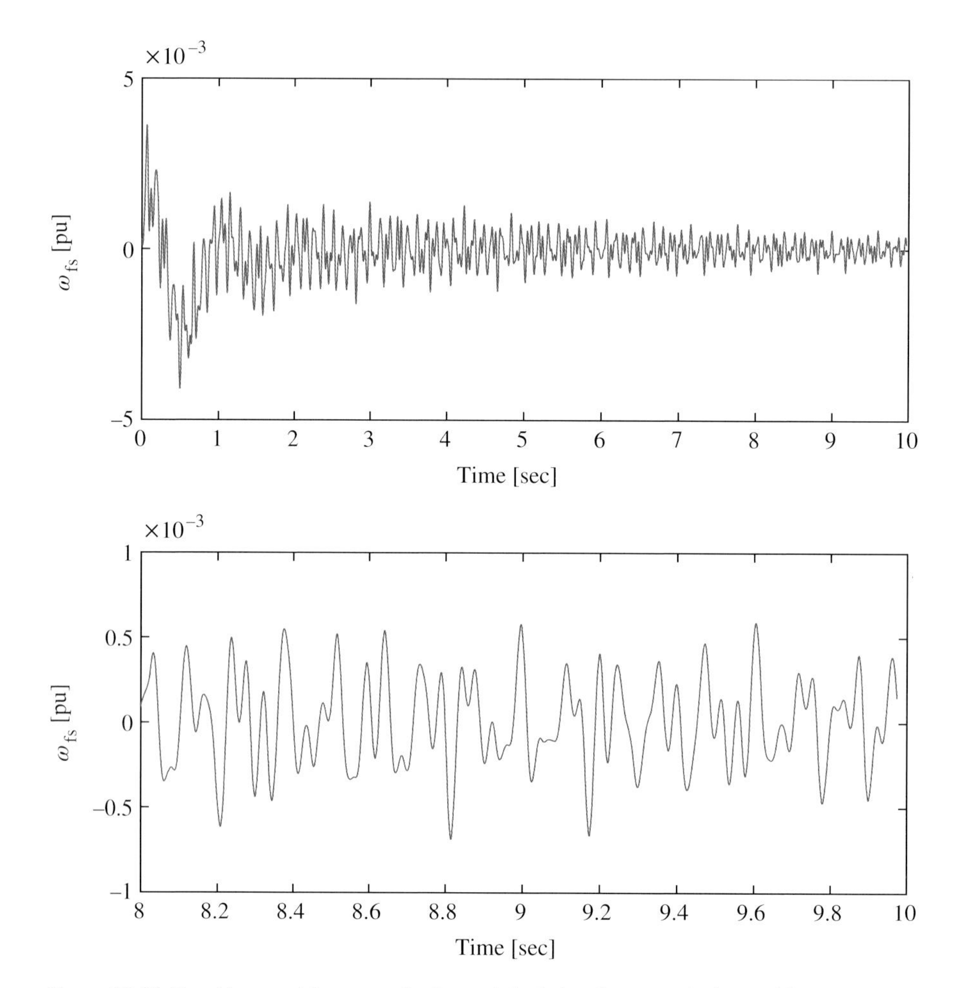

Figure 12.35 Top: Measured front standard speed deviation from nominal speed from a synchronization test [154]; bottom: enlarged plot.

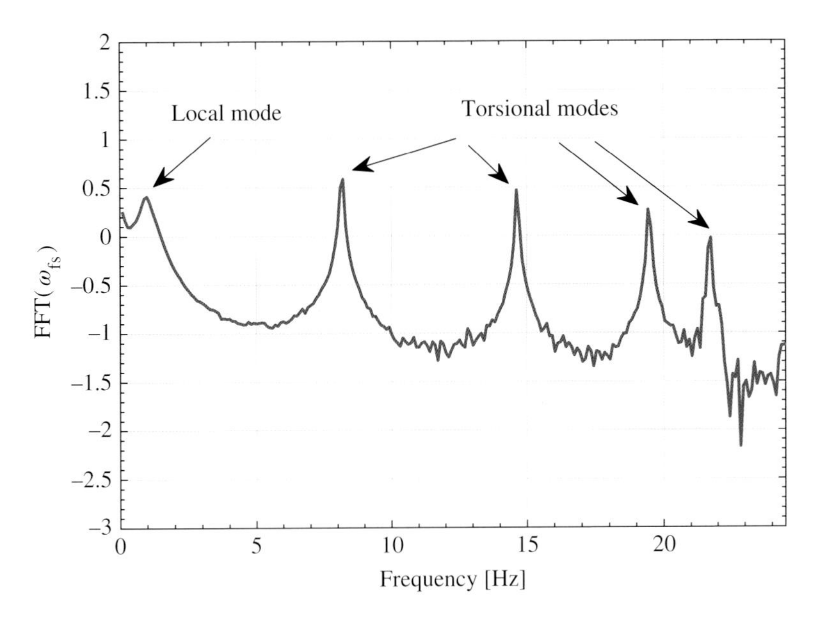

Figure 12.36 Spectrum of the front standard speed in Figure 12.35 [154].

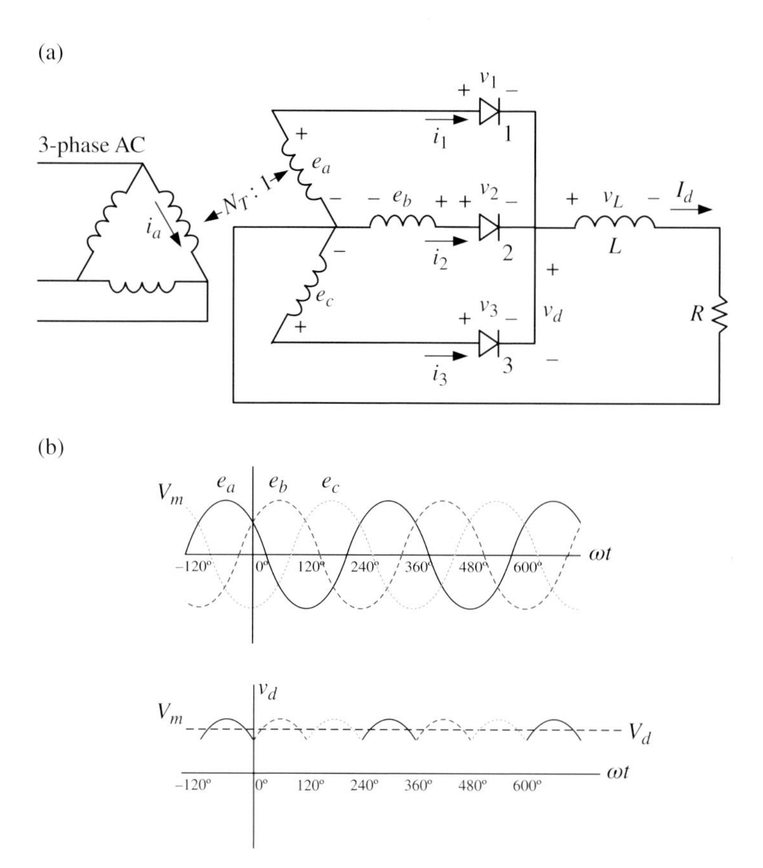

Figure 13.6 Three-phase one-way rectifier: (a) circuit and (b) voltage waveforms.

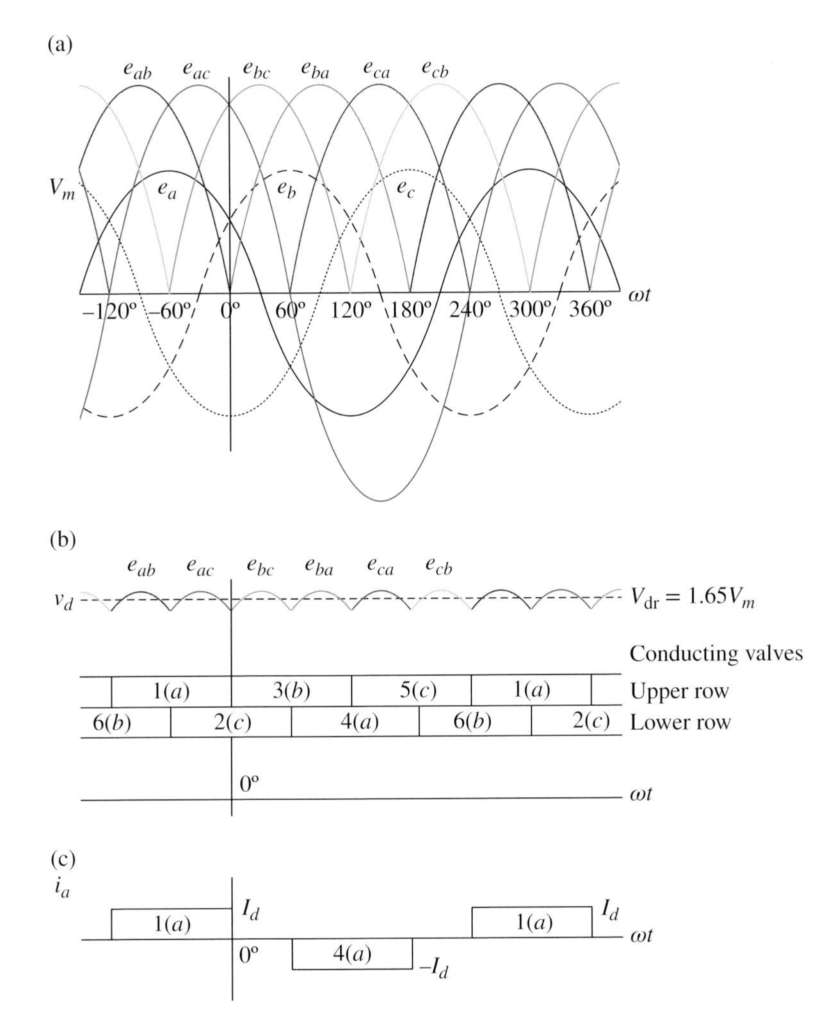

Figure 13.8 Three-phase full-wave bridge rectifier: (a) voltage waveforms, (b) DC voltage waveform and conducting valves, and (c) current in phase a.

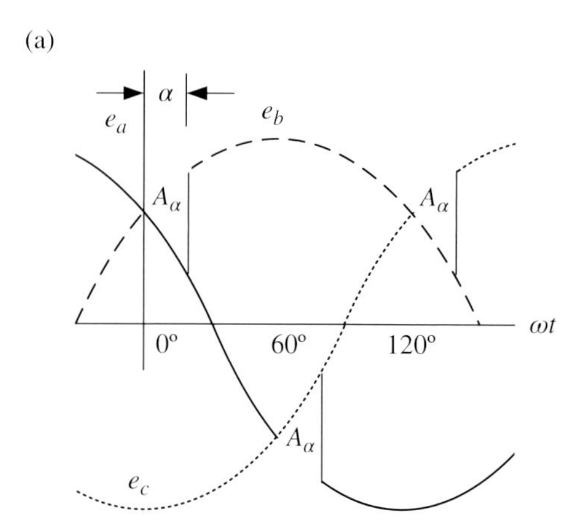

(b)

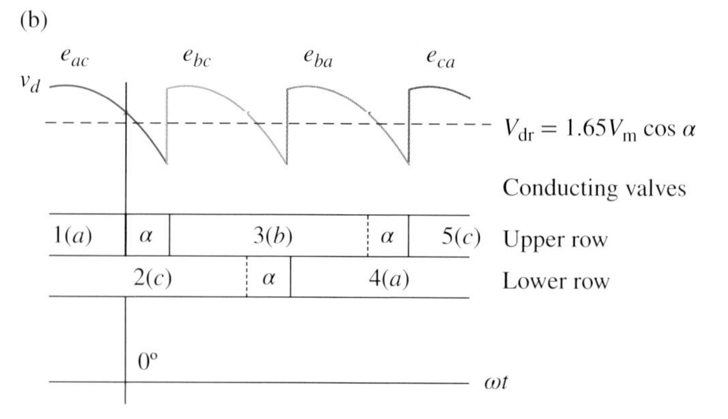

Figure 13.9 Three-phase full-wave bridge converter voltage and current waveforms with ignition delays: (a) voltage waveforms and (b) DC voltage waveform and conducting valves.

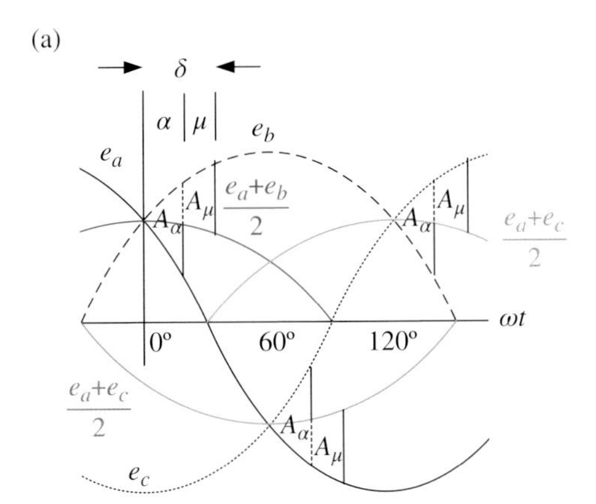

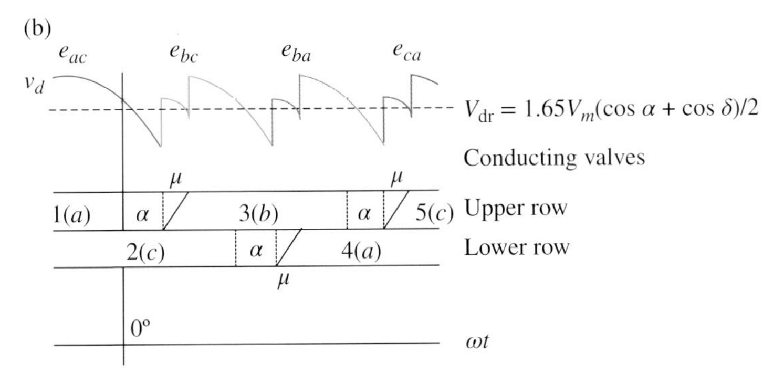

Figure 13.12 Voltage reduction during communtation: (a) voltage waveforms and (b) DC voltage waveform and conducting valves.

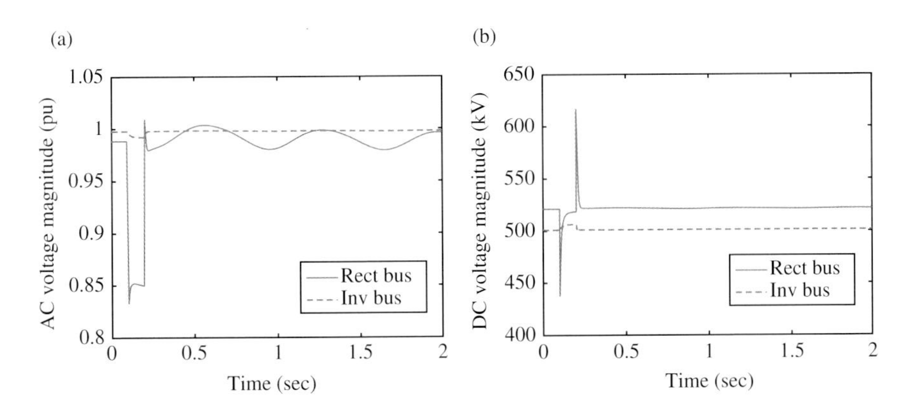

Figure 13.30 Disturbance response for Example 13.4: (a) AC voltages and (b) DC voltages.

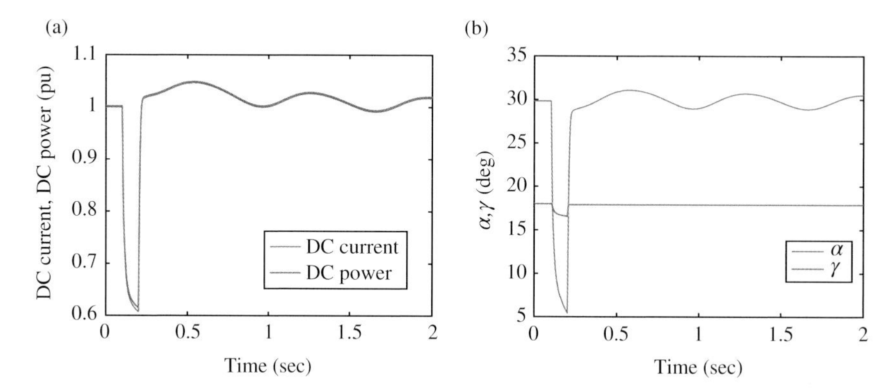

Figure 13.31 Disturbance response for Example 13.4: (a) DC current and power, and (b) firing and extinction angles.

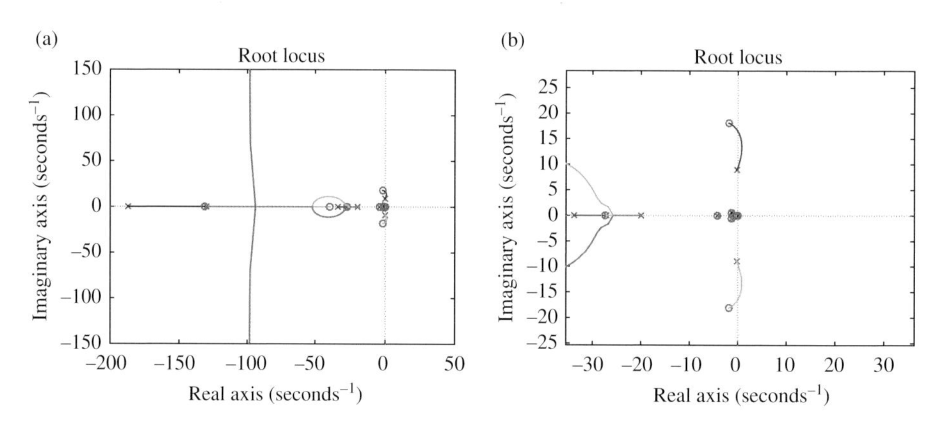

Figure 13.33 (a) Root-locus plot of uncompensated system and (b) enlarged plot around the swing mode.

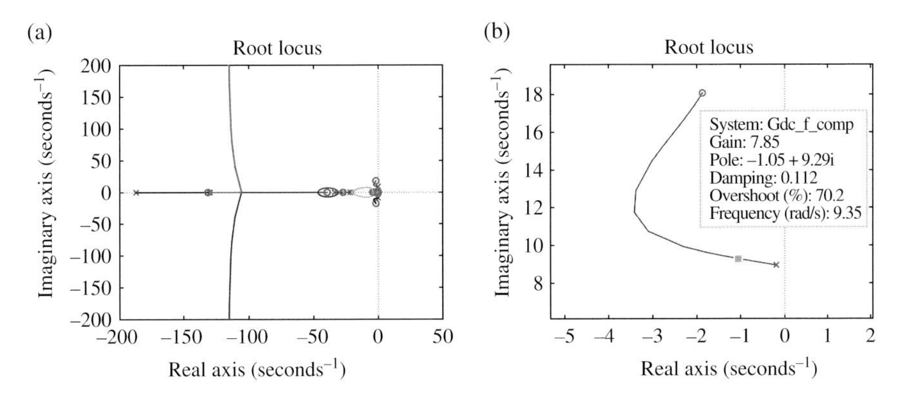

Figure 13.34 (a) Root-locus plot of compensated system and (b) enlarged plot around the swing mode.

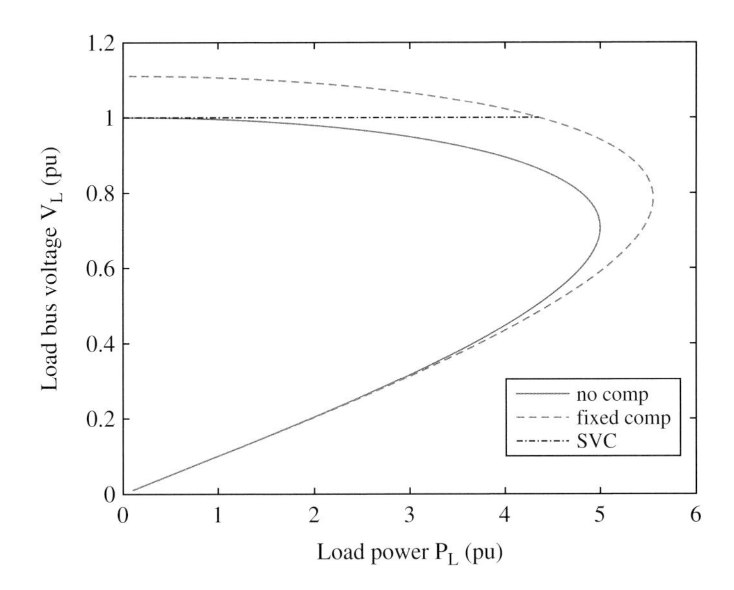

Figure 14.4 *PV* curves for the stiff-source-to-load system with no compensation, fixed capacitor, and SVC at the load bus.

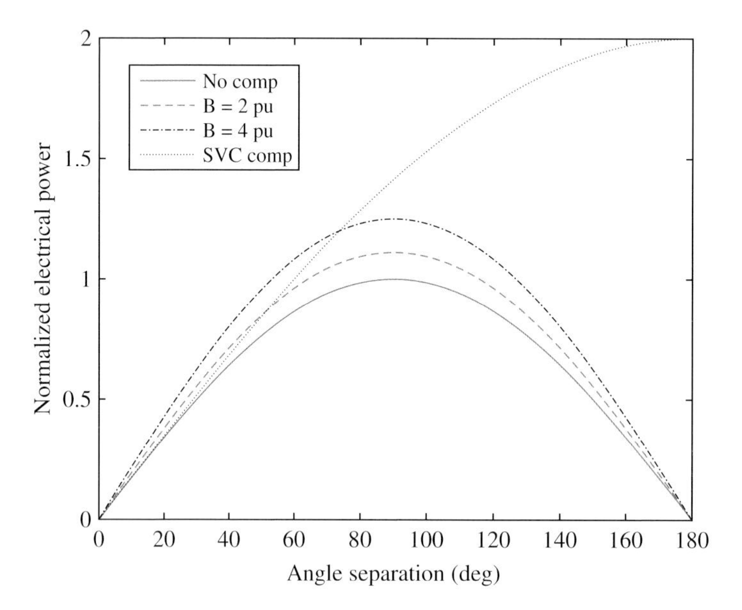

Figure 14.6 $P\delta$-curves for systems in Figure 14.5 with $X = 0.2$ pu and $V_m = 1$ pu on 100 MVA base - electrical power is normalized with respect to P_o.

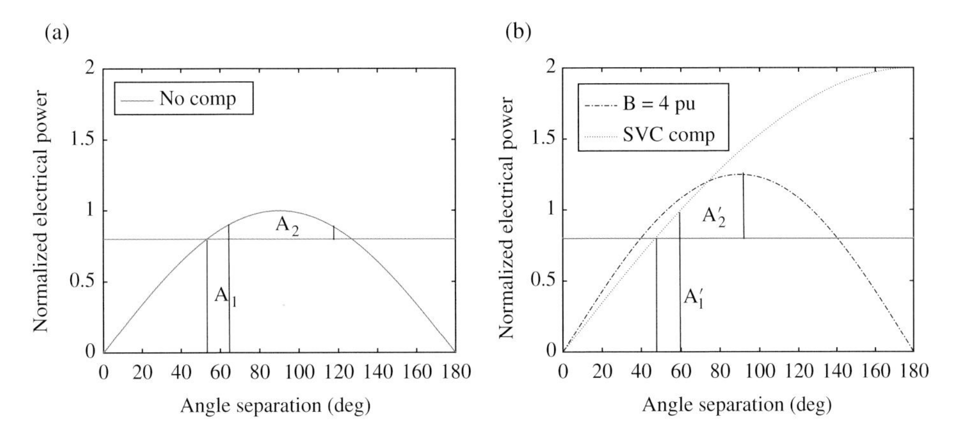

Figure 14.7 Equal area criterion applied to the $P\delta$ curves: (a) no compensation and (b) SVC compensation.

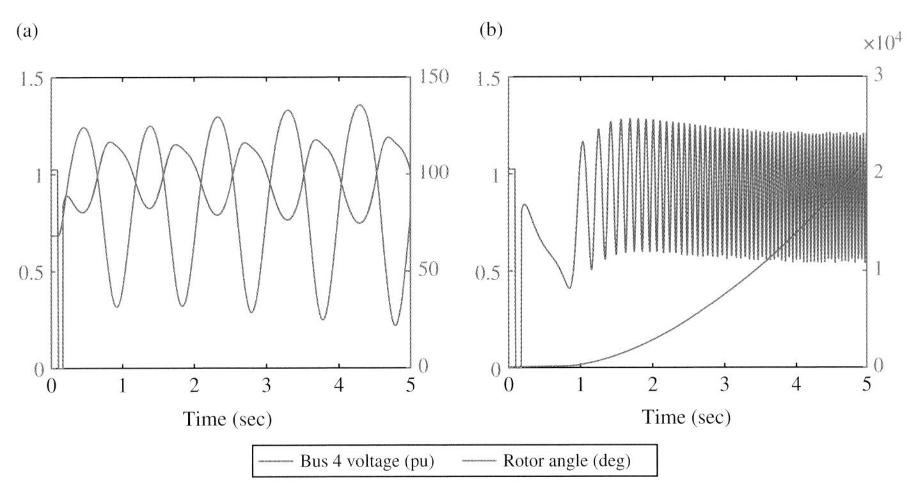

Figure 14.15 Bus 4 voltage and rotor angle response for system without SVC: (a) 4-cycle fault clearing and (b) 5-cycle clearing.

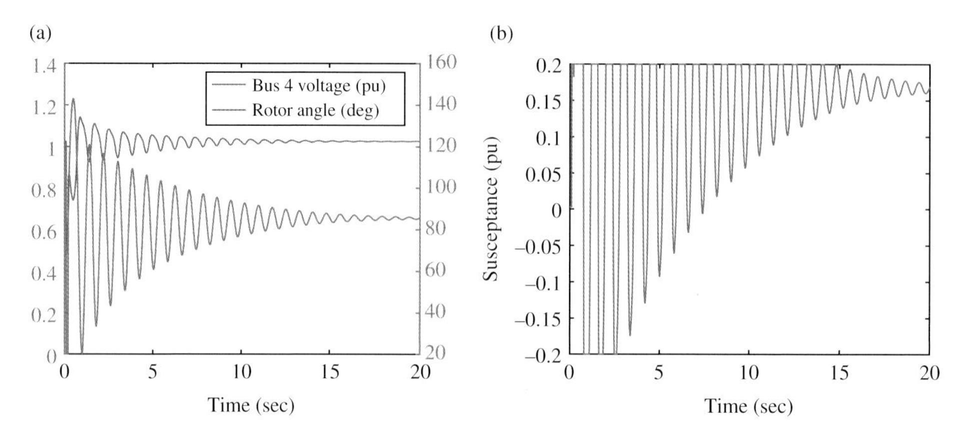

Figure 14.16 Time response of system with SVC for 6-cycle fault clearing: (a) Bus 4 voltage and (b) SVC susceptance response to Bus 4 voltage.

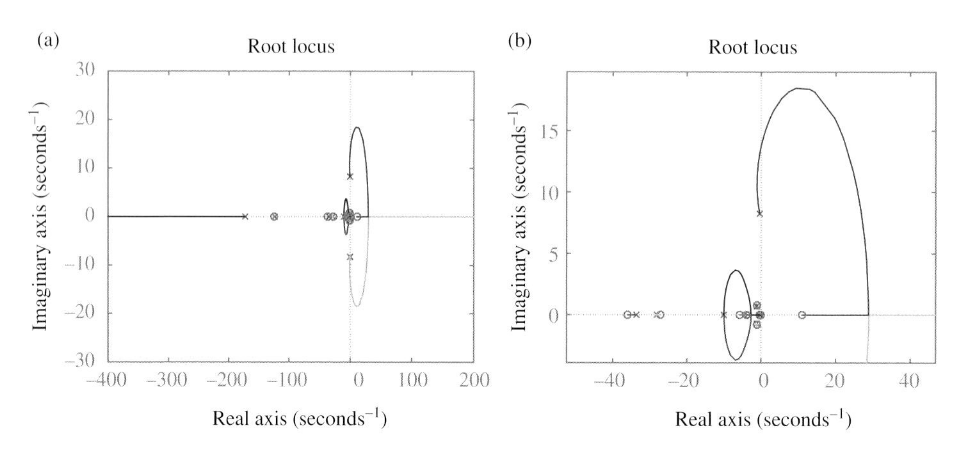

Figure 14.18 (a) Root-locus plot of uncompensated system and (b) enlarged plot around the swing mode.

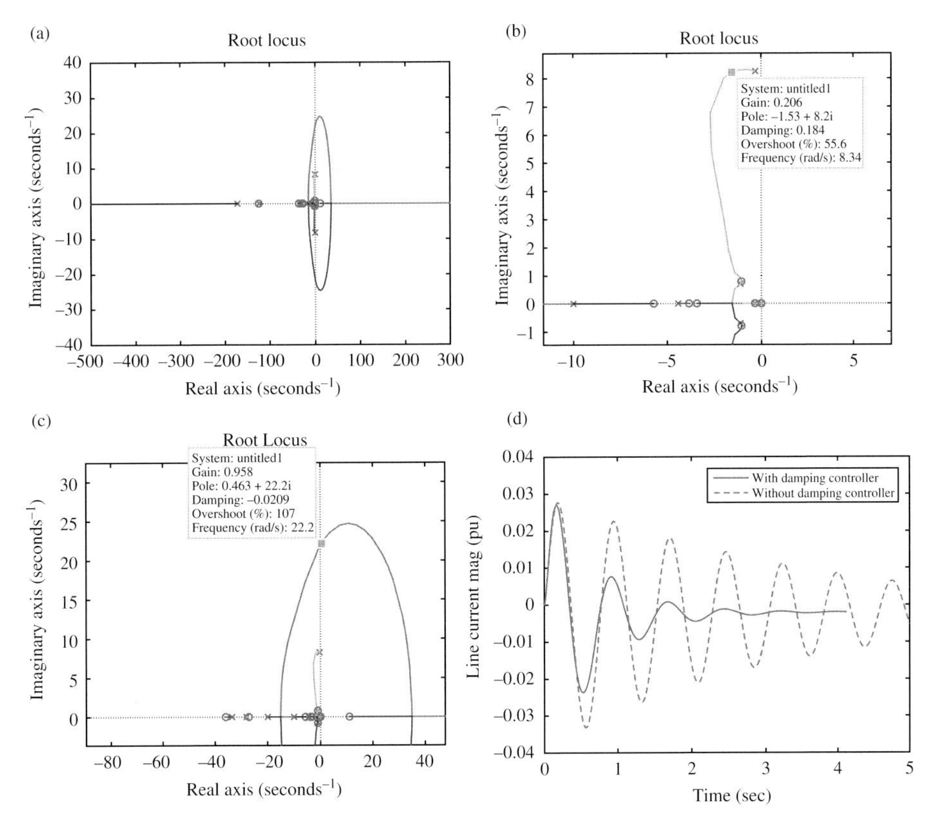

Figure 14.19 (a) Root-locus plot of compensated system, (b) enlarged plot around the swing mode, (c) unstable gain, and (d) comparison of compensated and uncompensated systems subject to a 0.05 pu step increase in the excitation system voltage reference.

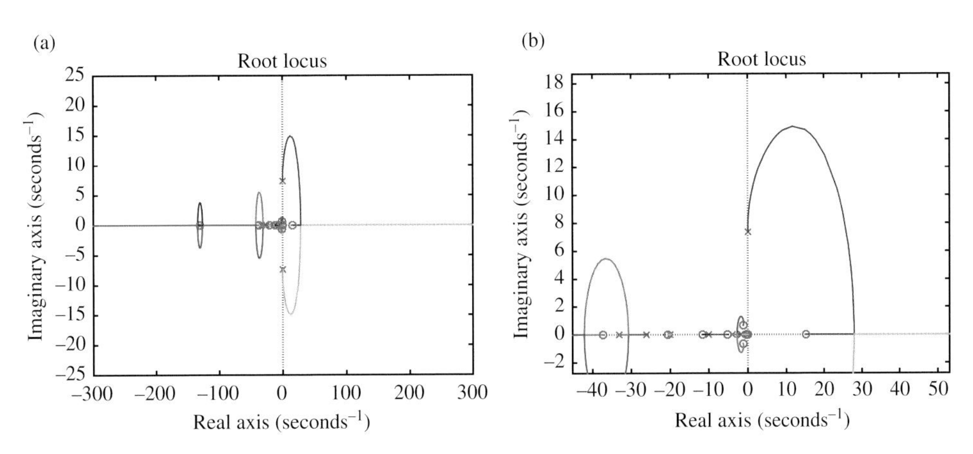

Figure 14.30 (a) Root-locus plot of uncompensated system and (b) zoomed-in plot around the swing mode.

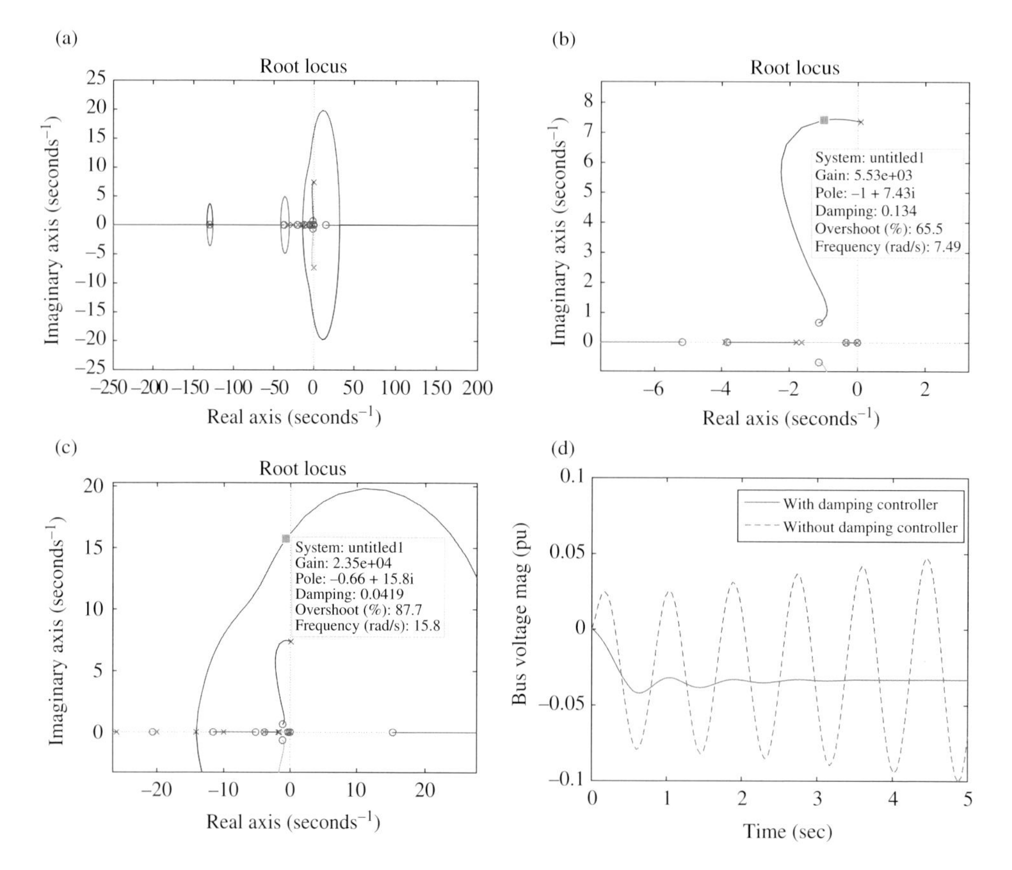

Figure 14.31 (a) Root-locus plot of compensated system, (b) zoomed-in plot around the swing mode, (c) unstable gain, and (d) comparison of compensated and uncompensated systems subject to a 0.05 pu step in excitation system voltage reference.

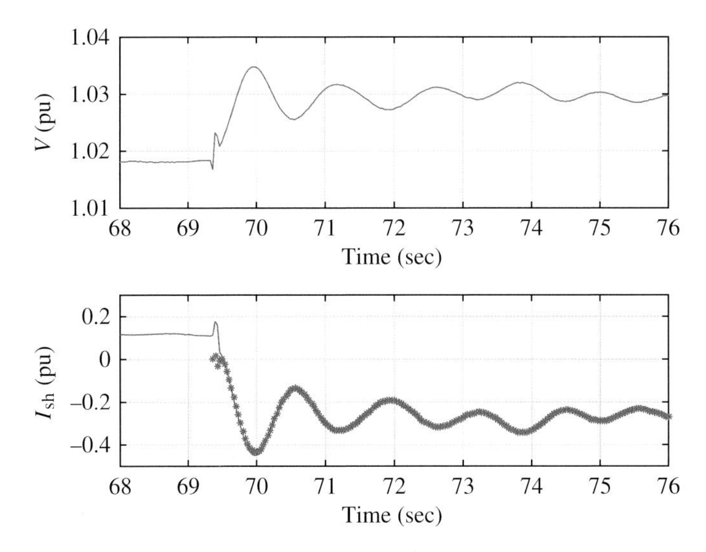

Figure 14.43 STATCOM disturbance response: (upper plot) measured high-side bus voltage magnitude, and (lower plot) measured STATCOM reactive current injection (solid curve): capacitive current is positive and inductive current is negative. The lower plot also shows the approximation achieved by the estimated transfer function (dotted curve).

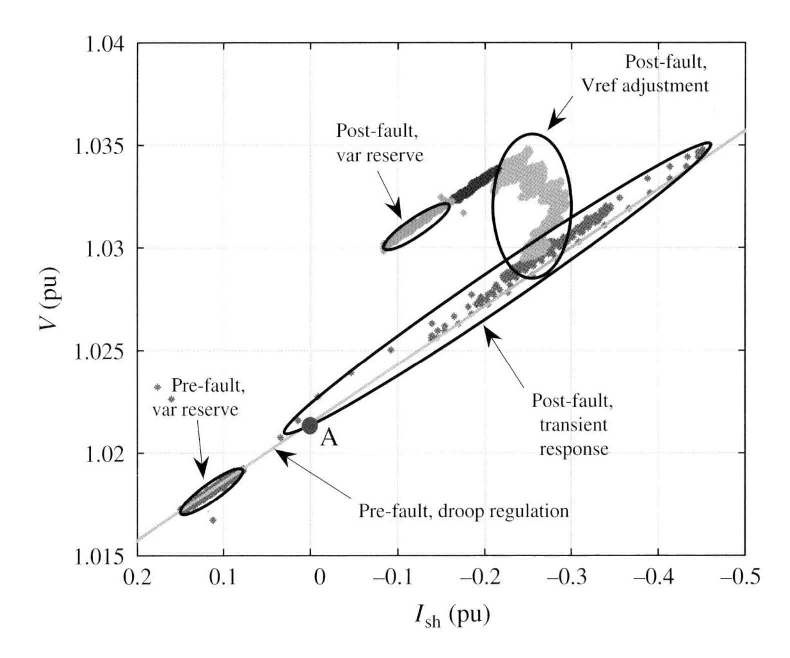

Figure 14.44 Pre-fault and post-fault STATCOM droop characteristics.

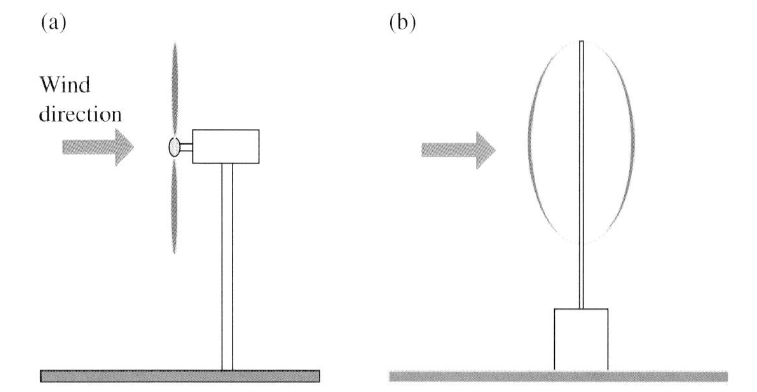

Figure 15.1 Wind energy conversion: (a) horizontal axis wind turbine (HAWT) and (b) vertical axis wind turbine (VAWT).

Figure 15.2 Horizontal axis wind turbine (courtesy of GE Renewable Energy).

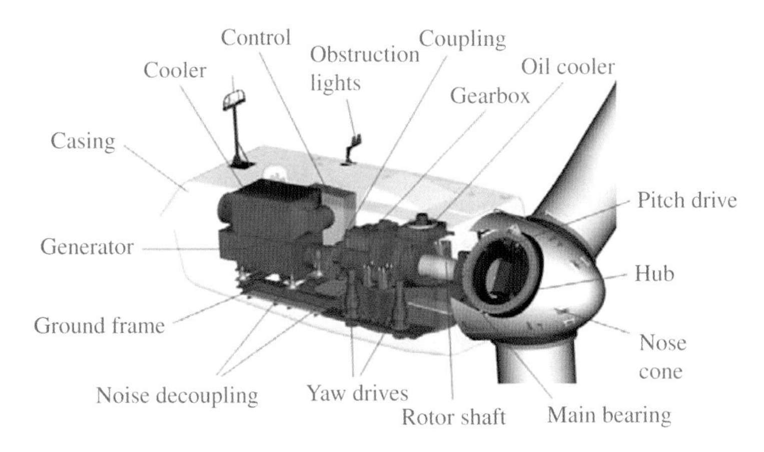

Figure 15.3 Nacelle (courtesy of GE Renewable Energy).

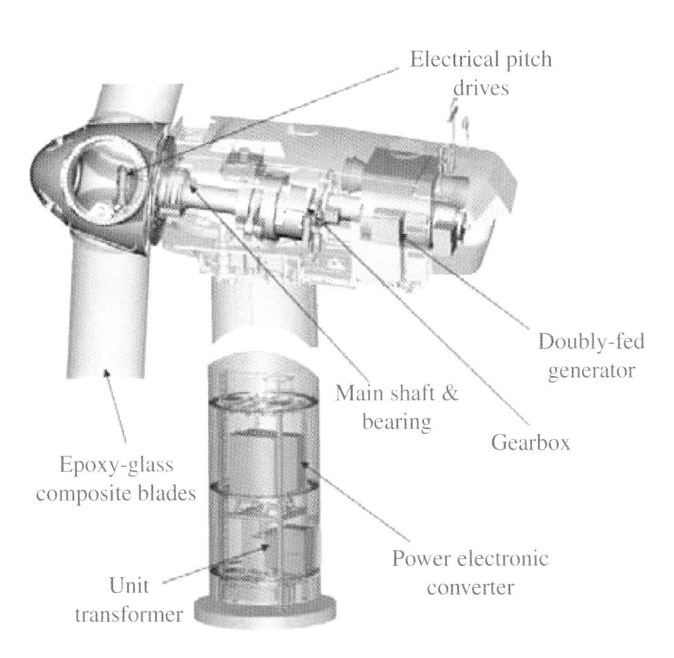

Figure 15.4 Tower and nacelle (courtesy of GE Renewable Energy).

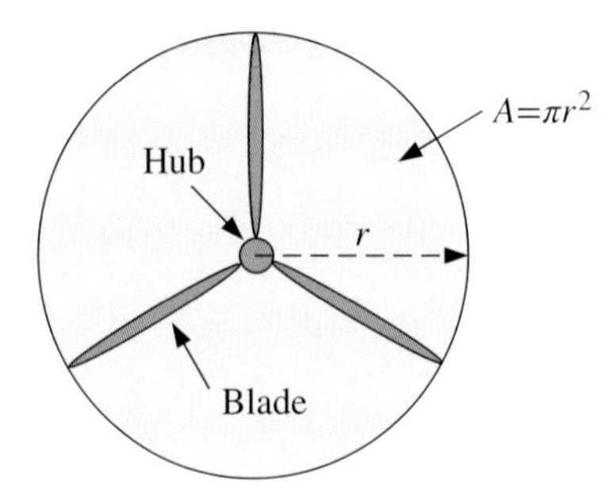

Figure 15.5 Area swept by wind turbine blades.

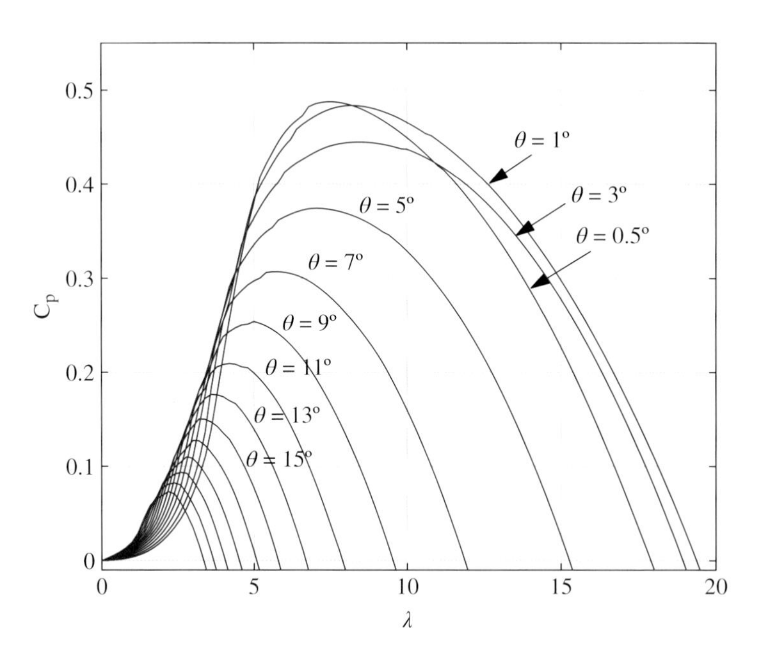

Figure 15.9 Power coefficient versus wind speed at different pitch angles (courtesy of GE Renewable Energy).

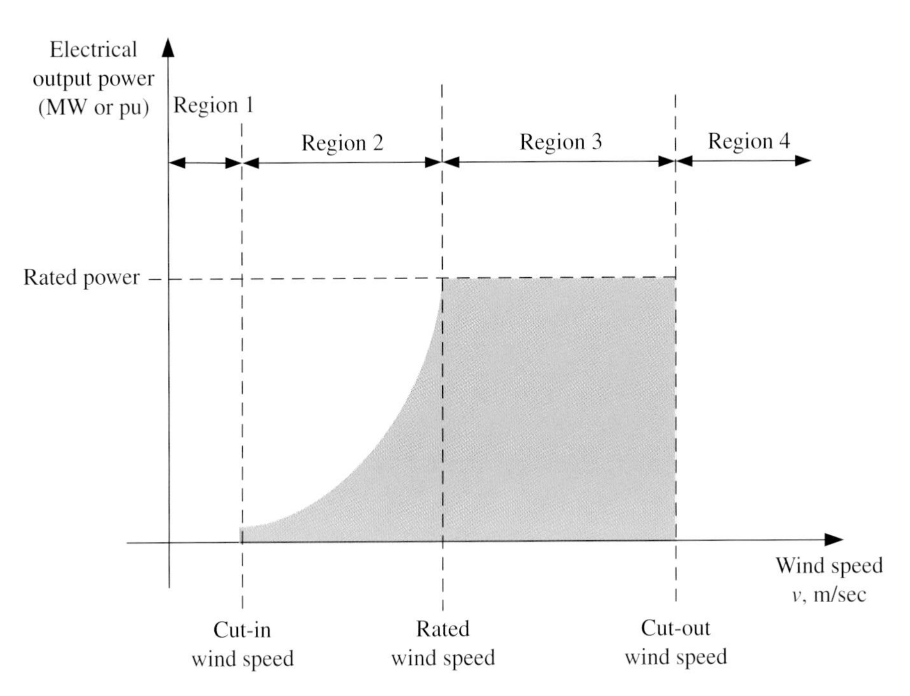

Figure 15.10 Power curve and four regions of operation.

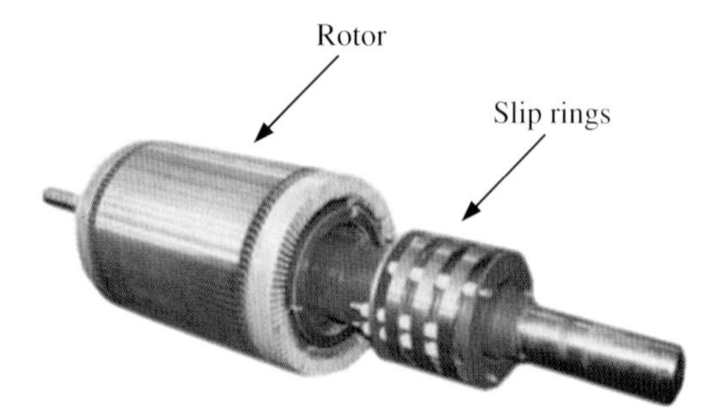

Figure 15.20 Rotor with slip rings (courtesy of GE Renewable Energy).

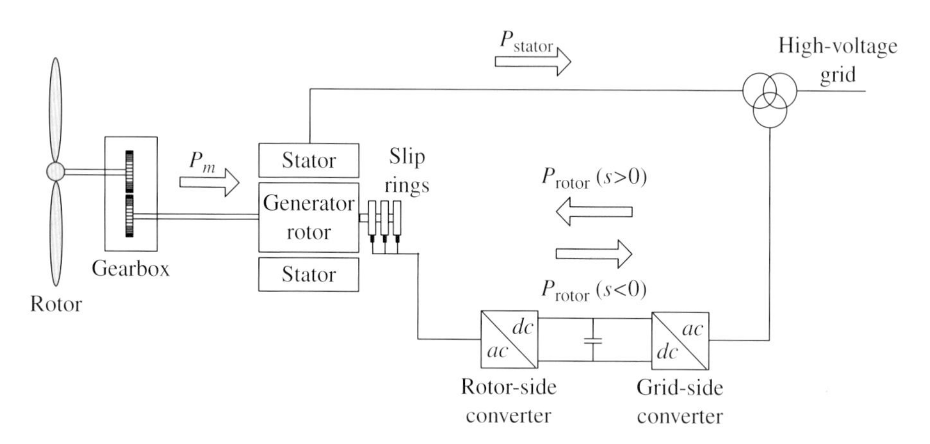

Figure 15.21 DFIG configuration and active power flows.

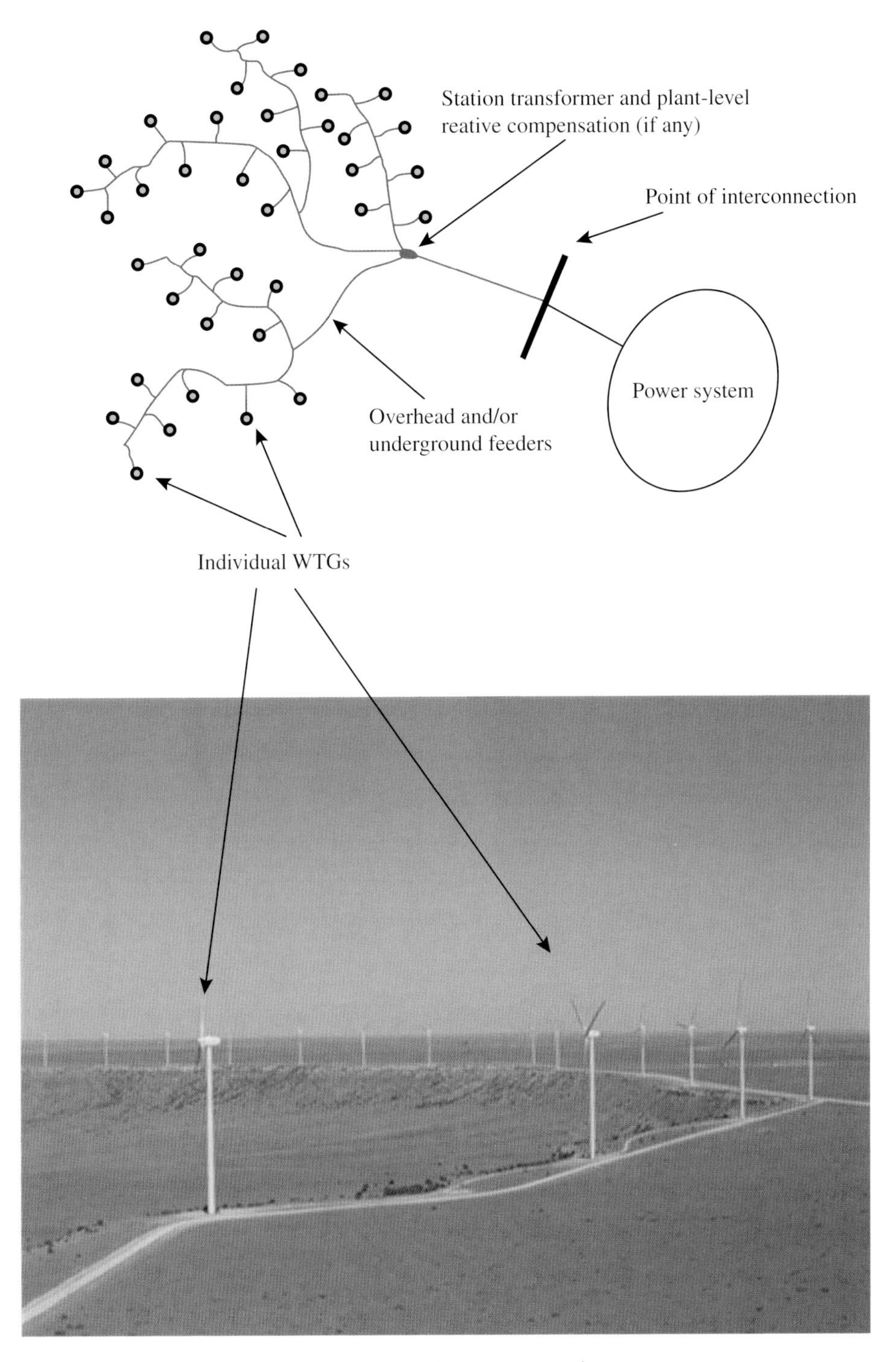

Figure 15.25 Components of a WPP (courtesy of GE Renewable Energy).

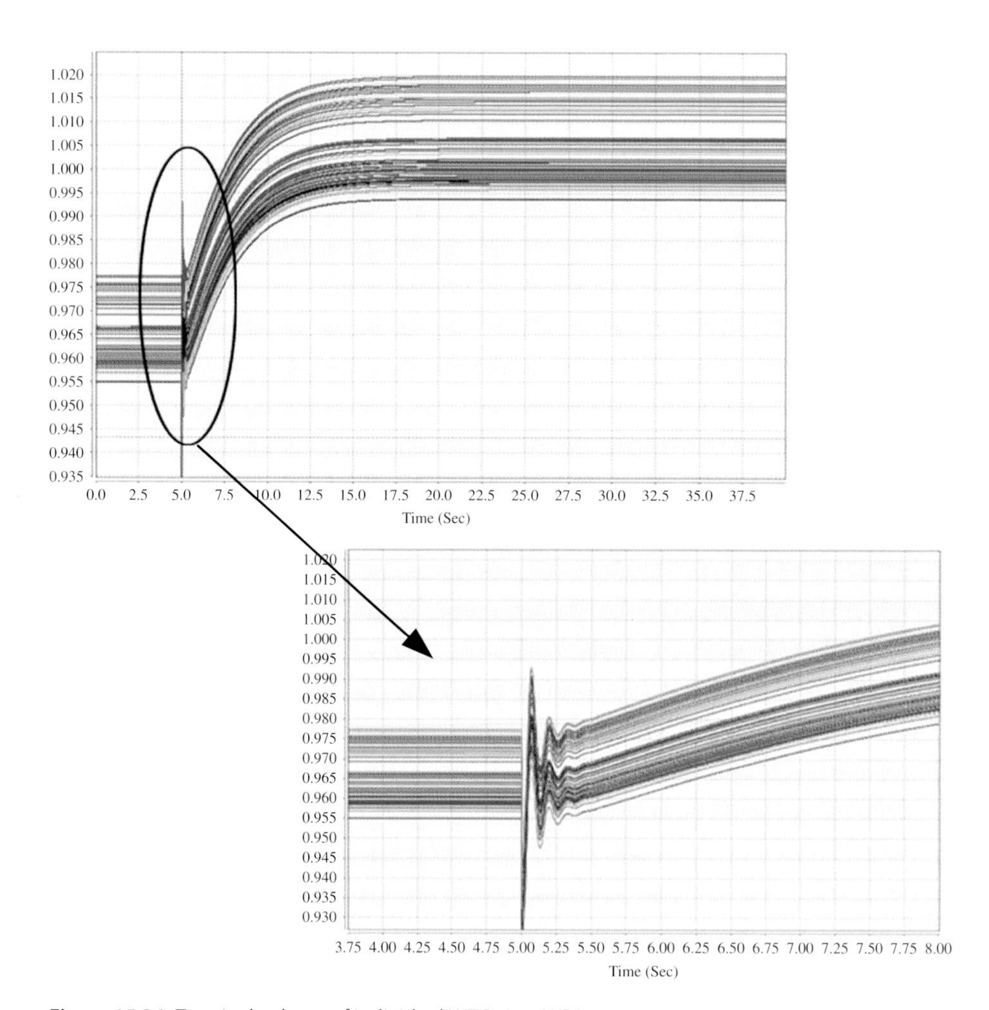

Figure 15.26 Terminal voltage of individual WTGs in a WPP.

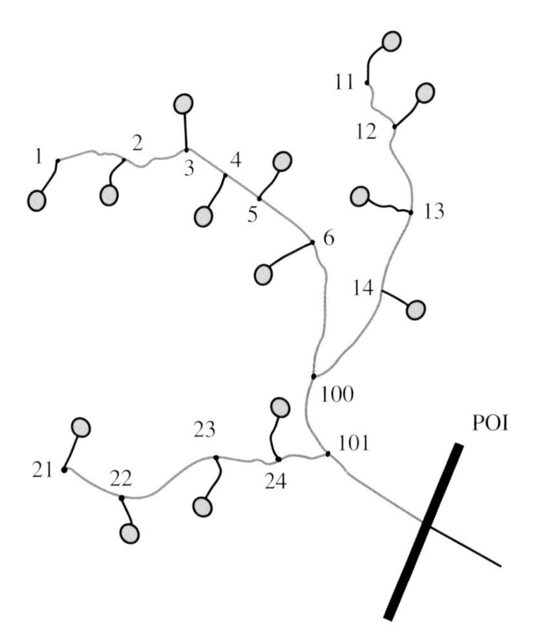

Figure 15.30 Wind power plant.

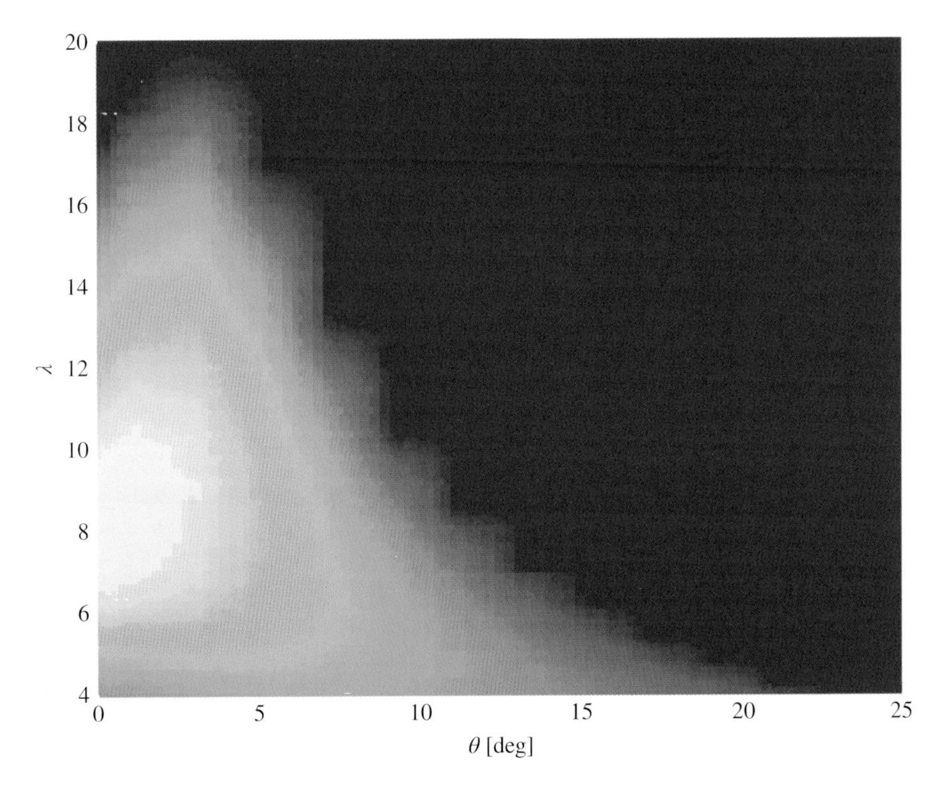

Figure 15.31 C_p versus θ and λ.

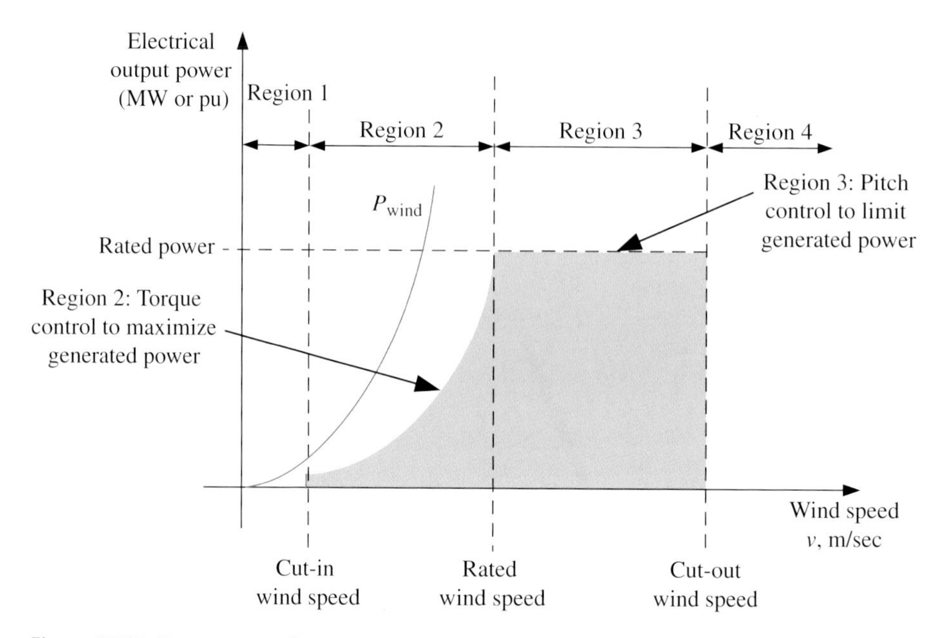

Figure 15.33 Power vs speed curve.

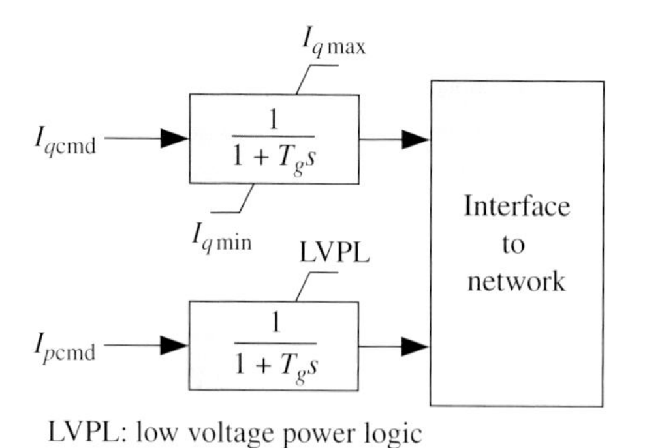

Figure 15.35 Generator/converter model.

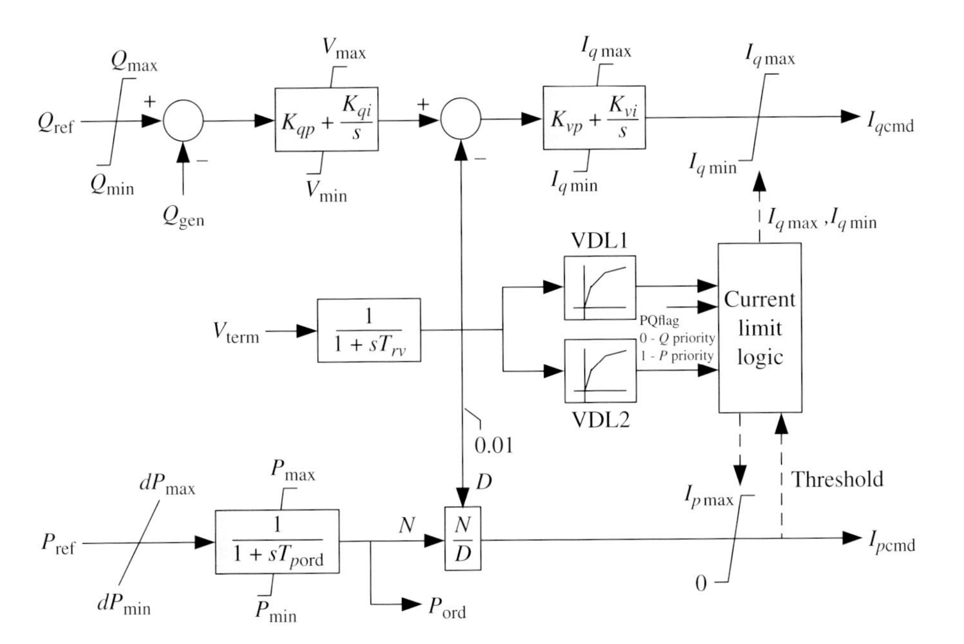

Figure 15.36 Electrical control model.

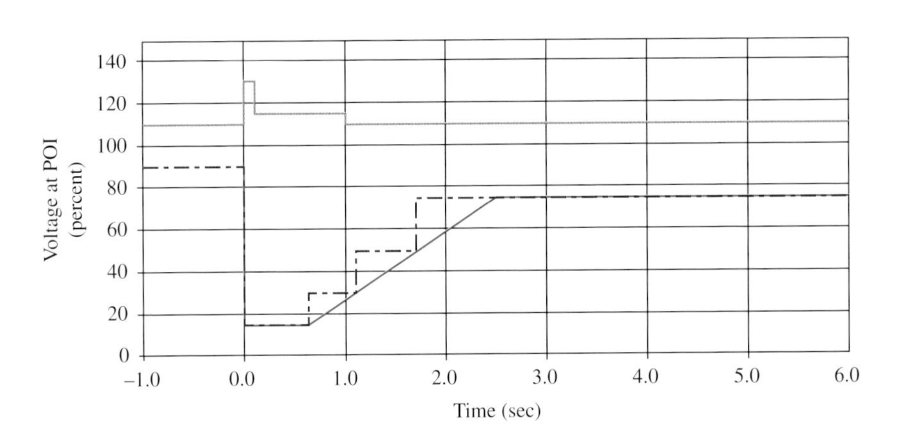

Figure 15.37 LVRT specifications [217].

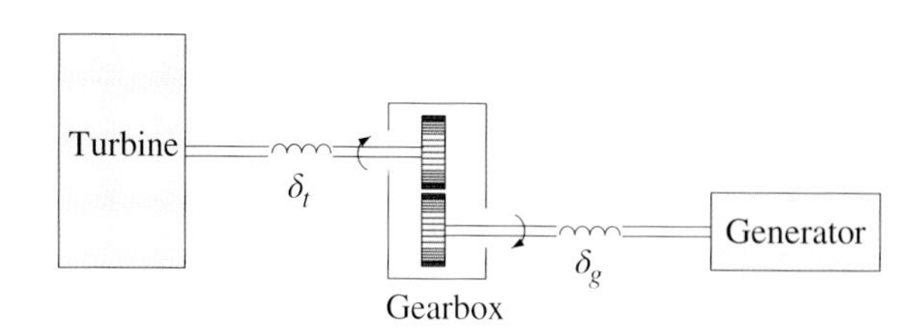

Figure 15.38 Two-mass turbine-generator model.

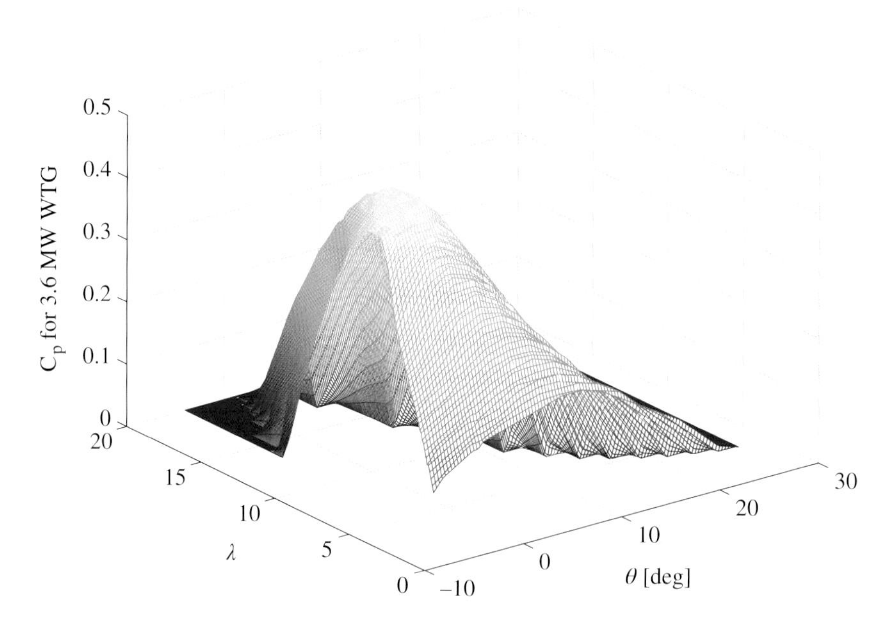

Figure 15.44 Three-dimensional C_p surface.

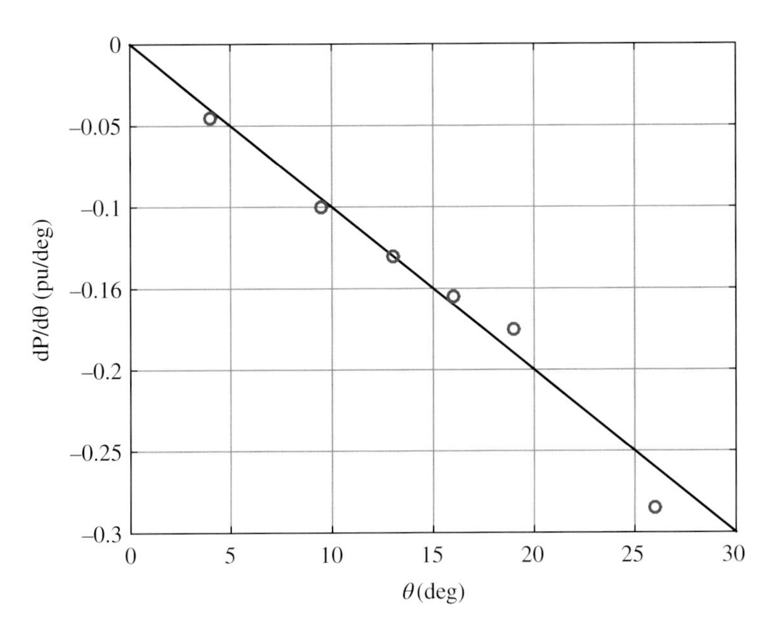

Figure 15.46 $dP/d\theta$ versus θ.

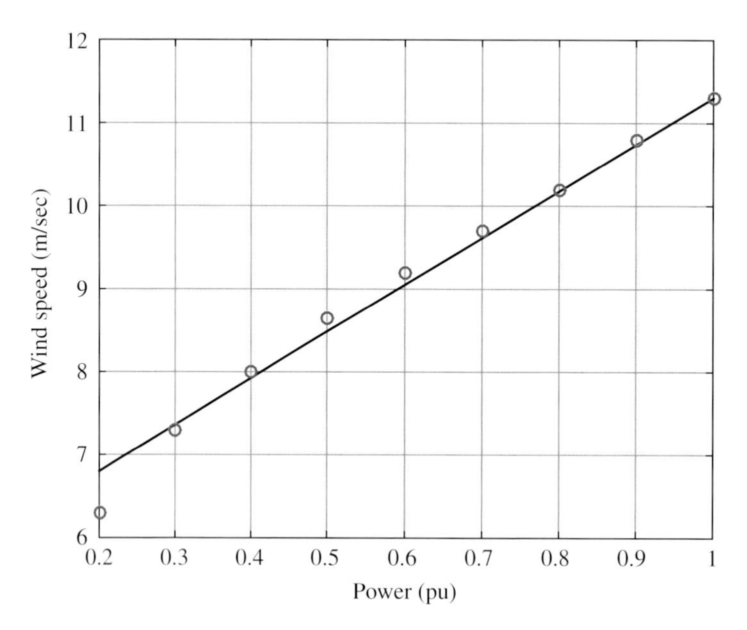

Figure 15.47 Wind speed versus generated power.

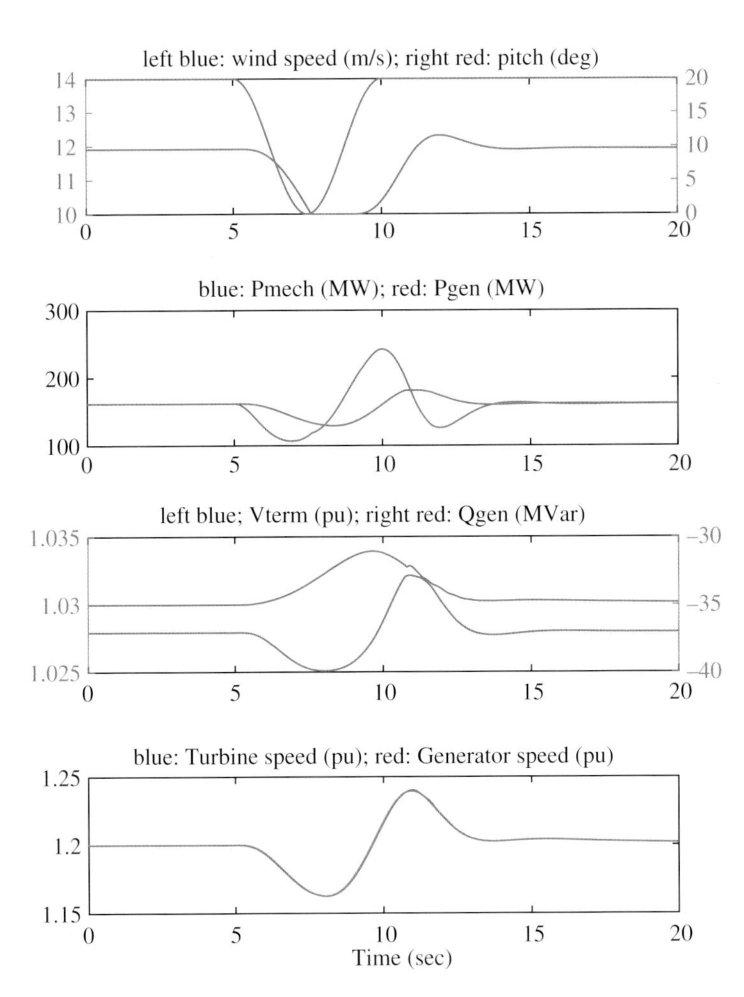

Figure 15.50 Simulation results for the wind gust.

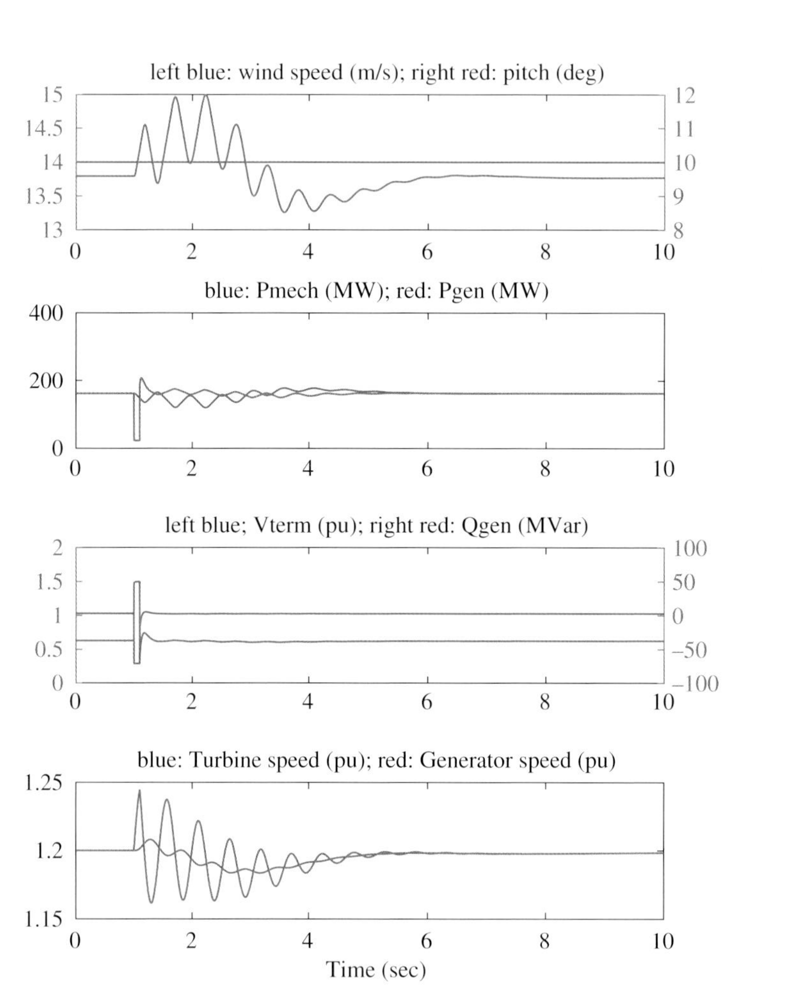

Figure 15.51 Simulation results for the fault condition.

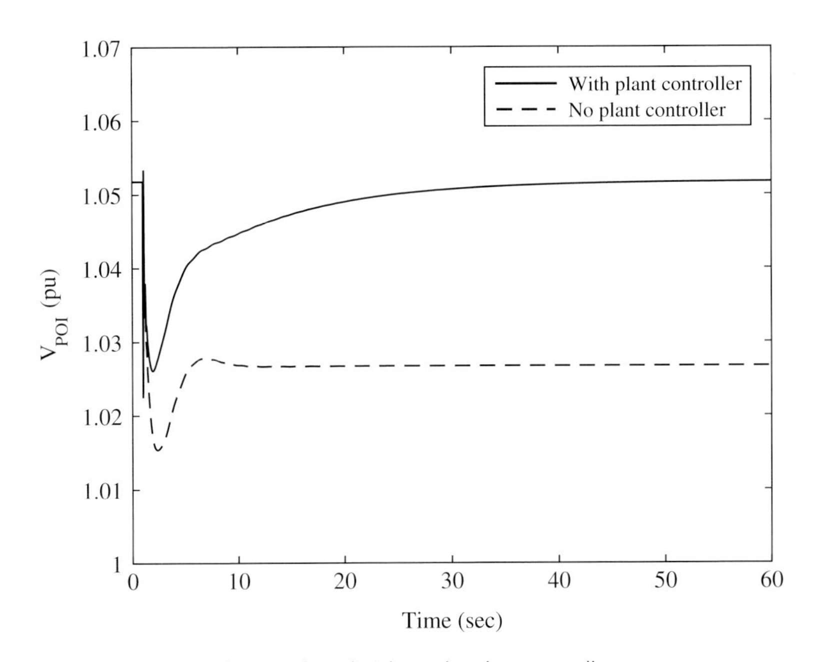

Figure 15.54 POI voltage with and without the plant controller.

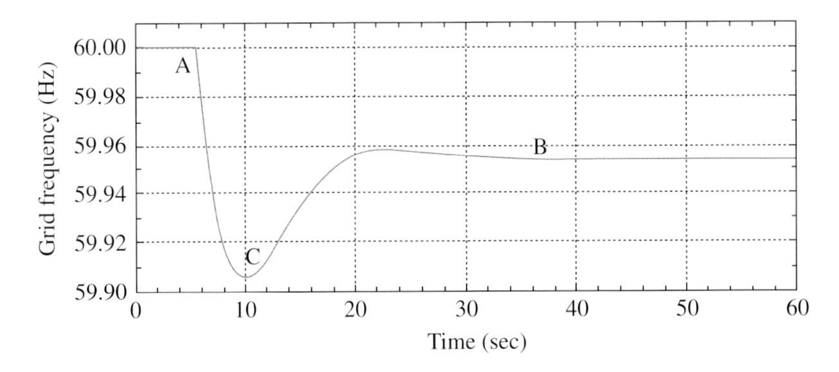

Figure 15.55 Typical frequency response of a power system for a loss of generation event.

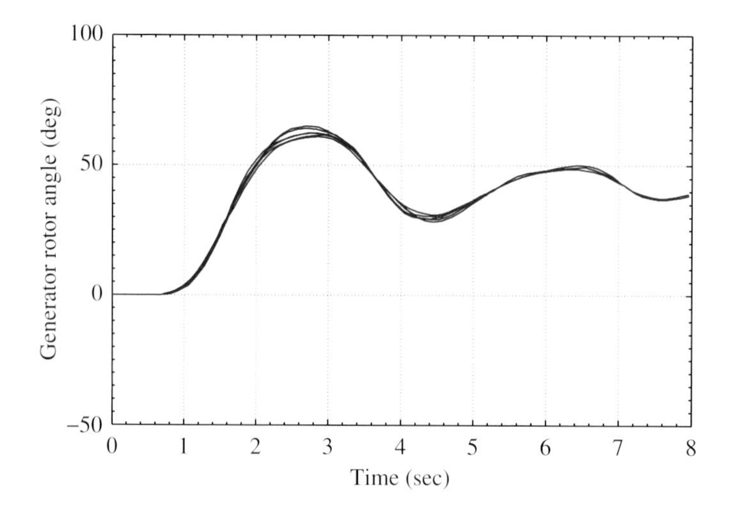

Figure 16.3 Swing curves of a group of 21 generators in Arizona [231]

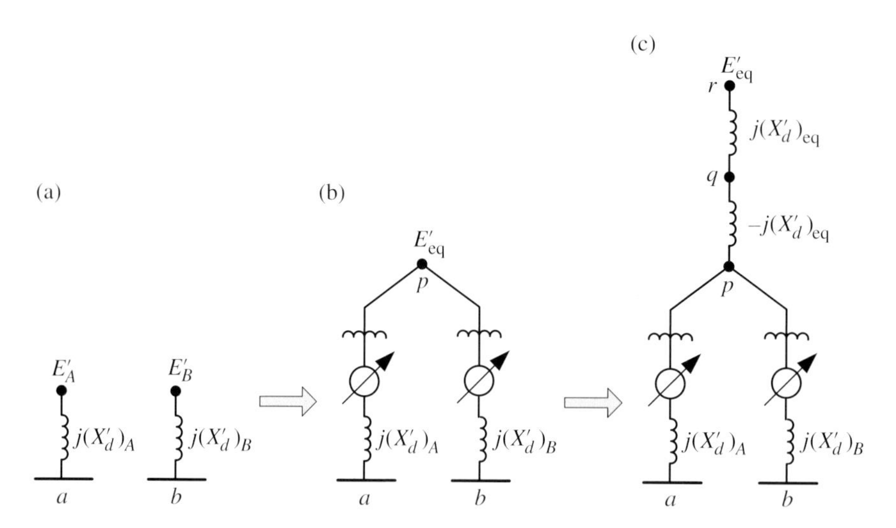

Figure 16.11 Inertial aggregation: (a) original network, (b) internal nodes of generators connected, and (c) internal node of the aggregate generator created.

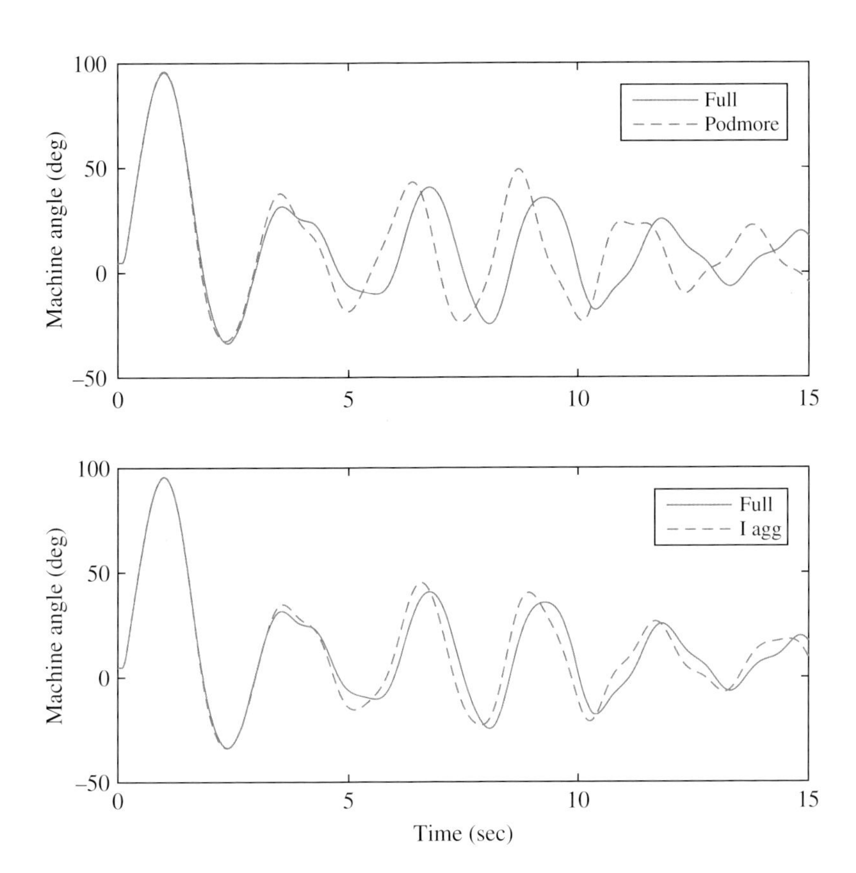

Figure 16.15 Time response of Machine 8 for the Medway disturbance.

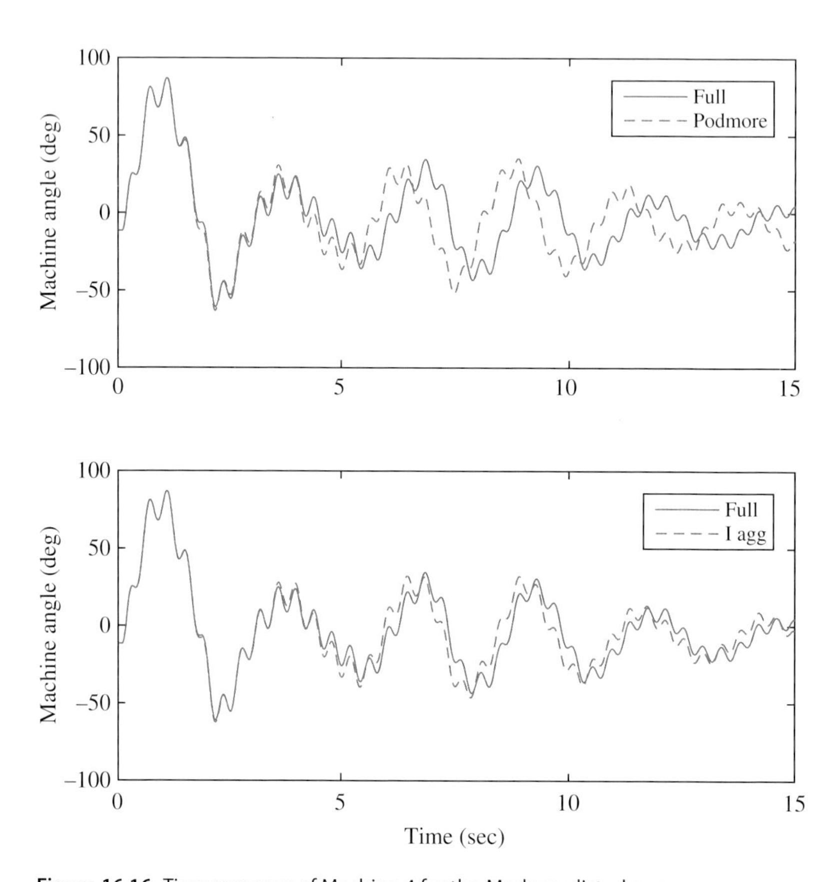

Figure 16.16 Time response of Machine 4 for the Medway disturbance